ABOUT THE COVER

The Liberty Bell appears on the cover of this textbook because the crack is internationally known and recognized more than any other fracture. However, there is considerable confusion as to the history of the Bell and how it gained such worldwide recognition. The following is intended to highlight major events in the Bell's existence from the casting foundry to the present.

To commemorate the fiftieth anniversary of the granting of William Penn's Charter of Liberties, the Pennsylvania Assembly purchased a bell for the Statehouse. Since there were no qualified bell foundries in the region, the bell was cast at the Whitechapel Foundry in London, England. The inscription on the bell was to read "Proclaim liberty through all the land unto all the inhabitants thereof," (Leviticus 25:10). On its completion, the bell was shipped to Philadelphia and placed in the Statehouse belfry. To the dismay of all, the bell cracked the first time it was struck. John Pass and Charles Stow, two area residents, agreed to recast the bell in time for the Charter of Liberty's jubilee celebration. After adjusting the alloy chemistry and recasting the bell twice, these amateur bell founders produced a bell with an acceptable tone. For their services, Pass and Stow were paid $295.25 and given a free advertisement: note their names on the shoulder of the bell.

Not leaving anything to chance, the Pennsylvania Assembly commissioned a second bell from the Whitechapel Foundry, which arrived from England when Pass and Stow had completed the third casting of the original bell. What were they to do with two bells? It was ultimately decided that the original bell (also known as the Liberty Bell) be used for grand occasions such as convening townsfolk for the first public reading of the Declaration of Independence and the second Whitechapel bell be used as the town's clockbell.

During the Revolutionary War, the Liberty Bell was taken to Allentown, Pennsylvania, to safeguard it from the advancing British armies. The city fathers were less concerned with protecting an American historical treasure (the bell had no historical value at that time) than with preventing the British from melting such bells to produce new artillery pieces. Cannon metal (also known as Admiralty bronze) contains 88% copper and 12% tin whereas bell metal contains roughly twice as much tin. After the bell was returned to Philadelphia in 1778, it continued to ring until 1835 when it cracked while tolling the funeral of Chief Justice Marshall. (The second Whitechapel bell was given to a church in 1828, before being destroyed in a fire.) After grinding the mating surfaces of the crack to prevent them from rubbing together, the Liberty Bell was struck once again in 1846 to celebrate Washington's birthday. After ringing for several hours, the original crack extended into the shoulder region. Since that time, the bell has effectively remained silent.

After being sent on a series of national tours, beginning with a trip to New Orleans in 1885, the Liberty Bell has become a symbol of American independence. It is now on permanent exhibition in the historical section of Philadelphia.

FOURTH EDITION

DEFORMATION AND FRACTURE MECHANICS OF ENGINEERING MATERIALS

FOURTH EDITION

DEFORMATION AND FRACTURE MECHANICS OF ENGINEERING MATERIALS

RICHARD W. HERTZBERG

New Jersey Zinc Professor
of Materials Science and Engineering and
Director, Mechanical Behavior Laboratory,
Materials Research Center,
Lehigh University

JOHN WILEY & SONS, INC.

New York   Chichester   Brisbane   Toronto   Singapore

Cover Photo	The Liberty Bell/Bettmann Archive
Acquisitions Editor	Cliff Robichaud
Assistant Editor	Catherine Beckham
Marketing Manager	Debra Riegert
Production Editor	Ken Santor
Manufacturing Coordinator	Dorothy Sinclair
Illustration Coordinator	Anna Melhorn

This book was set in 10.5/12.5 Times Roman by Ruttle, Shaw & Wetherill and printed and bound by Hamilton Printing. The cover was printed by Phoenix Color.

Recognizing the importance of preserving what has been written, it is a policy of John Wiley & Sons, Inc. to have books of enduring value published in the United States printed on acid-free paper, and we exert our best efforts to that end.

The paper on this book was manufactured by a mill whose forest management programs include sustained yield harvesting of its timberlands. Sustained yield harvesting principles ensure that the number of trees cut each year does not exceed the amount of new growth.

Library of Congress Cataloging-in-Publication Data

Hertzberg, Richard W., 1937–
    Deformation and fracture mechanics of engineering materials /
  Richard W. Hertzberg.
      p.   cm.
    Includes bibliographical references and index.
    ISBN 0-471-01214-9 (cloth)
    1. Deformations (Mechanics)   2. Fracture mechanics. I. Title.

  TA417.6.H46   1995
620.1′123—dc20
            95-35234
            CIP

Printed in the United States of America

10 9 8 7 6 5 4 3 2 1

To my wife Linda and my children Michelle, Jason, and Nicholas.

PREFACE TO THE FOURTH EDITION

The addition of several new topics and the updating of numerous segments in both sections of the book enables the fourth edition of this text to continue to serve both as a textbook and as a reference volume. These changes reflect recent developments in our understanding of deformation and fracture processes in structural materials. More than 130 additional references have been added to the text, raising the total to more than 1200. Twenty new example problems and 50 additional figures have been distributed throughout the manuscript to enable the reader to more readily understand the subject matter. Approximately 40 additional homework problems have been added at the end of most chapters; the solutions manual has been updated and is available to qualified users.

A new section on the isostress analysis, which characterizes the strength and stiffness of composites, has been added to complement the isostrain analysis that was introduced in the third edition. To establish the groundwork for expanded coverage of the fracture studies of ceramics, a new section on the modulus of rupture has been added. For students who possess a limited exposure to strength of materials concepts, the stress analysis of thin-walled cylinders has been included near the end of Chapter 1. This discussion will help the reader in later chapters when dealing with fracture problems associated with thin-walled pressure vessels; these include circumstances surrounding leak-before-break criteria. In connection with crystallographic textures, a new section on plastic anisotropy has been added to Chapter 3. A new section on creep fracture micromechanisms has been added at the conclusion of Chapter 5.

Two new analytical sections have been added to Chapter 7 that deal with the Weibull analysis of the statistical nature of fracture and the generation of thermal stresses and thermal shock-induced fracture. Several new example problems have been added to the important discussions of fracture mechanics in Chapter 8. The discussions dealing with the fracture toughness of metals, ceramics, and polymers have been updated and compared with respect to the controlling crack tip shielding mechanisms. Recent observations regarding environmental threshold test procedures are included in Chapter 11.

Discussion of "safe-life," "fail-safe," and "retirement-for-cause" fatigue life design procedures have been added to Chapters 12 and 13. Also, a new section on procedures for the avoidance of fatigue damage, including the development of favorable residual compressive stresses and pretensioning of load-bearing members, completes the discussion of Chapter 12. New topics added to Chapter 13 include: (1) calculation and prediction of FCP data based on fundamental crystal properties; (2) load interaction-induced macroscopic fracture surface appearance; and (3) updated discussion dealing with the fatigue crack propagation response of polymers and ceramics. Finally, a new failure analysis has been added in Chapter 14 along with additional discussions on the fracture surface appearance of metal alloys under block-loading and load-shedding conditions, respectively.

A number of individuals provided assistance with the completion of the book. Some contributed glossy prints of new figures, and others offered constructive criticism of new sections. To these individuals I express my sincere gratitude. Special thanks are extended to the author's son, J. L. Hertzberg, for drafting Section 5.7 and for providing several new figures. The author greatly appreciates the help of Drs. H. M. Chan, M. P. Harmer, and D. B. Williams for reviewing new sections of the book. I appreciate the help of several of my former students, H. Azimi, T. Pecorini, C. Ragazzo, and T. Clark, for use of their recent research findings and for their useful comments. I am grateful to the students in my recent graduate classes who reviewed several new sections of the book. Mr. Ken Marschall gratefully extended permission for my use of his artistic rendition of the sunken remains of the *Titanic*. The author also appreciates the editorial and production staffs at John Wiley and to M. Mattie, S. Siegler, and S. Coe from Lehigh University for their help with completion of the manuscript.

My wife Linda continues to be patient and understanding each time I revise this book. To her, I extend my love and appreciation.

Richard W. Hertzberg

PREFACE TO THE THIRD EDITION

In recognition of rapidly changing developments in the field of mechanical behavior of solids, the third edition of this book covers many new topics with significantly broadened attention given to all structural materials, including metals, polymers, ceramics, and their respective composites. Most notably, a new chapter on the strengthening mechanisms in metals has been added along with extensive discussion of composites, which is treated and cross-referenced in eight chapters throughout the book. In addition, numerous topics introduced in earlier editions have been updated, with every chapter undergoing substantial change. A solutions manual for the chapter problems is available to qualified users. As before, special attention is given to providing the reader with many references for further study. A new supplementary reading list is also added to the end of several chapters. The current edition contains over 1100 references; of the more than 200 new listings, over 70% represent citations published since the completion of the second edition in 1983. As such, this book should be viewed not only as a textbook for university study but also as a reference volume for professional use.

A major new section on composite materials was added to Chapter 1 and includes such topics as the isostrain analysis and the influence of fiber aspect ratio and orientation on composite strength. Chapter 1 now also includes additional tables of mechanical property data for ceramics and composites, an expanded discussion of strain-rate effects on strength, and a section on classical failure theories (Tresca, Von Mises, etc.). The topic of partial dislocations has been moved from Chapter 3 to Chapter 2 and a new section on superlattice dislocations has been added to the discussion. Chapters 3 and 4 from the second edition have been combined into one chapter dealing with the slip and twinning response of crystalline solids; the topics of crystallographic texturing and twinning have been condensed.

Chapter 4 represents a new addition to the book and is concerned with the strengthening mechanisms in metals. Topics in this chapter include strain (work) hardening, grain-boundary strengthening, solid solution, precipitation, dispersion and martensitic strengthening, and metal-matrix composite strengthening. The section in Chapter 5 on materials for elevated temperature use has been expanded significantly to include a discussion of recent developments in cast and powder-produced conventional alloys, oxide-dispersion-strengthened mechanically alloyed systems, composites, and high-temperature coatings. Chapter 6 has been updated and includes new discussions on designing with plastics (i.e., isochronous and isometric stress–strain curves), polymer composites, and expanded coverage of the strengthening of polymers.

Numerous changes and additions were also made to the second half of the book, which examines the fracture mechanics of engineering materials. New sections on fracture mechanisms in composites and quantitative fractography have been added to Chapter 7 and the section on microvoid coalescence was updated. Chapter 8 contains new discussion of fracture toughness measurements in ceramics based on microhardness indentation test methods. J_{IC} test procedures are updated. A discussion of impact testing in polymers has been added to Chapter 9.

Chapter 10 now begins with expanded coverage of generic toughening mechanisms found in metals, plastics, and ceramics, and their respective composites. The section on mechanical fibering has been moved from Chapter 4 to Chapter 10 and is included with consideration of fracture toughness anisotropy. Sections on toughness in various material groups have been streamlined and updated and now include treatment of the toughness of composites. New sections on liquid metal embrittlement and EAC of polymeric solids have been added to Chapter 11 and recent K_{IEAC} test methods are presented.

The subject of fatigue has been updated and expanded in Chapters 12 and 13, and sections on the fatigue of composites have been added to both chapters. Other new sections in Chapter 12 include discussions of the Bauschinger effect, rainflow-counting methods to account for damage during random cyclic loading, and expanded coverage of the macrofractography of fatigue fracture surfaces. Parts of Chapter 13 have been condensed to achieve a better balance of topic coverage and other sections have been updated (especially the topics concerning the growth of short cracks and the influence of load interactions on FCP); a new section on the fatigue of ceramics has been added. Finally, additional failure analysis case histories are briefly discussed and other topics are updated in Chapter 14.

The many changes and additions made in the third edition of this book benefited greatly from the considerable support of many individuals who critically evaluated the strengths and weaknesses of the previous edition, reviewed sections of new material, and provided advanced copies of their manuscripts and original photographs. These individuals include R. Queeney, N. S. Stoloff, C. E. Price, L. L. Clements, D. B. Williams, G. A. Miller, R. Landgraf, R. Jaccard, J. A.Manson, E. E. Underwood, W. A. Herman, T. Pecorini, T. Clark, R. Marissen, S. Suresh, R. Arsenault, A. Yee, R. Benn, J. Weber, E. Thompson, A. G. Evans, R. O. Ritchie, B. Smith, A. Benscoter, R. Bucci, T. Crooker, W. Hoffelner, and J. Mecholsky. My apologies to those whom I may have inadvertently failed to mention.

The manuscript was typed with great care and dedication by Betty Zdinak. Additional production assistance was rendered by Andrea Weiss, William Herman, Sharon Siegler and Kenneth Vecchio—I thank them all.

My wife Linda and children, Michelle and Jason, persevered once again. Their patience and understanding was exceptional and greatly appreciated.

Richard W. Hertzberg

PREFACE TO THE SECOND EDITION

Since the completion of the first edition of this book, several subject areas have witnessed significant growth and maturation. These topics include fracture analysis utilizing J integral procedures, metallurgical and short crack aspects of fatigue crack propagation, and fracture processes associated with polymeric solids and ceramics. Since these topics are of major interest to working engineers and academicians, a modification to the first edition was both timely and necessary. This second edition also incorporates numerous suggestions contributed by teaching colleagues, based on their considerable classroom experiences.

Chapters 1 to 4 contain modest changes designed to enhance the reader's understanding of selected topics. For example, Chapter 1 contains a discussion of the relationship between the modulus of elasticity and atomic bonding forces. The different origins of annealing and deformation twins are described in Chapter 4. The topics of superplasticity and deformation maps are updated, and there is an expanded discussion of materials for elevated temperature use. More recent findings pertaining to crazing in polymers are considered in Chapter 6. The section in this chapter on polymer toughness has been expanded and moved to Chapter 10.

The discussion on fracture mechanics of engineering materials has undergone the most extensive revision. Chapter 7 now includes a section on macroscopic features of fracture surfaces (chevron markings), which had appeared in Chapter 14 of the first edition. Chapter 7 also contains new sections on the fracture mechanisms in polymers and ceramics. Chapter 8 has undergone significant change. There is expanded coverage of stress intensity factors (including material originally found in Chapter 14 of the first edition), leak-before-break failure criterion, and updated discussion of K_{IC} test procedures. Appendix B has been added to provide the reader with stress intensity factor formulas for common K_{IC} test specimens. The chapter concludes with a considerable discussion of the J integral and its applicability in the elastic-plastic analysis of fracture toughness.

Chapter 9 has been streamlined with the deletion of some material and the addition of updated correlations between fracture toughness and Charpy energy values. Chapter 10 contains new sections on optimizing fracture toughness in ceramics and polymers. Metallurgical embrittlement phenomena, formerly considered in Chapter 11 of the first edition, are now examined in Chapter 10. Chapter 11 contains an expanded discussion of hydrogen embrittlement and stress corrosion cracking models.

Several new sections have been added to Chapter 12, including more detailed discussions of fatigue crack initiation and fatigue life estimations for notched components. A table of fatigue data for representative metal alloys has been added. Certain topics about the fatigue life of engineering plastics have been moved from Chapter

13 to Chapter 12. New discussions of fatigue crack propagation test techniques, including computer control procedures, have been added to Chapter 13. This chapter also contains new examples of fatigue life calculations and an updated discussion of crack closure phenomena. A major new section has been added that deals with fatigue threshold testing and the consequence of short crack lengths on fatigue crack propagation (FCP) rates. Sections on load interaction effects, corrosion fatigue, and polymer fatigue have been updated. The section on microstructural aspects of FCP has been expanded greatly to reflect significant developments in this subject area during the past few years. Three new case histories involving fracture mechanics have been added to Chapter 14 along with a new section on typical defects found in engineering structures. Finally, Appendix A has been expanded to include sections on fracture surface preservation and cleaning techniques.

The total number of problems at the end of each chapter has been increased by more than 50%. Seventy new figures and 280 additional references have also been added. About 70% of these references were published between 1977 and the present.

In an effort to update the book to the greatest extent possible, I sought the advice of numerous scientists and made extensive use of recent manuscripts that they provided for my study. Many also took the time to review pertinent sections of the book and offered constructive criticism. To these colleagues, I offer my sincerest thanks. These individuals include G. R. Yoder, P. C. Paris, J. D. Landes, K. Friedrich, J. J. Mecholsky, C. M. Rimnac, M. T. Hahn, R. S. Vecchio, C. Newton, C. J. McMahon, Jr., F. P. Ford, R. P. Wei, N. E. Dowling, R. Jaccard, R. J. Stofanak, N. Fleck, R. J. Bucci, R. O. Ritchie, S. Suresh, K. Tanaka, R. A. Smith, R. P. Gangloff, R. D. Zipp, N. S. Cheruvu, J. A. Manson, G. P. Conard, II, J. F. Throop, J. H. Underwood, J. E. Srawley, D. Hull, M. F. Ashby, D. Porter, V. Hanes, and F. D. Lemkey. My apologies to those whom I may have inadvertently failed to mention.

The manuscript was typed with considerable care by Louise Valkenburg, and Andrea Weiss provided excellent photographic assistance—I thank them both.

My wife, Linda, and my children, Michelle Ilyce and Jason Lyle, were very patient once again. Their sacrifices were great during this writing effort. I regret my periodic absence from family life, and I will try to make up for the lost time.

Richard W. Hertzberg

PREFACE TO THE FIRST EDITION

This book discusses macroscopic and microscopic aspects of the mechanical behavior of metals, ceramics, and polymers and emphasizes recent developments in materials science and fracture mechanics. The material is suitable for advanced undergraduate courses in metallurgy and materials, mechanical engineering, and civil engineering where a combined materials-fracture mechanics approach is stressed. The book also will be useful to working engineers who want to learn more about mechanical metallurgy and, particularly, the fracture-mechanics approach to the fracture of solids. I have assumed that readers have had previous training in strength of materials and basic calculus, and that they have been introduced to metallurgical principles including crystal structure.

My objective is to make the reader aware of several viewpoints held by engineers and materials scientists who are active in the field of mechanical metallurgy: the crystal physics approach and the role of dislocations in controlling mechanical properties; the classical metallurgical approach, which stresses the relationship between microstructure and properties; and the fracture mechanics approach which describes the relationship between material toughness, design stress, and allowable flaw size. The application of each viewpoint in the analysis of certain mechanical responses of solids is illustrated, with the hope that the reader will soon recognize which approach might best explain a given set of data or a particular service failure. I think that unless a proper perspective is gained regarding the limits of applicability of the atomistic, microstructural, and continuum viewpoints the reader will become too involved in the fine points of a concept that may prove to be irrelevant to the problem at hand. Indeed, this is vital to a successful failure analysis.

The book is divided into two sections. Section One is devoted to a study of the deformation of solids. Here, emphasis is placed on the role of microstructure, crystallography, and dislocations in explaining material behavior. Section Two, the larger section, deals with the application of fracture mechanics principles to the subject of the fracture in solids. Although familiarity with some topics discussed in Section One will be useful to the reader, the information is not critical to an understanding of Section Two. Therefore, the reader who wishes to focus on the subject of fracture can proceed from the introductory chapter on tensile behavior of solids (Chapter 1) directly to Section Two.

Chapter 1 examines the different macroscopic mechanical responses of metals, ceramics, and polymers in relation to their respective tensile stress-strain response. Chapters 2 to 5 constitute a closely related unit that deals with the deformation of crystalline solids. The elements of dislocation theory, discussed in Chapter 2, are source material for the discussion of slip and structure-property relationships in Chapters 3 to 5. A detailed treatment of the crystallography of twinning is presented in Chapter 4 with an analysis of the cold-worked structure of crystalline solids in terms of mechanical fibering and preferred crystallographic orientations. To acquaint the reader with the multidisciplinary character of time-dependent, high-temperature creep processes in crystalline solids, the topics discussed in Chapter 5 include empirical creep strain relationships with time, temperature, and stress; parametric time-temperature relationships, such as the Larson-Miller parameter used in engineering materials design; and evaluation of creep strain dependence on such material properties as diffusivity, melting point, activation energy, grain size, crystal structure, and elastic modulus. Superplasticity and deformation mechanism maps are also considered. Section One concludes with a discussion of deformation in polymeric materials. Here, again, the mechanical response of these materials is discussed both in terms of their continuum response (as described, for example, with linear viscoelastic relationships and mathematical analogs) and in terms of materials science considerations, involving such topics as the effect of structure on energy damping spectra and the micromechanisms of deformation in amorphous and crystalline polymers.

The subject of fracture is introduced in Section Two by a general overview, ranging from the continuum studies of Leonardo da Vinci in the fifteenth century to current fractographic examinations that employ sophisticated transmission and scanning electron microscopes. The importance of the stress intensity factor and the fracture mechanics approach in analyzing the fracture of solids is developed in Chapter 8 and is compared with the older transition temperature approach to engineering design (Chapter 9). From this macroscopic viewpoint, the emphasis shifts in Chapters 10 and 11 to a consideration of the role of microstructural variables in determining material fracture toughness and embrittlement susceptibility. Both environmental embrittlement (such as stress corrosion cracking and liquid metal and hydrogen embrittlement) and intrinsic material embrittlement (such as temper, irradiation, and 300°C embrittlement) are described. The fatigue of solids is discussed at length in Chapters 12 and 13, and cyclic stress life, cyclic strain life, fatigue crack propagation philosophies, and test data are given. In the final chapter, actual service failures are examined to demonstrate the importance of applying fracture mechanics principles in failure analysis. Several bridge, aircraft, and generator rotor shaft failures are analyzed. In addition, a checklist of information needed to best analyze a service failure is provided for use by the reader. This final chapter can be studied as a unit or as a source for specific case histories that may be considered when a particular point is introduced in an earlier chapter.

A number of scientific colleagues and former students provided valuable assistance in the planning and preparation of the book. Since a complete listing of them would be too lengthy and vulnerable to inadvertent omissions, they cannot be cited individually. I thank those who provided original prints of their previously published photographs that enhance the technical quality of this book. I am grateful to my colleagues at Lehigh University for their many contributions and, most especially, to P. C. Paris, D. A. Thomas, Y. T. Chou, J. A. Manson, M. R. Notis, N. Zettlemoyer, T. Smith, W. J. Mills, S. Siegler, B. Hayes, W. Walthier, and M. Skibo. The considerable care and exactness shown by Mrs. L. Valkenburg in typing the manuscript is deeply appreciated. I also thank the Alcoa Foundation and the Department of Metallurgy and Materials Science at Lehigh University for their financial support during the preparation of this manuscript.

Finally, this volume, which is the most significant project in my teaching career, could not have been attempted or completed without the understanding and patience of my wife, Linda, and my children, Michelle Ilyce and Jason Lyle. Their sacrifices were great; my gratitude is profound.

Richard W. Hertzberg

July 1976

ABOUT THE AUTHOR

Richard W. Hertzberg received his B.S. cum laude in Mechanical Engineering from the City College New York, his M.S. in Metallurgy from M.I.T. and his Ph.D. in Metallurgical Engineering from Lehigh University. A recipient of two Alcoa Foundation Awards of Outstanding Research Achievement, co-recipient of Lehigh University's Award of Outstanding Research, recipient of Lehigh University's College of Engineering Teaching Excellence Award, and co-recipient of Lehigh University's award in Recognition of outstanding contributions to the University, Dr. Hertzberg has served as Research Scientist for the United Aircraft Corporation Research Labs, and Visiting Professor at the Federal Institute of Technology, Lausanne, Switzerland. As an active member of several engineering societies, he has been elected as a Fellow of the American Society for Metals. He has authored approximately 220 scholarly articles, co-authored *Fatigue of Engineering Plastics* (Academic Press, 1980), and completed the fourth edition of *Deformation and Fracture Mechanics of Engineering Materials*. Dr. Hertzberg has also been an invited lecturer in the United States, Asia and Europe, and has served as a consultant to government and industry. Currently, he is New Jersey Zinc Professor of Materials Science and Engineering and Director of the Mechanical Behavior Laboratory of the Materials Research Center at Lehigh University.

CONTENTS

SECTION ONE

DEFORMATION OF ENGINEERING MATERIALS

TENSILE RESPONSE OF MATERIALS

The tensile test is the experimental test method most widely employed to characterize the mechanical properties of materials. From any complete test record, one can obtain important information concerning the material's elastic properties, the character and extent of plastic deformation, yield and tensile strengths, and toughness. That so much information can be obtained from one test justifies its extensive use in engineering materials research. To provide a framework for the varied response to tensile loading in load-bearing materials, several stress–strain plots reflecting different deformation characteristics will be introduced in this chapter.

1.1 DEFINITION OF STRESS AND STRAIN

Before discussing engineering material stress–strain response, it is appropriate to define the terms, stress and strain. This may be done in two generally accepted forms. The first definitions, used extensively in engineering practice, are

$$\sigma_{eng} = \text{engineering stress} = \frac{\text{load}}{\text{initial cross-sectional area}} = \frac{P}{A_0} \qquad (1\text{-}1a)$$

$$\epsilon_{eng} = \text{engineering strain} = \frac{\text{change in length}}{\text{initial length}} = \frac{l_f - l_0}{l_0} \qquad (1\text{-}1b)$$

where l_i = final gage length
l_0 = initial gage length

Alternatively, stress and strain may be defined by

$$\sigma_{true} = \text{true stress} = \frac{\text{load}}{\text{instantaneous cross-sectional area}} = \frac{P}{A_i} \qquad (1\text{-}2a)$$

$$\epsilon_{true} = \text{true strain} = \ln \frac{\text{final length}}{\text{initial length}} = \ln \frac{l_f}{l_0} \qquad (1\text{-}2b)$$

The fundamental distinction concerning the definitions for true stress and strain is recognition of the interrelation between gage length and diameter changes associated with plastic deformation. That is, since plastic deformation is a constant-volume process such that

$$A_1 l_1 = A_2 l_2 = \text{constant} \tag{1-3}$$

any extension of the original gage length would produce a corresponding contraction of the gage diameter. For example, if a 25-mm (1-in.)*-long sample were to extend uniformly by 2.5 mm owing to a tensile load P, the real or *true* stress would have to be higher than that computed by the *engineering* stress formulation. Since $l_2/l_1 = 1.1$, from Eq. 1-3 $A_1/A_2 = 1.1$, so that $A_2 = A_1/1.1$. The *true* stress is then shown to be $\sigma_{\text{true}} = 1.1 P/A_1$ and is larger than the *engineering* value.

By combining Eqs. 1-1a, 1-2a, and 1-3, the relationship between true and engineering stresses is shown to be

$$\sigma_{\text{true}} = \frac{P}{A_0} (l_i/l_0) = \sigma_{\text{eng}}(l_i/l_0) = \sigma_{\text{eng}} (1 + \epsilon_{\text{eng}}) \tag{1-4}$$

By combining Eqs. 1-1b and 1-2b, true and engineering strains may be related by

$$\epsilon_{\text{true}} = \ln(\epsilon_{\text{eng}} + 1) \tag{1-5}$$

The need to define true strain as in Eq. 1-4 stems from the fact that the actual strain at any given time depends on the instantaneous gage length l_i. Consequently, a fixed Δl displacement will result in a decreasing amount of incremental strain, since the gage length at any given time, l_i, will increase with each additional Δl increment. Furthermore, it should be possible to define the strain given to a rod by considering the total change in length of the rod as having taken place in either one step or any number of discrete steps. Stated mathematically, $\Sigma_n \epsilon_n = \epsilon_T$. As a simple example, take the case of a wire drawn in two steps with an intermediate annealing treatment. On the basis of *engineering* strain, the two deformation strains would be $(l_1 - l_0)/l_0$ and $(l_2 - l_1)/l_1$. Adding these two increments does *not* yield a final strain of $(l_2 - l_0)/l_0$. On the other hand, a summation of *true* strains does lead to the correct result. Therefore

$$\ln\frac{l_1}{l_0} + \ln\frac{l_2}{l_1} = \ln\frac{l_2}{l_0} = \epsilon_{\text{true total}}$$

EXAMPLE 1.1

A 25-cm (10-in.)†-long rod with a diameter of 0.25 cm is loaded with a 4500-newton (1012-lb)‡ weight. If the diameter decreases to 0.22 cm, compute the following:

---

* To convert from inches to millimeters, multiply by 25.4.

† To convert from inches to centimeters, multiply by 2.54.

‡ To convert from pounds to newtons, multiply by 4.448.

(a) The final length of the rod:
Since $A_1 l_1 = A_2 l_2$ (from Eq. 1-3)

$$l_2 = \frac{A_1}{A_2} l_1 = \frac{\frac{\pi}{4}(0.25)^2}{\frac{\pi}{4}(0.22)^2}(25)$$

$$l_2 = 32.3 \text{ cm}$$

(b) The true stress and true strain at this load:

$$\sigma_{true} = \frac{P}{A_i}$$

$$= \frac{4500}{(\pi/4)(2.2 \times 10^{-3})^2}$$

$$\sigma_{true} = 1185 \text{ MPa } (172{,}000 \text{ psi})^*$$

$$\epsilon_{true} = \ln \frac{l_f}{l_0}$$

$$= \ln \frac{32.3}{25}$$

$$\epsilon_{true} = 0.256 \text{ or } 25.6\%$$

(c) The engineering stress and strain at this load:

$$\sigma_{eng} = \frac{P}{A_0}$$

$$= \frac{4500}{\frac{\pi}{4}(2.5 \times 10^{-3})^2}$$

$$\sigma_{eng} = 917 \text{ MPa}$$

$$\epsilon_{eng} = \frac{l_f - l_0}{l_0}$$

$$= \frac{32.3 - 25}{25}$$

$$\epsilon_{eng} = 0.292 \text{ or } 29.2\%$$

The use of true strains offers an additional convenience when considering the constant-volume plastic deformation process in that $\epsilon_x + \epsilon_y + \epsilon_z = 0$. In contrast, we find a less convenient relationship, $(1 + \epsilon_x)(1 + \epsilon_y)(1 + \epsilon_z) = 1$, for the case of engineering strains.

---

* To convert from psi to pascals, multiply by 6.895×10^3.

1.2   STRESS–STRAIN CURVES

1.2.1   Elastic Response: Type 1

Over 300 years ago Robert Hooke reported in his classic paper "Of Spring" the following observations:[1]

> Take a wire string of 20 or 30 or 40 feet long and fasten the upper part . . . to a nail, and to the other end fasten a scale to receive the weights. Then with a pair of compasses [measure] the distance [from] the bottom of the scale [to] the ground or floor beneath. Then put . . . weights into the . . . scale and measure the several stretchings of the said string and set them down. Then compare the several stretchings of the . . . string and you will find that they will always bear the same proportions one to the other that the weights do that made them.

This observation may be described mathematically by the equation for an elastic spring:

$$F = kx \tag{1-6}$$

where F = applied force
  x = associated displacement
  k = proportionality factor often referred to as the spring constant

When the force acts on a cross-sectional area A and the displacement x related to some reference gage length l, Eq. 1-6 may be rewritten as

$$\sigma = E\epsilon \tag{1-7}$$

where $\sigma = F/A$ = stress
  $\epsilon = x/l$ = strain
  E = proportionality constant (often referred to as Young's modulus or the modulus of elasticity)

Equation 1-7—called Hooke's law—describes a material condition where stresses and strains are proportional to one another, leading to a stress–strain response shown in Fig. 1.1. A wide range of values of the modulus of elasticity for many materials is shown in Table 1.1. The major reason for these large property variations is related to differences in the strength of the interatomic forces between adjacent atoms or ions. To illustrate this fact, let us consider how the potential energy E between two adjacent particles changes with their distance of separation x (Fig. 1.2a). The equilibrium distance of particle separation x_0, corresponding to a minimum in potential energy, is associated with a balance of the energies of repulsion and attraction between two adjacent atoms or ions. The form of this relationship is given by E $= -\alpha/x^m + \beta/x^n$, where $-\alpha/x^m$ and β/x^n correspond to the energies of attraction and repulsion, respectively, and $n>m$. At x_0, the force ($F = dE/dx$) acting on the particles is equal to zero (Fig. 1.2b). The first derivative of the force with respect to distance of particle

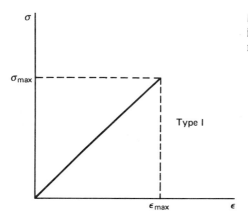

FIGURE 1.1   Type I stress–strain behavior revealing completely elastic material response.

TABLE 1.1a   Elastic Properties of Engineering Materials[a]

Material at 20°C	E (GPa)	G (GPa)	ν
Metals			
Aluminum	70.3	26.1	0.345
Cadmium	49.9	19.2	0.300
Chromium	279.1	115.4	0.210
Copper	129.8	48.3	0.343
Gold	78.0	27.0	0.44
Iron	211.4	81.6	0.293
Magnesium	44.7	17.3	0.291
Nickel	199.5	76.0	0.312
Niobium	104.9	37.5	0.397
Silver	82.7	30.3	0.367
Tantalum	185.7	69.2	0.342
Titanium	115.7	43.8	0.321
Tungsten	411.0	160.6	0.280
Vanadium	127.6	46.7	0.365
Other Materials			
Aluminum oxide (fully dense)	~415	—	—
Diamond	~965	—	—
Glass (heavy flint)	80.1	31.5	0.27
Nylon 66	1.2–2.9	—	—
Polycarbonate	2.4	—	—
Polyethylene (high density)	0.4–1.3	—	—
Poly(methyl methacrylate)	2.4–3.4	—	—
Polypropylene	1.1–1.6	—	—
Polystyrene	2.7–4.2	—	—
Quartz (fused)	73.1	31.2	0.170
Silicon carbide	~470	—	—
Tungsten carbide	534.4	219.0	0.22

[a]G. W. C. Kaye and T. H. Laby, *Tables of Physical and Chemical Constants,* 14th ed., Longman, London, 1973, p. 31.

TABLE 1.1b   Elastic Properties of Engineering Materials[a]

Material at 68°F	E (10^6 psi)	G (10^6 psi)	v
Metals			
Aluminum	10.2	3.8	0.345
Cadmium	7.2	2.8	0.300
Chromium	40.5	16.7	0.210
Copper	18.8	7.0	0.343
Gold	11.3	3.9	0.44
Iron	30.6	11.8	0.293
Magnesium	6.5	2.5	0.291
Nickel	28.9	11.0	0.312
Niobium	15.2	5.4	0.397
Silver	12.0	4.4	0.367
Tantalum	26.9	10.0	0.342
Titanium	16.8	6.35	0.321
Tungsten	59.6	23.3	0.280
Vanadium	18.5	6.8	0.365
Other Materials			
Aluminum oxide (fully dense)	~60	—	—
Diamond	~140	—	—
Glass (heavy flint)	11.6	4.6	0.27
Nylon 66	0.17	—	—
Polycarbonate	0.35	—	—
Polyethylene (high density)	0.058–0.19	—	—
Poly(methyl methacrylate)	0.35–0.49	—	—
Polypropylene	0.16–0.39	—	—
Polystyrene	0.39–0.61	—	—
Quartz (fused)	10.6	4.5	0.170
Silicon carbide	~68	—	—
Tungsten carbide	77.5	31.8	0.22

[a]G. W. C. Kaye and T. H. Laby, *Tables of Physical and Chemical Constants*, 14th ed., Longman, London, 1973, p. 31.

separation, dF/dx (i.e., d^2E/dx^2), then describes the stiffness or relative resistance to separation of the two particles. As such, dF/dx is analogous to the Young's modulus quantity given in Eq. 1-7. A simple analysis of bonding forces shows that the elastic stiffness is proportional to $1/x_0^n$. Examples of the strong dependence of elastic stiffness on x_0 for alkali metals are shown in Fig. 1.2c.

From the above discussion, it follows that values of E for metals and ceramics should decrease with increasing temperature. This is related to the fact that the distance of atom or ion separation increases with temperature (i.e., materials expand when heated). Note the dotted line in Fig. 1.2a, which corresponds to the locus of values of x_0 at temperatures above absolute zero. The loss of material stiffness with increasing temperature is gradual, with only a small percent decrease occurring for a 100°C (180°F) temperature change (Fig. 1.2d). Since E depends on the strength of the interatomic forces that vary with the type of bonding found in a given material, it is relatively insensitive to changes in microstructure. As a result, while heat treatment and minor alloying additions may cause the strength of a steel alloy to change from

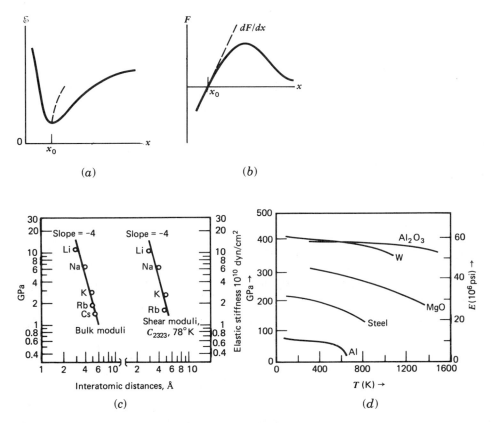

FIGURE 1.2 Dependence of elastic stiffness on interatomic spacing: (*a*) Potential energy versus interatomic spacing; (*b*) Force versus interatomic spacing; (*c*) Elastic stiffness of alkali metals versus interatomic spacing. (From J. J. Gilman, *Micromechanics of Flow in Solids*, McGraw-Hill, New York, 1969, with permission.); (*d*) Variation of Young's modulus with temperature in selected metals and ceramics. (From K. M. Ralls, T. H. Courtney, and J. Wulff, *Introduction to Materials Science and Engineering*, Wiley, 1976, with permission.)

210 to 2400 MPa, the modulus of elasticity of both materials remains relatively unchanged—about 200 to 210 GPa.

It is found that if the loads are removed from the tensile sample before the point of fracture, the corresponding strain will retrace itself along the same linear plot back to zero. The *reversible* nature of strain in this portion of the σ–ε curve is a basic element of elastic strains in any material, whether it is capable of much larger total strain or not. When a material is characterized by such a stress–strain curve and exhibits no plastic deformation, there is great concern for its ability to resist brittle (low energy) premature fracture. This point is treated extensively in Chapters 7–9. Typical materials that behave in this manner include glasses, rocks, many ceramics, heavily cross-linked polymers, and some metals at low temperature. Although these materials are not suitable for engineering applications involving tensile loading, they may be used with considerable success in situations involving compression loads for which the material exhibits much greater resistance to fracture. It is not uncommon to find the compressive strength of a brittle solid to be several times greater than the tensile value. Concrete is an excellent example of an industrial material used extensively in compression but not in tension. When tensile loads are unavoidable, the concrete is reinforced by the addition of steel bars that assume the tensile stresses.

Before proceeding with additional discussion of Hooke's law, it should be noted that elastomers also exhibit elastic, though nonlinear, behavior in nature. A brief discussion of the response of this material is found in Section 1.2.5. Anelastic deformation (time-dependent-reversible strain) is discussed in Section 6.3.

1.2.1.1 Generalized Hooke's Law

Hooke's law can be generalized to account for multiaxial loading conditions as well as material anisotropy. Regarding the former, readers should recall from their studies of the strength of materials that a stress in one direction (say the Y direction) will cause not only a strain in the Y direction but in the X and Z directions as well. Hence

$$\epsilon_{yy} = \frac{\sigma_{yy}}{E} \tag{1-8a}$$

$$\epsilon_{xx} = \epsilon_{zz} = -\frac{\nu\sigma_{yy}}{E} \tag{1-8b}$$

where σ_{yy} = stress acting normal to Y plane and in Y direction

$\epsilon_{xx}, \epsilon_{yy}, \epsilon_{zz}$ = corresponding strains in orthogonal directions

ν = Poisson's ratio $\left(= -\dfrac{\epsilon_{xx}}{\epsilon_{yy}} \right)$

E = modulus of elasticity

From Fig. 1.3, typical normal and shear strain components may be given by

$$\epsilon_{xx} = \frac{\partial u}{\partial x} \tag{1-9a}$$

$$\epsilon_{yy} = \frac{\partial v}{\partial y} \tag{1-9b}$$

$$\gamma_{xy} = \tan\alpha + \tan\beta = \frac{\partial v}{\partial x} + \frac{\partial u}{\partial y} \tag{1-9c}$$

with the other normal and shear strains defined in similar fashion. When multiaxial stresses are applied, the total strain in any given direction is the sum of all strains resulting from each normal and shear stress component. For the case of an isotropic material

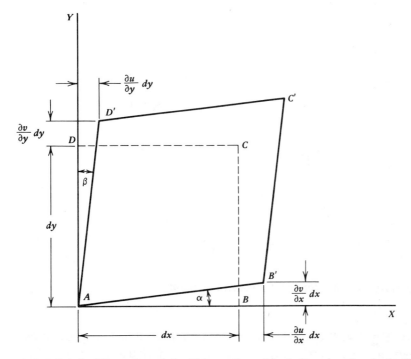

FIGURE 1.3 Distortion of the Z face of a cubical element. The dashed lines indicate the unstrained position of the cube.

$$\epsilon_{xx} = \frac{\sigma_{xx} - v(\sigma_{yy} + \sigma_{zz})}{E}$$

$$\epsilon_{yy} = \frac{\sigma_{yy} - v(\sigma_{xx} + \sigma_{zz})}{E}$$

$$\epsilon_{zz} = \frac{\sigma_{zz} - v(\sigma_{xx} + \sigma_{yy})}{E} \qquad \text{(1-10)}$$

$$\gamma_{xy} = \frac{\tau_{xy}}{G}$$

$$\gamma_{yz} = \frac{\tau_{yz}}{G}$$

$$\gamma_{xz} = \frac{\tau_{xz}}{G}$$

where τ_{ij} = stress acting on I plane and in J direction
G = shear modulus

The situation is complicated greatly when the material is anisotropic wherein the elastic constants vary as a function of crystallographic orientation. Since this is the case for practically all crystalline solids, it is important to consider the general loading

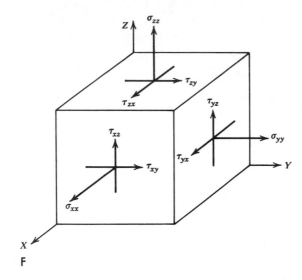

FIGURE 1.4   Stress components acting on a volume element.

conditon shown in Fig. 1.4. We see that there are three normal and six shear stress components. However, since $\tau_{yx} = \tau_{xy}$, $\tau_{yz} = \tau_{zy}$, and $\tau_{xz} = \tau_{zx}$ (so as to avoid rotation of the cube), only six independent stress components remain that determine the strains of the body. The strains in each direction may be given by

$$
\begin{aligned}
\epsilon_{xx} &= s_{11}\sigma_{xx} + s_{12}\sigma_{yy} + s_{13}\sigma_{zz} + s_{14}\tau_{yz} + s_{15}\tau_{zx} + s_{16}\tau_{xy} \\
\epsilon_{yy} &= s_{21}\sigma_{xx} + s_{22}\sigma_{yy} + s_{23}\sigma_{zz} + s_{24}\tau_{yz} + s_{25}\tau_{zx} + s_{26}\tau_{xy} \\
\epsilon_{zz} &= s_{31}\sigma_{xx} + s_{32}\sigma_{yy} + s_{33}\sigma_{zz} + s_{34}\tau_{yz} + s_{35}\tau_{zx} + s_{36}\tau_{xy} \\
\epsilon_{xy} &= s_{41}\sigma_{xx} + s_{42}\sigma_{yy} + s_{43}\sigma_{zz} + s_{44}\tau_{yz} + s_{45}\tau_{zx} + s_{46}\tau_{xy} \\
\epsilon_{yz} &= s_{51}\sigma_{xx} + s_{52}\sigma_{yy} + s_{53}\sigma_{zz} + s_{54}\tau_{yz} + s_{55}\tau_{zx} + s_{56}\tau_{xy} \\
\epsilon_{zx} &= s_{61}\sigma_{xx} + s_{62}\sigma_{yy} + s_{63}\sigma_{zz} + s_{64}\tau_{yz} + s_{65}\tau_{zx} + s_{66}\tau_{xy}
\end{aligned}
\tag{1-11}
$$

where s_{ij} = elastic compliances. Solving for stresses instead, we have

$$
\begin{aligned}
\sigma_{xx} &= c_{11}\epsilon_{xx} + c_{12}\epsilon_{yy} + c_{13}\epsilon_{zz} + c_{14}\gamma_{yz} + c_{15}\gamma_{zx} + c_{16}\gamma_{xy} \\
\sigma_{yy} &= c_{21}\epsilon_{xx} + c_{22}\epsilon_{yy} + c_{23}\epsilon_{zz} + c_{24}\gamma_{yz} + c_{25}\gamma_{zx} + c_{26}\gamma_{xy} \\
\sigma_{zz} &= c_{31}\epsilon_{xx} + c_{32}\epsilon_{yy} + c_{33}\epsilon_{zz} + c_{34}\gamma_{yz} + c_{35}\gamma_{zx} + c_{36}\gamma_{xy} \\
\tau_{yz} &= c_{41}\epsilon_{xx} + c_{42}\epsilon_{yy} + c_{43}\epsilon_{zz} + c_{44}\gamma_{yz} + c_{45}\gamma_{zx} + c_{46}\gamma_{xy} \\
\tau_{zx} &= c_{51}\epsilon_{xx} + c_{52}\epsilon_{yy} + c_{53}\epsilon_{zz} + c_{54}\gamma_{yz} + c_{55}\gamma_{zx} + c_{56}\gamma_{xy} \\
\tau_{xy} &= c_{61}\epsilon_{xx} + c_{62}\epsilon_{yy} + c_{63}\epsilon_{zz} + c_{64}\gamma_{yz} + c_{65}\gamma_{zx} + c_{66}\gamma_{xy}
\end{aligned}
\tag{1-12}
$$

where c_{ij} = elastic stiffnesses.

The reversibility of elastic strains leads to the fact that $s_{ij} = s_{ji}$ and $c_{ij} = c_{ji}$, which reduces the number of independent matrial constants from 36 to 21. As a result of symmetry considerations, the number of independent constants decreases further, with nine constants required to describe the elastic response of an orthorhombic crystal, five for hexagonal, and only three for cubic crystals. For the latter, the elastic compliance matrix reduces to

$$s_{ij} = \begin{matrix} s_{11} & s_{12} & s_{12} & 0 & 0 & 0 \\ s_{12} & s_{11} & s_{12} & 0 & 0 & 0 \\ s_{12} & s_{12} & s_{11} & 0 & 0 & 0 \\ 0 & 0 & 0 & s_{44} & 0 & 0 \\ 0 & 0 & 0 & 0 & s_{44} & 0 \\ 0 & 0 & 0 & 0 & 0 & s_{44} \end{matrix} \tag{1-13}$$

It can be shown for the case of cubic crystals that the modulus of elasticity in any given direction may be given by Eq. 1-14 in terms of these three independent elastic constants and the direction cosines of the crystallographic direction under study:

$$\frac{1}{E} = s_{11} - 2[(s_{11} - s_{12}) - \tfrac{1}{2}s_{44}](l_1^2 l_2^2 + l_2^2 l_3^2 + l_1^2 l_3^2) \tag{1-14}$$

where l_1, l_2, l_3 = direction cosines. Note that the elastic modulus for a given cubic material depends only on the magnitude of the direction cosines, with values for the principal crystallographic directions in the cubic lattice being given in Table 1.2. For example, the modulus in the $\langle 100 \rangle$ direction is given by $1/s_{11}$, since $\Sigma l_i^2 l_j^2 = 0$. By comparison, $\Sigma l_i^2 l_j^2 = \tfrac{1}{3}$ (the maximum value) in the $\langle 111 \rangle$ direction so that $1/E = s_{11} - \tfrac{2}{3}[(s_{11} - s_{12}) - \tfrac{1}{2}s_{44}]$. Depending on whether $(s_{11} - s_{12})$ is larger or smaller than $\tfrac{1}{2}s_{44}$, the modulus of elasticity may be greatest in either the $\langle 111 \rangle$ or $\langle 100 \rangle$ direction (see Problem 1.9). By comparison, the modulus in the $\langle 110 \rangle$ direction is in good agreement with the average value of E for a polycrystalline sample of the same material (see Example 1.2 and Problem 1.10).

The elastic constants for several materials are given in Table 1.3, and their relative elastic anisotropy is tabulated in Table 1.4. Note the large anisotropy exhibited by many of these crystals as compared with the isotropic behavior of tungsten for which $(s_{11} - s_{12}) = \tfrac{1}{2}s_{44}$. Owing to this equality, the modulus of elasticity in tungsten is independent of the direction cosines (Eq. 1-14).

EXAMPLE 1.2

Compute the modulus of elasticity for tungsten and iron in the $\langle 110 \rangle$ direction. From Tables 1.2 and 1.3 we obtain the necessary information regarding elastic compliance

TABLE 1.2   Direction Cosines for Principal Directions in Cubic Lattice

Direction	l_1	l_2	l_3
$\langle 100 \rangle$	1	0	0
$\langle 110 \rangle$	$1/\sqrt{2}$	$1/\sqrt{2}$	0
$\langle 111 \rangle$	$1/\sqrt{3}$	$1/\sqrt{3}$	$1/\sqrt{3}$

TABLE 1.3 Stiffness and Compliance Constants for Selected Crystals[a]

Material	$(10^{10}$ Pa$)$			$(10^{-11}$ Pa$^{-1})$		
	c_{11}	c_{12}	c_{44}	s_{11}	s_{12}	s_{44}
Cubic						
Aluminum	10.82	6.13	2.85	1.57	-0.57	3.51
Copper	16.84	12.14	7.54	1.50	-0.63	1.33
Gold	18.60	15.70	4.20	2.33	-1.07	2.38
Iron	23.70	14.10	11.60	0.80	-0.28	0.86
Lithium fluoride	11.2	4.56	6.32	1.16	-0.34	1.58
Magnesium oxide	29.3	9.2	15.5	0.401	-0.096	0.648
Molybdenum[b]	46.0	17.6	11.0	0.28	-0.08	0.91
Nickel	24.65	14.73	12.47	0.73	-0.27	0.80
Sodium chloride[b]	4.87	1.26	1.27	2.29	-0.47	7.85
Spinel (MgAl$_2$O$_4$)	27.9	15.3	15.3	0.585	-0.208	0.654
Titanium carbide[b]	51.3	10.6	17.8	0.21	-0.036	0.561
Tungsten	50.1	19.8	15.14	0.26	-0.07	0.66
Zinc sulfide	10.79	7.22	4.12	2.0	-0.802	2.43

	c_{11}	c_{12}	c_{13}	c_{33}	c_{44}	s_{11}	s_{12}	s_{13}	s_{33}	s_{44}
Hexagonal										
Cadmium	12.10	4.81	4.42	5.13	1.85	1.23	-0.15	-0.93	3.55	5.40
Cobalt	30.70	16.50	10.30	35.81	7.53	0.47	-0.23	-0.07	0.32	1.32
Magnesium	5.97	2.62	2.17	6.17	1.64	2.20	-0.79	-0.50	1.97	6.10
Titanium	16.0	9.0	6.6	18.1	4.65	0.97	-0.47	-0.18	0.69	2.15
Zinc	16.10	3.42	5.01	6.10	3.83	0.84	0.05	-0.73	2.84	2.61

[a] Data adapted from H. B. Huntington, *Solid State Physics*, Vol. 7, Academic, New York, 1958, p. 213, and K. H. Hellwege, *Elastic, Piezoelectric and Related Constants of Crystals*, Springer-Verlag, Berlin, 1969, p. 3.
[b] Note that $E_{100} > E_{111}$.

values and direction cosines. The modulus of elasticity in the $\langle 110 \rangle$ direction is then determined from Eq. 1-14.

For tungsten

$$\frac{1}{E_{110}} = 0.26 - 2\{[0.26 - (-0.07)] - \tfrac{1}{2}(0.66)\}(\tfrac{1}{4})$$
$$= 0.26 - (0)(\tfrac{1}{4})$$

TABLE 1.4a Elastic Anisotropy of Selected Materials

Metal	Relative Degree of Anisotropy $\left[\dfrac{2(s_{11} - s_{12})}{s_{44}} \right]$	E_{111} (GPa)	E_{100} (GPa)	$\left[\dfrac{E_{111}}{E_{100}} \right]$
Aluminum	1.219	76.1	63.7	1.19
Copper	3.203	191.1	66.7	2.87
Gold	2.857	116.7	42.9	2.72
Iron	2.512	272.7	125.0	2.18
Magnesium oxide	1.534	350.1	249.4	1.404
Spinel (MgAl$_2$O$_4$)	2.425	364.5	170.0	2.133
Titanium carbide	0.877	429.2	476.2	0.901
Tungsten	1.000	384.6	384.6	1.00

TABLE 1.4b   Elastic Anisotropy of Selected Materials

Metal	Relative Degree of Anisotropy $\left[\dfrac{2(s_{11} - s_{12})}{s_{44}} \right]$	E_{111} (10^6 psi)	E_{100} (10^6 psi)	$\left[\dfrac{E_{111}}{E_{100}} \right]$
Aluminum	1.219	11.0	9.2	1.19
Copper	3.203	27.7	9.7	2.87
Gold	2.857	16.9	6.2	2.72
Iron	2.512	39.6	18.1	2.18
Magnesium oxide	1.534	50.8	36.2	1.404
Spinel ($MgAl_2O_4$)	2.425	52.9	24.8	2.133
Titanium carbide	0.877	62.2	69.1	0.901
Tungsten	1.000	55.8	55.8	1.00

Therefore,

$$E_{110} = 384.6 \text{ GPa}$$

which is the same value given in Table 1.4 for E_{111} and E_{100}.

For iron

$$\frac{1}{E_{110}} = 0.80 - 2\{[0.80 - (-0.28)] - \tfrac{1}{2}(0.86)\}(\tfrac{1}{4})$$

$$E_{110} = 210.5 \text{ GPa}$$

Note that $E_{111} > E_{110} > E_{100}$ and that E_{110} is in good agreement with the average value of E for a polycrystalline sample (Table 1.1).

For the case of tungsten and any other isotropic material, then,

$$\begin{aligned}
\epsilon_{xx} &= s_{11}\sigma_{xx} + s_{12}\sigma_{yy} + s_{12}\sigma_{zz} \\
\epsilon_{yy} &= s_{12}\sigma_{xx} + s_{11}\sigma_{yy} + s_{12}\sigma_{zz} \\
\epsilon_{zz} &= s_{12}\sigma_{xx} + s_{12}\sigma_{yy} + s_{11}\sigma_{zz} \\
\gamma_{xy} &= s_{44}\tau_{xy} = 2(s_{11} - s_{12})\tau_{xy} \\
\gamma_{xz} &= s_{44}\tau_{xz} = 2(s_{11} - s_{12})\tau_{xz} \\
\gamma_{yz} &= s_{44}\tau_{yz} = 2(s_{11} - s_{12})\tau_{yz}
\end{aligned} \tag{1-15}$$

If we compare Eqs. 1-15 with 1-10, the elastic constants s_{ij} may be described in terms of the familiar strength of materials elastic constants. Therefore

$$s_{11} = \frac{1}{E} \tag{1-16a}$$

$$s_{12} = -\frac{\nu}{E} \tag{1-16b}$$

$$s_{44} = 2(s_{11} - s_{12}) = \frac{1}{G} \tag{1-16c}$$

Finally, from Eq. 1-16,

$$\frac{1}{G} = 2(s_{11} - s_{12}) = 2\left(\frac{1}{E} + \frac{v}{E}\right) = \frac{2}{E}(1 + v)$$

and

$$G = \frac{E}{2(1 + v)} \tag{1-16d}$$

For the case of hexagonal crystals, the matrix in Eq. 1-11 reduces to

$$\frac{1}{E} = s_{11}(1 - l_3^2)^2 + s_{33}l_3^4 + (2s_{13} + s_{44})l_3^2(1 - l_3^2) \tag{1-17}$$

where l_1, l_2, l_3 are direction cosines for directions in the hexagonal unit cell. From Eq. 1-17 note that in hexagonal crystals E depends only on the direction cosine l_3, which lies normal to the basal plane. Consequently, the modulus of elasticity in hexagonal crystals is isotropic everywhere in the basal plane.

1.2.1.2 Resiliency

The resilience of a material is a measure of the amount of energy that can be absorbed under elastic loading conditions and which is released completely when the loads are removed. From this definition, resilience may be measured from the area under the curve in Fig. 1.1:

$$\text{resilience} = \tfrac{1}{2}\sigma_{max}\epsilon_{max} \tag{1-18}$$

where σ_{max} = maximum stress for elastic conditions
$\quad\quad\ \epsilon_{max}$ = elastic strain limit

From Eq. 1-7

$$\text{resilience} = \frac{\sigma^2_{max}}{2E} \tag{1-19}$$

Should an engineering design require a material that allows only for elastic response with large energy absorption (such as in the case of a mechanical spring), the appropriate material to choose would be one possessing a high yield strength but low modulus of elasticity.

1.2.2 Elastic–Homogeneous Plastic Response: Type II

When a material has the capacity for plastic deformation—irreversible flow—the stress–strain curve often assumes the shape of Curve II (Fig. 1.5). Here we see the same elastic region at small strains but now find a smooth parabolic portion of the curve, which is associated with homogeneous plastic deformation processes, such as

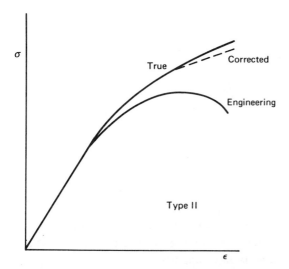

FIGURE 1.5 Type II stress–strain behavior revealing elastic behavior followed by a region of homogeneous plastic deformation. Data are plotted on the basis of engineering and true stress–strain definitions.

the irreversible movement of dislocations in metals, ceramics, and crystalline polymers, and a number of possible deformation mechanisms in amorphous polymers. That the curve continues to rise to a maximum stress level reflects an increasing resistance on the part of the material to further plastic deformation—a process known as strain hardening. The portion of the true stress–strain curve (from the onset of yielding to the maximum load) may be described empirically by the relationship generally attributed to Hollomon:[2]

$$\sigma = K\epsilon^n \qquad (1\text{-}20)$$

where σ = true stress
  ϵ = true plastic strain
  n = strain-hardening coefficient
  K = material constant, defined as the true stress at a true strain of 1.0

However, Bülfinger[3] proposed a similar parabolic relationship between stress and strain almost 200 years earlier. The magnitude of the strain-hardening coefficient reflects the ability of the material to resist further deformation. In the limit, n may be equal to unity, which represents ideally elastic behavior, or equal to zero, which represents an ideally plastic material. Selected values of strain-hardening coefficients for some engineering metal alloys are given in Table 1.5. (Note that n values are sensitive to thermomechanical treatment; they are generally larger for materials in the annealed condition and smaller in the cold-worked state.)

Such data may be derived by plotting true stress and associated true strain values on log-log paper. If Eq. 1-20 was absolutely correct, a straight line should result with a slope equal to n. However, this is found not always to be the case and reflects the

TABLE 1.5   Selected Strain-Hardening
Coefficients

Material	Strain-Hardening Coefficient, n
Stainless steel	0.45–0.55
Brass	0.35–0.4
Copper	0.3–0.35
Aluminum	0.15–0.25
Iron	0.05–0.15

fact that this relationship is only an empirical approximation.[4] (When a nonlinear log–log plot does result for a given material, the strain-hardening coefficient is often defined at a particular strain value.) In general, n increases with decreasing strength level and with decreasing mobility of certain dislocations in the crystalline lattice. More will be said about this material property in Chapter 3.

1.2.2.1   Strength Levels in Materials

It has become common practice to define several strength levels that characterize the material's tensile response. The *proportional limit* is that stress level below which stress is proportional to strain according to Eq. 1-7. The *elastic limit* defines that stress level below which the deformation strains are fully reversible. In most engineering materials, these two quantities are essentially equal. However, it is possible for a metal to exhibit nonlinear but elastic behavior. For example, very high-strength filamentary particles—often called whiskers—can exhibit elastic strains in excess of 2%. In this range of very large elastic strains, the modulus of elasticity reveals its weak dependence on strain—something that is completely obscured when strains are very small. For example, note that dF/dx in Fig. 1.2b decreases with increasing distance of particle separation. As such, Hooke's law (Eq. 1-7) represents an empirical relationship, albeit a good one at small strains. Consequently, the elastic limit is approximately equal to the proportional limit and may be slightly higher in some cases.

A much more important material property is the *yield strength*—a stress level related to the onset of irreversible plastic deformation. This quantity is difficult to define, since the point where plastic flow begins will depend on the sensitivity of the displacement transducer. The more sensitive the gage, the lower the stress level where plastic flow is found. In recent years, special capacitance strain gages have been used to measure strains in the range of 10^{-6}. In fact, a number of studies[5] dealing with the mechanical behavior of materials in the microstrain region have been undertaken as a result of this breakthrough in instrumentation. These investigations have shown, for example, that plastic deformation—the irreversible movement of dislocations—occurs at stress levels many times lower than the conventionally determined engineering yield strength. To arrive at a uniformly accepted method for determination of the yield strength, therefore, a standard test procedure[6] (ASTM Standard E 8-69) has been adopted. The yield-strength value is obtained in the following manner: (1) Determine the engineering stress–strain curve, such as the one shown in Fig. 1.6; (2) construct a line parallel to the elastic portion of the σ–ε curve but offset from the origin by a certain amount (the generally accepted offset is 0.002 or 0.2% strain); and (3) define

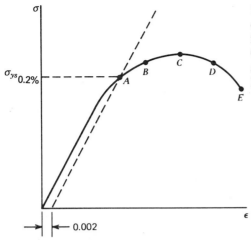

FIGURE 1.6 Engineering stress–strain curve. Tensile yield strength is defined at intersection of stress–strain curve and 0.2% offset line. Points *A*, *B*, *C*, *D*, and *E* are the arbitrary stress levels discussed in Fig. 1.7.

the yield strength at the intersection of the σ–ϵ curve and the offset line. As cited above, this value is usually defined as the 0.2% offset yield strength. The *ultimate tensile strength* is defined as the maximum load divided by the initial cross-sectional area, while the *true fracture stress* is the load at fracture divided by the final cross-sectional area with correction made for any localized deformation (necking) in the final fracture region (see Section 1.2.2.2). A compilation of tensile properties for a number of engineering materials is given in Tables 1.6 and 1.7.*

1.2.2.2 Plastic Instability and Necking

The true and engineering stress–strain plots from a tensile test reveal basic differences, as shown in Fig. 1.5. While the engineering curve reaches a maximum at maximum load and decreases thereafter to fracture, the true curve rises continually to failure. The inflection in the engineering curve is due to the onset of localized plastic flow and the manner in which engineering stress is defined. To understand this, consider for a moment the following sequence of events. When the stress reaches a critical level, plastic deformation will occur at the weakest part of the test sample, somewhere along the gage length. This local extension under tensile loading will cause a simultaneous area constriction so that the true local stress is higher at this location than anywhere else along the gage length. Consequently, all additional deformation would be expected to concentrate in this most highly stressed region. Such is the case in an ideally plastic material. For all other materials, however, this localized plastic deformation strain hardens the material, thereby making it more resistant to further damage. At this point, the applied stress must be increased to produce additional plastic deformation at the second weakest position along the gage length. Here again the material strain hardens and the process continues. On a macroscopic scale, the gage length extends uniformly in concert with a uniform reduction in cross-sectional area.

---

*Computerized data bases for material properties are presently being standardized [(e.g., see J. G. Kaufman, *Stand. News ASTM* **15**(3), 38 (1987)].

TABLE 1.6a   Tensile Properties for Selected Engineering Materials[a]

Material	Treatment	Yield Strength (MPa)	Tensile Strength (MPa)	Elongation in 5-cm Gage (%)	Reduction in Area (1.28-cm diameter) (%)
Steel Alloys					
1015	As-rolled	315	420	39	61
1050	"	415	725	20	40
1080	"	585	965	12	17
1340	Q + T (205°C)	1590	1810	11	35
1340	" (425°C)	1150	1260	14	51
1340	" (650°C)	620	800	22	66
4340	" (205°C)	1675	1875	10	38
4340	" (425°C)	1365	1470	10	44
4340	" (650°C)	855	965	19	60
301	Annealed plate	275	725	55	—
304	" "	240	565	60	—
310	" "	310	655	50	—
316	" "	250	565	55	—
403	Annealed bar	275	515	35	—
410	" "	275	515	35	—
431	" "	655	860	20	—
AFC-77	Variable	560–1605	835–2140	10–26	32–74
PH 15-7Mo	"	380–1450	895–1515	2–35	—
Titanium Alloys					
Ti-5Al-2.5Sn	Annealed	805	860	16	40
Ti-8Al-1Mo-1V	Duplex annealed	950	1000	15	28
Ti-6Al-4V	Annealed	925	995	14	30
Ti-13V-11Cr-3Al	Solution + age	1205	1275	8	—
Magnesium Alloys					
AZ31B	Annealed	103–125	220	9–12	—
AZ80A	Extruded bar	185–195	290–295	4–9	—
ZK60A	Artificially aged	215–260	295–315	4–6	—
Aluminum Alloys					
2219	-T31, -T351	250	360	17	—
2024	-T3	345	485	18	—
2024	-T6, -T651	395	475	10	—
2014	-T6, -T651	415	485	13	—
6061	-T4, -T451	145	240	23	—
7049	-T73	475	530	11	—
7075	-T6	505	570	11	—
7075	-T73	415	505	11	—
7178	-T6	540	605	11	—
Plastics					
ABS	Medium impact	—	46	6–14	—
Acetal	Homopolymer	—	69	25–75	—
Poly(tetra-fluorethylene)	—	—	14–48	100–450	—
Poly(vinylidene fluoride)	—	—	35–48	100–300	—
Nylon 66	—	—	59–83	60–300	—
Polycarbonate	—	—	55–69	130	—
Polyethylene	Low density	—	7–21	50–800	—
Polystyrene	—	—	41–54	1.5–2.4	—
Polysulfone	—	69	—	50–1000	—

[a] *Datebook 1974, Metal Progress* (mid-June 1974).

TABLE 1.6b   Tensile Properties for Selected Engineering Materials[a]

Material	Treatment	Yield Strength (ksi)	Tensile Strength (ksi)	Elongation in 2-in. Gage (%)	Reduction in Area (0.505-in. diameter) (%)
Steel Alloys					
1015	As rolled	46	61	39	61
1050	"	60	105	20	40
1080	"	85	140	12	17
1340	Q + T (400°F)	230	260	11	35
1340	" (800°F)	167	183	14	51
1340	" (1200°F)	90	116	22	66
4340	" (400°F)	243	272	10	38
4340	" (800°F)	198	213	10	44
4340	" (1200°F)	124	140	19	60
301	Annealed plate	40	105	55	—
304	" "	35	82	60	—
310	" "	45	95	50	—
316	" "	36	82	55	—
403	Annealed bar	40	75	35	—
410	" "	40	75	35	—
431	" "	95	125	20	—
AFC-77	Variable	81–233	121–310	10–26	32–74
PH 15-7Mo	"	55–210	130–220	2–35	—
Titanium Alloys					
Ti-5Al-2.5Sn	Annealed	117	125	16	40
Ti-8Al-1Mo-1V	Duplex annealed	138	145	15	28
Ti-6Al-4V	Annealed	134	144	14	30
Ti-13V-11Cr-3Al	Solution + age	175	185	8	—
Magnesium Alloys					
AZ31B	Annealed	15–18	32	9–12	—
AZ80A	Extruded bar	27–28	42–43	4–9	—
ZK60A	Artificially aged	31–38	43–46	4–6	—
Aluminum Alloys					
2219	-T31, -T351	36	52	17	—
2024	-T3	50	70	18	—
2024	-T6, -T651	57	69	10	—
2014	-T6, -T651	60	70	13	—
6061	-T4, -T451	21	35	23	—
7049	-T73	69	77	11	—
7075	-T6	73	83	11	—
7075	-T73	60	73	11	—
7178	-T6	78	88	11	—
Plastics					
ABS	Medium impact	—	6.8	6–14	—
Acetal	Homopolymer	—	10	25–75	—
Poly(tetra-fluorethylene)	—	—	2–7	100–450	—
Poly(vinylidene fluoride)	—	—	5.1–7	100–300	—
Nylon 66	—	—	8.6–12	60–300	—
Polycarbonate	—	—	8–10	130	—
Polyethylene	Low density	—	1–3	50–800	—
Polystyrene	—	—	6–9	1.5–2.4	—
Polysulfone	—	10	—	50–100	—

[a] *Databook 1974, Metal Progress* (mid-June 1974).

TABLE 1.7   Strength Properties of Selected Ceramics[a]

Material	Compressive Strength [MPa (ksi)]	Tensile Strength [MPa (ksi)]	Flexural Strength [MPa (ksi)]	Modulus of Elasticity [GPa (10^6 psi)]
Alumina (85% dense)	1620 (235)	125 (18)	295 (42.5)	220 (32)
Alumina (99.8% dense)	2760 (400)	205 (30)	345 (60)	385 (56)
Alumina silicate	275 (40)	17 (2.5)	62 (9)	55 (8)
Transformation toughened zirconia	1760 (255)	350 (51)	635 (92)	200 (29)
Partially stabilized zirconia +9% MgO	1860 (270)	—	690 (100)	205 (30)
Cast Si_3N_4	138 (20)	24 (3.5)	69 (10)	115 (17)
Hot-pressed Si_3N_4	3450 (500)	—	860 (125)	—

[a] *Guide to Engineering Materials,* Vol. 1(1), ASM, Metals Park, OH, 1986, pp. 16, 64, 65.

(Recall that plastic deformation is a constant-volume process.) With increasing load, a point is reached where the strain-hardening capacity of the material is exhausted and the nth local area contraction is no longer balanced by a corresponding increase in material strength. At this maximum load, further plastic deformation is localized in the necked region, since the stress increases continually with areal contraction even though the applied load is decreasing as a result of elastic unloading in the test bar outside the necked area. Eventually the neck will fail. Since engineering stress is based on A_0, the decreasing load on the sample after the neck has formed will result in the computation of a decreasing stress. By comparison, the decreasing load value is more than offset by the decrease in instantaneous cross-sectional area such that the true stress continues to rise to failure even after the onset of necking.

1.2.2.3   Strain Distribution in Tensile Specimen

The total strain distribution along the specimen gage length is shown schematically in Fig. 1.7 for various stress levels as indicated on the engineering stress–strain curve (Fig. 1.6). Owing to the variation of elongation along the gage length of the tensile specimen, researchers occasionally report both the total strain, $(l_f - l_0)/l_0$ or $\ln(l_f/l_0)$, and the uniform strain, which is related to the elongation just prior to local necking (Curve C in Fig. 1.7). It should be emphasized that the total strain reported for a given test result will depend on the gage length of the test bar. From Fig. 1.7, it is clear that as the gage length decreases, the elongation involved in the necking process becomes increasingly more dominant. Consequently, total strain values will increase the shorter the gage length. For this reason both specimen size and total strain data should be reported. ASTM has standardized specimen dimensions to minimize variability in test data resulting from such geometrical considerations. As noted in Table 1.8, the gage length to diameter ratio is standardized to a value of about 4.

1.2.2.4   Extent of Uniform Strain

From the standpoint of material usage in an engineering component, it is desirable to maximize the extent of uniform elongation prior to the onset of localized necking. It may be shown that the amount of uniform strain is related to the magnitude of the strain-hardening exponent.

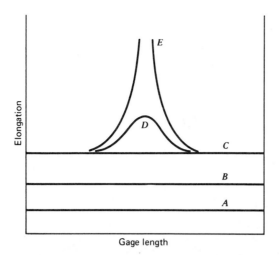

FIGURE 1.7 **Schematic representation of specimen elongation along the gage length. Uniform extension occurs up to the onset of necking (C). Additional displacements are localized in necked region (D and E). Stress levels A, B, C, D, and E are identified in Fig. 1.6.**

$$P = \sigma A \qquad\qquad (1\text{-}21)$$
$$dP = \sigma dA + A d\sigma$$

Recalling that necking occurs at maximum load

$$dP = 0$$

so that

$$\frac{d\sigma}{\sigma} = -\frac{dA}{A}$$

TABLE 1.8   Round Tension Test Specimen Dimensions[6]

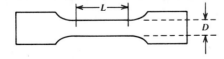

Diameter (D)		Gage Length (L)	
mm	(in.)	mm	(in.)
12.5	(0.5)	50	(2.0)
8.75	(0.345)	35	(1.375)
6.25	(0.25)	25	(1.0)
4.0	(0.16)	16	(0.63)
2.50	(0.1)	10	(0.394)

From Eq. 1-5

$$Adl + ldA = 0$$
$$-\frac{dA}{A} = \frac{dl}{l}$$

Since $dl/l \equiv d\epsilon$, we find

$$\sigma = \frac{d\sigma}{d\epsilon} \tag{1-22}$$

By using the Hollomon relation (Eq. 1-20)

$$K\epsilon^n = Kn\epsilon^{n-1}$$

Therefore

$$n = \epsilon \tag{1-23}$$

The true plastic strain at necking instability is numerically equal to the strain-hardening coefficient.

In addition to the necking strains, a triaxial stress state exists in the vicinity of the neck (Fig. 1.8). The radial (σ_r) and transverse (σ_t) stresses that are induced are developed as a result of a Poisson effect. In effect, the more highly stressed material within the neck wishes to pull in to accommodate the large local extensions (Fig. 1.7). Since the material immediately adjacent to the necked area experiences a much lower stress level, these regions will resist such contractions by exerting induced tensile stresses that act to retard deepening of the neck. Consequently, the triaxial stress field acts to plastically constrain the material from deforming in the reduced area. To provide for such plastic flow, the axial stress must be increased. The stress values recorded on the true stress–strain curve (Fig. 1.5) after the onset of necking reflect the higher axial stresses necessitated by the triaxial stress condition. In terms of the

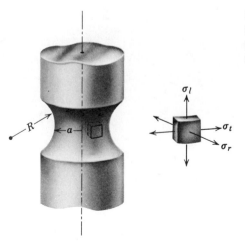

FIGURE 1.8 Triaxial tension stress distribution within necked region which acts to constrain additional deformation in the neck.

radius of curvature of the neck contour R and the radius of the minimum cross-sectional area a, Bridgman was able to correct the applied axial stress (σ_{app}) to determine the true stress (σ_{true}) that would be necessary to deform the material were it not for the presence of the neck. The corrected true stress–strain curve shown in Fig. 1.5 may be determined from the Bridgman[7] relation

$$\frac{\sigma_{true}}{\sigma_{app}} = \frac{1}{(1 + 2R/a)[\ln(1 + a/2R)]} \quad (1\text{-}24)$$

It is seen from this formula that the stress necessary to produce a given level of plastic deformation will increase with increasing notch root acuity for a given notch depth.

At some critical point, the triaxial tensile stress condition within the necked region causes small particles within the microstructure to either fracture or separate from the matrix. The resulting microvoids then undergo a period of growth and eventual coalescence, producing an internal, disk-shaped crack oriented normal to the applied stress axis. Final fracture then occurs by a shearing-off process along a conical surface oriented 45° to the stress axis. This entire process produces the classical cup–cone fracture surface appearance shown in Fig. 1.9. Sometimes the circular region in the middle of the sample (called the fibrous zone) is generated entirely by slow, stable crack growth, while the smooth shear walls are formed at final failure. Usually the fibrous zone contains a series of circumferential ridges reflecting slight undulations in the stable crack propagation direction. However, test conditions can be altered to suppress the extent of the slow, stable crack growth region; instead, the crack continues to grow on the same plane but in unstable fashion at a much faster rate. This new region, defined as the radial zone, contains radial markings (Fig. 1.10) often associated with the fracture of oriented inclusions in test bars prepared from rod stock. (More will be said of this fracture detail in Chapter 14.) The relative amount of fibrous,

FIGURE 1.9 Typical cup–cone fracture appearance of unnotched tensile bar: (*a*) cup portion; (*b*) cone portion. (Courtesy of Richard Sopko, Lehigh University.)

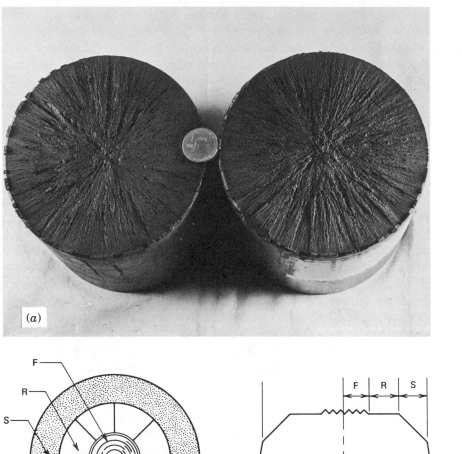

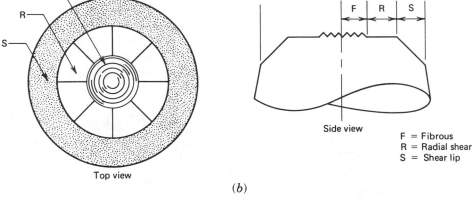

FIGURE 1.10 **Extent of fibrous, radial, and shear lip zones:** (*a*) **macrofractograph (courtesy of Richard Sopko, Lehigh University);** (*b*) **schema showing zone location. (After Larson and Carr[8]; reprinted by permission of the American Society for Metals, Metals Park, OH.)**

radial, and shear lip fracture zones has been found to depend on the strength of the material and the test temperature[8] (Fig. 1.11). Since the internal fracture process depends on plastic constraint resulting from the tensile triaxiality within the neck, the crack nucleation process could be suppressed by introducing hydrostatic pressure. Indeed, Bridgman[9] demonstrated that when a sufficiently large hydrostatic pressure is applied, necking can proceed uninterrupted almost to where the sample draws down to a point (Fig. 1.12).

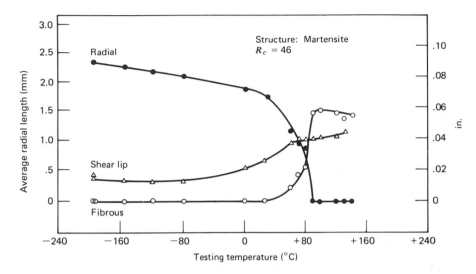

FIGURE 1.11 Effect of test temperature on relative size of fracture zones for AISI 4340 steel heat treated to R_c 46. (After Larson and Carr[8]; reprinted by permission of the American Society for Metals, Metals Park, OH.)

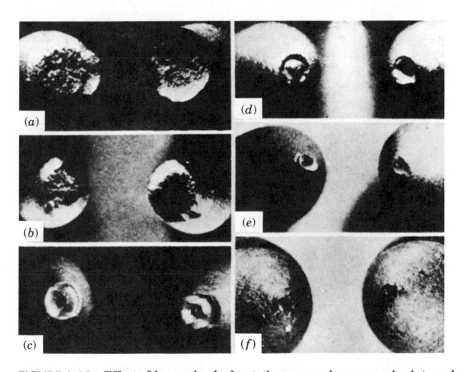

FIGURE 1.12 Effect of increasing hydrostatic pressure in suppressing internal void formation within necked region. (*a*) Atmospheric pressure, 10×; (*b*) 235-MPa hydrostatic pressure, 10×; (*c*) 1000 MPa, 12×; (*d*) 1290 MPa, 12×; (*e*) 1850 MPa, 12×; and (*f*) 2680 MPa, 18×. (After Bridgman[9]; reprinted by permission of the American Society for Metals, Metals Park, OH.)

1.2.2.5 Toughness

Another important material characteristic is its resistance to fracture (measured in units of energy). We may define a brittle material as one absorbing little energy, while a tough material would require a large expenditure of energy in the fracture process. For an unnotched tensile bar, the energy to break may be estimated from the area under the stress–strain curve.

$$\text{energy/volume} = \int_0^{\epsilon_f} \sigma d\epsilon \tag{1-25}$$

Maximum toughness, therefore, is achieved with an optimum combination of strength and ductility; neither high strength (e.g., glass) nor exceptional ductility (e.g., taffy) alone provides for large fracture energy absorption (Fig. 1.13). Material toughness will be considered in much greater detail in Chapters 7–11.

1.2.3 Elastic–Heterogeneous Plastic Response: Type III

Occasionally, a test specimen will produce a stress–strain curve that exhibits a series of serrations that are superimposed on the parabolic portion of Fig. 1.5 after the normal range of elastic response. Such behavior, shown in Fig. 1.14, reflects non-uniform or heterogeneous deformation within the material. Serrated stress–strain response is known to occur under at least two different conditions. When hexagonal close-packed metals are tested over a relatively wide temperature range, they tend to deform plastically by a combination of slip along glide planes and twinning in discrete zones within the specimen. (The nature of plastic deformation in metals is discussed in greater detail in Chapters 2–4.) When twinning occurs, extension of the gage length proceeds in discrete bursts that are associated with twin band nucleation and growth. Often, these bursts of deformation are associated with audible clicks emitted from within the sample. Whenever the instantaneous strain rate in the specimen exceeds the rate of motion of the test machine crosshead, a load drop will occur. A similar stress–strain response is found in body-centered-cubic metals tested at low tempera-

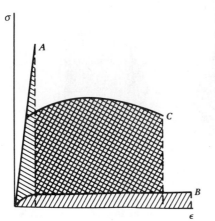

FIGURE 1.13 Stress–strain curves for strong material with little plastic flow capacity, A; low-strength but high-ductility material, B; and material with optimum combination of strength and ductility for maximum toughness, C.

Alternative login procedures and
help with logging in
Register for Athens

Current Search Results

Start a new search

psychology

⦿ Title ⦾ Title and Text

SEARCH

Advanced Search | Help

Empin...
psycho
Author(s
Publicati
Publicati
Source: ▮

Empiric
psycho
Author(s
Publicat
Publicat
Source:

Empiric
sympto
and car
Author(s
Publicat
Publicat
Source:

Empiric
psych
childre
Author(
Publica
Publica
Source

http://search.library.nhs.uk/nhs_sse/zengine?VD

658.5

620.084P

Admes in CE

CPD.

620·112 6

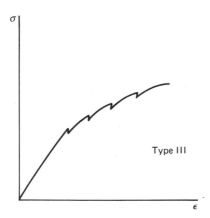

FIGURE 1.14 Type III stress–strain behavior reflecting elastic behavior followed by heterogeneous plastic flow. The latter can be caused by twin controlled deformation or solute atom–dislocation interactions.

tures and in face-centered-cubic metals tested at both low temperatures and high strain rates.

Serrated stress–strain curves are also encountered in body-centered-cubic iron alloys containing carbon in solid solution and in dilute solid solutions of aluminum. It has been argued that the Portevin-Le Chatelier effect (serrated σ–ϵ curve) is due to solute atom or vacancy interactions with lattice dislocations[10] (see Section 4.3.1). When a sufficiently large stress is applied, dislocations can break free from solute clusters and cause a load drop. If the solute atoms can diffuse quickly enough to retrap these dislocations, then more load must be applied once again to continue the deformation process.

1.2.4 Elastic–Heterogeneous Plastic–Homogeneous Plastic Response: Type IV

In many body-centered-cubic iron based alloys and some nonferrous alloys, a relatively narrow region of heterogeneous plastic deformation (with a range of approximately 1 to 3% strain) separates the elastic region from the homogeneous plastic flow portion of the stress–strain curve (Fig. 1.15). This segment of the curve is caused by

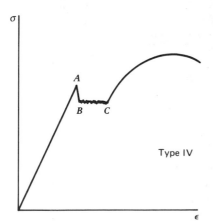

FIGURE 1.15 Type IV stress–strain behavior exhibiting a narrow heterogeneous deformation region between initial elastic and final homogeneous flow regions. Onset of local yielding occurs at upper yield point *A* with corresponding load drop to *B* defined as the lower yield point. After passage of Lüder band throughout the gage section, homogeneous deformation commences at *C*.

a dislocation–solute atom interaction related to that mentioned in the previous section. After being loaded elastically to A, defined as the upper yield point, the material is observed to develop a local deformation band (Fig. 1.16); the sudden onset of plastic deformation associated with this Lüder band is responsible for the initial load drop to B, defined as the lower yield point. Since the upper yield point is very sensitive to minor stress concentrations, alignment of the specimen in the test grips, and other related factors, measured values reflect considerable scatter. For this reason, the yield strength of materials exhibiting Type IV behavior (Fig. 1.15) is usually reported as the lower yield-point value. The remainder of the heterogeneous segment of deformation is consumed in the passage of the Lüder band across the entire gage section. (Occasionally more than one band may propagate simultaneously during this period.) When deformation has spread to all parts of the gage length, the material then continues to deform in a homogeneous manner, similar to that described for Type II behavior. Yield points are also found in ionic and covalent materials and are discussed more fully in Section 4.3.1.

1.2.5 Elastic–Heterogeneous Plastic–Homogeneous Plastic Response: Type V

Type V behavior may be found in the deformation of some crystalline polymers. While possessing an upper yield point and subsequent load drop (Fig. 1.17a) similar

(a)

(b)

FIGURE 1.16 (a) Concentrated deformation (Lüder) bands formed in plain carbon steel test sample. The band will grow across the gage section before homogeneous deformation develops at point C from Fig. 1.15. (b) Lüder band development from weld-related residual tensile stresses. (Photo courtesy P. Keating.)

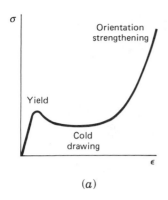
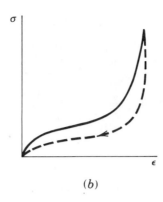

(a)   (b)

FIGURE 1.17   (*a*) Type V stress–strain behavior usually found in crystalline polymers. Minimum in curve reflects cold drawing and competition between breakdown of initial structure and its subsequent reorganization into a highly oriented strong material. (*b*) Rubber elasticity possesses no load drop. Unloading curve returns to origin.

to that observed in Type IV response, the final deformation stage is decidedly different. Meinel and Peterlin[11] have rationalized the shape of the Type V curve as reflecting the competition of two events. Initially, yielding occurs along with a breakdown of the original crystalline structure within the polymer; this produces the initial load drop and a general whitening of the sample in the region of greatest deformation. If failure does not occur soon after the necking process begins, continuing strains will lead to a reorganization of the broken-down structure into new, highly oriented, and strong units. As more and more of this new structure is produced, the polymer offers more resistance to deformation—it strain hardens—and the stress–strain curve begins to rise once again. Strain hardening continues by molecular alignment (sometimes called cold drawing) up to the fracture stress. As the cold-drawn regions become more highly oriented, the milky white regions associated with initial crystallite breakdown gradually become clear once again (Fig. 1.18). Deformation of crystalline polymers is discussed further in Chapter 6.

1.2.6   Rubber Elasticity

The tensile response of elastomers bears some resemblance to Type V behavior, though differences are to be noted. Elastomers do not exhibit a load drop at intermediate strains (the slope of the stress–strain curve is always positive for these materials). In addition, most strains in elastomers are fully reversible, though nonlinear. Hence, no permanent deformation would remain after the load was removed (Fig. 1.17*b*). Overall, elastomer or rubber elasticity is distinguished by two basic characteristics: very large nonlinear elastic strains (often in excess of 100%) and elastic moduli that increase with increasing temperature. The latter response is opposite that found in other materials (including rigid polymers). Rubber elasticity is related primarily to the straightening of amorphous polymer chains from their curled positions into partially extended conformations. As a result, the extent of elastic deformation is great, while the elastic moduli are very low because of the small contribution of

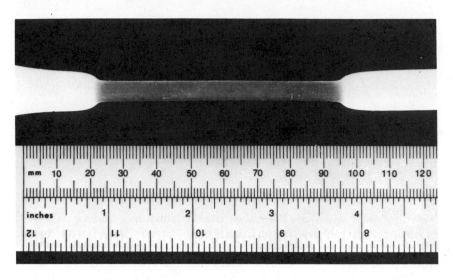

FIGURE 1.18 Cold drawing in polypropylene, which produces greater optical transparency in gage section as a result of enhanced molecular alignment.

actual polymer chain stretching. That is, a curled chain of length l is extended so that its end-to-end length approaches l with little additional chain lengthening attributed to the more difficult covalent bond extension mode. The straightening of the chains is responsible for the rapidly increasing apparent hardening of the material at large strains (Fig. 1.17). When the applied loads are relaxed, the chains return to a curled position, indicating the latter conformation to be preferred.

By simple application of the first and second laws of thermodynamics, it is possible to demonstrate that the elastic modulus of rubber should increase with increasing temperature. From the first law of thermodynamics

$$dU = \partial Q + \partial W \tag{1-26}$$

where dU = change in internal energy
∂Q = change in heat absorbed or released
∂W = work done on the system

For a reversible process, the second law of thermodynamics gives

$$dQ = TdS \tag{1-27}$$

where T = temperature
dS = change in entropy

If an elastomeric rod of length l is extended by an amount dl owing to a tensile force F, the work ∂W done on the rod is Fdl. Combining Eqs. 1-26 and 1-27 with the expression for ∂W gives

$$dU = TdS + Fdl \tag{1-28}$$

At constant temperature

$$F = \left(\frac{\partial U}{\partial l}\right)_T - T\left(\frac{\partial S}{\partial l}\right)_T \tag{1-29}$$

where $\left(\dfrac{\partial U}{\partial l}\right)_T = $ related to the strain energy associated with the application of a load

$\left(\dfrac{\partial S}{\partial l}\right)_T = $ related to the change in entropy or order of the rod as it is stretched

Since the chains prefer a random curled configuration, their initial degree of order is low and their entropy high. (Because of the very high degree of order of atoms in metals and ceramics, their entropy term by comparison is negligible.) However, when a tensile load is applied, the entropy decreases as the chains become straightened and aligned. As a consequence, $(\partial S/\partial l)_T$ is negative. The force required to extend the elastomer rod, therefore, increases with increasing temperature.

Other fundamental studies have shown that the shear modulus of rubber also increases with temperature in that

$$G = NkT \tag{1-30}$$

where N is the density of network crosslinks and k is Boltzmann's constant. As expected, rubber stiffness increases with increasing crosslink density and corresponding decrease in the molecular weight of chain segments between crosslinks (Mc). Regarding the latter, the shear modulus of rubber is found to vary inversely with Mc. Interestingly, rubbers are distinguished in that their elastic modulii can be predicted from molecular structural details.

1.2.7   Composite Materials Tensile Response*

Referring again to Fig. 1.13, we see that strong but brittle materials (A) can withstand large stresses prior to failure but possess limited ductility; soft, ductile materials (C) exhibit considerable plastic flow but little load-bearing capacity. Though only a relatively few materials exhibit both exceptional strength and ductility, a growing number of hybrid or composite materials have been developed to utilize the respective superior properties of the constituents of the composite material. For example, certain engineering plastics that possess considerable ductility are being reinforced with high-strength glass and aramid fibers to produce composite materials that possess both high strength and adequate ductility (see Table 1.9); such materials are challenging metal alloys for use in numerous components. Presently, manufacturers are making major substitutions of polymeric-based matrix composites in such items as automotive body frames and hood and door panels, aircraft wings, boat hulls, and sporting equipment. In parallel fashion, metal–matrix composites, reinforced with silicon carbide, silicon nitride, and/or alumina fibers, are being considered for use in fossil-fuel engine components such as turbochargers.

---

* See end of chapter for selected references.

TABLE 1.9 Tensile Properties of Selected Fibrous Composites and Reinforcements[a–d]

Materials	Modulus of Elasticity [GPa (10^6 psi)]	Tensile Strength [GPa (ksi)]	Elongation (%)
Composite			
Nylon 66 + 25 v/o carbon fibers	14 (2)	≈0.2 (29)	2.2
Epoxy resin + 60 v/o carbon fibers	220 (32)	1.4 (200)	0.8
Polyester resin + 50 v/o aligned glass fibers	38 (5.5)	0.75 (110)	1.8
Polyester resin + 20 v/o random glass fibers	8.5 (1.2)	0.11 (16)	2
Epoxy + 50 v/o boron fibers[b]	200 (29.2)	1.4 (200)	—
Epoxy + 72 v/o S-glass[b]	60.7 (8.8)	1.3 (187)	—
2024Al + 25 v/o SiC[d]	124–172 (18–25)	0.53–0.64 (77–93)	≤1
Reinforcement			
Al_2O_3 whiskers[a]	415–485 (60–70)	7.0–21.0 (1000–3000)	—
Aramid (Kevlar 49)[c]	125 (18)	2.8–3.6 (400–520)	2–3
Boron	380 (55)	3.4 (500)	—
Carbon fiber,[c] Type I	390 (57)	2.2 (320)	0.5
Carbon fiber,[c] Type II	250 (36)	2.7 (390)	1.0
E glass[c]	76 (11)	1.4–2.5 (200–360)	2–3
S glass[a]	85 (12)	4.5 (650)	—
SiC whiskers[a]	485 (70)	20.7 (3000)	—
Si_3N_4 whiskers[a]	380 (55)	1.4 (200)	—

[a] Z. D. Jastrzebski, *The Nature and Properties of Engineering Materials*, Wiley, New York, 1977, p. 546.

[b] *Guide to Engineering Materials,* Vol. 1(1) ASM, Metals Park, OH, 1986, p. 10.

[c] D. Hull, *An Introduction to Composite Materials*, Cambridge Univ. Press, Cambridge, England, 1981.

[d] A. P. Divecha, C. R. Crowe, and S. G. Fishman, *Failure Modes in Composites IV,* J. A. Cornei and F. W. Crossman, Eds., AIME, 1979, p. 406.

The stress–strain response of a given composite material reflects the properties of the individual constituents comprising the two-phase mixture. As such, the stress–strain curve may reveal hybrid response (e.g., a mixture of Type I and Type II or Type V tensile performance). To illustrate, glass fibers, which possess strengths in excess of 700 MPa but provide limited ductility, will exhibit a Type I stress–strain response. In sharp contrast, nylon 66 (a typical engineering thermoplastic) would reveal a modified Type V response, based on tensile strength and elongation values of 59–83 MPa and 60–300%, respectively. From Table 1.9, we see that when 25 volume percent (25 v/o) of glass fibers are added to a nylon 66 matrix, material strength and stiffness increases by factors of 3 and 7, respectively, whereas ductility decreases by a factor of 30. What would be the stress–strain response of a glass-fiber-reinforced nylon 66 matrix composite?

Before analyzing the stress–strain response of such a composite material, it is appropriate to consider the respective functions of both the matrix and reinforcing phases in the composite. The many discrete fibers, filaments, or platelets are supposed to carry much of the load applied to the composite. The fact that there are many discrete fibers in a given composite provides redundancy to the structure and precludes catastrophic fracture if one fiber were to contain a defect and, therefore, fracture prematurely. The matrix phase serves to isolate the fibers from one another and to protect the fiber surface from damage. Of considerable importance, the matrix transmits the applied loads to the fiber through localized shear stresses acting along the fiber–matrix interface.

1.2.7.1   Isostrain Analysis

The stress–strain response of a composite material depends on the respective properties of the matrix and reinforcing phases, their relative volume fraction, the absolute length of the fibers, and the orientation of the fibers relative to the applied stress direction. We begin our analysis of the strength of a reinforced composite by first assuming that the fibers are continuous (i.e., they extend the entire length of the sample), possess uniform strength, and are oriented parallel to the applied stress direction. If the fibers are properly bonded to the matrix and both phases behave elastically, the load applied to the composite in the direction of the fiber axes will be distributed such that

$$P_c = P_f + P_m \tag{1-31}$$

where $P_{c,f,m}$ = load carried by the composite, fiber, and matrix, respectively

To satisfy compatibility considerations, the strains experienced by the two phases must be similar so that $\epsilon_c = \epsilon_m = \epsilon_f$. From Eqs. 1-1 and 1-31, this *isostrain* condition leads to

$$\sigma_c A_c = \sigma_f A_f + \sigma_m A_m \tag{1-32}$$

where $\sigma_{c,f,m}$ = strength of composite, fibers, and matrix, respectively
$A_{c,f,m}$ = cross-sectional area of composite, fibers, and matrix, respectively

Since the area fraction of a continuous phase is equivalent to the volume fraction of the phase ($V_{f,m}$)

$$\frac{A_f}{A_c} = V_f \text{ and } \frac{A_m}{A_c} = V_m \tag{1-33}$$

Therefore,

$$\sigma_c = \sigma_f V_f + \sigma_m V_m = \sigma_f V_f + \sigma_m (1 - V_f) \tag{1-34}$$

From Hooke's law (Eq. 1-7),

$$\sigma_c = E_f \epsilon_f V_f + E_m \epsilon_m V_m \tag{1-35}$$

and

$$E_c = E_f V_f + E_m V_m \tag{1-36}$$

Also

$$\frac{P_f}{P_m} = \frac{E_f \, \epsilon_f \, V_f}{E_m \, \epsilon_m \, V_m} = \frac{E_f \, V_f}{E_m \, V_m} \tag{1-37}$$

From Eq. 1-37, the load distributed to the fibers and matrix of the composite depends on the respective moduli and volume fractions of the two phases. Therefore, the fibers will assume much of the applied load when the ratios of fiber–matrix elastic moduli and volume fractions of the two phases are large. Assuming that the fibers are of uniform strength and fail when the applied stress exceeds their strength limit, the composite will exhibit Type I behavior (provided the matrix remains elastic) and differ from the stress–strain response of the fiber phase only in terms of the modulus (Fig. 1.19a).

When the matrix is permitted to deform plastically, the strength of the composite can no longer be described by Eq. 1-34, but rather may be approximated by

$$\sigma_c = \sigma_f V_f + \sigma'_m(1 - V_f) \tag{1-38}$$

where σ'_m = stress acting on the matrix phase at ϵ_f, the strain to fracture of the fiber.

The two linear regions of the Type I stress–strain curve (Fig. 1.19b) correspond to the elastic response of both phases (Eq. 1-34) (Region *OA*) and the elastic fiber–plastic matrix response (Eq. 1-38) (Region *AB*), respectively.

If V_f, is less than some critical value $V_{f\text{crit}}$, the strength of the composite is actually less than that of the matrix alone (Fig. 1.20). This occurs because the matrix is capable of carrying a greater load than the fibers. In effect, the presence of a subcritical amount of a reinforcing phase reduces the overall load-bearing capacity of the matrix.

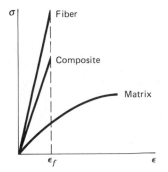 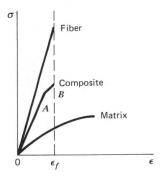

FIGURE 1.19 **Stress–strain response of fiber, matrix, and composite with (*a*) both phases elastic and (*b*) elastic fiber with plastically deformed matrix at fracture.**

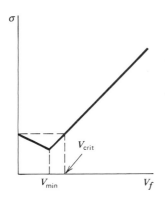

FIGURE 1.20   Composite strength versus volume fraction of reinforcing phase. Note critical volume fraction needed for reinforcement of matrix.

When fibers are discontinuous (i.e., their length is less than that of the component) and/or possess variable strength, the maximum strength potential of the composite is lower than the ideal value given by Eq. 1-35. An analysis of stress distributions within the composite reveals that the applied load is transmitted from the matrix to the fibers by shear stresses acting along the fiber–matrix interface in the vicinity of the fiber ends. For the case of a circular fiber with radius r, these shear stresses produce an axial stress along the fiber given by

$$\sigma_{zz}\pi r^2 = \tau_{rz}2\pi rz \tag{1-39}$$

where σ_{zz} = axial stress along the fiber length

τ_{rz} = shear stress acting along the fiber–matrix interface at the ends of the fiber

r = fiber radius

z = shear stress transfer length (distance from each fiber end)

Upon rearranging Eq. 1-39

$$\sigma_{zz} = \frac{2\tau_{rz}z}{r} \tag{1-40}$$

Note that the axial stress σ_{zz} is zero at the end of the fiber where $z = 0$ and increases with increasing transfer length.* When $\sigma_{zz} = \sigma_f$, the fiber will either fracture or deform plastically. At this point $z = l_c/2$ where $l_c/2$ is defined as the critical transfer length (Fig. 1.21). Upon substitution in Eq. 1-40,

$$\frac{l_c}{d} = \frac{\sigma_f}{2\tau_{rz}} \tag{1-41}$$

---

* This theoretical analysis for the axial stress distribution along the length of the fiber (see H. L. Cox, *Brit. J. Appl. Phys.*, **3**, 72 [1952]), was recently confirmed for the polydiacetylene fiber-epoxy matrix model composite system in experiments conducted by Robinson *et al* (see I. M. Robinson, R. J. Young, C. Galiotis and D. N. Batchelder, *J. Mater. Sci.*, **22**, 3942 [1987]).

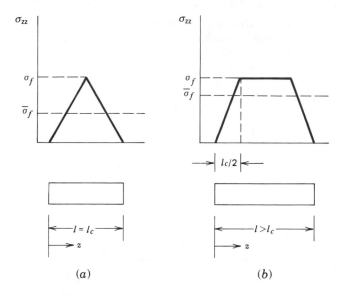

(a)   (b)

FIGURE 1.21   **Axial stress distribution along fiber length when fiber length is (a) equal to and (b) greater than critical length for reinforcement.**

To achieve a fiber stress of σ_f at the center of the fiber, the critical aspect ratio l_c/d must increase with increasing fiber strength and decreasing shear stress.* For example, when the composite temperature is increased, τ_{rz} tends to decrease faster than σ_f with the result that l_c/d increases at high temperatures. If $l<l_c$, then maximum strength potential cannot be reached.

From Fig. 1.21a, the average axial stress on the fiber is given by

$$\overline{\sigma}_f = \frac{\sigma_f(l_c/2)}{l_c} = \sigma_f/2 \tag{1-42}$$

When $z>l_c/2$, the maximum axial stress remains constant but now extends along the midregion of the fiber except near the fiber ends where σ_{zz} approaches zero (Fig. 1.21b). For this condition (assuming that the axial stress increases linearly from each end)

$$\overline{\sigma}_f = \frac{\sigma_f l - \sigma_f(l_c/2)}{l} = \sigma_f\left(1 - \frac{l_c}{2l}\right) \tag{1-43}$$

From Eq. 1-43 we see that the average strength of the fiber $\overline{\sigma}_f$ increases with increasing fiber length and approaches σ_f when $l\gg l_c$. When $\overline{\sigma}_f$ is substituted for σ_f in Eq. 1-38,

$$\sigma_c = \sigma_f\left(1 - \frac{l_c}{2l}\right)V_f + \sigma'_m(1 - V_f) \tag{1-44}$$

---

* For short-fiber composites, strength and stiffness properties depend critically on the integrity of the fiber–matrix interface (e.g., see M. R. Piggott, *Polym. Compos.*, **3**(4), 179 [1982]).

which describes the strength of a composite containing discontinuous elastic fibers embedded within a plastically deformed matrix phase.

It should be recognized that the many fibers in a particular composite do not possess the same strength level; rather, the strength of a given fiber falls within a statistical distribution of fiber strengths. As a result, the weaker fibers fracture prematurely, and the associated load is transferred to the broken fiber segments and to the remaining unbroken fibers. With repeated localized fiber fracture events, the effective l/d ratio decreases along with the composite strength (recall Eq. 1-44). At the same time, the stiffness of the composite decreases as noted by a decrease in the secondary slope (AB) of the stress–strain curve shown in Fig. 1.19b.

1.2.7.2 Influence of Fiber Orientation

In addition to the factors just described, the strength of a composite varies with fiber orientation. Due to processing variations, fibers may not always be oriented parallel to the intended stress axis. From Fig. 1.22, the cross-sectional area of the plane normal to the fiber axes is given by

$$A_\phi = \frac{A_0}{\cos\phi} \tag{1-45}$$

With the resolved load in the direction of the fiber axis (P_f) computed to be

$$P_f = P_0 \cos\phi \tag{1-46}$$

it follows that the stress acting parallel to the fibers is

$$\sigma_c = \frac{P_0 \cos\phi}{A_0/\cos\phi} = \sigma_0 \cos^2\phi \tag{1-47}$$

Note that the axial load-bearing capacity of the composite with off-axis fibers is less than that associated with a composite that contains fibers aligned parallel to the axis of the bar. For the case of off-axis loading, macroscopic shear stresses are developed in the matrix parallel to the fibers. From Fig. 1.22, the resolved load parallel to the fibers is again given by Eq. 1-46, with the shear surface area A_s, described by

$$A_s = \frac{A_0}{\sin\phi} \tag{1-48}$$

The shear stress τ_m acting in the matrix parallel to the axes of the fibers is then found to be

$$\tau_m = \sigma_0 \sin\phi \cos\phi \tag{1-49}$$

Finally, fracture can occur by separation transverse to the fiber length with

$$\sigma_N = \frac{P_0 \sin\phi}{A_0/\sin\phi} = \sigma_0 \sin^2\phi \tag{1-50}$$

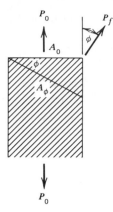

FIGURE 1.22   Applied load resolved in the direction of fiber orientation.

Whether shear fracture in the matrix takes place rather than fiber fracture depends on the angle of misorientation ϕ and the relative strengths of the matrix, interface, and fibers. For example, equating Eqs. 1-47 and 1-49 with respect to σ_0, shear failure will occur when

$$\tan\phi > \frac{\tau_m}{\sigma_c} \tag{1-51}$$

From Eq. 1-51, shear failure will occur at small misorientations when the matrix shear strength is small relative to that of the fiber breaking strength (e.g., see Fig. 1.23); it

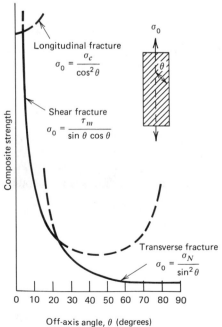

FIGURE 1.23   Competitive fracture processes depend on fiber misorientation.

follows that shear failure parallel to the fiber axis is more likely to occur in off-axis composites at elevated temperatures since τ_m decreases more rapidly than σ_c with increasing temperature. Finally, transverse tensile fracture perpendicular to the fiber axes will occur at large angles of fiber misorientation.

1.2.7.3 Isostress Analysis

When the applied stress direction is perpendicular to the fiber axes or reinforcing plates, the matrix and reinforcing phases are described as being in series with one another. Here the loads in the two phases are equal such that

$$P_c = P_f = P_m \tag{1-52}$$

where $P_{c,f,m}$ = load carried by the composite, fiber, and matrix, respectively. Unlike the isostrain analysis condition, displacements in the two phases under the *isostress* condition are additive with the total composite strain being the weighted sum of strains in the matrix and reinforcing phases. Accordingly,

$$\epsilon_c = V_f\epsilon_f + V_m\epsilon_m = V_f\epsilon_f + (1 - V_f)\epsilon_m \tag{1-53}$$

By combining Eq. 1-53 with Hooke's law (Eq. 1-7) for the fibers and matrix components, one finds that the composite modulus in the direction normal to the reinforcing phase is

$$E_c = \frac{E_fE_m}{V_fE_m + (1 - V_f)E_f} \tag{1-54}$$

As shown for the case of a glass-fiber reinforced epoxy composite, it is obvious that composite modulii increase linearly with fiber-volume fraction under isostrain conditions, whereas E_c increases nonlinearly for isostress loading conditions (Fig. 1.24)[12]. By examining modulii values (see Tables 1.1 and 1.9) for typical polymer matrices ($E \approx 2.5$ GPa) and glass fibers ($E \approx 76$ GPa), Eq. 1-54 can be approximated by

$$E_c \approx \frac{E_m}{(1 - V_f)E_f} \tag{1-55}$$

We conclude that the elastic modulus of a composite under isostress conditions is strongly dependent on the stiffness of the matrix, unlike the isostrain case where fiber stiffness dominates E_c (recall Eq. 1-36). Accordingly, the isostrain composite modulus is consistently larger than that associated with loading under isostress conditions (Fig. 1.24).

EXAMPLE 1.3

For the case of an epoxy + 70 v/o S-glass long fiber composite, what is the elastic modulus of the composite both parallel and perpendicular to the axis of the fibers? (Assume that $E_{epoxy} = 3$ GPa.)

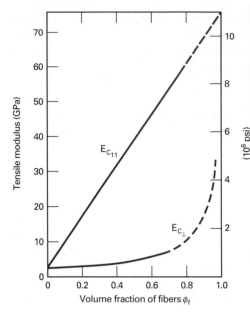

FIGURE 1.24 Tensile modulus in aligned glass-fiber reinforced epoxy resin as predicted from Eqs. 1-36 and 1-54[12]. N. G. McCrum, C. P. Buckley, and C. B. Bucknall, *Principles of Polymer Engineering*, Oxford Science Pub., Oxford, U.K. (1988). Reprinted by permission of Oxford University Press.

Using Eq. 1-36 and mechanical property data from Tables 1.1 and 1.9,

$$E_{c_\parallel} = E_f V_f + E_m V_m$$
$$\text{so that } E_{c_\parallel} = 85 \times 10^9 (0.7) + 3 \times 10^9 (0.3)$$
$$\text{Therefore, } E_{c_\parallel} = 60.4 \text{ GPa}$$

Notice good agreement between this value and the elastic modulus for a 72 v/o S-glass + epoxy matrix composite (see Table 1.9).

For the case of loading perpendicular to the S-glass fibers, we see from Eq. 1-54 that

$$E_{c_\perp} = \frac{85 \times 10^9 (3 \times 10^9)}{0.7(3 \times 10^9) + 0.3(85 \times 10^9)}$$
$$E_{c_\perp} = 9.24 \text{ GPa}$$

By comparison, Eq. 1-55 reveals the elastic modulus to be

$$E_{c_\perp} \approx \frac{3 \times 10^9}{0.3} \approx 10 \text{ GPa}$$

which is in relatively good agreement with the initial computation for $E_{c_\perp}$.

1.3 MODULUS OF RUPTURE

Though tensile testing methods are routinely used to determine the mechanical properties of metallic and polymeric materials, the strength characteristics of ceramic compounds are typically determined by flexural test methods. This arises from the fact that ceramics usually behave elastically and display essentially no plastic deformation (i.e., they exhibit Type I stress–strain behavior). As such, the mechanical properties of these materials are very sensitive to the presence of complex sample shapes that introduce stress concentrations. (Such is the case with threaded grips that are machined into tensile bars.) By contrast, bend bars have a smooth configuration, are easy to machine and test, and require simple load fixtures. The three-point and four-point methods represent two common loading configurations (see Fig. 1.25).

As the reader may recall from his or her Strength of Materials or Mechanics courses, the flexural stress in a bend bar is given by

$$\sigma = \frac{Mc}{I} \tag{1-56}$$

where M is the bending moment, c is the distance from the neutral axis to the outermost fiber surface, and I is the moment of inertia of the bar's cross section. For a rectangular configuration,

$$I = \frac{bh^3}{12} \tag{1-57}$$

where b and h are the beam width and height, respectively. It follows from Eq. 1.57 that for three-point loading, the bending moment increases linearly from either end of the beam to a maximum value at the midspan location, given by

$$\sigma_{3\text{–pt.}} = \frac{3PL}{2bh^2} \tag{1-58}$$

The maximum stress at fracture is referred to as the bend strength or the ''modulus of rupture.'' The reader should recognize that the maximum flexural stresses (both tensile or compressive) given by Eq. 1-58 are experienced only at the beam's midspan and outermost surfaces. Accordingly, the tensile (and compressive) bending stresses decrease linearly to zero from the bar's midspan to the outside loading points and also from the outermost surfaces to zero everywhere along the length of the bar at the neutral axis.

For the four-point loading configuration, the bending moment increases linearly from either loading point at the ends of the beam to a constant maximum value within the region bounded by the interior loading points. Here the flexural stress is given by

$$\sigma_{4\text{–pt.}} = \frac{3Pa}{bh^2} \tag{1-59}$$

where a is the distance from the exterior to interior loading points. It is clear from Fig. 1.25 that the volume of material experiencing maximum stress is considerably greater for the case of four-point loading than for three-point loading. Accordingly, there is a higher probability of locating a defect in the high-stress region of a four-point bend sample and that the associated modulus of rupture values will be lower

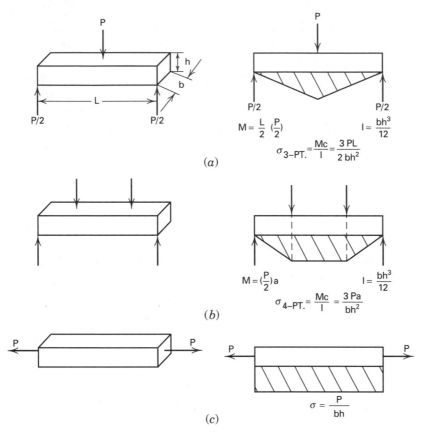

FIGURE 1.25 Bending and tensile stress formulii and distribution for (*a*) three-point; (*b*) four-point bending; and (*c*) uniaxial tensile loading of a rectangular bar.

than for the three-point bending configuration. Furthermore, since the maximum stress level in a tensile bar is experienced across the entire gage section and along the entire gage length, it follows that tensile strength values will be lower than modulus of rupture values for either bend bar. This is particularly true when a material contains a nonuniform flaw distribution, such as that found in ceramics (see Tables 1.7 and 1.10).

As such, the breaking strength or modulus of rupture of a ceramic sample depends on the sample size, type of loading, and flaw distribution. The reader will find additional discussion of the influence of sample size on the mechanical properties of solids in Section 7.3.1.

EXAMPLE 1.4

A 50-mm-long rod of Si_3N_4 has a rectangular cross section with width and depth dimensions of 6 mm and 3 mm, respectively. When tested in 3-pt. bending, the rod fails with an applied load of 670 N. If the rod were tested in tension, the breaking

TABLE 1.10   Room-Temperature Strengths of Ceramic Compounds[a]

Material	Modulus of Rupture* MPa (ksi)	Tensile Strength MPa (ksi)
Al_2O_3 (0-2% porosity)	350–580 (50–80)	200–310 (30–45)
Sintered BeO (3.5% porosity)	172–275 (25–40)	90–133 (13–20)
Sintered stabilized ZrO_2(<5% porosity)	138–240 (20–35)	138 (20)
Hot-pressed Si_3N_4 (<1% porosity)	620–965 (90–140)	350–580 (50–80)
Fused SiO_2	110 (16)	69 (10)
Hot-pressed TiC (<2% porosity)	275–450 (40–65)	240–275 (35–40)

[a] D. W. Richerson, *Modern Ceramic Engineering*, Marcel Dekker, Inc. New York (1992).
* Values corresponding to three- and four-point bending samples.

load would be 10 kN. What are the modulus of rupture and tensile strength properties for this ceramic and how well do these values agree with one another? Explain any property differences.

From Eq. 1-58, the modulus of rupture for the Si_3N_4 rod is

$$\sigma = \frac{3PL}{2bh^2}$$

$$\text{Modulus of Rupture} = \frac{3(670)(50 \times 10^{-3})}{2(6 \times 10^{-3})(3 \times 10^{-3})^2} = 930 \text{MPa}$$

If the rod were pulled in tension and the load to break equal to 10 kN, then the tensile strength would be

$$\frac{P}{A} = \frac{10 \times 10^3}{(6 \times 10^{-3})(3 \times 10^{-3})} = 556 \text{ MPa}$$

We see that the modulus of rupture (bend strength) is considerably greater than the tensile strength for this material. Typically, this is the case for brittle solids, which occurs since the mechanical properties of such materials are extremely sensitive to the presence of defects in the sample. Correspondingly, the properties of a brittle solid depend strongly on the existence of a defect in the region of the highest stress level. Since the maximum stress is experienced across the entire cross section of a tensile bar but is restricted to the surface layer beneath the center load point of a 3-pt. bend bar, it follows that there is a lower probability of finding a defect in the peak stress zone of the bend bar than in the tensile sample. Accordingly, the modulus of rupture in a brittle solid is higher than its corresponding tensile strength value.

1.4 THIN-WALLED PRESSURE VESSELS

The description of stresses in a *thin*-walled pressure vessel is introduced here to establish the basis for subsequent fatigue and fracture analyses that are discussed in later chapters. Consider a cylindrical vessel section of length, L, internal diameter, D, and wall thickness, t, that is subjected to a uniform gas or fluid pressure, p (Fig. 1.26a). By examining a free-body diagram of the lower half of the cylinder (Fig. 1.26b), one sees that the summation of forces acting normal to the midplane is given by

$$\left[\sum Y = 0\right] \quad F = pDL = 2P \tag{1-60}$$

or

$$P = \frac{pDL}{2} \tag{1-61}$$

The tangential or "hoop" stress, σ_t, acting on the wall thickness is then found to be

$$\sigma_t = \frac{P}{A} = \frac{pDL}{2Lt} = \frac{pD}{2t} \tag{1-62}$$

or

$$\sigma_t = \frac{pr}{t} \tag{1-63}$$

where r is the vessel radius. For the case of thin-walled cylinders, where $r/t \geq 10$, Eq. 1-63 describes the hoop stress at all locations through the wall thickness. (The reader is referred to Strength of Materials texts for the more complex analysis of stresses in *thick*-walled cylinders where hoop and radial stresses are found to vary with location through the wall thickness.)

A second free-body diagram to account for cylindrical stresses in the longitudinal direction is shown in Fig. 1.26c. Here we see the bursting force across the end of the cylinder is resisted by the tearing force P acting over the vessel circumference. In this instance, the sum of forces acting along the axis of the cylinder is

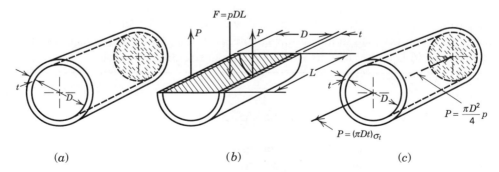

(a)   (b)   (c)

FIGURE 1.26   Thin-walled pressure vessel. (*a*) Overall shape; (*b*) free-body diagram of diametral section; and (*c*) free-body diagram of transverse section.

$$\frac{\pi D^2 p}{4} = P \qquad (1\text{-}64)$$

The cross-sectional area of the cylinder wall is characterized by the product of its wall thickness and the mean circumference [i.e., $\pi(D + t)t$]. For thin-walled pressure vessels where $D \gg t$, the cylindrical cross-sectional area may be approximated by πDt. Therefore, the longitudinal stress in the cylinder is given by

$$\sigma_l = \frac{\pi D^2 p}{4\pi Dt} = \frac{pD}{4t} \qquad (1\text{-}65)$$

By comparing Eqs. 1-62 and 1-65, one finds that the tangential or hoop stress is twice that in the longitudinal direction. Therefore, vessel failure is likely to occur along a longitudinal plane oriented normal to the transverse or hoop stress direction.

EXAMPLE 1.5

A cylindrical pressure vessel is fabricated by joining together 100-cm-diameter wide rings of 1015 as-rolled steel with a series of circumferential welds. The tank contains gas under a pressure of 15 MPa. If the strength of each weldment is 90% that of the base plate, where is failure most likely to occur and what is the minimum required thickness to ensure that the operating stress is no greater than 50% of the material's yield strength?

From Table 1.6, the yield strength of the 1015 alloy base plate is 315 MPa with the weldment strength estimated to be 283.5 MPa. For a design stress/yield strength ratio of 0.5, the hoop stress is computed to be

$$\sigma_l = \frac{pr}{t}$$

so that

$$0.5(315 \times 10^6) = \frac{15 \times 10^6 (50 \times 10^{-2})}{t}$$

$$\therefore t = 4.76 \text{ cm}$$

For the stresses in the longitudinal direction,

$$\sigma = \frac{pr}{2t}$$

so that

$$0.5(283.5 \times 10^6) = \frac{15 \times 10^6 (50 \times 10^{-2})}{2t}$$

$$\therefore t = 2.65 \text{ cm}$$

Therefore, the vessel must have a wall thickness of at least 4.76 cm and any significant overpressurization will cause failure along a longitudinal plane normal to the hoop stress direction and not as a result of longitudinal stresses acting across the weaker circumferential weldments.

1.5 TEMPERATURE AND STRAIN-RATE EFFECTS ON TENSILE BEHAVIOR

Brief mention was made in Section 1.2.2.4 of a temperature-induced transition in macroscopic fracture surface appearance. Since this transition most often parallels important changes in the strength and ductility of the material, some additional discussion is indicated. It is known that the general flow curve for a given material will decrease with increasing temperature T and decreasing strain rate $\dot{\epsilon}$ (Fig. 1.27). The magnitude of these changes varies with the material; body-centered-cubic metals (e.g., iron, chromium, molybdenum, and tungsten), and ceramic materials are much more sensitive to T and $\dot{\epsilon}$, than face-centered-cubic metals (e.g., aluminum, copper, gold, and nickel), with polymeric solids being especially sensitive. Over the years, a number of investigators have sought to define the overall response of a material in terms of some generalized equation of state reflecting the dependence of true stress on strain, strain rate, and temperature. The relationship between true stress and strain rate is of the same form as noted in Eq. 1-20 and given by

$$\sigma = K\dot{\epsilon}^m \tag{1-66}$$

where m = strain-rate sensitivity factor
$\dot{\epsilon}$ = strain rate
K = material constant
σ = true stress

For most metals m is low and varies between 0.02 and 0.2. Under certain conditions wherein $m>0.3$, a given material may exhibit a significant degree of strain-rate sensitivity in association with superplastic deformation behavior (see Section 5.4). In the limit where $m = 1$, the stress–strain-rate material response is analogous to that associated with Newtonian viscous flow (see Section 6.3.1).

Depending on the nature of the test or service condition, strain rates may vary by more than a dozen orders of magnitude. At low strain rates, below about 10^{-3} s^{-1}, material behavior is characterized by its creep and stress rupture response (see Chapter 5). At strain rates between 10^3 and 10^5 s^{-1}, the material experiences impact conditions and may fail with reduced fracture energy (Chapter 9). Ballistic conditions occur with strain rates above 10^5 s^{-1} and involve the shock-wave–material interactions associated with such circumstances as projectile impact, high-energy explosions, and meteorite impact with spacecraft (see Section 10.2).

Attempts have been made to characterize material properties in terms of parameters that include both test temperature and strain rate. For example, on the basis of simple rate theory, Bennett and Sinclair[13] proposed that the yield strength of iron and other body-centered-cubic transition metals be described in terms of a rate–temperature

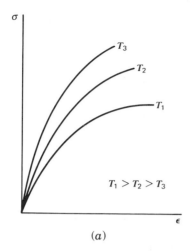
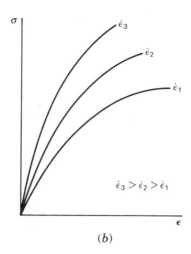

(a)   (b)

FIGURE 1.27   **Yield strength change as a function of (*a*) temperature and (*b*) strain rate.**

parameter $T\ln(A/\dot{\epsilon})$, where A is a frequency factor with an approximate value of 10^8/sec for these materials. As seen in Fig. 1.28, the parameter provides good correlation for the case of seven steels. While the lower strength steels reveal a somewhat larger yield-strength sensitivity to T and $\dot{\epsilon}$ than do the stronger alloys at low $T\ln(A/\dot{\epsilon})$ levels, the seven curves are remarkably similar, reflecting comparable *absolute* changes in yield strength with temperature and strain-rate variations. It is important to recognize, however, that the *relative* change in yield strength with $T\ln(A/\dot{\epsilon})$ is much greater in the lower strength alloys.

1.6   FAILURE THEORIES

Having characterized the varied tensile behavior of engineering solids, it is appropriate to comment briefly on how a material's tensile response may be used to predict failure in a component that experiences multiaxial loading. Several "failure theories" have been identified during the past 200 years and are based on the concept that the stress condition responsible for failure in a standard tensile bar was also responsible for failure in a component containing combined loads. The maximum normal stress theory is perhaps the simplest failure theory; it describes failure as occurring when the maximum tensile or compressive stress exceeds the uniaxial tensile or compressive strength, respectively, of the material. This theory is generally suitable for brittle materials, such as those exhibiting Type I stress–strain behavior.

The Tresca theory describes failure as taking place when the maximum shear stress exceeds the shear strength associated with yielding in the uniaxial tension test:

$$\tau_{max} = \frac{\sigma_1 - \sigma_3}{2} \tag{1-67}$$

where τ_{max} = maximum shear stress

$\sigma_{1,3}$ = largest and smallest principal tensile stresses, respectively

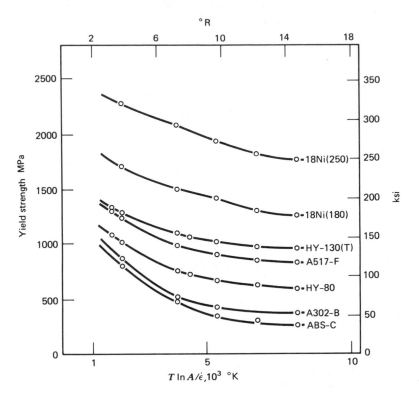

FIGURE 1.28   Yield strength for seven steels in terms of the Bennett-Sinclair parameter, $T\ln A/\dot{\epsilon}$.[14] **(Reprinted with permission from A. K. Shoemaker and S. I. Rolfe, *Engineering Fracture Mechanics*, 2, 319, © Pergamon Press, Elmsford, NY (1971.)**

For uniaxial loading conditions ($\sigma_2 = \sigma_3 = 0$), the shear yield strength is found to be equal to one-half the yield strength in tension.

The Tresca theory provides a reasonable description of yielding in ductile materials (e.g., those characterized by Type II stress–strain behavior), though the maximum distortion energy or Von Mises' yield criterion is preferred because it is based on better correlation with actual test data. In this theory, yielding is assumed to occur when the distortional energy in the tensile test associated wih a *shape* change is equal to the distortional energy in the component that experiences multiaxial loading. (The elastic energy associated with a *volume* change is not included in this analysis.) To characterize the distortional energy ($\propto \sigma^2/E$) within a specimen or component that experiences multiaxial loading, it is convenient to consider an equivalent stress, σ_e, given by

$$\sigma_e = \frac{\sqrt{2}}{2}[(\sigma_2 - \sigma_1)^2 + (\sigma_3 - \sigma_1)^2 + (\sigma_3 - \sigma_2)^2]^{1/2} \qquad (1\text{-}68)$$

where $\sigma_1, \sigma_2, \sigma_3$ = principal stresses

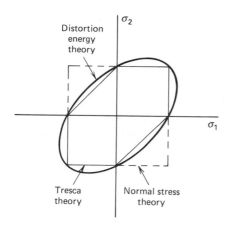

FIGURE 1.29 Failure envelopes for normal stress, Tresca, and distortion energy theories. Failure occurs when stress combinations fall outside the envelope for applicable theory.

Alternatively, the equivalent stress can be described in terms of the tensile and shear stresses acting on three orthogonal planes wherein

$$\sigma_e = \frac{\sqrt{2}}{2}[(\sigma_x - \sigma_y)^2 + (\sigma_y - \sigma_z)^2 + (\sigma_z - \sigma_x)^2 + 6(\tau_{xy}^2 + \tau_{yz}^2 + \tau_{xz}^2)]^{1/2} \quad (1\text{-}69)$$

If σ_e equals the yield strength from a uniaxial tensile test, yielding in the multiaxially loaded sample is predicted.

The two-dimensional yield loci for the normal stress, shear stress, and distortion-energy theories are shown in Fig. 1.29. Note that all three failure theories predict the same yielding conditions under uniaxial ($\sigma_1 = \sigma_1$; $\sigma_2 = \sigma_3 = 0$) and balanced biaxial ($\sigma_1 = \sigma_2$; $\sigma_3 = 0$) loading conditions; however, different failure conditions are predicted for conditions of pure shear ($\sigma_1 = -\sigma_2$; $\sigma_3 = 0$) with the distortion-energy theory predicting yielding when the applied stress is $0.577\sigma_{ys}$. Most data conform to predictions of the distortion-energy theory. For a more detailed discussion of such classical failure theories, see the text by Juvinall.[15]

REFERENCES

1. R. Hooke, "Of Spring," 1678, as discussed in S. P. Timoshenko, *History of the Strength of Materials*, McGraw-Hill, New York, 1953, p. 18.
2. J. H. Hollomon, *Trans. AIME* **162**, 268 (1945).
3. G. B. Bülfinger, *Comm. Acad. Petrop.* **4**, 164 (1735).
4. A. W. Bowen and P. G. Partridge, *J. Phys. D: Appl. Phys.* **7**, 969 (1974).
5. C. J. McMahon, Jr., Ed., *Advances in Materials Research*, Vol. 2, *Microplasticity*, Interscience, New York, 1968.
6. 1971 Annual Book of ASTM Standards, ASTM, Standard E8-69, p. 205.
7. P. W. Bridgman, *Trans. ASM* **32**, 553 (1944).
8. F. R. Larson and F. L. Carr, *Trans. ASM* **55**, 599 (1962).
9. P. W. Bridgman, *Fracturing of Metals*, ASM, Metals Park, OH, 1948, p. 246.

10. A. H. Cottrell, *Vacancies and Other Point Defects in Metals and Alloys*, Institute of Metals, London, 1958, p. 1.

11. G. Meinel and A. Peterlin, *J. Polym. Sci., Part A-2* **9,** 67 (1971).

12. N. G. McCrum, C. P. Buckley, and C. B. Bucknall, *Principles of Polymer Engineering*, Oxford Science Pub., Oxford, U.K. (1988).

13. P. E. Bennett and G. M. Sinclair, *Trans. ASME, J. Basic Eng.* **88,** 518 (1966).

14. A. K. Shoemaker and S. T. Rolfe, *Eng. Fract. Mech.* **2,** 319 (1971).

15. R. C. Juvinall, *Fundamentals of Machine Component Design,* John Wiley & Sons, New York, 1983.

FURTHER READINGS

L. J. Broutman and R. H. Krock, *Modern Composite Materials*, Addison-Wesley, Reading, MA, 1967.

J. A. Cornie and F. W. Crossman, *Failure Modes in Composites IV,* Metallurgical Society, AIME, New York, 1979.

D. Hull, *An Introduction to Composite Materials*, Cambridge Univ. Press, Cambridge, 1981.

R. M. Jones, *Mechanics of Composite Materials,* McGraw-Hill, New York, 1975.

Various ASTM publications including ASTM STP 617, 674, 676, 704, 734, 749, 768, 772, 775, 787, 893.

PROBLEMS

1.1 A rod is found to creep at a fixed rate over a period of 10,000 hr when loaded with a 1000-N weight. If the initial diameter of the rod is 10 mm and its initial length 200 mm, it is found that the steady-state-creep strain rate is 10^{-5} hr^{-1}. Calculate the following:

(a) The final length of the rod after 100 hr, 10,000 hr.

(b) The engineering and true strains after these two time periods.

(c) The engineering and true stresses after these two time periods.

1.2 Calculate the energy absorbed in fracturing a tensile specimen when the material obeys the Hollomon parabolic stress–strain relation.

1.3 A 200-mm-long rod with a diameter of 2.5 mm is loaded with a 2000-N weight. If the diameter decreases to 2.2 mm, compute the following:

(a) The final length of the rod.

(b) The *true* stress and *true* strain at this load.

(c) The *engineering* stress and strain at this load.

1.4 Calculate the elastic moduli for sodium chloride and nickel in the $\langle 100 \rangle$ and $\langle 111 \rangle$ directions. Compare the anisotropy in these two materials.

1.5 Calculate the elastic modulus for gold in the $\langle 110 \rangle$ direction and compare your results with the E_{100} and E_{111} values reported in Table 1.4 and the polycrystalline isotropic value given in Table 1.1.

1.6   Define true strain in terms of engineering strain.

1.7   Speculate on the relation between the modulus of elasticity of a group of crystalline solids and their respective melting points. Also relate the modulus of elasticity to the respective coefficients of thermal expansion.

1.8   A 5-cm-long circular rod of 1080 as-rolled steel (diameter = 1.28 cm) is loaded to failure in tension. What was the load necessary to break the sample? If 80% of the total elongation was uniform in character prior to the onset of localized deformation, compute the true stress at the point of incipient necking.

1.9   Calculate E_{111} and E_{100} for LiF and Mo. Describe the relative elastic anisotropy of the two materials.

1.10   Compute the modulus of elasticity for copper and aluminum single crystals in the ⟨100⟩, ⟨110⟩, and ⟨111⟩ directions. Compare these values with the modulus reported for polycrystalline samples of these two materials.

1.11   A platform is suspended by two parallel rods, as shown in the sketch, with each rod being 1.28 cm in diameter. Rod A is manufactured from 4340 steel $[Q + T (650°)]$ (E = 210 GPa, σ_{ys} = 855 MPa); rod B is made from 7075-T6 aluminum alloy (E = 70 GPa, σ_{ys} = 505 MPa).

  (a)   What uniform load can be applied to the platform before yielding will occur?

  (b)   Which rod will be the first to yield?
      (*Hint:* Both rods experience the same elastic strain.)

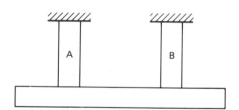

1.12   A 100-mm-long rod of 1340 steel $[Q + T (205°C)]$ is subjected to a load of 50,000 N. If the diameter of the rod is 10 mm, what strain would the rod experience? What strain would remain if the load were removed?

1.13   An annealed-steel tensile specimen (E = 210 GPa, σ_{ys} = 425 MPa) has a 1.28-cm minimum diameter and a 5-cm gage length. Maximum load is reached at 66,700 N, and fracture occurs at 44,500 N.

  (a)   What is the material's tensile strength?

  (b)   Why does fracture occur at a lower load than the maximum load?

  (c)   What does the specimen gage section look like an instant before failure?

  (d)   Compute the strain and describe its character when a tensile stress of 100 MPa is applied.

1.14   A 5-cm-long circular rod of 1015 as-rolled steel (diameter = 1.28 cm) is loaded in tension to failure.

    (a)   What load was necessary to break the sample?

    (b)   If 60% of the total elongation was uniform in character prior to the onset of localized deformation, compute the true stress at the point of incipient necking.

1.15   A tensile specimen (2.5-cm-long by 0.25-cm-diameter) is stretched uniformly to 3 cm, where it begins to neck under a load of 1400 N. Calculate the engineering and true stresses and strains at necking.

1.16   Derive a simple expression for the toughness of a material in terms of the true stress, true strain, and strain-hardening coefficient.

1.17   A specimen is placed in a creep machine and subjected to a load of 5000 N. The initial length of the circular rod was 200 mm, and its initial diameter was 5 mm. After several hundred hours, the length of the rod was found to be 230 mm.

    (a)   What was the rod diameter at this time?

    (b)   What were the true and engineering strains at this time?

    (c)   What were the true and engineering stresses at this time?

    (d)   If the test were to be conducted under conditions of constant true stress, would the load on the sample remain the same or change and, if so, by how much?

1.18   For three BCC metals, tungsten, molybdenum, and iron, compute the elastic moduli in the $\langle 100 \rangle$ and $\langle 111 \rangle$ directions. Compare the anisotropy in these three metals.

1.19   You are given a 135-cm-long length of high-strength wire with a circular cross-sectional diameter of 0.5 cm. The wire has the following properties: $E = 216$ GPa; $\epsilon_f = 0.02$. You are to manufacture a composite rod 15 cm long and having a cross-sectional area of 6.5 cm^2. The 135-cm-long high-strength wire is to be used as the high-strength constituent of the composite. The matrix phase will consist of a polymer resin that is to be cured after the high-strength wires have been positioned. Ignoring the strength contribution of the polymer resin matrix and any residual thermal shrinkage stresses, and assuming an ideal bond between the matrix and the reinforcing phase, calculate the strongest composite that can be made from the resin and length of wire provided.

1.20   A 5-cm-long, 1.28-cm-diameter rod of aluminum alloy 2024-T3 is tested in tension to failure. The yield and tensile strengths were found to be 345 and 485 MPa, respectively, and the total elongation was 18%. Calculate the load at yielding and the load at fracture. Assuming that necking occurs when the specimen has elongated uniformly by 15%, calculate the instantaneous diameter at the onset of necking and true stress at the onset of necking.

1.21   For a composite containing 60 v/o continuous aramid fibers in an epoxy matrix, compute the maximum theoretical composite modulus. Would the modulus of the composite be higher if 40 v/o carbon fibers (Type I) were used as the reinforcing phase instead of the aramid fibers?

1.22   A 10-cm long and 2-cm diameter rod of 403 stainless steel was pulled to fracture. The mechanical properties of this alloy are reported as follows:

> yield strength: 275 MPa
>
> tensile strength: 515 MPa
>
> total elongation: 35%

(a)   If the rod experienced 28% uniform strain, calculate the load at the onset of necking.

(b)   Calculate the true strain and true stress at the onset of necking.

1.23   From Table 1.9, we see that the stiffness of Nylon 66 + 25v/o carbon fibers is 14 GPa, whereas the stiffness of an epoxy resin + 60v/o carbon fibers is 220 GPa. If the elastic modulus of carbon fibers is 390 GPa, speculate on the nature of the two composites in question in terms of fiber length, aspect ratio, and fiber orientation.

1.24   A 10-cm-long cylindrical rod of steel yields at a stress level of 500 MPa with the application of a load of 39.3 kN.

(a)   What is the original diameter of the rod?

(b)   If the true stress at necking is 617 MPa in association with a load of 41 kN, what is the rod length and true strain at this stress level?

1.25   A cylindrical rod of steel is deformed elastically in tension to a load of 49 kN. If the original rod length and diameter are 25-cm and 15-mm, respectively, determine the rod length and diameter under load, assuming that the material possesses the following properties: $E = 205$ GPa, $\mu = 0.25$.

1.26   A 2024-T3 aluminum alloy possesses yield and tensile strengths of 345 and 485 MPa, respectively (scc Table 1.6). An unknown load causes a 400-mm long and 15-mm-diameter rod to extend by 4 mm. If possible, calculate the load in question, given that the modulus of elasticity of the alloy is 70 GPa. If not, why can't this be done?

1.27   A 40-cm-diameter pipe is used to carry a pressure of 20 MPa. Compute:

(a)   the lightest and

(b)   the cheapest pipe per unit length, based on the two possible alloy choices.

	2024-T3	**1015 steel**
σ_{ys} (MPa)	345	315
ρ (g/cm^3)	2.7	7.9
Cost ($/Kg)	2	0.5

1.28   A 40-cm-diameter pipe is fabricated from a polyester-glass composite with continuous glass fibers wrapped around the pipe circumference. What is the minimum thickness required to withstand an internal pressure of 4 MPa when $V_f = 0.5$, $\sigma_{\parallel} = 750$MPa, and $\sigma_{\perp} = 25$MPa?

1.29   A 150-mm-long and 15-mm-diameter rod-shaped component is to experience a 35kN load and is neither to yield nor to undergo a diameter reduction of more than 1.2×10^{-2} mm. Which of the four alloys listed below will satisfy these requirements?

	σ_{ys}	σ_{ts}	E	μ
301 Stainless Steel	275	725	205	0.3
Ti-5Al-2.5Sn	805	850	116	0.321
AZ31B Mg	103-125	220	45	0.291
2219-T31 Al	250	360	70	0.345

1.30   A 7178-T6 aluminum alloy is to be used to make a thin-walled pressure vessel. If the diameter is 40-cm, what wall thickness is required to ensure that a pressure of 50 MPa will result in a maximum stress no greater than 50% of the alloy's yield strength?

1.31   For the same rod of Si_3N_4 as discussed in Example 1.4, a load of 1050 N is necessary to cause fracture when the sample is tested in 4-pt. bending with the interior loads being located at L/4 and 3L/4, respectively. What is the modulus of rupture under this test condition and how does this value compare with the answers given for the case of 3-pt. bending and uniaxial tension?

CHAPTER 2

ELEMENTS OF DISLOCATION THEORY

2.1 STRENGTH OF A PERFECT CRYSTAL

In Chapter 1, the elastic limit of a given material was defined as that stress level above which strains are irreversible. The objective of this chapter is to consider the manner by which permanent deformations are generated and to estimate the magnitude of stresses necessary for such movement. To begin, consider atom movements along a particular crystallographic plane leading to the displacement of the upper half of a cube relative to the bottom (Fig. 2.1). If atom A is to move to position B, atom B to position C, etc., a *simultaneous* translation of all atoms on the slip plane must occur. Since this would involve simultaneous rupture of all the interatomic bonds acting across the slip plane (e.g., bonds A–A', B–B', C–C', etc.), the necessary stress would have to be large. From Fig. 2.2a we see that the equilibrium atom positions within the crystalline lattice are located at P and R, with a separation of b units. Midway between P and R the energy is a maximum at Q, which represents a metastable equilibrium position. The exact shape of the energy curve shown in Fig. 2.2a depends on the nature of the interatomic bonds. Since this is not known precisely, a sinusoidal waveform is assumed for simplicity in this analysis. Locating the atoms on the slip plane anywhere other than at an equilibrium position, such as P and R, requires a force defined by the slope of the energy curve (Fig. 2.2b) at that position where

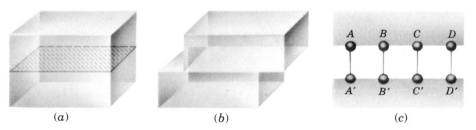

| (a) | (b) | (c) |

FIGURE 2.1 Movement of a solid cube along a particular slip plane. (*a*) Undeformed cube with anticipated slip plane; (*b*) slipped cube revealing relative translation of part of cube; (*c*) atom position showing bonds across slip plane (A–A', B–B', C–C', D–D').

57

$F \equiv - \; dE/dx$. For example, to place an atom between P and Q (that is, $0 < x < b/2$), a force acting to the right is required to counteract the tendency for the atom to move back to the equilibrium site at P. Note that between P and Q, the slope of the energy curve is everywhere positive so that the force curve is also positive from $0 \le x \le b/2$. When $b/2 < x < b$, the atom wants to slide into its new equilibrium position at R (that is, the energy decreases continually from Q to R). To prevent this, a force acting to the left is needed to keep the atom stationary at some location between Q and R. This force is in an opposite direction to that needed between P and Q and is, therefore, negative. Note that in the region between Q and R, the slope of the energy curve is negative. Therefore, the corresponding portion of the force curve in this same region must also be negative.

From the above discussion it is clear that the shear stress necessary to move the atoms on the slip plane varies periodically from zero at P, Q, and R to a maximum value at $b/4$ and $3b/4$. Therefore, the shear stress may be expressed in the following form (from an analysis due to Frenkel[1]), based on an assumed sinusoidal variation in energy throughout the lattice:

$$\tau = \tau_m \sin \frac{2\pi x}{b} \qquad (2\text{-}1)$$

where τ = applied shear stress
  τ_m = maximum theoretical strength of crystal
  x = distance atoms are moved
  b = distance between equilibrium positions

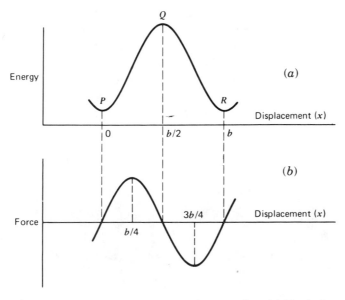

FIGURE 2.2   **The periodic nature of a lattice. (***a***) Variation of energy with atom position in lattice. Preferred atom sites are at *P* and *R*, associated with minimum energy; (***b***) variation in force acting on atoms throughout lattice. Force is zero at equilibrium site positions and maximum at *b*/4, 3*b*/4, 5*b*/4, ..., (2*n* − 1)*b*/4.**

Plastic flow (that is, irreversible deformation) will then occur when the upper part of the cube (Fig. 2.1) is translated a distance greater than $b/4$ because of an applied shear stress τ_m.

For elastic strains, the shear stress may also be defined by Hooke's law

$$\tau = G\gamma \qquad (2-2)$$

The shear strain may be approximated for small values by

$$\gamma \approx \frac{x}{a} \qquad (2-3)$$

where a = distance between slip planes. Combining Eqs. 2-1, 2-2, and 2-3 with sin $2\pi x/b$ approximated by $2\pi x/b$ for small strains gives

$$G\frac{x}{a} \approx \tau_m \frac{2\pi x}{b} \qquad (2-4)$$

Upon rearranging

$$\tau_m \approx \frac{Gb}{2\pi a} \qquad (2-5)$$

For most crystals b is of the same order as a, so Eq. 2-5 may be rewritten in the form

$$\tau_m \approx \frac{G}{2\pi} \qquad (2-6)$$

Because of the approximations made in this analysis, especially with regard to the form of the energy-displacement curve, the magnitude of the theoretical shear strength τ_m from Eq. 2-6 is of an approximate nature. More realistic estimates place τ_m in the range of $G/30$. Nevertheless, it is instructive to compare theoretical strength values calculated with Eq. 2-6 with experimentally determined shear strengths for single crystals of various materials. From Table 2.1, it is immediately obvious that very large discrepancies exist between theoretical and experimental values for all materials tabulated. Without question, the lack of precision regarding computations based on Eq. 2-6 is *not* responsible for these large errors. Rather, the discrepancies must be accounted for in a different manner.

2.2 THE NEED FOR LATTICE IMPERFECTIONS: DISLOCATIONS

In 1934 Taylor, Orowan, and Polanyi postulated independently the existence of a lattice defect that would allow the cube in Fig. 2.1 to slip at much lower stress levels.[3,4] By introducing an extra half plane of atoms into the lattice (Fig. 2.3), they

TABLE 2.1   Theoretical and Experimental Yield Strengths in Various Materials[2]

Material	$G/2\pi$		Experimental Yield Strength		
	GPa	10^6 psi	MPa	psi	τ_m/τ_{exp}
Silver	12.6	1.83	0.37	55	~3 × 10^4
Aluminum	11.3	1.64	0.78	115	~1 × 10^4
Copper	19.6	2.84	0.49	70	~4 × 10^4
Nickel	32	4.64	3.2–7.35	465–1,065	~1 × 10^4
Iron	33.9	4.92	27.5	3,990	~1 × 10^3
Molybdenum	54.1	7.85	71.6	10,385	~8 × 10^2
Niobium	16.6	2.41	33.3	4,830	~5 × 10^2
Cadmium	9.9	1.44	0.57	85	~2 × 10^4
Magnesium (basal slip)	7	1.02	0.39	55	~2 × 10^4
Magnesium (prism slip)	7	1.02	39.2	5,685	~2 × 10^4
Titanium (prism slip)	16.9	2.45	13.7	1,985	~1 × 10^3
Beryllium (basal slip)	49.3	7.15	1.37	200	~4 × 10^4
Beryllium (prism slip)	49.3	7.15	52	7,540	~1 × 10^3

were able to show that atom bond breakage on the slip plane could be restricted to the immediate vicinity of the bottom edge of the half plane (called the dislocation line). As the dislocation line moves through the crystal, bond breakage across the slip plane occurs *consecutively* rather than *simultaneously* as was necessary in the perfect lattice (Fig. 2.1). The consecutive nature of bond breakage is shown in Fig. 2.4 where the extra half plane is shown at different locations during its movement through the crystal. The end result of the movement of this half plane is the same as shown in Fig. 2.1—the upper half of the cube has been translated relative to the bottom half by an amount equal to the distance between equilibrium atomic positions **b.** The major difference is the fact that it takes much less energy to break one bond at a time than all the bonds at once. This concept is analogous to moving a large floor rug across

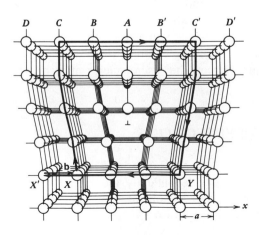

FIGURE 2.3   Lattice defect caused by introduction of an extra half plane of atoms, *A*. Note symmetrical displacement of planes *B*, *B'*, *C*, *C'*, etc. The dislocation line is defined as the edge of the half plane, *A*. The Burgers circuit *XCC'YX'* contains a closure failure *X'X*. (From Guy,[5] *Elements of Physical Metallurgy*, 2nd ed., Addison-Wesley, Reading, MA, 1959.)

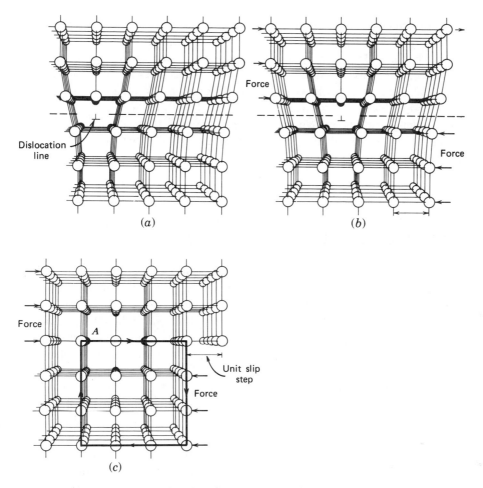

Dislocation line

Force

Force

(a)

(b)

Force

A

Unit slip step

Force

(c)

FIGURE 2.4 Successive positions of dislocation as it moves through crystal. Note that the final offset of crystal resulting from the passage of dislocation is the same as the simultaneous movement of the entire crystal. Note the perfect Burgers circuit in (c). (From Guy,[5] *Elements of Physical Metallurgy*, 2d ed., Addison-Wesley, Reading, MA, 1959.)

the room. If you have ever tried to grab the edge of the rug and pull it to a new position, you know that it is nearly impossible to move a rug in this manner. In this case, the "theoretical shear stress" necessary to move the rug is strongly dependent on the frictional forces between the rug and the floor. If you persisted in your task you probably discovered that the rug could be moved quite easily in several stages by first creating a series of buckles at the edge of the rug and then propagating them, one at a time, across the rug by shuffling your feet behind each buckle. In this way you were able to move the rug by increments equal to the size of the buckle. Since the only part of the rug to move at any given time was the buckled segment, there was no need to overcome the frictional forces acting on the whole rug. Since the lattice dislocation is a similar work-saving "device," one may reconcile the large errors between theoretical and experimental yield strengths (Table 2.1) by assuming the presence of dislocations in the crystals that were examined.

Before we begin to deal with the specific character of dislocations, it is natural to wonder whether the analysis leading to Eq. 2-6 is correct after all. What is needed, of course, are test data for crystals possessing *no* dislocations. Fortunately, such perfect crystals—called whiskers—have been prepared in the laboratory. The strengths of these extraordinary crystals, shown in Table 2.2, are seen to be in close agreement with theoretical maximum values computed from Eq. 2-6. On this basis, the Frenkel analysis is verified.

2.3 LATTICE RESISTANCE TO DISLOCATION MOVEMENT: THE PEIERLS STRESS

From Fig. 2.3 it is clear that the insertion of the extra half plane of atoms has perturbed the lattice and caused atoms to be pushed aside laterally, particularly in the upper half of the crystal. For example, atoms along planes B and C are displaced to the left, while atoms in planes B' and C' are displaced to the right. Since the forces acting on these groups of atoms are equal and of opposite sign (that is, pairing atoms in plane B with those in B' and those in C with atoms in plane C'), movement of the extra half plane A either to the left or right would be met by self-balancing forces on the other atoms within the distorted region. On this basis, the force necessary to move a dislocation would be zero. However, Cottrell[4] pointed out that although the above situation should prevail when the dislocation occurs in a symmetrical position (such as the one shown for plane A in Fig. 2.3), it would not hold true when the dislocation passes through nonsymmetrical positions. Consequently, some force is necessary to move the dislocation through the lattice. An important characteristic of this force (called the Peierls-Nabarro or Peierls force) is that its magnitude varies periodically as the dislocation moves through the lattice.

It is known that the magnitude of the Peierls force depends to a large extent on (1) the width of the dislocation W, which represents a measure of the distance over which the lattice is distorted because of the presence of the dislocation (Fig. 2.5), and (2)

TABLE 2.2   Theoretical and Experimental Strengths of Dislocation-Free Crystal (Whiskers)[6]

Material	Theoretical Strength ($G/2\pi$)		Experimental Strength		
	GPa	10^6 psi	GPa	10^6 psi	Error
Copper	19.1	2.77	3.0	0.44	~6
Nickel	33.4	4.84	3.9	0.57	~8.5
Iron	31.8	4.61	13	1.89	~2.5
B_4C	71.6	10.4	6.7	0.98	~10.5
SiC	132.1	19.2	11	1.60	~12
Al_2O_3	65.3	9.47	19	2.76	~3.5
C	156.0	22.6	21	3.05	~7

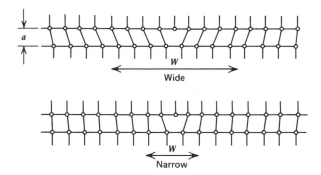

FIGURE 2.5 **Characteristic width of an edge dislocation that affects the Peierls-Nabarro stress.**[7] (Reproduced by courtesy of the Council of the Institution of Mechanical Engineers from *The Properties of Materials at High Rates of Strain* by A. H. Cottrell.)

the distance between similar planes a. The Peierls stress has been shown to depend on W and b in the form

$$\tau_{p-n} \propto G e^{-2\pi W/b} \tag{2-7}$$

where $\quad W = a/1 - v$

From Eq. 2-7, the Peierls stress for a given plane is seen to decrease with increasing distance between like planes. Since the distance between planes varies inversely with their atomic density, slip is preferred on closely packed planes. In addition, the Peierls stress depends on the dislocation width, which is dependent on atomic structure and the nature of the atomic bonding forces. For example, when the bonding forces are spherical in distribution and act along the line of centers between atoms, the dislocation width is large. Since this type of bonding is found in close-packed structures, it is seen that the Peierls stress in face-centered-cubic and close-packed hexagonal crystals is low. By contrast, when bonding forces are highly directional (as in the case of covalent. ionic, and body-centered-cubic crystals), the dislocation width is narrow and the Peierls stress correspondingly large. Although several attempts have been made to compute precisely the magnitude of the Peierls stress in a given lattice, considerable difficulties arise because the exact shape of the force–displacement curve is unknown. Foreman, Jaswon, and Wood[8] showed that when the amplitude of an assumed sinusoidal force–displacement law was reduced by half, the width of the dislocation increased fourfold. This, in turn, had the effect of reducing the computed Peierls stress value by more than *six orders of magnitude*. Until the force–displacement relation between atoms can be defined more precisely, the magnitude of the Peierls stress in crystals can be described only in qualitative terms.

2.3.1 Peierls Stress Temperature Sensitivity

One such qualitative characteristic of the Peierls stress relates to the temperature sensitivity of the yield strength. Since the Peierls stress depends on the short-range

TABLE 2.3   Relation between Dislocation Width and Yield-Strength Temperature Sensitivity

Material	Crystal Type	Dislocation Width	Peierls Stress	Yield-Strength Temperature Sensitivity
Metal	FCC	Wide	Very small	Negligible
Metal	BCC	Narrow	Moderate	Strong
Ceramic	Ionic	Narrow	Large	Strong
Ceramic	Covalent	Very narrow	Very large	Strong

stress field of the dislocation core, it is sensitive to the thermal energy in the lattice and, hence, to the test temperature. At low temperatures, where thermal enhancement of dislocation motion is limited, the Peierls stress is large. In crystals that have wide dislocations, however, the increase in Peierls stress with decreasing temperature is insignificant, since the Peierls stress is negligible to begin with.

Accordingly, there is limited yield strength-temperature dependence in FCC metals such as aluminum, copper, and austenitic stainless steel alloys. The situation is quite different in crystals that contain narrow dislocations. Although the Peierls stress may be small at elevated temperatures, it rises rapidly with decreasing temperature and represents a large component of the yield strength in the low-temperature regime. Recall from Fig. 1.28 that yield strengths in several BCC steels increase sharply with decreasing temperature. The yield strength-temperature sensitivity of several engineering materials is shown in Table 2.3. The large Peierls stress in ceramic materials is partly responsible for their limited ductility at low and moderate temperatures. However, the Peierls stress decreases rapidly with increasing temperature, thereby enhancing plastic deformation processes in these materials at high temperatures.

2.3.2   Effect of Dislocation Orientation on Peierls Stress

The Peierls stress as described above represents an upper bound to the stress necessary to move a dislocation through a crystal. In fact, dislocations will seldom lie completely along directions of lowest energy, or energy valleys, within the lattice. Rather, the dislocation line will contain bends or kinks that lie across energy peaks at some angle (Fig. 2.6a). The angle θ that the kink makes with the rest of the dislocation line, as well as its length l, is a direct consequence of the balance between two competing factors. On one hand, the dislocation will prefer to lie along the energy valleys such that the kink length is minimized and the kink angle maximized (90° (Fig. 2.6b)). (It should be noted that to create such a sharp kink angle will increase the energy of the dislocation, since the energy of any curved segment of a dislocation line increases with decreasing radius of curvature.) On the other hand, the dislocation line tries to be as short as possible to minimize its self-energy and in the limit would prefer a straight-line configuration (Fig. 2.6c; see also Section 2.6). The degree to which the kinked dislocation line approaches either extreme will depend strongly on ΔE, the amplitude of the periodic energy change along the crystal. When this amplitude is large, the dislocation line will prefer to lie along energy troughs such that short sharp kinks will be formed. Alternatively, when ΔE is small, long undulating kinks (more like gradual bends) will be observed.

The relative ease of movement of both the dislocation line segments lying along

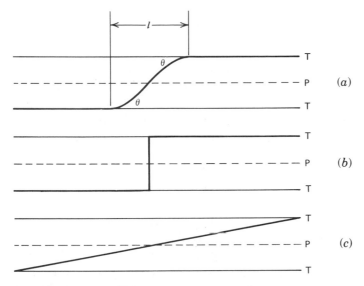

FIGURE 2.6   Position of dislocation line containing kinks with respect to energy troughs within lattice. (*a*) Typical configuration showing kink of length *l* with angle θ between kink segment and segment lying along energy trough; (*b*) sharp kink formed when magnitude of energy fluctuation in lattice is large. In this case *l*→0 and θ→90°; (*c*) broad kink formed when energy fluctuation in lattice is small. Here *l*→∞ and θ becomes very small.

energy troughs and the kinked sections, respectively, is now considered. Since the kinked sections are located across higher energy portions of the crystal, they can move more easily than the line segments along the energy troughs, which must overcome the maximum energy barrier if they are to move. Upon application of a shear stress, the kinked segments shown in Fig. 2.6*a* move to the left or right (depending on the sense of the stress), which in effect allows the entire dislocation line to move in a perpendicular direction from one energy trough to the adjacent one. The lateral movement of such a kink may be likened to the motion of a whip that has been snapped. Consequently, the introduction of kinks into dislocations eases their movement through the lattice.

It may be concluded, then, that the lattice resistance to the movement of a dislocation depends on both the magnitude of the Peierls stress and the orientation of the dislocation line within the periodically varying energy field in the lattice. Since both factors will depend on ΔE, which depends on the force–displacement relation between atoms, the importance of the latter relation is emphasized. Unfortunately, current lack of specific knowledge concerning the force law severely hampers quantitative treatments of dislocation–lattice interactions.

2.4   CHARACTERISTICS OF DISLOCATIONS

The reader should understand certain fundamental characteristics of dislocations. A dislocation is a lattice *line* defect that defines the boundary between slipped and unslipped portions of the crystal. Two basically different dislocations can be identified. The *edge* dislocation is defined by the edge of the extra half plane of atoms shown in Fig. 2.3. Note how this extra half plane is wedged into the top half of the crystal. As a result, the upper part of the crystal is compressed on either side of the half plane, while the region below the dislocation experiences considerable dilatation. By convention, the bottom edge of the half plane shown in Fig. 2.3 is defined as a *positive* edge dislocation. Had the extra half plane been introduced into the lower half of the crystal, the regions of localized compression and dilatation would be reversed and the dislocation line defined as a *negative* edge dislocation. Clearly, if a crystal contained both positive and negative edge dislocations lying on the same plane, their combination would result in mutual annihilation and the elimination of two high-energy regions of lattice distortion.

The movement of an edge dislocation and its role in the plastic deformation process may be understood more clearly by considering its Burgers circuit. (The Burgers circuit is a series of atom-to-atom steps along lattice vectors that generate a closed loop about any location in the lattice.) In a perfect lattice (Fig. 2.4c), the Burgers circuit beginning at A and progressing an equal and opposite number of lattice vectors in the horizontal and vertical directions will return to its starting position. When this occurs, the lattice contained within the circuit is considered perfect. When an edge dislocation is present in the lattice, the circuit does not close (Fig. 2.3). The vector needed to close the Burgers circuit $(X'X)$ is called the Burgers vector **b** of the dislocation and represents both the magnitude and direction of slip of the dislocation.

Another important feature of **b** is its orientation relative to the dislocation line. For the edge dislocation, **b** is oriented normal to the line defect. Ordinarily, plastic flow via edge dislocation movement is restricted to that one plane defined by the dislocation line and its Burgers vector. Such *conservative* motion will occur with the edge dislocation moving in the same direction as **b** (i.e., the direction of slip). From Section 2.3, the planes on which dislocations move are usually those of greatest separation and atomic density.

It is possible for an edge dislocation to undertake *nonconservative* motion, that is, movement out of its normal glide plane. This can occur by removal of a row of atoms, such as by the diffusion of lattice vacancies to the bottom of the extra half plane (Fig. 2.7). In this manner, the dislocation *climbs* from one plane to another where conservative glide may occur once again. Since vacancy diffusion is a thermally activated process, dislocation climb becomes an important process only at elevated temperatures above about one-half the melting point of the material. This mechanism will be discussed again in Chapter 5.

The other line defect, called the *screw* dislocation, is defined by the line *AB* in Fig. 2.8, the latter being generated by displacement of one part of the crystal relative to the other. The Burgers circuit about the screw dislocation assumes the shape of a helix, very much like a spiral staircase, wherein a 360° rotation produces a translation equal to one lattice vector in a direction parallel to the dislocation line *AB*. A right-handed screw dislocation is defined when a clockwise, 360° rotation causes the helix to advance one lattice vector. The same advance resulting from a 360° counterclockwise rotation is a left-handed screw dislocation. Thus, the screw dislocation and its Burgers vector are mutually parallel, unlike the orthogonal relationship found for the edge dislocation. Note that while the slip direction is again parallel to **b** as was found

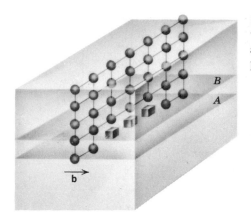

FIGURE 2.7 Dislocation climb involving vacancy (□) diffusion to edge dislocation allowing its movement to climb from plane A to plane B.

for the edge dislocation, the direction of movement of the screw dislocation is perpendicular to **b.** To better visualize this fact, take a piece of paper and tear it partway across its width. Note that the movement of your hands (the shear direction parallel to the Burgers vector) is perpendicular to the movement of the terminal point (the screw dislocation) of the tear.

Unlike the edge dislocation, a unique slip plane cannot be identified for the screw dislocation. Rather, an infinite number of potential slip planes may be defined, since the dislocation line and Burgers vector are parallel to one another. In fact, the movement of a screw dislocation is confined to those sets of planes that possess a low Peierls-Nabarro stress. Even so, the screw dislocation possesses greater mobility than the edge dislocation in moving through the lattice.

The movement of the screw dislocation from one slip plane to another takes place by a process known as *cross-slip* and may be understood by examining Fig. 2.9. At the onset of plastic deformation, the screw dislocation XY is seen to be moving on plane A (Fig. 2.9a). If continued movement on this plane is impeded by some obstacle, such as a precipitate particle, the screw dislocation can cross over to another equivalent plane, such as B, and continue its movement (Fig. 2.9b). Since the Burgers vector is unchanged, slip continues to occur in the same direction, though on a different plane. Movement of the screw dislocation may continue on plane B or return to plane A by a second cross-slip process (Fig. 2.9c). A summary of the basic differences between edge and screw dislocations is presented in Table 2.4.

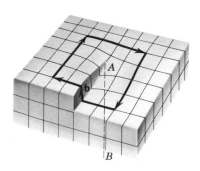

FIGURE 2.8 Screw dislocation AB resulting from displacement of one part of crystal relative to the other. Note that AB is parallel to b.

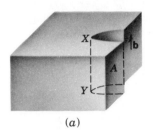 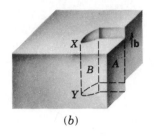 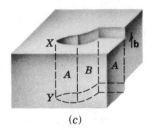

(*a*)   (*b*)   (*c*)

FIGURE 2.9 **Cross-slip of a screw dislocation *XY* from (*a*) plane *A* to (*b*) plane *B* to (*c*) plane *A*. Slip always occurs in direction of Burgers vector b.**

Since many dislocations in a crystalline solid are curved like the one shown in Fig. 2.10*a*, they take on aspects of both edge and screw dislocations. With **b** the same along the entire length of the dislocation, the dislocation is seen to be pure screw at *A* and pure edge at *B*. The reader should verify this by constructing a Burgers circuit around the dislocation at *A* and *B*. It follows that at all points between *A* and *B* the dislocation possesses both edge and screw components. For this reason, *AB* is called a mixed dislocation. Another example of a mixed dislocation is the dislocation loop. From Fig. 2.10*b* the loop is seen to be pure positive edge at *A* and pure negative edge at *B* while being pure right-handed screw at *D* and pure left-handed screw at *C*. Everywhere else the loop contains both edge and screw components. When a shear stress is applied parallel to **b**, we see from Table 2.4 that the loop will expand radially.

Dislocations can terminate at a free surface or at a grain boundary but never within the crystal. Consequently, dislocations either must form closed loops or networks with branches that terminate at the surface (Fig. 2.11). A basic characteristic of a network junction point or node involving at least three dislocation branches is that the sum of the Burgers vectors is zero:

$$\mathbf{b_1} + \mathbf{b_2} + \mathbf{b_3} = 0 \tag{2-8}$$

Furthermore, when these dislocations are of the same sense

$$\mathbf{b_1} = \mathbf{b_2} + \mathbf{b_3} \tag{2-9}$$

and the Burgers vector of one dislocation is equal to the sum of the other two Burgers vectors. This holds when two dislocations combine to form a third or when one dislocation dissociates into two additional dislocations.

TABLE 2.4 Characteristics of Dislocations

	Type of Dislocation	
Dislocation Characteristic	**Edge**	**Screw**
Slip direction	Parallel to **b**	Parallel to **b**
Relation between dislocation line and **b**	Perpendicular	Parallel
Direction of dislocation line movement relative to **b**	Parallel	Perpendicular
Process by which dislocations may leave the glide plane	Nonconservative climb	Cross-slip

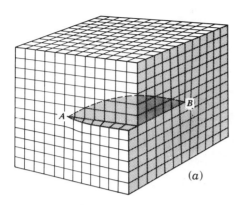
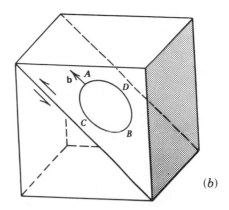

FIGURE 2.10 Curved dislocations containing edge and screw components. (*a*) Disloca-tion *AB* is pure screw at *A* and pure edge at *B*; (*b*) dislocation loop that grows out radially with shear stress applied parallel to b.

2.5 OBSERVATION OF DISLOCATIONS

Much effort has been devoted to the direct examination of dislocations within the crystal. One successful technique involves chemical or electrolytic etching of polished free surfaces. By carefully controlling the strength of the etchant, the high-energy dislocation cores exposed at the surface are attacked preferentially with respect to other regions on the polished surface. The result is the formation of numerous etch pits, each corresponding to one dislocation (Fig. 2.12). This technique has been used to study the effect of applied stress on dislocation velocity. Notable experiments by Johnston and Gilman[9] have identified the stress-induced, time-dependent change in

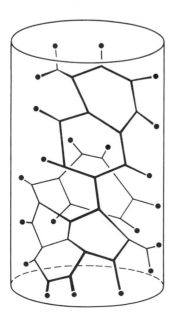

FIGURE 2.11 Network arrangement of dislocations in crystal. Dislocations can terminate only at a node, in a loop, or at a grain boundary or free surface.[7] (Repro-duced by courtesy of the Council of the Institution of Mechanical Engineers from *The Properties of Materials at High Rates of Strain*, by A. H. Cottrell.)

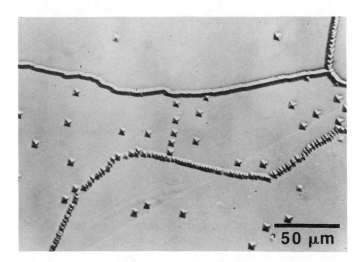

FIGURE 2.12   Etch pits on polished surface of lithium fluoride, each associated with an individual dislocation. The etch pit lineage indicates alignment of many dislocations in the form of low-angle boundaries (see Section 2.6). (From Gilman and Johnston;[18] reprinted by permission of General Electric Co.)

dislocation position by repeated etching, with each new etch pit representing the new location of the moving dislocation.

The most widely used technique for the study of dislocations involves their direct examination in the transmission electron microscope. Since electrons have little penetrating power, the specimens used for such studies are very thin films—only about 0.1 to 0.2 μm thick. Because dislocations are lattice defects, their presence perturbs the path of the diffracted electron beam relative to its path in a perfect crystal. As a result, various images are produced, depending on the prevailing diffraction conditions. Often, dislocations appear as single dark lines like those shown in planar array in Fig. 2.13*a*. Each dislocation lies along a particular plane and extends from the top to the bottom of the foil (Fig. 2.13*b*). As a result, the viewer sees only the projected length of the dislocation line, with the actual length being dependent on the foil thickness and angle of the plane containing the dislocations. Much progress was made about 35 years ago toward the understanding of electron diffraction images and the identification of numerous dislocation configurations. Some important publications in this area are cited at the end of the chapter.[11–17]

2.6   ELASTIC PROPERTIES OF DISLOCATIONS

As might be expected, there is an elastic stress field associated with the distorted lattice surrounding a dislocation. It is easy to describe the stresses developed around a screw dislocation (Fig. 2.14*a*). By rolling the cylindrical element out flat (Fig. 2.14*b*), the shear strain $\gamma_{\theta z}$ is seen in polar coordinates to be

$$\gamma_{\theta z} = \frac{b}{2\pi r} \qquad (2\text{-}10)$$

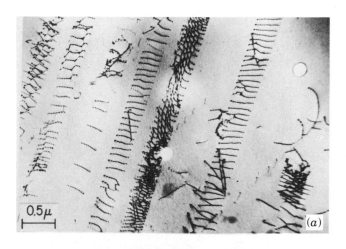

FIGURE 2.13 Observation of individual dislocations in thin foil. (*a*) Planar arrays of dislocations in 18Cr–8Ni stainless steels (from Michalak,[10] *Metals Handbook,* Vol. 8, copyright American Society for Metals, Metals Park, OH, 1973; used with permission); (*b*) diagram showing position of dislocations on the guide plane in the foil (after Hull[11]).

From Hooke's law, the corresponding stress is

$$\tau_{\theta z} = G\gamma_{\theta z} = \tau_{z\theta} = \frac{Gb}{2\pi r} \tag{2-11}$$

Since displacements are generated only in the z direction, the other stress components are zero. Equation 2-11 shows that the stress $\tau_{\theta z}$ becomes infinitely large as r approaches zero. Since this is unreasonable, there exists a limiting distance r_0 from the dislocation center (estimated to be 5 to 10 Å) within which Eq. 2-11 is no longer applicable. For the rectangular coordinates shown in Fig. 2.14b, the shear stresses surrounding the screw dislocation can also be given by

$$\tau_{xz} = \tau_{zx} = -\frac{Gb}{2\pi}\frac{y}{x^2 + y^2}$$
$$\tau_{yz} = \tau_{zy} = \frac{Gb}{2\pi}\frac{x}{x^2 + y^2} \tag{2-12}$$

Again all other stresses are zero.

The stress field surrounding an edge dislocation is more complicated, since both hydrostatic and shear stress components are present. In rectangular coordinates these stresses are given by[4]

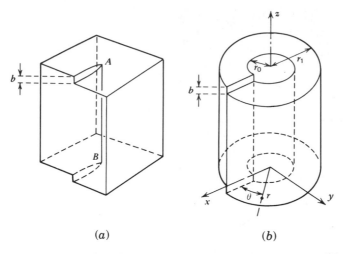

(a)   (b)

FIGURE 2.14 **Elastic distortions surrounding screw dislocation. (From Hull,[11] reprinted with permission from Hull,** *Introduction to Dislocations*, **Pergamon Press, Elmsford, NY, 1965.)**

$$\sigma_{xx} = -\frac{Gby}{2\pi(1-\nu)}\frac{(3x^2+y^2)}{(x^2+y^2)^2}$$

$$\sigma_{yy} = +\frac{Gby}{2\pi(1-\nu)}\frac{(x^2-y^2)}{(x^2+y^2)^2}$$

$$\tau_{xy} = \tau_{yx} = \frac{Gbx}{2\pi(1-\nu)}\frac{(x^2-y^2)}{(x^2+y^2)^2} \qquad (2\text{-}13)$$

$$\sigma_{zz} = \nu\,(\sigma_{xx}+\sigma_{yy})$$

$$\tau_{xz} = \tau_{zx} = \tau_{yz} = \tau_{zy} = 0$$

where ν = Poisson's ratio, Comparing Eq. 2-13 with Fig. 2.15, we find a region of pure compression directly above the edge dislocation $(X = 0)$ and pure tension below the bottom edge of the extra half plane. Along the slip plane $(Y = 0)$ the stress is pure shear. For all other positions surrounding the dislocation, the stress field is found to contain compressive and/or tensile components as well as a shear component.

The elastic strain energy is another elastic property of a dislocation. For the simple case of the screw dislocation, this quantity may be given by

$$E_{\text{screw}} = \frac{1}{2}\int_{r_0}^{r_1} \tau_{\theta z}\, b\, dr \qquad (2\text{-}14)$$

Note that the energy is defined for the region outside the core of the dislocation r_0 to the outer boundaries of the crystal r_1 (see Fig. 2.14). Combining Eqs. 2-11 and 2-14, we get

$$E_{\text{screw}} = \frac{1}{2}\int_{r_0}^{r_1} \frac{Gb^2}{2\pi}\frac{dr}{r} = \frac{Gb^2}{4\pi}\ln\frac{r_1}{r_0} \qquad (2\text{-}15)$$

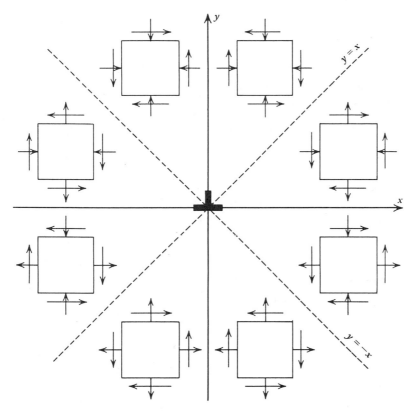

FIGURE 2.15   Elastic stress field surrounding an edge dislocation. (From Read,[3] *Dislocations in Crystals;* copyright McGraw-Hill Book Co., New York, © 1953. Used with permission of McGraw-Hill Book Company.)

The elastic energy of the edge dislocation is slightly larger and given by

$$E_{\text{edge}} = \frac{Gb^2}{4\pi(1-v)} \ln \frac{r_1}{r_0} \qquad (2\text{-}16)$$

Since a general dislocation contains both edge and screw components, its energy is intermediate to the limiting values given by Eqs. 2-15 and 2-16. For purposes of our discussion it is sufficient to note that

$$E = \alpha Gb^2 \qquad (2\text{-}17)$$

where E = energy of any dislocation
    α = geometrical factor with α taken between 0.5 and 1.0

A particularly important consequence of Eqs. 2-15 to 2-17 is that slip will usually occur in close-packed directions so as to minimize the Burgers vectors of the dislocation. The preferred slip directions in major crystal types are given in Chapter 3. Equations 2-15 to 2-17 also allow one to determine whether or not a particular dislocation reaction will occur. From Eq. 2-9, such a reaction will be favored when

$$b_1{}^2 > b_2{}^2 + b_3{}^2$$

(neglecting possible anisotropy effects associated with G).

Two other elastic properties of a dislocation are its line tension and the force needed to move the dislocation through the lattice. The line tension T is described in terms of its energy per unit length and is given by

$$T \propto Gb^2 \tag{2-18}$$

The line tension acts to straighten a dislocation line to minimize its length, thereby lowering the overall energy of the crystal (see Fig. 2.6c). Consequently, it is necessary to apply a stress τ so that the dislocation line remains curved. This stress is shown to increase with increasing line tension T and decreasing radius of curvature R where

$$\tau \propto \frac{T}{bR} \tag{2-19}$$

Combining Eqs. 2-18 and 2-19 we find that

$$\tau \propto \frac{Gb}{R} \tag{2-20}$$

This relationship will be referred to in Section 2.10.

Finally, the force acting on a dislocation is found to depend on the intrinsic resistance to dislocation movement through the lattice, the Peierls-Nabarro stress (Section 2.3), and interactions with other dislocations. As shown by Read[3] for the case of parallel dislocations, screw dislocations will always repel one another when the Burgers vectors of both dislocations are of the same sign; they will always attract one another when the signs of the Burgers vectors are opposite. In either case, the magnitude of the force is inversely proportional to the distance between the two dislocations. The force between two edge dislocations is complicated by a reversal in sign when the horizontal distance between two dislocations becomes less than the vertical distance between the two parallel slip planes (Fig. 2.16). Consequently, like edge dislocations are attracted to one another when $x<y$. As a result, like edge dislocations can form stable arrays of dislocations located vertically above one another in the form of simple tilt boundaries (Fig. 2.12).

2.7 PARTIAL DISLOCATIONS

As noted in the previous section, the likelihood that dislocation b_1 will dissociate into two dislocations b_2 and b_3, often referred to as Shockley partial dislocations, depends on whether the sum of the elastic energies of partial dislocations b_2 and b_3 is lower than the elastic energy associated with dislocation b_1. From Eq. 2-17, the dissociation will occur when

$$Gb_1{}^2 > Gb_2{}^2 + Gb_3{}^2$$

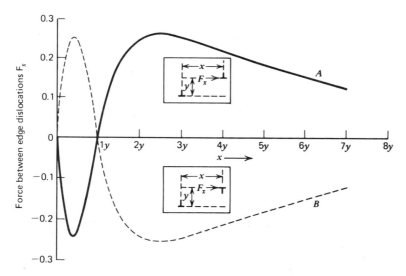

FIGURE 2.16 Force between parallel edge dislocations. Curve A corresponds to dislocations of the same sign. Force reversal when $X < Y$ causes like dislocations to become aligned as in a tilt boundary. Curve B corresponds to dislocations of opposite sign. Unit of force F_x is $Gb^2/2\pi(1 - v) \cdot x(x^2 - y^2)/(x^2 + y^2)^{2}.^{7}$ **(From A. H. Cottrell,** *The Properties of Materials at High Rates of Strain,* **Institute of Mechanical Engineering, London, 1957.)**

Using the FCC lattice as a model, it can be shown that the whole dislocation $\mathbf{b}_1$, oriented in the $\langle 110 \rangle$ close-packed direction, can dissociate into two dislocations of type $\langle 112 \rangle$. We see from Fig. 2.17a that the motion of the atoms on the slip plane is from A to B to C rather than directly in the close-packed direction AC. That is, the whole dislocation AC dissociates into two partial dislocations, AB and BC. For example

$$\frac{a}{2}[\bar{1}01] \rightarrow \frac{a}{6}[\bar{2}11] + \frac{a}{6}[\bar{1}\bar{1}2]$$

From Eq. 2-17

$$\frac{a^2}{4}(1 + 1) > \frac{a^2}{36}[4 + 1 + 1] + \frac{a^2}{36}[1 + 1 + 4]$$

$$\frac{a^2}{2} > \frac{a^2}{3}$$

(For simplicity, any anisotropy in elastic shear modulus has been ignored.) Therefore, the dislocation reaction will proceed in the direction indicated. It is possible to sense this dislocation reaction in a tactile way. If you were to hold a sheet of close-packed Ping-Pong balls (glued together, of course) in one hand and then slide it across a second sheet of balls parallel to one of the close-packed directions, you will note that

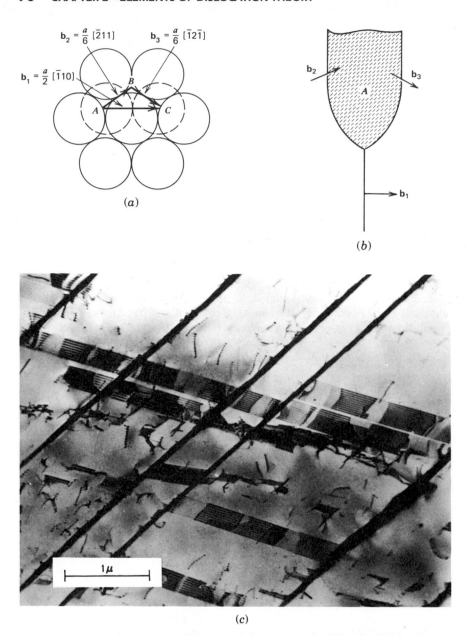

FIGURE 2.17 (*a*) Path of whole and partial (Shockley) dislocations; (*b*) Shockley b_2 and b_3 surrounding stacking fault region *A*; (*c*) Long stacking fault ribbons (bands of closely spaced lines) in low SFE 18Cr–8Ni stainless steel. Faults are bounded at ends by partial dislocations. Thin black bands are mechanical twins. (After Michelak;[10] reprinted with permission from *Metals Handbook*, Vol. 8, American Society for Metals, Metals Park, OH, © 1973.)

the sheets prefer to slide past one another in zigzag fashion along the troughs between the balls on the second sheet. In the FCC lattice, these troughs are parallel to $\langle 112 \rangle$ directions. Because of the reduction in strain energy and the fact that the partials have similar vector components, these partials will tend to repel one another and move apart. The extent of separation, denoted by area A in Fig. 2.17b, will depend on the nature of the change in stacking sequence that occurs between $\mathbf{b}_2$ and $\mathbf{b}_3$. Movement of these partial dislocations produces a change in the stacking sequence from the FCC type—ABCABCABC—to include a local perturbation involving the formation of a layer of HCP material—ABCBCABC. Examples of stacking faults are shown in Fig. 2.17c. For an FCC crystal, the layer of HCP material that is introduced will elevate the total energy of the system. Therefore, the equilibrium distance of separation of two partials reflects a balance of the net repulsive force between the two partial dislocations containing Burgers vector components of the same sign and the energy of the associated stacking fault. According to Cottrell,[4] this separation distance varies inversely with the stacking fault energy (SFE) and may be given by

$$d = \frac{G(\mathbf{b}_2\mathbf{b}_3)}{2\pi\gamma} \tag{2-21}$$

where
d = partial dislocation separation
$\mathbf{b}_2, \mathbf{b}_3$ = partial dislocation Burgers vectors
G = shear modulus
γ = stacking fault energy

The SFE of alloy crystals depends on their composition, and comparative values for pure metals also differ. Typical values for different elements and alloys are given in Table 2.5. For the case of copper-based alloys, Thornton el al.[19] showed SFE to be strongly affected by the material's electron/atom ratio. They found that when $e/a > 1.1$, the stacking fault energy usually decreased to below 20 mJ/m^2.

2.7.1 Movement of Partial Dislocations

The movement of the two Shockley partial dislocations is restricted to the plane of the fault, since movement of either partial on a different plane would involve ener-

TABLE 2.5 Selected Stacking Fault
Energies for FCC Metals

Metal	Stacking Fault Energy (mJ/m^2 = ergs/cm^2)
Brass	<10
Stainless steel	<10
Ag	~25
Au	~75
Cu	~90
Ni	~200
Al	~250

 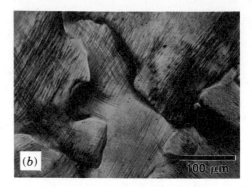

FIGURE 2.18 **Photomicrographs revealing slip character. (*a*) Planar glide in low stacking fault energy material; (*b*) wavy glide in high stacking fault energy material.**

getically unfavorable atomic movements. Therefore, cross-slip of an extended screw dislocation around obstacles is not permitted without thermally activated processes and, as a consequence, the slip offsets seen on a polished surface will be straight (Fig. 2.18*a*). Such is the case for a material of low stacking fault energy and widely separated partial dislocations. This type of dislocation movement is called *planar glide*. By the application of a suitably large stress, however, it is possible to squeeze the partial dislocations together against a barrier to form a whole dislocation. If this recombined dislocation is of the screw type, it may cross-slip (recall Fig. 2.9). As you might imagine, the stress necessary to recombine the partial dislocations will depend on the equilibrium distance of separation of the partials, which in turn depends on the magnitude of the stacking fault energy (Eq. 2-21). For materials with low stacking fault energy, partial dislocation separation is large (on the order of 10 to 20*b*) and the force necessary for recombination is large. Conversely, little stress is necessary to recombine partial dislocations in a high stacking fault energy material where partial dislocation separation is small (on the order of 1*b* or less). When cross-slip is easy, slip offsets on a polished surface take on a wavy pattern (Fig. 2.18*b*), and this deformation is called *wavy glide*.

One major implication of the dependence of cross-slip on stacking fault energy is the dominant role the latter plays in determining the strain-hardening characteristics of a material. When the stacking fault energy is low, cross-slip is restricted so that barriers to dislocation movement remain effective to higher stress levels than in material of higher stacking fault energy. That is to say, the low stacking-fault-energy material strain hardens to a greater extent. It is then possible to relate strain-hardening coefficients (Table 1.5) with stacking fault energy values (Table 2.5) as shown in Table 2.6. Note that the strain-hardening coefficient increases with decreasing stacking fault energy while the slip character changes from a wavy to a planar mode.

2.8 SUPERLATTICE DISLOCATIONS

The character of the deformation process is altered if dislocations are forced to move through a lattice containing long-range order. For example, consider the case for the

TABLE 2.6 Slip Character and Strain-Hardening Coefficients for Several Metals

Metal	Stacking Fault Energy (mJ/m^2)	Strain-Hardening Coefficient	Slip Character
Stainless steel	<10	~0.45	Planar
Cu	~90	~0.3	Planar/wavy
Al	~250	~0.15	Wavy

addition of aluminum atoms to a nickel crystal. Under certain compositional and thermal history conditions, an intermetallic compound Ni_3Al may form, with the aluminum atoms being located at the eight corner positions of the unit cell and the nickel atoms being located at the six cube faces (Fig. 2.19). Note that the ordered Ni_3Al phase is of the FCC type and contains four atoms (three nickel and one aluminum) per unit cell. The passage of a dislocation through half of a spherical particle containing this superlattice (consisting of Ni and Al atoms in specific lattice sites) generates an unfavorable rearrangement of the aluminum and nickel atoms, as shown in Fig. 2.20a. We see that the Ni (open circles) and Al (solid circles) atoms are opposite one another along that part of the slip plane that was traversed by the dislocation. This arrangement of nonpreferred atom pairs on the slip plane creates an antiphase domain boundary (APB). Since there is an additional energy associated with the APB, which depends on the degree of order in the lattice, dislocation motion is restricted. However, if a second identical dislocation were to sweep across the same plane, atomic disorder would be eliminated with atoms again assuming their preferred positions in the lattice (Fig. 2.20b). Note that the right side of the particle is still disordered since the second dislocation has passed through only half of the particle. The equilibrium distance separating these two dislocations (referred to as a superlattice dislocation) reflects a balance between an attractive force associated with minimization of APB energy and a repulsive force due to the stress fields of identical dislocations (recall Eq. 2-21). An example of superlattice dislocations (i.e., dislocation pairs) in Ni_3Al is shown in Fig. 2.21.

2.9 DISLOCATION–DISLOCATION INTERACTIONS

Since dislocations of the same sign will repel one another and not coalesce, they will tend to pile up (each with a unit Burgers vector) against a barrier on the slip plane

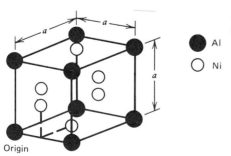

Origin

FIGURE 2.19 Nickel and aluminum atom locations in ordered Ni_3Al phase.

● Al
○ Ni

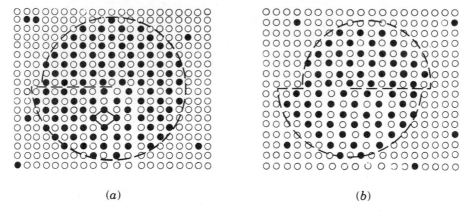

(a)   (b)

FIGURE 2.20 **(100) planar view of spherical Ni$_3$Al particle in Ni lattice. (a) Initial superlattice dislocation disorders atom pairs along slipped portion of glide plane. Note orientation of cube face in Ni$_3$Al particle. (b) Passage of second superlattice dislocation reorders Ni$_3$Al lattice. Dotted horizontal line corresponds to APB. Nickel (○) and aluminum (●) atom locations noted. (From Gleiter and Hornbogen.[20])**

(Fig. 2.22). As one might expect, a large stress concentration is developed at the leading edge of the pileup, which can lead to premature fracture in certain materials (see Chapter 10 for further discussion of this point).

Equally important are the intersections of dislocations moving on different slip planes. Two different edge dislocation interactions are shown in Fig. 2.23. In the first case, where the Burgers vectors of the two dislocations are at right angles, the intersection of dislocation AB leads to a simple lengthening of dislocation XY. On the other hand, dislocation XY with a Burgers vector $\mathbf{b}_1$ cuts dislocation AB, producing a jog PP' that has a length equal to that of $\mathbf{b}_1$. The Burgers vector of PP', however, is $\mathbf{b}_2$—the same as dislocation AB to which it belongs. Since $\mathbf{b}_2$ and PP' are normal to one another, PP' is of the edge type. It may be shown that this jog does not impede the movement of the dislocation AB. When the Burgers vectors of the edge dislocations are parallel, the jogs produced are different in character. As shown in Fig. 2.23b, dislocation XY with its Burgers vector $\mathbf{b}_1$ produces a jog PP' in dislocation AB. Since

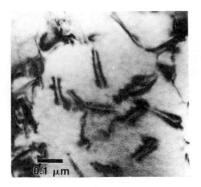

FIGURE 2.21 **Superlattice dislocation pairs in fully ordered Ni$_3$Al. (Photo courtesy M. Khobaib.[21])**

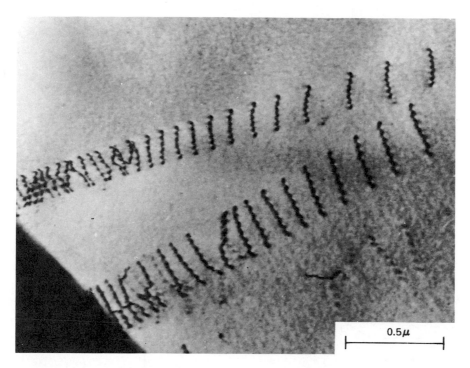

FIGURE 2.22 Dislocation pileups on two systems against a grain boundary in 309 stainless steel ($\gamma = 35$ mJ/m^2). (Courtesy of Anthony Thompson, Carnegie-Mellon University.)

the Burgers vector $\mathbf{b}_2$ in dislocation AB is parallel to the jog PP', the jog is of the screw type. Similarly, the jog QQ' in dislocation XY is found also to be of the screw type. The screw jogs PP' and QQ' have greater mobility than the edge dislocations to which they belong. Consequently, their presence does not impede the overall motion of the dislocation. In summary, jogs generated in edge dislocations will not affect the movement of the dislocation.

The same cannot be said for intersections involving screw dislocations. As illustrated in Fig. 2.24a the intersection of an edge and screw dislocation will produce a jog PP' in the edge dislocation AB and another jog QQ' in the screw dislocation XY. Since each jog assumes the same Burgers vector as its dislocation, it may be seen that PP' and QQ' are both edge jogs. From the above discussion, PP' will not impede the motion of dislocation AB whereas QQ' will restrict the movement of the screw dislocation XY. The same can be said for the edge type jogs PP' and QQ' found in the screw dislocations AB and XY, respectively, shown in Fig. 2.24b. The restriction placed on the mobility of the screw dislocations is caused by the inability of the edge jog to move on any plane other than that defined by the jog QQ' and $\mathbf{b}_2$ (i.e., the plane $QQ'YZ$; see Fig. 2.25). Consequently, when a shear stress is applied parallel to $\mathbf{b}_2$, the screw segments XQ and $Q'Y$ will produce displacements parallel to $\mathbf{b}_2$, while the screw dislocation lines move to DE and FG, respectively. The only way that the edge jog QQ' can follow along plane $EFQ'Q$ is by nonconservative motion involving

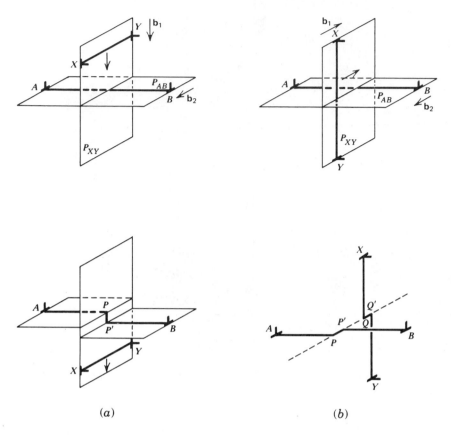

(a)   (b)

FIGURE 2.23 **Intersection of two edge dislocations.** (*a*) **Burgers vectors are at right angles and produce an edge jog** *PP′* **in dislocations** *AB*. **(From Read,[3]** *Dislocations in Crystals;* **copyright McGraw-Hill Book Co., New York. © 1953. Used with permission of McGraw-Hill Book Company.)** (*b*) **Burgers vectors are parallel and produce two screw jogs** *PP′* **and** *QQ′*. **(From Hull;[11] reprinted with permission from Hull,** *Introduction to Dislocations*, **Pergamon Press, Elmsford, NY, 1965.)**

vacancy-assisted dislocation climb. As shown schematically in Fig. 2.26 for the case of small jogs with heights of one or two atom spacings, the screw dislocation first bows out under application of a shear stress and then moves farther only by dragging along the edge jogs, which leave behind a trail of vacancies. When the jog height is greater as a result of multiple dislocation–dislocation intersections (e.g., about 50 to 100 Å in silicon iron), too many vacancies would be required for climb of the jog. As a result, long-edge dislocation segments (called dipoles) are left behind as the screw segments of the dislocation advance through the crystal (Fig. 2.27*a*). When the jog height is even larger (e.g., greater than 200 Å in silicon iron), the screw segments *XP* and *P′Y* move independently of one another (Fig. 2.27*b*). Examples of the three height categories of edge jogs in screw dislocations are shown in Fig. 2.28 for the case of silicon iron.

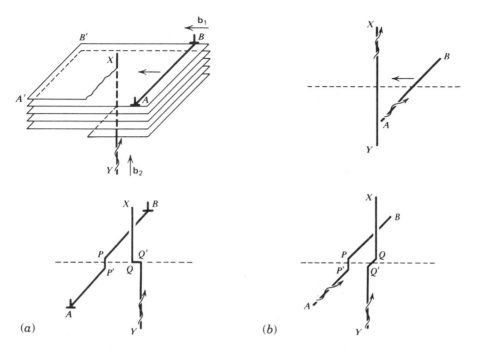

FIGURE 2.24   Intersection of screw dislocation *XY* with (*a*) edge dislocations *AB* to form two edge jogs *PP'* and *QQ'*. (From Read,[3] *Dislocations in Crystals;* McGraw-Hill Book Co., New York, © 1953. Used with permission of McGraw-Hill Book Company.) (*b*) Another screw dislocation *AB* which forms two edge jogs *PP'* and *QQ'*. (From Hull[11]; reprinted with permission from Hull, *Introduction to Dislocations*, Pergamon Press, Elmsford, NY, 1965.) Edge jogs in screw dislocations impede their motion.

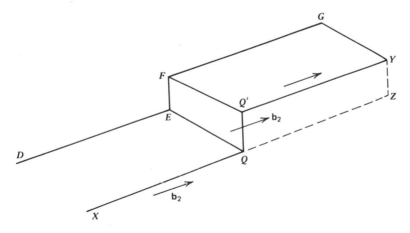

FIGURE 2.25   Screw dislocation *XY* containing an edge jog *QQ'* which can move conservatively on plane *QQ'YZ* but nonconservatively on plane *EFQ'Q* when screw components *XQ* and *Q'Y* move to *DE* and *FG*, respectively.

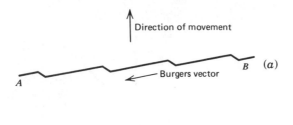

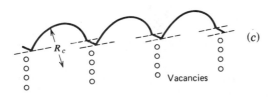

FIGURE 2.26 **Detailed movement of jogged screw dislocation.** (*a*) **Jogged dislocation under zero stress;** (*b*) **applied shear stress causes screw component to bow out between edge jogs;** (*c*) **edge jogs follow screw segments by nonconservative climb, leaving behind a trail of vacancies. (From Hull;[11] reprinted with permission from Hull,** *Introduction to Dislocations,* **Pergamon Press, Elmsford, NY, 1965.)**

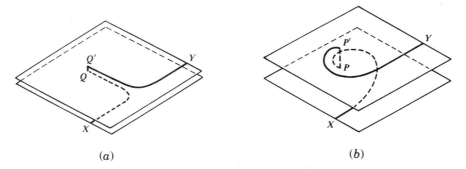

FIGURE 2.27 **Effect of jog height on screw dislocation mobility.** (*a*) **Intermediate jog height** *QQ′* **causes long-edge segments (dipoles) to form as screw segments glide through crystal;** (*b*) **large jog height** *PP′* **allows screw segments** *XP* **and** *YP′* **to move independently of one another. (From Gilman and Johnston;[22] reprinted with permission of the authors and Academic Press, Inc., New York.)**

FIGURE 2.28 Dislocations in silicon–iron thin film. Note dipole trails at *A*, pinched off dipoles at *B*, and independent dislocation movement at the large jog at *C*. (From Low and Turkalo;[23] reprinted with permission from Low, *Acta Met.* 10 (1962), Pergamon Press, Elmsford, NY.)

2.10 DISLOCATION MULTIPLICATION

Since slip offsets are clearly visible in a light microscope (e.g., see Fig. 2.18), they must be in the range of 1 μm in height. Since the typical Burgers vector for a dislocation is on the order of 2 to 3 × 10⁻⁸ cm, there is a requirement for approximately 10⁴ dislocations on each slip plane to create the slip step. That so many dislocations of the same sign should lie on the same plane *before* the crystal is stressed is highly unlikely. A possible alternative explanation is that additional dislocations must have been generated during deformation. This view is supported by the observations of electron microscopists who have found the dislocation densities in thin metal films to increase from $10^4 - 10^5$ to $10^{11} - 10^{12}$ dislocations/cm² as one proceeds from the annealed to heavily cold-worked state.

A widely accepted mechanism for dislocation generation is based on the Frank-Read source. In this model, a segment of a dislocation line is considered to be pinned either by foreign atoms or particles, or by interactions with other dislocations (Fig. 2.29). If a shear stress is applied to the crystal, the segment *AB* will bow out with a radius given by

$$\tau \propto \frac{Gb}{R} \tag{2-20}$$

Dislocation bowing will increase with increasing applied stress, while the radius of curvature decreases to the point where *R* equals half the pinned segment length *l* (Fig.

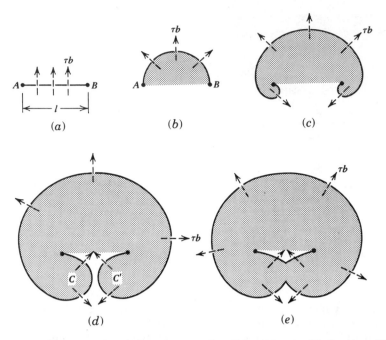

FIGURE 2.29   **Frank-Read source for dislocation multiplication. Slipped area is shaded. Loop instability point is reached when shear stress $\tau \approx Gb/l$. (From Read;[3] *Dislocations in Crystals*, McGraw-Hill Book Co., New York, © 1953. Used with permission of McGraw-Hill Book Company.)**

2.29*b*). At this point, the loop becomes unstable and begins to bend around itself (Fig. 2.29*c*). The stress necessary to produce this instability is given by

$$\tau \approx \frac{GB}{l} \qquad (2\text{-}22)$$

where l = distance between pinning points. Finally, the loop pinches off at C and C', since these two regions correspond to screw dislocations of opposite signs (Fig. 2.29*d*). After this has occurred, the loop and cusp ACB straighten, leaving the same segment AB as before but with an additional loop containing the same Burgers vector as the original segment (Fig. 2.29*e*). Upon further application of the stress, the segment AB can bow again to form a second loop while the initial loop moves out radially. With continued application of the shear stress, this source can generate an unlimited number of dislocations. In reality, however, the source is eventually shut down by back stresses produced by the pileup of dislocation loops against unyielding obstacles. The photograph shown in Fig. 2.30*a* represents a classic illustration of a Frank-Read source in a silicon crystal.

Another closely related dislocation generation mechanism has been suggested by Koehler[25] and modified by Low and Guard.[26] The basic feature of this model is that through the process of cross-slip, a screw dislocation can generate additional Frank-Read sources. From Fig. 2.30*b*, we see that a screw dislocation segment has cross-

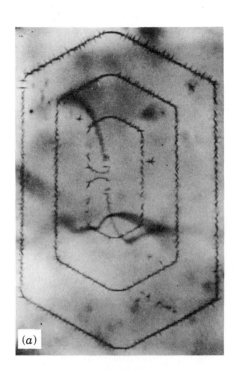

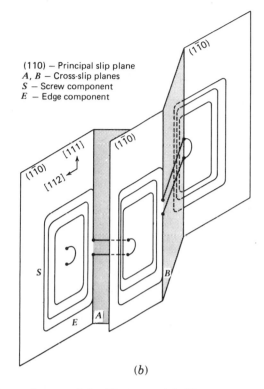

(110) — Principal slip plane
A, B — Cross-slip planes
S — Screw component
E — Edge component

FIGURE 2.30   Frank-Read sources. (a) Photomicrograph in silicon crystal. (From Dash;[24] reprinted with permission of General Electric Co.) (b) Dislocation multiplication by double cross-slip mechanism. (From Low and Guard;[26] reprinted with permission from Low, *Acta Met*. 7 (1959), Pergamon Press, Elmsford, NY.)

slipped twice to resume movement on a plane parallel to the initial slip plane. Note the additional dislocation loops that may be generated by this process.

REFERENCES

1. J. Frenkel, *Z. Phys*. **37,** 572 (1926).
2. W. J. McG. Tegart, *Elements of Mechanical Metallurgy*, Macmillan, New York, 1966.
3. W. T. Read, Jr., *Dislocations in Crystals*, McGraw-Hill, New York, 1953.
4. A. H. Cottrell, *Dislocations and Plastic Flow in Crystals*, Clarendon Press, Oxford, 1953.
5. A. G. Guy, *Elements of Physical Metallurgy*, 2d ed., Addison-Wesley, Reading, MA, 1959.
6. W. H. Sutton, B. W. Rosen, and D. G. Flom, *SPE J*. **72,** 1203 (1964).
7. A. H. Cottrell, *The Properties of Materials at High Rates of Strain*, Institute of Mechanical Engineering, London, 1957.
8. A. J. Forman, M. A. Jaswon, and J. K. Wood, *Proc. Phys. Soc. A* **64,** 156 (1951).

9.   W. G. Johnston and J. J. Gilman, *J. Appl. Phys.* **30,** 129 (1959).
10. J. T. Michalak, *Metals Handbook*, Vol. 8, ASM, Metals Park, OH, 1973, p. 218.
11. D. Hull, *Introduction to Dislocations,* Pergamon, Oxford, 1965.
12. P. B. Hirsch, *J. Inst. Met.* **87,** 406 (1959).
13. P. B. Hirsch, *Metall. Rev.* **4,** 101 (1959).
14. A. Howie, *Metall. Rev.* **6,** 467 (1961).
15. P. Kelly and J. Nutting, *J. Inst. Met.* **87,** 385 (1959).
16. P. B. Hirsch, *Prog. Met. Phys.* **6,** 236 (1956).
17. J. B. Newkirk and J. H. Wernick, eds., *Direct Observations of Imperfections in Crystals*, Interscience, New York, 1962.
18. J. J. Gilman and W. G. Johnston, in *Dislocations and Mechanical Properties of Crystals*, J. C. Fisher, W. G. Johnston, R. Thomson, and T. Vreeland, Jr., Eds., Wiley, New York, 1957, p. 116.
19. P. R. Thornton, T. E. Mitchell, and P. B. Hirsch, *Philos. Mag.* **7,** 1349 (1962).
20. H. Gleiter and E. Hornbogen, *Mater. Sci. Eng.* **2,** 285 (1967/68).
21. K. Khobaib, private communication.
22. J. J. Gilman and W. G. Johnston, *Solid State Phys.* **13,** 147 (1962).
23. J. R. Low and A. M. Turkalo, *Acta Met.* **10,** 215 (1962).
24. W. C. Dash, *Dislocations and Mechanical Properties of Crystals*, J. C. Fisher, Ed., Wiley, New York, 1957.
25. J. S. Koehler, *Phys. Rev.* **86,** 52 (1952).
26. J. R. Low and R. W. Guard, *Acta Met.* **7,** 171 (1959).

PROBLEMS

2.1 Demonstrate mathematically that dislocations at the head of a pileup will not combine to form a super dislocation with a Burgers vector of nb where $n = 2$, 3, 4, ..., n.

2.2 Consider the following face-centered-cubic dislocation reaction:

$$\frac{a}{2}[110] \rightarrow \frac{a}{6}[21\bar{1}] + \frac{a}{6}[121]$$

   (a) Prove that the reaction will occur.

   (b) What kind of dislocations are the $(a/6)\langle 121 \rangle$?

   (c) What kind of crystal imperfection results from this dislocation reaction?

   (d) What determines the distance of separation of the $(a/6)[21\bar{1}]$ and the $(a/6)[121]$ dislocations?

2.3 Distinguish between climb and cross-slip and discuss the role of stacking fault energy with regard to the latter.

2.4 Discuss the nature of the Peierls stress with regard to a dislocation and describe the role of the Peierls stress in determining the preferred slip plane in a crystal and the yield-strength temperature dependence of the crystal.

2.5 Why do dislocation loops tend to be circular? Why, then, are they angular for silicon as shown in Fig. 2.30a?

CHAPTER 3

SLIP AND TWINNING IN CRYSTALLINE SOLIDS

3.1 SLIP

We saw in the previous chapter that plastic deformation occurs primarily by sliding along certain planes with one part of a crystal moving relative to another. This blocklike nature of slip produces crystal offsets (called slip steps) in amounts given by multiples of the unit dislocation displacement vector **b** (see Fig. 3.1). To minimize the Peierls stress, slip occurs predominantly on crystallographic planes of maximum atomic density. In addition, slip will occur in the close-packed direction, which represents the shortest distance between two equilibrium atom postions and, hence, the lowest energy direction.

3.1.1 Crystallography of Slip

As shown in Fig. 3.2, the dominant slip systems (combinations of slip planes and directions) vary with the material's crystal lattice, since the respective atomic density of planes and directions are different. For the case of face-centered-cubic (FCC) crystals, slip occurs most often on $\{111\}$ octahedral planes and in $\langle 110 \rangle$ directions that are parallel to cube face diagonals. In all, there are 12 such slip systems (four $\{111\}$ planes and three $\langle 110 \rangle$ slip directions for each $\{111\}$ plane). Other FCC slip systems have been found but will not be considered here since they are activated only by unusual test conditions.

In body-centered-cubic (BCC) crystals, slip occurs in the $\langle 111 \rangle$ cube diagonal direction and on $\{110\}$ dodecahedral planes. Slip may occur on $\{112\}$ and $\{123\}$ planes as well. A total of 48 possible slip systems can be identified, based on combinations of these three slip planes and the common $\langle 111 \rangle$ slip direction. The fourfold greater number of slip systems in BCC as compared to FCC crystals does not mean that the former lattice provides more ductility; in fact, the reverse is true because FCC crystals have a much lower Peierls-Nabarro stress and contain more mobile dislocations.

Prediction of the preferred slip systems in hexagonal close-packed (HCP) materials is not an easy task. On the basis of the Peierls stress argument wherein the most densely packed planes of greater separation would be the preferred slip planes, one would expect the active slip planes in hexagonal crystals to vary with the *c/a* ratio

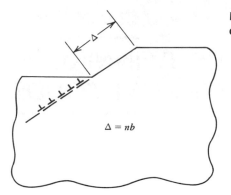

FIGURE 3.1   Diagram showing slip offset due to n dislocations leaving the crystal.

$$\Delta = nb$$

(see Fig. 3.2c). That is, if the c/a ratio is less than that for ideal packing (1.633), then the prism planes become atomically more dense relative to the basal plane. For this case, $\{10\bar{1}0\}$ prism slip would be preferred in the $\langle11\bar{2}0\rangle$ close-packed direction (Fig. 3.2). This has been found true for the case of zirconium and titanium but not for cobalt, magnesium, or beryllium (Table 3.1). Researchers have sought with little success other explanations to account for the observed slip behavior of these three metals.[1] On the other hand, when $c/a>1.633$, the basal plane should be the preferred slip plane as shown for the case of zinc and cadmium.

Thus far, my discussion has focused on the importance of the relative atomic density of crystallographic planes and directions in deciding whether a particular plane and direction combination could serve as a potential slip system. Certain slip-system combinations of ceramic crystals seem reasonable on the basis of density considerations, but are negated by the effects of strong directional bonding in covalent crystals or electrostatic interactions in ionic crystals. Since such atomic movements would be energetically unfavorable, the number of potential slip systems in these materials is restricted as is their overall ductility (except at relatively high test temperatures). A compilation of reported slip systems in selected ceramic crystals is given in Table 3.2.

The ductility of a material depends also on its ability to withstand a general homogeneous strain involving an arbitrary shape change of the crystal. Von Mises[2] showed this to be possible when five independent slip systems are activated. If we

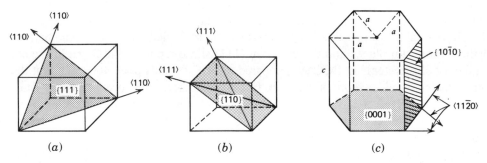

FIGURE 3.2   Diagram showing predominant slip systems in (*a*) FCC; (*b*) BCC; and (*c*) HCP crystals.

TABLE 3.1 Observed Dominant Slip Planes in Hexagonal Crystals

Metal	c/a Ratio	Observed Slip Plane
Be	1.568	$\{0001\}$
Ti	1.587	$\{10\bar{1}0\}$
Zr	1.593	$\{10\bar{1}0\}$
Mg	1.623	$\{0001\}$
Co	1.623	$\{0001\}$
Zn	1.856	$\{0001\}$
Cd	1.886	$\{0001\}$

allow one slip system to account for each of the six independent components of strain (Fig. 3.3), a total of six such systems would seem to be indicated; however, plastic deformation is a constant-volume process where $\epsilon_{xx} + \epsilon_{yy} + \epsilon_{zz} = 0$, thereby reducing to five the number of independent slip systems. An independent slip system is defined as one producing a crystal shape change that cannot be reproduced by any combination of other slip systems. On this basis, Taylor[3] showed that for the 12 possible $\{111\}$ $\langle 110 \rangle$ slip systems in FCC crystals, only five are independent. Furthermore, Taylor found there to be 384 different combinations of five slip systems that could produce a given strain, the activated combination being the one for which the sum of the glide shears is a minimum. Likewise, Groves and Kelly[4] found 384 combinations of five sets of $\{110\}$ $\langle 111 \rangle$ slip systems to account for slip in BCC metals. Since slip in BCC can occur also on $\{112\}$ $\langle 111 \rangle$ and $\{123\}$ $\langle 111 \rangle$ systems, the total number of combinations of five independent slip systems becomes incredibly large. Chin and coworkers[5,6] have applied computer techniques to identify the preferred slip-system combinations for the case of BCC metals.

Difficulties arise when one seeks five independent slip systems in the hexagonal materials. Of the three possible $\{0001\}$ $\langle 11\bar{2}0 \rangle$ slip systems, only two are independent.[4] Similarly, only two independent $\{0010\}$ $\langle 11\bar{2}0 \rangle$ slip systems can be identified from the three possible prism slip systems. Although four independent pyramidal $\{00\bar{1}1\}$ $\langle 11\bar{2}0 \rangle$ slip systems may be identified from a total of six such systems, the resulting deformations can be produced by simultaneous operation of the two independent basal and prism slip systems, respectively. Consequently, a fifth independent slip system is still needed. Besides some deformation twinning (see Section 3.2.6), additional non-

TABLE 3.2 Observed Slip Systems in Selected Ceramics[8]

Material	Structure Type	Preferred Slip System
C, Ge, Si	Diamond cubic	$\{111\}$ $\langle \bar{1}10 \rangle$
NaCl, LiF, MgO	Rock salt	$\{110\}$ $\langle 1\bar{1}0 \rangle$
CsCl	Cesium chloride	$\{110\}$ $\langle 001 \rangle$
CaF_2, UO_2, ThO_2	Fluorite	$\{001\}$ $\langle 1\bar{1}0 \rangle$
TiO_2	Rutile	$\{101\}$ $\langle 10\bar{1} \rangle$
$MgAl_2O_4$	Spinel	$\{111\}$ $\langle \bar{1}10 \rangle$
Al_2O_3	Hexagonal	$\{0001\}$ $\langle 11\bar{2}0 \rangle$

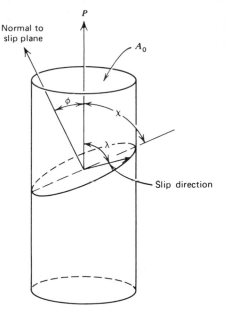

FIGURE 3.3 Diagram showing orientation of slip plane and slip direction in crystal relative to the loading axis.

basal slip with a c-axis Burgers vector component is necessary to explain the observed ductility in hexagonal engineering alloys.[7]

3.1.2 Geometry of Slip

It has been shown that the onset of plastic deformation in a single crystal takes place when the shear stress acting on the incipient slip plane and in the slip direction reaches a critical value. From Fig. 3.3, we see that the cross-sectional area of the slip plane is given by

$$A_{\text{slip plane}} = \frac{A_0}{\cos\phi} \tag{3-1}$$

where A_0 = cross-sectional area of single crystal rod
  ϕ = angle between the rod axis and the normal to the slip plane

Furthermore, the load on this plane resolved in the slip direction is given by

$$P_{\text{resolved}} = P\cos\lambda \tag{3-2}$$

where P = axial load
  λ = angle between load axis and slip direction

By combining Eq. 3-1 and 3-2 the resolved shear stress acting on the slip system is

$$\tau_{\text{RSS}} = \frac{P}{A} \cos\phi\cos\lambda \tag{3-3}$$

where $\cos\phi\cos\lambda$ represents an orientation factor (often referred to as the Schmid factor). Plastic deformation will occur when the resolved shear stress τ_{RSS} reaches a critical value τ_{CRSS}, which represents the yield strength of the single crystal. From Eq. 3-3, we see that yielding will occur on the slip system possessing the greatest Schmid factor. Consequently, if only a few systems are available, such as in the case of basal slip in zinc and cadmium, the necessary load for yielding can vary dramatically with the relative orientation of the slip system (i.e., the Schmid factor).[9,10] For example, the axial stress necessary for yielding anthracene crystals varies dramatically with crystal orientation (Fig. 3.4), while the critical resolved shear stress is unchanged.[11] (Note that the curve drawn in Fig. 3.4b was computed from Eq. 3-3 using a value of 137 kPa.)

Furthermore, the stress normal to the slip plane

$$\sigma_n = \frac{P}{A}\cos^2\phi \qquad (3\text{-}4)$$

can vary considerably without affecting the onset of yielding. For example, Andrade and Roscoe[10] found for cadmium that τ_{CRSS} varied within 2% for all crystal orientations examined, while the normal stress σ_n experienced a 20-fold change.

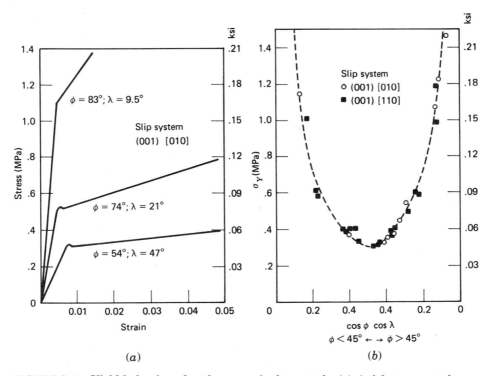

FIGURE 3.4   Yield behavior of anthracene single crystals. (a) Axial stress–strain curves for crystals possessing different orientations relative to the loading axis; (b) axial stress for many crystals plotted versus respective Schmid factors. Dotted curve represents relation given by Eq. 3-3 where $\tau_{crss} = 137$ kPa. (After Robinson and Scott;[11] reprinted with permission from Robinson, *Acta Met.* 15 (1967), Pergamon Press, Elmsford, NY.)

EXAMPLE 3.1

Three cylindrically-shaped tensile samples, each 12-mm in diameter, were machined from three different spherically-shaped single crystals. Samples A and B yielded with applied loads of 77.1 and 56 N, respectively. Does the difference in load level indicate that the crystals possessed different strength levels? Also, what load level would be necessary to cause Sample C to deform and what is the controlling stress for yielding?

The fact that different load levels were needed to cause yielding in Samples A and B *may* indicate that the materials in the two rods possessed different properties. Then, again, the materials may have been identical but, instead, machined at arbitrarily different orientations from the three spherically shaped single crystals. To resolve this issue, additional information is needed. Specifically, it is necessary to determine the crystallographic orientation of the three single crystals, relative to their respective loading directions. X-ray diffraction studies determined that the angles between the tensile axis and both slip plane normals and slip directions are

	$\emptyset$	λ	P
SAMPLE A	70.5	29	77.1
SAMPLE B	64	23	56
SAMPLE C	13	78	?

From Eq. 3-3, the resolved shear stresses for yielding in Samples A and B are

$$\tau_A = \frac{P}{A}\cos\phi\cos\lambda = \frac{77.1}{\pi(6\times10^{-3})^2}(\cos70.5)(\cos29) = 199{,}029 \text{ Pa}$$

$$\tau_B = \frac{P}{A}\cos\phi\cos\lambda = \frac{56}{\pi(6\times10^{-3})^2}(\cos64)(\cos23) = 199{,}803 \text{ Pa}$$

By contrast, the stress normal to the slip plane is given from Eq. 3-4 to be

$$\sigma_A = \frac{P}{A}\cos^2\phi = \frac{77.1}{\pi(6\times10^{-3})^2}\cos^2(70.5) = 75{,}961 \text{ Pa}$$

$$\sigma_B = \frac{P}{A}\cos^2\phi = \frac{56}{\pi(6\times10^{-3})^2}\cos^2(64) = 95{,}152 \text{ Pa}$$

We conclude that the strengths of the two samples are similar (approximately 199 kPa) and that yielding is controlled by the critical resolved shear stress acting on the slip system rather than the stress acting normal to the slip plane. It follows that the load needed to deform Sample C will generate a shear stress of 199 kPa on the active slip system.
Therefore,

$$\tau_C = \frac{P}{A}\cos\phi\cos\lambda = \frac{P}{\pi(6\times10^{-3})^2}(\cos13)(\cos78) \approx 199{,}000 \text{ Pa}$$

$$\therefore P = 111 \text{ N}$$

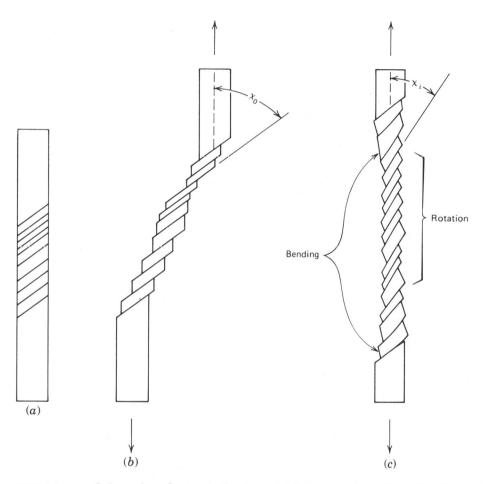

FIGURE 3.5 Orientation of crystal slip plane. (*a*) Prior to deformation; (*b*) after deformation without grip constraint where crystal segments move relative to one another but with no slip rotation; (*c*) after deformation with grip constraint revealing slip plane rotation in gage section (note $\chi_i < \chi_0$).

Let us now consider what happens to the single crystal once it begins to yield. From Fig. 3.5 we see that slip on planes oriented χ degrees away from the tensile axis can occur in two ways. First, the planes can simply slide over one another without changing their relative orientation to the load axis. This would be analogous to offsetting groups of playing cards on a table. Since such lateral movement of the crystal planes in a tensile bar would be forbidden by lateral constraints imposed by the specimen grips, the slip planes are forced to rotate with $\chi_i < \chi_0$. X-ray diffraction studies have shown that crystal planes undergo pure rotation in the middle of the gage length but experience simultaneous rotation and bending near the end grips. If we focus attention on the simpler midregion of the sample, it can be shown that the reorientation of the slip plane varies directly with the change in length of the specimen gage length according to the relationship[12]

$$\frac{L_i}{L_0} = \frac{\sin\chi_0}{\sin\chi_i} \tag{3-5}$$

where L_0, L_i = gage length before and after plastic flow, respectively

χ_0, χ_i = angle between slip plane and stress axis before and after plastic flow, respectively

(Note that $\chi + \phi = 90°$. However $\lambda + \phi = 90°$ *only* when the two vectors are coplanar.) By analogy, the deformation-induced rotation of slip planes is similar to the rotation of individual venetian blind slats—the more you pull on the cord, the more the individual slats deflect.

From the work of Schmid and Boas,[12] when $\chi_0 = \lambda_0$, the shear strain γ, after a given amount of extension, is found to be

$$\gamma = \frac{1}{\sin\chi_0}\left\{\left[\left(\frac{L_i}{L_0}\right)^2 - \sin^2\lambda_0\right]^{1/2} - \cos\lambda_0\right\} \tag{3-6}$$

Note that γ is determined by the initial orientation of the glide elements and by the amount of extension. Furthermore, Eq. 3-6 is valid when only one slip system is active, since multiple slip involves an undefined amount of crystal rotation from each system. The resolved shear stress is given by

$$\tau = \frac{P}{A}\sin\chi_0\left[1 - \frac{\sin^2\lambda_0}{(L_i/L_0)}\right]^{1/2} \tag{3-7}$$

For a detailed discussion of other relationships involving the shear stresses and strains in single crystals, see Schmid and Boas.[12]

It is instructive to trace the path of rotation of the slip plane. This is accomplished most readily with the aid of a stereographic projection.* For the purpose of this discussion, some basic understanding of this method is desirable. For the crystal block shown in Fig. 3.6a, imagine that a normal to each plane is extended to intersect an imaginary reference sphere that surrounds the block. Now place a sheet of paper (called the projection plane) tangent to the sphere. Next, take a position at the other end of the sphere diameter, which is oriented normal to the projection plane. From this position (called the point of projection) drawn lines through the points on the reference sphere and continue on to the projection plane (Fig. 3.6b). The points on the projection plane then reflect the relative position of various planes (or plane normals) with planar angle relationships faithfully reproduced. For convenience, standard stereographic projections are used to portray the relative positions of major planes, such as those shown in Fig. 3.7. Since a cubic crystal is highly symmetrical, the relative orientation of a crystal can be given with respect to any triangle within the stereographic projection. As a result, attention is usually focused on the central

---

* See B. D. Cullity, *Elements of X-Ray Diffraction*, Addison-Wesley, Reading, MA, 1956, for a treatment of stereographic projection.

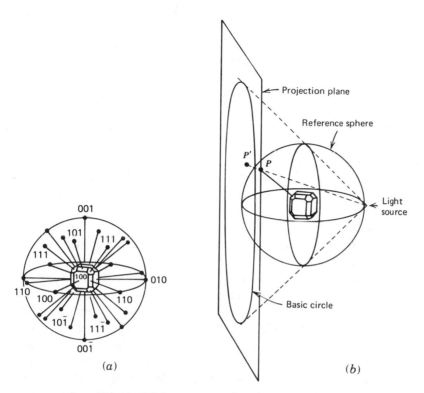

FIGURE 3.6 Geometric constructions to develop a stereographic projection. (*a*) Intersection of plane normals or reference sphere. (After C. W. Bunn, *Chemical Crystallography*, Clarendon Press, Oxford, 1946, p. 30.) (*b*) Projection of poles on the reference sphere to the projection plane. (After N. H. Polakowski and E. J. Ripling, *Strength and Structure of Engineering Materials*, p. 83. Reprinted by permission of Prentice-Hall, Inc., Englewood Cliffs, NJ, © 1966.)

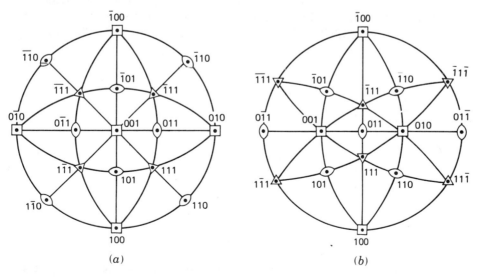

FIGURE 3.7 Standard stereographic projections for cubic crystals: (*a*) (001) and (*b*) (011) projections.

section of the projection. In Fig. 3.8, for example, we see the axis of a rod in terms of its angular relationship with the (001), (011), and ($\bar{1}$11) planes, respectively. That is, P is the normal to the plane lying perpendicular to the rod axis. When this rod is stressed to τ_{CRSS}, the crystal will yield on that slip system possessing the greatest Schmid factor and begin to rotate. For all orientations within triangle I (sometimes referred to as the standard triangle), the (111) [$\bar{1}$01] slip system possesses the greatest Schmid factor and will be the first to operate.[13,14] The rotation occurs along a great circle (corresponding to the trace of a plane on the reference sphere that passes through the center of the sphere) of the stereographic projection and toward the [$\bar{1}$01] slip direction. For simplicity, it is easier to consider rotation of the stress axis relative to the crystal than vice versa, so that P is seen to move toward the [$\bar{1}$01] pole. As the crystal rotates, λ will decrease while ϕ increases. In situations where $\lambda_0 > 45° > \phi_0$, rotation of the crystal will bring about an increase in the Schmid factor, since both λ_i and ϕ_i would approach 45°. As a result, yielding can continue at a lower load and the crystal is said to have undergone *geometrical softening*. Conversely, when $\phi_0 > 45° > \lambda_0$, crystal rotation will bring about a reduction in the Schmid factor, thereby increasing the load necessary for further deformation on the initial slip system. Bear in mind that this *geometrical hardening* is distinct from strain hardening which involves dislocation–dislocation interactions (see Section 4.1). Geometrical hardening continues as the crystal axis moves toward the [001]–[$\bar{1}$11] tie line. As soon as the relative crystal orientation crosses over into the adjacent triangle II, the Schmid factor on the primary slip system becomes less than that associated with the ($\bar{1}$1 1) [011] system, the latter being the slip system that would have operated had the crystal been oriented initially within triangle II. This newly activated slip system (the conjugate system) now causes the crystal to rotate along a different great circle toward the [011] direction of the conjugate slip system. Shortly, however, this movement returns the axis of the crystal to within the bounds of triangle I where primary slip resumes. The ultimate effect of this jockeying back and forth between primary and conjugate slip systems is the movement of the crystal axis along the [001]–[$\bar{1}$11] tie line to a location where further crystal rotations in either slip direction occur along the same great circle. This point is reached when the load axis is parallel to the [$\bar{1}$12] direction.

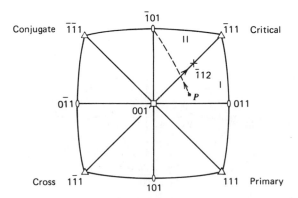

FIGURE 3.8   (001) stereographic projection showing lattice rotation for FCC crystals during tensile elongation.

The above analysis reflects the classic geometrical arguments proposed originally by Taylor and Elam.[13,14] In reality, some alloy crystals exhibit "overshooting," wherein the primary slip system continues to operate well into triangle II even though the Schmid factor of the conjugate system is greater. Similarly, the conjugate system, once activated, may continue to operate into triangle I (Fig. 3.9). Koehler[15] proposed that overshooting was caused by weakening of the primary system by passage of dislocations that destroyed precipitates and other solute atom clusters. Consequently, he argued that slip would be easier if continued on the softened primary plane. Alternatively, Piercy et al.[16] argued that overshooting resulted from a "latent hardening" process involving increased resistance to conjugate slip movement resulting from the dislocation debris found on the already activated primary system. That is, for slip to occur on the conjugate system, dislocations on this plane would have to cut across many dislocations lying on the primary plane. By comparison, then, Koehler[15] argued that overshooting resulted from a relative weakening of the primary plane while Piercy et al.[16] argued that the conjugate plane was strengthened relative to the primary plane by a latent hardening mechanism. By careful experimentation, the latent hardening theory was proven correct.

From Fig. 3.8, two other slip systems can be identified. These are denoted as the cross-slip system $(1\bar{1}1)$ $[\bar{1}01]$ and the critical slip system $(\bar{1}11)$ $[0\bar{1}1]$. The critical system is not encountered very often; the cross-slip system is the system involving the movement of screw dislocations that have cross-slipped out of the primary slip plane. Note that the slip direction is the same in this case.

3.1.3 Crystallographic Textures (Preferred Orientations)

From the previous section, it should not be surprising to find individual grains in a polycrystalline aggregate undergoing similar reorientation. As might be expected, lattice reorientation in a given grain is impeded by constraints introduced by contiguous grains, making the development of crystallographic textures in polycrystalline aggregates a complex process. In addition, the preferred orientation is found to depend on a number of additional variables, such as the composition and crystal structure of the metal and the nature, extent, and temperature of the plastic deformation process.[17] As a result, the texture developed by a metal usually is not complete, but instead may be described by the strength of one orientation component relative to another.

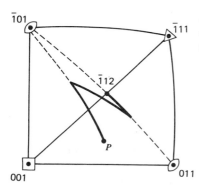

FIGURE 3.9 Lattice rotation of FCC crystal involving "overshoot" of primary and conjugate slip systems.

Crystallographic textures are portrayed frequently by the pole figure, which is essentially a stereographic projection showing the distribution of *one* particular set of {*hkl*} poles in orientation space. That is, X-ray diffractometer conditions are fixed for a particular diffraction angle and X-ray wavelength so that the distribution of one set of {*hkl*} poles in the polycrystalline sample can be monitored. To illustrate, consider the single-crystal orientation responsible for the (100) stereographic projection shown in Fig. 3.7a. The (100) pole figure for this crystal would reveal (100) diffraction spots at the north, south, east, and west poles and at the center of the projection (Fig. 3.10a). No information concerning the location of {110}, {111}, or {*hkl*} poles is collected, since diffraction conditions for these planes are not met. Their location would have to be surmised based on the position of the (100) poles and the known angular relation between the {100} and {*hkl*} poles. It is possible, of course, to change diffraction conditions to ''see'' the location of these other poles but then the {100} poles would ''disappear'' from the {*hkl*} pole figure. Figure 3.10b shows the same crystal as in Fig. 3.10a, but with its orientation portrayed by a (110) pole figure. It is important to appreciate that although these two pole figures look different, they convey the same information—the orientation of the crystal. By analogy, different {*hkl*} pole figures represent different languages by which the same thought (the preferred orientation) is conveyed.

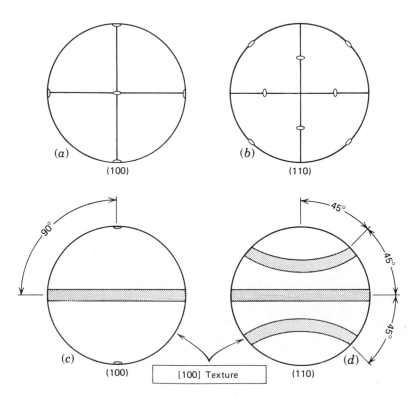

FIGURE 3.10 Pole figures depicting orientation of metals. (a) (100) pole figure for crystal orientation shown in Fig. 3.7a; (b) (110) pole figure for same orientation; (c) (100) pole figure for [100] wire texture; (d) (110) pole figure for [100] wire texture. Note rotational symmetry in (c) and (d).

When wires or rods are produced, such as by drawing or swaging, a uniaxial preferred orientation may develop in the drawing direction, with other crystallographic poles distributed symmetrically about the wire axis. For a [100] crystallographic wire texture, such as heavily deformed silver wire, the texture is given by Fig. 3.10c as portrayed by a (100) pole figure. Note the presence of {100} poles at the north (and south) pole of the projection corresponding to the drawing direction and the smearing out of the other {100} poles across the equator, the latter reflecting the rotational symmetry found in wire textures. The same texture is shown in Fig. 3.10d via a (110) pole figure. Here the rotational symmetry of the wire texture is again evident while the [100] wire texture must be inferred from the relative position of the {110} poles. Naturally, a (111) pole figure would present yet another interpretation of the same [100] wire texture. (The reader is advised to sketch the (111) pole figure for the [100] wire texture for his or her edification.) Typical wire textures for a number of FCC metals and alloys are given in Fig. 3.11, where the variation of texture with stacking fault energy (SFE) is shown clearly.[18] The explanation for the SFE dependence of texture transition and for the reversal in texture at very low stacking fault energies has been the subject of considerable debate. Cross-slip,[19−21] mechanical twinning,[22,23] overshooting,[24,25] and extensive movement of Shockley partial dislocation[26] mechanisms have been proposed as possible contributing factors toward development of both wire and sheet textures.

For the case of BCC metals, the wire texture is uncomplicated and found to be [110]. In HCP metals, texture is found to vary with the c/a ratio. When $c/a < 1.633$, a [10$\bar{1}$0] texture may be developed with the basal plane lying parallel to the rod axis.[17] By contrast, texture development is more complex when $c/a > 1.633$.

Although wire textures may be defined by one component—the direction parallel

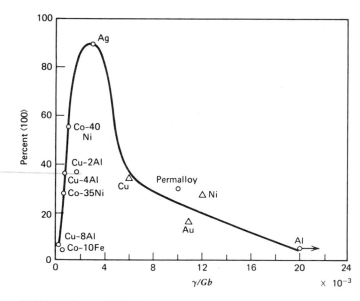

FIGURE 3.11 Relation between strength of [100] component in FCC wire texture and stacking fault energy parameter, γ/Gb. (After English and Chin;[18] reprinted with permission from Chin, *Acta Met.* **13** (1965), Pergamon Press, Elmsford, NY.)

to the wire axis—sheet textures are given by both the crystallographic plane oriented parallel to the rolling plane and the crystallographic direction found parallel to the rolling direction. Hence, rolling textures are described by the notation (hkl) $[uvw]$, where (hkl) corresponds to the plane parallel to the rolling plane and $[uvw]$ to the direction parallel to the rolling direction. Rolling textures are very complex, with several different components often existing simultaneously (Table 3.3).

The (110) $[\bar{1}12]$ texture is often referred to as the *brass* or *silver* texture, typical of FCC materials that possess low stacking fault energy. In copper, nickel, and aluminum, which possess intermediate and high SFE, respectively, the major textural components are (123) $[\bar{4}12]$, (146) $[\bar{2}11]$, and (112) $[\bar{1}\bar{1}1]$. This more complicated preferred orientation is called the *copper* texture. Since the SFE for an alloy depends on solute content (i.e., changes in the electron to atom ratio), the texture of a metal can change from copper to brass type with increasing alloy additions. Some researchers have argued that the importance of SFE in controlling the type of deformation texture is related to the relative ease by which cross-slip occurs, the brass texture being generated when cross-slip is more difficult. Others have suggested that the brass texture develops when mechanical twinning or deformation faulting is relatively easy. For this reason, the preferred orientation should also be sensitive to the temperature of deformation, since cross-slip, mechanical twinning, and faulting are thermally dependent processes. In studying the rolling texture in high-purity silver, Hu and Cline[27] found that cold rolling at 0°C produced a typical $\{110\}$ $\langle211\rangle$ *brass* or *silver* texture. However, when the silver was rolled at 200°C, near $\{123\}$ $\langle412\rangle$ and $\{146\}$ $\langle211\rangle$ components were observed, reflecting a *copper* type texture. A similar *brass*-to *copper*-type texture transition was found when 18-8 stainless steel was rolled at 200 and 800°C, respectively.[28] Conversely, a reverse *copper* to *brass* texture transition was realized for copper when the rolling temperature was reduced from ambient to $-196°C$.[29] Finally, by combining the effects of alloy content and deformation temperature on SFE, Smallman and Green[19] demonstrated for the silver–aluminum alloy that the *brass* to *copper* rolling texture transition temperature increased with decreasing initial stacking fault energy.

TABLE 3.3   Typical Rolling Textures in Selected Engineering Alloys [17]

Alloy	Rolling Texture
FCC	
Brass, silver, stainless steel	$(110)[\bar{1}12] + (110)[001]$
Copper, nickel, aluminum	$(123)[\bar{4}12] + (146)[\bar{2}11] + (112)[11\bar{1}]$
BCC	
Iron, tungsten, molybdenum, tantalum, niobium	$(001)[\bar{1}10]$ to $(111)[\bar{1}10] + (112)[\bar{1}10]$ to $(111)[\bar{2}11]$
HCP	
Magnesium, cobalt $(c/a \approx 1.633)$	$(001)[2\bar{1}\bar{1}0]$
Zinc, cadmium $c/a > 1.633$	(0001) plane tilted $\pm 20-25°$ from rolling plane about a $[10\bar{1}0]$ transverse direction axis
Titanium, zirconium, beryllium $c/a < 1.633$	(0001) plane tilted $\pm 30-40°$ from rolling plane about a $[10\bar{1}0]$ rolling direction axis

3.1.3.1 Plastic Anisotropy

It follows from the previous discussion that when a sheet or rod contains a preferred crystallographic orientation, the ability of the material to deform in an isotropic manner is altered. For example, assume that a sheet of α-titanium possesses an idealized texture with the (0001) basal planes oriented parallel to the plane of the sheet and $\langle 1\bar{2}10 \rangle$ directions aligned parallel to the rolling directions.[30–32] Slip can occur on (0001), $(10\bar{1}0)$, and $(10\bar{1}1)$ planes but only in the $\langle 11\bar{2}0 \rangle$ close-packed directions; therefore, no sheet thinning can occur in association with these slip systems. Recall that plastic deformation is a constant volume process (Eq. 1-3) where

$$\epsilon_l = (\epsilon_w + \epsilon_t) \tag{3-8a}$$

Since $\epsilon_t = 0$,

$$\epsilon_l = -\epsilon_w \tag{3-8b}$$

Therefore, tensile strains in a coupon prepared from a textured sheet (Fig. 3.12a) would be balanced only by a reduction in sample width. Accordingly,

$$\epsilon_l = \ln(l/l_o) = -\ln(w/w_o) \tag{3-8c}$$

whereas

$$\epsilon_t = \ln(t/t_o) = 0$$

A useful parameter to quantify the amount of plastic strain anisotropy in a sheet is identified by R where

$$R = \frac{\epsilon_w}{\epsilon_t} \tag{3-9a}$$

Since R and elastic modulus values typically vary within the plane of the textured sheet, it is common to describe an average R-value, $\bar{R}$, where

$$\bar{R} = \frac{R_0 + 2R_{45} + R_{90}}{4} \tag{3.9b}$$

with the subscripts corresponding to the orientation within the sheet.

For the ideal texture described previously, $\bar{R} = \infty$; alternatively, when no texture exists and the material behaves in an isotropic manner, $\bar{R} = 1$. For realistic crystallographic textures, such as in α-titanium alloys where basal planes are tilted at $\pm$ 30°–40° from the rolling plane (Table 3.3), $\bar{R}$ values of 3–7 are typically experienced; the higher the value of $\bar{R}$, the greater the sheet's resistance to thinning and the higher the material's yield strength under through-thickness compression or balanced biaxial tensile loading conditions (Fig. 3.12b). Note that the existence of texture and its influence on yield strength is obscured under uniaxial loading conditions and of limited importance in pure shear (i.e., where $\sigma_x = -\sigma_y$).

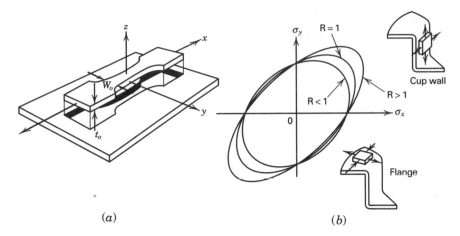

(a)   (b)

FIGURE 3.12   (*a*) Tensile coupon and dimensions as cut from sheet stock.[30] (Walter A. Backofen, Ed., et al, *Fundamentals of Deformation Processing* (Syracuse, N.Y.: Syracuse University Press, 1964). By permission of the publisher.) (*b*) Yield loci for textured material.[31] When $R > 1$, material exhibits thinning resistance and high strength under biaxial tension; when $R > 1$, the material displays easy thinning and low biaxial tensile strength. (W.F. Hosford and R.M. Caddell, *Metal Forming Mechanics and Metallurgy*, 2nd Ed. (1993). Reprinted by permission of Prentice-Hall, Inc., Englewood Cliffs, N.J.)

The influence of plastic anisotropy on metal forming is demonstrated by the deep drawing of flat sheets into cartridge cases, bathtubs, brass flashlight cases, and automobile panels. In this process, a circular sheet of metal is clamped over a die opening and then pressed through the die with a punch (Fig. 3.13). The load from the punch is transmitted along the sidewall of the cup to the flange area where most of the deformation takes place. Within the flange area, the stress state approaches that of pure shear, corresponding to tension in the radial direction and compression in the circumferential direction (Fig. 3.12*b*-Diagram A). By contrast, a plane strain biaxial tension condition exists in the cup wall (Fig. 3.12*b*-Diagram B). Failure occurs by localized necking within a narrow ring of material in the cup wall just above the radius of the punch. Analysis of this forming process reveals that the upperbound theoretical limiting drawing ratio (LDR) is estimated to be[32]

$$\text{LDR} \approx \left(\frac{D_0}{D_p}\right)_{\max} \approx e^{\eta} \qquad (3\text{-}10)$$

where D_0 and D_p are the original sheet and final cup diameters, respectively, and η is a parameter that accounts for frictional losses in the drawing process. For ideal efficiency, $\eta = 1$ and LDR ≈ 2.7. Typically, however, $\eta \approx 0.74$ to 0.79; hence, LDR ≈ 2.1 to 2.2.

The limiting drawing ratio can be increased—to permit the drawing of deeper cups—by restricting the material's ability to thin in the critical zone near the bottom of the cup wall. This can be achieved by strengthening the sheet in the thickness

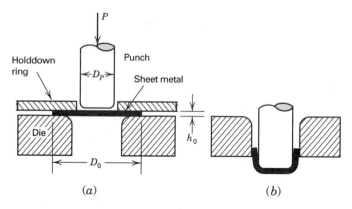

FIGURE 3.13 Illustration revealing deep drawing of a cylindrical cup (*a*) before and (*b*) after drawing. (Adapted from G. Dieter, *Mechanical Metallurgy*, 3rd Ed., 1986, with permission of McGraw-Hill, Inc.)

direction through the development of a crystallographic texture ($R > 1$) that limits deformation under the plane strain biaxial tension conditions experienced in the cup wall (recall Fig. 3.12*b*-Diagram B). Notice that LDR increases for several metal alloys with average plastic strain ratio, $\bar{R}$ (Fig. 3.14). For further discussion of the influence of texture on metal forming, the reader is referred to texts by Hosford and Caddell,[31] and Dieter.[33]

3.2 DEFORMATION TWINNING

As was noted in Section 3.1.1, the simultaneous operation of at least five independent slip systems is required to maintain continuity at grain boundaries in a polycrystalline solid. Failure to do so will lead to premature fracture. If a crystal possesses an

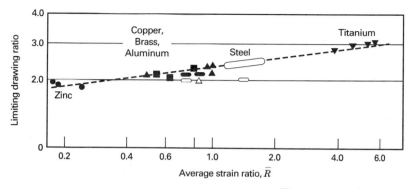

FIGURE 3.14 Influence of average strain ratio, $\bar{R}$, on limiting drawing ratio for several sheet metal alloys.[33] (M. Atkinson, *Sheet Metal Industries*, **44**, 167 (1967) with permission.)

insufficient number of independent slip systems, twin modes may be activated in some metals to provide the additional deformation mechanisms necessary to bring about an arbitrary shape change.

3.2.1 Comparison of Slip and Twinning Deformations

The most obvious difference between a slipped versus a twinned crystal is the shape change resulting from these deformations. Whereas slip involves a simple translation across a slip plane such that one rigid portion of the solid moves relative to the other, the twinned body undergoes a shape change (Fig. 3.15).

According to Bilby and Crocker,[35] "A deformation twin is a region of a crystalline body which had undergone a homogeneous shape deformation in such a way that the resulting product structure is identical with that of the parent, but oriented differently." As pointed out in Chapter 2, dislocation movement associated with slip will take place in multiples of the unit displacement—the dislocation Burgers vector. By contrast, the shape change found in the twinned solid results from atom movements taking place on all planes in fractional amounts within the twin. In fact, we see from Fig. 3.15c that the displacement in any plane within the twin is directly proportional to its distance from the twin–matrix boundary. Upon closer examination of these twinning displacements in a simple cubic lattice, it is seen that the twinning process has effected a rotation of the lattice such that the atom positions in the twin represent a mirror image of those in the untwinned material (Fig. 3.16). By contrast, slip occurs by

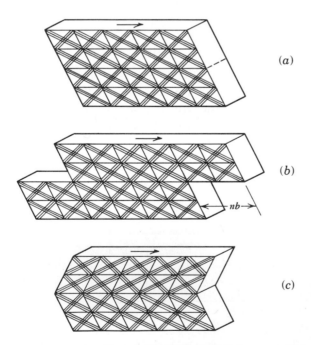

(a)

(b)

(c)

FIGURE 3.15 Shape change in solid cube caused by plastic deformation. (a) Undistorted cube; (b) slipped cube with offsets nb; (c) twinned cube revealing reorientation within twin. Displacements are proportional to distance from twin plane.

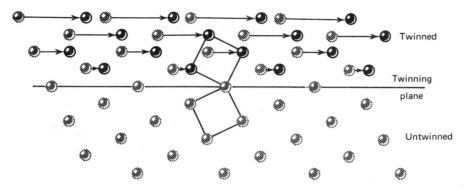

FIGURE 3.16 **Twinning on the (120) plane in a simple cubic crystal. Gray circles represent original atom positions. Black circles are final atom positions.**

translations along widely spaced planes in whole multiples of the displacement vector, so that the relative orientation of different regions in the slipped cube remains unchanged.

The differences associated with these deformation mechanisms are revealed when one examines the deformed surface of a prepolished sample (Fig. 3.17). Offsets due to slip are revealed as straight or wavy lines (depending on the stacking fault energy of the material) with no change in contrast noted on either side of the slip offset. Twin bands do exhibit a change in contrast, since the associated lattice reorientation within the twin causes the incident light to be reflected away from the objective lens of the microscope. After repolishing and etching the sample, only twin band markings persist, since they were associated with a reorientation of the lattice (Figs. 3.17*b* and 3.17*d*).

Before proceeding further, it is appropriate to distinguish between *deformation* twins (Figs. 3.17*a* and 3.17*b* and *annealing* twins (Figs. 3.17*c* and 3.17*d*). The deformation twins in the zinc specimen were generated as a result of plastic deformation, where the annealing twins in the brass sample *preexisted* plastic deformation. The annealing twins were formed instead during prior heat treatment of the brass in association with recrystallization and growth of new grains. During the formation of a new packing order in the new crystals, the emerging grains could have encountered packing sequence defects in the original grains (such as stacking faults); this interaction would result in the formation of annealing twins.

For example, the error indicated by the vertical line in the following planar packing sequence—*ABCABℂBACBA*—constitutes a twin boundary. Note also that the two planes on either side of the twin plane are similar: This arrangement constitutes a stacking fault (i.e., *BℂB*). Without the preexistence of stacking faults in the old grains, annealing twins are unlikely to form. Hence, annealing twins are rarely seen in aluminum, which has a high stacking fault energy (low stacking fault probability). Conversely, annealing twins are observed readily in brass, which has a low stacking fault energy (high stacking fault probability). Since the stacking fault probability also depends on the extent of deformation, the number of annealing twins found in a given material should increase with increasing prior cold work. As such, the number of

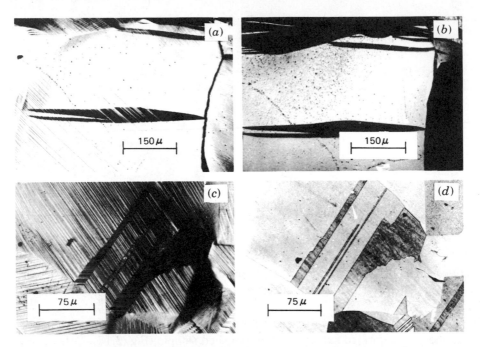

FIGURE 3.17 Surface markings resulting from plastic deformation. (*a*) Prepolished and deformed zinc revealing slip lines (upper left to lower right markings) and twin bands (large horizontal band); (*b*) same as (*a*) but repolished and etched to show only twin bands; (*c*) prepolished and deformed brass revealing straight slip lines (reflecting low stacking fault energy) and preexisting annealing twins; (*d*) same as (*c*) but repolished and etched to show only annealing twins.

annealing twins found in a recrystallized material provides a clue as to the deformation history of the material.

3.2.2 Geometry of Twin Formation[36]

Consider the growth of a twin over the upper half of a crystalline unit sphere. Any point on the sphere will be translated from coordinates X, Y, Z to X', Y', Z', where $X = X'$, $Z = Z'$ and $Y' = Y + SZ$ (Fig. 3.18). Since S represents the magnitude of the shear strain, we see that the shear displacement on any plane is directly proportional to the distance from the twinning plane (called the composition plane). Therefore, the equation for the distorted sphere is given by

$$X'^2 + Y'^2 + Z'^2 = 1 = X^2 + Y^2 + 2SZY + S^2Z^2 + Z^2 = 1 \qquad (3\text{-}11)$$

or

$$X^2 + Y^2 + 2SZY + Z^2(S^2 + 1) = 1 \qquad (3\text{-}12)$$

which defines a quadric surface. Specifically, the distorted sphere forms an ellipsoid whose major axis is inclined to η_1 by an angle ϕ. It is clear from Fig. 3.18 that most

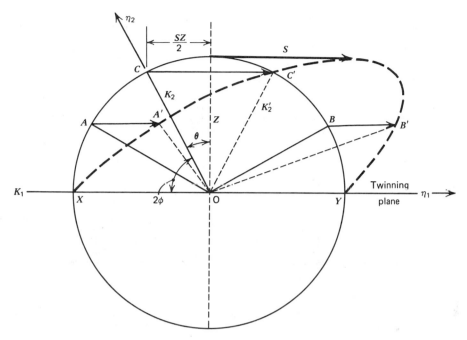

FIGURE 3.18   Crystal sphere distorted to that of an ellipsoid. Undistorted planes are K_1 and K_2, separated by angle 2ϕ. Note foreshortening of plane OA after twinning, while plane OB is extended.

planes contained within the sphere are either foreshortened or extended. For example, consider the movement of points A and B, which are translated by the twinning deformation to A' and B', respectively. If AO and BO represent the traces of two different planes, it is clear that AO has been foreshortened ($A'O$) while BO has been stretched ($B'O$). Only two planes remain undistorted after the twin shear has been completed. The first is the composition plane, designated as K_1; the direction of the shear is given by η_1. The second undistorted plane is the one shown in profile by the line OC. (Note that $OC = OC'$.) This plane is designated as the K_2 plane, where η_2 is defined by the line of intersection of the K_2 plane and the plane of shear (the plane of this page). The final position of this second undistorted plane is designated as the K_2' plane. Therefore, all planes located between X and C will be compressed, while all planes located between C and Y will be extended. Typical values for K_1, K_2, η_1, and η_2 are shown in Table 3.4 and discussed in the following sections. By definition,[39] when K_1 and η_2 are rational and K_2 and η_1 are not, we speak of this twin as being of the *first kind*. The orientation change resulting from this twin can be accounted for by reflection in the K_1 plane or by a 180° rotation about the normal to K_1. When K_2 and η_1 are rational but K_1 and η_2 are not, the twin is of the *second kind*. The twin orientation in this case may be achieved either by a 180° rotation about η_1 or by reflection in the plane normal to η_1. When all twin elements are rational, the twin is designated as *compound*. This occurs often in crystals possessing high symmetry (such as most metals), where the reflection and rotation operations are equivalent.

The magnitude of the shear strain S in the unit sphere is given by the angle 2ϕ between the two undistorted planes K_1 and K_2. From Fig. 3.18

TABLE 3.4 Observed Twin Elements in Metals[37,38]

Metal	Crystal Structure	c/a Ratio	K_1	K_2	η_1	η_2	S	$(l' - l/l)_{max}$
Al, Cu, Au, Ni, Ag, γ-Fe	FCC		$\{111\}$	$\{11\bar{1}\}$	$\langle 11\bar{2}\rangle$	$\langle 112\rangle$	0.707	41.4%
αFe	BCC		$\{112\}$	$\{\bar{1}\bar{1}2\}$	$\langle\bar{1}\bar{1}1\rangle$	$\langle 111\rangle$	0.707	41.4
Cd	HCP	1.886	$\{10\bar{1}2\}$	$\{\bar{1}012\}$	$\langle 10\bar{1}\bar{1}\rangle$	$\langle 10\bar{1}1\rangle$	0.17	8.9
Zn	HCP	1.856	$\{10\bar{1}2\}$	$\{\bar{1}012\}$	$\langle 10\bar{1}\bar{1}\rangle$	$\langle 10\bar{1}1\rangle$	0.139	7.2
Mg	HCP	1.624	$\{10\bar{1}2\}$	$\{\bar{1}012\}$	$\langle 10\bar{1}\bar{1}\rangle$	$\langle 10\bar{1}1\rangle$	0.131	6.8
			$\{11\bar{2}1\}$	$\{0001\}$	$\langle 11\bar{2}6\rangle$	$\langle 11\bar{2}0\rangle$	0.64	37.0
Zr	HCP	1.589	$\{10\bar{1}2\}$	$\{\bar{1}012\}$	$\langle 10\bar{1}\bar{1}\rangle$	$\langle 10\bar{1}1\rangle$	0.167	8.7
			$\{11\bar{2}1\}$	$\{0001\}$	$\langle 11\bar{2}6\rangle$	$\langle 11\bar{2}0\rangle$	0.63	36.3
			$\{11\bar{2}2\}$	$\{11\bar{2}4\}$	$\langle 11\bar{2}3\rangle$	$\langle 22\bar{4}3\rangle$	0.225	11.9
Ti	HCP	1.587	$\{10\bar{1}2\}$	$\{\bar{1}012\}$	$\langle 10\bar{1}\bar{1}\rangle$	$\langle 10\bar{1}1\rangle$	0.167	8.7
			$\{11\bar{2}1\}$	$\{0001\}$	$\langle 11\bar{2}6\rangle$	$\langle 11\bar{2}0\rangle$	0.638	36.9
			$\{11\bar{2}2\}$	$\{11\bar{2}4\}$	$\langle 11\bar{2}3\rangle$	$\langle 22\bar{4}3\rangle$	0.255	11.9
Be	HCP	1.568	$\{10\bar{1}2\}$	$\{\bar{1}012\}$	$\langle 10\bar{1}\bar{1}\rangle$	$\langle 10\bar{1}1\rangle$	0.199	10.4

$$\tan\theta = SZ/2/Z = \frac{S}{2} \tag{3-13}$$

Since

$$\theta + 2\phi = 90°$$

$$\cot 2\phi = \frac{S}{2} \tag{3-14}$$

3.2.3 Elongation Potential of Twin Deformation

Hall[36] has shown that the total deformation strain to be expected from a completely twinned crystal may be given by

$$\frac{l'}{l} = [1 + S\tan\chi]^{1/2} \tag{3-15}$$

where l, l' = initial and final lengths, respectively

$$\tan\chi = \frac{S \pm \sqrt{S^2 + 4}}{2}$$

From Eq. 3-15, the maximum potential elongation of the metals shown in Table 3.4 is quite small, particularly in HCP crystals, which undergo $\{10\bar{1}2\}$ type twinning. Although the twinning reaction contributes little to the total elongation of the sample, the rotation of the crystal within the twin serves mainly to reorient the slip planes so that they might experience a higher resolved shear stress and thereby contribute more deformation by slip processes.

3.2.4 Twin Shape

From the above geometrical analysis, one would assume twinned regions to be bounded by two parallel composition planes representing the two twin–matrix coherent interfaces. In practice, twins are often found to be lens-shaped, so that the interface must consist of both coherent and noncoherent segments. These noncoherent portions of the interface can be described in terms of particular dislocation arrays (Fig. 3.19). Mahajan and Williams[40] have reviewed the literature and found that twin formation has been rationalized both in terms of heterogeneous nucleation at some dislocation arrangement or by homogeneous nucleation in a region of high stress concentration. It is worth noting that dislocations are also needed to account for the requirement of a much lower stress to move a twin boundary than the theoretically expected value.

Cahn[39] postulated that the lens angle β should increase with decreasing shear strain. Since the magnitude of β controls the permissible thickness of the lens, Cahn's postulate correctly predicts the empirical fact that twin thickness increases with decreasing shear strain. More recently, Friedel[41] also concluded that the optimum lens thickness to length ratio should increase with decreasing shear strain. Since twin formation involves discontinuous deformations, some type of lattice accommodation is necessary along the perimeter of the twin lens. When the parent lattice possesses limited ductility, the lens angle β is kept small and the strain discontinuity accommodated by crack formation.[42] At the other extreme, lattice plane bending and/or slip may be introduced to "smear out" the strain discontinuity resulting from the twin. If the crystal is able to slip readily, the lens angle β can increase, thereby enabling the twin to thicken. Therefore, we find that in ductile crystals, the thickness of deformation twins increases with decreasing twin shear strain (Fig. 3.20). From this discussion, it follows that the twins seen in the brass sample in Figs. 3.17*a* and *b* were of the *annealing* type since the deformation twin strain ($S = 0.707$) in this material would have produced thin deformation twins similar to those shown in Fig. 3.20*a*.

3.2.5 Stress Requirements for Twinning

It is now generally known that the twin initiation stress is much greater than the stress needed to propagate a preexistent twin. Similarly, the *nucleation* of a twin is associated with a sudden load drop (responsible for serrated stress–strain curves as shown in Fig. 1.14), while the *growth* of a twin exhibits smoother loading behavior. Both

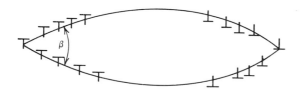

FIGURE 3.19 Diagram of lens-shaped twin with dislocations to accommodate noncoherent twin-matrix interface regions. Lens angle β increases with decreasing twin shear and increasing ability of matrix to accommodate the twin strain concentration.

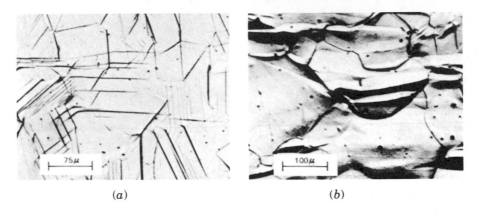

FIGURE 3.20 **Prepolished and subsequently deformed surfaces revealing shape of twins.** *(a)* **Narrow deformation twins in α-Fe** $(S = 0.707)$**; *(b)* broad deformation twins in Mg** $(S = 0.131)$**. (After Eckelmeyer and Hertzberg[43]; American Society for Metals, Metals Park, OH, © 1970.)**

observations point to the need for a large stress concentration to nucleate the twin, although the stress concentration apparently is not needed for twin growth. The prerequisite of a stress concentration to initiate plastic deformation via a twinning mode raises serious doubts regarding the existence of a critical resolved shear stress (CRSS) for twinning, as a counterpart to the well-documented CRSS for slip. Furthermore, the large degree of scatter in reported ''CRSS for twinning'' in various materials raises additional doubts as to its existence.

3.2.6 Twinning in HCP Crystals

Among the three major unit cells found in metals and their alloys, twinning is most prevalent in HCP materials. Over a broad temperature range, twinning and slip are highly competitive deformation processes. We saw from Section 3.1.1 that regardless of the *c/a* ratio (that is, whether basal or prism slip was preferred), an insufficient number of independent slip systems can operate to satisfy the von Mises requirement.[44] Since alloys of magnesium, titanium, and zinc are known to possess reasonable ductility, some other deformation mechanisms must be operative. While combinations of basal, prism, and pyramidal slip do not provide the necessary five independent slip systems necessary for an arbitrary shape change in a polycrystalline material, deformation twinning often is necessary to satisfy von Mises' requirement.

Twinning in HCP metals and alloys has been observed on a number of different planes (Table 3.4). One twin mode common to many HCP metals is that involving $\{10\bar{1}2\}$ planes. One of three possible sets of these planes is shown in Fig. 3.21a. Activation of one particular set will depend on the respective Schmid factors. Naturally, twinning will occur on the $\{10\bar{1}2\}$ plane and $\langle\bar{1}011\rangle$ direction that experiences the highest resolved shear stress. For additional clarification, the angular relationships

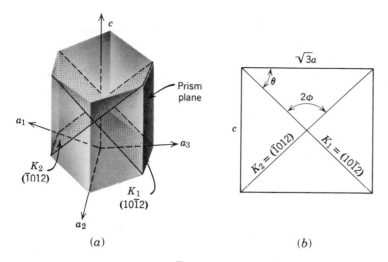

FIGURE 3.21   One set of $\{10\bar{1}2\}$ type K_1 and K_2 planes in HCP crystal. (*a*) **Inclined view of prism, basal and two undistorted planes;** (*b*) **important planes viewed on edge along a_2 direction.**

between the undistorted $\{10\bar{1}2\}$ planes and the prism and basal planes are shown in Fig. 3.21*b*. From Fig. 3.21*b*

$$\tan\theta = \frac{c}{\sqrt{3}a} \tag{3.16}$$

Since $2\phi + 2\theta = 180°$ and $\tan 2\phi = 2/S$

$$\tan 2\phi = 2/S = \tan(180 - 2\theta) \tag{3-17}$$

With trigonometric identities it may be shown that

$$S = \frac{\tan^2\theta - 1}{\tan\theta} \tag{3-18}$$

Combining Eqs. 3-16 and 3-18 and rearranging we find

$$S = [(c/a)^2 - 3]\frac{\sqrt{3}a}{3c} \tag{3-19}$$

From Eq. 3.19, it is seen that the sense of the twin deformation is opposite for HCP metals exhibiting c/a ratios $\gtrless\sqrt{3}$. When $c/a = \sqrt{3}$, the analysis predicts that $S = 0$ and that twinning would not occur by the $\{10\bar{1}2\}$ mode. Stoloff and Gensamer[45] have verified this in a magnesium crystal alloyed with cadmium to produce a c/a ratio of $\sqrt{3}$. The reversal in sense of the twin deformation is seen when the responses of

TABLE 3.5 Interplanar Angles in Beryllium and Zinc

	$\{10\bar{1}2\}$–$\{0001\}$	$\{10\bar{1}2\}$–$\{\bar{1}012\}$	$\{10\bar{1}2\}$–$\{10\bar{1}0\}$
Beryllium	42°10′	84°20′	47°50′
Zinc	46°59′	86°02′	43°01′

beryllium ($c/a = 1.568$) and zinc ($c/a = 1.856$) are compared using strain ellipsoid diagrams. The relevant interplanar angles in each metal are determined by

$$\cos\theta = \frac{h_1h_2 + k_1k_2 + \frac{1}{2}(h_1k_2 + h_2k_1) + \frac{3a^2}{4c^2}l_1l_2}{\left[\left(h_1^2 + k_1^2 + h_1k_1 + \frac{3a^2}{4c^2}l_1^2\right)\left(h_2^2 + k_2^2 + h_2k_2 + \frac{3a^2}{4c^2}l_2^2\right)\right]^{1/2}} \tag{3-20}$$

and are given in Table 3.5.

For the case of beryllium, the basal plane bisects the acute angle separating the $\{10\bar{1}2\}$ planes, and the prism plane bisects its supplement. In addition, the prism plane may be positioned simply by the fact that it must lie 90° away from the basal plane. From Fig. 3.22, we see that the twinning process in beryllium involves compression of the basal plane and tension of the prism plane. Consequently, if a single crystal were oriented with the basal plane parallel to the loading direction, the crystal would twin if the loads were compressive but not if the loads were tensile. The crystal would be able to twin in tension only if the basal plane were oriented perpendicularly to the loading axis.

The situation is completely opposite for zinc. Here, because the prism plane bisects the acute angle between K_1 and K_2, zinc will twin when the applied stress causes compression of the prism or extension of the basal plane (Fig. 3.23). The response of any HCP metal that twins by the $\{10\bar{1}2\}$ mode is summarized in Fig. 3.24. When $c/a < \sqrt{3}$, twinning will occur if compressive loads are applied parallel to the basal plane or tensile loads applied parallel to the prism planes. The opposite is true for the case of $c/a > \sqrt{3}$ where twinning occurs when the prism plane is compressed or the basal plane extended.

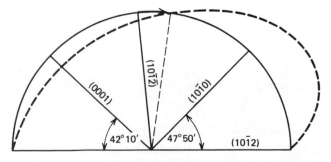

FIGURE 3.22 Strain ellipsoid for beryllium revealing twin-related foreshortening of basal plane and extension of prism plane. Twinning by $\{10\bar{1}2\}$ mode will occur when compression is applied parallel to basal plane or tension applied parallel to prism plane.

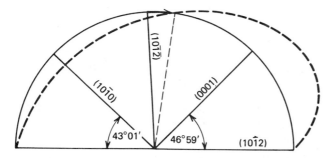

FIGURE 3.23 Strain ellipsoid for zinc revealing twin-related foreshortening of prism plane and extension of basal plane. Twinning by $\{10\bar{1}2\}$ mode will occur when compression is applied parallel to prism plane or tension applied parallel to basal plane.

The other HCP twin modes shown in Table 3.4 may operate under certain conditions; however, they are generally not preferred since the strain energy of the twin increases with S.[2] Therefore, if the resolved shear stress for given K_1 and η_1 twin elements is sufficient, twinning will occur via the mode possessing the lowest shear strain. As might be expected, there is competition not only between twin modes but also between slip and twinning as the dominant deformation mechanism under specific test conditions. Reed-Hill[37] examined the likelihood of either prism slip or $\{10\bar{1}2\}$ twinning in zirconium and found these mechanisms to be complementary (Fig. 3.25). As such, slip will occur on a viable slip system if the resolved shear stress is high enough; twinning will occur if the resolved shear stress along the K_1 and η_1 twin elements is high enough *and* the direction of loading consistent with the twinning process. For example, Fig. 3.25 shows that the highest shear stress along the K_1 and η_1 elements, corresponding to the largest orientation factor, *may* generate twinning when the angle between the applied stress axis and pole of the basal plane is zero degrees. In this situation, twinning will occur *only* when the applied stress is tensile; when this stress is compressive in nature, the resolved shear stress will be the same, but no twinning will occur.

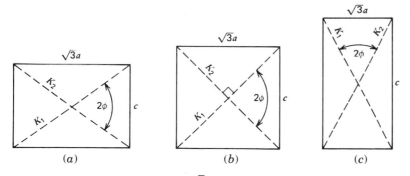

FIGURE 3.24 Conditions for $\{10\bar{1}2\}$ twinning in hexagonal crystals. (*a*) When $c/a < \sqrt{3}$, twinning results when the basal plane is compressed or the $\{10\bar{1}0\}$ planes stretched; (*b*) when $c/a = \sqrt{3}$, no twinning by $\{10\bar{1}2\}$ mode occurs; (*c*) when $c/a > \sqrt{3}$, twinning occurs when the prism planes are compressed or the basal planes stretched.

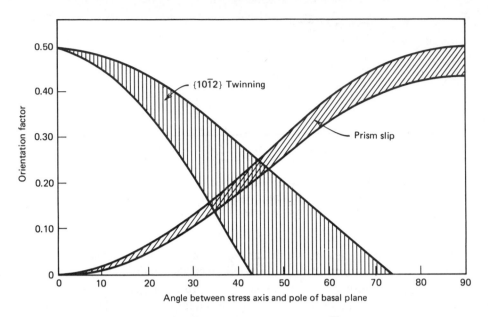

FIGURE 3.25 Competitive aspects of prism slip and {10$\bar{1}$2} twinning in zirconium. (After Reed-Hill[37]; reprinted with permission from Gordon and Breach Science Publishers, New York.)

3.2.7 Twinning in BCC and FCC Crystals

Twinning in BCC materials has been examined most closely for the case of ferritic steels, because of their engineering significance. Twin formation in steels (called Neumann bands) occurs most readily under high strain rate and/or low-temperature test conditions. The twin plane is found to be of type {112}, with the shearing direction parallel to [$\bar{1}\bar{1}$1]. What is intriguing again is the fact that twinning will depend on the direction of shear; twinning will occur in the [$\bar{1}\bar{1}$1] direction but not in the opposite [11$\bar{1}$] direction.[46] Deformation twinning is found least frequently in FCC metals except under cryogenic temperature conditions, extremely high strain rates, and in certain alloys. Since the twin elements {111}, {11$\bar{1}$}, ⟨11$\bar{2}$⟩, and ⟨112⟩ produce a large twin strain (0.707), it would appear that slip processes are more highly favored (i.e., partial dislocation motion along close-packed planes) than twin-related movements. By comparison, it should be pointed out that while *deformation* twinning is found only under extreme conditions, some FCC alloys may exhibit many *annealing* twins. As discussed in Section 3.2.1, these twins result from accidents associated with the growth of recrystallized grains from previously deformed material possessing a high density of stacking faults.

REFERENCES

1. P. G. Partridge, *Met. Rev.* **12,** 118 (1967).
2. R. von Mises, *Z. Ang. Math. Mech.* **8,** 161 (1928).

3. G. I. Taylor, *J. Inst. Met.* **62,** 307 (1938).
4. G. W. Groves and A. Kelly, *Philos. Mag.* **8,** 877 (1963).
5. G. Y. Chin and W. L. Mammel, *Trans. Met. Soc. AIME* **239,** 1400 (1967).
6. G. Y. Chin, W. L. Mammel, and M. T. Dolan, *Trans. Met. Soc. AIME* **245,** 383 (1969).
7. W. J. McG. Tegart, *Philos. Mag.* **9,** 339 (1964).
8. A. Kelly and G. W. Groves, *Crystallography and Crystal Defects,* Addison-Wesley, Reading, MA, 1970, p. 175.
9. D. C. Jillson, *Trans. Met. Soc. AIME* **188,** 1129 (1950).
10. E. N. Andrade and R. Roscoe, *Proc. R. Soc.* **49,** 166 (1937).
11. P. M. Robinson and H. G. Scott, *Acta Met.* **15,** 1581 (1967).
12. E. Schmid and W. Boas, *Plasticity of Crystals,* Hughes, London, 1950, p. 55.
13. G. I. Taylor and C. F. Elam, *Proc. R. Soc. London Ser. A* **108,** 28 (1925).
14. C. F. Elam, *Distortion of Metal Crystals,* Oxford Univ. Press, London, 1935.
15. J. S. Koehler, *Acta Met.* **1,** 508, (1953).
16. G. R. Piercy, R. W. Cahn, and A. H. Cottrell, *Acta Met.* **3,** 331 (1955).
17. H. Hu, *Texture* **1**(4), 233 (1974).
18. A. T. English and G. Y. Chin, *Acta Met.* **13,** 1013 (1965).
19. R. E. Smallman and D. Green, *Acta Met.* **12,** 145 (1964).
20. I. L. Dillamore and W. T. Roberts, *Acta Met.* **12,** 281 (1964).
21. N. Brown, *Trans. Met. Soc. AIME* **221,** 236 (1961).
22. J. S. Kallend and G. J. Davies, *Texture* **1,** 51 (1972).
23. G. Wassermann, *Z. Met.* **54,** 61 (1963).
24. E. A. Calnan, *Acta Met.* **2,** 865 (1954).
25. J. F. W. Bishop, *J. Mech. Phys. Sol.* **3,** 130 (1954).
26. H. Hu, R. S. Cline, and S. R. Goodman, *Recrystallization Grain Growth and Textures,* ASM, Metals Park, OH, 1965, p. 295.
27. H. Hu and R. S. Cline, *J. Appl. Phys.* **32,** 760 (1961).
28. S. R. Goodman and H. Hu, *Trans. Met. Soc. AIME* **230,** 1413 (1964).
29. H. Hu and S. R. Goodman, *Trans. Met. Soc. AIME* **227,** 627 (1963).
30. W. F. Hosford and W. A. Backofen, *Fundamentals of Deformation Processing,* Syracuse University Press, Syracuse, New York (1964).
31. W. F. Hosford and R. M. Caddell, *Metal Forming, Mechanics and Metallurgy,* 2d. ed. PTR Prentice Hall, Englewood Cliffs, N. J. (1993).
32. W. A. Backofen, *J. Mech. Phys. Solids.* **14,** 233 (1966).
33. G. Dieter, *Mechanical Metallurgy, 3d ed.* McGraw-Hill, New York, (1987).
34. M. Atkinson, *Sheet Met. Ind.* **44,** 167 (1967).
35. B. A. Bilby and A. G. Crocker, *Proc. R. Soc. London Ser. A* **288,** 240 (1965).
36. E. O. Hall, *Twinning and Diffusionless Transformations in Metals,* Butterworth, London, 1954.
37. R. E. Reed-Hill, *Deformation Twinning,* Gordon & Breach, New York, 1964, p. 295.
38. P. G. Partridge, *Met. Rev.* **12,** 169 (1967).
39. R. W. Cahn, *Adv. Phys.* **3,** 363 (1954).
40. S. Mahajan and D. F. Williams, *Int. Metall. Rev.* **18,** 43 (1973).
41. J. Friedel, *Dislocations,* Pergamon, Oxford, 1964.
42. D. Hull, *Fracture of Solids,* Interscience, New York, 1963, p. 417.
43. K. E. Eckelmeyer and R. W. Hertzberg, *Met. Trans.* **1,** 3411 (1970).

44. R. von Mises, *Z. Ang. Math. Mech.* **8,** 161 (1928).
45. N. S. Stoloff and M. Gensamer, *Trans. Met. Soc. AIME* **227,** 70 (1963).
46. R. Clark and G. B. Craig, *Prog. Met. Phys.* **3,** 115 (1952).

PROBLEMS

3.1 From the work of D. C. Jillson, *Trans. AIME* **188,** 1129 (1950), the following data were taken relating to the deformation of zinc single crystals.

ϕ	λ	F (newtons)
83.5	18	203.1
70.5	29	77.1
60	30.5	51.7
50	40	45.1
29	62.5	54.9
13	78	109.0
4	86	318.5

The crystals have a normal cross-sectional area of 122×10^{-6} m^2.
ϕ = angle between loading axis and normal to slip plane
λ = angle between loading axis and slip direction
F = force acting on crystal when yielding begins

(a) Name the slip system for this material.

(b) Calculate the resolved shear τ_{RSS} and normal σ_n stresses acting on the slip plane when yielding begins.

(c) From your calculations, does τ_{RSS} or σ_n control yielding?

(d) Plot on graph paper the Schmid factor versus the normal stress P/A acting on the rod.

3.2 What would be the stress–strain response of an Fe–C steel reloaded bar had the second loading taken place several weeks later or had the bar been given a moderate temperature aging treatment prior to being reloaded.

3.3 Prove to yourself that the *c/a* ratio for ideal packing in the case of a hexagonal closed-packed structure is 1.633.

3.4 What crystallographic factors usually determine the planes and directions on which slip occurs? How does one of these factors determine the temperature sensitivity of the yield strength? (Refer to Chapter 2.)

3.5 Calculate the Schmid factor for an FCC single-crystal rod oriented with the $\langle 100 \rangle$ direction parallel to the loading axis.

3.6 From the work of Andrade and Roscoe, *Proc. Phys. Soc., London,* **49,** 152 (1937), the following data were taken relating to the deformation of cadmium single crystals.

A (mm²)	Force (g)	λ	φ
0.172	53.2	80	10.1
0.1835	21.8	54.2	38.7
0.181	24.5	30.5	61.4
0.185	35.2	35.2	70.6
0.191	37.2	23.3	72.3
0.181	41.0	20.1	75.4
0.179	46.0	11.8	78.2

The values correspond to those described in Problem 3.1.

(a) What is the probable slip system?

(b) Calculate the resolved shear τ_{RSS} and normal σ_n stresses acting on the slip plane when yielding begins.

(c) From your calculations, does τ_{RSS} or σ_n control yielding?

(d) Plot on graph paper the Schmid factor versus the normal stress P/A acting on the rod.

3.7 Two rods of zinc ($c/a = 1.856$) were loaded in tension to the point where plastic deformation occurred at the same critical stress level. Yielding was found to occur at different load levels, however. Examination of the 1.25-cm rods revealed them to have been oriented differently with respect to the direction of loading, as noted in the following table:

	φ	λ
Rod A	83.5	18
Rod B	13	78

where ϕ = angle between loading axis and normal to slip plane
λ = angle between loading axis and slip direction

(a) What is the most probable slip system for this material?

(b) If Rod A yielded with a load of 203 N, what would have been the load at yield for Rod B?

3.8 A single-crystal rod of FCC nickel is oriented with the [100] direction parallel to the rod axis.

(a) Name the slip system involved in the plastic flow of nickel.

(b) How many such slip systems are in a position to be activated at the same time when the load is applied parallel to this crystallographic direction?

(c) What is the Schmid factor for this slip system? (The angles between the {100} and {110} and {100} and {111} planes are 45 and 54.7°, respectively.)

3.9 Draw the strain ellipsoid for magnesium ($c/a = 1.623$), locate the major planes, and describe those conditions that would produce twinning in the metal for the case where K_1 and K_2 are of the $\{10\bar{1}2\}$ type. Confirm the shear strain value given in Table 3.4.

3.10   Draw the (111) pole figure for the [100] wire texture in silver and for the [110] wire texture in iron wires.

3.11   From Figure 3.11, we see that copper exhibits a mixed [100] + [111] wire texture. Draw both (100) and (111) pole figures to portray this duplex texture.

3.12   The tensile strength for cold-rolled magnesium alloy AZ31B plate is approximately 160 MPa for specimens tested either parallel or perpendicular to the rolling direction. When similarly oriented specimens are compressed, the yield strength is only 90 MPa. Why? (*Hint:* Consider the possible deformation mechanisms available in the magnesium alloy and any crystallographic texture that might exist in the wrought plate.)

3.13   For the same magnesium alloy discussed in Problem 12, what would happen to a test bar loaded in simple bending?

3.14   An HCP alloy, known as Hertzalloy 200, has a *c/a* ratio of 1.600.

   (a)   What is the most probable slip system?

   (b)   For each of the following diagrams, determine whether slip will occur and whether twinning will occur (consider only {10$\bar{1}$2} twinning).

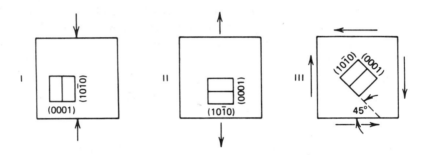

3.15   Three different investigators are each given a cube-shaped single crystal of an HCP metal. It is not known whether the cubes are of the same material. The investigators are to cut out tensile and compression specimens and measure the loads necessary for yielding. In addition, they are to observe the nature of deformation in each case. A summary of their results is given in the accompanying table.

   From these results, two investigators concluded that three different materials were involved in the test program. The third investigator claimed that the materials were the same.

   (a)   Who was correct?

   (b)   Could both conclusions be correct?

   (c)   If the materials were the same, what additional information would have to be known and how could it be used to support the claim of similar materials?

	Investigator I	Investigator II	Investigator III
Tension Test	Low loads for yielding. Only slip observed.	Low loads for yielding. Little slip and much twinning.	High loads for yielding. Little slip and no twinning.
Compression Test	Low loads for yielding. Only slip mechanism.	High loads for yielding. Little slip and no twinning.	Low loads for yielding. Little slip and much twinning.

3.16 If a single crystal of zinc (1.856) is oriented with the c axis parallel to the direction of loading, would the crystal be expected to slip and/or twin if the load was tensile in character? What would be your answer if the load was compressive?

3.17 Examine the change in orientation of the slip traces in the brass and zinc samples as they passed through twinned regions (Fig. 3.17a and 3.17c). For each case, determine whether slip or twinning occurred first.

3.18 Determine when and if twinning will occur in the following HCP crystals. Define the loading direction and sense with respect to specific crystallographic planes. Assume the twinning will involve {1012}-type planes.

	Interplane Angles		
Metal	{1012}–{0001}	{1012}–{1012}	{1012}–{1010}
A	47	86	43
B	42	84	48
C	45	90	45

3.19 Single crystals of magnesium are prepared in the form of rods that are tested in tension. The six test results are listed below where

ϕ = angle between stress axis (parallel to rod axis) and normal to basal plane;

λ = angle between stress axis and slip direction

σ_A = applied stress at onset of yielding defined by applied load divided by cross-sectional area of rod.

ϕ	λ	σ_A (MPa)
82	11	2.74
63	A	1.06
52	41	0.82
33	50	0.69
29	B	0.96
13	77	C

> (a) What is the probable slip system for the magnesium crystals?
>
> (b) Compute the resolved shear stresses and normal stresses acting on the slip system in each crystal.
>
> (c) Which stress controls the yielding process?
>
> (d) Based on the information from (b), provide estimates for the data corresponding to *A, B,* and *C.*

3.20 A cylindrical single crystal rod of zirconium is loaded in compression normal to one of its prism planes.

> (a) What load is necessary to deform the metal if the crystal diameter is 1.2-cm and the yield strength is 400 kPa?
>
> (b) Speculate as to whether this load would change if the direction of loading were reversed.

3.21 A square-shaped single rod of a Ni-base superalloy possesses a yield strength of 600 MPa. If the [100] direction of the crystal is parallel to the direction of loading, what was the rod width if yielding occurred with the application of a load of 146.8 kN?

CHAPTER 4

STRENGTHENING MECHANISMS IN METALS

We are now in a position to examine the various ways by which a metal may be strengthened. As was discussed in Chapter 1, metals (as well as plastics) can be strengthened by the addition of high-strength fibers. In a sense, such strengthening can be viewed as being *extrinsic* in nature since the load on the matrix is transferred to the high-strength fibers. As such, the *intrinsic* resistance to deformation in the matrix is not changed. It should be noted that *intrinsic* strengthening in metals, per se, relates to processes by which dislocation motion is restricted within the lattice. A number of strengthening mechanisms have been identified and include dislocation interactions with other dislocations (strain hardening), grain boundaries, solute atoms (solid solution strengthening), precipitates (precipitation hardening), and dispersoids (dispersion strengthening). These topics, along with additional discussion of metal–matrix composites, will be addressed in this chapter.

4.1 STRAIN (WORK) HARDENING

Strain hardening (also referred to as work hardening or cold working) dates back to the Bronze Age and is perhaps the first widely used strengthening mechanism for metals. Artisans hammered and bent metals to desired shapes and achieved superior strength in the process. Typical cold-worked commercial products that find use today include cold-drawn piano wire and cold-rolled sheet metal. Strain hardening results from a dramatic increase in the number of dislocation–dislocation interactions and which reduces dislocation mobility. As a result, larger stresses must be applied in order that additional deformation may take place. It is interesting to note that the strength of a metal approaches extremely high levels when there are either no dislocations present (recall Eq. 2-6) or when the number of dislocations is extremely high ($\gtrsim 10^{10}/cm^2$); low strength levels correspond to the presence of moderate numbers of dislocations ($\sim 10^3 - 10^5/cm^2$) (Fig. 4.1).

To characterize more clearly the strain-hardening behavior of metal crystals, it is helpful to examine the stress–strain response of single crystals. From Fig. 4.2, the

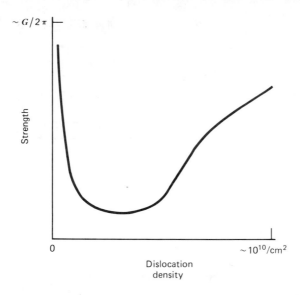

FIGURE 4.1 **Strength of metal crystals as a function of dislocation density.**

resolved shear stress–shear strain curve is seen to contain several distinct regions: an initial region of elastic response where the resolved shear stress is less than τ_{CRSS}; Stage I, a region of easy glide; Stage II, a region of linear hardening; and Stage III, a region of dynamic recovery or parabolic hardening. The latter three regions involve different aspects of the plastic deformation process for a given crystal. It is known that the extent of Stages I, II, and III depends on such factors as the test temperature, crystal purity, initial dislocation density, and initial crystal orientation.[1,2] It should be noted that Stage III closely resembles the stress–strain response of the polycrystal form of the same material.

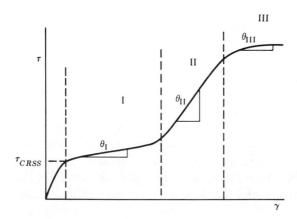

FIGURE 4.2 **Shear stress–strain curve for single crystal revealing elastic behavior when $\tau < \tau_{crss}$ and Stage I, II, III plastic response when $\tau > \tau_{crss}$. θ_I, θ_{II}, θ_{III} measure the strain hardening rate in each region.**

A number of theories have been proposed to explain the strain-hardening process in crystals, including the reason for the dramatic changes in strain-hardening rate associated with the three stages of plastic deformation. An extensive literature[3] has developed regarding these theories, all of which have focused on some of the dislocation interaction mechanisms described in the previous chapter. Seeger[4] and Friedel,[5] for example, argued that rapid strain hardening in Stage II resulted from extensive formation of dislocation pileups at strong obstacles such as Cottrell-Lomer locks.[6,7] The latter represents a sessile (nonmobile) dislocation that impedes the motion of other dislocations on their respective slip planes. An example of such a barrier is given by

$$\frac{a}{2}[01\bar{1}] + \frac{a}{2}[101] = \frac{a}{2}[110] \qquad (4\text{-}1)$$

The $[01\bar{1}]$ and $[101]$ dislocations, which move along their slip planes, (111) and $(11\bar{1})$, respectively, join to produce the sessile dislocation [110], which cannot move along either plane. Note that this dislocation reaction is permissible since the total elastic energy is reduced (recall Eq. 2-17). Mott[8] proposed that heavily jogged dislocations produced by dislocation–dislocation interactions (see Section 2.9) would be more resistant to movement, thereby enhancing the hardening rate. Unfortunately, a certain degree of confusion has arisen in this field because of the varying importance of certain dislocation interactions in different alloy crystals. One wonders then why the three distinct stages of deformation are so reproducible from one material to another and why the work hardening coefficient θ_{II} associated with Stage II deformation is almost universally constant at $G/300$. For these reasons, the ''mesh length'' theory of strain hardening proposed by Kuhlmann-Wilsdorf[9,10] is appealing pedagogically, since it does not depend on any specific dislocation model that might be appropriate for one material but not for another. Her theory may be summarized as follows: In Stage I a heterogeneous distribution of low-density dislocations exists in the crystal. Since these dislocations can move along their slip planes with little interference from other dislocations, the strain hardening rate θ_I is low. The easy glide region (Stage I) is considered to end when a fairly uniform dislocation distribution of moderate density is developed but not necessarily in lockstep with the onset of conjugate slip where a marked increase in dislocation–dislocation interactions would be expected. At this point Kuhlmann-Wilsdorf theorizes the existence of a quasi-uniform dislocation array with clusters of dislocations surrounding cells of relatively low dislocation density (Fig. 4.3a). It is believed that such cell structures represent a minimum energy and, hence, preferred dislocation configuration within the crystal.[11] Studies have shown that high stacking fault energy metals (e.g., aluminum) exhibit cell walls that are narrower and cell interiors that are more dislocation-free than in lower stacking fault energy metals (e.g., copper) (Fig. 4.3b). (In very low stacking fault energy metals (e.g., Cu–7%Al) the crystal substructure is characterized by dislocation planar arrays, consistent with the tendency for these materials to exhibit restricted cross slip (Fig. 4.3c)).

The stress necessary for further plastic deformation is then seen to depend on the

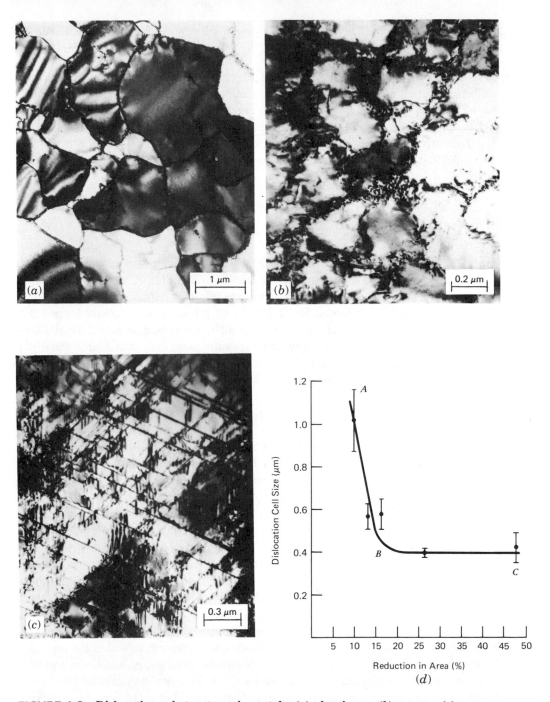

FIGURE 4.3 Dislocation substructures in metals: (*a*) aluminum; (*b*) copper; (*c*) copper–7% aluminum. (Photographs courtesy K. S. Vecchio.) (*d*) Variation in dislocation cell size with percentage reduction of area in polycrystalline niobium steel alloy.[12] (With permission.)

mean free dislocation length $\bar{l}$ in a manner similar to that necessary for the activation of a Frank–Read source where

$$\tau \propto \frac{Gb}{\bar{l}} \tag{4-2}$$

Since the dislocation density is proportional to $1/\bar{l}^2$, Eq. 4-2 may be written in the form

$$\Delta\tau \propto Gb \sqrt{\rho} \tag{4-3}$$

where ρ = dislocation density
    $\Delta\tau$ = incremental shear stress necessary to overcome dislocation barriers

This relationship has been verified experimentally for an impressive number of materials[13] and represents a necessary requirement for any strain hardening theory. With increasing plastic deformation, ρ increases resulting in a decrease in the mean free dislocation length $\bar{l}$. From Eqs. 4-2 and 4-3, the stress necessary for further deformation then increases. Kuhlmann-Wilsdorf suggests[9] that there is a continued reduction in cell size and an associated increase in flow stress throughout the linear hardening region. In other words, the *character* of the dislocation distribution remains unchanged, only the *scale* of the distribution changes (see region AB in Fig. 4.3d). With further deformation, the number of free dislocations within the cell interior decreases to the point where glide dislocations can move relatively unimpeded from one cell wall to another. Since the formation of new cell walls (and hence a reduction in $\bar{l}$) is believed to depend on such interactions, a point would be reached where the cell size $\bar{l}$ would stabilize or at best decrease slowly with further deformation. According to Kuhlmann-Wilsdorf,[10] this condition signals the onset of Stage III and a lower strain-hardening rate, since $\bar{l}$ would not decrease further. Recently Bassin and Klassen[12] provided experimental confirmation that Stage III behavior corresponds to strain levels where $\bar{l}$ remains constant (see region BC in Fig. 4.3d). Of particular note, the data reported in Fig. 4.3d are measurements taken from a polycrystalline niobium steel alloy; as such, the mesh length theory of strain hardening is applicable for both single-crystal and polycrystalline commercial alloys.

Stacking fault energy is considered to be important to the onset of Stage III. Seeger[4] has argued that Stage III begins when dislocations can cross-slip around their barriers, a view initially supported by Kuhlmann-Wilsdorf. From Seeger's point of view, Stage III would occur sooner for high stacking fault energy materials since cross-slip would be activated at a lower stress. Conversely, a low stacking fault energy material, such as brass, would require a larger stress necessary to force the widely separated partial dislocations to recombine and hence cross-slip. More recently, Kuhlmann-Wilsdorf[10,11] suggested that the mesh length theory could also explain the sensitivity of τ_{III} to stacking fault energy by proposing that enhanced cross-slip associated with a high value of stacking fault energy would accelerate the dislocation rearrangement process. Consequently, $\bar{l}$ would become stablized at a lower stress level. Setting aside for the moment the question of the correctness of the Seeger versus Kuhlmann-

Wilsdorf interpretations, it is sufficient for us to note that both theories account for the inverse dependence of τ_{III} on stacking fault energy.

In discussing the deformation structure of metals, it is important to keep in mind the temperature of the operation. It is known that the highly oriented grain structure in a wrought product, which has a very high dislocation density (10^{11} to 10^{13} dislocations/cm^2), remains stable only when the combination of stored strain energy (related to the dislocation substructure) and thermal energy (determined by the deformation temperature) is below a certain level. If not, the microstructure becomes unstable and new strain-free equiaxed grains are formed by combined recovery, recrystallization, and grain growth processes. These new grains will have a much lower dislocation density (in the range of 10^4 to 10^6 dislocations/cm^2). When mechanical deformation at a given temperature causes the microstructure to recrystallize spontaneously, the material is said to have been *hot worked*. If the microstructure were stable at that temperature, the metal experienced *cold working*. The temperature at which metals undergo hot working varies widely from one alloy to another but is generally found to occur at about one-third the absolute melting temperature. Accordingly, lead is hot worked at room temperature, while tungsten may be cold worked at 1500°C.

Before concluding the discussion of single-crystal stress–strain curves, it is appropriate to consider whether one can relate qualitative and quantitative aspects of the stress–strain response of single-crystal and polycrystalline specimens of the same material. For one thing, the early stages of single-crystal deformation would not be expected in a polycrystalline sample because of the large number of slip systems that would operate (especially near grain boundary regions) and interact with one another. Consequently, the tensile stress–strain response of the polycrystalline sample is found to be similar only to the Stage III single-crystal shear stress–strain plot. A number of attempts have been made to relate these two stress–strain curves. From Eq. 3-3

$$\sigma = \frac{P}{A} = \tau \frac{1}{\cos\phi\cos\lambda} = \tau M \qquad (4\text{-}4)$$

where $M = 1/(\cos\phi\cos\lambda)$

Assuming the individual grains in a polycrystalline aggregate to be randomly oriented, M would vary with each grain such that some average orientation factor $\overline{M}$ would have to be defined. Since there are 384 combinations of the five necessary slip systems to accomplish an arbitrary shape change, $\overline{M}$ is not easy to compute. From Section 3.1, Taylor[14] determined the preferred combination to be the one for which the sum of the glide shears was minimized. As a result it may be shown[15] that

$$\epsilon = \frac{\gamma}{\overline{M}} \qquad (4\text{-}5)$$

By combining Eq. 4-4 and 4-5 it is seen that

$$\frac{d\sigma}{d\epsilon} = \overline{M}^2 \frac{d\tau}{d\gamma} \qquad (4\text{-}6)$$

For the case of {111}⟨110⟩ slip in FCC metals and {110}⟨111⟩ slip in BCC metals, Taylor[14] and Groves and Kelly[16] showed $\overline{M}$ equal to 3.07. Subsequently, Chin et al.[17,18] analyzed the more difficult case of {110}⟨111⟩ + {112}⟨111⟩ + {123}⟨111⟩ slip in BCC crystals and found $\overline{M}$ = 2.75. In either case, one can see from Eq. 4-6 that the strain-hardening rate of a polycrystalline material is many times greater than its single-crystal counterpart.

4.2 GRAIN-BOUNDARY STRENGTHENING

The presence of grain boundaries has an additional effect on the deformation behavior of a material by serving as an effective barrier to the movement of glide dislocations. From the work of Petch[19] and Hall,[20] the yield strength of a polycrystalline material could be given by

$$\sigma_{ys} = \sigma_i + k_y d^{-1/2} \tag{4-7}$$

where σ_{ys} = yield strength of polycrystalline sample
σ_i = overall resistance of lattice to dislocation movement
k_y = "locking parameter," which measures relative hardening contribution of grain boundaries
d = grain size

The derivation of Eq. 4-7 can be traced to the work of Eshelby et al.[21] The number of dislocations that can occupy a distance L between the dislocation source and the grain boundary is given by

$$n = \frac{\alpha \tau_s d}{Gb} \tag{4-8}$$

where n = number of dislocations in the pileup
α = constant
τ_s = average resolved shear stress in the slip plane
d = grain diameter
G = shear modulus
b = Burgers vector

The stress acting on the lead dislocation is found to be n times greater than τ_s. When this local stress exceeds a critical value τ_c, the blocked dislocations are able to glide past the grain boundary. Hence

$$\tau_c = n\tau_s = \frac{\alpha \tau_s^2 d}{Gb} \tag{4-9}$$

Since the resolved shear stress τ_s is equal to the applied stress τ less the frictional stress τ_i associated with intrinsic lattice resistance to dislocation motion, Eq. 4-9 may be rewritten as

$$\tau_c = \frac{\alpha(\tau - \tau_i)^2 d}{Gb} \qquad (4\text{-}10)$$

After rearranging,

$$\tau = \tau_i + k_y d^{-1/2} \qquad (4\text{-}11)$$

which is the shear stress form of Eq. 4-7. The Petch–Hall relation also characterizes alloy yield strength in terms of other microstructural parameters such as the pearlite lamellae spacing and martensite packet size (see Section 4.6). It is readily seen that grain refinement techniques (e.g., normalizing alloy steels) provide additional barriers to dislocation movement and enhance the yield strength. As will be shown in Chapter 10, improved toughness also results from grain refinement.

Conrad[22] has demonstrated clearly that σ_i may be separated into two components: σ_{ST}, which is not temperature sensitive but structure sensitive where dislocation–dislocation, dislocation–precipitate, and dislocation–solute atom interactions are important; and σ_T, which is strongly temperature sensitive and related to the Peierls stress. The yield strength of a material may then be given by

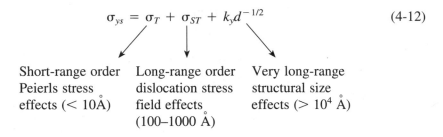

$$\sigma_{ys} = \sigma_T + \sigma_{ST} + k_y d^{-1/2} \qquad (4\text{-}12)$$

Short-range order Peierls stress effects ($< 10\overset{\circ}{A}$)	Long-range order dislocation stress field effects (100–1000 $\overset{\circ}{A}$)	Very long-range structural size effects ($> 10^4$ $\overset{\circ}{A}$)

Note that the overall yield strength of a material depends on both short- and long-range stress field interactions with moving dislocations.

The universal use of the Petch–Hall relation to characterize the behavior of metal alloys should be viewed with caution since other equations can sometimes better describe the observed strength–microstructural size relation.[23] In addition, extrapolation of Eq. 4-7 to extremely small grain sizes leads to the prediction of unrealistic yield strength levels that approach theoretical levels. Finally, there is growing consensus that grain boundary-induced dislocation pileups may not be responsible for the yield-strength–microstructural size relation described above. Instead, recent thought focuses on the important role of the grain boundary as a source for dislocations, with the yield strength being given by[24]

$$\tau = \tau_i + \alpha Gb \sqrt{\rho} \qquad (4\text{-}13)$$

Li[24] theorized that dislocations were generated at grain-boundary ledges and noted that the dislocation density ρ was inversely proportional to the grain size. Consequently, Eq. 4-13 has the same form as Eq. 4-11.

4.3 SOLID SOLUTION STRENGTHENING

Up to this point, we have considered strain-hardening and grain-boundary strengthening mechanisms that would be operative in both pure metals and alloys. When two or more elements are combined such that a single-phase microstructure is retained, various elastic, electrical, and chemical interactions take place between the stress fields of the solute atoms and the dislocations present in the lattice.[25-28] Of these, elastic interactions are believed to be most important and will be the focus of our discussion.

With reference to the stress fields surrounding both edge and screw dislocations, we see from Figs. 2.14 and 2.15 that shear stresses are associated with a screw dislocation, whereas both shear and hydrostatic stress fields surround an edge dislocation. Regarding the latter, one finds that the edge dislocation is surrounded by combined shear/hydrostatic stress fields at all locations except along the Y and X axes. Along the Y axis, the stress field is one of hydrostatic compression above the dislocation line and of hydrostatic tension below the dislocation line. This should be intuitively obvious to the reader since the extra half plane associated with the edge dislocation is squeezed into the top half of the crystal and, as such, acts to dilate the bottom half of the crystal, much as an axe blade splits open a log of wood. Along the X axis, no hydrostatic stresses are present and the stresses are pure shear in nature.

When the shear stress fields associated with both edge and screw dislocations are resolved into their normal stress components (Fig. 4.4), note that the absolute magnitude of the shear stress is equal to the normal stress; of importance, however, is the fact that the sign of the normal stress is reversed along the $\pm 45°$ directions. It follows that the shear stress field surrounding a screw dislocation is distortional (i.e., stretched in one direction and compressed in the other), whereas the edge dislocation contains both distortional and dilatational components.

The potential interaction between an edge or screw dislocation with a solute atom depends on the stress field associated with the solute atom. For example, if an atom of chromium were to substitute for an atom of FCC nickel or BCC iron, the host lattices would experience a symmetrical (hydrostatic) misfit stress associated with differences in size between solute and solvent atoms.[29] Lattice distortion would be felt equally in all directions, with the strengthening contribution being proportional to the magnitude of the misfit ϵ such that

$$\epsilon \propto \frac{da}{dc} \qquad (4\text{-}14)$$

where a = lattice parameter
c = solute concentration

The hydrostatic stress field of a substitutional solute atom interacts with the hydrostatic stress field associated with edge dislocations but not with the distortional stress field surrounding screw dislocations in the lattice. The level of hardening also depends on how much the local modulus G of the crystal was altered as a function of solute content (i.e., $G^{-1}(dG/dc)$.

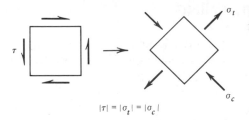

FIGURE 4.4 Resolution of shear stress field into normal stress components.

$|\tau| = |\sigma_t| = |\sigma_c|$

A much greater solute atom–dislocation interaction occurs when the stress field associated with the solute atom interacts with both edge and screw dislocations. The stress fields associated with the four lattice defects shown in Fig. 4.5 satisfy this requirement in that they are nonsymmetrical and, as such, will interact with the nonsymmetrical stress components of both edge and screw dislocations. The defect type shown in Fig. 4.5a identifies one of the octahedral interstitial sites within the BCC iron lattice where carbon and/or nitrogen atoms are located. The size of this octahedral interstitial site along any edge in the BCC lattice (or its equivalent location in the middle of each cube face) is not symmetrical and provides insufficient room for carbon and nitrogen atoms in the $\langle 100 \rangle$ direction[30]; this arises from the fact that the site size is 0.38 and 1.56 Å in the $\langle 100 \rangle$ and $\langle 110 \rangle$ directions, respectively, whereas the diameter of the carbon atom is 1.54 Å. Theoretical consideratons as well as experimental findings have shown that steel alloy strength increases rapidly at small carbon concentrations with a relationship of the form $\tau \propto \sqrt{c}$ (e.g., see Fig. 4.6). Such alloy strengthening is of great commercial interest to the steel industry. The insufficient amount of space available for the carbon atom in the BCC lattice also accounts for the very limited solid solubility of carbon in BCC iron (approximately 0.02%)

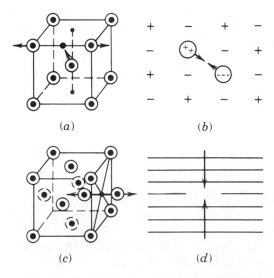

FIGURE 4.5 Nonsymmetrical stress fields in crystals. (*a*) Octahedral interstitial site in BCC crystal ($\langle 100 \rangle$ anisotropy); (*b*) divalent ion-vacancy pair ($\langle 110 \rangle$ anisotropy); (*c*) interstitial pair in FCC crystal ($\langle 100 \rangle$ anisotropy); (*d*) vacancy disk ($\langle 111 \rangle$ anisotropy).

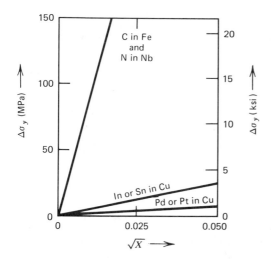

FIGURE 4.6 **Alloy strength dependence on solute content. Greater strengthening associated with nonsymmetrical defect sites. (Reprinted with permission from K. M. Ralls, T. H. Courtney and J. Wulff, *Introduction to Materials Science and Engineering*, Wiley, New York (1976).**

and leads to the development of a body-centered-*tetragonal* lattice in high carbon martensite rather than the body-centered-*cubic* crystal form for pure iron. It should be noted that the octahedral interstitial site in FCC iron is symmetrical and provides space for an atom whose diameter is as great as 1.02 Å. Since the extent of lattice distortion in the FCC lattice is much less than that found in the BCC form, the strengthening contribution of carbon in FCC iron (i.e., austenite) is low. (At the same time, the solubility limit of carbon in FCC iron is in excess of 2%—more than 100 times greater than that associated with carbon in the BCC ferrite phase.) To summarize, the strengthening potential for carbon in FCC iron is much less than that for carbon in BCC iron since the strain field surrounding the interstitial atom site is symmetrical in the FCC lattice; solute atom interaction with screw dislocations is then much weaker than for the placement of carbon atoms in the nonsymmetrical interstitial sites in the BCC lattice.

Other nonsymmetrical defects are shown in Fig. 4.5. The substitution of a divalent ion in a monovalent crystal requires that two monovalent ions be replaced by a single divalent ion; this is necessary to maintain charge balance. The divalent ion and the associated vacancy have an affinity for one another which establishes a nonsymmetrical stress field in the $\langle 110 \rangle$ direction (Fig. 4.5*b*). Interstitial atom pairs such as those resulting from irradiation damage in an FCC crystal produce a stress field in the $\langle 100 \rangle$ direction (Fig. 4.5*c*). Finally, the collapsed vacancy disk in an FCC lattice produces a dislocation loop with asymmetry in the $\langle 111 \rangle$ direction (Fig. 4.5*d*).

From the above discussion, it is seen that the relative strengthening potential for a given solute atom is determined by the nature of the stress field associated with the solute atom. When the stress field is symmetrical, the solute atom interacts only with the edge dislocation and solid solution strengthening is limited. Examples of such symmetrical defects are shown in Table 4.1. In sharp contrast, when the stress field

TABLE 4.1   Dislocation–Solute Interaction Potential[25]

Material	Defect	Hardening Effect $\dfrac{d\tau}{dc}$ as $f(G)$
Symmetrical Defects		
Al	Substitutional atom	$G/10$
Cu	Substitutional atom	$G/20$
Fe	Substitutional atom	$G/16$
Ni	Interstitial carbon	$G/10$
Nb	Substitutional atom	$G/10$
NaCl	Monovalent substitutional ion	$G/100$
Nonsymmetrical Defects		
Al	Vacancy disk (quenched)	$2G$
Cu	Interstitial Cu (irradiation)	$9G$
Fe	Interstitial carbon	$5G$
LiF	Interstitial fluorine (irradiation)	$5G$
NaCl	Divalent substitutional ion	$2G$

surrounding the solute atom is nonsymmetrical in character, the solute atom interacts strongly with both edge and screw dislocations; in this instance, the magnitude of solid solution strengthening is much greater (Table 4.1). *Note that the degree of solid solution strengthening depends on whether the solute atom possesses a symmetrical or nonsymmetrical stress field and not whether it is of the substitutional or interstitial type.* Examples of solid solution strengthening in both symmetrical (Pd or Pt in Cu) and asymmetrical distortional stress fields (C in Fe and N in Nb) are shown in Fig. 4.6. Finally, it is interesting to note that the addition of a given amount of solute atoms to the host metal may, in some instances, lead to solid solution hardening at one temperature and *softening* at another.[31,32] It has been suggested that this contrasting response is due to complex temperature-dependent interactions of screw dislocations with Peierls and solute misfit strain fields.

4.3.1   Yield-Point Phenomenon and Strain Aging

We are now in a position to rationalize the discrete load drops associated with Type III and Type IV stress–strain behavior (recall Chapter 1). As we noted in the previous section, carbon and nitrogen atoms possess a strong attraction for both edge and screw dislocations within the BCC iron lattice; accordingly, a solute ''atmosphere'' is formed around each dislocation core. Since these dislocations are ''pinned'' by such solute atmospheres, dislocation motion is severely restricted until a sufficiently high stress (the upper yield point) is applied to enable dislocation to rip free and move through the lattice. According to current theory,[29,33] these unpinned dislocations multiply rapidly by a multiple-cross-slip mechanism (Section 2.10). As a result, the number of mobile dislocations increases sharply, yielding becomes easier, and the load necessary for continued deformation decreases to the level associated with the lower yield point. As additional regions (i.e., Lüder bands) deform in this manner, the stress level remains relatively constant until essentially all dislocations have broken free from their respective solute atom clusters. At this point continued deformation takes place

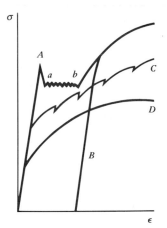

FIGURE 4.7 **Stress–strain curves influenced by discontinuous yielding. Curve *A:* Type IV behavior; Curve *B:* Type IV material response after reloading; Curve *C:* Type III behavior associated with dislocation–solute atom interactions leading to heterogeneous plastic deformation; Curve *D:* Type II behavior associated with homogeneous plastic deformation.**

by homogeneous plastic flow (Fig. 4.7, curve *A*). Furthermore, if the test was interrupted after completion of the Lüder strain region (*ab*) and the load removed and then immediately reapplied, the subsequent stress–strain curve would not display any yield point (see Fig. 4.7, curve *B*).

Although this explanation for yield-point phenomenon may be appropriate for iron single crystals containing small solute additions of interstitial carbon and nitrogen, it does not explain similar yield-point behavior in other material such as silicon, germanium, and lithium fluoride. Johnston[34] and Hahn[35] have proposed that yield-point behavior in these crystals is related to an initially low mobile dislocation density and a low dislocation-velocity stress sensitivity. Regarding the latter, studies by Stein and Low,[36] Gilman and Johnston,[37,38] and others demonstrated that the dislocation velocity v depends on the resolved shear stress as given by Eq. 4-15

$$v = \left(\frac{\tau}{D}\right)^m \qquad (4\text{-}15)$$

where v = dislocation velocity
    τ = applied resolved shear stress
    D, m = material properties

Defining the plastic strain rate by

$$\dot{\epsilon}_p \propto Nbv \qquad (4\text{-}16)$$

where $\dot{\epsilon}_p$ = plastic strain rate
    N = number of dislocations per unit area free to move about and multiply
    b = Burgers vector
    v = dislocation velocity

Johnston[34] argued that when the initial mobile dislocation density in these materials is low, the plastic strain rate would be less than the rate of movement of the test machine crosshead and little overall plastic deformation would be detected. At higher stress levels, the dislocations would be moving at a higher velocity and also begin to multiply rapidly such that the total plastic strain rate would then exceed the rate of crosshead movement. To balance the two rates, the dislocation velocity would have to decrease. From Eq. 4-15, this may be accomplished by a drop in stress, the magnitude of which would depend on the stress-sensitivity parameter m. If m were very small (less than 20 as in the case of covalent- and ionic-bonded materials as well as in some BCC metals), then a large drop in load would be required to reduce the dislocation velocity by the necessary amount. If m were large (greater than 100 to 200 as found for FCC metal crystals), only a small load drop would be required to effect a substantial change in dislocation velocity. The severity of the yield drop is depicted in Fig. 4.8a for a range of dislocation velocity stress sensitivity values. Note the magnitude of the yield drop increasing with decreasing m. If there are many free dislocations present at the outset of the test, they may multiply more gradually at lower stress levels, precluding the occurrence of a sudden avalanche of dislocation generation at higher stress levels. The corresponding decrease in magnitude of the yield drop with increasing initial mobile dislocation density is shown in Fig. 4.8b. From the above discussion, a yield point is pronounced in crystals that (1) contain few mobile dislocations at the beginning of the test, (2) have the potential for rapid dislocation multiplication with increasing plastic strain, and (3) exhibit relatively low dislocation-velocity stress sensitivity. Since many ionic- and covalent-bonded crystals possess these characteristics,[15] yield points are predicted and found experimentally in these materials.

For the case of carbon- and nitrogen-locked dislocations in iron, dislocation mo-

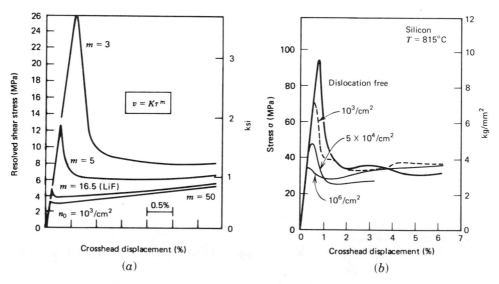

FIGURE 4.8 (a) Effect of stress sensitivity m in LiF. (From Johnston;[34] reprinted with permission of American Institute of Physics, New York.) (b) Effect of initial mobile dislocation density n_o in Si on severity of yield drop. (From Patel and Chaudhuri;[39] reprinted with permission of American Institute of Physics, New York.)

bility is essentially zero prior to the upper yield point where dislocations are finally able to tear away from interstitial atmospheres. It is theorized that the unpinning of some dislocations, their rapid multiplication, and weak velocity stress sensitivity (i.e., low m value) all contribute to the development of a yield point in engineering iron alloys. By contrast, most FCC metals have an initially high mobile dislocation density and a very high dislocation-velocity stress sensitivity, thereby making a yield drop an unlikely event in most of these materials.

The serrated character of Type III stress–strain behavior in plain carbon steel alloys can also be explained in terms of dislocation–solute atom interactions. After dislocations have ripped away from their solute atmospheres, homogeneous plastic flow occurs (e.g., curve A in Fig. 4.7) unless the tensile test was conducted at moderately elevated temperatures (e.g., approximately 200°C); for this condition, enhanced diffusion would enable carbon and/or nitrogen atmospheres to reform, thereby repinning dislocations. The applied stress would then have to increase further to enable dislocations to again break free from their solute clusters. So long as the diffusion rate for the solute atoms is equal to or slightly greater than the rate of plastic deformation, dislocations will alternately break free from solute atmospheres and then be repinned; this produces the serrated stress–strain curve shown in Fig. 4.7 (curve C). If the test temperature was much higher, homogeneous dislocation flow would take place since solute atmosphere formation would no longer be favored; accordingly, the stress–strain curve would be smooth (Fig. 4.7, curve D).

4.4 PRECIPITATION HARDENING

4.4.1 Microstructural Characteristics

When the solute concentration in an alloy exceeds the limits of solubility for the matrix phase, equilibrium conditions dictate the nucleation and growth of second-phase particles, provided that suitable thermal conditions are present. From Fig. 4.9,

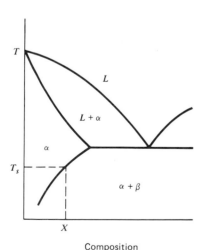

FIGURE 4.9 Portion of equilibrium phase diagram showing alloy composition X and associated solvus temperature T_s.

which shows a portion of an equilibrium phase diagram, we see that for an alloy of composition X, a single phase α is predicted at temperatures above T_s whereas two phases, α and β, are stable below the solvus line. When such an alloy is heated into the single-phase field (called a solution treatment) and then rapidly quenched, the resulting microstructure contains only supersaturated solid solution α even though the phase diagram predicts a two-phase mixture; the absence of the β phase is attributed to insufficient atomic diffusion. If this alloy is heated to an intermediate temperature (called the aging temperature) below the solvus temperature, diffusional processes are enhanced and result in the precipitation of β particles either within α grains or at their respective grain boundaries.

The onset of precipitation depends strongly on the aging temperature itself (Fig. 4.10). At temperatures approaching the solvus temperature, there is little driving force for the precipitation process, even though diffusion kinetics are rapid. Alternatively, precipitation of the second phase proceeds slowly at temperatures well below T_s despite the large driving force for nucleation of the second phase; in this instance, diffusional processes are restricted. An optimal temperature for rapid precipitation is then identified at an intermediate temperature corresponding to an ideal combination of particle nucleation and growth rates.

The development of the two-phase mixture can most generally be described as taking place in three stages. After an incubation period, clusters of solute atoms form and second-phase particles nucleate and begin to grow either homogeneously within the host grains or heterogeneously along host grain-boundary sites. During the second stage of aging, particle nucleation continues along with the growth of existing precipitates; these processes continue until the equilibrium volume fraction of the second phase has been reached. In the third and final stage of aging, these second-phase particles coarsen, with larger particles growing at the expense of smaller ones. This process, referred to as Ostwald ripening, is diffusion-driven so as to reduce the total amount of interfacial area between the two phases.

For reasons to be addressed shortly, the precipitation of second-phase particles throughout the matrix increases the difficulty of dislocation motion through the lattice. (Conversely, little strengthening has been attributed to the presence of grain-boundary

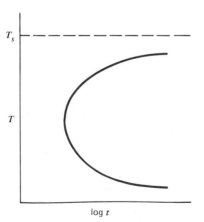

FIGURE 4.10 Precipitation rate is maximized at intermediate aging temperatures.

precipitates.) Typically, the hardness and strength of the alloy increases initially with time (and particle size) but may then decrease with further aging (Fig. 4.11). The strength and sense of the strength–time slope ($d\tau/dt$) depends on four major factors: the volume fraction, distribution, the nature of the precipitate, and the nature of the interphase boundary. Surely, were all things to remain constant, the resistance to dislocation motion through the lattice would be expected to increase with increasing volume fraction of the dislocation barrier (i.e., the precipitate). Accordingly, the first two stages of aging generally contribute to increased strengthening with time and/or particle dimension (i.e., positive $d\tau/dt$). On the other hand, Ostwald ripening, corresponding to long aging times and/or the growth of large second-phase particles, leads to negative $d\tau/dt$ conditions (see curves B and C in Fig. 4.11).

Whether the dislocation cuts through or avoids the precipitate depends on the structure of the second phase and the nature of the particle–matrix interface. The interface between the two phases may be coherent, which implies good registry between the two lattices. A dislocation moving through one phase would then be expected to pass readily from the matrix lattice into that of the precipitate. Such a coherent interface possesses a low surface energy. At the same time, however, lattice misfit (related to the difference in lattice parameters between the two phases) leads to the development of elastic strain fields surrounding the coherent phase boundary. Researchers have found that the shape of the precipitate particles depends on the degree of misfit. For example, when the misfit strain is small, spherical particles are formed such as in the case of the Al–Li binary alloy (Fig. 4.12a). When such particles grow in size and/or when a large misfit is developed, cuboidal particles are formed as in nickel superalloys (Fig. 4.12b). With increasing particle size and/or misfit strain, the microstructure reveals aligned cubes or rodlike particles.[40] As these small coherent precipitates grow with time, their interfaces may become semicoherent, with the increased lattice misfit between the two phases being accommodated by the development of interface dislocations, which bring the two lattices back into registry. At

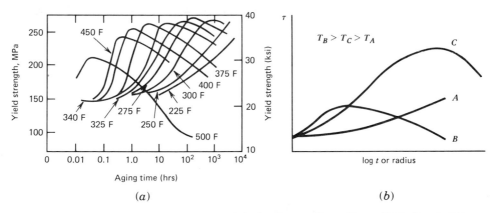

FIGURE 4.11 (*a*) **Aging curves in 6061-T4 aluminum alloy. (From J. E. Hatch, Ed.,** *Aluminum Properties and Physical Metallurgy,* **ASM, Metals Park, OH, 1984, p. 178; with permission.)** (*b*) **Schematic representation of aging process at low (*A*), high (*B*), and intermediate (*C*) temperatures.**

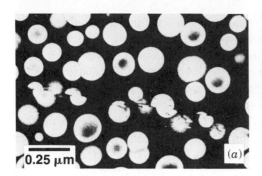 

FIGURE 4.12   **Precipitate morphology dependence on degree of lattice misfit. (*a*) Low misfit spherical particles in Al-Li alloy. (Courtesy S. Baumann.) (*b*) Moderate misfit cuboidal particles in Ni–Al alloy. Both are sheared by dislocations.**

this stage, misfit energy decreases markedly, whereas surface energy increases to a significant degree. Finally, in the latter stages of aging associated with the development of coarse particles, the interface between the two phases may break down completely and become incoherent; the surface energy associated with this interphase boundary is then increased whereas its strain field is essentially eliminated.

4.4.2   Dislocation–Particle Interactions

Why do the strength–time plots, as shown in Fig. 4.11, vary with aging temperature? Why does alloy strength increase with time (particle size) and then decrease after maximum strength has been achieved? The answers to these questions involve assessment of several dislocation–particle interactions that depend on whether dislocations are able to cut through precipitate particles or, instead, are forced to loop around them. When particle cutting occurs, hardening depends to some extent on the relative importance of elastic interactions between the dislocations and the precipitates. As previously noted, differences in lattice parameter between the host and precipitate phase will produce misfit strains that slow the movement of dislocations through the host lattice. Researchers[40,41] have found the strengthening contribution of misfit hardening to be

$$\tau \propto G\epsilon^{3/2}(rf)^{1/2} \qquad (4\text{-}17)$$

where ϵ = misfit strain (proportional to difference in lattice parameter of the two phases)
   r = particle radius
   f = volume fraction of precipitated second phase
   G = shear modulus

For many nickel-based superalloys, however, metallurgists tinker with alloy composition to limit misfit strains so as to maintain coherency for larger precipitates. As a result, the strengthening contribution of lattice misfit in these alloys is relatively minor.[42] On the other hand, low misfit strains minimize Ostwald ripening, which leads to enhanced creep resistance. Other elastic interactions include those associated with differences in shear modulus and stacking fault energy between the two phases. Here, again, for a number of important precipitation-hardened commercial alloys, the strengthening contribution of these factors is relatively small.

A second group of dislocation–particle cutting interactions involves energy storing mechanisms associated with the generation of new interphase boundary and anti-phase-domain boundary area. For example, when dislocations cut through a particle, additional precipitate–matrix interfacial area is created, which increases the overall energy of the lattice (recall Fig. 4.12); since the interfacial energy of coherent precipitates is small, this hardening mechanism contributes little to the strength of alloys that contain low misfit precipitates. On the other hand, if the precipitate has an ordered lattice (such as $CuAl_2$ particles in an aluminum alloy or Ni_3Al (γ') precipitates in a nickel-based superalloy) the passage of a single dislocation destroys the periodicity of the superlattice (recall Section 2.8) and contributes markedly to alloy strengthening. Though the passage of a second identical dislocation on the same slip plane reorders the particle, an anti-phase-domain boundary (APB) is formed between the two super-lattice dislocations (recall Section 2.8). Since the APB energy is roughly ten times greater than interphase boundary energy in nickel superalloys[43] and in Al–Li alloys, the strengthening contribution due to APB formation in these systems is significant. Gleiter and Hornbogen[40–44] and Ham[45] reported the strengthening contribution associated with this mechanism to be of the form

$$\tau \propto \gamma^{3/2}(rf/G)^{1/2} \qquad (4\text{-}18)$$

where γ = APB energy

It is interesting to note that the passage of superlattice dislocation pairs through a precipitate particle effectively reduces the length of the ordered path for subsequent dislocation pairs; for this reason, dislocation movement within microstructures containing ordered precipitates is of a heterogeneous nature and typically involves the activity of relatively few slip planes associated with large slip steps.

If the misfit strain is large, the interface incoherent, or the average particle separation above a certain critical value, dislocations are unable to cut through the precipitate; instead they loop around individual particles as shown in Fig. 4.13. (In some alloy systems, both particle cutting and looping can occur simultaneously.) Note the strong similarity of such dislocation looping with the Frank–Read mechanism for dislocation multiplication (Fig. 2.29). The stress necessary for the dislocation to loop around the precipitate is the same as that given for activation of the Frank–Read source, where l is the distance between the particles:[46]

$$\tau = Gb/l \qquad (2\text{-}22)$$

With the passage of subsequent dislocations, the *effective* distance between two adjacent precipitates l' decreases with the increasing number of dislocation loops surrounding the particles. As such, the dislocation looping mechanism provides a measure of strain hardening.[47] For a given volume fraction of second-phase particles, l increases as the precipitates grow larger with further aging. Consequently, the stress necessary for dislocations to loop around precipitates should decrease with increasing particle

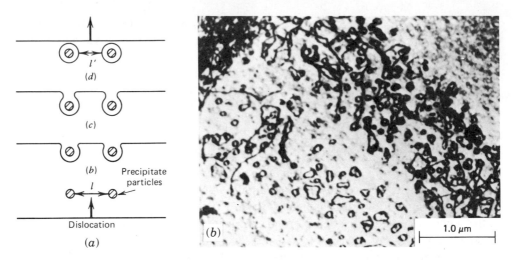

FIGURE 4.13 Dislocation looping around particles. (*a*) Schema revealing reduced "effective" particle spacing with looping; (*b*) looping in Al–Li alloy. (Photo courtesy S. Baumann.)

size. It is important to note that dislocation looping (often referred to as Orowan looping) is controlled by the spacing between particles and not by the nature of the particle itself. Furthermore, such slip activity is of a more homogeneous nature than that described above for the case of deformation by particle cutting.

The complex interaction between this group of strengthening mechanisms is responsible for the strength–time relations shown in Fig. 4.11, which may be summarized in the following manner. After solution treatment and subsequent quenching, the alloy experiences the greatest *potential* for solid solution strengthening, since the greatest amount of solute is present in the host matrix. If the solute possesses a nonsymmetrical stress field, solid solution strengthening would be great, as in the case of as-quenched carbon martensite (see Section 4.6). In sharp contrast, were the solute to possess a symmetrical stress field (such as in aluminum- and nickel-based alloys), limited solid solution strengthening would be expected. With aging, and the associated precipitation of second-phase particles, the solute level in the host matrix would decrease along with the solid solution strengthening component. This relatively small reduction in absolute strength in aluminum- and nickel-based alloys is more than compensated for by several precipitation hardening mechanisms, such as dislocation interactions with precipitate misfit strain fields, particle cutting, and elastic modulus interaction effects. As noted above, the extent of such hardening *increases* with time (i.e., particle size). With further aging, leading to the loss of coherency, or a wide interparticle spacing, dislocation looping around the particles takes place. Alloy strength then *decreases* with further aging time and/or particle size. The overall aging response of the alloy may then be characterized by the attainment of maximum strength at intermediate aging times and particle dimensions. A schematic representation of the interplay between these hardening processes is shown in Fig. 4.14 and bears close resemblance to the experimental results given in Fig. 4.11. Regarding the latter figure, "underaged" conditions are associated with curve *A* and the left portions of curves *B* and *C*; "overaging" corresponds to aging times greater than those needed

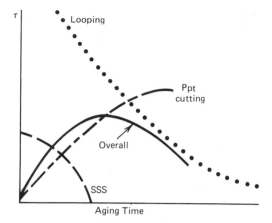

FIGURE 4.14   Schematic representation showing role of major hardening mechanisms in development of overall hardening response.

for peak strengthening. It should be noted that with the exception of the γ' phase in nickel superalloys, most homogeneous precipitates in other alloy systems are metastable.

4.5   DISPERSION STRENGTHENING

Alloys can also be strengthened by the addition of oxide particles that obstruct dislocation motion. By adding Al_2O_3 flakes and ThO_2 particles to aluminum and nickel matrices, respectively, these metals achieve attractive strength properties at temperatures approaching their melting points.[48,49] As expected, alloy strength increases with oxide volume fraction and decreasing particle spacing.[50,51] Since the strengthening potential for dislocation looping around noncoherent particles is less than that associated with particle cutting processes, such alloys are not among the strongest structural materials. On the other hand, the microstructures of dispersion-hardened alloys are more stable than those associated with precipitation-hardened alloys, thereby making them more suitable for load-bearing applications at elevated temperatures.

Further improvements in oxide-dispersion-strengthened (ODS) alloys have resulted from the use of a "mechanical alloying" process.[52] Mixtures of powder particles of different constituents are blended together in a dry, high-energy ball mill. The discrete particles are repeatedly welded together, fractured, and rewelded. Such intimate mechanical mixing leads to the formation of particles with a homogeneous phase distribution that is extremely fine grained and heavily cold worked. The powders are then hot compacted, hot extruded, and/or hot rolled to produce materials with attractive mechanical properties.[53,54] By choosing a matrix alloy of virtually any composition, it is possible to tailor a material to meet a wide range of property requirements. The compositions of three commercial nickel-based ODS alloys are given in Table 4.2. These alloys contain Cr in solid solution for elevated temperature corrosion resistance and Y_2O_3 for dispersion hardening. Alloy MA754 contains a mixture of yttrium oxides

TABLE 4.2   Chemical Composition of Mechanically Alloyed ODS Superalloys

Alloy	Ni	Cr	Al	Ti	Y_2O_3	W	Mo	Ta
Inconel MA 754	bal	20	0.3	0.5	0.6	—	—	—
Inconel MA 6000	bal	15	4.5	2.5	1.1	4	2	2
Alloy 51	bal	9.5	8.5	—	1.1	6.6	3.4	—

and yttria aluminates in a size range from 5 to 100 nm; these fine particles have a planar spacing of approximately 0.1 μm and constitute about 1 v/o of the alloy.[55] Inconel MA 6000 contains approximately 7 w/o Al + Ti, which introduces the precipitation-hardening γ' phase to the nickel matrix. These alloys are being selected for gas turbine vanes, turbine blades and sheets for use in oxidizing/corrosive atmospheres. The elevated temperature strength and rupture behavior of ODS alloys are discussed in the next chapter. For the present, it is timely to examine the rupture strength of ODS alloys versus precipitation-hardened nickel-based superalloys (see Fig. 5.34). At 750°C, the dispersion-strengthened ODS alloys are seen to exhibit lower strength levels than the precipitation-hardened superalloys. However, at temperatures in excess of 900 to 950°C, the more stable dispersion-strengthened ODS alloys are found to possess superior strengths than in the precipitation strengthened alloys; in the latter instance, the precipitate particles have begun to coarsen and/or redissolve in the matrix in this temperature range.

4.6   STRENGTHENING OF STEEL ALLOYS

A brief discussion of the strengthening mechanisms associated with steel alloys is appropriate since this class of materials is of major commercial importance. In addition, different steel alloys derive their strength from various combinations of the strengthening mechanisms considered thus far; as such, an analysis of the strength of steel alloys provides pertinent examples of these strengthening mechanisms. Several review articles[56-58] concerning the strength of steel alloys point to the fact that some or all of the major strengthening mechanisms that we have studied (i.e., solid solution strengthening, strain hardening, grain-boundary hardening, and precipitation and dispersion strengthening) are operative in each alloy system, depending on the character of the transformation product(s). The four major microstructural features found in steel alloys are ferrite, pearlite, bainite or lath martensite, and plate martensite (Fig. 4.15); the associated microstructural features are represented in Fig. 4.16. Clearly, ferrite (single-phase solid solution of iron) exhibits the simplest microstructure, with the principle strengthening mechanism corresponding to grain refinement[59]

$$\sigma_{ys} = \sigma_i + k_y d^{-1/2} \qquad (4\text{-}7)$$

where d = ferrite grain size

Values of σ_i and k_y are found to vary with alloy content and test temperature (Fig. 4.17). Note the stronger dislocation locking tendency for nitrogen solute additions relative to that of carbon atoms. In pearlitic steels, the interlamellar spacing (S)

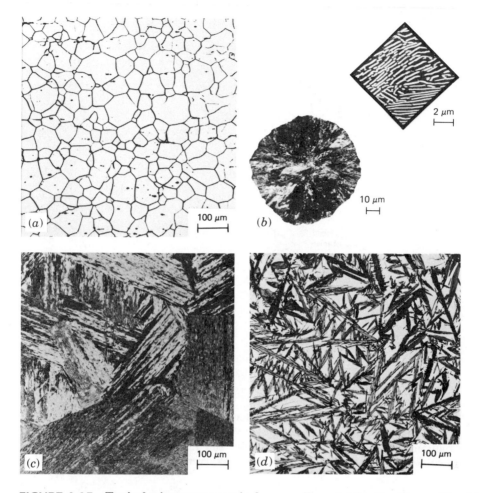

FIGURE 4.15   Typical microstructures in ferrous alloys: (*a*) ferrite; (*b*) pearlite; (*c*) lath martensite; (*d*) plate (acicular) martensite. (Courtesy A. Benscoter and J. Ciulik.)

between the ferrite and iron carbide lamellae dominates alloy strength; little influence on alloy strength is noted by changes in austenite (FCC form of iron) grain size and nodule diameter.[58] Langford[60] has shown that the strength of pearlitic steels may be described by

$$\sigma_{ys} = \sigma_i + k_1 S^{-1/2} + k_2 S^{-1} \tag{4-19}$$

where S = interlamellar spacing

k_1, k_2 = constants

σ_i = resistance of lattice to dislocation movement

The specific influence of S on alloy strength (i.e., $S^{-1/2}$ versus S^{-1} dependence) varies with the dominant hardening mechanism, which, in turn, reflects a change in the controlling free path for movement of dislocations; an $S^{-1/2}$ dependence of σ_{ys} cor-

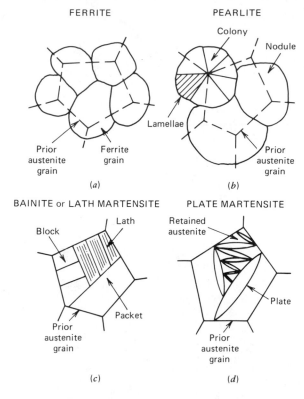

FIGURE 4.16 **Schematic representation of microconstituents shown in Fig. 4.15.[58] (Reprinted with permission from the proceedings of an International Conference on Phase Transformations in Ferrous Alloys, A. R. Marder and J. I. Goldstein, eds., 1984, The Metallurgical Society, 420 Commonwealth Drive, Warrendale, PA., 15086.)**

responds to processes associated with the formation of dislocation pile-ups (recall Eq. 4-7). When strength is controlled by the work required for the generation of dislocations, an S^{-1} lamellae size–strength dependence develops (recall Eq. 4-2).

When asked which strengthening mechanism controls the mechanical properties of martensitic steels, the reader would not err by replying, "All of the above." Indeed, the high strength of martensite draws upon several mechanisms, with solid solution strengthening exerting the greatest influence (recall Fig. 4.6). To illustrate, lath martensite contains up to 0.6 wt% carbon and possesses boundary obstacles (e.g., packet boundaries) along with a highly dislocated substructure ($>10^{10}$ dislocations/cm^2). Norstrom[61] has proposed a comprehensive relation to describe the yield strength of lath martensite

$$\sigma_{ys} = \sigma_i + k\sqrt{c} + k_y d^{-1/2} + \alpha Gb\sqrt{\rho} \qquad (4\text{-}20)$$

Peierls   Solid solution   Boundary   Strain
stress   strengthening   hardening   hardening

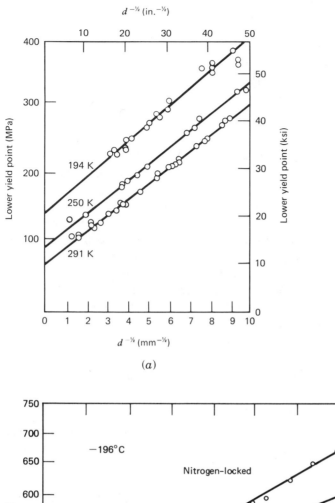

(a)

(b)

FIGURE 4.17   (a) Influence of grain size on yield strength in ferritic steel; (b) grain-size dependence of lower yield point in steel reflecting greater dislocation locking k_y with nitrogen interstitial.[59] (Reprinted with permission from MIT Press, Cambridge, MA.)

where σ_i = resistance of lattice to dislocation movement

  k, α = constants

  c = solute atom concentration

  k_y = locking parameter

  d = packet size (recall Fig. 4.16)

  G = shear modulus

  b = Burgers vector

  ρ = dislocation density

Furthermore, a dispersion-hardening component is introduced with the precipitation of iron carbide particles during tempering. It should be noted, however, that total alloy strength decreases with tempering since the solid solution strengthening contribution is correspondingly reduced. For this reason, carbon martensites are strongest after rapid quenching from the austenite region, but lose strength with tempering. Such behavior contrasts markedly with precipitation-hardening alloys (recall Section 4.4), which are soft upon quenching from the solution treatment zone but then strengthen with aging. Maraging steels are alloys that conform to the latter strengthening type since precipitation of fine second-phase particles is responsible for strengthening in such alloys; correspondingly, extremely low carbon levels in these alloys precludes significant solid solution strengthening.

4.7   METAL–MATRIX COMPOSITE STRENGTHENING

4.7.1   Whisker-Reinforced Composites

Thus far, we have examined several intrinsic strengthening mechanisms in metal alloys. As mentioned in Section 1.2.7, alloys may also be strengthened extrinsically through the addition of aligned high-strength fibers such as carbon, aramid (e.g., Kevlar), and boron, and reinforcement may be achieved with Al_2O_3 or SiC whiskers. Some metal–matrix composites have been fabricated by liquid infiltration of the matrix around the fibers; other composite systems have been prepared by extrusion of hot compacted matrix powders and high-strength whiskers. Such powder metallurgy composites, consisting of aluminum alloys reinforced with SiC whiskers, have attracted considerable attention. Several investigators[62–65] have noted that the strength of Al–SiC composites exceeds that predicted from conventional composite theory (recall Section 1.2.7). Transmission electron microscope studies have determined that these enhanced strength levels are attributed to the presence of relatively high dislocation densities in the aluminum alloy matrices examined thus far (Fig. 4.18). Arsenault and coworkers[63,64] theorized and subsequently confirmed that such high dislocation-density levels resulted from the large difference (10:1) in coefficients of thermal expansion between the aluminum alloy matrix and the SiC whiskers. Accordingly, when the composite is cooled from elevated temperatures, the misfit strains that develop are relieved by the generation of dislocations at the ends of the SiC whiskers. Recent studies have also shown that this increased dislocation density accelerates the aging process within the aluminum alloy matrix by shortening the aging time to achieve maximum strength relative to that associated with the unreinforced matrix alloy.[62,65] These preliminary findings are of major significance since computations of composite

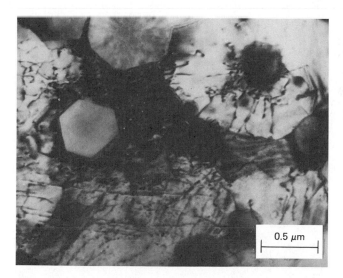

FIGURE 4.18 High dislocation density on 6061 aluminum alloy reinforced with 20 v/o SiC[63]. (Reprinted with permission from M. Vogelsang, R. J. Arsenault, and R. M. Fisher, *Metallurgical Transactions,* 17A, 379 (1986).

strength must account for alterations in matrix properties due to the presence of the reinforcing phase. As such, the high-strength phase serves to strengthen the matrix both *extrinsically* by load transfer to the fibers/whiskers and *intrinsically* by increasing the dislocation density.

4.7.2 Eutectic Composites

A different group of whisker-reinforced metal–matrix composites can be cast directly from the melt by unidirectional solidification of certain alloys of eutectic composition. The reader should recognize that when such an alloy is cooled from the melt, the liquid will undergo, at the eutectic temperature, a first-order transformation to two distinctly different solid phases at a fixed temperature. If the solidification of an ingot of eutectic composition is carried out under a steep axial thermal gradient, it is possible to establish a planar solid–liquid interface separating the molten and solid portions of the ingot. Such conditions may be achieved by the gradual withdrawal of an ingot from a furnace such that uniaxial heat flow conditions are established. When this occurs, the two solid phases produced from the eutectic reaction deposit at the liquid–solid interface and grow parallel to the direction of movement of this reaction front. Consequently, two phases are formed and oriented parallel to the direction of solidification. The resulting aligned microstructures of such eutectic alloys may take several forms,[66], but often assume a ''normal'' configuration of many fine fibers embedded within a continuous matrix (Fig. 4.19*a,b*) or parallel lamellae of each phase (4.19*c*).

Cooksey et al.[67] and Jackson and Hunt[68] have shown that the type of morphology observed in a given ''normal'' eutectic microstructure will depend upon the relative volume fraction of each phase. After examination of many eutectic systems, it is now known that the rod morphology will prevail when one phase is present in amounts less than $1/\pi$ of the total volume. Alternatively, when the minor phase constitutes

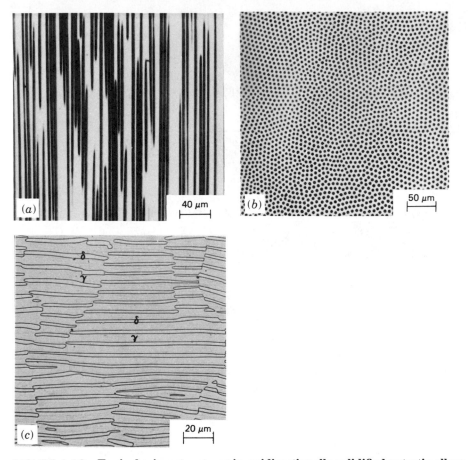

FIGURE 4.19 **Typical microstructures in unidirectionally solidified eutectic alloys. (***a***) Longitudinal and (***b***) transverse sections of MnSb whiskers embedded in an Sb matrix. (After Jackson.[69]) (***c***) Transverse section of the Ni–Ni$_3$Nb eutectic alloy revealing lamellar microstructure.**

more than $1/\pi$ of the total volume, a lamellar structure consisting of alternate platelets of the two solid phases is preferred.

Recognizing that the volume fraction of each phase is determined from the phase diagram and that the microstructural morphology is controlled in large part by the relative volume fraction of each constituent, Tiller[70] further showed that the spacing (and, hence, the size of each particle) of the interpenetrating phases could be controlled by the velocity of the solid–liquid interface.

$$\lambda^2 R = \text{constant} \qquad (4\text{-}21)$$

where λ = interparticle spacing
         R = solidification rate

In this manner, two-phase composite structures of given relative volume fraction can be produced with a range of particle sizes.

While the two phases are clearly aligned, the associated alloy cannot be classified as a composite material until it can be demonstrated that the microcrystals are capable of supporting high stress levels and arc adequately bonded to the matrix. These two critical steps required for composite manufacture by directional solidification of eutectic alloys were revealed in studies of the Cu–Cr system.[71,72] Lemkey and Kraft[71] showed that extracted Cr whiskers of submicron diametral size could withstand elastic stresses in excess of 7 GPa. In a closely related study, a colleague and I demonstrated that the Cu–Cr system exhibited excellent bonding between the Cr whiskers and the Cu matrix.[72] Unfortunately, no reinforcement in this alloy system was achieved since the volume fraction of Cr whiskers was less than the critical amount necessary for composite strengthening (recall Fig. 1.20).

The development of a reinforced composite system by unidirectional solidification of a eutectic alloy was confirmed initially with the fibrous Al–Al_3Ni and lamellar Al–$CuAl_2$ systems.[73,74] The results depicted in Fig. 4.20 for the Al–Al_3Ni system are in close agreement with composite theory since empirical strength matched closely the theoretical expectation. Also, ultimate composite strain was found equal to the strain limit of the Al_3Ni whiskers; therefore, when the whiskers fractured, the entire composite failed. Subsequent studies of nickel and cobalt binary and pseudo-binary eutectic systems have shown these alloys to possess excellent strength and creep resistance[75] (see Section 5.7). Were it not for the relatively time-consuming casting procedures required to produce the desired microstructures in directionally solidified eutectic composites, these materials would be more widely used.

While the formation of high-strength crystalline fibers or filaments directly from the melt is an intriguing process, it is by no means a unique event in nature; rather, the materials scientist must yield to the long-recorded activities of arachnids and silkworms. For example, researchers have determined that spider silk is generated by the drawing of an amorphous protein liquid from various glands, which then converts quickly to a highly oriented, very long crystalline filament with a diameter on the order of several hundred angstroms. The highly oriented and crystalline morphology

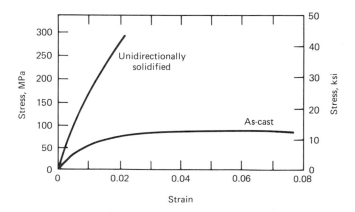

FIGURE 4.20   Stress–strain curves for unidirectionally solidified and as-cast Al–Al_3Ni eutectic microstructures.[74] (Reprinted with permission from Transactions of the Metallurgical Society, 233, 334 (1965), a publication of the Metallurgical Society, Warrendale, PA., 15086.)

of these filaments is believed responsible for their reported strengths in excess of 690 MPa.[76] In one intriguing investigation, Lucas[76] showed spider silk to have twice the tenacity, defined in grams/denier^2, of a 2070-MPa steel wire and four times the extension at break.

Such high strengths along with the apparent abundant supply of spider silk have prompted enterprising individuals to seek commercial markets for the product of our arachnid friends. In one such feasibility study in 1709, several pairs of stockings and gloves were woven from spider silk and presented to the French Academy of Science for consideration.[77] It would appear that the silkworm has emerged the clear victor over the spider in the battle for the silk market.

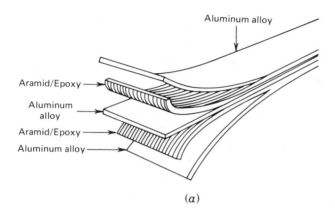

(a)

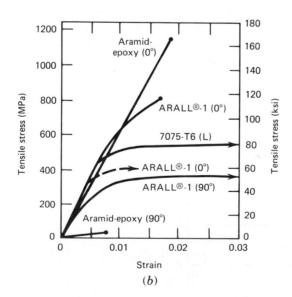

(b)

FIGURE 4.21   (a) **Layup of ARALL hybrid composite, consisting of alternate layers of aluminum alloy and aramid/epoxy laminates.** (b) **Tensile stress–strain curves for ARALL and constituent layers as function of loading angle relative to fiber axis. Compression test results shown with dashed curve. (From R. J. Bucci, Alcoa Technical Center, with permission.)**

4.7.3 Laminated Composites

A number of laminated composites have been used in various engineering components, with contiguous plies being joined together by such methods as diffusion bonding or with the use of adhesives. A new hybrid composite has attracted considerable attention in Europe and in the United States as a possible candidate material for aircraft structural components.[78–80] This material consists of thin aluminum alloy sheets (typically 2024-T3 or 7075-T6) and epoxy/aramid composite layers that are adhesively bonded together (Fig. 4.21a). The material, called ARALL (aramid aluminum laminate), can be machined and formed into useful shapes. Also of potentially great interest is the fact that the aluminum alloy outer layers provide impact resistance and damage detectability. Furthermore, cracks that may initiate in the aluminum alloy surface layers are arrested when the crack front encounters the epoxy/aramid fiber layer. As discussed in Chapter 13, this greatly extends overall fatigue lifetime for components fabricated with this unusual material.

Panels of ARALL typically contain three layers of aluminum alloy (approximately 0.3 to 0.5 mm thick) which sandwich two epoxy/aramid fiber adhesive layers, each 0.25 mm thick. Since the volume fraction of the aligned aramid fibers in the epoxy resin is between 40 and 50 v/o, the overall volume fraction of aramid fibers in the hybrid composite is approximately 15%. Typical stress–strain curves for ARALL along with its constituent layers are shown in Fig. 4.21b as a function of loading direction relative to the alignment direction of the aramid fibers.[80] As with other reinforced composites, the compressive strength of ARALL is inferior to that in tension, owing to buckling of the high-strength fibers.

REFERENCES

1. J. Garstone and R. W. K. Honeycombe, *Dislocations and Mechanical Properties of Crystals,* J. Fisher, Ed., Wiley, New York, 1957, p. 391.
2. L. M. Clarebrough and M. E. Hargreaves, *Progress in Materials Science,* Vol. 8, Pergamon, London, 1959, p. 1.
3. J. P. Hirth and J. Weertman, Eds., *Work Hardening,* Gordon & Breach, New York, 1968.
4. A. Seeger, *Dislocations and Mechanical Properties of Crystals,* J. C. Fisher, Ed., Wiley, New York, 1957, p. 243.
5. J. Friedel, *Philos. Mag.* **46,** 1169 (1955).
6. W. M. Lomer, *Philos. Mag.* **42,** 1327 (1951).
7. A. H. Cottrell, *Philos. Mag.* **43,** 645 (1952).
8. N. F. Mott, *Trans. Met. Soc. AIME* **218,** 962 (1960).
9. D. Kuhlmann-Wilsdorf, *Trans. Met. Soc. AIME* **224,** 1047 (1962).
10. D. Kuhlmann-Wilsdorf, *Work Hardening,* J. P. Hirth and J. Weertman, Eds., Gordon & Breach, New York, 1968, p. 97.
11. D. Kuhlmann-Wilsdorf, *Work Hardening in Tension and Fatigue,* A. W. Thompson, Ed., AIME, New York, 1977, p. 1.
12. M. N. Bassin and R. J. Klassen, *Mater, Sci. Eng.* **81,** 163 (1986).
13. H. M. Otte and J. J. Hren, *Exp. Mech.* **6,** 177 (1966).
14. G. I. Taylor, *J. Inst. Met.* **62,** 307 (1938).

15. W. J. McG. Tegart, *Elements of Mechanical Metallurgy,* Macmillan, New York, 1966.
16. G. W. Groves and A. Kelly, *Philos. Mag.* **8,** (1963).
17. G. Y. Chin and W. L. Mammel, *Trans. Met. Soc. AIME* **239,** 1400 (1967).
18. G. Y. Chin, W. L. Mammel, and M. T. Dolan, *Trans. Met. Soc. AIME* **245,** 383 (1969).
19. N. J. Petch, *JISI* **173,** 25 (1953).
20. E. O. Hall, *Proc. Phys. Soc. B* **64,** 747 (1951).
21. J. D. Eshelby, F. C. Frank, and F. R. N. Nabarro, *Philos. Mag.* **42,** 351 (1951).
22. H. Conrad, *JISI* **198,** 364 (1961).
23. R. W. Armstrong, Y. T. Chou, R. A. Fisher, and N. Lovat, *Philos. Mag.* **14,** 943 (1966).
24. J. C. M. Li, *Trans. Met. Soc. AIME* **227,** 239 (1963).
25. R. L. Fleischer, *The Strengthening of Metals,* D. Peckner, Ed., Reinhold, New York, 1964, p. 93.
26. P. Haasen, *Physical Metallurgy,* R. M. Cahn and P. Haasen, Eds., Vol. 2, Chap. 21, North-Holland, Amsterdam, 1983, p. 1341.
27. K. R. Evans, *Treatise on Materials Science and Technology,* H. Herman, Ed., Vol. 4, Academic, New York, 1974, p. 113.
28. R. L. Fleischer, *Acta Metall.* **11,** 203 (1963).
29. A. H. Cottrell, Report on the Conference on Strength of Solids, The Physical Society, London, 1948, p. 30.
30. G. K. Williamson and R. E. Smallman, *Acta Crystallogr.* **6,** 361 (1953).
31. A. Sato and M. Meshii, *Acta Metall.* **21,** 753 (1973).
32. D. J. Quesnel, A. Sato, and M. Meshii, *Mater. Sci. Eng.* **18,** 199 (1975).
33. H. Conrad, *JISI* **198,** 364 (1961).
34. W. G. Johnston, *J. Appl. Phys.* **33,** 2716 (1962).
35. G. T. Hahn, *Acta Metall.* **10,** 727 (1962).
36. D. F. Stein and J. R. Low, Jr., *J. Appl. Phys.* **31,** 362 (1960).
37. J. J. Gilman and W. G. Johnston, *J. Appl. Phys.* **31,** 687 (1960).
38. W. G. Johnston and J. J. Gilman, *J. Appl. Phys.* **30,** 129 (1959).
39. J. R. Patel and A. R. Chaudhuri, *J. Appl. Phys.* **34,** 2788 (1963).
40. H. Gleiter and E. Hornbogen, *Mater. Sci. Eng.* **2,** 284 (1967/68).
41. L. M. Brown and R. K. Ham, *Strengthening Methods in Crystals,* A. Kelly and R. B. Nicholson, Eds., Applied Science, London, 1971, p. 9.
42. V. Gerald and H. Haberkorn, *Phys. Status Solidi* **16,** 675 (1966).
43. E. Nembach and G. Neite, *Prog. Mater. Sci.* **29**(3), 177 (1985).
44. H. Gleiter and E. Hornbogen, *Phys. Status Solidi,* **12,** 235 (1965).
45. R. K. Ham, *Trans. Japan Inst. Met.* **9** (supplement), 52 (1968).
46. E. Orowan, Discussions in *Symposium on Internal Stresses in Metals and Alloys,* Institute of Metals, London, 451 (1948).
47. J. C. Fisher, E. W. Hart, and R. H. Pry, *Acta Metall.* **1,** 336 (1953).
48. I. Irmann, *Metallurgia* **49,** 125 (1952).
49. G. B. Alexander, U. S. Patent No. 2,972,529, Feb. 21, 1961.
50. E. Gregory and N. J. Grant, *Trans. AIME* **200,** 247 (1954).
51. F. V. Lenel, A. B. Backensto, Jr., and M. V. Rose, *Trans. AIME* **209,** 124 (1957).
52. J. S. Benjamin, *Met. Trans.* **1,** 2943 (1970).
53. R. C. Benn, L. R. Curwick, and G. A. J. Hack, *Powder Metall.* **24,** 191 (1981).
54. R. Sunderesan and F. H. Froes, *J. Metals.* **39**(8), 22 (1987).

55. T. E. Howson, J. E. Stulga, and J. K. Tien, *Met. Trans.* **11A,** 1599 (1980).
56. E. Hornbogen, Strengthening Mechanisms in Steel, in *Steel-Strengthening Mechanisms,* Climax Molybdenum Co., Zurich, 1969, p. 1.
57. F. B. Pickering, *Physical Metallurgy and the Design of Steels,* Applied Science, London, 1978, Chap. 1.
58. A. R. Marder and J. I. Goldstein, Eds., *Phase Transformations in Ferrous Alloys,* AIME, Warrendale, PA, 1984.
59. N. J. Petch, *Fracture,* Proceedings, Swampscott Conference, Wiley, New York, 1959, p. 54.
60. G. Langford, *Met. Trans.* **8,** 861 (1977).
61. L. A. Norstrom, *Scand. J. Metall.* **5,** 159 (1976).
62. T. G. Nieh and R. F. Karlak, *Scripta Met.* **18,** (1984).
63. M. Vogelsang, R. J. Arsenault, and R. M. Fisher, *Met. Trans.* **17A,** 379 (1986).
64. R. J. Arsenault and R. M. Fisher, *Scripta Met.* **17,** 67 (1983).
65. T. Christman and S. Suresh, Brown University Report NSF-ENG-8451092/1/87, June 1987.
66. L. M. Hogan, R. W. Kraft, and F. D. Lemkey, *Advances in Materials Research,* Vol. 5, Wiley, New York, 1971, p. 83.
67. D. J. S. Cookey, D. Munson, M. P. Wilkinson, and A. Hellawell, *Philos. Mag.* **10,** 745 (1964).
68. K. A. Jackson and J. D. Hunt, *Trans. AIME* **236,** 1129 (1966).
69. M. R. Jackson, M. S. Thesis, Lehigh University, 1967.
70. W. A. Tiller, *Liquid Metals and Solidification,* ASM, Cleveland, OH, 1958, p. 276.
71. F. D. Lemkey and R. W. Kraft, *Rev. Sci. Inst.* **33,** 846 (1962).
72. R. W. Hertzberg and R. W. Kraft, *Trans. AIME* **227,** 580 (1963).
73. R. W. Hertzberg, F. D. Lemkey, and J. A. Ford, *Trans. AIME* **233,** 342 (1965).
74. F. D. Lemkey, R. W. Hertzberg, and J. A. Ford, *Trans. AIME* **233,** 334 (1965).
75. F. D. Lemkey, *Proceedings, MRS Conference 1982, CISC IV,* Vol. 12, F. D. Lemkey, H. E. Cline, and M. McLean, Eds., Elsevier Science, Amsterdam, 1982.
76. F. Lucas, *Discovery* **25,** 20 (1964).
77. W. J. Gertsch, *American Spiders,* Van Nostrand, New York, 1949.
78. R. Marissen and L. B. Vogelesang, Int. SAMPE Meeting, January 1981, Cannes, France.
79. R. Marissen, DFVLR-FB 84-37, Institute Für Werkstoff-Forschung, Koln, Germany, 1984.
80. R. J. Bucci, L. N. Mueller, R. W. Schultz, and J. L. Prohaska, 32nd Int. SAMPE Meeting, April 1987, Anaheim, CA.

PROBLEMS

4.1 An alloy of unknown composition was heated to temperature T_1 and then quenched rapidly to room temperature T_3. After quenching, hardness readings for the alloy were low. The material was then reheated to an intermediate temperature T_2 for varying lengths of time. Room temperature hardness was found to increase progressively to a maximum before decreasing with longer exposures to the T_2 exposure. Speculate on the nature of the phase diagram corresponding to this alloy, the approximate composition of the alloy, and the location of the temperatures mentioned above. Speculate on the strengthening mechanisms responsible for this hardness–time profile.

4.2 A bar of BCC iron containing carbon and nitrogen solute additions is stressed to the point where general yielding occurs and the test interrupted prior to necking. Describe the stress–strain response phenomenologically and in terms of dislocation dynamics. Now immediately reload the test bar into the plastic strain range and describe the new stress–strain response. Explain any different response in terms of dislocation dynamics.

4.3 The lattice parameters in nickel and Ni_3Al are 3.52×10^{-10} m and 3.567×10^{-10} m, respectively. What is the lattice misfit between these phases and its associated influence on precipitation hardening? What change in misfit hardening would occur if the lattice parameter of nickel was increased to 3.525×10^{-10} m due to the addition of 50 a/o Cr?

4.4 **(a)** Tungsten wire, wrapped in a drum, is to be introduced into a copper matrix so as to produce a long square rod with a cross-sectional area of 6.25 cm^2. If the tungsten wire diameter is 0.75 cm, what is the maximum stiffness (Young's modulus of elasticity E) of a tungsten–copper composite that can be fabricated? (The copper is melted and then poured into a mold that contains the aligned tungsten wires.) $E_w = 411$ GPa, $E_{Cu} = 130$ GPa.

(b) Within the elastic regime, what is the load distribution on the tungsten wire and copper matrix?

4.5 **(a)** The radius of the iron atom in the BCC lattice is 1.24×10^{-10} m. Compute the size of the octahedral interstitial site and compare with the size of a carbon atom, which has a radius of 0.77×10^{-10} m. Repeat the calculation for FCC iron lattice where the atomic radius is 1.27×10^{-10} m.

(b) From your answers in (a), comment on the degree of carbon solubility and solid solution strengthening potential in the two iron lattices.

4.6 Fig. 4.20 shows that the strength of the Al-Al_3Ni eutectic composite increases dramatically to approximately 310 MPa when unidirectional solidification aligns the Al_3Ni whiskers parallel to the stress axis. Assuming that the whisker l/d ratio is large, estimate the matrix contribution to overall composite strengthening if the alloy contains 10 v/o Al_3Ni and the whisker elastic modulus and fracture strain are 138 GPa and 0.02, respectively. The composite fracture strain is also 0.02, at which point the Al matrix experiences plastic flow.

4.7 Maraging steels are relatively soft upon quenching from the austenitizing temperature range but strengthen greatly following exposure to a reheating treatment at intermediate temperature. Given that the carbon level of such steels is typically less than 0.03%, whereas the alloy contains additions of Ni, Mo, and Ti, speculate as to the probable strengthening mechanism that controls the strength of this class of alloys.

CHAPTER 5

HIGH-TEMPERATURE DEFORMATION RESPONSE OF CRYSTALLINE SOLIDS

Brief mention was made in Section 1.3 of the effect of strain rate and temperature on the mechanical response of engineering materials. It was shown that tensile strength increased with increasing strain rate and decreasing temperature. In Section 2.3.1, the temperature sensitivity of strength in crystalline solids was related to the role played by the Peierls-Nabarro stress in resisting dislocation movement through a given lattice. In addition, the potential importance of temperature in controlling crystalline deformation through thermally activated edge dislocation climb was mentioned in Section 2.4. In the next two chapters, temperature and strain rate effects on mechanical properties are explored more extensively. Particular attention is given to the time- and temperature-dependent deformation characteristics of solids.

5.1 CREEP OF SOLIDS: AN OVERVIEW

For the most part, our discussions of deformation in solids thus far have been limited to the instantaneous deformation response—elastic or plastic—to the application of a load. A time-dependent change in the observed strain adds a new dimension to the problem. As shown in Fig. 5.1, after a load has been applied, the strain increases with time until failure finally occurs. For convenience, researchers have subdivided the creep curve into three regimes, based on the similar response of many materials. After the initial instantaneous strain ϵ_0, materials often undergo a period of transient response where the strain rate $d\epsilon/dt$ decreases with time to a minimum steady-state value that persists for a substantial portion of the material's life. Appropriately, these two regions are referred to in the literature as the transient or primary creep stage and the steady-state creep stage, respectively. Final failure with a rupture life t_R then comes soon after the creep rate increases during the third, or tertiary, stage of creep.

It is generally believed that the varying creep response of a material (Fig. 5.1) reflects a continually changing interaction between strain hardening and softening (recovery) processes, which strongly affect the overall strain rate of the material at a

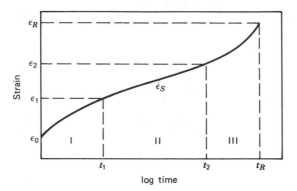

FIGURE 5.1   Typical creep curve showing three regions in strain–time–space.

given temperature and stress. Strain hardening at elevated temperatures is believed to involve subgrain formation associated with the rearrangement of dislocations,[1] while thermally activated cross-slip and edge dislocation climb represent two dominant recovery processes (see Chapter 2). It is logical to conclude that the decrease in strain rate in Stage I must be related to substructure changes that increase overall resistance to dislocation motion. Correspondingly, the constant strain rate in Stage II would indicate a stable substructure and a dynamic balance between hardening and softening processes. Indeed, Barrett et al.[2] verified that the substructure in Fe–3Si was invariant during Stage II. At high stress and/or temperature levels, the balance between hardening and softening processes is lost, and the accelerating creep rate in the tertiary stage is dominated by a number of weakening metallurgical instabilities. Among these microstructural changes are localized necking, corrosion, intercrystalline fracture, microvoid formation, precipitation of brittle second-phase particles, and resolution of second phases that originally contributed toward strengthening of the alloy. In addition, the strain-hardened grains may recrystallize and thereby further destroy the balance between material hardening and softening processes.

The engineering creep strain curve shown schematically in Fig. 5.1 reflects the material response under constant tensile loading conditions and represents a convenient method by which most elevated temperature tests are conducted. However, from Eq. 1-2a, the true stress increases with increasing tensile strain. As a result, a comparable true creep strain–time curve should differ significantly if the test is conducted under constant *stress* rather than constant *load* conditions (Fig. 5.2). This is especially true for Stage II and III behavior. As a general rule, data being generated for engineering purposes are obtained from constant load tests, while more fundamental studies involving the formulation of mathematical creep theories should involve constant stress testing. In the latter instance, the load on the sample is lowered progressively with decreasing specimen cross-sectional area. This is done either manually or by the incorporation of automatic load-shedding devices in the creep stand load train.

The creep response of materials depends on a large number of material and external variables. Certain material factors are considered in more detail later in this chapter. For the present, attention will be given to the two dominant external variables—stress

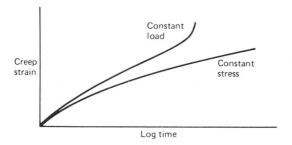

FIGURE 5.2 **Creep curves produced under constant load and constant stress conditions.**

and test temperature—and how they affect the shape of the creep–time curve. Certainly, environment represents another external variable because of the importance of corrosion and oxidation in the fracture process. Unfortunately, consideration of this variable is not within the scope of this book.

The effect of temperature and stress on the minimum creep rate and rupture life are the two most commonly reported data for a creep or creep rupture test, although different material parameters are sometimes reported.[3,4] The rupture life at a given temperature and stress is obtained when it is necessary to evaluate the response of a material for use in a short-life situation, such as for a rocket engine nozzle ($t_R \approx 100$ s) or a turbine blade in a military aircraft engine ($t_R \approx 100$ hr). In such short-life situations, the dominant question is whether the component will or will not fail, rather than by how much it will deform. As a result, the details of the creep–time curve are not of central importance to the engineering problem. For this reason, *creep rupture* tests usually provide only one datum—the rupture lift t_R. Rupture life information is sometimes used in the design of engineering components that will have a service life up to 10^5 hr. An example of such data is given in Fig. 5.3 for the high-temperature, iron-based alloy, S590. As expected, the rupture life t_R is seen to decrease with increasing test temperature and stress. When preparing this plot, Grant and Bucklin[5] chose to separate the data for a given temperature into several discrete regimes. This was done to emphasize the presence of several metallurgical instabilities that they identified metallographically and which they believed to be responsible for the change in slope of the $\log\sigma$–$\log t_R$ curve.

For long-life material applications, such as in a nuclear power plant designed to operate for several decades, component failure obviously is out of the question. However, it is equally important that the component not creep excessively. For long-life applications, the minimum creep rate represents the key material response for a given stress and test temperature. To obtain this information, *creep* tests are performed into Stage II, where the steady-state creep rate $\dot{\epsilon}_s$ can be determined with precision. Therefore, the *creep* test focuses on the early deformation stages of creep and is seldom carried to the point of fracture. As one might expect, the accuracy of $\dot{\epsilon}_s$ increases with the length of time the specimen experiences Stage II deformation. Consequently, $\dot{\epsilon}_s$ values obtained during instrumented creep rupture tests are not very accurate because of the inherently short time associated with the creep rupture test.

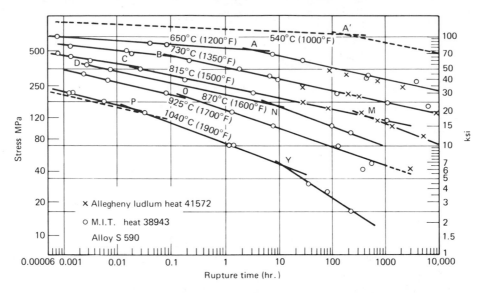

FIGURE 5.3 **Stress–rupture life plot at several test temperatures for iron-based alloy S-590. (From N. J. Grant and A. G. Bucklin, copyright American Society for Metals, Metals Park, OH, © 1950.)**

The magnitude of $\dot{\epsilon}_s$ often depends strongly on stress. As a result, steady-state creep rate data are usually plotted against applied stress, as shown in Fig. 5.4. The significance of the $\dot{\epsilon}_s$ differences between α − and γ − iron[6] at the allotropic transformation temperature is discussed in Section 5.2.

Since the creep and creep rupture tests are similar (though defined over different stress and temperature regimes), it would seem reasonable to assume the existence of certain relations among various components of the creep curve (Fig. 5.1). In his text, Garofalo[3] summarized a number of log–log relations between t_R and other quantities, such as $t_2 − t_1$, t_2, and the steady-state creep rate $\dot{\epsilon}_s$. Regarding the latter, Monkman and Grant[7] identified an empirical relation between t_R and $\dot{\epsilon}_s$ with the form

$$\log t_R + m \log \dot{\epsilon}_s = B \qquad (5\text{-}1)$$

where t_R = rupture life
$\dot{\epsilon}_s$ = steady-state creep rate
m, B = constants

For a number of aluminum, copper, titanium, iron, and nickel base alloys, Monkman and Grant found $0.77 < m < 0.93$ and $0.48 < B < 1.3$. To a first approximation, then, the rupture life was found to be inversely proportional to $\dot{\epsilon}_s$. This would allow t_R to be estimated as soon as $\dot{\epsilon}_s$ was determined. Of course, the magnitude of t_R could be estimated from Eq. 5-1 only after the validity of the relation for the material in question was established and the two constants identified.

A number of other empirical relations have been proposed to relate the primary creep strain to time at stress and temperature. Garofalo[3] summarized the work of

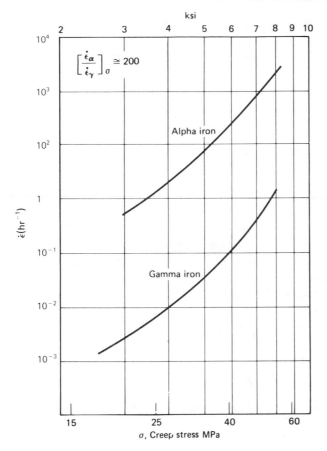

FIGURE 5.4 **Stress–steady-state creep rate for α- and γ-iron at 910°C. (From O. D. Sherby and J. L. Lytton;[6] reprinted with permission of the American Institute of Mining, Metallurgical and Petroleum Engineers, 1956.)**

others and showed that for low temperatures ($0.05 < T_h^* < 0.3$) and small strains, a number of materials exhibit *logarithmic creep*:

$$\epsilon_t \propto \ln t \qquad (5\text{-}2)$$

where ϵ_t = true strain
       t = creep time

In the range $0.2 < T_h < 0.7$ another relation has been employed with the form

$$\epsilon_t = \epsilon_{0_t} + \beta t^m \qquad (5\text{-}3)$$

where ϵ_{0_t} = instantaneous true strain accompanying application of the load
       β, m = time-independent constants

---

* T_h represents the homologous temperature—the ratio of ambient to melting point temperatures (absolute units).

Creep response in materials according to Eq. 5-3 is often referred to in the literature as *parabolic creep* or β *flow*. Since $0 < m < 1$ in transient creep, both Eqs. 5-2 and 5-3 reflect a decreasing strain rate with time. The strain rate $\dot{\epsilon}$ can be derived from Eqs. 5-2 and 5-3 with the form

$$\dot{\epsilon} \propto t^{-n} \tag{5-4}$$

as suggested by Cottrell,[8] where

$$\dot{\epsilon} = \text{strain rate}$$
$$t = \text{time}$$
$$n = \text{constant.}$$

It is generally found that n decreases with increasing stress and temperature. At low temperatures when $n = 1$, Eq. 5-4 describes logarithmic creep (see Eq. 5-2). In the parabolic creep regime at higher temperatures, $m = 1 - n$. To provide a transition from Stage I to Stage II creep, another term $\dot{\epsilon}_s t$, has to be added to Eq. 5-3 to account for the steady-state creep rate in Stage II. Hence

$$\epsilon_t = \epsilon_{0_t} + \beta t^m + \dot{\epsilon}_s t \tag{5-5}$$

where $\dot{\epsilon}_s$ = steady-state creep rate in Stage II reflecting a balance between strain hardening and recovery processes.

When $m = \frac{1}{3}$, Eq. 5-5 reduces to the relation originally proposed by Andrade[9] in 1910.

5.2 TEMPERATURE–STRESS–STRAIN-RATE RELATIONS

Since the creep life and total elongation of a material depends strongly on the magnitude of the steady-state creep rate $\dot{\epsilon}_s$ (Eqs. 5-1 and 5-5), much effort has been given to the identification of those variables that strongly affect $\dot{\epsilon}_s$. As mentioned in Section 5.1, the external variables, temperature and stress, exert a strong influence along with a number of material variables. Hence the steady-state creep rate may be given by

$$\dot{\epsilon}_s = f(T, \sigma, \epsilon, m_1, m_2) \tag{5-6}$$

where T = absolute temperature
  σ = applied tensile stress
  ϵ = creep strain
  m_1 = various intrinsic lattice properties, such as the elastic modulus G and the crystal structure
  m_2 = various metallurgical factors, such as grain and subgrain size, stacking fault energy, and thermomechanical history

It is important to recognize that m_2 also depends on T, σ, and ϵ. For example, subgrain diameter decreases markedly with increasing stress. Consequently, there exists a subtle but important problem of separating the effect of the major test variables on the structure from the deformation process itself that controls the creep rate. Dorn, Sherby, and coworkers[10–13] suggested that where $T_h > 0.5$ for the steady-state condition, the structure could be defined by relating the creep strain to a parameter θ

$$\epsilon = f(\theta) \tag{5-7}$$

where $\theta = te^{-\Delta H/RT}$ described as the temperature-compensated time parameter
$\quad\quad t = $ time
$\quad\Delta H = $ activation energy for the rate-controlling process
$\quad\quad T = $ absolute temperature
$\quad\quad R = $ gas constant

The activation energy ΔH, shown schematically in Fig. 5.5 represents the energy barrier to be overcome so that an atom might move from A to the lower energy location at B. Upon differentiating Eq. 5-7 with respect to time, one finds

$$Z = f(\epsilon) = \dot{\epsilon}e^{\Delta H/RT} \tag{5-8}$$

which describes the strain-rate–temperature relation for a given stable structure and applied stress. When the rate process is given by the minimum creep rate $\dot{\epsilon}_s$ and its logarithm plotted against $1/T$, a series of parallel straight lines for different stress levels is predicted from Eq. 5-8 (Fig. 5.6). The slope of these lines, $\Delta H/2.3R$, then defines the activation energy for the controlling creep process. The fact that the isostress lines were straight in Fig. 5.6 suggests that only one process had controlled creep in the TiO_2 single crystals throughout the stress and temperature range examined. Were different mechanisms to control the creep rate at different temperatures, the $\log\dot{\epsilon}_s$ vs. $1/T$ plots would be nonlinear. When multiple creep mechanisms are present and act in a concurrent and dependent manner, the slowest mechanism would control $\dot{\epsilon}_s$. The overall strain rate would take the form

$$\frac{1}{\dot{\epsilon}_T} = \frac{1}{\dot{\epsilon}_1} + \frac{1}{\dot{\epsilon}_2} + \frac{1}{\dot{\epsilon}_3} + \dots + \frac{1}{\dot{\epsilon}_n} \tag{5-9}$$

where $\quad\dot{\epsilon}_T = $ overall creep rate
$\quad\dot{\epsilon}_{1,2,3,\dots,n} = $ creep rates associated with n mechanisms

For the simple case where only two mechanisms act interdependently

$$\dot{\epsilon}_T = \frac{\dot{\epsilon}_1\dot{\epsilon}_2}{\dot{\epsilon}_1 + \dot{\epsilon}_2} \tag{5-10}$$

Conversely, if the n mechanisms were to act independently of one another, the fastest one would control. For this case, $\dot{\epsilon}_T$ would be given by

$$\dot{\epsilon}_T = \dot{\epsilon}_1 + \dot{\epsilon}_2 + \dot{\epsilon}_3 + \dots + \dot{\epsilon}_n \tag{5-11}$$

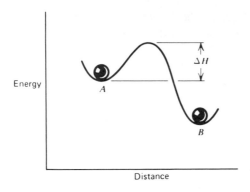

FIGURE 5.5   Diagram revealing significance of activation energy required in moving an atom from *A* to *B*.

To determine the activation energy for creep over a small temperature interval, where the controlling mechanism would not be expected to vary, researchers often make use of the temperature differential creep test method. After a given amount of strain at temperature T_1, the temperature is changed abruptly to T_2, which may be slightly above or below T_1. The difference in the steady-state creep rate associated with T_1 and T_2 is then recorded (Fig. 5.7). If the stress is held constant and the assumption made that the small change in temperature does not change the alloy

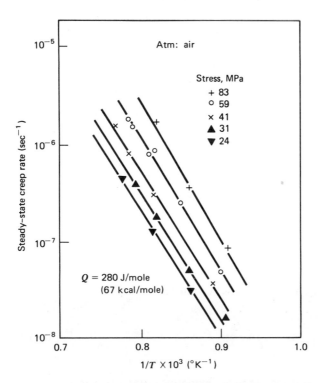

FIGURE 5.6   Log steady-state creep rate versus reciprocal of absolute temperature for rutile (TiO$_2$) at various stress levels. (From W. M. Hirthe and J. O. Brittain;[14] reprinted with permission from the American Ceramic Society, copyright © 1963.)

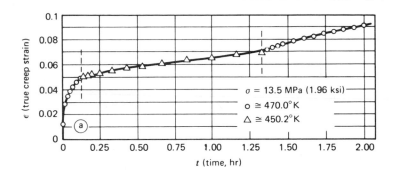

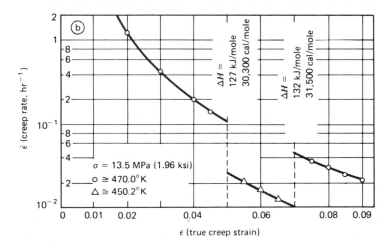

FIGURE 5.7 **Incremental step test involving slight change in test temperature to produce change in steady-state creep rate in aluminum. (From J. E. Dorn, *Creep and Recovery*, reprinted with permission from American Society for Metals, Metals Park, OH, copyright © 1957.)**

structure, then Z is assumed constant. From Eq. 5-8 the activation energy for creep may then be calculated by

$$\Delta H_C = \frac{R\ln \dot{\epsilon}_1/\dot{\epsilon}_2}{1/T_2 \ - \ 1/T_1} \tag{5-12}$$

where ΔH_C = activation energy for creep
         $\dot{\epsilon}_1, \dot{\epsilon}_2$ = creep rates at T_1 and T_2, respectively

This value of ΔH_C should correspond to the activation energy determined by a data analysis like that shown in Fig. 5.6, as long as the same mechanism controls the creep process over the expanded temperature range in the latter instance. As shown in Fig. 5.8, this is not always the case. The activation energy for creep in aluminum is seen to increase with increasing temperature up to $T_h \approx 0.5$, whereupon ΔH_C remains

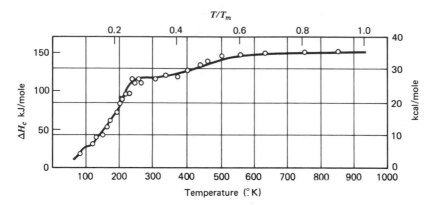

FIGURE 5.8 **Variation of apparent activation energy for creep in aluminum as a function of temperature. (From O. D. Sherby, J. L. Lytton, and J. E. Dorn,[13] reprinted with permission from Sherby and Pergamon Press, Elmsford, NY, 1957.)**

constant up to the melting point. Similar results have been found in other metals.[15] It would appear that different processes were rate controlling over the test temperature range.[13] Furthermore, it should be recognized that ΔH_C may represent some average activation energy reflecting the integrated effect of several mechanisms operating simultaneously and interdependently (see Section 5.3).

Dorn,[12] Garofalo,[3] and Weertman[16] have compiled a considerable body of data to demonstrate that at $T_h \geqslant 0.5$, ΔH_C is most often equal in magnitude to ΔH_{SD}, the activation energy for self-diffusion (Fig. 5.9); this fact strongly suggests the latter to be the creep rate-controlling process in this temperature regime. While the approximate equality between ΔH_C and ΔH_{SD} seems to hold for many metals and ceramics at temperatures equal to and greater than half the melting point, some exceptions do exist, particularly for the case of intermetallic and nonmetallic compounds. It is found that small departures from stoichiometry of these compounds have a pronounced effect on ΔH_C, which in turn affects the creep rate. For example, a reduction in oxygen content in rutile from TiO_2 to $TiO_{1.99}$ causes a reduction in ΔH_C from about 280 to 120 kJ/mol (67–29 kcal/mol)* with an associated 100-fold increase in $\dot{\epsilon}_s$.[14] For the more general case, however, the creep process is found to be controlled by the diffusivity of the material

$$D = D_0 e^{-\Delta H_{SD}/RT} \tag{5-13}$$

where $\quad D$ = diffusivity, cm^2/s
$\qquad D_0$ = diffusivity constant$\approx 1\ cm^2/s$
$\qquad \Delta H_{SD}$ = activation energy, J/mol
$\qquad R$ = gas constant, J/K
$\qquad T$ = absolute temperature, K

---

* To convert from kcal to kJ, multiply by 4.19.

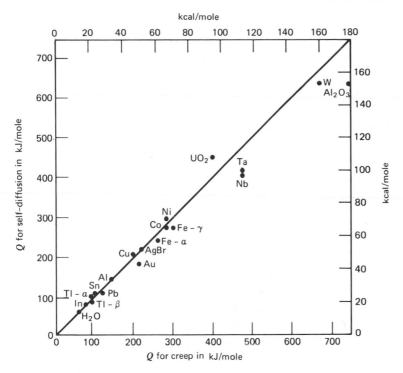

FIGURE 5.9 Correlation between activation energy for self-diffusion and creep in numerous metals and ceramics. (From J. Weertman,[16] reprinted with permission from American Society for Metals, Metals Park, OH, copyright © 1968.)

$$D = D_0 e^{-(K_0 + V)T_m/T} \qquad (5\text{-}14)$$

K_0 = dependent on the crystal structure and equal to 14 for BCC lattice, 17 for FCC and HCP lattices, and 21 for diamond-cubic lattice

V = valence of the material

T_m = absolute melting temperature

The constants K_0 are estimates associated with an assumed diffusivity constant ≈ 1 cm^2/s.

By combining Eqs. 5-13 and 5-14

$$\Delta H_{SD} = RT_m(K_0 + V) \qquad (5\text{-}15)$$

we see that the activation energy for self-diffusion increases (corresponding to a reduction in D) with increasing melting point, valence, packing density, and degree of covalency. Consequently, although refractory metals with high melting points, such as tungsten, molybdenum, and chromium, seem to hold promise as candidates for high-temperature service, their performance in high-temperature applications is adversely affected by their open BCC lattice, which enhances diffusion rates. From Eq.

5-15, ceramics are identified as the best high-temperature materials because of their high melting point and the covalent bonding that often exists.

It is important to recognize that creep rates for all materials cannot be normalized on the basis of D alone because other test variables affect the creep process in different materials. For example, Barrett and coworkers[19] noted the important influence of elastic modulus on the creep rate and on determination of the true activation energy for creep. A semi-empirical relationship with the form

$$\frac{\dot{\epsilon}_s kT}{DGb} = A \left(\frac{\sigma}{G}\right)^n \tag{5-16}$$

has been proposed[1] to account for other factors where

$$\dot{\epsilon} = \text{steady-state creep rate}$$
$$k = \text{Boltzman's constant}$$
$$T = \text{absolute temperature}$$
$$D = \text{diffusivity}$$
$$G = \text{shear modulus}$$
$$b = \text{Burgers vector}$$
$$\sigma = \text{applied stress}$$
$$A,n = \text{material constants}$$

By combining Eqs. 5-8 and 5-13, the steady-state creep rate at different temperatures can be normalized with respect to D to produce a single curve, as shown in Fig. 5.10. This is an important finding since it allows one to conveniently portray a great deal of data for a given material. For example, we see from a reexamination of Fig. 5.4 that at the allotropic transformation temperature, the creep rate in γ-iron (FCC lattice) is found to be approximately 200 times slower than that experienced by α-iron (BCC lattice).[6] This substantial difference is traced directly to the 350-fold lower diffusivity in the close-packed FCC lattice in γ-iron. Similar findings were reviewed by Sherby and Burke[17] for the allotropic transformation from HCP to BCC in thallium. Therefore, it is appropriate to briefly consider those factors that strongly influence the magnitude of D. Sherby and Simnad[18] reported an empirical correlation showing D to be a function of the type of lattice, the valence, and the absolute melting point of the material.

Here again we see that creep is assumed to be diffusion controlled. Even after normalizing creep data with Eq. 5-16, a three-decade scatter band still exists for the various metals shown in Fig. 5.11. While some of this difference might be attributable to actual test scatter or relatively imprecise high-temperature measurements of D and G, other as yet unaccounted for variables most likely will account for the remaining inexactness. For example, there appears to be a trend toward higher creep rates in FCC metals and alloys possessing high stacking fault energy (SFE). Whether the SFE variable should be incorporated into either A or n is the subject of current discussion.[20–22] The role of substructure on A and n must also be identified more precisely.

One important factor in Eq. 5-16 is the stress dependency of the steady-state creep

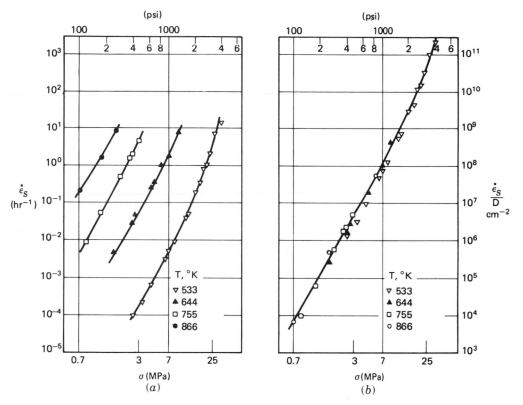

FIGURE 5.10 Creep data in aluminum. (*a*) Stress versus steady-state creep rate $\dot\epsilon_s$ at various test temperatures; (*b*) data normalized by plotting stress versus $\dot\epsilon_s$ divided by the diffusion coefficient. (From O. D. Sherby and P. M. Burke;[17] reprinted with permission from Sherby and Pergamon Press, Elmsford, NY, 1968.)

rate. It is now generally recognized that $\dot\epsilon_s$ varies directly with σ at low stresses and temperatures near the melting point. At intermediate to high stresses and at temperatures above $0.5T_m$, where the thermally activated creep process is dominated by the activation energy for self-diffusion, $\dot\epsilon_s \propto \sigma^{4-5}$ (so-called power law creep). It should be noted that this stress dependency holds for pure metals and their solid solutions. Much stronger stress dependencies of $\dot\epsilon_s$ and t_R have been reported in oxide-dispersion-strengthened superalloys (see Section 5.7). At very high stress levels $\dot\epsilon_s \propto e^{\alpha\sigma}$. Garofalo[23] showed that power law and exponential creep represented limiting cases for a general empirical relationship

$$\dot\epsilon_s \propto (\sinh \alpha\sigma)^n \tag{5-17}$$

Equation 5-17 reduces to power law creep when $\alpha\sigma < 0.8$, but approximates exponential creep when $\alpha\sigma > 1.2$. An explanation for the changing stress dependence of $\dot\epsilon_s$ in several operative deformation mechanisms is discussed in the next section.

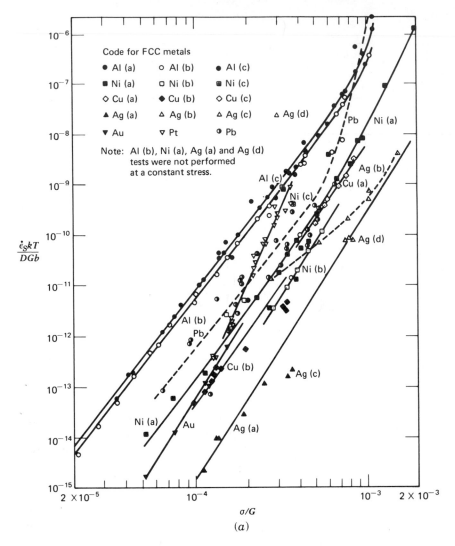

FIGURE 5.11 Creep data in metals. (*a*) Data for FCC metals; materials with high-stacking fault energy tend to have higher steady-state creep rates. (*b*) Data for BCC metals. (From A. K. Mukherjee, J. E. Bird, and J. E. Dorn[1]; copyright American Society for Metals, Metals Park, OH, © 1969.)

5.3 DEFORMATION MECHANISMS

At low temperatures relative to the melting point of crystalline solids, the dominant deformation mechanisms are slip and twinning (Chapter 3). However, at intermediate and high temperatures, other mechanisms become increasingly important and dominate material response under certain conditions. It is with regard to these additional deformation modes that attention will now be focused.

Over the years a number of theories have been proposed to account for the creep

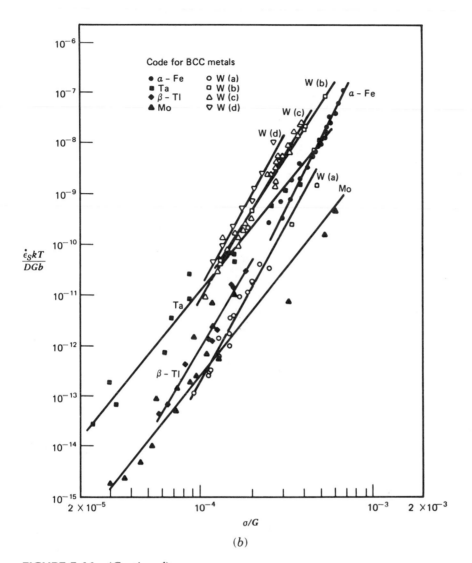

$$\frac{\dot{\epsilon}_s kT}{DGb}$$

(b)

FIGURE 5.11 (*Continued*)

data trends discussed in the previous sections. In fact, the empirical form of Eq. 5-16 takes account of mathematical formulations for several proposed creep mechanisms. At low stresses and high temperatures, where the creep rate varies with applied stress, Nabarro[24] and Herring[25] theorized that the creep process was controlled by stress-directed atomic diffusion. Such *diffusional creep* is believed to involve the migration of vacancies along a gradient from grain boundaries experiencing tensile stresses to boundaries undergoing compression (Fig. 5.12); simultaneously atoms would be moving in the opposite direction, leading to elongation of the grains and the test bar. This gradient is produced by a stress-induced decrease in energy to create vacancies when tensile stresses are present and a corresponding energy increase for vacancy formation

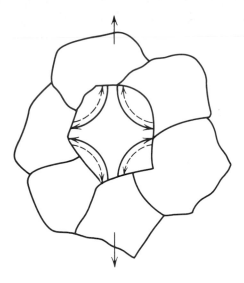

FIGURE 5.12 **Stress-directed flow of vacancies (solid lines) from tensile to compressive grain boundaries and corresponding reverse flow of atoms or ions (dashed lines).**

along compressed grain boundaries. Nabarro-Herring creep can be described by Eq. 5-16 when $A \approx 7 \, (b/d^2)$ (d = grain diameter) and $n = 1$, such that[21]

$$\dot{\epsilon}_s \approx \frac{7\sigma D_v b^3}{kTd^2} \tag{5-18}$$

where D_v = volume diffusivity through the grain interior

As expected, $\dot{\epsilon}_s$ is seen to increase with increasing number of grain boundaries (i.e., smaller grain size). A closely related *diffusional creep* process described by Coble[26] involves atomic or ionic diffusion along grain boundaries. Setting $A \approx 50(b/d)^3$ and $n = 1$, Eq. 5-16 reduces to the Coble relationship

$$\dot{\epsilon}_s \approx \frac{50\sigma D_{gb} b^4}{kTd^3} \tag{5-19}$$

(Note that Coble creep is even more sensitive to grain size than is Nabarro-Herring creep.) In complex alloys and compounds there is a problem in deciding which particular atom or ion species controls the diffusional process and along what path such diffusion takes place. This is usually determined from similitude arguments. That is, if ΔH_C is numerically equal to ΔH_{SD} for element A along a particular diffusion path, then it is presumed that the self-diffusion of element A had controlled the creep process.

At intermediate to high stress levels and test temperatures above $0.5T_m$, creep deformation is believed to be controlled by diffusion-controlled movement of dislocations. Several of these theories have been evaluated by Mukherjee et al.,[1] with the Weertman[16,27] model being found to suffer from the least number of handicaps and found capable of predicting best the experimental creep results described in Section 5.2. Weertman proposed that creep in the above-mentioned stress and temperature regime was controlled by edge dislocation climb away from dislocation barriers. Again using Eq. 5-16 as the basis for comparison, Bird et al.[21] showed that when A is constant and $n \approx 5$, *dislocation creep* involving the climb of edge dislocations could be estimated by

$$\dot{\epsilon}_s \approx \frac{ADGb}{kT} \left(\frac{\sigma}{G}\right)^5 \tag{5-20}$$

It should be noted that in many creep situations, the dislocation creep process dominates the elevated temperature $(T \geq 0.5T_m)$ response of engineering alloys.

The actual Weertman relationship expresses the shear strain rate $\dot{\gamma}_s$ in terms of the shear stress τ by

$$\dot{\gamma}_s \propto \tau^2 \sinh \tau^{2.5} \tag{5-21}$$

As such, the transition from power law to exponential creep mentioned earlier is readily predicted from Eq. 5-21. Weertman[27] theorized that the onset of exponential creep ($\dot{\epsilon}_s \propto e^{\alpha\sigma}$) at high stress levels was related to accelerated diffusion, because of an excess vacancy concentration brought about by dislocation–dislocation interactions.

Another high-temperature deformation mechanism involves grain-boundary sliding. The problem in dealing with grain-boundary sliding, however, is that it does not represent an independent deformation mechanism; it must be accommodated by other deformation modes. For example, consider the shear-induced displacement of the two grains in Fig. 5.13a. At sufficiently high temperatures, the local grain-boundary stress fields can cause diffusion of atoms from the compression region BC to the tensile region AB by either a Nabarro-Herring or Coble process. As might be expected, the

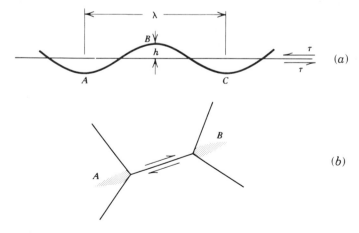

FIGURE 5.13   **Accommodation mechanisms for grain-boundary sliding. (*a*) Shear along boundary accommodated by diffusional flow of vacancies of region *AB* to *BC*; (*b*) grain-boundary sliding accommodated by dislocation climb within contiguous grains *A* and *B*.**

rate of sliding should depend strongly on the shape of the boundary. Raj and Ashby[28] demonstrated that $\dot{\epsilon}_s$ increased rapidly as the ratio of perturbation period λ to perturbation height h increased. Furthermore, when λ is small and the temperature relatively low, diffusion is found to be controlled by a grain-boundary path. On the other hand, when λ is large and the temperature relatively high, volume diffusion controls the grain-boundary sliding process.[28] Consequently, grain-boundary sliding may be accommodated by diffusional flow, which is found to depend on both the temperature and the grain-boundary morphology. For this case, the sliding rate would be directly proportional to stress (see Eqs. 5-18 and 5-19). By examining this problem from a different perspective, one finds that Nabarro-Herring and Coble creep models are themselves dependent on grain-boundary sliding! From Fig. 5.14, note that the stress-directed diffusion of atoms from compression to tension grain boundaries causes the grain boundaries to separate from one another (Fig. 5.14b). Grain-boundary sliding is needed, therefore, to maintain grain contiguity during diffusional flow processes (Fig. 5.14c).[28-30] On the basis of this finding, Raj and Ashby concluded that Nabarro-Herring and Coble diffusional creep mechanisms were "identical with grain-boundary sliding with diffusional accommodation."[28]

For the internal boundary shown in Fig. 5.13b, grain-boundary sliding could be accommodated by dislocation creep within grains A and B. Matlock and Nix[31] examined this condition for several metals and found that the grain-boundary-sliding strain-rate contribution was proportional to σ^{n-1}, where n is the exponent associated with the dislocation creep mechanism ($\approx 4 - 5$). Unfortunately this stress sensitivity does not agree with any presently known theoretical predictions.

It is apparent from the above discussion that these high-temperature deformation mechanisms all depend on atom or ion diffusion but differ in their sensitivity to other variables such as G, d, and σ. As such, a particular strengthening mechanism may strengthen a material *only* with regard to a particular deformation mechanism but not another. For example, an increase in alloy grain size will suppress Nabarro-Herring and Coble creep along with grain boundary sliding, but will not substantially change

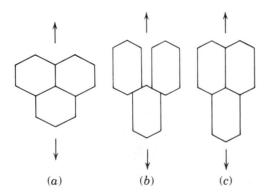

(a) (b) (c)

FIGURE 5.14 Stress-induced diffusional flow elongates grains and could lead to grain separation (b), but is accommodated by grain-boundary sliding which brings grains together (c).

the dislocation climb process.[1] As a result, the rate-controlling creep deformation process would shift from one mechanism to another. Consequently, marked improvement in alloy performance requires simultaneous suppression of several deformation mechanisms. This point is considered further in Section 5.5.

5.4   SUPERPLASTICITY

As we have just seen, fine-grained structures are to be avoided in high-temperature, load-bearing components since this would bring about an increase in creep strains resulting from Nabarro-Herring, Coble, and grain-boundary-sliding creep mechanisms. In fact, recent metallurgical developments[33] reveal improved creep response in alloys possessing either no grain boundaries (i.e., single-crystal alloys) or highly elongated boundaries (produced by unidirectional solidification) oriented parallel to the major stress axis. Where the opposite of creep resistance (i.e., easy flow) is required, such as in hot-forming processes, fine-grained structures are preferred. Some such materials are known to possess *superplastic* behavior[34] with total strains in excess of 1000% (Fig. 5.15). These large strains, generated at low stress levels, drastically improve the formability of certain alloys.

By expressing the flow stress–strain-rate relation (Eq. 1-52) in the form

$$\sigma = \frac{F}{A} = K\dot{\epsilon}^m \tag{5-22}$$

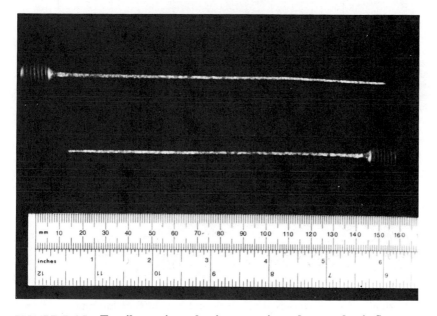

FIGURE 5.15   Tensile specimen having experienced superplastic flow.

where F = applied force

A = cross-sectional area

K = constant

$$\dot\epsilon = \frac{1}{l}\frac{dl}{dt} = -\frac{1}{A}\frac{dA}{dt}$$

m = strain-rate sensitivity factor

superplasticity is found when m is large[34–36] and approaches unity. Figures 5.16 and 5.17 show the normalized stress–strain-rate relation for loading in the superplastic region. After substituting for $\dot\epsilon$ and rearranging, the change in cross-sectional area with time, dA/dt, is given by

$$\frac{-dA}{dt} = \frac{F^{(1/m)}}{K}\left[\frac{1}{A^{(1-m/m)}}\right] \tag{5-23}$$

In the limit, as the rate sensitivity factor m approaches unity, note that dA/dt depends only on the applied force and is independent of any irregularies in the specimen cross-sectional area, such as incipient necks and machine tool marks which are maintained but not worsened. That is, the sample undergoes extensive deformation without pronounced necking.

Superplastic behavior has been reported in numerous metals, alloys, and ceramics[34] and associated in all cases with (1) a fine grain size (on the order of 1–10 μm), (2)

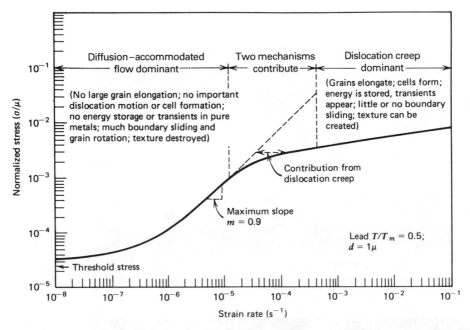

FIGURE 5.16   Normalized stress versus strain rate plot in lead showing intermediate region associated with superplastic behavior. (From M. F. Ashby and R. A. Verall;[37] reprinted with permission from Ashby and Pergamon Press, Elmsford, NY, 1973.)

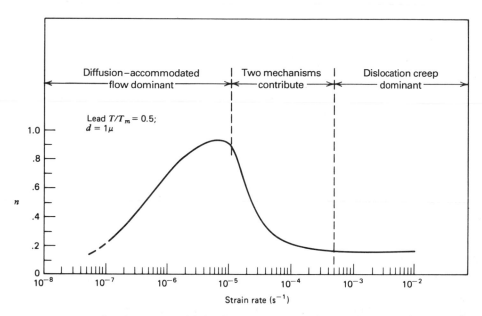

FIGURE 5.17 **Strain-rate sensitivity factor versus strain rate in lead. (From M. F. Ashby and R. A. Verall;[37] reprinted with permission from Ashby and Pergamon Press, Elmsford, NY, 1973.)**

deformation temperature $> 0.5 \, T_m$, and (3) a strain-rate sensitivity factor $m > 0.3$. The strain-rate range associated with superplastic behavior has been shown to increase with decreasing grain size and increasing temperature, as shown schematically in Fig. 5.18. There has been considerable debate, however, regarding the mechanisms responsible for the superplastic process. Avery and Backofen[36] originally proposed that a combination of deformation mechanisms involving Nabarro-Herring diffusional flow at low stress levels and dislocation climb at higher stresses were rate controlling. The applicability of the Nabarro-Herring creep model in the low stress regime has been questioned, based on experimental findings and theoretical considerations. First, it is generally found that m is of the order 0.5 rather than unity, the latter being associated with Nabarro-Herring creep. Furthermore, Nabarro-Herring creep would lead to the formation of elongated grains proportional in length to the entire sample. To the contrary, equiaxed grain structures are preserved during superplastic flow. More recent theories have focused with greater success on grain-boundary-sliding arguments, with diffusion-controlled accommodation[37-39] as the operative deformation mechanism associated with superplasticity at low stress levels.

As mentioned above, the formability of a material is enhanced greatly when in the superplastic state, while forming stresses are reduced substantially. To this end, grain refinement is highly desirable. Grain sizes on the order of 1–3 μm are commonly needed to attain superplastic behavior. The alert reader will immediately recognize, however, that once an alloy is rendered superplastic through a grain-refinement treatment, it no longer possesses the optimum grain size for high-temperature load applications. To resolve this dichotomy, researchers are currently seeking to develop duplex

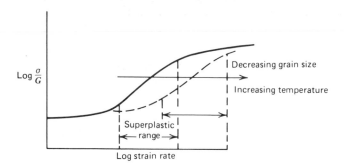

FIGURE 5.18 Temperature- and grain-size-induced shift in strain-rate range associated with superplastic behavior.

heat treatments to optimize both hot-forming and load-bearing properties of an alloy. For example, a nickel-based superalloy to be used in a gas turbine engine may first receive a grain-refining heat treatment to provide superplastic response during a forging operation. Once the alloy has been formed into the desired component, it is given another heat treatment to coarsen the grains so as to suppress Nabarro-Herring, Coble, and grain-boundary-sliding creep processes during high-temperature service conditions. For reviews of the superplasticity literature, see the papers by Edington et al.[40] and Taplin et al.[41] along with an analysis of current problems in our understanding of superplasticity.[42] Several additional articles pertaining to the mechanical, microstructural, and fracture processes in superplastically formed materials are recommended for the reader's attention.[43] Commercial applications of superplasticity are described by Hubert and Kay[44] (also see Section 5.7).

5.5 DEFORMATION-MECHANISM MAPS

It is important for the materials scientist and the practicing engineer to identify the deformation mechanisms that dominate a material's performance under a particular set of boundary conditions. This can be accomplished by solving the various constitutive equations for each deformation mechanism (e.g., Eqs. 5-16 to 5-20) and recognizing their respective interdependence or independence (Eqs. 5-9 and 5-11). Solutions to these equations reveal over which range of test variables a particular mechanism is rate controlling. Ashby and coworkers[45–47] have displayed such results pictorially in the form of maps in stress–temperature space based on the original suggestion by Weertman.[16] Typical deformation-mechanism maps for pure silver and germanium are shown in Fig. 5.19, where most of the high-temperature deformation mechanisms discussed in Section 5.3 (as well as pure glide) are shown. Each mechanism is rate controlling within its stress–temperature boundaries. Consistent with the previous discussion, dislocation creep is seen to dominate the creep process in both materials at relatively high stresses and homologous temperatures above 0.5. For the FCC metal, diffusional creep by either Nabarro-Herring or Coble mechanisms dominates at high temperatures but lower stress levels. The virtual absence of these two

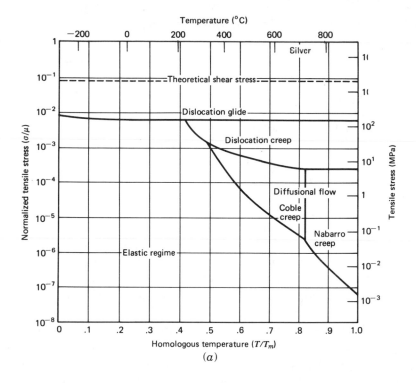

(a)

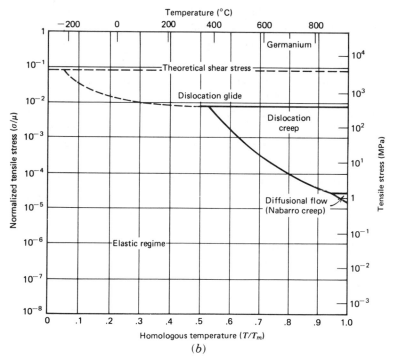

(b)

FIGURE 5.19 Deformation mechanism map for *(a)* pure silver and *(b)* germanium, showing stress–temperature space where different deformation mechanisms are rate controlling. Grain size in both materials is 32 μm. Elastic boundaries determined at a strain rate of 10^{-8}/s. (From M. F. Ashby[45]; reprinted with permission from Ashby and Pergamon Press, Elmsford, NY, 1972.)

diffusional flow mechanisms in covalently bonded diamond-cubic germanium is traced to its larger activation energy for self-diffusion and associated lower diffusivity. The boundaries separating each deformation field are defined by equating the appropriate constitutive equations (Eqs. 5-16 to 5-20) and solving for stress as a function of temperature. This amounts to the boundary lines representing combinations of stress and temperature, wherein the respective strain rates from the two deformation mechanisms are equal. Triple points in the deformation map occur when a particular stress and temperature produce equal strain rates from three mechanisms.

The maps shown in Fig. 5.19 do not portray a grain-boundary-sliding region, since uncertainties exist regarding the appropriate constitutive equation for this mechanism (see discussion, Section 5.3). Recent studies[47] have shown, however, that the dislocation creep field can be subdivided with a grain-boundary-sliding contribution existing at the lower stress levels associated with lower creep strain rates. Regarding the latter point, it is desirable to portray on the deformation map the strain rate associated with a particular stress–temperature condition, regardless of the rate-controlling mechanism. This may be accomplished by plotting the diagram contours of isostrain rate lines calculated from the constitutive equations. Examples of such modified maps are given in Fig. 5.20 for pure nickel prepared with two different grain sizes. These maps allow one to pick any two of the three major variables—stress, strain rate, and temperature—which then identifies the third variable as well as the dominant deformation mechanism. This is particularly useful in identifying the location of testing domains (such as creep and tensile tests) relative to the stress–temperature–strain-rate domains experienced by the material (e.g., hot-working, hot torsion, and geological processes) (Fig. 5.21). Note that in most instances, the laboratory test domains do not conform to the material's application experience. Certainly a better correspondence would be more desirable.

There are two additional points to be made regarding Fig. 5.20. First, the dislocation climb field has been divided into low- and high-temperature segments, corresponding to dislocation climb controlled by dislocation core and lattice diffusion, respectively. Furthermore, since Coble creep involves grain boundary diffusion, three diffusion paths are represented on these maps. Second, a large change in grain size in pure nickel drastically shifts the isostrain rate contours and displaces the deformation field boundaries. For example, at $T_h = 0.5$ and a strain rate of 10^{-9}/s, a 100-fold decrease in grain size causes the creep rate-controlling process to shift from low-temperature dislocation creep to Coble creep. Furthermore, the stress necessary to produce this strain rate decreases by almost three orders of magnitude! Both the expansion of the Coble creep regime and the much lower stress needed to produce a given strain rate reflect the strong inverse dependence of grain size on the rate of this mechanism (Eq. 5-19). The Nabarro-Herring creep domain also expands for the same reason (Eq. 5-18). Since grain size effects on deformation maps are large, some researchers[29,48] have further modified the maps to include grain size as one of the dominant variables along with stress and isostrain rate contour lines. The diagrams, such as the one shown in Fig. 5.22, portray the deformation field boundaries at a fixed temperature, where the grain-size dependence of each deformation mechanism is clearly indicated. (Note the lack of grain-size dependence in the dislocation creep region.)

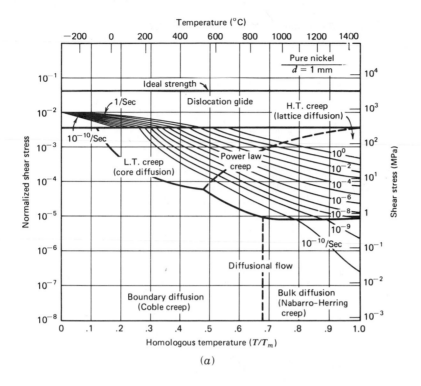

(a)

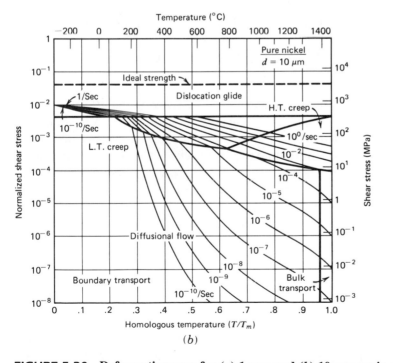

(b)

FIGURE 5.20 Deformation map for (a) 1-mm and (b) 10-μm grain-size nickel. Iso-strain rate lines superimposed on map. Dislocation climb region divided into low-temperature (core diffusion) and high-temperature (volume diffusion) regions. Note lower strain rates in more coarsely grained material. (M. F. Ashby[46]; reprinted with permission of the Institute of Metals.)

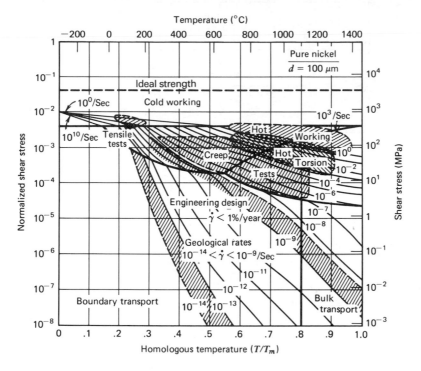

FIGURE 5.21 Deformation map for 100-μm nickel showing laboratory test regimes relative to deformation fields experienced by the material. (From M. F. Ashby;[46] reprinted with permission of the Institute for Metals.)

Figure 5.23 provides one final map comparison by showing the effect of nickel-based superalloy (MAR-M200) multiple strengthening mechanisms in shrinking the dislocation climb domain relative to that associated with pure nickel. In addition, the creep strain rates in the stress–temperature region associated with gas turbine material applications are reduced substantially. By combining alloying additions *and* grain coarsening, the isostrain rate contours are further displaced, thereby providing additional creep resistance to the material.[33] In summary, it must be recognized that displacement of a particular boundary resulting from some specific strengthening mechanism does not in itself eliminate an engineering design problem. It may simply shift the rate-controlling deformation process to another mechanism. The materials designer then must suppress the strain rate of the new rate-controlling process with a different flow attenuation mechanism. As such, the multiple strengthening mechanisms built into high-temperature alloys are designed to counteract simultaneously a number of deformation mechanisms much in the same manner as an all-purpose antibiotic attacks a number of bacterial infections that may assault living organisms.

Recent studies involving deformation maps have focused on new mechanism portrayal methods.[49–52] For example, deformation maps have been constructed as a function of the creep rate $\dot{\epsilon}$ versus T_m/T as compared with normalized stress σ/G versus T/T_m diagrams (e.g., Figs. 5.19–5.21). Furthermore, three-dimensional maps have been

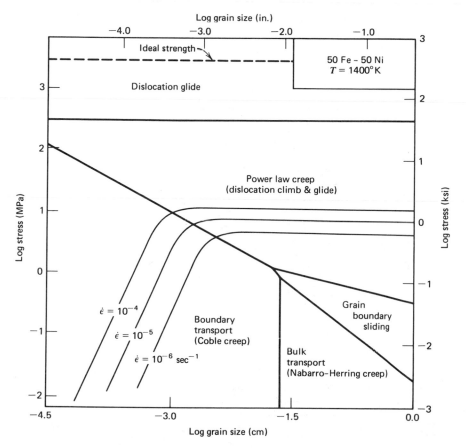

FIGURE 5.22   Deformation map for 50Fe–50Ni in stress–grain-size space at a temperature of 1400 K. Note inclusion of grain-boundary sliding field. (Courtesy of Michael R. Notis, Lehigh University.)

developed using coordinates of $\dot{\epsilon}$, T_m/T, and d/b or σ/G, T_m/T, and d/b where d is the grain size and b the atomic diameter[51,52]; as before, these maps identify those regions associated with a dominant deformation mechanism. For example, Fig. 5.24 reveals the individual regions corresponding to six different deformation mechanisms in a high stacking fault energy FCC alloy.[52] Oikawa suggested that $\dot{\epsilon}$-based diagrams are useful in defining strain-rate conditions associated with easier hot working. On the other hand, σ/G-based diagrams are useful in describing conditions associated with higher creep resistance.

Proceeding in another direction, Ashby and coworkers[53–55] have constructed fracture mechanism maps wherein the conditions for various failure mechanisms are defined. Thus far, fracture mechanism maps have been compiled for various FCC, BCC, and HCP metals and alloys and ceramics.[53–56] For a more detailed study of fracture micromechanisms in metals, ceramics, and engineering plastics, see Sections 7.7, 13.3, and 13.8.

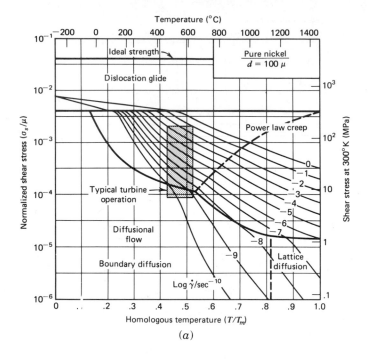

(a)

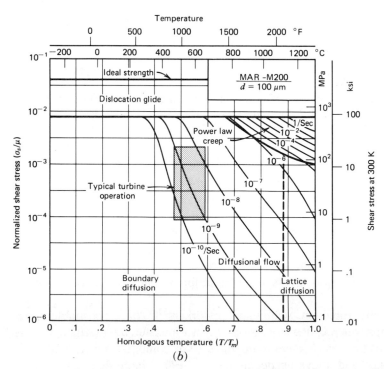

(b)

FIGURE 5.23 Deformation map for (*a*) nickel (100 μm), (*b*) MAR-M200 nickel-based allow (100 μm), and (*c*) MAR-M200 (1 cm). Creep rate is suppressed by multiple strengthening mechanisms and grain coarsening. (From M. F. Ashby[46]; reprinted with permission from the Institute of Metals.)

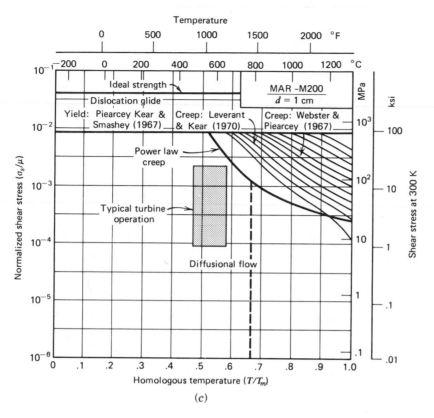

FIGURE 5.23 (*Continued*)

5.6 PARAMETRIC RELATIONS: EXTRAPOLATION PROCEDURES FOR CREEP RUPTURE DATA

It goes without saying that an engineering alloy will not be used for a given elevated temperature application without first obtaining a profile of the material's response under these test conditions. Although this presents no difficulty in short-life situations, such as for the rocket engineer nozzle or military gas turbine blade, the problem becomes monumental when data are to be collected for prolonged elevated temperature exposures, such as those encountered in a nuclear power plant. If the component in question is to withstand 30 or 40 years of uninterrupted service, should there not be data available to properly design the part? If this were done, however, final design decisions concerning material selection would have to wait until all creep tests were concluded. Not only would the laboratory costs of such a test program be prohibitively expensive, but all plant construction would have to cease and the economies of the world would stagnate. In addition, while such tests were being conducted, superior alloys most probably would have been developed to replace those originally selected. Assuming that some of these new alloys were to replace the older alloys in the component manufacture, a new series of long-time tests would have to be initiated. Obviously, nothing would ever be built!

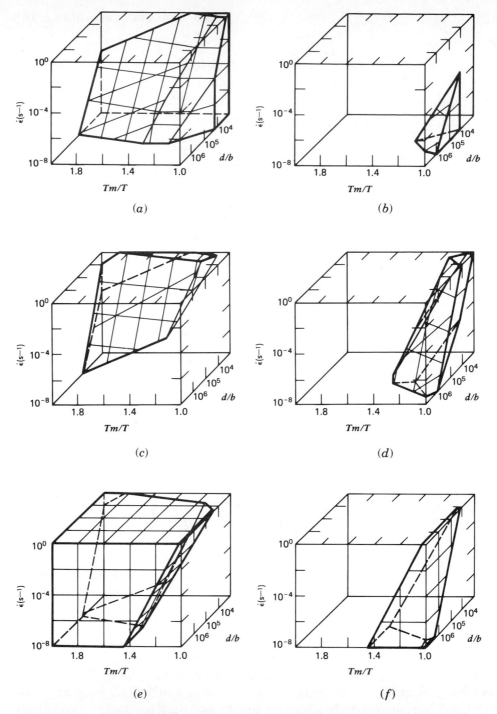

FIGURE 5.24 Three-dimensional deformation mechanism maps for a high stacking fault energy FCC metal. Each map reveals the conditions associated with a specific deformation mechanism. (*a*) Coble creep; (*b*) Nabarro-Herring creep; (*c*) grain-boundary sliding controlled by grain-boundary diffusion; (*d*) grain-boundary sliding controlled by lattice diffusion; (*e*) power-law creep controlled by dislocation–core diffusion; and (*f*) power-law creep controlled by lattice diffusion. (From Oikawa[52]; with permission from Pineridge Press Ltd.)

The practical alternative, therefore, is to perform certain creep and/or creep rupture tests covering a convenient range of stress and temperature and then to *extrapolate* the data to the time–temperature–stress regime of interest. A considerable body of literature has been developed that examines parametric relations (of which there are over 30) intended to allow one to extrapolate experimental data beyond the limits of convenient laboratory practice. A textbook[4] on the subject has even been written. Although it is beyond the scope of this book to consider many of these relations to any great length, it is appropriate to consider two of the more widely accepted parameters.

The Larson-Miller parameter is, perhaps, most widely used. Larson and Miller[57] correctly surmised creep to be thermally activated with the creep rate described by an Arrhenius-type expression of the form

$$r = Ae^{-\Delta H/RT} \tag{5-24}$$

where
r = creep process rate
ΔH = activation energy for the creep process
T = absolute temperature
R = gas constant
A = constant

Equation 5-24 also can be written as

$$\ln r = \ln A - \frac{\Delta H}{RT} \tag{5-25}$$

After rearranging and multiplying by T, Eq. 5-25 becomes

$$\Delta H/R = T(\ln A - \ln r) \tag{5-26}$$

Since $r \propto (1/t)$ (also suggested by Eq. 5-1), Eq. 5-24 can be written as

$$\frac{1}{t} = A'e^{-\Delta H/RT} \tag{5-27}$$

Therefore,

$$-\ln t = \ln A' - \frac{\Delta H}{RT} \tag{5-28}$$

and after rearranging Eq. 5-28, multiplying by T, and converting $\ln t$ to $\log t$

$$\Delta H/R = T(C + \log t) \tag{5-29}$$

which represents the most widely used form of the Larson-Miller relation. Assuming ΔH to be independent of applied stress and temperature (not always true as demonstrated earlier) the material is thought to exhibit a particular Larson-Miller parameter

$[T(C + \log t)]$ for a given applied stress. That is to say, the rupture life of a sample at a given stress level will vary with test temperature in such a way that the Larson-Miller parameter $T(C + \log t)$ remains unchanged. For example, if the test temperature for a particular material with $C = 20$ were increased from 800 to 1000°C, the rupture life would decrease from an arbitrary value of 100 hr at 800°C to 0.035 hr at 1000°C. The value of this parametric relation is shown by examining the creep rupture data in Fig. 5.25, which are the very same data used in Fig. 5.3. The normalization potential of the Larson-Miller parameter for this material is immediately obvious. Furthermore, long-time rupture life for a given material can be estimated by extrapolating high-temperature, short rupture life response toward the more time-consuming low-temperature, long rupture life regime. It is generally found that such extrapolations to longer time conditions are reasonably accurate at higher stress levels because a smaller degree of uncertainty is associated with this portion of the Larson-Miller plot. In-

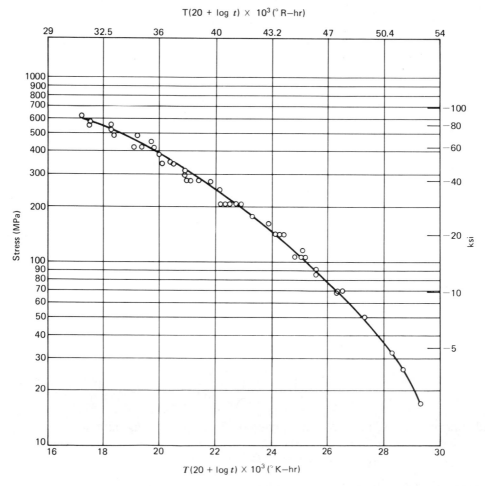

FIGURE 5.25 **Larson-Miller plot showing S-590 iron-based alloy data presented in Fig. 5.3.**

creased extrapolation error is found at lower stress levels where experimental scatter is greater. A comparison between predicted and experimentally determined rupture lives will be considered later in this section.

The magnitude of C for each material may be determined from a minimum of two sets of time and temperature data. Again, assuming $\Delta H/R$ to be invariant and rearranging Eq. 5-29

$$C = \frac{T_2 \log t_2 - T_1 \log t_1}{T_1 - T_2}$$

(5-30)

It is also possible to determine C graphically based on a rearrangement of Eq. 5-29 where

$$\log t = - C + \frac{\text{constant}}{T}$$

(5-31)

When experimental creep rupture data are plotted as shown in Fig. 5.26, the intersection of the different stress curves at $1/T = 0$ defines the value of C. It is important to note that not all creep rupture data give the same trends found in Fig. 5.26. For example, isostress lines may be parallel, as shown in Fig. 5.6, for the case of rutile (TiO_2) and other ceramics and metals. Representative values of C for selected materials[57] are given in Table 5.1. For convenience, the constant is sometimes not determined experimentally but instead assumed equal to 20. Note that the magnitude of the material constant C does not depend on the temperature scale but only on units of time. (Since practically all data reported in the literature give both the material constant C and the rupture life in more convenient units of hours rather than in seconds—the recommended SI unit for time—test results in this section will be described in units of hours.)

In addition to being used for the extrapolation of data, the Larson-Miller parameter also serves as a figure of merit against which the elevated temperature response of

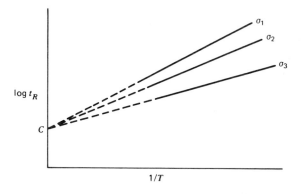

FIGURE 5.26   Convergence of isostress lines in plot of $\log t_R$ versus $1/T$ to determine magnitude of constant C in Larson-Miller parameter.

TABLE 5.1 **Material Constants for Selected Alloys**[57]

Alloy	C	
	Time, hr	Time, s
Low carbon steel	18	21.5
Carbon moly steel	19	22.5
18-8 stainless steel	18	21.5
18-8 Mo stainless steel	17	20.5
$2\frac{1}{4}$ Cr–1 Mo steel	23	26.5
S-590 alloy	20	23.5
Haynes Stellite No. 34	20	23.5
Titanium D9	20	23.5
Cr–Mo–Ti–B steel	22	25.5

different materials may be compared (e.g., in the case of alloy development studies). For example, when the curves for two materials with the same constant C are coincident, the materials obviously possess the same creep rupture behavior (Fig. 5.27a). The same conclusion does not follow, however, when the coincident curves result from materials with different values of C (Fig. 5.27b). When $C_A < C_B$, material A would be the stronger of the two. (For the same parameter P, and at the same test temperature, $\log t_{R_A}$ for alloy A would have to be greater than $\log t_{R_B}$ since $C_B > C_A$.) A direct comparison of material behavior is evident when C is the same but the parametric curves are distinct from one another (Fig. 5.27c). Here alloy A is clearly the superior material. While such alloy comparisons for specified conditions of stress and temperature are possible using the Larson-Miller parameter (and other parameters as well), it should be understood that such parameters provide little insight into the mechanisms responsible for the creep response in a particular time–temperature regime. This is done more successfully by examining deformation maps (Section 5.4).

The Sherby-Dorn (SD) parameter $\theta = t_R e^{-\Delta H/RT}$ (where $t = t_R$) described in Eq. 5-7 has been used to compare creep rupture data for different alloys much in the same manner as the Larson-Miller (LM) parameter. Reasonably good results have been obtained with this parameter in correlating high-temperature data of relatively pure metals[10] (Fig. 5.28). The reader should recognize that if the Sherby-Dorn parameter does apply for a given material, then when θ is constant, a plot of the logarithm of rupture life against $1/T$ should yield a series of straight lines corresponding to different

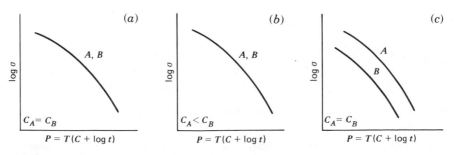

FIGURE 5.27 **Parametric comparison of alloy behavior.** (a) Alloy A = Alloy B; (b) and (c) Alloy A superior to Alloy B.

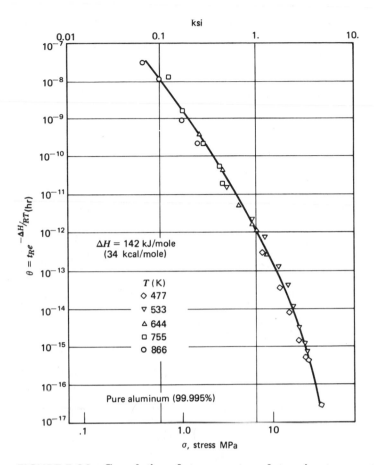

FIGURE 5.28 **Correlation of stress rupture data using temperature-compensated time parameter $\theta = t_{Re}^{-\Delta H/RT}$ for pure aluminum. (From J. E. Dorn,[12] *Creep and Recovery*; reprinted with permission from American Society for Metals, Metals Park, OH, copyright © 1957.)**

stress levels. This is contrary to the response predicted by the Larson-Miller parameter, where the isostress lines converge when $1/T = 0$. The choice of the LM or SD parameters to evaluate a material's creep rupture response would obviously depend on whether the isostress lines converge to a common point or are parallel. In fact, the choice of a particular parameter (recall that over 30 exist) to correlate creep data for a specific alloy is a very tricky matter. Some parameters seem to provide better correlations than others for one material but not another. This may be readily seen by considering Goldhoff's tabulated results[58] for 19 different alloys (Table 5.2). Shown here are root-mean-square (RMS) values reflecting the accuracy of the LM, SD, and other parameters in predicting creep rupture life. The RMS value is defined as

$$\text{RMS} = \left[\frac{\sum (\log \text{ actual time to rupture} - \log \text{ predicted time to rupture})^2}{\text{number of long-time data points}} \right]^{1/2}$$

(5-32)

TABLE 5.2 Comparative RMS Values Reflecting Accuracy of Different Time–Temperature Parameters[58]

Data Set	Alloy	Data Points		LM^a	MH^b	SD^c	MS^d	Best TTP^e	MCM^f
		Short-Time	Long-Time						
1	Al 1100-0	53	11	0.347	0.377	0.308	0.488	0.308	0.260
2	Al 5454-0	68	7	0.099	0.166	0.143	0.287	0.099	0.081
4	Carbon steel	18	8	0.456	0.313	0.415	0.396	0.313	0.084
5	Cr-Mo steel	23	10	0.152	0.102	0.056	0.191	0.056	0.122
6	Cr-Mo-V steel	17	9	0.389	0.091	0.162	0.477	0.091	0.102
7A	304 stainless steel	33	19	0.375	0.207	0.185	0.309	0.185	0.194
7B	304 stainless steel	41	11	0.454	0.167	0.272	0.292	0.167	0.179
8	304 stainless steel	26	13	0.334	0.349	0.237	0.457	0.237	0.228
9	316 stainless steel	28	10	0.244	0.296	0.212	0.323	0.212	0.073
11A	347 stainless steel	18	24	0.368	0.203	0.298	0.265	0.203	0.123
11B	347 stainless steel	31	13	0.291	0.173	0.267	0.211	0.173	0.107
12	A-286	19	5	0.097	0.338	0.089	0.111	0.089	0.220
13	Inco 625	78	21	0.343	0.283	0.337	0.329	0.283	0.317
14	Inco 718	17	9	0.104	0.565	0.110	0.100	0.100	0.084
15	René 41	26	11	0.106	0.144	0.139	0.113	0.106	0.131
16	Astroloy®	21	12	0.302	0.343	0.231	0.264	0.231	0.107
17A	Udimet 500	65	38	0.252	0.342	0.316	0.348	0.252	0.268
17B	Udimet 500	93	12	0.111	1.057	0.247	0.173	0.111	0.124
18A	L-605	51	49	0.319	0.652	0.420	0.261	0.261	0.247
18B	L-605	76	28	0.374	0.641	0.460	0.305	0.305	0.290
19	Al 6061-T651	74	25	0.361	0.382	0.217	0.473	0.217	0.311
Average of above 21 data sets				0.280	0.342	0.244	0.294	0.190	0.174
Average excluding B data sets				0.273	0.303	0.228	0.305	0.191	0.174

^{a} Larson-Miller parameter.
^{b} Manson-Haferd parameter.
^{c} Sherby-Dorn parameter.
^{d} Manson-Succop parameter.
^{e} Time-temperature parameter.
^{f} Minimum commitment method.

Note that for some metals, either the LM or SD parameter represented the best time–temperature parameter (TTP) of the four examined by Goldhoff and predicted actual test results most correctly. Alternatively, these two parameters provided poor correlations when compared to other parameters for different materials; the use of the LM or SD parameters in evaluating these alloys led to significant error in the prediction of actual rupture life.

The inconsistency with which a particular TTP predicts actual creep rupture life for different alloys represents a severe shortcoming of the parametric approach to creep design. These deficiencies may be traced in part to some of the assumptions underlying each parameter. For example, the LM and SD parameters are based on the assumption that the activation energy for the creep process is not a function of stress and temperature. Clearly, the test results shown in Fig. 5.8 and the extended discussion in Section 5.3 discredit this supposition. (Recall, however, that when $T \geq 0.5T_m$, the activation energy for creep is essentially constant and equivalent to the activation energy for self-diffusion.) Furthermore, none of the TTP make provision for metallurgical instabilities.

Attempts are being made to standardize creep data parametric analysis procedures through the establishment of required guidelines by which an investigator arrives at the selection of a particular TTP. In this regard, the minimum commitment method (MCM)[59,60] holds considerable promise in that it presumes initially a very general time–temperature–stress relation. The precise form is obtained on the basis of actual test data. As such, the MCM can lead to the selection of a standard parametric relation, such as LM or SD, or it may define a new parameter that can reflect the possible existence of metallurgical instabilities. Note the reduced RMS values for the MCM method as compared to the LM, SD, or the other two TTP evaluated by Goldhoff (Table 5.2).

Another method, referred to as the graphical optimization procedure (GOP), also has been used to improve the accuracy of life predictions based on various extrapolation procedures.[61,62] To illustrate this point, Woodford employed the GOP to demonstrate that the material constant C used in the Larson-Miller parameter was a function of rupture life. For example, he found for the case of IN718 nickel-based alloy that C varied from 27.1 at short lives to 20 at a 10,000 rupture hour.[61,62] By utilizing the correct time-dependent value of C in the Larson-Miller formula, less scatter was observed in the data normalization procedure.

5.7 MATERIALS FOR ELEVATED TEMPERATURE USE

From the previous discussions, a material suitable for high-temperature service should possess a high melting point and modulus of elasticity, and low diffusivity. In addition, such materials must possess a combination of superior creep strength, thermal fatigue resistance, and oxidation and hot corrosion resistance. As a result, alloy development has focused primarily on nickel- and cobalt-based superalloys, with earlier iron-based alloys being replaced because of their relatively low melting point and high diffusivity.[63–68] These high-temperature alloys have been produced by several methods including casting, mechanical forming, powder metallurgy, directional solidification of columnar and single crystals, and mechanical alloying.

For the case of nickel-based superalloys, constituent elements are introduced to enhance solid solution properties, as precipitate and carbide formers, and as grain-boundary and free surface stabilizers.[69] Tungsten (W), molybdenum (Mo), and titanium (Ti) are very effective solid solution strengtheners; W and Mo also serve to lower the diffusion coefficient of the alloy. (There is a general inverse relation between the melting point and alloy diffusivity.) Though the incremental influence of chromium (Cr) on solid solution strengthening is small (i.e., $d\tau/dc$ is low), the overall solid solution strengthening potential of Cr in nickel (Ni) alloys is large since large amounts of Cr can be dissolved in the Ni matrix. Cobalt (Co) provides relatively little solid solution strengthening but serves to enhance the stability of the submicron-size $Ni_3(Al,X)$ (γ') precipitates within the nickel solid solution (γ) matrix (Fig. 5.29a). Within the γ' phase, X corresponds to the presence of Ti, niobium (Nb), or tantalum (Ta). The difficulty of dislocation motion through the ordered γ' particles in these alloys is responsible for their high creep strength at elevated temperatures. Of particular note, the γ' phase exhibits unusual behavior in that strength *increases* by three- to sixfold with increasing temperature from ambient to approximately 700°C.[70–72]

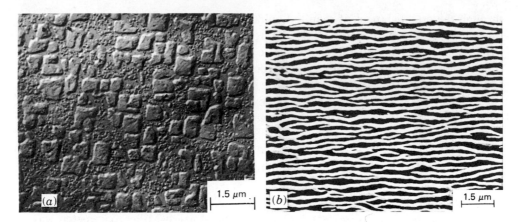

FIGURE 5.29 Electron micrographs revealing Ni₃Al precipitates (γ') in a nickel solid solution (γ) matrix. (*a*) Cubic form in MAR M-200. (*b*) Rafted morphology in Ni-14.3Mo-6Ta-5.8Al (Alloy 143). Tensile stress axis is in vertical direction and parallel to [001] direction. Creep tested with 210 MPa at 1040C[73]. (Courtesy E. Thompson.)

Also noteworthy is the fact that γ' precipitates in single-crystal alloys tend to coarsen under stress at 1000°C and form thin parallel platelike arrays that are oriented normal to the applied stress axis (Fig. 5.29*b*). Recent studies have confirmed that alloy creep resistance is enhanced by the development of this "rafted" microstructure[73,74]; it is believed that the absence of dislocation climb around the γ' particles, due to their lenticular shape, forces dislocations to cut across the ordered γ' phase. As noted in Section 4.4.2, this dislocation path enhances the alloy's resistance to plastic flow.

The presence of carbides along grain boundaries in polycrystalline alloys serves to restrict grain-boundary sliding and migration. Carbide formers such as W, Mo, Nb, Ta, Ti, Cr, and vanadium (V) lead to the formation of M_7C_3, $M_{23}C_6$, M_6C, and MC, with MC carbides being most stable (e.g., TiC). When Cr levels are relatively high, $Cr_{23}C_6$ particles are formed.

Surface stabilizers include Cr, Al, boron (B), zirconium (Zr), and hafnium (Hf). The presence of Cr in solid solution allows for the formation of Cr_2O_3, which reduces the rate of oxidation and hot corrosion. Aluminum contributes to improved oxidation resistance and resistance to oxide spalling. Finally, B, Zr, and Hf are added to impart improved hot strength, hot ductility, and rupture life.[75] Cobalt-based alloys derive their strength from a combination of solid solution hardening and carbide dispersion strengthening. The mechanical properties of representative nickel-based and cobalt-based alloys are given in Table 5.3; references 63 to 68 provide additional information concerning these materials.

Recent efforts to improve the high-temperature performance of superalloys have tended more toward optimizing component design and making use of advanced processing techniques rather than tinkering with alloy chemistry.[76] For example, when inlet guide vanes and first-stage turbine blades of the gas turbine engine are air cooled via internal channels, the gas turbine inlet temperature can be increased markedly with a concomitant improvement in engine operating efficiency. Several processing techniques have been developed and applied to the manufacture of gas turbine com-

TABLE 5.3   Mechanical Properties of Selected Superalloys

Alloy Designation	Yield Strength [MPa(ksi)]			100-hr Rupture Strength [MPa(ksi)]		1000-hr Rupture Strength [MPa(ksi)]	
	21°C (70°F)	760°C (1400°F)	982°C (1800°F)	760°C (1400°F)	982°C (1800°F)	760°C (1400°F)	982°C (1800°F)
Cast Alloys							
B1900	825 (120)	808 (117)	415 (60)	505 (73)[a]	170 (25)	380 (55)[a]	105 (15)
IN-100	850 (123)	860 (125)	370 (54)	625 (91)	170 (25)	515 (75)	105 (15)
MAR-M-200	840 (122)	840 (122)	470 (68)	635 (92)	179 (26)	580 (84)	130 (18.5)
MAR-M-200(DS)[b]	860 (125)	925 (134)	620 (90)	725 (105)	200 (29)	660 (96)	140 (20)
TRW-NASA VI A	940 (136)	945 (137)	520 (75)	725 (105)[c]	215 (31)	585 (85)	140 (20)
MAR-M 509	570 (83)	365 (53)	180 (26)	345 (50)	105 (15)	260 (38)	79 (11.5)
Wrought Alloys							
Astroloy	1050 (152)	910 (132)	275 (40)	540 (78)	105 (15)	430 (62)	55 (8.0)
Hastelloy X	360 (52)	260 (38)	110 (16)	145 (21)	26 (3.8)	100 (15)	14 (2.0)
Waspaloy	795 (115)	675 (98)	140 (20)	415 (60)	45 (6.5)	290 (42)	— (—)
ODS Alloys[d]							
MA 6000	1069 (155)	781 (113)	344 (50)	485 (70)	210 (30)	410 (59)	180 (26)
Alloy 51	903 (131)	972 (141)	517 (75)	600 (87)	221 (32)	469 (68)	186 (27)

[a] Data correspond to 816°C (1500°F).

[b] Directionally solidified.

[c] Extrapolated values.

[d] Data courtesy Inco Alloys Inc.

ponents. One such technique involves the directional solidification of conventional superalloys to produce either highly elongated grain boundaries or single-crystal components (Fig. 5.30). Helical molds are used to cast single-crystal turbine blades; multiple grains form initially and grow into the helical section of the mold. The faster growing $\langle 100 \rangle$-oriented grains then crowd out other grains until a single $\langle 100 \rangle$ grain is left to fill the mold cavity.[77–79] Current sophisticated mold designs now allow for the simultaneous growth of two turbine blades from the same single crystal.[79] The alignment of airfoils (turbine blades) along the $\langle 100 \rangle$ axis parallel to the centrifugal stress direction allows for a 40% reduction in the elastic modulus and associated lower plastic strain range during thermal fatigue cycling; a 6- to 10-fold improvement in thermal fatigue resistance is thus achieved. Since grain boundaries are eliminated, their influence on grain-boundary sliding, cavitation, and cracking is obviated.[77,78] Furthermore, it is no longer necessary to add such elements as hafnium, boron, carbon, and zirconium for the purpose of improving grain-boundary hot strength and ductility.[80] Without these elements, the incipient melting temperature of the alloy is increased by approximately 120°C and the alloy chemistry simplified. The development of cast superalloy turbine blades is shown in Fig. 5.31a; the relative ranking of the rupture lifetime for equiaxed and columnar polycrystalline alloys is compared with that of single-crystal alloys in Fig. 5.31b. By applying unidirectional solidifcation to alloys of eutectic composition, it has been possible to produce eutectic composite alloys possessing properties superior to those found in conventional superalloys[81] (Fig. 5.32). A number of these alloys contain a γ/γ' matrix that is reinforced with high-strength whiskers of a third phase; these strong filamentary particles are oriented

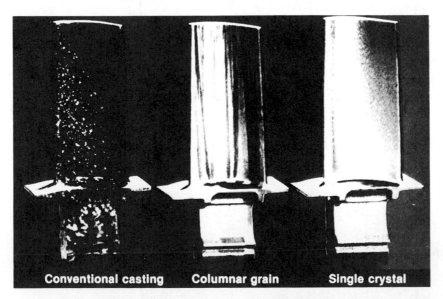

FIGURE 5.30 Conventional and directional solidification used to prepare gas turbine blades with equiaxed, columnar, and single-crystal morphologies. (F. L. VerSnyder and E. R. Thompson, *Alloys for the 80's,* R. Q. Barr, Ed., Climax Molybdenum Co., 1980, p. 69; with permission.)

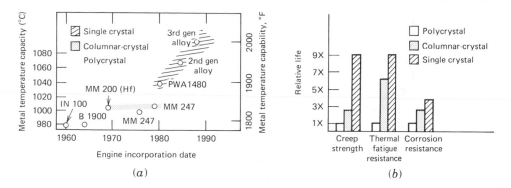

(a)   (b)

FIGURE 5.31 (*a*) Development of turbine blade temperature capability. (*b*) Comparative high temperature strength and corrosion resistance of equiaxed, columnar, and single-crystal superalloys.[79] (Reprinted with permission from *Journal of Metals*, **39**(7), 11 (1987), a publication of the Metallurgical Society, Warrendale, PA. 15086.)

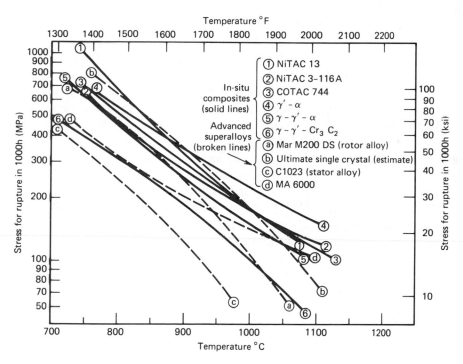

FIGURE 5.32 1000-hr strength as a function of temperature in eutectic superalloys and conventional directionally solidified single-crystal and oxide-dispersion-strengthened superalloys. In situ (eutectic) composites reveal generally superior stress rupture behavior. (From Lemkey[81]; reprinted by permission of the publisher from F. D. Lemkey, *Proceedings,* MRS Conference, CISC IV, Vol. 12, F. D. Lemkey, H. E. Cline, and M. McLean, Eds., copyright by Elsevier Science Publishing Co., Inc., Amsterdam, © 1982.)

parallel to the maximum stress direction. Athough the properties of these alloys are very good, the allowable solidification rates for their manufacture are much lower than those permissible in the manufacture of directionally solidified columnar or single-crystal microstructures. One is then faced with a trade-off between the superior properties of eutectic composites and their higher manufacturing costs.

Another new fabrication technique involves forging under superplastic conditions.[82] In this process, the material is first hot extruded just below the γ' solvus temperature, which causes the material to undergo spontaneous recrystallization. Since the γ' precipitates in the nickel solid solution matrix tend to restrict grain growth, the recrystallized grain diameter remains relatively stable in the size range of 1 to 5 μm. The part is then forged isothermally at a strain rate that enables the material to deform superplastically (recall Section 5.4). At this point, the superplastically formed component is solution treated to increase the grain size for the purpose of enhancing creep strength. The material is then quenched and aged to optimize the γ/γ' microstructure and the associated set of mechanical properties. One major advantge of superplastic forging is its ability to produce a part closer to its final dimensions, thereby reducing final machining costs.

Superalloys can also be fabricated from powders produced by vacuum spray atomization of liquid or by solid-state mechanical alloying techniques (recall Section 4.5). Powders may then be placed in a container that is a geometrically larger version of the final component shape. The can is then heated under vaccum and hydrostatically compressed to yield a fully dense component with dimensions close to the design values. The microstructure of hot isostatically pressed (HIP) Astroloy superalloy is shown in Fig. 5.33a.[83] Note the persistence of the necklace of prior particle boundary borides, carbides, and oxides that surround the atomized powder particles. Hot isostatic pressing is also being used to heal defects in conventionally cast parts and to heal certain defects in parts that experience creep damage in service.

With significant additions of γ' formers, such as Al and Ti, mechanically alloyed oxide-dispersion-strengthened (MA/ODS) products possess attractive strength levels over a broad temperature range.[84,85] Two such alloys are MA6000 and Alloy 51, which contain approximately 55 v/o and 75 v/o γ', respectively (Fig. 5.33b).[84,85] The 1000-hr rupture strength (normalized with respect to density) of these alloys and others is shown in Fig. 5.34 as a function of temperature. As expected, directionally solidified (DS MAR-M200) and single-crystal (PWA 1480) cast alloys are superior to the two mechanically alloyed products at temperatures up to 900°C with the relative rankings being reversed above this temperature. At high temperatures near the γ' solvus temperature, the γ' particles that dominate the precipitation hardening process tend to coarsen and/or go back into solution. The superiority of MA materials relative to that of directionally solidified and single-crystal cast alloys at temperatures in excess of 900°C is due to the oxide-dispersion-strengthening influence of the Y_2O_3 particles that remain in the microstructure and do not coarsen to any significant degree.

Recent attention has focused on the unusual creep rate and rupture-life stress dependence of ODS alloys. Whereas most pure metals and associated solid solutions reveal a σ^{4-5} dependence of $\dot{\epsilon}$ (recall Eq. 5-15 and 5-20), the steady-state creep rate in ODS alloys exhibits a stress dependency of 20 or more.[70,84,86] Furthermore, the apparent activation energy for the creep process is found to be two to three times

 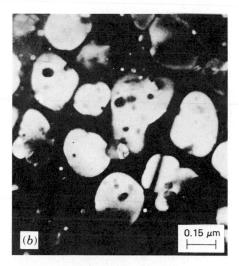

FIGURE 5.33   Transmission electron micrographs of P/M nickel-based alloys. (*a*) Microstructure of HIP'd Astroloy superalloy. Note persistent necklaces of prior particle boundary borides, carbides, and oxides.[83] (Reprinted with permission from J. S. Crompton and R. W. Hertzberg, *J. Mater Sci.*, 21, 3445 (1986), Chapman & Hall Pub.) (*b*) Microstructure of MA 6000 showing γ' precipitates (large light areas) and Y_2O_3 dispersoids (small dark regions). [(Photo courtesy W. Hoffelner from W. Hoffelner and R. F. Singer, Metallurgical Transactions 16A, 393 (1985).)]

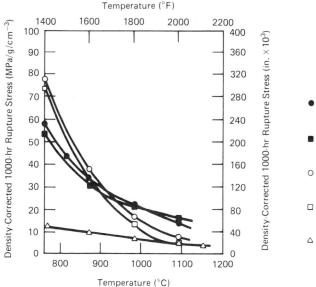

FIGURE 5.34   Comparison of 1000-hr rupture strength (density corrected) in directionally solidified and oxide-dispersion-strengthened nickel-based superalloys.[85] Note superior properties of ODS alloys at temperatures above 900°C. (Reprinted with permission from S. K. Kang and R. C. Benn, *Metallurgical Transactions,* 16A, 1285 (1985).)

greater than the activation energy for self-diffusion. Tien and coworkers[70,86] have suggested that these apparent differences in creep response can be rationalized by considering creep to be dominated by an *effective* stress rather than the applied stress; the effective stress is defined as the applied stress minus a back stress that reflects dislocation interactions with Y_2O_3 dispersion strengthening particles. When the applied stress level is replaced by the effective stress value in Eq. 5-20, the stress dependency of $\dot{\epsilon}_s$ and the apparent activation energy for creep are found to be similar to those values corresponding to pure metals (i.e., $n \sim 4$–5 and $\Delta H_c \sim \Delta H_{SD}$).

In corresponding fashion, the rupture life of ODS alloys can reveal a very strong applied stress dependency and an *upward* slope change with increasing rupture lifetime, opposite to that observed in many other alloys (e.g., recall Fig. 5.3). Figure 5.35 reveals that MA6000 and Alloy 51 exhibit two regions of behavior; Region I corresponds to high stress levels and intermediate temperatures and is dominated by the γ' precipitates. At higher temperatures, lower stress levels and longer times (Region II), stress rupture is dominated by the Y_2O_3 dispersoid phase. Note that ODS alloy MA754, which contains no γ' phase, does not exhibit Region I behavior; conversely, cast alloy IN939, which contains no dispersion strengthening phase, exhibits no Region II behavior. Recent studies have sought to clarify the nature of the dislocation–dispersoid particle interaction so as to better understand the unique phenomenological behavior of ODS alloys.[87]

In another recent thrust, researchers have focused attention on the development of a gas turbine engine using ceramic components. Since ceramics often possess higher melting points and moduli of elasticity and lower diffusivities than metal systems, they offer considerable potential in such applications. Unfortunately, ceramics suffer from low ductility and brittle behavior in tension (see Table 10.8). This serious problem must be resolved before the ceramic engine can become a reality. Progress toward this end is being made as discussed in Section 10.4.3.

Finally, fiber-reinforced superalloys are receiving increased attention as candidate materials for structural use at elevated temperatures. Tungsten fibers hold promise as

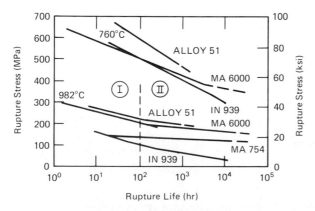

FIGURE 5.35 Stress rupture response of MA/ODS and cast nickel superalloys.[84] (From R. C. Benn and S. K. Kang, *Superalloys 1984,* American Society for Metals, Metals Park, OH, 1984, with permission.)

a suitable reinforcement for superalloys in that they possess superior high-temperature strength and creep resistance.[88] In addition, a good interface is developed between the superalloy matrix and the tungsten fibers without excessive surface reactions that degrade W-fiber mechanical properties. Preliminary studies have shown that operating temperatures of fiber-reinforced superalloys may be increased by 175°C over that of unreinforced superalloys.

Whatever the alloy or process used to fabricate superalloy parts, the high-temperature environments that are experienced demand that careful attention be given to the suppression of oxidation and corrosion damage. To this end, coatings such as MCrAl/Y (where M = Ni, Co, and Fe) may be placed on the component's exterior surface; surface coatings with such compositions promote the formation and retention of Al_2O_3, which serves as an effective barrier to the diffusion of oxygen into the component interior.[79] Unfortunately, these coatings tend to spall away during thermal cycling and must be stabilized. Other ceramics (e.g., ZrO_2) may serve as thermal barrier coatings (approximately 0.25 mm thick) that can reduce superalloy turbine blade surface temperatures by as much as 125–250°C. Here, too, the tendency for spallation due to thermally induced strains must be suppressed.

5.8 CREEP FRACTURE MICROMECHANISMS

In closing this chapter, it is appropriate to consider mechanisms associated with the fracture of materials at elevated temperatures. As noted in Chapter 3, slip and twinning deformation processes occur at high stresses and ambient temperatures and which lead to the *transgranular* fracture of polycrystalline materials. With increasing temperatures and relatively low stress levels, however, *intergranular* fracture generally dominates material response. This change in fracture path occurs since grain boundaries become weaker with respect to the matrix as temperature is increased. The transition temperature for this crack path change-over is often referred to as the "equicohesive temperature." It is generally recognized that intergranular fracture takes place by a combination of grain boundary sliding and grain boundary cavitation associated with stress concentrations or structural irregularities such as grain boundary ledges, triple points, and hard particles.[89]

Grain boundary sliding (GBS), generally thought to occur by grain boundary dislocation motion, becomes operational at temperatures greater than approximately $0.4T_h$ and contributes to both creep strain and intergranular fracture in polycrystals (Fig. 5.36a). As discussed in Section 5.3, the GBS process is accommodated by two major deformation processes; at high temperatures, diffusional creep (i.e., Nabarro-Herring and Coble creep) accommodates the sliding process, whereas at lower temperatures and elevated stresses, grain boundary sliding is controlled by dislocation creep, involving glide and climb of lattice dislocations. If neither diffusional mass transport or intragranular plastic flow is operable, grain boundary decohesion develops by the formation of a planar array of grain boundary cavities (Fig. 5.37a) that eventually coalesce to form grain boundary cracks (Fig. 5.37b); in addition, failure can occur by the formation of wedge cracks at grain boundary triple points (Fig. 5.37c).

The magnitude of the contribution of grain boundary sliding (GBS) to creep strain

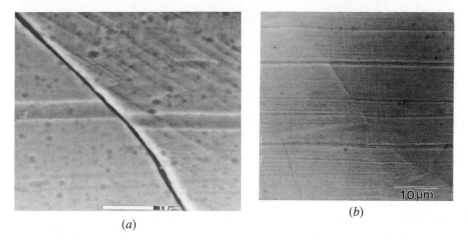

(a)   (b)

FIGURE 5.36   Influence of 300 wt-ppm carbon-doping of ultra-high purity Ni-16Cr-9Fe alloy after 20% elongation at 360°C in argon. (a) Clear evidence of grain boundary sliding in UHP alloy as noted by displacement of fiduciary line; (b) lack of grain boundary sliding when alloy contains 300 wt-ppm carbon in solid solution. (Courtesy Jason L. Hertzberg.)

is a strong function of the stress level and, accordingly, is inversely proportional to the minimum creep rate (Fig. 5.38).[90] At low strain rates (corresponding to low stress levels) and high temperatures, the rate of grain boundary sliding represents a significant portion of the overall creep rate; at high strain rates (i.e., high stress levels), deformation within the grains occurs at a much faster rate than sliding, thereby leading to a negligible contribution of GBS on total creep strain. While it is generally agreed that grain boundary particles can reduce the amount of boundary sliding by their pinning of grain boundaries, the effects of other changes in boundary structure (e.g., solute segregation, grain misorientation) on sliding and associated cracking propensity have recently been investigated.[91,92] Figure 5.36a illustrates typical grain boundary sliding in an ultra-high purity (UHP) Ni-16Cr-9Fe alloy. However, when 300 ppm C was added, the carbon in solid solution was found to limit both grain boundary sliding and cavitation, thereby suppressing intergranular cracking during slow strain rate testing at 360°C ($\sim.38T_h$)[92] (Fig. 5.36b).

Since cavity formation usually occurs by decohesion at grain boundary-particle interfaces, the cavity formation rate will depend on the extent of impurity segregation at such boundaries. In this regard, it has been shown that the critical radius, r_c, for stable cavity development is given by

$$r_c = 2\gamma_s/\sigma \tag{5-33}$$

where γ_s = grain boundary surface energy
σ = local tensile stress normal to the grain boundary

Hence, the stability of a newly formed pore is not guaranteed but, instead, depends on the pore radius, relative to that of the critical pore size, r_c. For example, if the pore

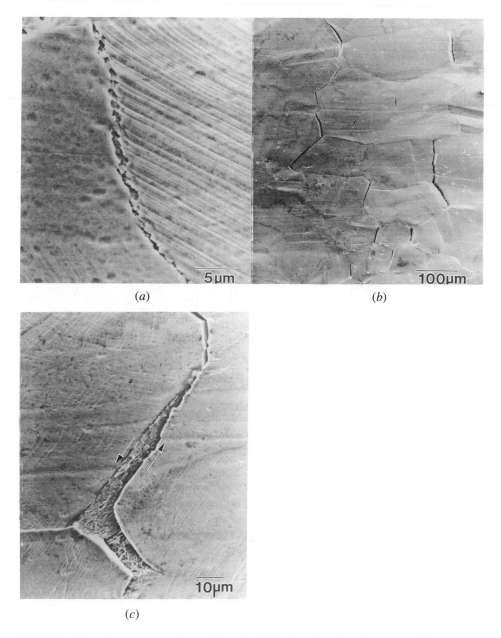

(a)

(b)

(c)

FIGURE 5.37   Cavitation and cracking in UHP Ni-16Cr-9Fe allow after 35% elongation at 360°C in argon. Initial strain rate was 3×10^{-7} sec^{-1}. (a) slip-boundary induced cavitation. (b) intergranular cracking in UHP alloy; (c) triple point cracking. Note involvement of grain boundary sliding (i.e., displacement of fudiciary markings) and grain boundary microvoid coalescence on new fracture surface. (Courtesy Jason L. Hertzberg.)

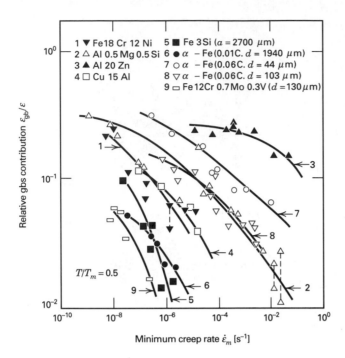

FIGURE 5.38 **Relative contribution of grain boundary sliding as a function of minimum creep rate. Abscissa can also be viewed as describing influence of increasing stress level.[90] (J. Cadek, *Creep in Metallic Materials* (Elsevier, Amsterdam: 1988). Reprinted by permission.)**

radius, r, is less than the critical value, r_c, the cavity surfaces will begin to sinter and close up. On the other hand, when the $r > r_c$, the pore is stable and will grow larger;[93,94] intergranular failure then results from grain boundary cavity growth and coalescence. The magnitude of the effect of impurity level on pore stability can depend upon impurity atom size, with interstitial impurities generally having a potentially greater stabilizing effect than substitutional atoms.[93,94] Furthermore, the cavity growth rate depends on grain boundary diffusivity which in turn varies with solute levels. For example, beneficial hafnium and boron solute segregation to grain boundaries in nickel-based alloys reduces grain boundary diffusion rates by an order of magnitude, thereby slowing cavitation growth kinetics.[95]

REFERENCES

1. A. K. Mukherjee, J. E. Bird, and J. E. Dorn, *Trans. ASM* **62,** 155 (1969).
2. C. R. Barrett, W. D. Nix, and O. D. Sherby, *Trans. ASM* **59,** 3 (1966).
3. F. Garofalo, *Fundamentals of Creep and Creep-Rupture in Metals*, Macmillan, New York, 1965.
4. J. B. Conway, *Stress-Rupture Parameters: Origin, Calculation and Use,* Gordon & Breach, New York, 1969.

5. N. J. Grant and A. G. Bucklin, *Trans. ASM* **42,** 720 (1950).
6. O. D. Sherby and J. L. Lytton, *Trans. AIME* **206,** 928 (1956).
7. F. C. Monkman and N. J. Grant, *Proc. ASTM* **56,** 593 (1956).
8. A. H. Cottrell, *J. Mech. Phys. Sol.* **1,** 53 (1952).
9. E. N. DaC. Andrade, *Proc. R. Soc. London Ser. A* **84,** 1 (1910).
10. R. L. Orr, O. D. Sherby, and J. E. Dorn, *Trans. ASM* **46,** 113 (1954).
11. O. D. Sherby, T. A. Trozera, and J. E. Dorn, *Trans. ASTM* **56,** 789 (1956).
12. J. E. Dorn, *Creep and Recovery,* ASM, Metals Park, OH, 1957, p. 255.
13. O. D. Sherby, J. L. Lytton, and J. E. Dorn, *Acta Met.* **5,** 219 (1957).
14. W. M. Hirthe and J. O. Brittain, *J. Am. Ceram. Soc.* **46**(9), 411 (1963).
15. S. L. Robinson and O. D. Sherby, *Acta Met.* **17,** 109 (1969).
16. J. Weertman, *Trans. ASM* **61,** 681 (1968).
17. O. D. Sherby and P. M. Burke, *Prog. Mater. Sci.* **13,** 325 (1968).
18. O. D. Sherby and M. T. Simnad, *Trans. ASM* **54,** 227 (1961).
19. C. R. Barrett, A. J. Ardell, and O. D. Sherby, *Trans. AIME* **230,** 200 (1964).
20. C. R. Barrett and O. D. Sherby, *Trans. AIME* **230,** 1322 (1964).
21. J. E. Bird, A. K. Mukherjee, and J. E. Dorn, *Quantitative Relation Between Properties and Microstructure,* Israel Universities Press, Haifa, Israel, 1969, p. 255.
22. H. J. Frost and M. F. Ashby, NTIS Report AD-769821, August 1973.
23. F. Garofalo, *Trans AIME* **227,** 351 (1963).
24. F. R. N. Nabarro, *Report of a Conference on the Strength of Solids*, Physical Society, London, 1948, p. 75.
25. C. Herring, *J. Appl. Phys.* **21,** 437 (1950).
26. R. L. Coble, *J. Appl. Phys.* **34,** 1679 (1963).
27. J. Weertman, *J. Appl. Phys.* **28,** 362 (1957).
28. R. Raj and M. F. Ashby, *Met. Trans.* **2,** 1113 (1971).
29. T. G. Langdon, *Deformation of Ceramic Materials,* R. C. Bradt and R. E. Tressler, Eds., Plenum, New York, 1975, p. 101.
30. L. M. Lifshitz, *Sov. Phys. JETP* **17,** 909 (1963).
31. D. K. Matlock and W. D. Nix, *Met Trans.* **5,** 961 (1974).
32. D. K. Matlock and W. D. Nix, *Met Trans.* **5,** 1401 (1974).
33. B. J. Piearcey and F. L. Versnyder, *Met. Prog.,* 66 (Nov. 1966).
34. R. H. Johnson, *Met. Mater.* **4**(9), 389 (1970).
35. W. A. Backofen, I. R. Turner, and D. H. Avery, *Trans. ASM* **57,** 981 (1964).
36. D. H. Avery and W. A. Backofen, *Trans. ASM* **58,** 551 (1965).
37. M. F. Ashby and R. A. Verall, *Acta Met.* **21,** 149 (1973).
38. T. H. Alden, *Acta Met.* **15,** 469 (1967).
39. T. H. Alden, *Trans. ASM* **61,** 559 (1968).
40. J. W. Edington, K. N. Melton, and C. P. Cutler, *Prog. Mater. Sci.* **21,** 61 (1976).
41. D. M. R. Taplin, G. L. Dunlop, and T. G. Langdon, *Annu. Rev. Mater. Sci.* **9,** 151 (1979).
42. T. G. Langdon, *Creep and Fracture of Engineering Materials and Structures,* B. Wilshire and D. R. J. Owen, Eds., Pineridge Press, Swansea, U.K., 1981, p. 141.
43. *Metallurgical Transactions* **13**(5), 689–744 (1982).
44. J. F. Hubert and R. C. Kay, *Met. Eng. Quart.* **13,** 1 (1973).
45. M. F. Ashby, *Acta Met.* **20,** 887 (1972).

46. M. F. Ashby, *The Microstructure and Design of Alloys,* Proceedings, Third International Conference on Strength of Metals and Alloys, Vol. 2, Cambridge, England, 1973, p. 8.

47. F. W. Crossman and M. F. Ashby, *Acta Met.* **23,** 425 (1975).

48. M. R. Notis, *Deformation of Ceramic Materials,* R. C. Bradt and R. E. Tressler, Eds., Plenum, New York, 1975, p. 1.

49. T. G. Langdon and F. A. Mohamed, *J. Mater. Sci.* **13,** 1282 (1978).

50. T. G. Langdon and F. A. Mohamed. *Mater. Sci. Eng.* **32,** 103 (1978).

51. H. Oikawa, *Scripta Met.* **13,** 701 (1979).

52. H. Oikawa, *Creep and Fracture of Engineering Materials and Structures*, B. Wilshire and D. R. J. Owen, Eds., Pineridge Press, Swansea, U.K. 1981, p. 113.

53. M. F. Ashby, *Fracture 1977*, Vol. 1, Waterloo, Canada, 1977, p. 1.

54. M. F. Ashby, C. Gandhi, and D. M. R. Taplin, *Acta Met.* **27,** 699 (1979).

55. C. Gandhi and M. F. Ashby, *Acta Met.* **27,** 1565 (1979).

56. Y. Krishna, M. Rao, V. Kutumba Rao, and P. Rama Rao, *Titanium 80,* H. Kimura and O. Izumi, Eds., AIME, Warrendale, PA, 1981, p. 1701.

57. F. R. Larson and J. Miller, *Trans. ASME,* **74,** 765 (1952).

58. R. M. Goldhoff, *J. Test. Eval.* **2**(5), 387 (1974).

59. S. S. Manson, *Time-Temperature Parameters for Creep-Rupture Analysis*, Publication No. D8-100, ASM, Metals Park, OH, 1968, p. 1.

60. S. S. Manson and C. R. Ensign, *NASA Tech. Memo TM X-52999,* NASA, Washington, DC, 1971.

61. D. A. Woodford, *Mater, Sci. Eng.* **15,** 69 (1974).

62. D. A. Woodford, *Creep and Fracture of Engineering Materials and Structures*, B. Wilshire and D. R. J. Owen, Eds., Pineridge Press, Swansea, U.K., 1981, p. 603.

63. C. T. Sims and W. C. Hagel, Eds., *Superalloys*, Wiley, New York, 1972.

64. B. H. Kear, D. R. Muzyka, J. K. Tien, and S. T. Wlodek, Eds., *Superalloys: Metallurgy and Manufacturer*, Claitors, Baton Rouge, LA, 1976.

65. J. K. Tien, S. T. Wlodek, H. Morrow III, M. Gell, and G. E. Mauer, Eds., *Superalloys 1980,* ASM, Metals Park, OH, 1980.

66. E. F. Bradley, Ed., *Source Book on Materials for Elevated Temperature Applications*, ASM, Metals Park, Oh, 1979.

67. *High Temperature High Strength Nickel Base Alloys*, International Nickel Co., 3d Ed., New York, 1977.

68. M. Gell, C. S. Kortovich, R. H. Bricknell, W. B. Kent and J. F. Radavich, Eds., *Superalloys 84*, AIME, Warrendale, PA, 1984, p. 357.

69. A. K. Jena and M. C. Chaturvedi, *J. Mater. Sci.* **19,** 3121 (1984).

70. R. R. Jensen and J. K. Tien, *Metallurgical Treatises,* J. K. Tien and J. F. Elliott, Eds., AIME, Warrendale, PA, 1981, p. 529.

71. N. S. Stoloff, *Strengthening Methods in Crystals*, A. Kelly and R. B. Nicholson, Eds., Wiley, New York, 1971, p. 193.

72. P. H. Thornton, R. G. Davies, and T. L. Johnston, *Met Trans.* **1,** 207 (1970).

73. E. R. Thompson, Private communication.

74. D. D. Pearson, F. D. Lemkey, and B. H. Kear, *Superalloys 1980*, ASM, Metals Park, OH, 1980, p. 513.

75. R. F. Decker and J. W. Freeman, *Trans. AIME* **218,** 277 (1961).

76. B. II. Kear and E. R. Thompson, *Science* **208,** 847 (1980).
77. B. H. Kear and B. J. Piearcey, *Trans. AIME* **238,** 1209 (1967).
78. F. L. VerSnyder and M. E. Shank, *Mater, Sci. Eng.* **6,** 213 (1970).
79. M. Gell, D. N. Duhl, D. K. Gupta, and K. D. Sheffler, *J. Met.* **39**(7), 11 (1987).
80. D. N. Duhl and C. P. Sullivan, *J. Met.* **23**(7), 38 (1971).
81. F. D. Lemkey, *Proceedings, MRS Conference 1982, CISC IV*, Vol. 12, F. D. Lemkey, H. E. Cline, and M. McLean, Eds., Elsevier Science, Amsterdam, 1982.
82. J. B. Moore and R. L. Athey, U.S. Patent 3,519,503, 1970.
83. J. S. Crompton and R. W. Hertzberg, *J. Mater. Sci.* **21,** 3445 (1986).
84. R. C. Benn and S. K. Kang, *Superalloys 1984,* ASM, Metals Park, OH, 1984, p. 319.
85. S. K. Kang and R. C. Benn, *Met. Trans.* **16A,** 1285 (1985).
86. T. E. Howson, J. E. Stulga, and J. K. Tien, *Met. Trans.* **11A,** 1599 (1980).
87. A. H. Cooper, V. C. Nardone, and J. K. Tien, *Superalloys 1984*, ASM, Metals Park, OH, 1984, p. 319.
88. D. W. Petrasek, D. L. McDanels, L. J. Westfall, and J. R. Stephans, *Metal Prog.* **130**(2), 27 (1986).
89. M. H. Yoo and H. Trinkhaus, *Metall. Trans.,* **14A,** 547 (1983).
90. J. Cadek, *Creep in Metallic Materials*, Elsevier, Amsterdam, (1988).
91. T. Watanabe, *Mater. Sci. & Eng.,* **A166,** 11 (1993).
92. V. Thaveeprungsriporn, T. M. Angeliu, D. J. Paraventi, J. L. Hertzberg and G. S. Was, *Proc. of 6th Int. Sym. on Env. Degrad. of Matl. in Nuclear Power Systems/Water Reactors*, R. E. Gold and E. P. Simonen, eds., TMS, Warrendale, PA, 721 (1993).
93. E. D. Hondros and M. P. Seah, *Metall. Trans.* **8A,** 1363 (1977).
94. M. P. Seah, *Phil Trans. Roy. Soc. London*, **295,** 265 (1980).
95. J. H. Schneibel. C. L. White, and R. A. Padgett, *Proc. 6th Int. Conf. Strength of Metals and Alloys, ICSMA6*, **2,** R. C. Gifkins, Ed., Pergamon Press, Oxford, 649 (1982).

PROBLEMS

5.1 **(a)** Two researchers have simultaneously and independently studied the activation energy for creep of Zeusalloy 300 (Tm.p. = 1000 K). Both used the Eyring rate equation to plot their data:

$$\dot{\epsilon}_s = Ae^{-\Delta H/RT}$$

where $\dot{\epsilon}_s$ = minimum creep rate, T = temperature, and ΔH = activation energy for the creep process. When their data are plotted on semilog paper ($\log \dot{\epsilon}$ vs. $1/T$) their results were different. Investigator I concluded from her work that one activation energy was controlling the creep behavior of the material. Investigator II, on the other hand, concluded that creep in Zeusalloy 300 was very complex since the activation energy was not constant throughout his test range. Which investigator is correct or are they both correct? Explain your answer.

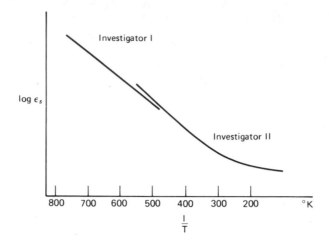

(b) How might the activation energy have been determined?

5.2 Why is the extrapolation of short-time creep rupture data to long times at the same temperatures extremely dangerous? Give a few examples to justify your answer.

5.3 For the following creep rupture data, construct a Larson-Miller plot (assuming $C = 20$). Determine the expected life for a sample tested at 650°C with a stress of 240 MPa, and at 870°C with a stress of 35 MPa. Compare these values with actual test results of 32,000 and 9000 hr, respectively.

Temp. (°C)	Stress (MPa)	Rupture Time (hr)	Temp. (°C)	Stress (MPa)	Rupture Time (hr)
650	480	22	815	140	29
650	480	40	815	140	45
650	480	65	815	140	65
650	450	75	815	120	90
650	380	210	815	120	115
650	345	2700	815	105	260
650	310	3500	815	105	360
705	310	275	815	105	1000
705	310	190	815	105	700
705	240	960	815	85	2500
705	205	2050	870	83	37
760	205	180	870	83	55
760	205	450	870	69	140
760	170	730	870	42	3200
760	140	2150	980	21	440
			1095	10	155

5.4 For the data given in the previous problem, what is the maximum operational temperature such that failure should not occur in 5000 hr at stress levels of 140 and 200 MPa, respectively.

5.5 The test results given below for René 41[58] are from a different alloy from that described in the previous two problems. From a Larson-Miller plot (assume $C = 20$) estimate the rupture life associated with 650°C exposure at 655 MPa and 870°C exposure at 115 MPa. Compare these values with actual test results of 3900 and 1770 hr, respectively.

Temp. (°C)	Stress (MPa)	Rupture Time (hr)	Temp. (°C)	Stress (MPa)	Rupture Time (hr)
650	1000	12	815	460	15
650	950	56	815	395	47
650	895	165	815	345	91
650	840	479	815	260	379
650	820	319	815	220	773
650	790	504	870	260	33
705	815	23	870	215	73
705	695	110	870	180	196
705	635	254	870	145	300
705	590	478	900	170	69
760	600	28	900	140	163
760	515	62	900	105	738
760	440	291			
760	415	390			

5.6 For a certain high-temperature alloy, failure was reported after 3500 hr at 650°C when subjected to a stress level of 310 MPa. If the same stress were applied at 705°C, how long would the sample be expected to last? Assume the material constant needed for this computation to be 20.

5.7 For a given stress level, by how much would the minimum creep rate in aluminum be expected to increase if the test temperature were increased from 550 to 750 K?

5.8 If the Larson-Miller parameter for a given elevated temperature alloy was found to be 26,000, by how much would the rupture life of a sample be estimated to decrease if the absolute temperature of the test were increased from 1100 to 1250 K? Assume that the Larson-Miller constant is equal to 20.

5.9 Gas turbine component A was originally designed to operate at 700°C and exhibited a stress rupture life of 800 h. Component B in the same section of the turbine was redesigned, thereby allowing its operating temperature to be raised to 725°C. Could component A be used at that temperature without modification so long as its stress rupture life exceeds 100 h? (Assume that the Larson-Miller constant for the material is equal to 20.)

CHAPTER 6

DEFORMATION RESPONSE OF ENGINEERING PLASTICS

Polymers exist in nature in such forms as wood, rubber, jute, hemp, cotton, silk, wool, hair, horn, and flesh. In addition, there are countless man-made polymeric products, such as synthetic fibers, engineering plastics, and artificial rubber. In certain respects, the deformation of polymeric solids bears strong resemblance to that of metals and ceramics: Polymers become increasingly deformable with increasing temperature, as witnessed by the onset of additional flow mechanisms (albeit, different ones for polymers). Also, the extent of polymer deformation is found to vary with time, temperature, stress, and microstructure consistent with parallel observations for fully crystalline solids. Furthermore, a time–temperature equivalence for polymer deformation is indicated, which is strongly reminiscent of the time–temperature parametric relations discussed in Section 5.6. Before beginning our discussion of polymer deformation, it is appropriate to describe basic features of the polymer structure that dominate flow and fracture properties. In preparing this section, several excellent books about polymers were consulted to which the reader is referred.[1-6] Additional reading material on polymers is cited at the end of the reference section.

6.1 POLYMER STRUCTURE: GENERAL REMARKS

A polymer is formed by the union of two or more structural units of a simple compound. In polymeric materials used in engineering applications, the number of such unions—known as the degree of polymerization (DP)—often exceeds many thousands. Two major polymerization processes are available to the polymer chemist: addition and condensation polymerization. To examine the former, consider the polymerization of ethylene C_2H_4, which contains a double carbon bond. If the necessary energy is introduced, the double bond may be broken, resulting in a molecule with two free radicals.

$$
\begin{array}{ccc}
\overset{\displaystyle H}{\underset{\displaystyle H}{|}} \quad \overset{\displaystyle H}{\underset{\displaystyle H}{|}} & & \overset{\displaystyle H}{\underset{\displaystyle H}{|}} \quad \overset{\displaystyle H}{\underset{\displaystyle H}{|}} \\
C = C + \text{Energy} \longrightarrow -C - C - + \text{Energy} & & (6\text{-}1)
\end{array}
$$

This activated molecule is now free to combine with other activated molecules to form very long chains. Since this process is exothermic, special care is needed to assure optimum temperature and pressure conditions for continuation of the polymerization process. Therefore, the initial energy input, usually in the form of heat, is needed only to initiate the reaction. To this end, the process is often aided by the addition to the monomer of an active chemical containing free radicals. These initiators break down the double bond and may then become attached to one end of the activated molecule; the other end grows by linking up with other activated molecules. The growth phase occurs very quickly, and the degree of polymerization exceeds several hundred in a fraction of a second. The key to *addition* polymerization is the availability of double or even triple carbon bonds in the monomer that may be broken if sufficient energy is provided. Examples of polymers prepared by addition polymerization include polypropylene, polystyrene, poly(vinyl chloride), and poly(methyl methacrylate).

Condensation polymerization proceeds by a different process and involves the chemical reaction between smaller molecules containing reactive groups with the associated elimination of a small molecular byproduct, usually water. For example, by combining an alcohol (R–OH) with an acid (R'–COOH)

$$R - OH + R' - COOH \longrightarrow R' - COO - R + H_2O \qquad (6\text{-}2)$$

The hydroxyl groups have reacted to allow the two molecules to join, leaving a single molecule of water as a reaction by-product. It should be recognized that the above reaction can proceed no further, since all active hydroxyl groups have been consumed. As in addition polymerization, the key to continued condensation polymerization is the availability of *multiple* active groups to react with other molecules. Here again the two molecules must be *bifunctional* or *trifunctional* to allow the activated molecules to react with *two* or *three* molecules and not just one as noted for the monofunctional alcohol–acid example cited above. For the case of nylon 66, this reaction takes the form

$$H - N - (CH_2)_6 - N - H + OH - \overset{\overset{\displaystyle O}{\|}}{C} - (CH_2)_4 - \overset{\overset{\displaystyle O}{\|}}{C} - OH \longrightarrow$$
$$\underset{\displaystyle H}{\overset{}{|}} \qquad \underset{\displaystyle H}{\overset{}{|}}$$

$$(6\text{-}3)$$

$$H - N - (CH_2)_6 - N - \overset{\overset{\displaystyle O}{\|}}{C} - (CH_2)_4 - \overset{\overset{\displaystyle O}{\|}}{C} - OH + H_2O$$
$$\underset{\displaystyle H}{\overset{}{|}} \qquad \underset{\displaystyle H}{\overset{}{|}}$$

Note that the new molecule retains H and OH groups that react shortly to continue the polymerization reaction. Consequently, long chains of nylon 66 can be formed with repeating units of

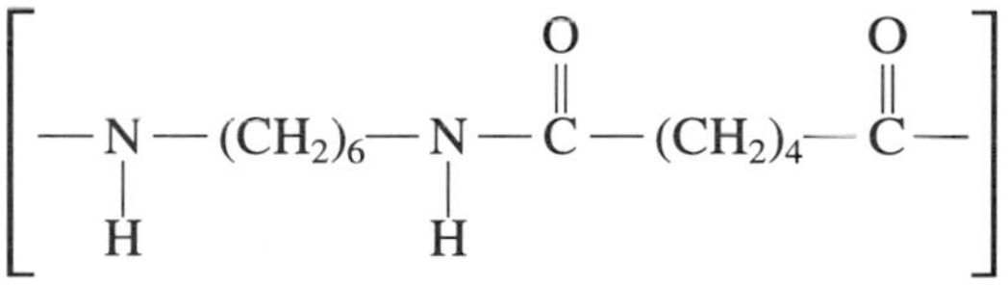

$$\left[\; -N-(CH_2)_6-N-\overset{\overset{\displaystyle O}{\|}}{C}-(CH_2)_4-\overset{\overset{\displaystyle O}{\|}}{C}- \right]$$
$$\quad\;\; \underset{H}{|} \qquad\qquad \underset{H}{|}$$

Polycarbonate and epoxy are other examples of condensation polymerization.

The length of a given polymer chain is determined by the statistical probability of a specific activated mer attaching itself to a particular chain. For example, a chain growing from one side (the other side being stabilized by the attachment of an initiator radical) can (1) continue to grow by the addition of successive activated mers, (2) cease to grow by the attachment of a free radical group to the free end of the chain, or (3) join with another growing chain at its free end, thus terminating the growth stage for both chains (Fig. 6.1). (Other interactions are possible but will not be considered in this book.) It is clear that some chains will be very short, while others might be very long. Consequently, the reader should appreciate one of the most distinctive characteristics of a polymeric solid: There is no unique chain length for a given polymer and no specific molecular weight (MW). Instead, there is a distribution of these values. Contrast this with metal and ceramic solids that exhibit a well-defined lattice parameter and unit cell density. An example of the molecular weight distribution (MWD) for all the chains in a polymer is shown in Fig. 6.2; in this case, a larger number of small chains exist relative to the very long chains. The MWD will vary with the nature of the monomer and the conditions of polymerization so as to be skewed to higher or lower MW and/or made narrower or broader. For example, when the processing temperature is high and/or large amounts of initiator are added to the melt, MW will be low, and vice versa.

Rather than referring to a molecular weight distribution curve to describe the character of a polymer, it is often more convenient to think in terms of an average molecular weight $\overline{M}$. Such a value can be described in a number of ways, but is usually described in terms of either the number of weight fraction of molecules of a given weight. The *number average* molecular weight $\overline{M}_n$ is defined by

$$\overline{M}_n = \frac{\displaystyle\sum_{i=1}^{\infty} N_i M_i}{\displaystyle\sum_{i=1}^{\infty} N_i} \tag{6-4}$$

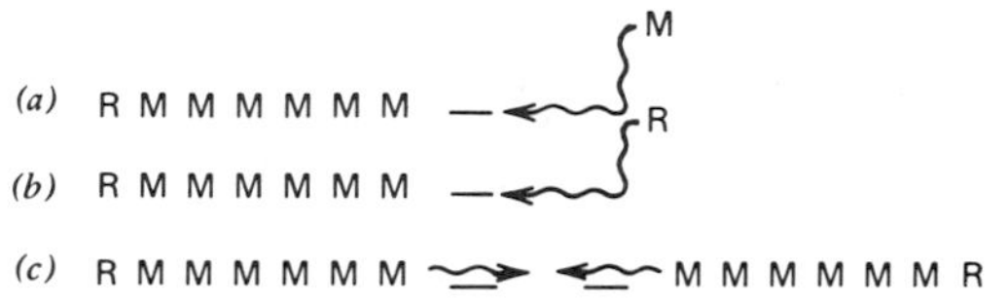

(a) R M M M M M M ⟶ ⟨M / R⟩

(b) R M M M M M M ⟶

(c) R M M M M M M ⟶ ⟵ M M M M M R

FIGURE 6.1 Different events in the polymerization process: (*a*) continuation of chain lengthening; (*b*) chain termination; (*c*) simultaneous termination of two chains.

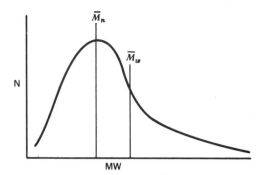

FIGURE 6.2 Molecular weight distribution showing location of average number $\overline{M}_n$ and average weight $\overline{M}_w$ molecular weights.

and represents the total weight of material divided by the total number of molecules. $\overline{M}_n$ emphasizes the importance of the smaller MW chains. The *weight average* molecular weight $\overline{M}_w$ is given by

$$\overline{M}_w = \frac{\sum\limits_{i=1}^{\infty} N_i M_i^2}{\sum\limits_{i=1}^{\infty} N_i M_i} \tag{6-5}$$

which reflects the weight of material of each size rather than their number. It is seen that $\overline{M}_w$ emphasizes relatively high MW fractions. As will be shown in later sections, MW exerts a very strong influence on a number of polymer physical and mechanical properties. The molecular weight distribution can be described by the ratio $\overline{M}_w/\overline{M}_n$. A narrow MWD prepared under carefully controlled conditions may have $\overline{M}_w/\overline{M}_n <$ 1.5, while a broad MWD would reveal $\overline{M}_w/\overline{M}_n$ in excess of 25.

We now look more closely at a segment of the polyethylene (PE) chain just described. In the fully extended conformation, the chain assumes a zigzag pattern, with the carbon–carbon bonds describing an angle of about 109° (Fig. 6.3). With the

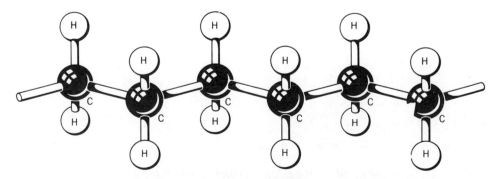

FIGURE 6.3 Extended chain of polyethylene showing coplanar zigzag arrangement of C–C bonds with hydrogen pairs located opposite one another.

zigzag carbon main chain atoms lying in the plane of this page, the two hydrogen atoms are disposed above and below the paper. The chain is truly three dimensional, though it is often represented schematically in two-dimensional space only. Adjacent pairs of hydrogen atoms are positioned relative to one another so as to minimize their steric hindrance. That is, as rotations occur about a C–C bond (permissible as long as the bond angle remains 109°), both favorable and unfavorable juxtapositions of the hydrogen atom pairs are experienced. This is perhaps more readily seen by examining the rotations about the C–C bond in ethane, C_2H_6 (Fig. 6.4), recalling that the hydrogen

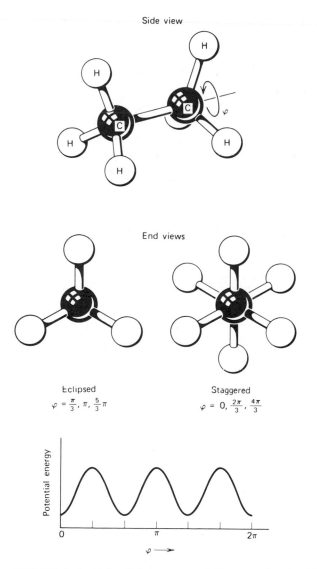

FIGURE 6.4 Potential energy variation associated with C–C bond rotation in ethane.[4] (T. Alfrey and E. F. Gurnee, *Organic Polymers*, 1967. Reprinted by permission of Prentice-Hall Inc., Englewood Cliffs, NJ.)

atoms do not lie in the plane of the page. We see that when the hydrogen atoms are located opposite one another, the potential energy of the system is maximized. Conversely, when they are staggered where $\phi = 0$, $2\pi/3$, and $4\pi/3$, the configuration has lowest potential energy. From this, it is seen that the facility by which C–C bond rotation occurs will depend on the magnitude of the energy barrier in going from one low-energy configuration to another. For the two pairs of adjacent hydrogen atoms in the PE chain, the lowest potential energy trough occurs when the hydrogen atom pair associated with one carbon atom is 180° away from its neighboring hydrogen pairs (Fig. 6.3).

As might be expected, the extent of rotational freedom about the C–C bond depends on the nature of side groups often substituted for hydrogen in the PE chain. When one hydrogen atom is replaced, we have a vinyl polymer. As shown in Table 6.1, a number of different atoms or groups can be added to form a variety of vinyl polymers, which all act to restrict C–C rotation to a greater or lesser degree. Generally, the bigger and bulkier the side group and the greater its polarity, the greater the resistance to rotation, since the peak-to-valley energy differences (Fig. 6.4) would be greater. Restrictions to such movement may also be effected by double carbon bonds in the main chain, which rotate with much greater difficulty. Furthermore, the main chain in some polymers may contain flat cyclic groups (such as a benzene ring) which prefer to lie parallel to one another. Consequently, C–C bond rotation would be made more difficult by their presence.

TABLE 6.1 Selected Vinyl Polymers

Repeat Unit	Polymer
H, H; —C—C—; H, H	Polyethylene
H, Cl; —C—C—; H, H	Poly(vinyl chloride)
H, F; —C—C—; H, H	Poly(vinyl fluoride)
H, CH$_3$; —C—C—; H, H	Polypropylene
H, (benzene ring); —C—C—; H, H	Polystyrene

Thus far, we have discussed the effect of side group size, shape, and polarity on main chain mobility. The *location* of these groups along the chain is also of critical importance, since it affects the relative packing efficiency of the polymer. It is seen from Fig. 6.5 that the side groups can be arranged either randomly along the chain, only on one side, or on alternate sides of the chain. These three configurations are termed atactic, isotactic, and syndiotactic, respectively. Atactic polymers with large side groups (e.g., polystyrene) have low packing efficiency, with the chains arranged in a random array. Consequently, polystyrene, poly(methyl methacrylate), and, to a large extent, poly(vinyl chloride) are amorphous. In a regular and symmetric polymer (e.g., polyethylene), the chains can be packed close together, resulting in a high degree of crystallinity. In fact, the density of a given polymer serves as a useful measure of crystallinity; the higher the density, the greater the degree of crystallinity.

For the data shown in Table 6.2, densities were varied by the amount of main chain branching produced during polymerization (Fig. 6.6). Extensive branching reduces the opportunity for closer packing, and little branching promotes the polymerization of higher density polyethylene. Polypropylene represents an example of a stereoregular polymer that has a high packing efficiency and resultant crystallinity. Although it is not stereoregular, the propensity for crystallinity in nylon 66 is enhanced by the highly polar nature of the nylon chain.

$$
\text{The } \underset{\underset{H}{|}}{N} - \overset{\overset{\displaystyle O}{\|}}{C} \text{ groups in adjacent chains have great affinity for one another with the}
$$

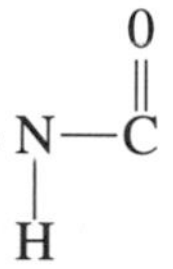

associated hydrogen bond providing additional cause for closer packing and chain alignment.

To summarize, the degree to which polymers will crystallize depends strongly on the polarity, symmetry, and stereoregularity of the chain and its tendency for branch-

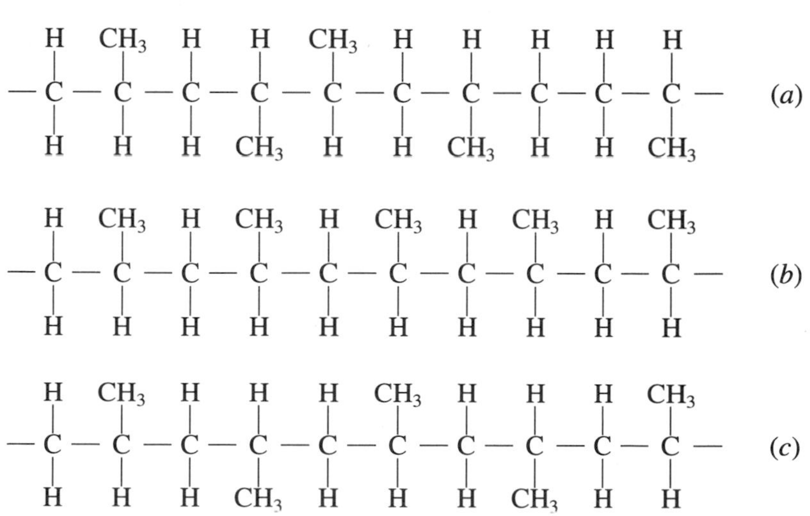

FIGURE 6.5 Location of side groups in polypropylene: (*a*) atactic; (*b*) isotactic; (*c*) syndiotactic.

TABLE 6.2 Relation between Density–Crystallinity and Ultimate Tensile Strength in Polyethylene[7]

Density (g/cm³)	Crystallinity (%)	Ultimate Tensile Strength	
		MPa	ksi
0.92	65	13.8	2
0.935	75	17.2	2.5
0.95	85	27.6	4
0.96	87	31.0	4.5
0.965	95	37.9	5.5

ing. The extent of crystallinity of several polymers is given in Table 6.3, along with other material characteristics. In general, higher levels of crystallinity yield greater stiffness, strength, thermal stability, and chemical resistance. Conversely, elongation and toughness are enhanced by reduced levels of crystallinity.

6.1.1 Morphology of Amorphous and Crystalline Polymers and Their Unoccupied Free Volume

At all temperatures above absolute zero, the existing thermal energy causes the polymer chains to vibrate and wriggle about. First, small-scale vibrations are permitted. Then, with increasing temperature, molecule segments begin to move more freely. Finally, at sufficiently high temperatures associated with the molten state, entire chains are free to move about. It is seen from Fig. 6.7 that these large-amplitude molecular vibrations cause the polymer to become less dense. If crystallization is likely for the type of polymer described in Section 6.1, the material undergoes upon cooling a first-order transformation at B associated with the melting point T_m. Heat of fusion is liberated and the specific volume drops abruptly to C. Further cooling involves ad-

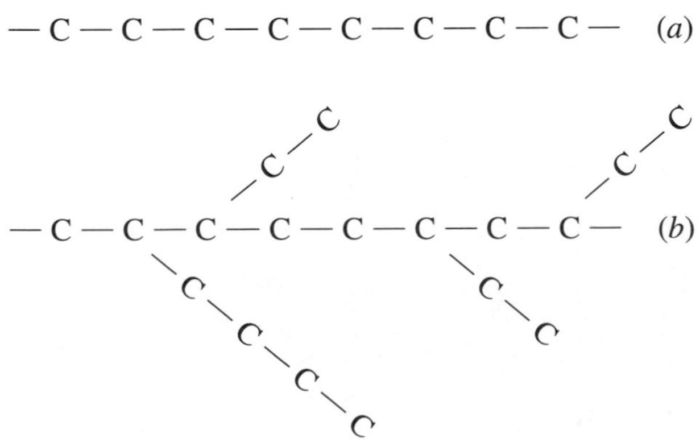

FIGURE 6.6 Degree of chain branching in polymeric solid.
a) Linear; *b)* Branched.

TABLE 6.3 Characteristics of Selected Polymers[2]

Material	Repeat Unit	Major Characteristics	Applications
Low-density polyethylene	$\left[\begin{array}{cc} H & H \\ -C-C- \\ H & H \end{array}\right]$	Considerable branching; 55–70% crystallinity; excellent insulator; relatively cheap	Film; moldings; cold water plumbing; squeeze bottles
Polypropylene	$\left[\begin{array}{cc} H & CH_3 \\ -C-C- \\ H & H \end{array}\right]$	Extent of crystallinity depends on stereoregularity; can be highly oriented to form integral hinge with extraordinary fatigue behavior.	Hinges; toys; fibers; pipe; sheet; wire covering
Acetal copolymer	$\left[\begin{array}{ccc} H & H & H \\ -C-O-C-C-O- \\ H & H & H \end{array}\right]$	Highly crystalline; thermally stable; excellent fatigue resistance	Speedometer gears; instrument housing; plumbing valves; glands; shower heads
Nylon 66	See p. 212	Excellent wear resistance; high strength and good toughness; used as plastic and fiber; highly crystalline; strong affinity for water	Gears and bearings; rollers; wheels; pulleys; power tool housings; light machinery components; fabric
Poly(tetrafluoroethylene) (Teflon)	$\left[\begin{array}{cc} F & F \\ -C-C- \\ F & F \end{array}\right]$	Extremely high MW; high crystallinity; extraordinary resistance to chemical attack; nonsticking	Coatings for cooking utensils; bearings and gaskets; pipe linings; insulating tape; nonstick, loadbearing pads
Poly(vinyl chloride)	$\left[\begin{array}{cc} H & Cl \\ -C-C- \\ H & H \end{array}\right]$	Primarily amorphous; variable properties through polymeric additions; fire self-extinguishing; fairly brittle when unplasticized; relatively cheap	Floor covering; film; handbags; water pipes; wiring insulation; decorative trim; toys; upholstery
Poly(methyl methacrylate) (Plexiglas)	$\left[\begin{array}{cc} H & CH_3 \\ -C-\!\!\!\!\!\!-C- \\ H & O=C-O-CH_3 \end{array}\right]$	Amorphous; brittle; general replacement for glass	Signs; canopies; windows; windshields; sanitary ware

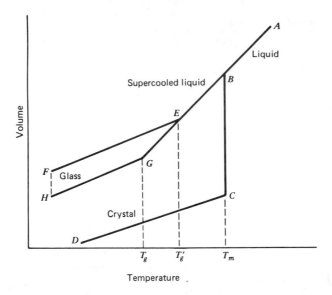

FIGURE 6.7   Change in volume as function of temperature. Crystalline melting point at $T_{m'}$, glass transition temperature at $T_{g'}$, and excess free volume FH.

ditional change in the specific volume (D) as molecular oscillations become increasingly restricted. When crystallization does not occur in the polymer, the liquid cools beyond T_m (location B) without event. However, a point is reached where molecular motions are highly restricted and the individual chains are no longer able to arrange themselves in equilibrium configurations within the supercooled liquid. Below this point (G) (called the glass transition temperature T_g), the material is relatively frozen into a glassy state. The change from a supercooled liquid to glass represents a second-order transformation that does not involve a discrete change in specific volume or internal heat. From Fig. 6.7, it is seen that the polymer in the amorphous state occupies more volume than in the crystalline form. This is to be expected, since higher density forms of a particular polymer are associated with greater crystallinity as a result of greater chain-packing efficiency (Table 6.2). The relative difference in chain-packing density can be described in terms of the fractional unoccupied (or free) volume given by Litt and Tobolsky[8] as

$$\bar{f} = \frac{v_a - v_c}{v_a} = 1.0 - \left(\frac{d_a}{d_c}\right) \tag{6-6}$$

where $\bar{f}$ = fractional unoccupied free volume
v_a, d_a = specific volume and density of amorphous phase
v_c, d_c = specific volume and density of crystalline phase

For many polymers, $0.01 < \bar{f} < 0.1$.

Since the glass transition occurs where molecular and segmental molecular motions are restricted, it is sensitive to cooling rate. Consequently, a polymer may not exist at its glassy equilibrium state. Instead, nonequilibrium cooling rates could preclude the attainment of the lowest possible free volume in the amorphous polymer. In Fig. 6.7, this would correspond to line *EF* with the glass transition temperature increasing to T_g'. Petrie[9] describes the difference between the equilibrium and actual glassy free volume as the *excess free volume* and postulates that this quantity is important in understanding the relation between polymer properties and their thermodynamic state. Note that the free volume will differ from one polymer to another; within the same polymer, the excess free volume is sensitive to thermal history.

The crystalline structure of polymers can be described by two factors: chain conformation and chain packing. The conformation of a chain relates to its geometrical shape. In polyethylene, the chains assume a zigzag pattern as noted above and pack flat against one another. This is not observed in polypropylene, which has a single large methyl group, and in poly(tetrafluoroethylene), which contains four large fluorine atoms. Instead, these zigzag molecules twist about their main chain axis to form a helix. In this manner, steric hindrance is reduced. It is interesting to note that poly(tetrafluoroethylene)'s extraordinary resistance to chemical attack, mentioned in Table 6.3, is believed partly attributable to the sheathing action of the fluorine atoms that cover the helical molecule.

It is currently believed that chain packing in the crystalline polymer is effected by repeated chain folding, such that highly ordered crystalline lamellae are formed. Two models involving extensive chain folding to account for the formation of crystalline lamellae are shown in Fig. 6.8, along with an electron micrograph of a polyethylene single crystal lamellae.[10,11] The thickness of these crystals is generally about 100 to 200 Å, while the planar dimensions can be measured in micrometers. Since chains are many times longer than the observed thickness of these lamellae, chain folding is required. Consequently, chains are seen to extend across the lamellae but reverse direction on reaching the crystallite boundary. In this manner, a chain is folded back on itself many times. The loose loop model depicting less perfect chain folding (some folds occurring beyond the nominal boundary of the lamellae; see Fig. 6.8*b*) is viewed by Clark[12] as being more realistic in describing the character of real crystalline polymers. In addition to loose loops and chain ends (cilia) that lie on the surface of the lamellae, there exist tie molecules that extend from one crystal to another[13] (Fig. 6.9). The latter provide mechanical strengthening to the crystalline aggregate as is discussed later. In the unoriented condition, crystalline polymers possess a spherulitic structure consisting of stacks of lamellae positioned along radial directions (Fig. 6.10). Since the extended chains within the lamellae are normal to the lamellae surface, the extended chains are positioned tangentially about the center of the spherulite. Finally, a new class of semicrystalline polymers has been developed recently that possesses the attractive high-temperature strength and chemical resistance of thermosets, but has the processing ease of thermoplastics. Examples of these resins include poly(ethylene terephthalate) (PET), poly(butyliene terephthalate) (PBT), poly(phenylene sulfide) (PPS), liquid-crystal polymers (LCP), and polyetheretherketone (PEEK). For example, PEEK

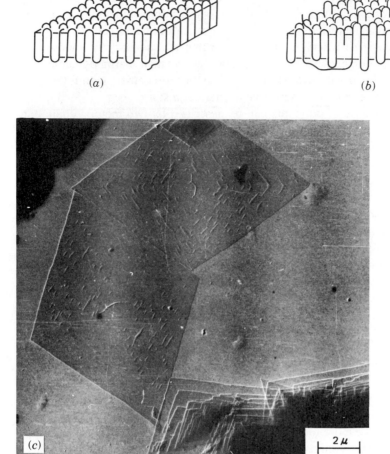

FIGURE 6.8 Crystalline lamellae in polymers. Models for (*a*) regular and (*b*) irregular chain folds, which produce thin crystallites.[10] (Reprinted with permission from *Chem. Eng. News,* 43(33), copyright © by the American Chemical Society.) (*c*) Photomicrograph showing single crystals in polyethylene.[11] (Reprinted with permission from John Wiley & Sons, Inc.)

(containing phenyl (C_6H_5) and carbonyl groups (CO) has high-temperature stability and chemical resistance as does epoxy, but is about ten times tougher than this thermoset material. The PEEK resin is being used to make bearings, electrical connectors, pump impellors, and coatings for cables.

For many years, the structure of amorphous polymers was presumed to consist of a collection of randomly coiled molecules surrounding a certain unoccupied volume. (A coiled molecule can be created by a random combination of C–C bond rotations along the backbone of the molecule.) Recent studies have suggested that this simple view is not correct. Instead, Geil and Yeh[14–16] have proposed that seemingly amorphous polymers actually contain small domains (about 30 to 100 Å in size) in which the molecules are aligned. More work is needed to identify the character of short-

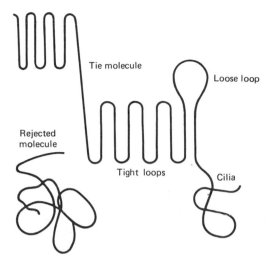

FIGURE 6.9 Schematic representation of chain-folded model containing tie molecules, loose loops, cilia (chain ends), and rejected molecules.[13] (By permission, from *Polymeric Materials,* copyright American Society for Metals, © 1975).

FIGURE 6.10 Sheaflike stacks of crystal lamellae in polychlorotrifluoroethylene which represent intersecting spherulites.[11] (Reprinted with permission from John Wiley & Sons, Inc.)

range order domains in "amorphous" polymers before the existence and role of such structural micro-details in determining properties can be proven convincingly.

6.2   POLYMER ADDITIONS

Up to this point, discussion has been confined to the structure of a pure material or homopolymer. A second monomer can be added to the starting feedstock such that the polymerization process produces a *copolymer*, designed to exhibit a certain set of properties. As shown in Fig. 6.11 polymer B can be added to the polymer A chain at random, in discrete blocks, or grafted to the side of a chain of polymer A. Two polymers also can be cross-linked within the presence of the other to form an inter-penetrating polymer network (IPN) with each polymer interlocking within the network of the other.[17,18] In this manner, the domain size of one polymer within the other is controlled by the distance between cross-links.

Often, commercial polymeric products contain a variety of additives that change the overall structure and associated properties. It is, therefore, appropriate to identify the major types of additives and their primary functions.

Pigments and Dyestuff. These materials are added to impart color to the polymer.

Stabilizers.   Stabilizers suppress molecular breakdown in the presence of heat, light, ozone, and oxygen. One form stabilizes the chain ends so the chains will not "unzip," thereby reversing the polymerization process. Other stabilizers act as anti-oxidants and antiozonants that are attached preferentially by O_2 and O_3 relative to the polymer chain.

Fillers.   Various ingredients are sometimes added to the polymer to enhance certain properties. For example, the additon of carbon black to automobile tires improves their strength and abrasion resistance. Fillers also serve to lower the volume cost of the polymer–filler aggregate, since the cost of the filler is almost always much lower than that of the polymer.

Plasticizers.   Plasticizers are high-boiling point, low-MW monomeric liquids that possess low volatility. They are added to a polymer to improve its processability and/or ductility. These changes arise for a number of reasons. Plasticizers add a low MW fraction to the melt, which broadens the MWD and shifts $\overline{M}$ to lower values. This enhances polymer processability. The liquid effectively shields chains from one another, thus decreasing their intermolecular attraction. Furthermore, by separating large chains, the liquids provide the chains with greater mobility for molecule segmental motion. The decrease in $\overline{M}$ and the lowering of intermolecular forces contribute toward improving polymer ductility and toughness. It should be recognized that these bene-

A A B A B B A A A B A B B A   (*a*)

A A A A B B B A A A B B B A A A A   (*b*)

A A A A A A A A A A A A   (*c*)
        B
        B
        B
        B
        B
        B

FIGURE 6.11   Conformations of a co-polymer: (*a*) random; (*b*) block; and (*c*) graft copolymer.

ficial changes occur while stiffness and maximum service temperature decrease. Consequently, the extent of polymer plasticization is determined by an optimization of processability, ductility, strength, and stiffness, and service temperature requirements. It is interesting to note that nylon 66 is inadvertently plasticized by the moisture it picks up from the atmosphere (Table 6.3).

Blowing Agents.   These substances are designed to decompose into gas bubbles within the polymer melt, producing stable holes. (In this way, expanded or foamed polymers are made.) The timing of this decomposition is critical. If the viscosity of the melt is too high, the bubbles will not form properly. If the viscosity is too low, the gas bubbles burst with the expanding polymer mass collapsing like an aborted soufflé.

Cross-linking Agents.   The basic difference between thermoplastic and thermosetting polymers lies in the nature of the intermolecular bond. In thermoplastic materials, these bonds are relatively weak and of the van der Waals type. Consequently, heating above T_g or T_m provides the thermal energy necessary for chains to move independently. This is not possible in thermosetting polymers, since the chains are rigidly joined with primary covalent bonds. Sulfur is a classic example of a cross-linking agent as used in the vulcanization of rubber.

From this very brief description of polymer additives, it is clear that a distinction should be made between a pure polymer and a polymer plus assorted additives; the latter is often referred to as a *plastic*, though rubbers also may be compounded. Although the terms *plastic* and *polymer* are often used synonymously in the literature, they truly represent basically different entities.

6.3 VISCOELASTIC RESPONSE OF POLYMERS AND THE ROLE OF STRUCTURE

The deformation response of many materials depends to varying degrees on both time-dependent and time-independent processes. For example, it was shown in Chapter 5 that when the test temperature is sufficiently high, a test bar would creep with time under a given load (Fig. 6.12a). Likewise, were the same bar to have been stretched to a certain length and then held firmly, the necessary stress to maintain the stretch would gradually relax (Fig. 6.12b). Such response is said to be *viscoelastic*. Since T_g and T_m of most polymeric materials are not much above ambient (and in fact may be lower as in the case of natural rubbers), these materials exhibit viscoelastic creep and relaxation phenomena at room temperature. When the elastic strains and viscous flow rate are small (approximately 1 to 2% and 0.1 s^{-1}, respectively), the viscoelastic strain may be approximated by

$$\epsilon = \sigma \cdot f(t) \tag{6-7}$$

That is, the stress–strain ratio is a function of time only. This response is called *linear* viscoelastic behavior and involves the simple addition of linear elastic and linear viscous (Newtonian) flow components. When the stress–strain ratio of a material varies with time and *stress*

$$\epsilon = g(\sigma, t) \tag{6-8}$$

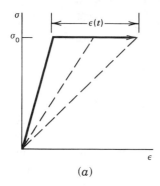
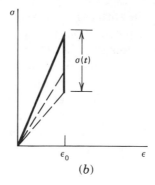

(a)   (b)

FIGURE 6.12   Time-dependent stress–strain response in polymers:
(a) creep and (b) relaxation behavior.

the viscoelastic response is nonlinear. A comparison of creep behavior between metals and polymers is summarized in Table 6.4.

On the basis of the simple creep test it is possible to define a creep modulus

$$E_c(t) = \frac{\sigma_0}{\epsilon(t)} \tag{6-9}$$

where $E_c(t)$ = creep modulus as a function of time

σ_0 = constant applied stress

$\epsilon(t)$ = time-dependent strain

Note different values of $E_c(t)$ in Fig. 6.12a. Likewise, in a stress relaxation test where the strain ϵ_0 is fixed and the associated stress is time dependent, a relaxation modulus $E_r(t)$ may be defined

$$E_r(t) = \frac{\sigma(t)}{\epsilon_0} \tag{6-10}$$

Figure 6.12b illustrates the change in E_r with time. These quantities can be plotted against log time to reveal their strong time dependence, as shown schematically in Fig. 6.13a for $E_r(t)$. (For small strains and up to moderate temperatures, corresponding to linear viscoelastic behavior, $E_r \approx E_c$.) It is clear that material behavior changes radically from one region to another. For very short times, the relaxation modulus approaches a maximum limiting value where the material exhibits glassy behavior

TABLE 6.4   Comparison of Creep Behavior in Metals and Polymers

Creep Behavior	Metals	Polymers
• Linear elastic	No	Sometimes
• Recoverable	No	Partially
• Temperature range	High temperatures above $\sim 0.2\, T_h$	All temperatures above $\sim -200°C$

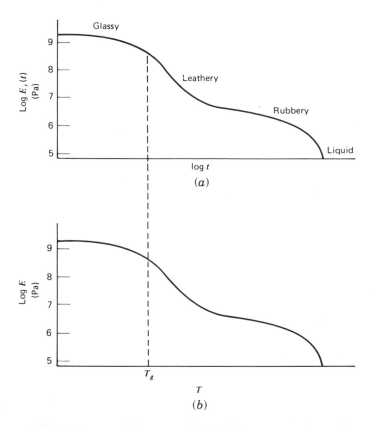

FIGURE 6.13 **Time–temperature dependence of elastic modulus in thermoplastic polymeric solids: (*a*) change in relaxation modulus $E_r(t)$ as function of time; (*b*) change in tensile modulus as function of temperature.**

associated with negligible molecule segmental motions. At longer times, the material experiences a transition to leathery behavior associated with the onset of short-range molecule segmental motions. At still longer times, complete molecule movements are experienced in the rubbery region associated with a further drop in the relaxation modulus. Beyond this point, liquid flow occurs. It is interesting to note that the same type of curve may be generated by plotting the modulus (from a simple tensile test) against test temperature (Fig. 6.13*b*). In this instance, the initial sharp decrease in E from its high value in the glassy state occurs at T_g. The shape of this curve can be modified by structural changes and polymer additions. For example, the entire curve is shifted downward and to the left as a result of plasticization (Fig. 6.14*a*). As $\overline{M}$ increases, the rubbery flow region is displaced to longer times (Fig. 6.14*b*), because molecular and segmental molecular movements are suppressed when chain entanglement is increased. Molecular weight has relatively little effect on the onset of the leathery region, since T_g is relatively independent of $\overline{M}$ except at low $\overline{M}$ values (Fig. 6.15). The effect of $\overline{M}$ on T_g is believed to be related to the chain ends.[1] Since the ends are freer to move about, they generate a greater than average amount of free volume. Adjacent chains are then freer to move about and contribute to greater

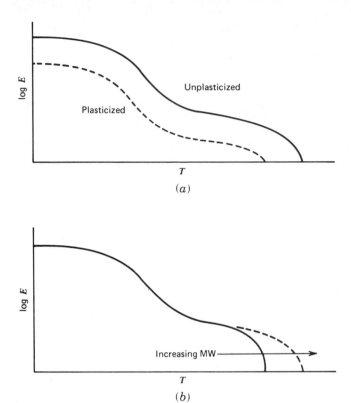

FIGURE 6.14   Effect of (*a*) plasticization and (*b*) molecular weight on elastic modulus as function of temperature.

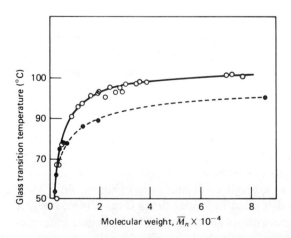

FIGURE 6.15   Glass transition temperature in PMMA (○) and polystyrene (●) as a function of $\overline{M}_n$.[19] (M. Miller, *The Structure of Polymers,* © 1966 by Litton Educational Publishing by permission of Van Nostrand Reinhold.)

mobility of the polymer. Since the chain ends are more sensitive to $\overline{M}_n$ than $\overline{M}_w$, T_g is best correlated with the former measure of molecular weight. The leathery region is greatly retarded by cross-linking, while the flow region is completely eliminated, the latter being characteristic of thermosetting polymers (Fig. 6.16).

The temperature–time (i.e., strain rate^{-1}) equivalence seen in Fig. 6.13 closely parallels similar observations made in Chapter 5. It is seen that the same modulus value can be obtained either at low temperatures and long times or at high test temperatures but short times. In fact, this equivalence is used to generate E_r versus log t curves as shown in Fig. 6.13a. The reader should appreciate that since such plots extend over 10 to 15 decades of time, they cannot be determined conveniently from direct laboratory measurements. Instead, relaxation data are obtained at different temperatures over a convenient time scale. Then, after choosing one temperature as the reference temperature, the remaining curves are shifted horizontally to longer or shorter times to generate a single master curve (Fig. 6.17). This approach was first introduced by Tobolsky and Andrews[22] and was further developed by Williams et al.[23] Assuming that the viscoelastic response of the material is to be controlled by a single function of temperature (i.e., a single rate-controlling mechanism), Williams et al.[23] developed a semiempirical relation for an amorphous material, giving the time shift factor a_T as

$$\log a_T = \log \frac{t_T}{t_{T_0}} = \frac{C_1(T - T_0)}{C_2 + T - T_0} \qquad (6\text{-}11)$$

where a_T = shift factor that is dependent on the difference between the reference and data temperatures $T - T_0$

t_T, t_{T_0} = time required to reach a specific E_r at temperatures T and T_0, respectively

C_1, C_2 = constants dependent on the choice of the reference temperature T_0

T = test temperatures where relaxation data were obtained, K

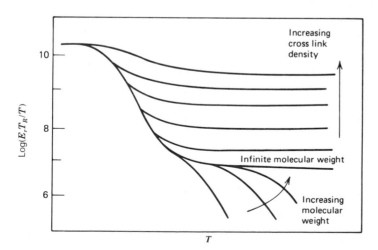

FIGURE 6.16 Effect of molecular weight and degree of cross-linking on relaxation modulus.[20] (Reprinted with permission from McGraw-Hill Book Company.)

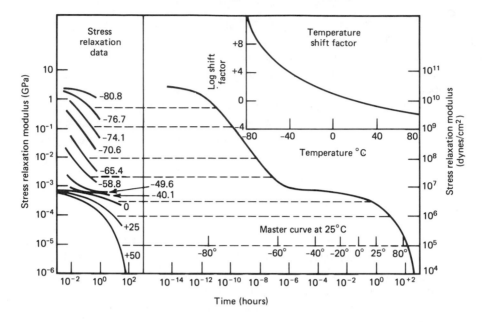

FIGURE 6.17 **Modulus–time master plot for polyisobutylene based on time–temperature superposition of data to a reference temperature of 25°C. (From Catsiff and Tobolsky,[21] with permission from John Wiley & Sons, Inc.)**

This relation is found to hold in the temperature range $T_g < T < T_g + 100$ K, but is sometimes used beyond these limits on an individual basis as long as time–temperature superposition still occurs. This would indicate that the same rate-controlling processes were still operative. Two reference temperatures are often used to normalize experimental data—T_g and $T_g + 50$ K—for which the constants C_1 and C_2 are given in Table 6.5.

The shift function may be used to normalize creep data,[4] enabling this information to be examined on a single master curve as well. Furthermore, by normalizing the creep strain results relative to the applied stress σ_0, the normalization of both axes converts individual creep–time plots into a master curve of creep compliance versus adjusted time (Fig. 6.18). These curves can be used to demonstrate the effect of MW and degree of cross-linking on polymer mechanical response much in the manner as the modulus relaxation results described in Figs. 6.14 and 6.16. Note that viscous flow is eliminated and the magnitude of the creep compliance reduced with increasing cross-linking in thermosetting polymers. For the thermoplastic materials, compliance decreases with increasing viscosity, usually the result of increased MW.

As previously noted (e.g., see Eqs. 6-9 and 6-10), the elastic modulus of engineering

TABLE 6.5   Constants for WLF Relationship

Reference Temperature	C_1	C_2
T_g	−17.44	51.6
$T_g + 50$ K	−8.86	101.6

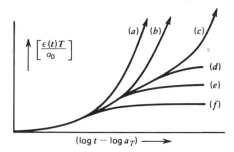

FIGURE 6.18 Master creep curve revealing effect of increasing MW ($a \rightarrow b \rightarrow c$) and degree of cross-linking ($d \rightarrow e \rightarrow f$) on creep strain. (T. Alfrey and E. F. Gurnee, *Organic Polymers,* © 1967. Reprinted by permission of Prentice-Hall Inc., Englewood Cliffs, NJ.)

plastics varies with time as a result of time-dependent deformation. For this reason, the designer of a plastic component must look beyond basic tensile data when computing the deformation response of a polymeric component. For example, if a designer were to limit component strain to less than some critical value ϵ_c, the maximum allowable stress would be given by $E\epsilon_c$ so long as the material behaved as an ideally elastic solid. Since most engineering plastics experience creep, the level of strain in the component would increase with time as noted in Fig. 6.19a. To account for this

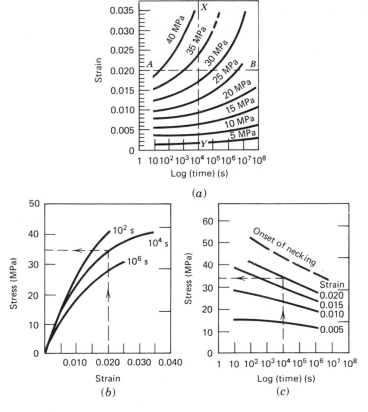

FIGURE 6.19 Creep response in PVC pipe resin at 20°C. (*a*) Creep curve; (*b*) isochronous stress–strain curves; (*c*) isometric stress–time curves.[2] (By permission of ICI Ltd.)

additional deformation, designers often make use of *isochronous* stress–strain curves which are derived from such creep data (e.g., see line *xy*). Figure 6.19*b* shows three isochronous stress–strain curves corresponding to loading times of 10^2, 10^4, and 10^6 s, respectively. To illustrate the use of these curves, we see that to limit the strain in a component to no more than 0.02 after 10^4 s, the allowable stress must not exceed 32 MPa.

The creep data shown in Fig. 6.19*a* can be analyzed in alternative fashion by considering the stress–time relation associated with various strain levels (e.g., line *AB*, Fig. 6.19*a*). The resulting *isometric* curves provide stress–time plots corresponding to different strain levels (Fig. 6.19*c*). For example, if a component were designed that would strain less than 0.02 after 10^4 s, the maximum permissible stress level would again be 32 MPa.

EXAMPLE 6.1

A PVC rod experiences a load of 500 N. An acceptable design calls for a maximum strain of 1% after one year of service. What is the minimum allowable rod diameter?

We will assume that the creep characteristics of the PVC pipe are identical to data shown in Fig. 6.19. Since one year is equal to 3.15×10^7 s, we see from Fig. 6.19*a* that an allowable strain of 1% would correspond to a stress of approximately 15 MPa. A similar result could have been identified with an isometric stress–time curve corresponding to 1% strain or with an isochronous stress–strain curve, corresponding to 3.15×10^7 s. The minimum rod diameter is then found to be

$$\sigma = \frac{P}{\frac{\pi}{4}d^2}$$

$$15 \times 10^6 = \frac{500}{\frac{\pi}{4}d^2}$$

$$\therefore d \sim 6.5 \text{ mm}$$

6.3.1   Mechanical Analogs

The linear viscoelastic response of polymeric solids has for many years been described by a number of mechanical models (Fig. 6.20). Many, including this author, have found that these models provide a useful physical picture of time-dependent deformation processes. The spring element (Fig. 6.20*a*) is intended to describe linear elastic behavior

$$\epsilon = \frac{\sigma}{E} \quad \text{and} \quad \gamma = \frac{\tau}{G} \qquad (1\text{-}7)$$

such that resulting strains are not a function of time. (The stress–strain–time diagram for the spring is shown in Fig. 6.21*a*.) Note the instantaneous strain upon application

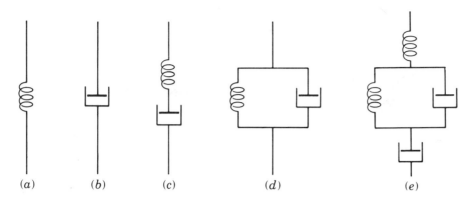

FIGURE 6.20 Mechanical analogs reflecting deformation processes in polymeric solids: (*a*) elastic; (*b*) pure viscous; (*c*) Maxwell model for viscoelastic flow; (*d*) Voigt model for viscoelastic flow; (*e*) four-element viscoelastic model.

of stress σ_0, no further extension with time, and full strain recovery when the stress is removed. The dashpot (a piston moving in a cylinder of viscous fluid) represents viscous flow (Fig. 6.20*b*).

$$\dot{\epsilon} = \frac{\sigma}{\eta} \quad \text{and} \quad \dot{\gamma} = \frac{\tau}{\eta} \tag{6-12}$$

where $\dot{\epsilon}, \dot{\gamma}$ = tensile and shear strain rates
   σ, τ = applied tensile and shear stresses
   η = fluid viscosity in units of stress–time

The viscosity η varies with temperature according to an Arrhenius-type relation

$$\eta = Ae^{\,\Delta H/RT} \tag{6-13}$$

where ΔH = viscous flow activation energy at a particular temperature
   T = absolute temperature

On the basis of time–temperature equivalence, η is seen, therefore, to depend strongly on time as well. For example, at $t = 0$ the viscosity will be extremely high, while at $t \rightarrow \infty$, η is small. The deformation response of a purely viscous element is shown in Fig. 6.21*b*. Upon loading ($t = 0$), the dashpot is infinitely rigid. Consequently, there is no instantaneous strain associated with σ_0 (the same holds when the stress is removed). With time, the viscous character of the dashpot element becomes evident as strains develop that are directly proportional to time. When the stress σ_0 is removed these strains remain. When the spring and dashpot are in series, as in Fig. 6.20*c* (called the Maxwell model), we are able to describe the mechanical response of a material possessing both elastic and viscous components. The stress–strain–time diagram for this model is shown in Fig. 6.21*c*. Note that all the elastic strains are recovered, but the viscous strains arising from creep of the dashpot remain. Since the elements are

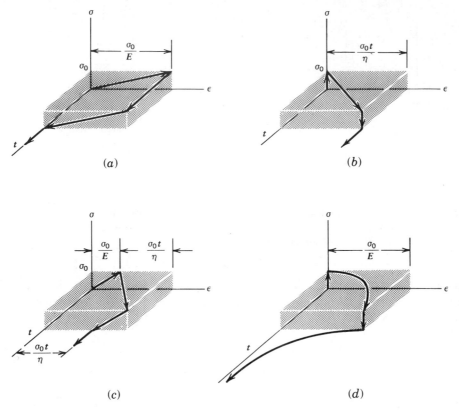

FIGURE 6.21 **Stress–strain–time diagrams for mechanical analogs; (a) simple spring; (b) simple dashpot; (c) Maxwell model; (d) Voigt model.**

in series, the stress on each is the same, and the total strain or strain rate is determined from the sum of the two components. Hence

$$\frac{d\epsilon}{dt} = \frac{\sigma}{\eta} + \frac{1}{E}\frac{d\sigma}{dt} \tag{6-14}$$

For stress relaxation conditions, $\epsilon = \epsilon_0$ and $d\epsilon/dt = 0$. Upon integration, Eq. 6-14 becomes

$$\sigma(t) = \sigma_0 e^{-Et/\eta} = \sigma_0 e^{-t/\mathcal{F}} \tag{6-15}$$

where $\mathcal{F} \equiv$ relaxation time defined by η/E. From Eq. 6-15, the extent of stress relaxation for a given material will depend on the relationship between $\mathcal{F}$ and t. When $t \gg \mathcal{F}$, there is time for viscous reactions to take place so that $\sigma(t)$ will drop rapidly. When $t \ll \mathcal{F}$, the material behaves elastically such that $\sigma(t) \approx \sigma_0$.

When the spring and dashpot elements are combined in parallel, as in Fig. 6.20d (the Voigt model), this unit predicts a different time-dependent deformation response. First, the strains in the two elements are equal, and the total stress on the pair is given by the sum of the two components

$$\epsilon_T = \epsilon_S = \epsilon_D \qquad (6\text{-}16)$$
$$\sigma_T = \sigma_S + \sigma_D$$

Therefore

$$\sigma_T(t) = E\epsilon + \eta \frac{d\epsilon}{dt} \qquad (6\text{-}17)$$

For a creep test, $\sigma_T(t) = \sigma_0$ and after integration

$$\epsilon(t) = \frac{\sigma_0}{E} \left(1 - e^{-t/F} \right) \qquad (6\text{-}18)$$

The strain experienced by the Voigt element is shown schematically in Fig. 6.21d. The absence of any instantaneous strain is predicted from Eq. 6-18 and is related in a physical sense to the infinite stiffness of the dashpot at $t = 0$. The creep strain is seen to rise quickly thereafter, but reach a limiting value σ_0/E associated with full extension of the spring under that stress. Upon unloading, the spring remains extended, but now exerts a negative stress on the dashpot. In this manner, the viscous strains are reversed, and in the limit when both spring and dashpot are unstressed, all the strains have been reversed. Consequently, the Maxwell and Voigt models describe different types of viscoelastic response. A somewhat more realistic description of polymer behavior is obtained with a four-element model consisting of Maxwell and Voigt models in series (Fig. 6.20e). By combining Eqs. 1-7, 6-12, and 6-18, it can be readily shown that the total strain experienced by this model may be given by

$$\epsilon(t) = \frac{\sigma}{E_1} + \frac{\sigma}{E_2} (1 - e^{-t/F}) + \frac{\sigma}{\eta_3} t \qquad (6\text{-}19)$$

which takes account of elastic, viscoelastic, and viscous strain components, respectively (Fig. 6.22). Even this model is overly simplistic with many additional elements often required to adequately represent mechanical behavior of a polymer. For example, such a model might include a series of Voigt elements, each describing the relaxation response of a different structural unit in the molecule.

Even so, the four-element model is useful in characterizing the response of different types of polymers. For example, a stiff and rigid material, such as a polyester thermoset resin, can be simulated by choosing stiff springs and high-viscosity dashpots. These elements would predict high stiffness and little time-dependent deformation, characteristic of a thermoset material. On the other hand, a soft and flexible material such as low-density polyethylene could be simulated by choosing low stiffness springs and dashpots with low viscosity levels. Accordingly, considerable time-dependent deformation would be predicted. Finally, the temperature dependence of the mechanical response of a polymer can be modeled by appropriate adjustment in dashpot and spring values (i.e., lower spring stiffness and dashpot viscosity levels for higher temperatures and vice versa for lower temperature conditions).

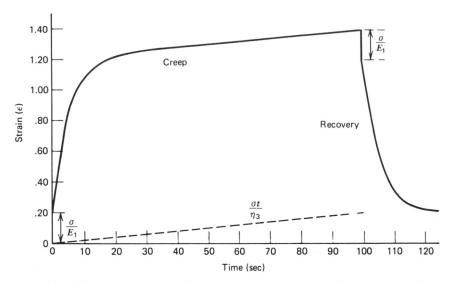

FIGURE 6.22 **Creep response of four-element model with $E_1 = 5 \times 10^2$ MPa, $E_2 = 10^2$ MPa, $\eta_2 = 5 \times 10^2$ MPa-sec, $\eta_3 = 50$ GPa-sec, and $\sigma = 100$ MPa.[6]** (L. Nielsen, *Mechanical Properties of Polymers,* © 1962 by Litton Educational Publishing Inc., reprinted by permission of Van Nostrand Reinhold.)

EXAMPLE 6.2

Let us examine the viscoelasticity of a soft and flexible material—cheese. This edible commodity is composed primarily of protein substances which are polymeric in nature. Sperling and coworkers* conducted experiments to examine the viscoelastic response of Velveeta® brand processed cheese. Such cheeses are plasticized or softened by the addition of water. A 15-cm-long block of this cheese, with cross-sectional dimensions of 4 cm × 6 cm, was supported in a slightly tilted holder and subjected to a compressive load of 4.9 N for approximately 2 h. The height of the cheese block was measured prior to loading and every 5 minutes thereafter. No additional displacement measurements were made after removal of the load. A duplicate experiment was conducted with a second cheese block under a compressive load of 6.85 N. The two creep curves from these experiments are illustrated below. With the exception of the unloading portion of the curve shown in Fig. 6.22, note the similarity in shape between the experimental Velveeta® creep curves and the computed curve for the stiffer polymer.

If we assume that the creep response of the Velveeta® cheese may be characterized by a four-element viscoelastic model (Fig. 6.20e), the strain–time plot is given by Eq. 6-19

$$\epsilon = \frac{\sigma}{E_1} + \frac{\sigma}{E_2}\left(1 - e^{-\left(\frac{E_2}{\eta_2}\right)t}\right) + \frac{\sigma}{\eta_3}(t)$$

* V. S. Chang, J. S. Guo, Y. P. Lee, and L. H. Sperling, *J. Chem. Ed.*, **63,** 1077 (1986).

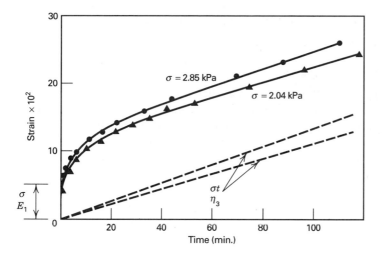

For the two experiments, the applied stress, σ, is equal to

$$\sigma = \frac{P}{A} = \frac{4.9}{(4 \times 10^{-2})(6 \times 10^{-2})} = 2.04 \text{ kPa}$$

Also

$$\sigma = \frac{P}{A} = \frac{6.85}{(4 \times 10^{-2})(6 \times 10^{-2})} = 2.85 \text{kPa}$$

As shown in Fig. 6.22, the elastic modulus, E_1, for the spring in series is determined by the strain at zero time (i.e., $E_1 = \sigma/\epsilon$). The viscosity, η_3, of the dashpot in series is determined from the slope of the linear portion of the creep curve at long times. Finally, the strain associated with the viscoelastic Voigt elements is obtained from the total strain less that associated with the spring and dashpot series elements. By simple curve fitting, the Voigt elements, E_2 and η_2, can then be determined. The constants for the four-element model are listed in the accompanying table. We see relatively good agreement between the two sets of values. As expected, the elastic and viscous elements for the processed cheese are much lower than those associated with the engineering polymer, described in Fig. 6.22.

Experimentally Determined Constants for Four-Element Viscoelastic Model of Velveeta® Cheese*

	4.8 Newtons	6.85 Newtons
E_1 (kPa)	4.88×10^4	5.18×10^4
E_2 (kPa)	2.82×10^4	4.24×10^4
η_2 (MPa-s)	1.52×10^7	1.78×10^7
η_2 (MPa-s)	1.00×10^8	1.21×10^8

* V. S. Chang, J. S. Guo, Y. P. Lee, and L. H. Sperling, *J. Chem. Ed.,* **63,** 1077 (1986).

6.3.2 Dynamic Mechanical Testing and Energy-Damping Spectra

Another method by which time-dependent moduli and energy-dissipative mechanisms are examined is through the use of dynamic test methods. These studies have proven to be extremely useful in identifying the major molecular relaxation at T_g as well as secondary relaxations below T_g. It is believed that such relaxations are associated with motions of specific structural units within the polymer molecule. Two basically different types of dynamic test equipment have been utilized by researchers. One type involves the free vibration of a sample, such as that which takes place in the torsion pendulum apparatus shown in Fig. 6.23. A specimen is rotated through a predeter-

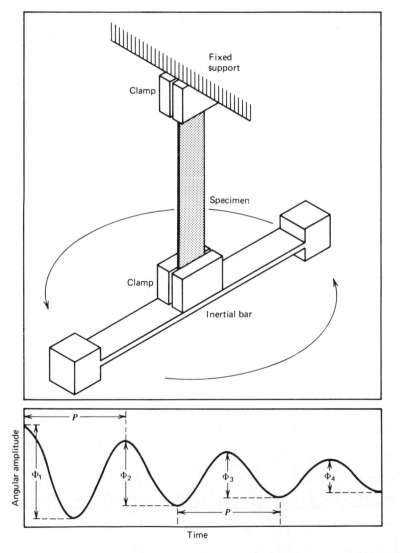

FIGURE 6.23 Simple torsion pendulum and amplitude–time curve for free decay of torsional oscillation.[6] (L. Nielsen, *Mechanical Properties of Polymers*, © 1962 by Litton Educational Publishing Inc., reprinted by permission of Van Nostrand Reinhold.)

mined angle and then released. This causes the sample to oscillate with decreasing amplitude resulting from various energy-dissipative mechanisms. The extent of mechanical damping is defined by the decrement in amplitude of successive oscillations as given by

$$\Delta = \ln \frac{A_1}{A_2} = \ln \frac{A_2}{A_3} = \ldots = \ln \frac{A_n}{A_{n+1}} \tag{6-20}$$

where Δ = log (base e) decrement which measures the amount of damping
A_1, A_2 = amplitude of successive oscillations of the freely vibrating sample

From these same observations, stiffness of the sample is determined from the period of oscillation P, the shear modulus G increasing with the inverse square of P.

The other type of dynamic instruments introduce to the sample a forced vibration at different set frequencies. The amount of damping is found by noting the extent to which the cyclic strain lags behind the applied stress wave. The relation between the instantaneous stress and strain values is shown in Fig. 6.24. Note that the strain vector ϵ_0 lags the stress vector σ_0 by the phase angle δ. It is instructive to resolve the stress vector into components both in phase and 90° out of phase with ϵ_0. These are given by

$$\sigma' = \sigma_0 \cos\delta \text{ (in-phase component)} \tag{6-21}$$
$$\sigma'' = \sigma_0 \sin\delta \text{ (out-of-phase component)}$$

The corresponding in-phase and out-of-phase moduli are determined directly from Eq. 6-21 when the two stress components are divided by ϵ_0. Hence

$$E' = \frac{\sigma'}{\epsilon_0} = \frac{\sigma_0}{\epsilon_0} \cos\delta = E^* \cos\delta$$
$$E'' = \frac{\sigma''}{\epsilon_0} = \frac{\sigma_0}{\epsilon_0} \sin\delta = E^* \sin\delta \tag{6-22}$$

where E^* = absolute modulus = $(E'^2 + E''^2)^{1/2}$. E' reflects the elastic response of the material, since the stress and strain components are in phase. This part of the

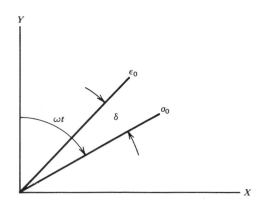

FIGURE 6.24   Forced vibration resulting in phase lag δ between applied stress σ_0 and corresponding strain ϵ_0.

strain energy, introduced to the system by the application of stress σ_0, is stored but then completely released when σ_0 is removed. Consequently, E' is often referred to as the *storage* modulus. E'', on the other hand, describes the strain energy that is completely dissipated (mostly in the form of heat) and for this reason is called the *loss* modulus. The relative amount of damping or energy loss in the material is given by the loss tangent, $\tan\delta$

$$\frac{E''}{E'} = \frac{E^*\sin\delta}{E^*\cos\delta} = \tan\delta \qquad (6\text{-}23)$$

By comparison,[6]

$$\frac{G''}{G'} \approx \frac{\Delta}{\pi} \qquad (6\text{-}24)$$

with the result that

$$\Delta \approx \pi\tan\delta \qquad (6\text{-}25)$$

When dynamic tests are conducted, the values of the storage and loss moduli and damping capacity are found to vary dramatically with temperature (Fig. 6.25). Note the correlation between the rapid drop in G', the rise in G'', and the corresponding damping maximum. The relaxation time associated with these changes (occurring in Fig. 6.13b in the vicinity of T_g) is considered to have an Arrhenius-type temperature dependence associated with a specific activation energy. In turn, the activation energy is then used to identify the molecular motion responsible for the change in dynamic behavior. Dynamic tests can be conducted either over a range of test temperatures at a constant frequency or at different frequencies for a constant temperature. Since the fixed frequency tests are usually more convenient to perform, most studies employ this procedure. Experiments of this type are now conducted routinely in many laboratories to characterize polymers with regard to effects of thermal history, degree of crystallinity, molecular orientation, polymer additions, molecular weight, plasticization, and other important variables. Consequently, the extant literature for such studies is enormous. Fortunately, a number of books and review articles have been prepared[13,24–28] on the subject to which the interested reader is referred. Within the scope of this book, we can only highlight some of the major findings.

When dynamic tests are performed over a sufficiently large temperature range, multiple secondary relaxation peaks are found in addition to the T_g peak shown in Fig. 6.25. Boyer[13] has summarized some of these data in the schematic form shown in Fig. 6.26. He noted that relaxation response in amorphous and semicrystalline polymers could be separated conveniently into four regions, as summarized in Table 6.6. Furthermore, crude temperature relations between various damping peaks were identified (e.g., $T_m \approx 1.5T_g$ and the $T < T_g$ transition (β) occurring at about $0.75T_g$). The dynamic mechanical spectra for a given material characterizes localized molecular movement such as small-scale segmental motions and side-chain group rotations. These transitions have been described by different activation energy levels that in-

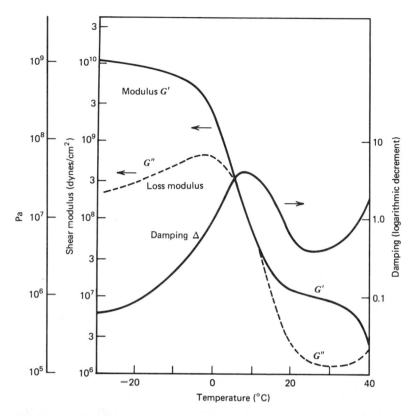

FIGURE 6.25 Dynamic mechanical response of un-cross-linked styrene and butadiene copolymer revealing temperature dependence of G', G", and Δ.[6] (L. Nielsen, *Mechanical Properties of Polymers,* © 1962 by Litton Educational Publishing Inc., reprinted by permission of Van Nostrand Reinhold.)

crease with increasing temperature of the transition and size of the side group responsible for the transition. Boyer[28] and Heijboer[24] have considered possible correlations between the size of the β peak and impact resistance (toughness).

6.4 DEFORMATION MECHANISMS IN CRYSTALLINE AND AMORPHOUS POLYMERS

We now consider the mechanisms by which crystalline and amorphous polymers deform. Unoriented crystalline materials are found to deform by a complex process involving initial breakdown and subsequent reorganization of crystalline regions.[29−31] After an initial stage of plastic deformation in the spherulites, the latter begin to break down. Lamellae packets oriented normal to the applied stress may separate along the amorphous boundary region between crystals, while others begin to rotate toward the stress axis (analogous to slip-plane rotation discussed in Section 3.1.2). The crystals themselves are now broken into smaller blocks, but the chains maintain their folded conformation. As this phase of the deformation process continues, these small bundles

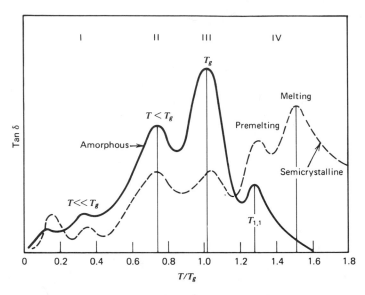

FIGURE 6.26 **Energy damping spectra for semicrystalline and amorphous polymers at various temperatures normalized to T_g. Several damping peaks are found for each material.[13] (By permission, from** *Polymeric Materials,* **copyright American Society for Metals, 1975.)**

become aligned in tandem along the drawing direction, forming long microfibrils (Fig. 6.27). Note that the extended chains within each bundle are positioned parallel to the draw axis along with a large number of fully extended tie molecules. Since many tandem blocks are torn from the same lamellae, they remain connected through a number of tie molecules created by unfolding chains from the original lamellae. The combination of many more fully extended tie molecules and the orientation of the bundles within each fibril contributes toward a rapid increase in strength and stiffness. By contrast, few primary bonds join blocks in adjacent microfibrils, except those representing tie molecules from the original lamellae (Fig. 6.27b). It is this initial spherulite structure breakdown, followed by microfibril formation, that gives rise to the substantial hardening associated with Type V stress–strain response discussed in Section 1.2.5. Continued deformation of the microfibrillar structure is extremely difficult because of the high strength of the individual microfibrils and the increasing extension of the interfibrillar tie molecules. These tie molecules become more ex-

TABLE 6.6 **Transition Regions in Polymers[13]**

Region	Temperature of Occurrence	Cause
I	$T \ll T_g$ (the γ peak)	Believed to be caused by movements of small groups involving only a few atoms
II	$T < T_g$ (the β peak)	Believed to be related to movement of 2–3 consecutive repeat units
III	T_g (the α peak)	Believed to be related to coordinated movements of 10–20 repeat units
IV	$T > T_g$	Large-scale molecular motions

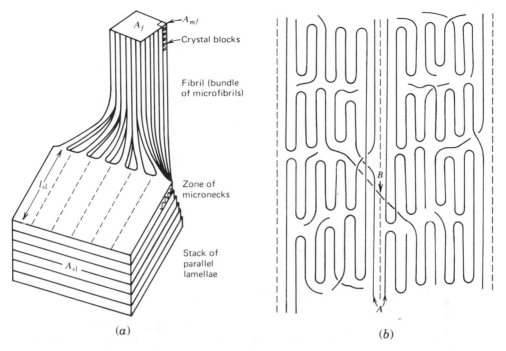

(a) (b)

FIGURE 6.27 (*a*) Model depicting transformation of a stack of parallel lamellae into a bundle of densely packed and aligned microfibrils.[30] Crystal blocks oriented as shown in *b*. (Reprinted from *Macromol. Chem.* **8**, 277 (1973), with permission.) (*b*) Alignment of crystal blocks in microfibrils. Intrafibrillar extended tie molecules shown at *A* with interfibrillar extended tie molecule at *B*.[31] (By permission from *Polymeric Materials*, copyright © American Society for Metals, Metals Park, OH, 1975.)

tended as a result of microfibril shear relative to one another. For additional comments regarding highly oriented microfibrils see Section 6.5.

Centuries ago, it was noted that pottery glazes tended to develop intricate and often beautiful arrays of fine cracks. The phenomenon was referred to as *crazing*. Analysis of crazes in glasses revealed that they usually developed on surfaces that were at right angles to the principal tensile axis. When the study and applications of polymers began to be pursued intensely, it was recognized that similarly oriented crazes in glassy polymers were not, in fact, simple cracks as in ceramics; rather, they consisted of expanded material containing oriented fibrils interspersed with small (100- to 200-Å) interconnected voids[25,32–41] (Fig. 6.28*a*). For several key reviews and current articles pertaining to craze formation and fracture, the reader is referred to references 42 to 52. The combination of fibrils (extending across the craze thickness) and interconnected microvoids contributes toward an overall weakening of the material, though *the craze is capable of supporting some reduced stress relative to that of the uncrazed matrix*. The latter point is proven conclusively by the load-bearing capability of samples containing crazes that extend completely across the sample.[25,42] Researchers have found that crazes tend to grow along planes normal to the principal tensile stress direction with little change in craze thickness being noted (Fig. 6.28*b*). The typical

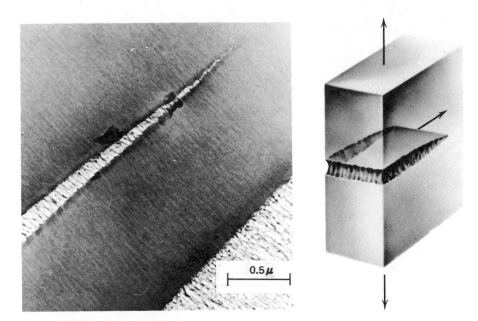

FIGURE 6.28 (*a*) **Crazes in polyphenylene oxide revealing interconnected microvoids and aligned fibrils.**[37] **(Reprinted with permission from the Polymer Chemistry Division of ACS.)** (*b*) **Sketch showing craze development normal to applied principal stress. Note slight surface dimpling along craze perimeter.**

craze thickness in glassy polymers is on the order of 5 μm or less, which corresponds in some cases to plastic strains in excess of 50%.[36] Since the lateral surface contraction associated with the craze (Fig. 6.28*b*) is negligible by comparison, it can only be concluded that the density of the craze should decrease. Indeed, Kambour[36] computed a craze density 40 to 60% of that associated with the uncrazed matrix. The presence of the interconnected void network within the craze is consistent with this finding. Since the refractive index of the craze is lower than that of the polymer, the craze is observed readily with the unaided eye, assuming of course that it is large enough to resolve.

Kausch[46] has pointed out that the initiation of a craze depends simultaneously on factors associated with the macroscopic state of stress and strain, the nature and distribution of heterogeneities within the solid, and the molecular behavior of the polymer for a given set of thermal and environmental conditions. As such, the polymer literature contains a number of models for craze formation. In one often cited model, Gent[53] proposed a theory in terms of stress-induced devitrification (change from a glassy to a weaker rubbery state) at a flaw, followed by cavitation resulting from the hydrostatic component of the stress. The following equation has been proposed for the critical stress required to develop (and permit growth of) a thin band of softened material (which then can undergo cavitation with deformation characteristic of crazing):[53]

$$\overline{\sigma}_c = [\beta(T_g - T) + P]/k \qquad (6\text{-}26)$$

where β is a coefficient relating the effect of an applied hydrostatic pressure P on the T_g, T is the ambient temperature, and k is the stress concentration factor. Thus, crazing should be favored by high temperatures and high stress concentrations, and restricted by applied hydrostatic pressure and the occurrence of creep or other types of flow that could decrease k. Regarding this point, Sternstein[54] found that under multiaxial stress conditions, glassy polymers could yield by two distinct mechanisms: normal yielding (crazing) and shear yielding (Fig. 6.29). For the biaxial stress case, shear yielding was found to depend on the mean normal stress with

$$\tau_{oct} = \tau_0 - \mu\sigma_m \qquad (6\text{-}27)$$

where τ_{oct} = octahedral yield stress
τ_0 = rate- and temperature-dependent octahedral yield stress for pure shear ($\sigma_m = 0$)
$\sigma_m = (\sigma_1 + \sigma_2)/2$
μ = material constant

Note that when $\sigma_m > 0$ (dilatation), τ_{oct} is reduced, with the result that shear yielding in pure shear is more difficult than in an ordinary uniaxial tensile test. Conversely, when $\sigma_m < 0$ (compressional), yielding is even more difficult than under pure shear conditions.

As mentioned previously, crazing develops on planes oriented normal to the principal tensile stress direction and is sensitive to the hydrostatic stress component as well. Sternstein combined these two observations and showed that

$$\sigma_b = |\sigma_1 - \sigma_2| = A(T) + \frac{B(T)}{I_1} \qquad (6\text{-}28)$$

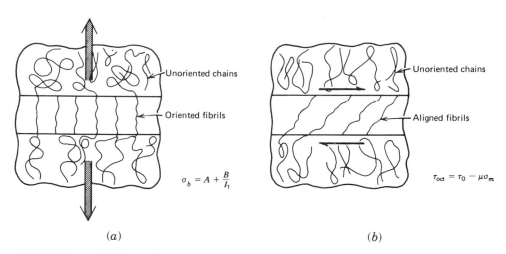

FIGURE 6.29 Deformation mechanisms in amorphous polymers: (*a*) normal yielding (crazing) and (*b*) shear yielding.

where σ_b = stress bias given by the difference in principal stresses σ_1 and σ_2

$A(T)$, $B(T)$ = temperature-dependent material constants

$I_1 = \sigma_1 + \sigma_2$, which must be positive to provide the necessary dilatation for craze formation

From the above, the stress bias necessary for crazing decreases with increasing dilatation I_1. Note that the deformation mechanism for uniaxial tension switches from crazing to shear yielding when hydrostatic pressure is raised.

The growth of a craze occurs by extension of the craze tip into uncrazed material. At the same time, the craze thickens by lengthening of the fibrils. Regarding the latter, Kramer and Lauterwasser[47] and Verhaulpen-Heymans[48] have shown that in monotonically loaded thin films, craze thickening occurs by drawing in new material from the craze flanks as opposed to molecular chain disentanglement of craze fibrils. The latter mechanism is believed to dominate craze thickening under cyclic loading conditions.[55]

Because of its intrinsic weakness, the craze is an ideal path for crack propagation. Craze breakdown mechanisms include:[46]

1. Chain slippage and disentanglement of molecular coils.

2. Rupture of primary bonds within the molecular entanglements.

3. Detachment at the craze–matrix interface.

A deformation map for PMMA is shown in Fig. 6.30 and characterizes the regions where particular deformation mechanisms are dominant.[56] Note the analogous form of this map to the ones described in Section 5.5. For a detailed discussion of the fracture surface micromorphology of crazes, see Section 7.7.4. While the presence of

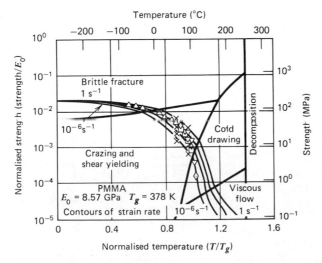

FIGURE 6.30 Deformation map for PMMA showing deformation regions as a function of normalized stress versus normalized temperature.[56] (Reprinted with permission from M. F. Ashby and D. R. H. Jones, *Engineering Materials 2*, Pergamon Press, 1986.)

the craze is considered undesirable from the standpoint of introducing a likely crack path, the localized process of fibril and void formation does absorb considerable strain energy. This is particularly true in the case of rubber-toughened polymers (see Section 10.4.4).

6.5 STRENGTHENING OF POLYMERS

A number of approaches may be taken to strengthen a polymer. Through changes in *chemistry* some polymers can be cross-linked to lock their molecules together in rigid fashion, thereby precluding viscous flow (Section 6.2). For example, a bowling ball contains a much higher cross-link density than a handball. Strengthening through changes in chemistry can also be brought about by the introduction of large side groups and intrachain groups which restrict C–C bond rotation (Section 6.1).

The superstructure or *architecture* of a polymer can be modified to effect dramatic changes in mechanical strength greater than those made possible by chemistry adjustments. First, mechanical properties are found to increase with molecular weight, and relations[57] often assume the forms

$$\text{mechanical property} = A - \frac{B}{\overline{M}} \qquad (6\text{-}29a)$$

or

$$\text{mechanical property} = C + \frac{D}{\overline{M}} \qquad (6\text{-}29b)$$

An example of such data[58] is shown in Fig. 6.31. It may be argued that as chain length increases beyond a critical length, the combined resistance to flow from chain entanglement and intermolecular attractions exceeds the strength of primary bonds, which can then be broken. Consequently, once the molecular weight exceeds a critical lower limit $\overline{M}_c$, entanglement and primary bond breakage occur. At this point the mechanical property becomes less sensitive to MW. On the basis of available data it is not clear whether mechanical properties correlate best with $\overline{M}_w$ or $\overline{M}_n$.[59] Studies have shown that polymer viscosity depends on MW. When MW $< M_c$, the viscosity η is proportional to MW. When MW $> M_c$, $\eta \propto \text{MW}^{3.5}$ (Fig. 6.32).[60]

Mechanical properties are improved most dramatically by molecular and molecular segment alignment parallel to the stress direction. This stands to reason, since the loads would then be borne by primary covalent bonds along the molecule rather than by weak van der Waals forces between molecules. Figure 6.33 illustrates the rapid increase in polymer stiffness with increasing fraction of covalent bonds aligned in the loading direction.[56] A similar curve would describe the strength of the polymer. Such orientation hardening is distinct from the strain-hardening phenomenon found in metals. In the latter case, strength is lost when the sample is annealed due to the annihilation of dislocations. The strength of oriented polymers is high even after annealing (below T_g) since the polymer chain orientation is retained. The alignment

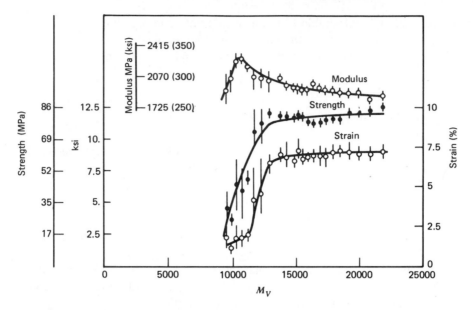

FIGURE 6.31   Mechanical properties in polycarbonate as a function of molecular weight.[58] (Reprinted with permission from John Wiley & Sons, Inc.)

of molecules in an amorphous polymer is described with the aid of Fig. 6.34. Thermal energy causes the molecules in the polymer at $T > T_g$ to vibrate with relative ease in random fashion, but below T_g the randomness is "frozen in." If the material were drawn quickly at a temperature not too far above T_g (say T_1), some chain alignment could be achieved and effectively "frozen in," provided the stretched polymer were to be quenched from that temperature. The resulting material would be stronger in the direction of drawing and correspondingly weakened in the lateral direction. An example of this anisotropy is shown in Fig. 6.35 for drawn polystyrene. Note that the strength anisotropy parallels a deformation mechanism transition from crazing to shear

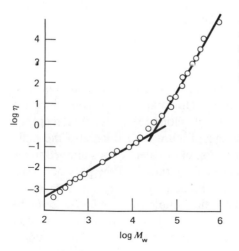

FIGURE 6.32   Dependence of melt viscosity on molecular weight in polydimethylsiloxane at 20°C.[60] (With permission, N. J. Mills, *Plastics: Microstructure, Properties and Applications*, Edward Arnold, London, 1986.)

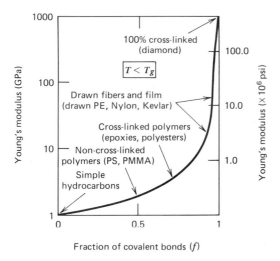

FIGURE 6.33 **Polymer stiffness dependence on fraction of covalent bonds in the load-ing direction.**[56] **(Reprinted with permission from M. F. Ashby and D. R. H. Jones,** *Engi-neering Materials 2,* **Pergamon Press, 1986.)**

yielding as the tensile axis approaches the draw direction. If the drawing were con-ducted slowly at T_1 or even at T_2, the elongation could be accommodated by viscous flow without producing chain alignment. Consequently, no strengthening would result.

It should be recognized that the oriented structure is unstable and will contract upon subsequent heating above T_g. By contrast, the polymer stretched at T_2 would be dimensionally stable, since it never departed from its preferred fully random state.

In semicrystalline polymers, crystallite alignment may be produced by cold drawing spherulitic material (Section 6.4) and by forcing or drawing liquid through a narrow orifice. During the past few years, attempts have been made to extend the practice of polymer chain orientation to its logical limit—the full extension of the molecule chain—with the potential of producing a very strong and stiff fiber.[61-63] Indeed, this has been partially accomplished. Highly oriented and extended commercial fibers, such as DuPont Kevlar, possess a tensile modulus two-thirds that of steel but with a much lower density. This is truly extraordinary, since commercial plastics generally

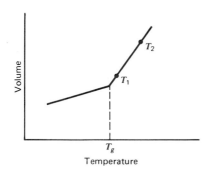

FIGURE 6.34 **Rapid drawing at T_1 followed by quenching can produce molecular alignment and polymer strengthening. Viscous flow during draw-ing at T_2 precludes such alignment.**

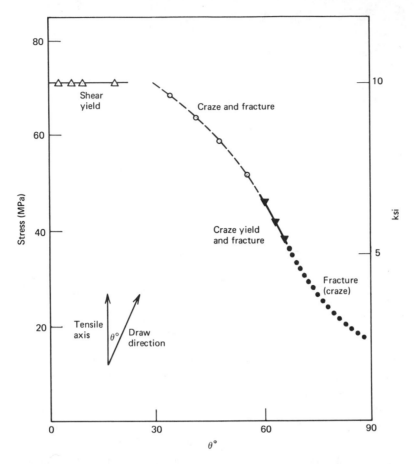

FIGURE 6.35 **Effect of orientation on tensile properties of polystyrene tested at 20°C, hot drawn to a draw ratio of 2.6. Note deformation mechanism transition with test direction.[41] (By permission, from *Polymeric Materials*, copyright American Society for Metals, 1975.)**

exhibit elastic moduli fully two orders of magnitude smaller than steel. By converting the folded chain conformation to a fully extended one, the applied stresses are sustained by the very strong main chain covalent bonds, which are less compliant than the much weaker intermolecular van der Waals forces. Highly oriented fibers have been produced both by cold forming and direct spinning from the melt.

From this brief discussion, it is clear that the elastic moduli of polymeric solids are very structure sensitive, unlike those found in metals and ceramics (see Chapter 1). Specifically, it has been found[62] that the modulus of a polymer increases with (1) thermodynamic stability of the main chain bonds, (2) percentage of crystallinity, (3) packing density of the chains, (4) chain orientation in the tensile direction, (5) chain end accommodation in the polymer crystal, and (6) a minimization of chain folding. There is great potential commercial use for these fibers, such as in tire cord and engineering composites, including sporting equipment.

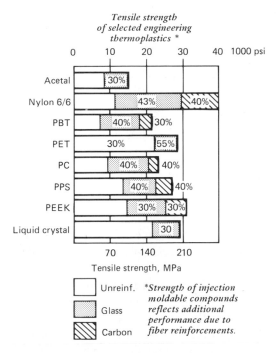

FIGURE 6.36 Tensile strength of selected engineering thermoplastics and their respective composites.[64] (Reprinted with permission from *Guide to Engineering Materials*, 2nd Ed.; ASM International, Metals Park, OH, 1987.)

6.5.1 Polymer Composites

As discussed in previous sections, thermoplastic resins exhibit creep or relaxation and loss of strength under long-term loading conditions. The addition of fibers to polymeric matrices (e.g., epoxy, polyester, and nylon 66) enhances polymer strength, stiffness, dimensional stability, and elevated temperature resistance at the expense of ductility. The change in tensile strength of selected polymers with the addition of glass and carbon fibers is shown in Fig. 6.36. Note the superior strengthening potential of graphite fibers relative to that of glass. Since the density of nonmetallic fibers is relatively low (see Table 6.7), the specific strength and stiffness of polymeric composites (σ_{TS}/ρ and E/ρ, where ρ = density) exceeds that of conventional metal alloys. For this reason, high-performance polymer composites are finding increasing use in

TABLE 6.7 Density of Selected Fibers and Matrices[56]

Fibers	Density (g/cm^3)	Matrices	Density (g/cm^3)
Carbon Type 1	1.95	Epoxies	1.2–1.4
Carbon Type 2	1.75	Polyesters	1.1–1.4
E glass	2.56	Nylon	1.1–1.2
Kevlar	1.45	Concrete/cement	2.4–2.5
SiC	2.5–3.2	Aluminum alloys	2.6–2.9
Al$_2$O$_3$	3.9	Steel alloys	7.8–8.1

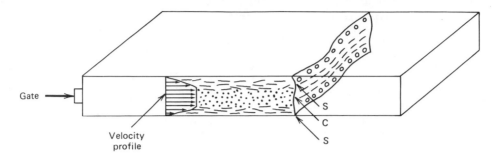

Gate

Velocity
profile

S

C

S

FIGURE 6.37 **Through-thickness fiber orientation in injection molded part.**

aircraft applications. (Manufacturers usually report fiber content in a composite by either weight or volume fraction. For the case of glass-reinforced plastics, the density of the glass is roughly twice that of the polymer matrix; consequently, a 30 w/o glass content corresponds to approximately 15 v/o.)

Polymer composites, containing glass and graphite fibers, have found extensive use in sporting equipment, such as vaulting poles, baseball bats, tennis rackets, skis and ski boots, fishing rods, golf clubs, and bicycle frames. These materials have reduced the weight of these items and otherwise optimized mechanical properties to enable athletes to achieve new world records; the dramatic increase in heights achieved during pole-vaulting competitions symbolizes progress in this regard.

The properties of fiber-reinforced plastics depend in a complex manner on processing history. For example, during injection molding, the chopped fibers fracture, with the result that many glass and carbon fibers have lengths in the range of 100 to 500 μm. In addition, fiber orientation differs throughout the thickness of the injection-molded part. This follows from the flow characteristics associated with injection molding.[65,66] Near the mold wall (S), the fibers tend to be aligned in the molding

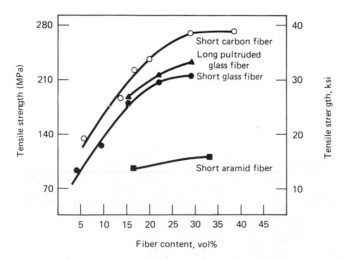

FIGURE 6.38 **Strength in nylon 66 composites as a function of fiber content, type, and length.**[67] **(Reprinted with permission from** *Advanced Materials and Processes,* **131(2), 57 (1987), ASM International, Metals Park, OH, 1987.)**

direction. In the interior core region (C), however, the fibers tend to be aligned parallel to the advancing liquid front (Fig. 6.37); as such, these fibers are nominally normal to the flow direction. Accordingly, the properties of an injection-molded component depend both on the relative thickness of the surface and core layers and the direction of loading. Likewise, extruded composites contain oriented fibers that were broken during the extrusion process into lengths smaller than their initial size. Note that fiber breakage during injection molding, extrusion, or any other process reduces the strength and stiffness of the composite below the theoretical potential for the composite (recall Eq. 1-44). Figure 6.38 shows the difference in strength of nylon 66 composites as a function of fiber length.

REFERENCES

1. R. D. Deanin, *Polymer Structure, Properties and Applications*, Cahners, Boston, MA, 1972.
2. R. M. Ogorkiewicz, Ed., *Engineering Properties of Thermoplastics*, Wiley-Interscience, London, 1970.
3. M. Kaufman, *Giant Molecules*, Doubleday, Garden City, NY, 1968.
4. T. Alfrey and E. F. Gurnee, *Organic Polymers*, Prentice-Hall, Englewood Cliffs, NJ, 1967.
5. S. L. Rosen, *Fundamental Principles of Polymeric Materials for Practicing Engineers,* Barnes & Noble, New York, 1971.
6. L. E. Nielsen, *Mechanical Properties of Polymers*, Reinhold, New York, 1962.
7. H. V. Boening, *Polyolefins: Structure and Properties,* Elsevier, Lausanne, 1966, p. 57.
8. M. H. Litt and A. V. Tobolsky, *J. Macromol. Sci. Phys. B* **1**(3), 433 (1967).
9. S. E. B. Petrie, *Polymeric Materials*, ASM, Metals Park, OH, 1975, p. 55.
10. P. H. Geil, *Chem. Eng. News* **43**(33), 72 (Aug. 16, 1965).
11. P. H. Geil, *Polymer Single Crystals,* Interscience, New York, 1963.
12. E. S. Clark, *Polymeric Materials*, ASM, Metals Park, OH, 1975, p. 1.
13. R. F. Boyer, *Polymeric Materials*, ASM, Metals Park, OH, 1975, p. 277.
14. P. H. Geil, *Polymeric Materials,* ASM, Metals Park, OH, 1975, p. 119.
15. G. S. Y. Yeh, *Crit. Rev. Macromol. Sci.* **1,** 173 (1972).
16. G. S. Y. Yeh and P. H. Geil, *J. Macromol. Sci. B* **1,** 235 (1967).
17. L. H. Sperling, *Interpenetrating Polymer Networks and Related Materials*, Plenum, New York, 1981.
18. L. H. Sperling, *Multicomponent Polymer Materials*, D. R. Paul and L. H. Sperling, Eds., *Adv. Chem. Ser. 211*, ACS, Washington, DC, 1986, p. 21.
19. M. L. Miller, *The Structure of Polymers*, Reinhold, New York, 1966.
20. F. Rodriguez, *Principles of Polymer Systems,* McGraw-Hill, New York, 1970.
21. E. Catsiff and A. V. Tobolsky, *J. Polym. Sci.* **19,** 111 (1956).
22. A. V. Tobolsky and R. D. Andrews, *J. Chem. Phys.* **13,** 3 (1945).
23. M. L. Williams, R. F. Landel, and J. D. Ferry, *J. Appl. Phys.* **26,** 359 (1955).
24. J. Heijboer, *J. Polym. Sci.* **16,** 3755 (1968).
25. R. P. Kambour and R. E. Robertson, *Polymer Science: A Materials Science Handbook*, Vol. 1, A. D. Jenkins, Ed., North Holland, 1972, p. 687.
26. N. G. McCrum, B. E. Read, and G. Williams, *Anelastic and Dielectric Effects in Polymeric Solids*, Wiley, London, 1967.

27. R. F. Boyer, *Rubber Chem. Tech.* **36,** 1303 (1963).

28. R. F. Boyer, *Poly. Eng. Sci.* **8**(3), 161 (1968).

29. A. Peterlin, *Advances in Polymer Science and Engineering*, K. D. Pae, D. R. Morrow, and Y. Chen, Eds., Plenum, New York, 1972, p. 1.

30. A. Peterlin, *Macromol. Chem.* **8,** 277 (1973).

31. A. Peterlin, *Polymeric Materials*, ASM, Metals Park, OH, 1975, p. 175.

32. B. Maxwell and L. F. Rahm, *Ind. End. Chem.* **41,** 1988 (1949).

33. J. A. Sauer, J. Marin, and C. C. Hsiao, *J. Appl. Phys.* **20,** 507 (1949).

34. C. C. Hsiao and J. A. Sauer, *J. Appl. Phys.* **21,** 1071 (1950).

35. J. A. Sauer and C. C. Hsiao, *Trans. ASME* **75,** 895 (1953).

36. R. P. Kambour, *Polymer* **5,** 143 (1964).

37. R. P. Kambour and A. S. Holick, *Polym. Prepr.* **10,** 1182 (1969).

38. R. P. Kambour, *J. Polym. Sci. Part A-2* **4,** 17 (1966).

39. J. P. Berry, *J. Polym. Sci.* **50,** 107 (1961).

40. J. P. Berry, *Fracture Processes in Polymeric Solids*, B. Rosen, Ed., Interscience, 1964, p. 157.

41. D. Hull, *Polymeric Materials*, ASM, Metals Park, OH, 1975, p. 487.

42. S. Rabinowitz and P. Beardmore, *CRC Crit. Rev. Macromol. Sci.* **1,** 1 (1972).

43. E. H. Andrews, *The Physics of Glassy Polymers*, R. N. Haward, Ed., Wiley, New York, 1973, p. 394.

44. R. P. Kambour, *J. Polym. Sci. Macromol. Rev.* **7,** 1 (1973).

45. E. J. Kramer, *Developments in Polymer Fracture*, E. H. Andrews, Ed., Applied Science Publ., London, 1979.

46. H. H. Kausch, *Polymer Fracture*, Springer-Verlag, Berlin, 1978.

47. E. J. Kramer and B. D. Lauterwasser, *Deformation Yield and Fracture of Polymers*, Plastics and Rubber Institute, London, 1979, p. 34-1.

48. N. Verheulpen-Heymans, *Polymer* **20,** 356 (1979).

49. A. S. Argon and J. G. Hannoosh, *Philos. Mag.* **36**(5), 1195 (1977).

50. J. A. Sauer and C. C. Chen, *Adv. Polym. Sci.* **52/53,** 169 (1983).

51. A. S. Argon, R. E. Cohen, O. S. Gebizlioglu, and C. E. Schwier, *Adv. Polym. Sci.* **52/53,** 276 (1983).

52. W. Döll, *Adv. Polym. Sci.* **52/53,** 105 (1983).

53. A. Gent, *J. Mater. Sci.* **5,** 925 (1970).

54. S. S. Sternstein, *Polymeric Materials*, ASM, Metals Park, OH, 1975, p. 369.

55. R. W. Hertzberg and J. A. Manson, *Fatigue of Engineering Plastics*, Academic, New York, 1980.

56. M. F. Ashby and D. R. H. Jones, *Engineering Materials 2*, Pergamon, Oxford, 1986.

57. P. J. Flory, *J. Am. Chem. Soc.* **67,** 2048 (1945).

58. J. H. Golden, B. L. Hammant, and E. A. Hazell, *J. Polym. Sci.* **2A,** 4787 (1974).

59. J. R. Martin, J. F. Johnson, and A. R. Cooper, *J. Macromol, Sci.—Rev. Macromol. Chem.* **C8**(1), 57 (1972).

60. N. J. Mills, *Plastics: Microstructure, Properties and Applications*, Edward Arnold, London, 1986.

61. F. C. Frank, *Proc. R. Soc. London, Ser. A* **319,** 127 (1970).

62. R. S. Porter, J. H. Southern, and N. Weeks, *Polym. Eng. Sci.* **15,** 213 (1975).

63. J. Preston, *Polym. Eng. Sci.* **15,** 199 (1975).

64. *Guide to Engineering Materials*, Vol. 2(1), ASM, Metals Park, OH, 1987.

65. Z. Tadmor, *J. Appl. Polym. Sci.* **18,** 1753 (1974).

66. S. S. Katti and J. M. Schultz, *Polym. Eng. Sci.* **22**(16), 1001 (1982).

67. *Advance Materials and Processing,* Vol. 131(2), ASM, Metals Park, OH, 1987, p. 59.

68. R. J. Crawford, *Plastics Engineering*, Pergamon, Oxford, 1981.

FURTHER READING

M. F. Ashby and D. R. H. Jones, *Engineering Materials 2*, Pergamon, Oxford, 1986. *Encyclopedia of Polymer Science and Engineering,* Wiley, New York, 1986. *Engineering Design with Plastics; Principles and Practice.* Plastice and Rubber Institute, London, 1981, p. 1982.

R. J. Crawford, *Plastics Engineering*, Pergamon, Oxford, 1981.

A. J. Kinloch and R. J. Young, *Fracture Behavior of Polymers,* Applied Science, New York, 1983.

N. G. McCrum, C. P. Buckley, and C. B. Bucknall, *Principles of Polymer Engineering*, Oxford Science Publishing, Oxford, 1988.

N. J. Mills, *Plastics Microstructure, Properties, and Applications*, Edward Arnold, London, 1986.

Modern Plastics Encyclopedia, McGraw-Hill, New York, 1986/87.

Structural Plastics Design Manual, FHWA-TS-79-203, U.S. Dept. of Transportation, 1979.

PROBLEMS

6.1 It is found that the viscosity of a liquid above T_g can be expressed by the relation

$$\log\frac{\eta_T}{\eta_{T_g}} = \frac{-17.44(T - T_g)}{51.6 + T - T_g}$$

which is similar to Eq. 6-11. If the viscosity of a material at T_g is found to be 10^4 GPa-sec, compute the viscosity of the liquid 5, 10, 50, and 100 K above T_g.

6.2 If it takes 300 s for the relaxation modulus to decay to a particular value at T_g, to what temperature must the material have to be raised to effect the same decay in 10 s?

6.3 Calculate the relaxation time for glass and comment on its propensity for stress relaxation at room temperature. $E \approx 70$ GPa and $\eta \approx 1 \times 10^{12}$ GPa-s (10^{22} poise).

6.4 For the four-element model shown in Fig. 6.20, compute the strain–time plot over the same time span when (a) all constants are the same except $\eta_3 = 10$ GPa-s, and (b) all constants are the same except $E_1 = 1$ GPa and $\eta_2 = 10$ GPa-s. Compare the three plots.

6.5 The deformation response of a certain polymer can be described by the Voigt model. If $E = 400$ MPa and $\eta = 2 \times 10^{12}$ MPa-s, compute the relaxation time. Compute $\epsilon(t)$ for times to $5\mathcal{T}$ when the steady stress is 10 MPa. How much creep strain takes place when $t = \mathcal{T}$ and when $t = \infty$?

6.6 For the material described in the previous problem, compare the fractional amount of total deformation that would occur if $t = \mathcal{T}$ when $\eta = 2 \times 10^{12}$ and 8×10^{12} MPa-s, respectively.

6.7 The molecular weight distribution was determined for a polyethylene sample with the following number fractions corresponding to particular molecular weight ranges found to be:

N_i	0.26	0.31	0.21	0.13	0.07	0.015	0.001
M_i	10^3	3×10^3	10^4	3×10^4	10^5	3×10^5	10^6

Compute M_N, M_W, and MWD as defined by M_W/M_N.

6.8 A 10-cm-long rod of PVC is used to connect two stiff plates that are a fixed distance apart. The fastening process causes the rod to extend by 0.15 cm. Given that the elastic modulus of PVC is 3 GPa and assuming that the material's creep behavior is similar to that characterized by the data shown in Fig. 6.19, estimate the initial stress on the rod and stress after 30 days.

6.9 A polypropylene pipe is to withstand an internal pressure of 0.5 MPa for a minimum of three years. If the pipe diameter is 100 mm, what is the minimum necessary wall thickness to ensure that the pipe will not experience a strain greater than 1.3%? To solve this problem, use the accompanying graph that reveals the room temperature creep response for polypropylene.[68] R. J. Craw-

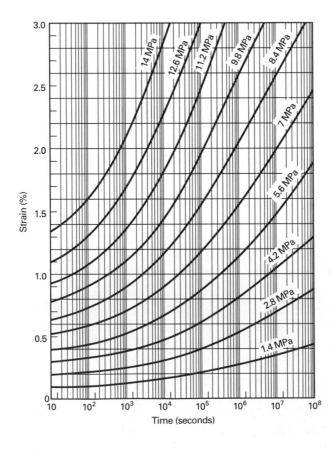

ford, *Plastics Engineering*, SPE, Brookfield Center, CT (1981). Reprinted by permission.

6.10 A 10-cm-long cylindrical rod of polypropylene is subjected to a tensile load of 550 N. If the maximum allowable strain of 2% is experienced no earlier than after four months of service, what is the minimum required rod diameter? Also, what is the rod diameter after the four-month service period?

6.11 A 200-mm-long polypropylene rod, with a rectangular cross-section that is 20 mm by 4 mm, is subjected to a tensile load of 300 N, directed along its length. If the rod extends by 0.5 mm after being under load for 100 s, determine the creep modulus.

6.12 The creep response of a polymer, corresponding to a load application for 100 s, may be described by either a Voigt or Maxwell model. The elastic and viscous elements of the Voigt model have properties of 2 GPa and 100×10^9 Ns/m^2, respectively. Given that the viscous element in the Maxwell model has a viscosity of 200×10^9 Ns/m^2, estimate the elastic constant of the spring in the Maxwell model if the two models predict the same level of creep strain.

SECTION TWO

FRACTURE MECHANICS OF ENGINEERING MATERIALS

CHAPTER 7

FRACTURE: AN OVERVIEW

7.1 INTRODUCTION

On January 15, 1919, something frightening happened on Commercial Street in Boston. A huge tank, 27 meters in diameter and about 15 meters high, fractured catastrophically, and over 7.5×10^6 liters (2×10^6 gallons) of molasses cascaded into the streets.

> *Without an instant's warning the top was blown into the air and the sides were burst apart. A city building nearby, where the employees were at lunch, collapsed burying a number of victims and a firehouse was crushed in by a section of the tank, killing and injuring a number of firemen.*[1]
>
> *On collapsing, a side of the tank was carried against one of the columns supporting the elevated structure [of the Boston Elevated Railway Co.]. This column was completely sheared off . . . and forced back under the structure . . . the track was pushed out of alignment and the superstructure dropped several feet. . . . Twelve persons lost their lives either by drowning in molasses, smothering, or by wreckage. Forty more were injured. Many horses belonging to the paving department were drowned, and others had to be shot.*[2]

The molasses tank failure dramatically highlights the necessity of understanding events that contribute to premature fracture of any engineering component. Other manufactured structures are susceptible to the same fate. For example, several bridges have fractured and collapsed in Belgium, Canada, Australia, Netherlands Antilles, and the United States during the past 50 years, resulting in the loss of many lives. Two such case histories are described in Section 14.5. In addition, numerous cargo ship failures have occurred, dating from World War II to the present (Fig. 7.1a). Subsequent studies concluded that these failures, which broke the vessels in two, were primarily attributable to the presence of stress concentrations in the ship superstructure and the ability of cracks to traverse welds that joined adjacent steel plates; the existence of faulty weldments and inferior steel quality were also cited as contributing factors in the fracture process.[3] A number of more recent oil cargo ship failures have resulted in extensive pollution of rich fishing grounds and coastal resort beach areas.

It is intriguing to note the similar fracture path of the cargo tanker (Fig. 7.1a) with that of the passenger liner *Titanic* which struck a large iceberg during its maiden

(a)

(b)

FIGURE 7.1 (a) Fractured T-2 tanker, the S. S. *Schenectady*, which failed in 1941.[3] (Reprinted with permission of Earl R. Parker, *Brittle Behavior of Engineering Structures,* National Academy of Sciences, National Research Council, Wiley, New York, 1957.); (b) bow portion of the Titanic. (Painting by Ken Marschall, based on photographs taken aboard Alvin, Angus, and Argo research vessels. (Courtesy Dr. Robert D. Ballard, *The Discovery of the Titanic.*)

voyage in 1912 and sank, causing the death of 1500 passengers and crew members (Fig. 7.1b). The remains of this vessel were first discovered in 1985 by Dr. Robert Ballard and co-workers aboard the Alvin, Angus, and Argo minisubmarine research vessels at a depth of 3.6-km beneath the surface of the Atlantic Ocean.* Gardze et al[4] speculated that the *Titanic*'s sinking was caused by a brittle fracture of the steel superstructure as a result of the ship having struck an iceberg in the North Atlantic Ocean. Indeed, Gannon[5] reported that a Charpy fracture test (see Chapter 9), conducted on a retrieved fragment of the ship's hull at $-1°C$, approximating the ambient water temperature at the time of the catastrophic event, confirmed that the hull was fabricated from brittle steel. This brittle condition was attributable to a high sulfur content (e.g., see Fig. 10.15) and/or to a high ductile–brittle transition temperature (see Chapter 9). Furthermore, the edges of the retrieved fragment appeared ". . . jagged, almost shattered. And the metal itself showed no evidence of bending."[5]

Additional support for the brittle metal fracture theory to account for the rapid sinking of the *Titanic* is based on Marschall's examination of a series of photographs of ship hull fragments taken from around the wreck by the Ballard research team. "The pieces look like cracked sections of an eggshell and the fractures seem to have paid no heed to rivets or plate boundaries."** The final separation of the bow and stern sections of the vessel is believed to have occurred in the following manner: When the bow section of the vessel struck the iceberg, it took on water and submerged, thereby lifting the stern above the water line. The overhanging stern section would then have created a maximum bending moment amidships which would have caused the vessel to tear apart, beginning at or near the top deck where the bending stresses were tensile. Correspondingly, note evidence for buckling along the side of the bow section near the lower portion of the vessel (Fig. 7.1b); such markings would be consistent with the existence of compressive bending stresses near the ship's keel.

Various aircraft and rockets also are not immune to periodic failure. The debris from a ruptured 660-cm-diameter rocket motor casing is shown in Fig. 14.14, and the failure is analyzed in Section 14.5. Additional fractures of domestic products are shown in Fig. 7.2.

It is quite apparent, then, that the subject of fracture in engineering components and structures is certainly a dynamic one, with new examples being provided continuously for evaluation. You might say that things are going wrong all the time. (In all seriousness, the reader should recognize that component failures are the exception and not the rule.)

7.2 THEORETICAL COHESIVE STRENGTH

Recall from Chapter 2 that the theoretical shear stress necessary to deform a perfect crystal was many orders of magnitude greater than values commonly found in engineering materials. It is appropriate now to consider how high the cohesive strength σ_c might be in an ideally perfect crystal. Again, using a simple sinusoidal force-

---

* Robert Ballard, *The Discovery of the Titanic*, Guild Pub., London, 1987.
** Ken Marschall, private communication.

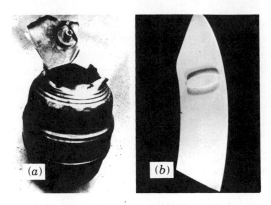

FIGURE 7.2   Fractured components and devices. (*a*) Ruptured beer barrel; (*b*) fractured toilet seat. (From R. W. Hertzberg et al., *Int. J. Fracture* 22 [1983].)

displacement law with a half period of $\lambda/2$, we see from Fig. 7.3 that the shape of the curve may be approximated by

$$\sigma = \sigma_c \sin \frac{\pi x}{\lambda/2} \qquad (7\text{-}1)$$

where σ reflects the tensile force necessary to pull atoms apart. For small atom displacements, Eq. 7-1 reduces to

$$\sigma = \sigma_c \frac{2\pi x}{\lambda} \qquad (7\text{-}2)$$

and the slope of the curve in this region becomes

$$\frac{d\sigma}{dx} = \frac{2\sigma_c \pi}{\lambda} \qquad (7\text{-}3)$$

FIGURE 7.3   Simplified force versus atom displacement relation.

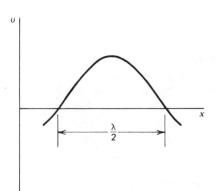

Since Hooke's law (Eq. 1-7) applies to this region as well, the slope of the curve also may be given by

$$E = \frac{\text{stress}}{\text{strain}} = \frac{\sigma}{x/a_0} \tag{7-4}$$

where a_0 = equilibrium atomic separation
E = modulus of elasticity

Upon differentiation

$$\frac{d\sigma}{dx} = E/a_0 \tag{7-5}$$

By combining Eqs. 7-3 and 7-5 and solving for σ_c

$$\sigma_c = \frac{E\lambda}{2\pi a_0} \tag{7-6}$$

If we let $a_0 \approx \lambda/2$, a reasonably accurate assumption, then

$$\sigma_c \approx E/\pi \tag{7-7}$$

It is apparent from Table 7.1 that some materials have the potential for withstanding extremely high stresses before fracture.

If the energetics of the fracture process are considered, the fracture work done per unit area during fracture is given by

$$\text{fracture work} = \int_0^{\lambda/2} \sigma_c \sin \frac{\pi x}{\lambda/2} \, dx = \sigma_c \frac{\lambda}{\pi} \tag{7-8}$$

If all this work is set equal to the energy required to form two new fracture surfaces 2γ we may substitute for λ and show from Eq. 7-6 that

$$\sigma_c = \sqrt{\frac{E\gamma}{a_0}} \tag{7-9}$$

TABLE 7.1 Maximum Strengths in Solids[6]

Material	σ_f		E		E/σ_f
	GPa	(psi $\times$ 10^6)	GPa	(psi $\times$ 10^6)	
Silica fibers	24.1	(3.5)	97.1	(14.1)	4
Iron whisker	13.1	(1.91)	295.2	(42.9)	23
Silicon whisker	6.47	(0.94)	165.7	(24.1)	26
Alumina whisker	15.2	(2.21)	496.2	(72.2)	33
Ausformed steel	3.14	(0.46)	200.1	(29.1)	64
Piano wire	2.75	(0.40)	200.1	(29.1)	73

EXAMPLE 7.1

What is the cohesive strength of fused silica? Using a value of 1750 mJ/m^2 for the estimated surface energy in fused silica, 1.6Å for the equilibrium Si–O atomic separation, and 69 to 76 GPa for the elastic modulus, the cohesive strength is found from Eq. 7-9 to be approximately 28 GPa, in agreement with experimental results from carefully prepared specimens (see Table 7.1). Note that a comparable value could have been computed from Eq. 7-7 had the modulus been the only known quantity.

Regardless of the equation used to obtain σ_c, the problem discussed in Chapter 2 reappears—it is necessary to explain not the great strength of solids, but their weakness.

7.3 DEFECT POPULATION IN SOLIDS

Materials possess low fracture strengths relative to their theoretical capacity because most materials deform plastically at much lower stress levels and eventually fail by an accumulation of this irreversible damage. In addition, materials contain defects that are microstructural in origin or introduced during the manufacturing process. These include porosity, shrinkage cavities, quench cracks, grinding and stamping marks (such as gouges, burns, tears, scratches, and cracks), seams, and weld-related cracks. Other microconstituents, such as inclusions, brittle second-phase particles, and grain-boundary films, can lead to crack formation if the applied stress level exceeds some critical level. An illustration of the role of preexisting defects in the failure of engineering structures is the Duplessis Bridge failure in Quebec, Canada, in 1951, which was traced to a preexistent crack in the steel superstructure.[3] In fact, when a crack that had been sighted before the collapse was examined, *paint* was found on the fracture surface near the crack origin. Certainly, this crack had to be present some time before failure.

In light of such findings, is it not reasonable and even conservative to assume that an engineering component will fail as a consequence of preexistent defects and that this hypothesis provides the basis for fracture control design planning? To a first approximation, then, the problem reduces to one of statistics. How many defects are present in the component or structure, how big are they, and where are they located with respect to the highly stressed portions of the part? Certainly, component size should have some bearing on the propensity for premature failure if for no other reason than the fact that larger pieces of material should contain more defects than smaller ones. Indeed, Leonardo da Vinci used the simple wire-testing apparatus shown in Fig. 7.4 500 years ago to demonstrate that short iron wires were stronger than long sections. In one of his manuscripts we find the following passage:[7]

> *The object of this test is to find the load an iron wire can carry. Attach an iron wire 2 braccia [about 1.3 m] long to something that will firmly support it, then attach a basket or any similar container to the wire and feed into the basket*

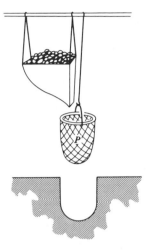

FIGURE 7.4 **Sketch from the notebook of Leonardo da Vinci illustrating tensile test apparatus for iron wires.**

some fine sand through a small hole placed at the end of a hopper. A spring is fixed so that it will close the hole as soon as the wire breaks. The basket is not upset while falling, since it falls through a very short distance. The weight of sand and the location of the fracture of the wire are to be recorded. The test is repeated several times to check the results. Then a wire of one-half the previous length is tested and the additional weight it carries is recorded; then a wire of one-fourth length is tested and so forth, noting each time the ultimate strength and the location of the fracture.

7.3.1 Statistical Nature of Fracture: Weibull Analysis

The fact that Leonardo repeated his experiments several times to verify his results reflected his concern with the statistical nature of the fracture event. In general, statistical variation in a given mechanical property value (e.g., fracture strength) depends on inherent measurement errors (including those resulting from variations in specimen alignment and test environment) and inherent property variations of the material. For purposes of this discussion, we will consider only the latter source of statistical variation in fracture properties. As such, the failure event depends on the probability that a flaw of a certain size and orientation is present when a specific stress is applied. For example, assume the existence of four different flaws as shown in Fig. 7.5. For argument's sake, let us further assume that lengths $A = a$ and $B = b$, and that each defect acts independently of the other three. It will be shown in Chapter 8 that if $\sigma\sqrt{B}$ reaches a critical value, failure will occur. Since $A < B$, failure will not occur in association with crack A. Furthermore, since cracks a and b are parallel to the direction of stress, failure will not occur either, even though length $b = B$. Consequently, the probability of failure will depend only on the existence of the B type crack.

Since a key element in the design and fabrication of a component pertains to its reliability (usually determined by safety and/or economic considerations), it is important that one knows the statistical probability of a given fracture event. For the case of the fracture of brittle wires, such as those studied by da Vinci, failure depends

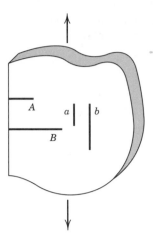

FIGURE 7.5 Arbitrary cracks in a solid body. Note that crack lengths $a = b$ and $A = B$.

on the length of the wire: The longer the wire, the greater the likelihood that a critical defect is present to cause failure. This weakest-link theory pertains to fracture being dependent on the volume of material in the wire where the volume is proportional to the length of the wire of a given diameter. Proceeding further, if we assume that more defects in a solid are found on the surface of the body rather than at the interior (e.g., due to mechanical handling-induced damage and/or environmental degradation), it follows that the probability of failure will depend on the overall surface area of a component.

For design purposes, it is often more important to determine the probability that a property of a component will exceed some specified threshold value than to identify a particular value such as the average tensile strength. Experience has shown that a normal (Gaussian) distribution of property values for a *ductile* material yields a reasonably accurate characterization of material behavior; the same cannot be said for the case of a *brittle* solid. Instead, other statistical theories (e.g., the Weibull analysis[8–10]) are necessary to account for the variability of strength and the probability of survival of a particular component as a function of its volume and the applied stress.

For the case of a brittle rod of length, L, subjected to a stress σ, the survival probability is given as $S(L)$. For another piece of the same rod with length, L', the survival probability at the same stress is $S(L')$. If one were to attach x pieces of this rod, the overall probability of survival at the same stress would be $S(L_o)^x$ where L_o is a unit length of wire. [The multiplicity of survival probabilites of x segments of the rod is analogous to the probability of obtaining a specific result from a series of coin tosses. Since the probability of flipping a "head" is $1/2$, it follows that the probability of tossing two consecutive "heads" is $1/4$. For three "heads," the probability would be $1/8$, or $(1/2)^3$.] If the flaw distribution in each volume is the same, then the probability of survival of the rod would be

$$S(V) = S(V_o)^x \tag{7-10}$$

where V_o is the unit volume of the solid. By taking logarithms and rearranging Eq. 7-10, we have

$$S(V) = \exp(x \ln[S(V_o)])$$ (7-11)

Weibull defined the risk of rupture R as

$$R = -x \ln[S(V_o)]$$ (7-12)

Therefore,

$$S(V) = \exp(-R)$$ (7-13)

For an infinitesimal volume dV, the risk of rupture, R, is shown to depend only on stress where

$$R = \int f(\sigma) \, dV$$ (7-14)

Weibull postulated that

$$R = \left(\frac{\sigma - \sigma_u}{\sigma_o} \right)^m$$ (7-15)

where σ = applied stress

σ_u = stress below which there is a zero probability of failure. This implies an upper limit to the size of defects present in the material. [For the case of fatigue in steel alloys, $\sigma_u = \sigma_{fat}$, where σ_{fat} represents the fatigue endurance limit (see Section 12.2). For brittle ceramics, $\sigma_u = 0$ since any tensile stress might cause failure. Failure would not be expected when $\sigma < 0$ since compressive stresses would act to close the crack.]

σ_o = a characteristic strength that is analogous to the mean strength of the material

m = Weibull modulus that characterizes the variability in the strength of the material and is analogous to the standard deviation of the material's strength.

Increasing m values reflect more homogeneous material behavior with strength levels for a given component being more predictable. In the limit where m approaches ∞, the probability of fracture is zero for all stress levels $< \sigma_o$. When $\sigma = \sigma_o$, the probability of fracture becomes unity. Conversely, when m approaches zero, R approaches unity and failure occurs with equal certainty at any stress level. For all other values ($0 < m < \infty$), the risk of failure also is a function of the Weibull modulus. For example, if $\sigma_u = 0$ and for the case of the average strength of a material (i.e., where the risk of failure is 0.5),

$$0.5 = (\sigma_{R=0.5} / \sigma_0)^m$$ (7-16)

It follows that the stress level corresponding to a risk of failure of 0.5 is given by

$$\sigma_{R=0.5} = (0.5)^{1/m} \sigma_o$$ (7-17)

When $m = 2$ or 20, $\sigma_{R=0.5}$ equals 0.71 σ_o and 0.97 σ_o, respectively. Note that the lower the value of m, the lower the allowable stress to ensure the same probability of failure. Furthermore, it is dangerous to assign the same factor of safety* to the average strengths in the two materials since the same probability of fracture of the two different brittle ceramics corresponds to different fractions of the average strength of the two materials. For example, when m approaches zero, the minimum strength approaches zero and the probability of fracture approaches unity for all stress levels. Conversely, when m approaches ∞, the scatter in strength values approach zero, $\sigma_{min} = \sigma_{avg}$, and $R = 0$ for $\sigma < \sigma_{avg}$.

7.3.1.1 Effect of Size on the Statistical Nature of Fracture
By combining Eqs. 7-13 to 7-16, the probability of survival is

$$S(V) = \exp\left\{ -V\left(\frac{\sigma - \sigma_u}{\sigma_o}\right)^m \right\} \tag{7-18}$$

Alternatively, the risk of failure is given by

$$R = 1 - \exp\left\{ -V\left(\frac{\sigma - \sigma_u}{\sigma_o}\right)^m \right\} \tag{7-19}$$

By taking double logarithms of Eq. 7-18 and rearranging, a linear relationship is found between the probability of survival of a component and the applied stress with the slope of the data plot being characterized by the Weibull modulus, m (Eq. 7-20).

$$\ln\ln\left(\frac{1}{S(V)}\right) = \ln V + m \ln(\sigma - \sigma_u) - m \ln \sigma_o \tag{7-20}$$

Tensile strength data[11] for 21 samples of SiC are shown in Table 7.2, where strength levels are ranked in ascending value. These results, assuming $\sigma_u = 0$, are plotted in Fig. 7.6, using Weibull-probability graph paper with $\ln\ln(1/S(V))$ as ordinate and $\ln \sigma$ as abscissa. The Weibull modulus is then given by the negative value of the slope of the plot and is 10 for this set of data. (The negative of the slope is used since $\ln\ln(1/S(V))$ increases as $S(V)$ decreases.)

---

* A factor of safety essentially provides a margin of error in design of a component. For example, by adding a factor of safety of two to the material's yield strength, the design stress is then set at 50% of the yield strength.

TABLE 7.2 Tensile-Strength Data for Self-Bonded Silicon Carbide[11]

Rank (1-21)	1	2	3	4	5	6	7
Strength (MPa)	232	252	256	274	282	285	286
	8	9	10	11	12	13	14
	289	294	308	311	314	316	324
	15	16	17	18	19	20	21
	334	337	339	341	365	379	382

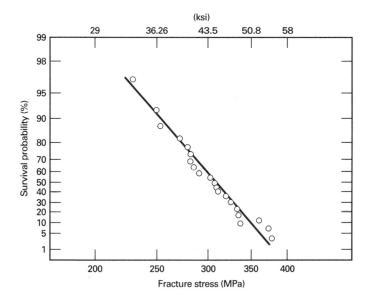

FIGURE 7.6 **Survival probability in SiC as a function of fracture stress,[11] according to Eq. 7-20 with $\sigma_u = 0$. Data listed in Table 7.2. Weibull modulus, $m = 10$. R. W. Davidge, *Mechanical Behavior of Ceramics*, Cambridge University Press, Cambridge (1979). Reprinted with the permission of Cambridge University Press.**

Equation 7-18 reveals that for the same probability of survival, the fracture strength of a material varies with the volume of the component. For example, if $\sigma_u = 0$ and $V_1 > V_2$, we find

$$V_1(\sigma_1)^m = V_2(\sigma_2)^m \qquad (7\text{-}21)$$

and

$$\sigma_1/\sigma_2 = (V_2/V_1)^{1/m} \qquad (7\text{-}22)$$

Clearly, for an equal probability of survival, the larger the volume of the component, the lower the stress necessary for fracture (see Fig. 7.7).

EXAMPLE 7.2

Two different brittle ceramic parts are being considered for use in a particular application. The Weibull moduli for the two materials are 2 and 10, respectively. If σ_A and σ_B represent the stresses necessary for the same probability of survival in laboratory specimens of materials A and B, respectively, by how much will the respective fracture stresses change if the volume of the full-scale components of each material are ten times larger than that of the laboratory specimens?

In each instance, the stress to produce the same probability of survival will vary with the specimen/component volume ratio (see Eq. 7-22). For the case of material B

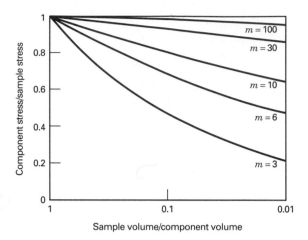

FIGURE 7.7   **Influence of component volume on allowable stress in a component to maintain a given probability of survival.**

with the higher Weibull modulus, the stress to generate the same probability of survival will decrease by a factor of 1.26 when the component volume is ten times greater than that of the laboratory specimen. Of greater concern, the stress necessary for the same probability of survival in material A will decrease by a factor of 3.16 when the volume changes by a factor of ten. Clearly, the material with a larger degree of scatter in mechanical properties (i.e., lower Weibull modulus) exhibits a more pronounced size effect.

The probability of survival will also depend on the existence of stress gradients and the associated volume of material that experiences the maximum stress in the component or specimen. For example, it is to be expected that the probability of fracture would be greater under a tensile rather than a flexural loading condition (recall Section 1.3). Clearly, a tensile load generates the same stress across a component's entire cross-sectional area whereas flexural loading generates a maximum stress only at the outermost fiber of the beam. By taking into consideration the respective volumes of the maximum tensile and flexural stresses, it can be shown that the same probability of survival would yield different fracture stress levels in the tensile versus the bend bar samples. For purposes of illustration, the stress ratio between bending and tensile loading for the same probability of survival is given by

$$\sigma_{3\text{-pt bend}}/\sigma_{\text{tensile}} = \{2(m + 1)^2\}^{1/m} \tag{7-23}$$

If the Weibull modulus of a given material were 2 versus 10, as in the previous illustrative problem, the ratio of flexural/tensile strengths for an equal probability of survival would be 4.24 and 1.73, respectively. The bend bar is seen to possess the greater strength since its volume under maximum stress is smaller than that of the tensile sample. This is especially true for cases where m values are low. Conversely, as m approaches ∞, the properties of the tensile and bend samples converge to the same value.

As we will see in Chapter 8, the probability argument is not the complete explanation of the size effect in fracture.

7.4   THE STRESS-CONCENTRATION FACTOR

By analyzing a plate containing an elliptical hole, Inglis[12] was able to show that the applied stress σ_a was magnified at the ends of the major axis of the ellipse (Fig. 7.8) so that

$$\frac{\sigma_{max}}{\sigma_a} = 1 + \frac{2a}{b} \tag{7-24}$$

where σ_{max} = maximum stress at the end of the major axis
  σ_a = applied stress applied normal to the major axis
  a = half major axis
  b = half minor axis

Since the radius of curvature ρ at the end of the ellipse is given by

$$\rho = \frac{b^2}{a} \tag{7-25}$$

Eqs. 7-24 and 7-25 may be combined so that

$$\sigma_{max} = \sigma_a(1 + 2\sqrt{a/\rho}) \tag{7-26}$$

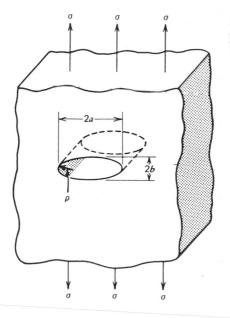

FIGURE 7.8   Elliptical hole in infinitely large panel produces stress concentration of $1 + 2a/b$.

In most cases $a \gg \rho$, therefore

$$\sigma_{max} \approx 2\sigma_a \sqrt{a/\rho} \qquad (7\text{-}27)$$

The term $2\sqrt{a/\rho}$ is defined as the stress-concentration factor k_t and describes the effect of crack geometry on the local crack-tip stress level. Many textbooks and standard handbooks describe stress concentrations in components with a wide range of crack configurations; a few are listed in the bibliography.[13,14] Although the exact formulations vary from one case to another, they all reflect the fact that k_t increases with increasing crack length and decreasing crack radius. Therefore, all cracks, if present, should be kept as small as possible. One way to accomplish this is by periodic inspection and replacement of components that possess cracks of dangerous length. Alternatively, once a crack has developed, the relative severity of the stress concentration can be reduced by drilling a hole through the crack tip. In this way, ρ is increased from a curvature associated with a natural, sharp crack tip to that of the hole radius. In fact, it was suggested by Wilfred Jordan that the crack in the Liberty Bell "might have been stopped by boring a very small hole through the metal a short distance beyond the former termination of the first crack. [However], it was never done, with the result that the original fracture, known to millions, has more than doubled in length and has extended up and around the shoulder of the Bell."[15]

The latter procedure is used occasionally in engineering practice today but should be employed with caution. Obviously, a crack is still present after drilling and may continue to grow beyond the hole after possible reinitiation. Also, field failures have occurred earlier than expected, simply because the blunting hole was introduced *behind* the crack tip, and the sharp crack-tip radius was not eliminated.

Stress concentrations may also be defined for component configuration changes, such as those associated with section size changes. As shown in Fig. 7.9, k_t increases for a number of design configurations whenever there is a large change in cross-sectional area and/or where the associated fillet radius is small. For every book written on the analysis of stress concentrations, no doubt many others could be filled with case histories of component failures attributable to either ignorance or neglect of these same factors.

The completely elastic stress–strain material response, denoted in Chapter 1 as Type I behavior, is affected greatly by the presence of stress concentrations within the sample. As shown schematically in Fig. 7.10, the maximum stress and strain level that any component may support decreases with increasing k_t. This explains, in part, why many ceramic materials are much weaker in tension than in compression. The higher the stress concentration, the easier it is to break the sample. This fact is put to good use regularly by glaziers, who first score and then bend glass plate to induce fracture with minimal effort and along a desired path.

Fortunately, stress concentrations in most materials will not result in the escalation of the local crack-tip stress to dangerously high levels. Instead, this potentially damaging stress elevation is avoided by plastic deformation processes in the highly stressed crack-tip region. As a result, the local stress does not greatly exceed the material's yield-strength level as the crack tip blunts, thereby reducing the severity of the stress concentration. The ability of a component to plastically deform in the vicinity of a crack tip is the saving grace of countless engineering structures. It should also be recognized that a component that contains a stress concentration may still perform in a satisfactory manner. This follows from the fact that the local stress at the crack tip is represented by the product of the applied stress and k_t values (Eq. 7-27); so long as the applied stress level is sufficiently low, $\sigma_a k_t$ values will remain comfortably below the local stress level necessary for fracture.

7.5 NOTCH STRENGTHENING

When an appreciable amount of plastic deformation is possible, an interesting turn of events may occur with regard to the fracture behavior of notched components. We saw in Chapter 1 that plastic constraint is developed in the necked region of a tensile bar as a result of a triaxial stress state; the unnecked regions of the sample experience a lower true stress than the necked section and, therefore, restrict the lateral contraction of the material in the neck. Similar stress conditions exist in the vicinity of a notch in a round bar. When the net section stress reaches the yield-strength level, the material in the reduced section attempts to stretch plastically in the direction parallel to the loading axis. Since conservation of volume is central to the plastic deformation process, the notch root material seeks to contract also, but is constrained by the bulk of the sample still experiencing an elastic stress. The development of tensile stresses in the other two principal directions— the constraining stresses—makes it necessary to raise the axial stress to initiate plastic deformation. The deeper the notch, the greater is the plastic constraint and the higher the axial stress must be to deform the sample. Consequently, the yield strength of a notched sample may be *greater* than the yield strength found in a smooth bar tensile test. The data shown in Table 7.3 demonstrate the "notch-strengthening" effect in 1018 steel bars, notched to reduce the cross-sectional area by up to 70%.

A laboratory demonstration is suggested to illustrate a seemingly contradictory test response in two different materials. First, austenitize and quench a high-strength steel, such as AISI 4340, to produce an untempered martensite structure, and then perform a series of notched tensile tests. You will note that the net section stress *decreases* with increasing notch depth because of the increasing magnitude of the stress-concentration factor. Now conduct notch tests with a ductile material such as a low-carbon steel or aluminum alloy. In this case, note that the net section stress will *increase* with increasing notch depth as a result of the increased plastic constraint. In this manner, you may prove to yourself that materials with limited deformation capacity will *notch weaken*, and highly ductile materials will *notch strengthen*.

EXAMPLE 7.3

Two 0.5-cm-diameter rods of 1020 steel ($\sigma_{ts} = 395$ MPa) are to be joined with a silver braze alloy 0.025 cm thick ($\sigma_{ys} = 145$ MPa) to produce one long rod. What will be the ultimate strength of this composite? The response of this bar may be equated to that of a notched rod of homogeneous material. In this instance, preferential yielding in the weaker braze material would be counterbalanced by a constraining

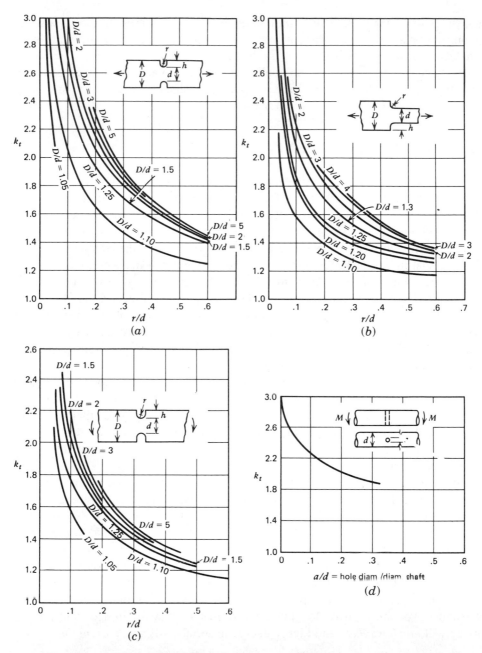

FIGURE 7.9 Selected stress-concentration factors:[16] (*a*) axial loading of notched bar; (*b*) axial loading of bar with fillet; (*c*) bending of notched bar; (*d*) bending of shaft with transverse hole; (*e*) axial loading of bar with transverse hole; (*f*) straight portion of shaft keyway in torsion. (From *Metals Engineering; Design* by American Society of Mechanical Engineers. Copyright © 1953 by the American Society of Mechanical Engineers. Used with permission of McGraw-Hill Book Company.)

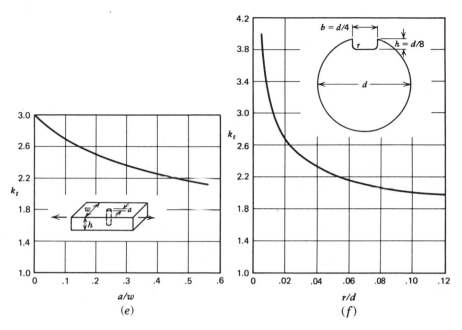

FIGURE 7.9 (*Continued*)

triaxial stress field similar to that found in a notched bar of homogeneous material. As such, the strength of the joint will depend to a great extent on the geometry of the joint. Specifically, it would be expected that braze joint constraint would increase with increasing rod diameter and decreasing joint thickness. The experimental results by Moffatt and Wulff[17] (Fig. 7.11) reflect the importance of these two geometrical variables on the composite strength $\bar{\sigma}$. Accordingly, $\bar{\sigma}$ is found to be approximately 345 MPa.

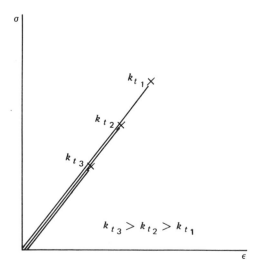

FIGURE 7.10 Effect of stress concentration k_t on allowable stress and strain in completely elastic material.

TABLE 7.3 Notch Strengthening
in 1018 Steel

Reduction of Area in Notched Sample	Yield Strength, Notched Bar Smooth Bar
0	1.00
20	1.22
30	1.36
40	1.45
50	1.64
60	1.85
70	2.00

Two factors need to be emphasized when discussing the observed notch-strengthening effect. First, even though the notched component may have a higher net section stress, it requires a *lower* load for fracture than does the smooth sample when based on the *gross* cross-sectional area. I trust that this should temper the enthusiasm of any overzealous student who might otherwise race about, hacksaw in hand, with the intent of "notch strengthening" all the bridges in town. Second, there is a limit to the amount of notch strengthening that a material may exhibit. From theory of plasticity considerations, it is shown that the net section stress in a deformable material may be elevated to $2\frac{1}{2}$ to 3 times the smooth bar yield-strength value. It would appear, then, that the brazed joint system described in the previous example represented an optimum matching of material properties. Using a higher strength steel with the same braze alloy would not have made the joint system stronger.

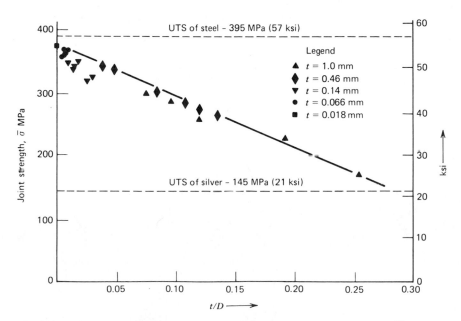

FIGURE 7.11 **Brazed joint strength as a function of joint geometry.**[17] **(Reprinted with permission from Metallurgical Society of AIME.)**

7.6   EXTERNAL VARIABLES AFFECTING FRACTURE

As mentioned in Section 7.4, the damaging effect of an existing stress concentration depends strongly on the material's ability to yield locally and thereby blunt the crack tip. Consequently, anything that affects the deformation capacity of the material will affect its fracture characteristics as well. Obviously, any metallurgical strengthening mechanism designed to increase yield strength will simultaneously suppress plastic deformation capacity and the ability of the material to blunt the crack tip. For any given material, there are, in addition, a trilogy of external test conditions that contribute to premature fracture: notches, reduced temperatures, and high strain rates (Fig. 7.12). The more the flow curve is raised for a given material, the more likely brittle fracture becomes. As we saw in Section 7.5, the presence of a notch acts to plastically constrain the material in the reduced section and serves to elevate the net section stress necessary for yielding. Likewise, lowering the test temperature and/or increasing strain rate will elevate the yield strength (the magnitude of this change depends on the material). For example, yield-strength temperature and strain-rate sensitivity varies with the Peierls stress component of the overall yield strength of the material (Section 2.3). In BCC alloys, such as ferritic steels, the Peierls stress increases rapidly with decreasing temperature, thereby causing these materials to exhibit a sharp rise in yield strength at low temperatures. This can and often does precipitate premature fracture in structures fabricated from these materials. In other materials, such as FCC alloys, the Peierls stress component is small. Consequently, yield-strength temperature sensitivity is small, and these materials may be used under cryogenic conditions, provided they possess other required mechanical properties.

7.6.1   Thermal Stresses and Thermal Shock-Induced Fracture

In addition to the influence of temperature on the deformation and fracture response of materials, temperature changes can generate residual stresses that ultimately may lead to the fracture of components. Scientists, engineers, and homemakers have long known that when hot ceramic components are cooled quickly, some will crack, whereas others will not. The development of thermally induced stresses and the

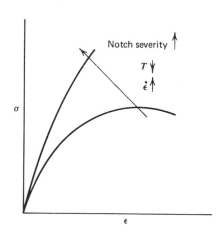

FIGURE 7.12   Effect of temperature, strain rate, and plastic constraint on flow curve.

potential for thermal shock-induced fracture have been the subject of study for at least 2200 years! Indeed, Roman historians reported that Hannibal's military engineers used thermal shock to fracture rocks that blocked the path of the Carthaginian army during its advance across the Alps in 218 BC. Two hundred years later, Livy[18] reported that ". . . It was necessary to remove a rock. Trees were felled and the wood piled onto it in a huge pyre which was lighted and burned fiercely with the help of a fortunate breeze. The hot stone was then drenched with vinegar [presumably sour wine] to disintegrate it and attacked with pickaxes." In all likelihood, the hot rocks were doused with water to cause them to crack. For example, Pliny[19] reported that ". . . if fire has not disintegrated a rock, the addition of water makes it split." Other more recent reports of thermal shock-induced fracture of rocks were cited by Agricola[20] and de Beer.[21]

The modern technical literature provides more rigorous analyses of thermal stresses and thermal shock resistance than that found in early Roman documents. Of importance, Hasselman and co-workers[22,23] developed a unified theory of thermal stress fracture that accounts for both the initiation and propagation of cracks in a brittle solid. For the case of fracture initiation, failure occurs when the thermal stress exceeds the material's fracture strength. In this instance, fracture may be characterized by a "maximum stress" failure theory (recall Section 1.4). For the case of failure under conditions of critical crack growth, the Griffith theory of failure accounts for the fracture event (see Section 8.1).

7.6.1.1 Initiation-Controlled Thermal Fracture

Let us first consider the case where initiation-controlled fracture by thermal cracking cannot be tolerated as in glasses, porcelain, electronic ceramics, and whiteware. If an unconstrained rod of length L is cooled from one temperature (T_1) to the other (T_2), it will contract by an amount

$$\Delta L = \alpha (T_1 - T_2)L \qquad (7\text{-}28)$$

where α = coefficient of thermal expansion (CTE). The CTE for a given material depends on the bond strength between atoms. As discussed in Section 1.2.1, increasing bond strength results in a decreasing interatomic distance of separation and a higher elastic stiffness. Likewise, increased interatomic bond strength (associated with higher melting points) leads to lower values of CTE[24] (Fig. 7.13). (For the case of ceramics, CTE decreases with both increasing bond strength and percentage of covalent bonding.) Additional CTE values are given in Table 7.4[25] for various metals, ceramics, and organic solids. (These data are approximate since CTE values tend to increase moderately with temperature.) Note that the highest CTE values are associated with organic solids, whereas ceramics display the lowest values of CTE. As previously discussed, these trends reflect the fact that ceramics and organic solids possess the highest and lowest interatomic bond strengths, respectively.

If the rod were constrained between two rigid walls, a thermal stress would be generated with the magnitude

$$\sigma = E \alpha (T_1 - T_2) \qquad (7\text{-}29)$$

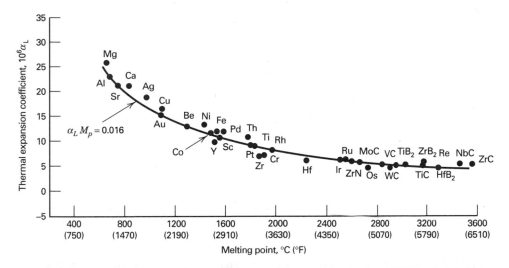

FIGURE 7.13 Relation between coefficient of thermal expansion (CTE) and melting point for metals, carbides, and borides with close-packed structures.[24] L. G. Van Vitert, H. M. O'Bryan, M. E. Lines, H. J. Guggenheim, and G. Zydzik, *Materials Research Bulletin*, **12**, 261 (1977). Reprinted by permission.

where E = modulus of elasticity. The in-plane thermal stress developed in a constrained thin slab is given by

$$\sigma = \frac{E \, \alpha \, \Delta T}{1 - \mu} \tag{7-30}$$

where μ = Poisson's ratio
$\Delta T = T_1 - T_2$

EXAMPLE 7.4

The temperature of a 10-cm-long rod of alumina (99.8% dense) decreases from 200 to 0°C. (a) What would be the change in length if the rod were unconstrained? (b) Would the rod survive this drop in temperature if it were fully constrained?

To solve this problem, we must first collect relevant material property data. From Tables 1.7 and 7.4, we find

$$\alpha = 6.7 \times 10^{-6}/°C$$
$$E = 385 \text{ GPa}$$
$$\sigma_{\tau s} = 205 \text{ MPa}$$

(a) When the rod is unconstrained, we see from Eq. 7-28 that the 200°C temperature change generates a length change of

$$\Delta L = 6.7 \times 10^{-6}(200)(10)$$
$$\Delta L = 0.0134 \text{ cm}$$

which corresponds to a thermal strain of 0.134%.

(b) If the rod were fully constrained, the thermal stress is computed from Eq. 7-29 where

$$\sigma = 385 \times 10^9 (6.7 \times 10^{-6})(200)$$
$$\sigma = 516 \text{ MPa}$$

Since the tensile strength of alumina is 205 MPa, the rod would have fractured. Indeed, the maximum allowable temperature rise for a constrained rod of alumina is calculated to be

$$\Delta T = \frac{205 \times 10^6}{385 \times 10^9 \ (6.7 \times 10^{-6})} = 79.5°C$$

TABLE 7.4 **Selected Coefficients of Thermal Expansion (CTE) for Metals, Ceramics, and Organic Solids $(10^{-7}/°C)$**[25]

Metals	
Invar	20
Molybdenum	52
Alloy 42 (*FeNi*)	60
Titanium	100
Iron	120
Gold	142
Nickel	130–150
Gold–tin eutectic	160
Copper and its alloys	160–180
Silver	190
Lead–tin eutectic	210
Aluminum and its alloys	220–250
5–95 tin–lead	280
Lead	290
Ceramics, semiconductors, etc.	
Silica glasses	5–10
Silicon carbide	26
Silicon (single crystal)	28
Alumina	67
Beryllia	80
Gold–silicon eutectic	130
Gold–germanium eutectic	130
Organics*	
Kevlar®	−20
Epoxy–glass (FR4)—horizontal	110–150
vertical	600–800
Polyimides	400–500
Polycarbonates	500–700
Epoxies	600–800
Polyurethanes	1800–2500
Sylgard®	3000
RTV	8000

* Below glass transition temperature

For the case of composite materials, thermal stresses are generated at temperatures other than that associated with fabrication; such stresses will vary directly with differences in coefficients of thermal expansion between the matrix (α_m) and reinforcing phases (α_r). To illustrate, when a thin coating is applied to an underlying substrate, Burgreen[26] showed that

$$\sigma = \frac{E \, \Delta\alpha \, \Delta T}{1 - \mu} \tag{7-31}$$

where $\Delta\alpha = |\alpha_m - \alpha_r|$. Whichever phase has the larger value of α will be the phase in tension. For example, thermal barrier coatings in gas turbine engine components develop residual compressive thermal stresses in the outer ceramic layer due to differences in CTE between the intermediate metallic bond—higher CTE—coating layer (typically (Ni/Co)CrAlY) and the outer ceramic—lower CTE—thermal barrier coating (typically Zr-O_2-based). Failure of the coating then occurs by delamination and subsequent spallation of the outer thermal barrier coating.[27] Such difficulties can be reduced by minimizing the CTE mismatch.

The stress levels given by Eqs. 7-29 to 7-31 correspond to maximum values for the case when the average temperature of the body after quenching is unchanged* and its surface temperature matches that of the quenching medium. The latter condition is approached only under extreme quenching conditions and in materials possessing a very low coefficient of thermal conductivity.

The magnitude of thermal stress more typically depends on the heat transfer coefficient (h) between the cooling fluid and the quenched solid, the material's coefficient of thermal conductivity (k), and the geometry of the component. It is customary to compute the magnitude of the thermal stress by including a nondimensional parameter [defined as the Biot modulus (β)] that incorporates these factors where

$$\beta = xh/k \tag{7-32}$$

where x = specimen dimension such as slab thickness or rod diameter. For large component dimensions, high heat transfer rates (corresponding to conductive rather than convective or radiative heat transfer), and/or low levels of thermal conductivity, β values are large ($\beta > 20$) and thermal stress levels approach those given by Eqs. 7-29 to 7-31.

For the case of more moderate quenching, Eq. 7-30 is rewritten in the form

$$\sigma = \frac{c\beta E\alpha(\Delta T)}{1 - \mu} \tag{7-33}$$

where c = function of specimen geometry. Jaeger[28] determined that values of $c\beta$ fall in the range $0 < c\beta \leq 1$. By rearranging Eq. 7-33 and combining with Eq. 7-32, the

---

* If the body of a component experiences a thermal gradient, there will exist a stress gradient from the surface to the core of the object.

material's initiation-controlled thermal shock resistance is defined by the temperature change that can be tolerated without fracture. Hence,

$$\Delta T = \frac{\sigma_f(1 - \mu)k}{E\alpha ch} \tag{7-34}$$

where σ_f = fracture strength. Note that a material's resistance to thermal stress-induced failure is enhanced by a high fracture strength and coefficient of thermal conductivity, and low values of the modulus of elasticity, coefficient of thermal expansion, and heat transfer rate. It is ironic that those materials that possess superior high temperature stiffness and resistance to environmental degradation are most susceptible to catastrophic thermal-shock failure.

Table 7.5[29] lists relevant thermal and mechanical property data for selected ceramics, where R' represents the measure of thermal fracture resistance, comparable to that expressed by Eq. 7-34. Higher values of R' reflect greater thermal shock resistance. The relative ranking of these materials must be viewed with caution, however, since property values such as α and k increase and decrease, respectively, with temperature. As a result, R' values for a given material will vary as a function of temperature. Consequently, it is not possible to provide a simple ranking of materials in terms of their resistance to thermal shock.

7.6.1.2 Propagation-Controlled Thermal Fracture

When crack initiation cannot be avoided as a result of severe thermal conditions, it becomes necessary to minimize crack extension. For example, it is important that thermal stress fluctuations do not lead to the spalling away of pieces of porous refractory bricks and tiles. Hasselman[22] determined that the extent of crack propagation is dependent on the available amount of elastic strain energy at fracture (proportional to σ^2/E) and inversely proportional to the material's fracture energy (γ) necessary for the creation of new crack surface. Correspondingly, he defined a thermal stress damage resistance parameter, R'''', so as to describe a material's resistance to crack instability with

$$R'''' = \frac{E\gamma}{\sigma_f^2}(1 - \mu) \tag{7 35}$$

where γ = surface fracture energy. Note that resistance to crack instability is maximized by high values of E and low fracture strength levels—just the opposite to that needed to limit crack initiation (recall Eq. 7-34).* Proceeding further, Hasselman[22] computed the temperature change necessary for unstable crack extension as a function of crack size. For the case of a constrained flat plate (Fig. 7.14), the critical temperature difference ΔT_c needed for simultaneous propagation of N circular and noninteracting cracks of equal diameter(l)/unit volume, is given by

---

* Eq. 7-35 loses relevancy as σ_f approaches zero.

TABLE 7.5 Mechanical and Thermal Properties of Selected Ceramics and Associated Thermal Shock Resistance Parameter [29]

Material	Bend strength σ (MPa)	Young's modulus E (GPa)	Poisson's ratio ν	Thermal expansion coefficient α, 0–1000 °C (10^{-6} K^{-1})	Thermal conductivity k at 500 °C (W m^{-1} K^{-1})	$R' = \dfrac{\sigma k(1-\nu)}{E\alpha}$ (kW m^{-1})
Hot-pressed Si$_3$N$_4$	850	310	0.27	3.2	17	11
Reaction-bonded Si$_3$N$_4$	240	220	0.27	3.2	15	3.7
Reaction-bonded SiC	500	410	0.24	4.3	84	18
Hot-pressed Al$_2$O$_3$	500	400	0.27	9.0	8	0.8
Hot-pressed BeO	200	400	0.34	8.5	63	2.4
Sintered WC (6% Co)	1400	600	0.26	4.9	86	30

$$\Delta T_c = \left(\frac{2\gamma}{\pi l \alpha^2 E}\right)^{0.5} (1 + 2\pi N l^2) \qquad (7\text{-}36)$$

Figure 7.15 shows the critical temperature difference (ΔT_c) necessary for crack instability as a function of initial crack length and crack density. For the case of short cracks, ΔT_c decreases with increasing crack length (i.e., the allowable temperature change to ensure crack stability decreases with increasing crack length) until a minimum is reached at $l = l_m$. At this point, l_m is found to be equal to $(6\pi N)^{-1/2}$. When a critical temperature difference is reached and the crack ($l < l_m$) propagates in an unstable manner, the strain energy release rate is greater than that necessary to create new fracture surface. The excess strain energy is then transformed into kinetic energy so as to drive the crack until the two energy terms are in balance and the crack arrests. Hasselman[22] determined that this energy balance corresponds to those arrested crack

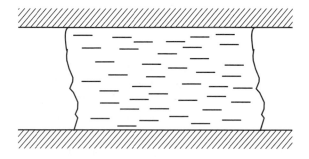

FIGURE 7.14 Crack distribution in rigidly constrained body.

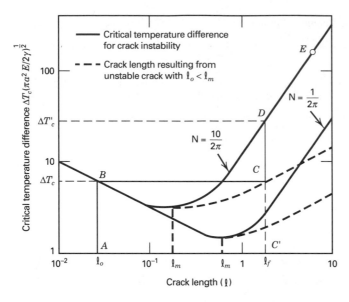

FIGURE 7.15 **Thermal shock resistance described in terms of critical temperature difference ΔT_c versus crack length as a function of crack density N.[32] D. R. Larson, J. A. Coppola, D. P. H. Hasselman, and R. C. Bradt,** *Journal of the American Ceramic Society,* **57, 417 (1974). Reprinted by permission.**

lengths defined by the dotted lines in Fig. 7.15. The two solid lines in Fig. 7.15 where $l > l_m$ represent values of ΔT_c associated with conditions for quasi-stable crack growth in brittle bodies containing two different crack densities. Note that thermal shock resistance is expected to increase with increasing microcrack density. Indeed, significantly less thermal shock damage was found in both BeO-matrix composites, containing 15 w/0 SiC dispersed particles, and in MgO-3v/0 W composites.[30,31] In both instances, three-dimensional networks of microcracks developed to relieve thermal stresses generated upon cooling from the fabrication temperature in association with the coefficient of thermal expansion mismatches in the two composites.

We see from Fig. 7.15 that for cracks of length l_0, instability occurs when the temperature difference increases to ΔT_c. As noted above, these cracks continue to grow unstably until they reach a size l_f and then arrest. The cracks remain stable until the temperature difference is increased by an additional amount equal to $\Delta T_{c'}$. For this crack length (either for short cracks which grew rapidly to length l_f or for long cracks that were originally of length l_f), the cracks then grow in a quasi-static manner with increasing increments of ΔT.

The changes in crack length and corresponding material strength with increasing temperature difference for stable and unstable crack extension is summarized in Fig. 7.16. For the case where $l_0 < l_m$, the crack length change with increasing ΔT is shown by the diagram in Fig. 7.16a. Note that crack instability occurs at point B (see also Fig. 7.15) when $\Delta T = \Delta T_c$. Strength of the solid correspondingly decreases abruptly to a lower level associated with C in Fig. 7.16b. At this point, crack length and residual strength remain fixed until a temperature differential $\Delta T_{c'}$ is reached whereupon stable crack extension resumes at D and strength decreases gradually. When l_0

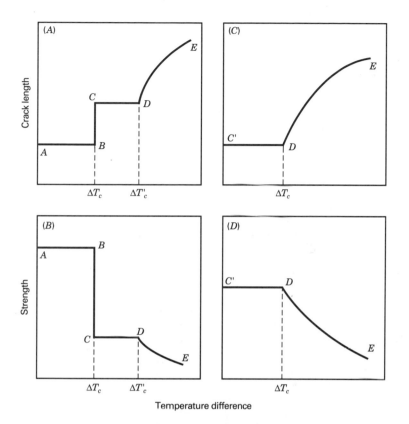

FIGURE 7.16 **Changes in crack length and corresponding strength as a function of temperature change for unstable (a, b) and stable (c, d) crack propagation response.[23] D. P. H. Hasselman and J. P. Singh, *Thermal Stresses*, R. B. Hetnaski, et al, Eds., Vol 1, Chap 4, 264 (1986). Reprinted by permission.**

$> l_m$, stable crack extension occurs at the critical temperature difference (Fig. 7.16c) and residual strength decreases gradually (Fig. 7.16d).

Representative tensile strength data as a function of ΔT are shown in Fig. 7.17a for the case of industrial-grade polycrystalline aluminum oxide. Note the strength plateau in the regime where $\Delta T_c \leq \Delta T \leq \Delta T_{c'}$ that is described by region CD in Fig. 7.16d. Alternatively, the strength-ΔT plot, shown in Fig. 7.17b for stable crack growth in a high-alumina refractory corresponds to material behavior shown in Fig. 7.16d. Note that the absence of unstable crack extension in the refractory material is associated with its lower strength level as compared with that found in the industrial grade aluminum oxide (recall Eq. 7-35).

The choice of a suitable brittle solid to maximize thermal shock resistance may then be characterized in two basically different ways:[34]

1. For relatively mild thermal environments, thermal shock fracture can be avoided by the selection of a material with high strength and thermal conductivity and low values of elastic modulus and thermal expansion.

2. For severe thermal fluctuations, optimal material properties should include low

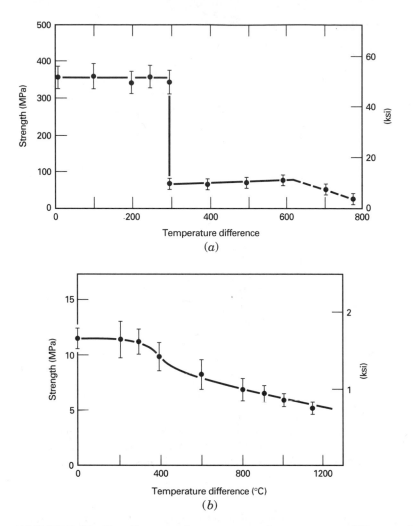

FIGURE 7.17 Tensile strength versus quench temperature difference for (a) industrial polycrystalline aluminum oxide;[33] and (b) high-alumina refractory.[32] D. P. H. Hasselman, *Journal of the American Ceramic Society*, 53, 490 (1979). Reprinted by permission. D. P. H. Hasselman and J. P. Singh, *Thermal Stresses*, R. B. Hetnaski, et al, Eds., Vol 1, Chap 4, 264 (1986). Reprinted by permission.

thermal expansion and elastic modulus and high fracture surface energy. In addition, the material should be designed to contain a high density of microcracks such that $l_0 > l_m$. The latter will ensure stable crack propagation. Note that the choice of a low strength ceramic in connection with its population of microcracks should be restricted to low load bearing applications.

7.7 NOMENCLATURE OF THE FRACTURE PROCESS

Many words and phrases have been used to characterize failure processes in engineering materials. Since these terms are born of different disciplines, each having its own relatively unique jargon, confusion exists along with some incorrect usage. In an attempt to simplify and clarify this situation, I have found it convenient to describe the fracture of a test sample or engineering component in terms of three general characteristics: energy of fracture, macroscopic fracture path and texture, and microscopic fracture mechanisms.

7.7.1 Energy to Break

As discussed in Chapter 1, the toughness of a given material is a measure of the energy absorbed before and during the fracture process. The area under the tensile stress–strain curve would provide a measure of toughness where

$$\text{energy/volume} = \int_0^{\epsilon_f} \sigma d\epsilon \qquad (1\text{-}25)$$

If the energy is high (such as for Curve C in Fig. 1.13), the material is said to be tough or possess high fracture toughness.

Conversely, if the energy is low, the material (e.g., Curves A and B in Fig. 1.13) is described as brittle. In notched samples, a determination of toughness is more complicated, and will be considered in more detail in Chapter 8. For the time being, the relative toughness or brittleness of a material may be estimated by noting the extent of plasticity surrounding the crack tip. Since the stress concentration at a crack tip will often elevate the applied stress above the level necessary for irreversible plastic deformation, a zone of plastically deformed material will be found at the crack tip, embedded within an elastically deformed media. Since much more energy is dissipated during plastic flow than during elastic deformation, the toughness of a notched sample should increase with the potential volume of the crack-tip plastic zone. As shown in Fig. 7.18, when the plastic zone is small just before failure, the

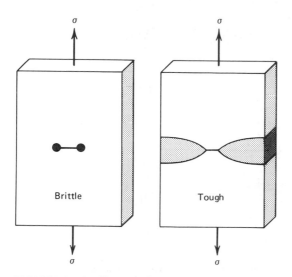

FIGURE 7.18   Extent of plastic zone development (shaded region) at fracture for brittle and tough fractures.

overall toughness level of the sample is low and the material is classified as *brittle*. On the other hand, were plasticity to extend far from the crack tip to encompass the specimen's unbroken ligament, the energy to break would be high and the material would be *tough*.

An additional measure of toughness in notched samples is that reflected by the notched to smooth bar tensile strength ratio. A low ratio reflects considerable notch sensitivity and low toughness, and a high ratio—say, greater than two—depicts high toughness behavior and minimal notch sensitivity. As will be discussed in Chapter 10, the notch test can reveal metallurgical embrittlement, with brittle behavior becoming apparent through changes in the notched to smooth bar tensile strength as a function of heat treatment.

7.7.2   Macroscopic Fracture Mode and Texture

The macroscopic crack path may also provide useful information about the toughness exhibited by a component prior to failure. This is particularly true in the case of sheet- or plate-type components where the level of toughness can be related to the relative amounts of flat and slant fracture (Fig. 7.19). For a given material, the fracture toughness values would be higher in the thin sample exhibiting a full slant fracture condition than the thick sample exhibiting a completely flat fracture mode. Correspondingly, samples with mixed-mode (i.e., part slant and part flat fracture) would

FIGURE 7.19   Fractured surfaces of aluminum test specimens revealing flat- and slant-type failure. Toughness level increases with increasing relative amount of slant fracture.

reflect an intermediate toughness level. The significance of the macroscopic fracture mode transition is discussed in greater detail in Sections 8.6 and 14.2.

As is obvious from Figs. 7.1 and 7.2, engineering service failures can generate large areas of fracture surface. Since a key element in analyzing a given failure lies in identifying the mechanism(s) by which a critical crack is developed, it is necessary for investigators to focus most of their attention on the small crack origin zone rather than the vast areas associated with rapid, unstable fracture. This is particularly important when an electron microscope fracture surface analysis is to be conducted. For example, a piecewise examination of the *entire* fracture surface of the T-2 tanker (Fig. 7.1*a*), a square millimeter at a time, would take years to complete and be extremely costly. Fortunately, the crack often leaves a series of fracture markings in its wake that may indicate the relative direction of crack motion. For example, the curved lines (called "chevron" markings) that seem to converge near midthickness of the fracture surface shown in Fig. 7.20 have been shown to point back toward the crack origin. Another example of chevron markings is shown in Fig. 8.17*c*. It is believed that within the material localized separations ahead of the crack grow back to meet the advancing crack front and form these curved tear lines. All one needs to do is follow the chevron arrow. When a crack initiates in a large component, it sometimes branches out in several directions as it runs through the structure. Although the chevron markings along each branch will point in different directions relative to the component geometry, it is important to recognize that the different sets of chevron markings all point in the same *relative* direction—back toward the origin (Fig. 7.21). Chevron markings grow out radially from an internally located origin, as shown in Fig. 7.22 for the case of a 6-cm-diameter steel (σ_{ys} = 550 MPa) reinforcing bar.

When we approach the task of describing the character of the fracture surface texture, we find a plethora of terms to excite a sympathetic image in the mind of the reader. What do you visualize when you hear the words "rock-candy-like," "woody," "glassy"? If you pictured fractured surfaces consisting of small shiny

FIGURE 7.20 Chevron markings curve in from the two surfaces and point back to the crack origin. (Courtesy of Roger Slutter, Lehigh University.)

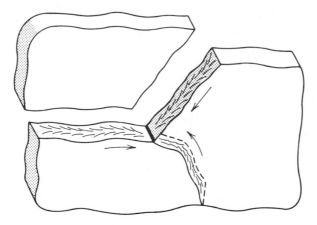

FIGURE 7.21 **Multiple chevron patterns emanating from crack origin. Each pattern points back to crack nucleation site.**

angular facets, a rough splintered surface, and a smooth featureless surface, respectively, the terms are apt descriptions of the fracture texture. As unscientific as these words may sound, they perform their descriptive function successfully. However, although such terms may create a useful mental picture of the macroscopic texture of the fracture surface, they do not provide a precise description of crack propagation through the material's microstructure.

7.7.3 Microscopic Fracture Mechanisms: Metals

As recently as 35 years ago, the light microscope was the tool most often used in the microscopic examination of fracture processes. Because of the very shallow depth of focus, examination of the fracture surface was not possible except at very low magnifications. Consequently, the fracture surface analysis procedure entailed the examination of a metallographic section containing a profile of the fracture surface. Using

FIGURE 7.22 **Chevron markings radiating from internal flaw in steel reinforcing bar. (Courtesy of Roger Slutter, Lehigh University.)**

this technique, it was possible to obtain important information about the fracture path. For example, by comparing the path of the fracture with the metallographic grain structure, it was possible to determine whether the failure was of a transcrystalline or intercrystalline nature (Fig. 7.23). This point is often more easily clarified when secondary cracks are present in the sectioned component, thereby revealing profiles of mating fracture surfaces. Since the condition of the profile edge is critical for proper failure analysis, precautions are often taken to preserve the sharpness of the fracture profile. To this end, fracture surfaces are usually plated with nickel to protect the specimen edge from rounding caused by the metallographic polishing procedure. The most widely used procedures for metallographic specimen preparation are described in a standard metallographic text.[35]

The understanding of fracture mechanisms in materials increased dramatically when the electron microscope was developed. Because its depth of field and resolution were superior to those of the light microscope, many topographical fracture surface features were observed for the first time. Many of these markings have since been applied to current theories of fracture. Until recently, much of the fractographic work had been conducted on transmission electron microscopes (TEM). Since the penetrating power of electrons is quite limited, fracture surface observations in a TEM require the preparation of a replica of the fracture surface that allows transmission of the high-energy electron beam.

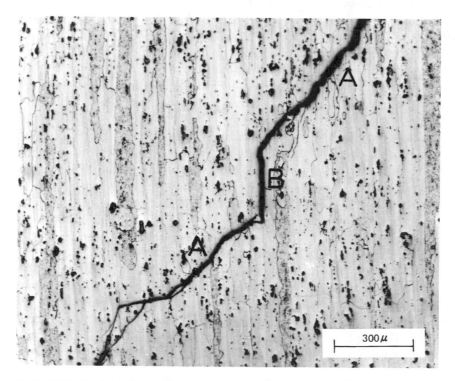

300 μ

FIGURE 7.23 Metallographic section revealing transcrystalline crack propagation at (A) and intercrystalline crack growth at (B).

During the past 20 years, however, encouraging progress also has been made in the utilization of scanning electron microscopy in failure analysis. A major advantage of the scanning electron microscope (SEM) for some examinations is that the fractured sample may be viewed directly in the instrument, thereby obviating the need for replica preparation. When it is not possible to cut the fractured component to fit into the viewing chamber, replicas must be used instead. It is anticipated that both instruments may soon be found in a typical laboratory committed to failure analysis.

Before one can proceed with an interpretation of fracture surface markings, it is necessary to review replication techniques and electron image contrast effects. To this end, the reader is referred to Appendix A. In addition, four fractography handbooks[36-39] are available that contain both discussions of techniques and thousands of electron fractographs. For the purposes of the present discussion, it is important only to recognize that the most commonly employed replication technique for the TEM leads to a reversal in the ''apparent'' fracture surface morphology. That is, electron images may suggest that the fracture surface consists of mounds or hillocks when in reality, it is composed of troughs or depressions. *Everything that looks up is really down and vice versa.* On the other hand, SEM images do not possess this height deception. For completeness and comparative purposes, most microscopic fracture mechanisms discussed in this book are described with both TEM and SEM electron images.

7.7.3.1 Microvoid Coalescence

Microvoid coalescence (MVC), observed in most metallic alloys and many engineering plastics, takes place by the nucleation of microvoids, followed by their growth and eventual coalescence. These mechanically induced micropores should not be confused with preexistent microporosity sometimes present as a result of casting or powder sintering procedures. The initiation stage has largely been attributed to either particle cracking or interfacial failure between an inclusion or precipitate particle and the surrounding matrix. Accordingly, the spacing between adjacent microvoids is closely related to the distance between inclusions. Where a given material contains more than one type of inclusion, associated with a bimodal size distribution, microvoids with different sizes are often found on the fracture surface. The criteria for void nucleation are complex and depend on several factors, including inclusion size, stress and strain levels, local deformation modes, and alloy purity.[40] Earlier computations have shown that most of the fracture energy associated with MVC is consumed during growth of the microvoids.[41] At least two growth mechanisms have been identified: (1) plastic flow of the matrix that surrounds the nucleation site, and (2) plastic flow enhanced by decohesion of small particles in the matrix. The final step of MVC that leads to final failure involves the coalescence of countless microvoids into large cracks. Often, this process occurs by the necking down of material ligaments located between adjacent microvoids, thereby leading to the impingement of the adjacent microvoids.[42] Coalescence may also proceed by linking together large microvoids with many smaller voids that form within strain-localized intense shear bands[43] (see Fig. 10.26).

The fracture surface appearance of microvoids depends on the state of stress.[44] Under simple uniaxial loading conditions, the microvoids will tend to form in association with fractured particles and/or interfaces and grow out in a plane generally normal to the stress axis. (This occurs in the fibrous zone of the cup–cone failure shown in Section 1.2.2.4.) The resulting micron-sized "equiaxed dimples" are generally spherical, as shown in Fig. 7.24a, b. Since the growth and coalescence of these voids involves a plastic deformation process, it is to be expected that total fracture energy should be related in some fashion to the size of these dimples. In fact, it has been shown in laboratory experiments that fracture energy does increase with increasing depth and width of the observed dimples.[45,46]

When failure is influenced by shear stresses, the voids that nucleate in the manner cited above grow and subsequently coalesce along planes of maximum shear stress. Consequently, those voids tend to be elongated and result in the formation of parabolic depressions on the fracture surface, as shown in Fig. 7.25a, b (such voids are found in the shear walls of the cup–cone failure). If one were to compare the orientation of these "elongated dimples" from matching fracture faces, one would find that the voids are elongated in the direction of the shear stresses and point in opposite directions on the two matching surfaces.

Finally, when the stress state is one of combined tension and bending, the resulting tearing process produces "elongated dimples," which can appear on gross planes normal to the direction of loading. The basic difference between these "elongated dimples" and those produced by shear is that the tear dimples point in the same direction on both halves of the fracture surface. It is important to note that these dimples point back toward the crack origin. Consequently, when viewing a replica that contains impressions of tear dimples, the dimples may be used to direct the viewer to the crack origin. A schematic diagram illustrating the effect of stress state on microvoid morphology is presented in Fig. 7.26.

It may be desirable to determine the chemical composition of the particle responsible for the initiation of the voids. By selected area diffraction in the TEM of particles extracted from replicas or by X-ray detector instrumentation in the SEM, it often is possible to identify the composition of particles responsible for microvoid initiation.

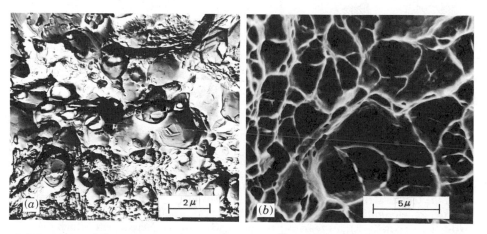

FIGURE 7.24 **Microvoid coalescence under tensile loading, which leads to "equiaxed dimple" morphology: (a) TEM fractograph shows "dimples" as mounds; (b) SEM fractograph shows "dimples" as true depressions.**

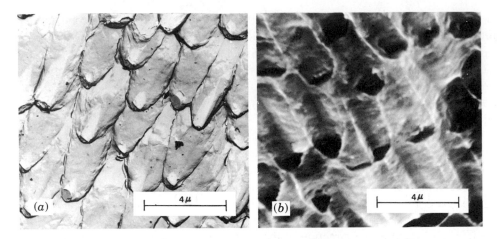

FIGURE 7.25 Microvoid coalescence under shear loading, which leads to "elongated dimple" morphology: (*a*) TEM fractograph shows "dimples" as raised parabolas; (*b*) SEM fractograph shows "dimples" as true elongated troughs.

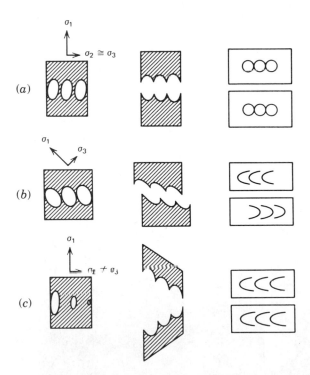

FIGURE 7.26 Diagrams illustrating the effect of three stress states on microvoid morphology: (*a*) tensile stresses produce equiaxed microvoids; (*b*) pure shear stresses generate microvoids elongated in the shearing direction (voids point in opposite directions on the two fracture surfaces); (*c*) tearing associated with nonuniform stress, which produces elongated dimples on both fracture surfaces that point back to crack origin.[44] (Reprinted with permission of the American Society of Mechanical Engineers.)

With this information, it may be possible to select a different heat treating procedure and/or select an alloy of higher purity so as to suppress the void formation initiation process.

7.7.3.2 Cleavage

The process of cleavage involves transcrystalline fracture along specific crystallographic planes and is usually associated with low-energy fracture. This mechanism is observed in BCC, HCP, and ionic and covalently bonded crystals, but occurs in FCC metals only when they are subjected to severe environmental conditions. Cleavage facets are typically flat, although they may reflect a parallel plateau and ledge morphology (Fig. 7.27a). Often these cleavage steps appear as "river patterns" wherein fine steps are seen to merge progressively into larger ones (Fig. 7.27b). It is generally believed that the "flow" of the "river pattern" is in the direction of microscopic crack propagation (from right to left in Fig. 7.27b). The sudden appearance of the "river pattern" in Fig. 7.27b was probably brought on by the movement of a cleavage crack across a high-angle grain boundary, where the splintering of the crack plane represents an accommodation process as the advancing crack reoriented in search of cleavage planes in the new grain. It is also possible that the cleavage crack traversed a low-angle twist boundary, and the cleavage steps were produced by the intersection of the cleavage crack with screw dislocations.[47]

In some materials, such as ferritic steel alloys, the temperature and strain-rate regime necessary for cleavage formation is similar to that required to activate deformation twinning (see Chapter 3). Fine-scale height elevations (so-called tongues) seen in Fig. 7.28a, b provide proof of deformation twinning during or immediately preceding failure. In BCC iron, etch pit studies have verified that these fracture surfaces consist of {100} cleavage facets and {112} tongues, the latter representing failure along twin matrix interfaces.

Little information may be obtained from cleavage facets for use in failure analyses. However, one may learn something about the phase responsible for failure by noting the shape of the facet and comparing it to the morphology of different phases in the

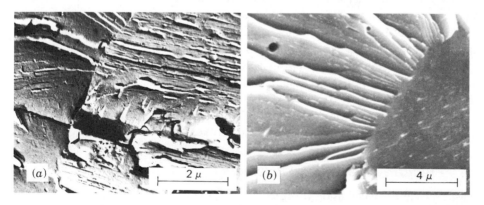

FIGURE 7.27 Cleavage fracture in a low-carbon steel. Note parallel plateau and ledge morphology and river patterns reflecting crack propagation along many parallel cleavage planes: (a) TEM; (b) SEM.

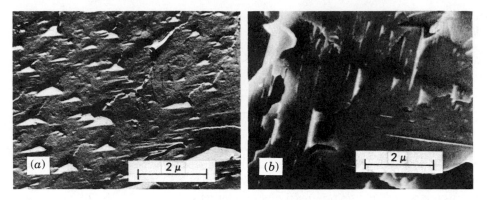

FIGURE 7.28   Cleavage facets revealing fine-scale height elevations caused by localized deflection of the cleavage crack along twin-matrix interfaces: (*a*) TEM; (*b*) SEM.

alloy. Furthermore, in materials that undergo a fracture mechanism transition (e.g., void coalescence to cleavage failure), it is possible to relate the presence of the cleavage mechanism to a general set of external conditions. In most mild steel alloys (which undergo the above fracture mechanism transition), the observation of cleavage indicates that the component was subjected to some combination of low-temperature, high-strain-rate, and/or a high tensile triaxial stress condition. This point is discussed further in Chapter 9.

7.7.3.3   Intergranular Fracture

Perhaps the most readily recognizable microscopic fracture mechanism is that of intergranular failure wherein the crack prefers to follow grain surfaces. The resulting fracture surface morphology (Fig. 7.29) immediately suggests the three-dimensional character of the grains that comprise the alloy microstructure. The occurrence of intergranular fracture can result from a number of processes. These include microvoid nucleation and coalescence at inclusions or second-phase particles located along grain bundaries; grain-boundary crack and cavity formations associated with elevated temperature stress rupture conditions; decohesion between contiguous grains due to the presence of impurity elements at grain boundaries and in association with aggressive atmospheres such as gaseous hydrogen and liquid metals (Sections 10.6 and 11.1); and stress corrosion cracking processes associated with chemical dissolution along grain boundaries. Additional discussion of intergranular fracture under cyclic loading conditions is found in Section 13.7. Also, if the material has an insufficient number of independent slip systems (see Chapter 3) to accommodate plastic deformation between contiguous grains, grain-boundary separation may occur.

7.7.4   Fracture Mechanisms: Polymers

As was the case for metal alloy fracture surfaces, the fracture surface micromorphology of engineering plastics reflects both the microstructure and the deformation mechanisms of these materials. Let us consider first the fracture surface appearance of an amorphous plastic.[48–52] As discussed in Section 6.4, deformation in many amorphous

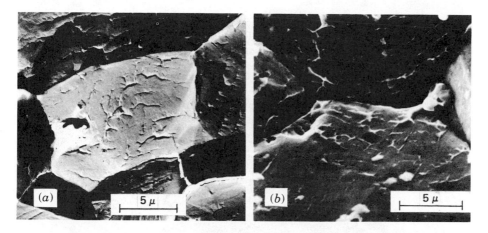

FIGURE 7.29 Intergranular fracture in maraging steel tested in the gaseous hydrogen environment: (*a*) TEM; (*b*) SEM.

polymers involves the formation of thin crazes that contain interconnected microvoids and polymer fibrils extended in the craze thickness direction (recall Fig. 6.28). Subsequent fracture then occurs usually in two stages, typified by either a mirrorlike or a misty macroscopic fracture surface appearance. The first zone, which is smooth and highly reflective, is appropriately referred to as the "mirror" area. The mirror region also exhibits colorful patterns that reflect the presence of a layer of material (craze matter) with a different refractive index on the fracture surface.[53-55] One color indicates the existence of a single craze with a uniform thickness; packets of multicolor fringes reflect the presence of a few craze layers at the fracture surface or a single craze with variable thickness.

In the mirror region, fracture progresses through precrazed matter in one of several ways. Initially, the crack grows by the nucleation and growth of conically shaped voids along the midplane of the craze (Figs. 7.30, Region A, and 7.31). When these pores are viewed with transmitted light, they appear as a series of concentric rings that correspond to changes in the crazed thickness on the fracture surface (Fig. 7.31*b*). As the crack velocity increases, the crack tends to grow alternately along one craze–matrix boundary interface and then the other (Fig. 7.30, Region B).[49,50] Consequently, the fracture surface contains islands or patches of craze matter attached to one-half of the fracture surface (Figs. 7.31*a* and 7.32*a*) which mate with detachment zones on

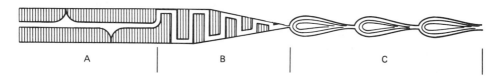

FIGURE 7.30 Model of crack advance in association with craze matter. Region A: crack advance by void formation through craze midplane. Region B: crack advance along alternate craze–matrix interfaces to form patch or mackerel patterns. Region C: crack advance through craze bundles to form hackle bands. (Adopted from Hull.[48])

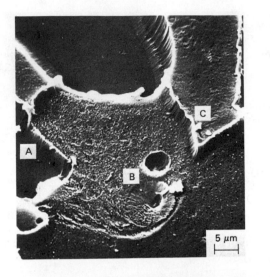

(a)

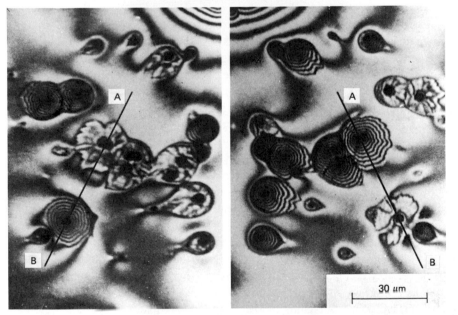

(b)

FIGURE 7.31 Void formation. (a) SEM image revealing two conical voids along mid-plane of craze (B). Craze–matrix detachment at A and C. Note fibril extension between AB and BC. (b) Transmitted light images revealing concentric rings on one surface that match with relatively featureless zones on the mating fracture surface. (From Hull,[48] "Nucleation and Propagation Processes in Fracture," *Polymeric Materials*, American Society for Metals, 1975, p. 524.)

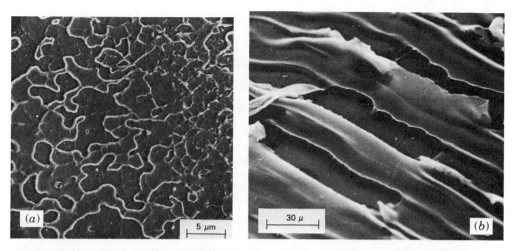

FIGURE 7.32 Crack advance from left to right along alternate craze–matrix interfaces in polystyrene. (*a*) Patch pattern. Note decreasing patch size in crack propagation direction. (Courtesy of Clare Rimnac, Lehigh University.) (*b*) Stereoscan micrograph showing bands of craze separated by areas where the craze has been stripped off. Mackerel pattern. Fully detached craze regions appear darker in this photograph. (Murray and Hull;[50] with permission from John Wiley & Sons, Inc.)

the mating surface. Note evidence for fibril elongation along the walls of the craze patches (Fig. 7.31*a*). These detached craze zones tend to decrease in size in the crack direction since crazes tend to get thinner toward the craze tip (Figs. 7.30, Region B, and 7.32*a*). Craze matter also can detach alternately from each craze–matrix interface to produce a series of fracture bands that are parallel to the advancing crack front (Fig. 7.32*b*). These markings are referred to in the literature as "mackerel" bands, since they are reminiscent of scale markings along the flanks of the mackerel fish.

During the terminal stage of fast fracture, the crack front outpaces the craze tip, with continued crack extension taking place through bundles of secondary crazes that form at the crack tip. The banded nature of this surface (referred to as "hackle" bands) suggests that the crack propagates through one bundle of crazes, at which point a new craze bundle is formed and the process is repeated (Figs. 7.30, Region C, and 7.33*a*).[51] A close examination of the hackle zone reveals the patch morphology described above (Fig. 7.33*b*). Since this part of the fracture surface is much rougher and contains bundles of crazes, it takes on a misty appearance. Parabolic-shaped voids are also found on some portions of the mist region. Similar to the elongated dimples associated with the tearing mode of fracture (Fig. 7.25), these parabolic-shaped voids are produced by the nucleation and growth of cavities ahead of the advancing crack front, with the extent of cavity elongation depending on the relative rates of cavity growth and crack-front advance. Figure 7.34 reveals several elongated tear dimples in ABS polymer which were nucleated at duplex butadiene–polystyrene particles (also see Fig. 10.40).

The fracture surface appearance of the semicrystalline polymers depends on the crack path with respect to underlying microstructural features. For example, a crack

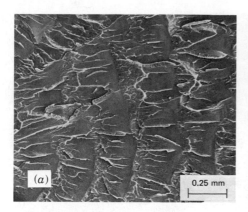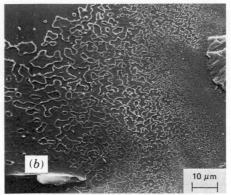

FIGURE 7.33   Banded hackle markings in fast fracture region. (*a*) Crack advances in jumps through craze bundles. (*b*) Patch appearance on hackle band surface. Crack propagation from left to right. (Courtesy of Clare Rimnac, Lehigh University.)

may choose an interspherulitic crack path or pass through the spherulite along a tangential or radial direction.[56] Four possible crack paths through or around a spherulite are shown in Figure 7.35*a*, with respect to the orientation of crystal lamellae in the spherulite (Fig. 7.35*b*). Examples of the fracture surface micromorphologies corresponding to these four paths in spherulitic polypropylene are shown in Fig. 7.35*c* for the case of a fast-running crack. A completely interspherulitic fracture path is observed when the crack velocity was low (Fig. 7.35*d*). The similarity in appearance between intergranular fracture in maraging steel (Fig. 7.29) and interspherulitic fracture in polypropylene is truly striking. It should be noted that fractographic evidence for transspherulitic or interspherulitic failure may be obscured by extensive prior deformation of the polymer, which distorts beyond recognition characteristic details of the underlying microstructure.

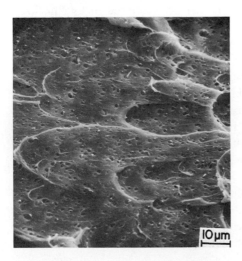

FIGURE 7.34   Tear dimples in Noryl polymer. Microvoid nucleation at butadiene–polystyrene duplex particles. (Courtesy of Clare Rimnac, Lehigh University.)

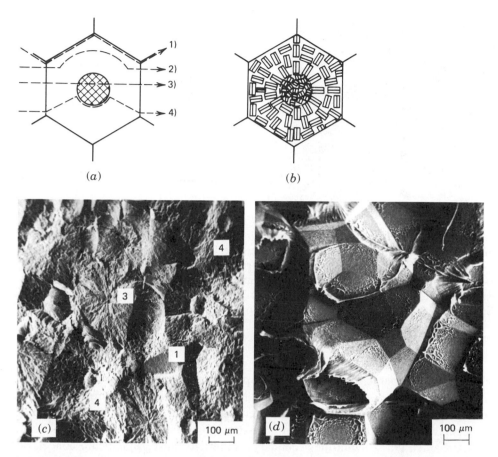

FIGURE 7.35 Fracture associated with spherulites in crystalline polymers. (*a*) Schema showing possible crack paths through a spherulite. (*b*) Orientation of crystal lamellae in spherulite. Lamellae are believed to be randomly oriented in core region, radially oriented in midregion, and tangentially oriented along surface of spherulite. (*c*) Fast running crack fracture surface in polypropylene revealing the four crack paths as outlined in (*a*). (*d*) Interspherulitic fracture in polypropylene associated with slow crack velocity. (From Friedrich;[56] reprinted with permission from *Fracture 1977*, Vol. 3, 1977, p. 1119, Pergamon Press.)

7.7.5 Fracture Surfaces of Ceramics

The fracture surfaces of brittle solids, such as glassy and crystalline ceramics, often reveal several characteristic regions as shown in Fig. 7.36.[57–65] Surrounding the crack origin is a mirror region associated with a highly reflective fracture surface. This smooth area is bordered by a misty region that contains small radial ridges associated with numerous microcracks. The mist region in turn is surrounded by an area that is rougher in appearance and contains larger secondary cracks. Depending on the size of the sample, this hackle region may be bounded by macroscopic crack branching. The occurrence of these different regions depends on such factors as the specimen

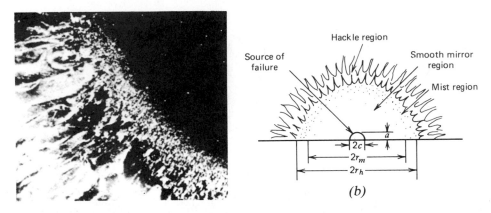

FIGURE 7.36   Fracture surface appearance in glassy ceramic revealing mirror, mist, and hackle regions, (*a*) Plate-glass fracture surface. Tensile fracture stress = 28.3 MPa, or 4.1 ksi. (From Orr.[64]) Arrow indicates approximate crack origin. (*b*) Schematic diagram showing different fracture regions and approximate textural detail. (From Mecholsky et al.[59] Copyright, American Society for Testing and Materials, Philadelphia, PA. Reprinted with permission.)

size and the material in question. For example, if the strength of the glass is low, it is not uncommon for the mirror zone to extend across the entire fracture surface.[63] In addition, Abdel-Latif et al[63] have called attention to the fact that multiple mist zones can occur within the hackle region in high-strength specimens of soda lime silicate and fused silica glasses with the secondary and tertiary mist bands being concentric with the initial mist region shown in Fig. 7.36. Different theories have been proposed to explain the onset of these different fracture zones, but the subject remains a topic of considerable controversy.[62,63] (Although the terms mirror, mist, and hackle are identical to those used in describing the fracture surface appearance of glassy polymers, the associated fracture micromechanisms are different; recall Section 7.7.4 and compare with nomenclature to describe fracture surface markings in ceramics.[66])

Of great significance, the radii of these various fracture regions have been shown to vary inversely with the fracture stress σ, with a relation of the form

$$\sigma(r_{m,h,cb})^{1/2} = M_{m,h,cb} \qquad (7\text{-}37)$$

where σ = fracture stress, in MPa
   $r_{m,h,cb}$ = radius from crack origin to mirror–mist, mist–hackle, and crack branching boundaries, respectively
   $M_{m,h,cb}$ = ''mirror constant'' corresponding to r_m, r_h and r_{cb}

Logarithmic plots of fracture stress as a function of the mist–hackle boundary radius are shown in Fig. 7.37 for several polycrystalline ceramic materials. Note the inverse square root dependence of r on the fracture stress and the fact that the mirror constant

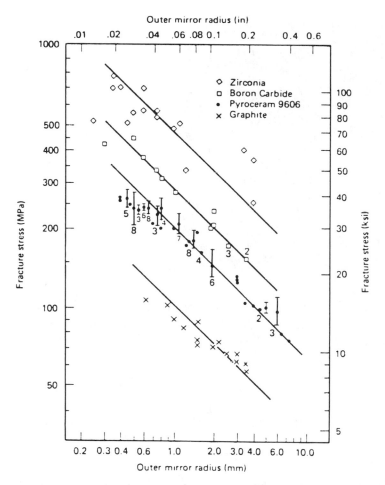

FIGURE 7.37 **Fracture stress versus mist–hackle boundary radius, r_h, for four polycrystalline ceramics. (From Mecholsky et al.,[65] *J. Mater. Sci.* 11, 1310 (1976), with permission from Chapman & Hall.)**

is different for the four materials shown. The relation between the mirror constant and other fracture properties is discussed in Section 10.4.3. For additional discussion of the fracture surface appearance of ceramics, see the extensive review by Rice and several other papers contained in that reference.[67]

EXAMPLE 7.5

The fracture surface of a glass-fiber reinforced polyethylene terephthalate (PET) composite reveals the presence of ruptured glass fibers. Estimate the stress at fracture for the fiber identified by ''A'' in the accompanying photograph (courtesy H. Azimi).

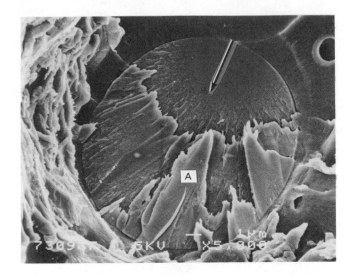

This broken fiber shows a clear example of a mirror-mist zone (see arrow) where the radius is measured to be 4.6 μm. Since the mirror constant for E-glass is approximately 2 MPa$\sqrt{\text{m}}$ (see Fig. 10.32), it follows from Eq. 7-14 that the stress to cause fracture of this fiber is

$$M_m = \sigma\sqrt{r_m}$$

$$2 \times 10^6 = \sigma\sqrt{4.6 \times 10^{-6}}$$

$$\therefore \sigma = 933 \text{ MPa}$$

This value is smaller than estimates for the tensile strength of E-glass (see Table 1.9) (i.e., 1400–2500 MPa) and may reflect fiber damage during manufacture of the composite. (For further analysis of fracture marking-inferred stress level determinations of embedded fibers, see A. C. Jaras, B. J. Norman, and S. C. Simmens, *J. Mater. Sci.,* **18,** 2459 (1983).

7.7.6 Fracture Mechanisms in Composites

It is not possible to describe in detail the fracture surface appearance of the large number of composites that are commercially available since their fracture micromorphologies vary markedly and depend on the fracture characteristics of the many reinforcements and matrices that are chosen. In addition, fracture surfaces and fracture paths are found to vary with the nature of the fiber–matrix interface and its dependence on the environment and temperature. Too many material/test conditions combinations—too many fractographs—too little available space in the text! Some comments are appropriate, however, regarding the fracture surface appearance of selected fibrous and particulate composites. Fracture of composites can occur in a cohesive (matrix phase adheres to the fiber) or adhesive (no matrix phase residue on fiber) manner.

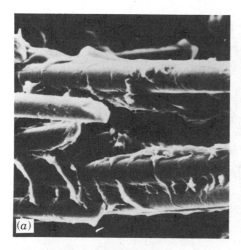 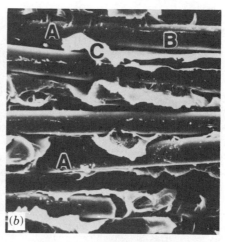

FIGURE 7.38   **Fracture surface appearance in short glass-fiber reinforced nylon 66. (a) Fast fracture appearance revealing matrix adhesion to glass fibers. (b) Fatigue fracture revealing fiber–matrix debonding (A and B) and matrix drawing (C). (Courtesy R. Lang.)**

Examples of these two different fracture paths are shown in Fig. 7.38 for a nylon 66–glass fiber composite.[68] The strong bond between fiber and matrix in this composite is confirmed by the cohesive failure produced under monotonic tensile loading conditions (Fig. 7.38a). Note that the glass fibers are coated with the nylon 66 matrix material. When the sample was cyclically loaded to produce a fatigue crack, fracture took place along the fiber–matrix interface (Fig. 7.38b). Adhesive failures are expected to prevail when the fiber–matrix interface is weak. However, interfacial failure does occur in fatigue, with monotonic loading at elevated temperatures and in the presence of a moist environment, even though the fiber–matrix interface is deemed to be strong.

An additional example of interfacial fracture between particle and matrix is shown in Fig. 7.39. Here we see the fatigue fracture surface of an epoxy resin containing hollow glass beads.[69] Figure 7.39a shows that some of the beads are retained on this half of the fracture surface, whereas other particles were either pulled out or shattered. Note the fracture step on the trailing surface of many glass beads, which represents the location where the crack front linked up after passing around opposite sides of the spherical glass particle. The silicon x-ray map shown in Fig. 7.39b clearly reveals the presence of both undamaged and fragmented hollow glass spheres on the fracture surface. Figure 7.40 reveals that fracture surface features in graphite–epoxy composites change with the mode of loading.[70] Under tensile loading conditions, interply failure reveals a cleavagelike appearance in the epoxy resin as evidenced by the presence of river patterns, which point in the crack propagation direction (Fig. 7.40a). When shear stresses dominate the fracture process, the epoxy resin reveals a series of hackle markings on the fracture surface, which presumably reflect the coalescence of many tension-induced microcracks inclined at an angle to the overall fracture plane (Fig. 7.40b). For more complete reviews of the fracture surface appearance in composites, see references 37, 70, and 71.

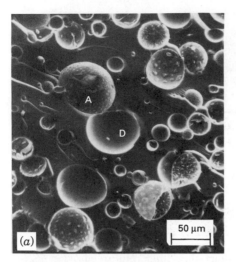
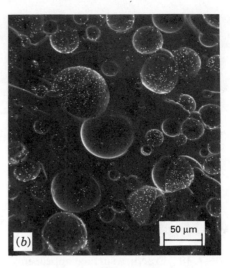

FIGURE 7.39 Fracture surface in hollow glass sphere-filled epoxy resin. (*a*) Glass particles located at A, detached at D, and cracked at C. (*b*) Silicon X-ray map showing location of glass spheres. (Courtesy M. Breslaur.)

7.7.7 Quantitative Fractography

Having identified several fractographic features in different engineering solids, a brief discussion of their size in relation to appropriate stress parameters is appropriate. To establish such relations requires precise measurement of these fracture surface details as typically observed in the electron microscope. Since the SEM provides a *projected* view of the fracture surface and TEM replicas collapse onto their support grids, the linear and areal dimensions derived from such observations are not accurate and provide essentially no quantitative information pertaining to the roughness (i.e., non-

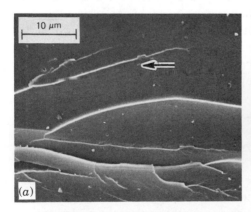
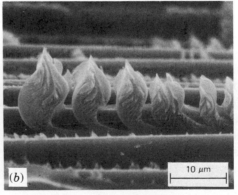

FIGURE 7.40 Fracture surface appearance in graphite fiber-reinforced epoxy. (*a*) Tensile failure revealing cleavagelike pattern. Arrow indicates crack direction. (*b*) Shear failure reveals presence of hackle markings in epoxy resin.[70] (Reprinted with permission. Copyright ASTM).

planar nature) of the fracture surface. Recent studies[68-77] have shown that the key to quantitative fractography lies in the determination of the true fracture surface area through stereological measurements. Several parameters have been defined and relations developed to describe the roughness of the fracture surface. The experimental procedure currently preferred involves taking successive vertical sections of the fracture surface to obtain crack profiles as a function of distance across the sample.[72] Digitizing image analysis systems are then used to trace the actual fracture surface profile with

$$R_L = \frac{L_t}{L_p} \tag{7-38}$$

where R_L = linear roughness parameter
L_t = true crack profile length
L_p = projected crack length

In turn, the surface roughness parameter R_s—a difficult quantity to measure directly—can be estimated from Eq. 7-39[74]

$$R_s = \frac{4}{\pi} (R_L - 1) + 1 \tag{7-39}$$

where $R_s = S_t/A_p$, the true surface area divided by its projected area.

Equations 7-38 and 7-39 have been used to obtain the true microvoid diameter and intergranular facet sizes in 4340 steel and Al–4Cu alloys, respectively.[78] In both instances, R_s values were in excess of 2.0, which means that the true surface areas of these fracture surface features actually were more than 100% larger than measurements based on SEM projected images. Clearly one must exercise considerable care when making such measurements.

Further efforts to quantify fracture surface markings have made use of fractal analysis.[73,79,80] Mathematical concepts of fractals recognize that the length of an irregular profile (e.g., a geographical coastline or fracture profile) increases as the size of the measurement unit decreases. The theoretical relation that characterizes this behavior is given by

$$L(\eta) = L_0 \eta^{-(D-1)} \tag{7-40}$$

where $L(\eta)$ = apparent length of profile as affected by the measuring unit
η = measurement unit
L_0 = constant with dimensions of length
D = fractal dimension

Accordingly, the apparent fracture dimension is observed to depend on the measurement unit η and the fractal dimension D. Recent studies have shown that D is not constant but varies somewhat as a function of microstructure and measurement unit.

Accordingly, a modified form of Eq. 7-40 has been proposed.[73] Of interest, Mecholsky and Passoja[81] have suggested that the fracture resistance of brittle ceramics may be characterized by a parameter that is related to the fractal dimension.

REFERENCES

1. *Scientific American* **120,** 99 (Feb. 1, 1919).
2. *Engineering News-Record* **82**(20), 974 (May 15, 1919).
3. E. R. Parker, *Brittle Behavior of Engineering Structures,* John Wiley & Sons, New York (1957).
4. W. Garzke, D. Yoerger, S. Harris, Society of Naval Architects and Marine Engineers, Centennial Meeting, New York, 15–1 September 15, 1993.
5. R. Gannon, *Popular Science,* **246**(2), 49 (1995).
6. W. J. McGregor Tegart, *Elements of Mechanical Metallurgy,* Macmillan, New York (1966).
7. W. B. Parsons, *Engineers and Engineering in the Renaissance,* Williams and Wilkins, Baltimore, 72 (1939).
8. W. Weibull, *Proceedings 151,* Stockholm: Royal Swedish Academy of Engineering Sciences (1939).
9. W. Weibull, *J. Appl. Mech.* **18**(3), 293 (1951).
10. W. Weibull, *J. Mech. Phys. Solids* **8,** 100 (1960).
11. R. W. Davidge, *Mechanical Behavior of Ceramics,* Cambridge Univ. Press, Cambridge (1979).
12. C. E. Inglis, *Proceedings, Institute of Naval Architects* **55,** 219 (1913).
13. R. E. Peterson, *Stress Concentration Design Factors,* John Wiley & Sons, New York, 1974.
14. H. Neuber, *Kerbspannungslehre,* Springer, Berlin; English translation available from Edwards Bros., Ann Arbor, MI, 1959.
15. W. Jordan, *Proceedings, American Numismatical and Antiquarian Society* **27,** 109 (1915).
16. *ASME Handbook—Metals Engineering—Design,* McGraw-Hill, New York, 1953.
17. W. Moffatt and J. Wulff, *J. Met.,* 440 (Apr. 1957).
18. Livy, xxi, 37.2.
19. Pliny, *Historia Naturalis,* xxiii. 27.57.
20. G. Agricola, *De Re Metallica,* translated from the first Latin edition of 1556 by H. C. Hoover and L. H. Hoover, Dover Publications, New York, 1950, p. 119.
21. G. de Beer, *Alps and Elephants—Hannibal's March,* Geoffrey Bles, London, 1955.
22. D. P. H. Hasselman, *J. Am. Ceram. Soc.* **52,** 600 (1969).
23. D. P. H. Hasselman, and J. P. Singh, *Thermal Stresses I,* Vol. 1, R. B. Hetnarski, Ed., North-Holland, New York, Chap. 4, 264 (1986).
24. L. G. Van Uitert et al., *Mater. Res. Bull.* **12,** 261 (1977).
25. *Microelectronics Packaging Handbook,* R. R. Tummala and E. J. Rymaszewski, Eds., Van Nostrand Reinhold, New York, 278 (1989).
26. D. Burgreen, *Elements of Thermal Stress Analysis,* C. P. Press, New York, 462 (1971).

27. R. A. Miller and C. E. Lowell, *Thin Solid Films* **95,** 265 (1982).
28. J. C. Jaeger, *Phil Mag.* **36,** 419 (1945).
29. R. W. Davidge, *Mechanical Behavior of Ceramics,* Cambridge University Press, Cambridge, 1979.
30. R. C. Rossi, *Ceram. Bull.* **48**(7), 736 (1969).
31. R. C. Rossi, *J. Amer. Ceram. Soc.* **52**(5), 290 (1969).
32. D. R. Larson, J. A. Coppola, D. P. H. Hasselman, and R. C. Bradt, *J. Amer. Ceram. Soc.* **57,** 417 (1974).
33. D. P. H. Hasselman, *J. Amer. Ceram. Soc.* **53,** 490 (1970).
34. D. P. H. Hasselman, *Ceramics in Severe Environments,* W. W. Kriegel and H. Palmour, III, Plenum Press, New York, 1971, p. 89.
35. G. L. Kehl, *Principles of Metallographic Laboratory Practice,* McGraw-Hill, New York, 1949.
36. A. Phillips, V. Kerlins, and B. V. Whiteson, *Electron Fractography Handbook,* AFML TDR-64-416, WPAFB, Ohio, 1965.
37. *Metals Handbook,* Vol. 12, American Society of Metals, Metals Park, OH, 1987.
38. L. Englel, H. Klingele, G. W. Ehrenstein, and H. Schaper, *An Atlas of Polymer Damage,* Prentice-Hall, Englewood Cliffs, NJ, 1981.
39. *Fractography Handbook,* Chubu Keiei Kaihatsu Center, Nagoya, Japan, 1985.
40. R. H. VanStone, T. B. Cox, J. R. Low, Jr., and J. A. Psioda, *Int. Met. Rev.* **30**(4), 157 (1985).
41. D. A. Shockey, L. Seaman, K. C. Dao, and D. R. Curan, Report NP-701-SR, EPRI, Palo Alto, CA, 1978.
42. A. H. Cottrell, *Fracture,* B. L. Averbach, Ed., Tech. Press of MIT, New York, 1959, Chap. 20, p. 20.
43. T. B. Cox and J. R. Low, Jr., *Met. Trans.* **5A,** 1457 (1974).
44. C. D. Beachem, *Trans. ASME J. Basic Eng. Ser. D,* **87,** 299 (1965).
45. A. J. Birkle, R. P. Wei, and G. E. Pellissier, *Trans. ASM* **59,** 981 (1966).
46. D. R. Passoja and D. C. Hill, *Met. Trans.* **5,** 1851 (1974).
47. J. J. Gilman, *Trans. Met. Soc. AIME* **212,** 310 (1958).
48. D. Hull, *Polymeric Materials,* ASM, Metals Park, OH, 1975, p. 487.
49. J. Murray and D. Hull, *Polymer* **10,** 451 (1969).
50. J. Murray and D. Hull, *J. Polym. Sci. A-2* **8,** 583 (1970).
51. D. Hull, *J. Mater. Sci.* **5,** 357 (1970).
52. R. P. Kusy and D. T. Turner, *Polymer* **18,** 391 (1977).
53. J. P. Berry, *Nature (London)* **185,** 91, (1960).
54. R. P. Kambour, *J. Polym. Sci. Part A* **3,** 1713 (1965).
55. R. P. Kambour, *J. Polym. Sci. Part A-2* **4,** 349 (1966).
56. K. Friedrich, *Fracture 1977,* Vol. 3, ICF4, Waterloo, Canada, 1977, p. 1119.
57. E. B. Shand, *J. Am. Ceram. Soc.* **37**(12), 572 (1954).
58. N. Terao, *J. Phys., Proc. Phys. Soc., Japan* **8,** 545 (1953).
59. J. J. Mecholsky, S. W. Freiman, and R. W. Rice, ASTM *STP 645,* 1978, p. 363.
60. J. J. Mecholsky and S. W. Freiman, ASTM *STP 678,* 1979, p. 136.
61. A. I. A. Abdel-Latif, R. C. Bradt, and R. E. Tressler, *Int. J. Fract.* **13**(31), 349 (1977).
62. J. J. Mecholsky and S. W. Freiman, ASTM *STP 733,* 1981, p. 246.
63. A. I. A. Abdel-Latif, R. C. Bradt, and R. E. Tressler, ASTM *STP 733,* 1981, p. 259.

64.   L. Orr, *Materials Research and Standards,* **12**(1), 21 (1971).
65.   J. J. Mecholsky, S. W. Freiman, and R. W. Rice, *J. Mater. Sci.* **11,** 1310 (1976).
66.   V. D. Frechette, ASTM *STP 827,* 1984, p. 104.
67.   R. W. Rice, ASTM, *STP 827,* 1984, p. 5.
68.   R. W. Lang, J. A. Manson, and R. W. Hertzberg, *Polym. Eng. Sci.* **22,** 982 (1982).
69.   M. Breslaur, private communication.
70.   B. W. Smith and R. A. Grove, ASTM *STP 948,* 1987, p. 154.
71.   *Fractography of Modern Engineering Materials: Composites and Metals,* ASTM *STP 948,* J. E. Masters and J. J. Au, Eds., Philadelphia, PA, 1987.
72.   E. E. Underwood and K. Banerji, *Metals Handbook*, Vol. 12 *Fractography*, 9th ed., ASM Int., Metals Park, OH, 1987, p, 193.
73.   E. E. Underwood and K. Banerji, *Mater. Sci. Eng.* **80,** 1 (1986).
74.   E. E. Underwood and K. Banerji, *Acta Stereol.* **2** (Suppl. 1), 75 (1983).
75.   E. E. Underwood, *Applied Metallography,* G. F. Vander Voort, Ed., Chapter 8, Van Nostrand Reinhold, New York, 1986, p. 101.
76.   M. Coster and J. L. Chermant, *Int. Met. Rev.* **28**(4), 228 (1983).
77.   S. M. Soudani, *Metallography* **7,** 271 (1974).
78.   E. E. Underwood, *J. Met.* **38,** 30 (1986).
79.   B. B. Mandelbrot, *The Fractal Geometry of Nature,* Witt. Freeman, San Francisco, 1982.
80.   E. E. Underwood and K. Banerji, *Metals Handbook*, Vol. 12, *Fractography,* 9th ed., ASM Int., Metals Park, OH, 1987, p. 211.
81.   J. J. Mecholsky and D. E. Passoja, *Fractal Aspects of Materials,* R. B. Leibowitz, B. B. Mandelbrot, and D. E. Passoja, Eds., Materials Research Society, Pittsburgh, PA, 1985, p. 117.

PROBLEMS

7.1   For a given engineering alloy, it is found that the notched tensile strength decreased with increasing notch depth (assuming a constant notch root radius) to a point beyond which σ_{net} began to increase. Explain this behavior.

7.2   Demonstrate to yourself that the intersection of a cleavage crack with a screw dislocation will produce a step on the fracture surface and that no step will result from intersection of the crack with an edge dislocation.

7.3   For most of the configurations shown in Fig. 7-9, k_t increases toward infinity as the notch root radius to minor diameter ratio approaches zero. In practice, such high k_t values are never experienced. Why?

7.4   Since cleavage in BCC crystals is preferred along (100) planes, consider what might happen to the fracture energy of a randomly oriented polycrystalline sample if the grain size were reduced severalfold. (Trace the crack path across the grain boundaries.)

7.5   Might the molasses tank that fractured in Boston have remained intact had it been erected in the tropical zone?

7.6   It has been shown that carefully prepared rods of silicon can withstand extremely high stresses before fracturing. When failure does occur, the sample explodes into powder. Why?

7.7   You are to examine the fracture surfaces from a metal that failed by microvoid coalescence and a glassy polymer that revealed a patch-type fracture morphology. From your knowledge of these fracture mechanisms, how would the mating fracture surfaces appear?

7.8   Estimate the stress necessary to cause fracture in boron carbide if a mirror–mist–hackle fracture surface morphology transition takes place and the mist–hackle boundary radius is measured to be 1 mm. Also, compute the "mirror constant" for this material.

7.9   A 1-cm-diameter rod of alumina was initially gripped between two rigid supports at a temperature of 25°C. Some time later when the temperature had decreased to 0°C, the supports were moved apart by an unknown amount, at which point the rod fractured. What load was necessary to break the rod? The relevant properties for the alumina rod are: $E = 385$ GPa, $\sigma_{ts} = 205$ MPa, $\alpha = 6.7 \times 10^{-6}/°C$.

7.10   A glass rod was attached between two rigid fixtures. A sudden drop in temperature caused the rod to fracture. The fracture surface revealed a mirror zone with an outer radius of 1 mm. Determine the magnitude of the temperature drop, given that the mirror constant and coefficient of thermal expansion for the glass are 2 MPa$\sqrt{m}$ and $7.5 \times 10^{-6}/°C$, respectively.

7.11   **(a)**   Describe the difference between Young's modulus and the Weibull modulus.

   (b)   How do you measure the two?

   (c)   Characterize two different materials if their elastic and Weibull modulii are 70 and 385 GPa and 60 and 4, respectively.

   (d)   For equal probabilities of failure in two materials, characterized by Weibull modulii of 3 and 30, how would the fracture stress change for each material if the volume of the test bar were to increase by an order of magnitude?

CHAPTER 8

ELEMENTS OF FRACTURE MECHANICS

As outlined in the previous chapter, the fracture behavior of a given structure or material will depend on stress level, presence of a flaw, material properties, and the mechanism(s) by which the fracture proceeds to completion. The purpose of this chapter is to develop quantitative relations between some of these factors. With knowledge of these relations, fracture phenomena may be better understood and design engineers more equipped to anticipate and thus prevent structural deficiencies. In addition to several sample problems given in this chapter, other real case histories are discussed in Chapter 14, which bear upon the material discussed below.

8.1 GRIFFITH CRACK THEORY

The quantitative relations that engineers and scientists use today in determining the fracture of cracked solids were initially stated some 75 years ago by A. A. Griffith.[1] Griffith noted that when a crack is introduced to a stressed plate of elastic material, a balance must be struck between the decrease in potential energy (related to the release of stored elastic energy and work done by movement of the external loads) and the increase in surface energy resulting from the presence of the crack. Likewise, an existing crack would grow by some increment if the necessary additional surface energy were supplied by the system. This "surface energy" arises from the fact that there is a nonequilibrium configuration of nearest neighbor atoms at any surface in a solid. For the configuration seen in Fig. 8.1, Griffith estimated the surface energy term to be the product of the total crack surface area $(2a \cdot 2 \cdot t)$, and the specific surface energy γ_s, which has units of energy/unit area. He then used the stress analysis of Inglis[2] for the case of an infinitely large plate containing an elliptical crack and computed the decrease in potential energy of the cracked plate to be $(\pi\sigma^2 a^2 t)/E$. Hence, the change in potential energy of the plate associated with the introduction of a crack may be given by

$$U - U_0 = -\frac{\pi\sigma^2 a^2 t}{E} + 4at\gamma_s \qquad (8\text{-}1)$$

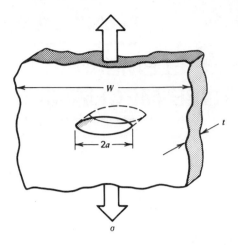

FIGURE 8.1 **Through-thickness crack in a large plate.**

where U = potential energy of body with crack
        U_0 = potential energy of body without crack
        σ = applied stress
        a = one-half crack length
        t = thickness
        E = modulus of elasticity
       γ_S = specific surface energy

By rewriting Eq. 8-1 in the form

$$U = 4at\gamma_S - \frac{\pi\sigma^2 a^2 t}{E} + U_0 \tag{8-2}$$

and determining the condition of equilibrium by differentiating the potential energy U with respect to the crack length and setting equal to zero

$$\frac{\partial U}{\partial a} = 4t\gamma_S - \frac{2\pi\sigma^2 at}{E} = 0 \tag{8-3}$$

($\partial U_0/\partial a = 0$, since U_0 accounts for the potential energy of the body without a crack and does not vary with crack length). Therefore

$$2\gamma_S = \frac{\pi\sigma^2 a}{E} \tag{8-4}$$

which represents the equilibrium condition.

     The nature of the equilibrium condition described by Eqs. 8-3 and 8-4 is determined by the second derivative, $\partial^2 U/\partial a^2$. Since

$$\frac{\partial^2 U}{\partial a^2} = -\frac{2\pi\sigma^2 t}{E} \tag{8-5}$$

and is negative, the equilibrium condition described by Eq. 8-3 is unstable, and the crack will always grow.

Griffith rewrote Eq. 8-4 in the form

$$\sigma = \sqrt{\frac{2E\gamma_S}{\pi a}} \qquad (8\text{-}6)$$

for the case of plane stress (biaxial stress conditions), and

$$\sigma = \sqrt{\frac{2E\gamma_S}{\pi a(1 - \nu^2)}} \qquad (8\text{-}7)$$

for the case of plane strain (triaxial stress conditions associated with the suppression of strains in one direction).

Since Poisson's ratio ν is approximately 0.25 to 0.33 for many materials, the difference in allowable stress level in a given material subjected to plane–strain or plane–stress conditions does not appear to be large. However, major differences do arise, as will be discussed in Section 8.6.

It is important to recognize that the Griffith relation was derived for an elastic material containing a very sharp crack. Although Eqs. 8-6 and 8-7 do not explicitly involve the crack-tip radius ρ, as was the case for the stress concentration in Eq. 7-27, the radius is assumed to be very sharp. As such, the Griffith relation, as written, should be considered necessary but not sufficient for failure. The crack–tip radius also would have to be atomically sharp to raise the local stress above the cohesive strength.

8.1.1 Verification of the Griffith Relation

The other half of Griffith's classic paper was devoted to experimentally confirming the accuracy of Eqs. 8-6 and 8-7. Thin round tubes and spherical bulbs of soda lime silica glass were deliberately scratched or cracked with a sharp instrument, annealed to eliminate any residual stresses associated with the cracking process, and fractured by internal pressure. By recording the crack size and stress at fracture for these glass samples, Griffith was able to compute values of $\sigma\sqrt{a}$ in the range of 0.25 to 0.28 MPa $\sqrt{m}$ (0.23 to 0.25 ksi $\sqrt{in.}$),* which correspond to values of $\sqrt{2\gamma_S E/\pi}$ (Eq. 8-6). From experimentally determined surface tension values γ_S for glass fibers between 745 and 1110°C, a room temperature value was obtained by extrapolation (risky business, but reasonable for a first approximation). By multiplying this value, 0.54 N/m, by the modulus of elasticity, 62 GPa, the value of $\sqrt{2\gamma_S E/\pi}$ determined from material properties was found to be 0.15 MPa $\sqrt{m}$. It should be recognized that the exceptional agreement between theoretical and experimental values may be somewhat fortuitous in light of some inaccuracies contained in the original development by Griffith.[1] For example, the second law of thermodynamics requires some inefficiency in transfer of energy from stored strain energy to surface energy. Also, recent estimates

---

* To convert from ksi $\sqrt{in.}$ to MPa $\sqrt{m}$ multiply by 1.099.

of the fracture resistance of soda lime glass are more than three times greater than that reported by Griffith (see Tables 10.8a and b). Nevertheless, the Griffith equation and its underlying premise are basically sound and represent a major contribution to the fracture literature.

Since plastic deformation processes in amorphous glasses are very limited, the difference in surface energy and fracture energy values is not expected to be great. This is not true for metals and polymers, where the fracture energy is found to be several orders of magnitude greater than the surface energy of a given material. Orowan[3] recognized this fact and suggested that Eq. 8-6 be modified to include the energy of plastic deformation in the fracture process so that

$$\sigma = \sqrt{\frac{2E(\gamma_S + \gamma_P)}{\pi a}} = \sqrt{\frac{2E\gamma_S}{\pi a}\left(1 + \frac{\gamma_P}{\gamma_S}\right)} \tag{8-8a}$$

where γ_P = plastic deformation energy and $\gamma_P \gg \gamma_S$.

Under these conditions

$$\sigma \approx \sqrt{\frac{2E\gamma_S}{\pi a}\left(\frac{\gamma_P}{\gamma_S}\right)} \tag{8-8b}$$

The applicability of Eqs. 8-6 or 8-8 in describing the fracture of real materials will depend on the sharpness of the crack and the relative amount of plastic deformation. The following relation reveals these two factors to be related. By combining Eqs. 7-9 and 7-27 and letting $\sigma_{max} = \sigma_c$, we see that the applied stress σ_a for fracture will be

$$\sigma_a = \frac{1}{2}\sqrt{\frac{E\gamma_S}{a}\left(\frac{\rho}{a_0}\right)} \text{ or } \sqrt{\frac{2E\gamma_S}{\pi a}\left(\frac{\pi\rho}{8a_0}\right)} \tag{8-8c}$$

The similarity between Eqs. 8-8b and 8-8c is obvious and suggests a correlation between γ_P/γ_S and $\pi\rho/8a_0$; that is, plastic deformation can be related to a blunting process at the crack tip—ρ will increase with γ_P. From Eqs. 8-8b and 8-8c, it is seen that the Griffith relation (Eq. 8-6) is valid for sharp cracks with a tip radius in the range of $(8/\pi)a_0$. Equation 8-6 is believed to be applicable also where $\rho < (8/\pi)a_0$, since it would be unreasonable to expect the fracture stress to approach zero as the crack root radius became infinitely small. When $\rho > (8/\pi)a_0$, Eqs. 8-8b or 8-8c would control the failure condition where plastic deformation processes are involved.

At the same time, Irwin[4] was also considering the application of Griffith's relation to the case of materials capable of plastic deformation. Instead of developing an explicit relation in terms of the energy sink terms, γ_S or $\gamma_S + \gamma_P$, Irwin chose to use the energy source term (i.e., the elastic energy per unit crack–length increment $\partial U/\partial a$). Denoting $\partial U/\partial a$ as $\mathcal{G}$, Irwin showed that

$$\sigma = \sqrt{\frac{E\mathcal{G}}{\pi a}} \tag{8-9}$$

which is one of the most important relations in the literature of fracture mechanics. By comparison of Eqs. 8-8 and 8-9, it is seen that

$$\mathcal{G} = 2(\gamma_S + \gamma_P) \tag{8-10}$$

At the point of instability, the elastic energy release rate $\mathcal{G}$ (also referred to as the crack driving force) reaches a critical value $\mathcal{G}_c$, whereupon fracture occurs. This critical elastic energy release rate may be interpreted as a material parameter and can be measured in the laboratory with sharply notched test specimens.

8.1.2 Energy Release Rate Analysis

In the previous section, the elastic energy release rate $\mathcal{G}$ was related to the release of strain energy and the work done by the boundary forces. The significance of these two terms will now be considered in greater detail. For an elastically loaded body containing a crack of length a (Fig. 8.2), the amount of stored elastic strain energy is given by

$$V = \frac{1}{2} P\delta \text{ or } \frac{1}{2} \frac{P^2}{M_1} \tag{8-11}$$

where V = stored strain energy
P = applied load
δ = load displacement
M_1 = body stiffness for crack length a

If the crack extends by an amount da, the necessary additional surface energy is obtained from the work done by the external body forces $P\,d\delta$ and the release of strain energy dV.[5] As a result

$$\mathcal{G} = \frac{dU}{da} = P\frac{d\delta}{da} - \frac{dV}{da} \tag{8-12}$$

with the stiffness of the body decreasing to M_2. Whether the body was rigidly gripped such that incremental crack growth would result in a load drop from P_1 to P_2 or whether the load was fixed such that crack extension would result in an increase in δ by an amount $d\delta$, the stiffness of the plate M would decrease. For the fixed grip case, both P and M would decrease but the ratio P/M would remain the same, since from Fig. 8.2

$$\delta_1 = \delta_2 = \frac{P_1}{M_1} = \frac{P_2}{M_2} \tag{8-13}$$

The elastic energy release rate would be

$$\left(\frac{\partial U}{\partial a}\right)_\delta = \frac{1}{2}\left[\frac{2P}{M}\frac{\partial P}{\partial a} + P^2\frac{\partial(1/M)}{\partial a}\right] \tag{8-14}$$

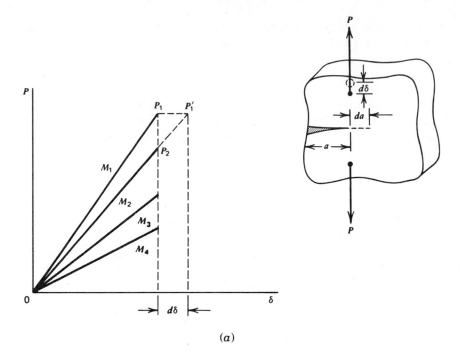

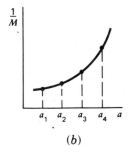

(b)

FIGURE 8.2 (a) **Load-deflection response of cracked plate such as shown in Fig. 8.1 for case where crack length increases by** da. OP_2 **corresponds to fixed grip condition while** OP'_1 **corresponds to fixed load cases.** (b) **Compliance dependence on crack length.**

By differentiating Eq. 8-13 to obtain

$$\frac{1}{M}\frac{\partial P}{\partial a} + P\frac{\partial(1/M)}{\partial a} = 0 \tag{8-15}$$

and substituting the result into Eq. 8-14

$$\left(\frac{\partial U}{\partial a}\right)_{\delta} = -\frac{1}{2}P^2\frac{\partial(1/M)}{\partial a} \tag{8-16a}$$

It may be shown[5] that under fixed load conditions

$$\left(\frac{\partial U}{\partial a}\right)_P = \frac{1}{2} P^2 \frac{\partial(1/M)}{\partial a} \tag{8-16b}$$

Note that in both conditions, the elastic energy release rate is the same (only the sign is reversed), reflecting the fact that $\mathcal{G}$ is independent of the type of load application (e.g., fixed grip, constant load, combinations of load change and displacement, and machine stiffness). At instability, then, the critical strain energy release rate is

$$\mathcal{G}_c = \frac{P_{max}^2}{2} \frac{\partial(1/M)}{\partial a} \tag{8-17}$$

where $1/M$ is the compliance of the cracked plate, which depends on the crack size. Once the compliance versus crack–length relation has been established for a given specimen configuration, $\mathcal{G}_c$ can be obtained by noting the load at fracture, provided the amount of plastic deformation at the crack tip is kept to a minimum. To illustrate, load–displacement plots corresponding to samples that contain cracks of different lengths are shown in Fig. 8.2a. Since compliance $(1/M)$ is given by δ/P, the crack–length dependence of $1/M$ takes the form given in Fig. 8.2b.

8.2   STRESS ANALYSIS OF CRACKS

The fracture of flawed components also may be analyzed by a stress analysis based on concepts of elastic theory. Using modifications of analytical methods described by Westergaard,[6] Irwin[7] published solutions for crack–tip stress distributions associated with the three major modes of loading shown in Fig. 8.3. Note that these modes involve different crack surface displacements.

Mode I.   Opening or tensile mode, where the crack surfaces move directly apart.

Mode II.   Sliding or in-plane shear mode, where the crack surfaces slide over one another in a direction perpendicular to the leading edge of the crack.

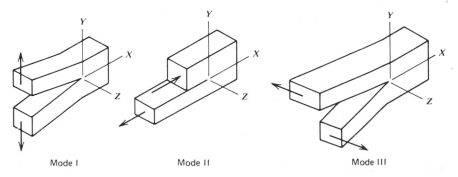

Mode I   Mode II   Mode III

FIGURE 8.3   Basic modes of loading involving different crack surface displacements.

Mode III. Tearing or antiplane shear mode, where the crack surfaces move relative to one another and parallel to the leading edge of the crack.

Mode I loading is encountered in the overwhelming majority of actual engineering situations involving cracked components. Consequently, considerable attention has been given to both analytical and experimental methods designed to quantify Mode I stress–crack–length relations. Mode II is found less frequently and is of little engineering importance. One example of mixed Mode I–II loading involves axial loading (in the *Y* direction) of a crack inclined as a result of rotation about the *Z* axis (Fig. 8.4). Even in this instance, analytical methods[8] show the Mode I contribution to dominate the crack–tip stress field when $\beta > 60°$. Mode III may be regarded as a pure shear problem such as that involving a notched round bar in torsion.

For the notation shown in Fig. 8.5, the crack–tip stresses are found to be

$$\sigma_y = \frac{K}{\sqrt{2\pi r}} \cos\frac{\theta}{2}\left(1 + \sin\frac{\theta}{2}\sin\frac{3\theta}{2}\right)$$

$$\sigma_x = \frac{K}{\sqrt{2\pi r}} \cos\frac{\theta}{2}\left(1 - \sin\frac{\theta}{2}\sin\frac{3\theta}{2}\right) \tag{8-18}$$

$$\tau_{xy} = \frac{K}{\sqrt{2\pi r}}\left(\sin\frac{\theta}{2}\cos\frac{\theta}{2}\cos\frac{3\theta}{2}\right)$$

It is apparent from Eq. 8-18 that these local stresses could rise to extremely high levels as *r* approaches zero. As pointed out earlier in the chapter, this circumstance is

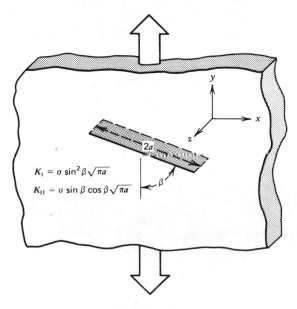

FIGURE 8.4 **Crack inclined β degrees about *z* axis. Mode I dominates when $\beta > 60°$.**

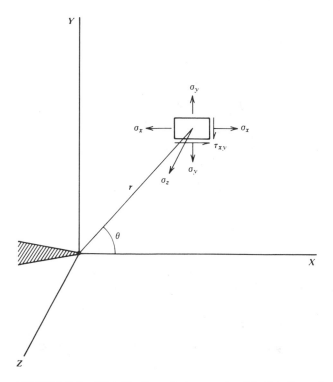

FIGURE 8.5   Distribution of stresses in vicinity of crack tip.

precluded by the onset of plastic deformation at the crack tip. Since this plastic enclave is embedded within a large elastic region of material and is acted upon by either biaxial ($\sigma_y + \sigma_x$) or triaxial ($\sigma_y + \sigma_x + \sigma_z$) stresses, the extent of plastic strain within this region is suppressed. For example, if a load were applied in the Y direction, the plastic zone would develop a positive strain ϵ_y and attempt to develop correspond- ing negative strains in the X and Z direction, thus achieving a constant volume condition required for a plastic deformation process ($\epsilon_y + \epsilon_z + \epsilon_x = 0$). However, σ_x acts to restrict the plastic zone contraction in the X direction, while the negative ϵ_z strain is counteracted by an induced tensile stress σ_z. Since there can be no stress normal to a free surface, the through-thickness stress σ_z must be zero at both surfaces but may attain a relatively large value at the midthickness plane. At one extreme, the case for a thin plate where σ_z cannot increase appreciably in the thickness direction, a condition of *plane stress* dominates, so

$$\sigma_z \approx 0 \tag{8-19}$$

In thick sections, however, a σ_z stress is developed, which creates a condition of triaxial tensile stresses acting at the crack tip and severely restricts straining in the z

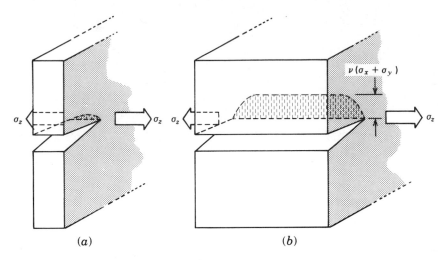

FIGURE 8.6   **Through-thickness stress σ_z in (a) thin sheets under plane-stress state and (b) thick plates under plane-strain conditions.**

direction. This condition of *plane strain* can be shown to develop a through-thickness stress

$$\sigma_z \approx \nu(\sigma_x + \sigma_y) \tag{8-20}$$

The distribution of σ_z stress through the plate thickness is sketched in Fig. 8.6 for conditions of plane stress and plane strain.

An important feature of Eq. 8-18 is the fact that the stress distribution around any crack in a structure is similar and depends only on the parameters r and θ. The difference between one cracked component and another lies in the magnitude of the stress field parameter K, defined as the *stress–intensity factor*. In essence, K serves as a scale factor to define the magnitude of the crack–tip stress field. From Irwin's paper we see that

$$K = f(\sigma, a) \tag{8-21}$$

where the functionality depends on the configuration of the cracked component and the manner in which the loads are applied. Many functions have been determined for various specimen configurations and are available from the fracture mechanics literature.[8-11] In recent years, stress–intensity factor functions have been determined by mathematical procedures other than the Airy stress function approach used by Westergaard. Several solutions are shown in Fig. 8.7 for both commonly encountered cracked component configurations and standard laboratory test sample shapes, where the function is defined by $Y(a/W)$. Consistent with Eq. 8-21, the stress–intensity factor is most often found to be a function of stress and crack length. (Situations where K is independent of crack length or varies inversely with a (e.g., Eq. 13-4) are reported elsewhere.[8-11]) Typical expressions for $Y(a/W)$, corresponding to some of the speci-

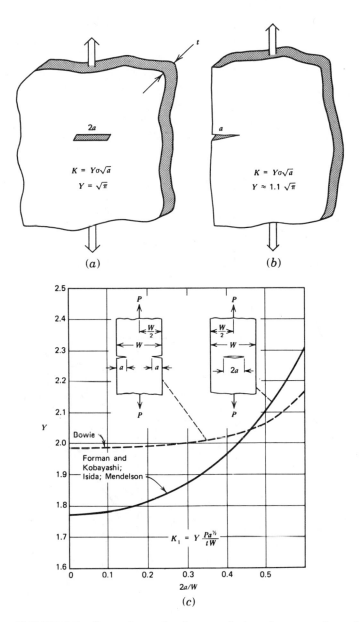

FIGURE 8.7 **Stress-intensity-factor solutions for several specimen configurations.**[12] **(Reprinted by permission of the American Society for Testing and Materials from copyright material.)**

men configurations shown in Fig. 8.7, are given in Appendix B. The solution for the center–cracked panel, however, deserves further comment here. Accurate stress–intensity–factor solutions in polynomial form for this configuration were reported by several investigators with the results shown graphically in Fig. 8.7c.[12] More recently, Feddersen[13] noted that these polynomial expressions could be described with equal precision but much more conveniently by a secant expression with the form

$$K = \sigma\sqrt{\pi a} \cdot \sqrt{\sec \pi a/W} \qquad (8\text{-}22)$$

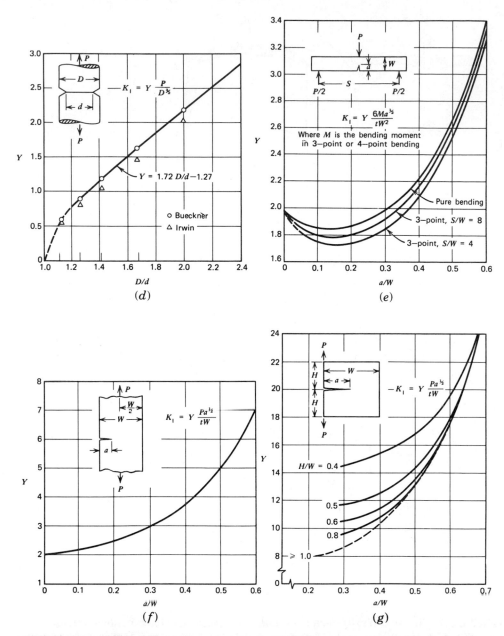

FIGURE 8.7 (*Continued*)

Equation 8-22 can be used also to describe the K conditions associated with a partial through-thickness flaw where a corresponds to the depth of penetration of the crack through the component wall thickness; W is replaced by a quantity equal to two times the thickness dimension. It is intriguing to note that Eq. 8-22 is an *empirical* expression. Perhaps some future fracture mechanics analyst will provide the solution leading to this exact expression.

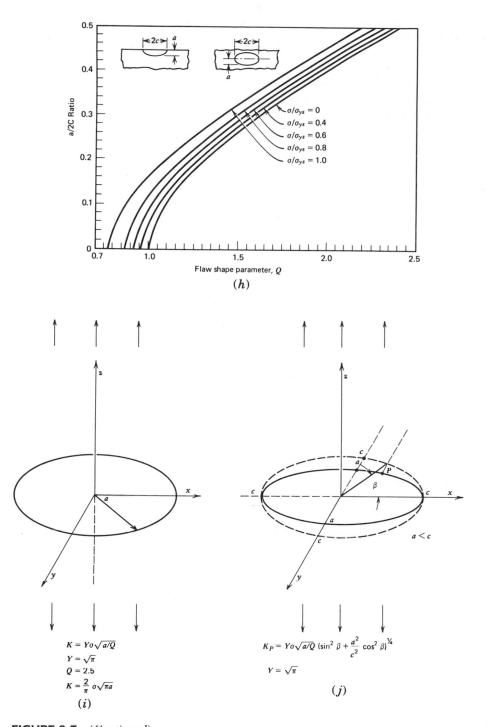

$$K = Y\sigma\sqrt{a/Q}$$
$$Y = \sqrt{\pi}$$
$$Q = 2.5$$
$$K = \frac{2}{\pi}\,\sigma\sqrt{\pi a}$$

(i)

$$K_P = Y\sigma\sqrt{a/Q}\left(\sin^2\beta + \frac{a^2}{c^2}\cos^2\beta\right)^{\frac{1}{4}}$$

$$Y = \sqrt{\pi}$$

(j)

FIGURE 8.7   (*Continued*)

EXAMPLE 8.1

You are asked to confirm that the fracture toughness of a particular steel alloy is approximately 60 MPa$\sqrt{\text{m}}$. Furthermore, the test is to be conducted with the smallest possible load cell, using either a 1-cm-thick, center-cracked panel (CCT) or compact [$C(T)$] specimen. Which specimen would you choose for this property verification?

The stress intensity relations for these two specimen configurations are given in Appendix B. For the CCT panel, $a = 2$ cm and $W = 10$ cm, whereas $a = 3$ cm and $W = 6$ cm for the $C(T)$ sample.

For the CCT panel,

$$K = \frac{P\sqrt{\pi a}}{BW} f(a/W)$$

where

$$f(a/W) = \sqrt{\sec \frac{\pi a}{W}}$$

Therefore, $K = 60$ MPa$\sqrt{\text{m}} = \dfrac{P\sqrt{\pi 0.02}}{0.01(0.1)} \sqrt{\sec \dfrac{\pi 0.02}{0.1}}$

Hence, $P = 4.35 \times 10^4$ Newtons.

For the $C(T)$ specimen,

$$K = \frac{P}{B\sqrt{W}} f(a/W)$$

where

$$f(a/W) = \frac{(2 + a/W)}{(1 - a/W)^{1.5}} [0.886 + 4.64a/W - 13.32(a/W)^2$$
$$+ 14.72(a/W)^3 - 5.6(a/W)^4]$$

By substitution, $P = 6.21 \times 10^3$ Newtons.

We see that the maximum load needed for the $C(T)$ sample is seven times smaller than that required for the CCT configuration. It is also worth noting that much less material is needed to prepare the $C(T)$ sample than for the CCT panel. Indeed, the $C(T)$ sample was developed by Westinghouse engineers to minimize the volume of radioactive material needed to determine the fracture toughness of neutron irradiated material.

There is an important connection between the stress–intensity correction factors given in Figs. 8.7h and 8.7i. Note that as $a/2c$ approaches 0.5, Q approaches 2.5, where $\sqrt{1/Q}$ equals $2/\pi$, which is the solution for an embedded circular flaw or a semicircular surface flaw. An embedded elliptical or semielliptical surface flaw will grow such that $a/2c$ always increases to a limiting value of 0.5. (For the semielliptical surface flaw, the equilibrium $a/2c$ is closer to 0.36, because an additional K correction associated with the free surface is present.) This results from a variation in K along the surface of the ellipse (Fig. 8.7j). When $\beta = 90°$, K is maximized but is smallest where $\beta = 0°$. As a result, the crack will always grow fastest where $\beta = 90°$ until the crack assumes a circular configuration. At this point, the K level is the same along the entire crack perimeter, with additional crack growth maintaining a circular shape. A corollary to this case is that elliptical flaws will grow first into a circular configuration before appreciably increasing the crack size in the direction of the major axis of the ellipse.

8.2.1   Multiplicity of Y Calibration Factors

It is important to recognize that the stress–intensity factor for a given flaw shape and loading configuration often may involve several Y calibration factors. For example, an alternative form for the stress–intensity–factor solution for the elliptically shaped crack (Fig. 8.7j) can be written as

$$K = Y_1 \cdot Y_2 \cdot Y_3 \sigma \sqrt{a} \tag{8-23}$$

where $Y_1 = \sqrt{\pi}$

$Y_2 = \sqrt{1/Q}$

$Y_3 = \left(\sin^2\beta + \frac{a^2}{c^2} \cos^2\beta \right)^{1/4}$

Furthermore, were this defect to grow to an appreciable size relative to that of the section thickness, then Eq. 8-23 also would have to include a finite width correction factor (e.g., $Y_4 = \sqrt{\sec(\pi a/2t)}$). Other stress–intensity–factor solutions involving multiple Y calibration factors are now estimated in the following three examples.

CASE 1
Crack Emanating from a Round Hole[8,14]
This configuration (Fig. 8.8a) is commonly found in engineering practice (especially in aircraft components, which contain many rivet holes) since cracks often emanate from regions of high stress concentration. For the case of a round hole, k_t equals 3 at position A and is minus 1 at B. For a shallow crack ($L \ll R$), the crack tip is embedded within this local stress concentration and may be considered to be a shallow surface flaw. The stress–intensity factor is then estimated to be

$$K \approx 1.12(3\sigma)\sqrt{\pi L} \; (L \ll R) \tag{8-24}$$

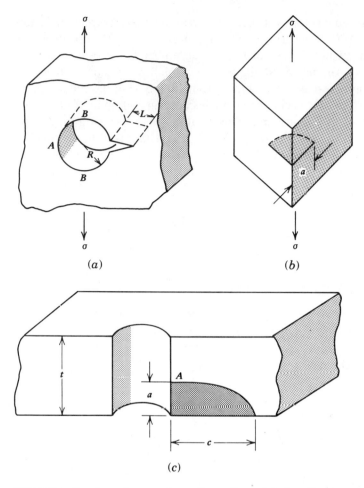

FIGURE 8.8 Complex crack configurations. (*a*) Crack emanating from hole; (*b*) circular corner crack; (*c*) elliptical corner crack emanating from through-thickness hole.

As one might expect, K drops quickly when L increases because the crack runs out of the high stress concentration region. Equation 8-24, therefore, represents an upper bound solution for this crack configuration. A lower bound solution may be estimated from conditions where $L > R$. Here we may estimate the crack length to be $L + 2R$. Hence

$$K \approx \sigma \sqrt{\pi\left(\frac{L + 2R}{2}\right)} \quad (L > R) \tag{8-25}$$

Both upper and lower bound solutions are given in Table 8.1 for several L/R ratios. Also shown are correction factors $F(L/R)$ for this crack configuration based on the solution by Bowie[14] where

$$K = F(L/R)\sigma\sqrt{\pi L} \tag{8-26a}$$

TABLE 8.1 Stress–Intensity Correction
Factors for a Single Crack Emanating
from a Hole

L/R	Eq. 8-24	Eq. 8-25	$F(L/R)$[14]
0.00	3.39	—	3.39
0.10	—	—	2.73
0.20	—	—	2.30
0.30	—	1.96	2.04
0.40	—	1.73	1.86
0.50	—	1.58	1.73
0.60	—	1.47	1.64
0.80	—	1.32	1.47
1.0	—	1.22	1.37
1.5	—	1.08	1.18
2.0	—	1.00	1.06
3.0	—	0.91	0.94
5.0	—	0.84	0.81
10.0	—	0.77	0.75
∞	—	0.707	0.707

Note the excellent agreement between Eqs. 8-24 and 8-25 at both $L/R \approx 0$ and $L/R > 1$ extremes. Paris and Sih[8] also tabulated the correction factors for the case involving cracks emanating from both sides of the hole.

The only difficulty in using the engineering approximations for K (i.e., Eqs. 8-24, 8-25) arises in the region $0 < L/R < 0.5$, where the stress concentration at the hole decays rapidly. Broek[15] considered this problem and concluded that the residual strength of cracked sheets and the crack propagation rate of cracks emanating from holes could be accounted for in reasonable fashion by considering the hole as being part of the crack (i.e., by using Eq. 8-25).

Case 2
Semicircular Corner Crack[9]
This configuration (Fig. 8.8b) involves geometries shown in Figs. 8.7b and 8.7i. Since the crack is circular and lies along two free surfaces, the prevailing stress–intensity level may be approximated by

$$K \approx (1.12)^2 \frac{2}{\pi} \sigma \sqrt{\pi a} \qquad (8\text{-}26b)$$

where $(1.12)^2$ represents two surface flaw corrections and $2/\pi$ represents the correction for a penny-shaped embedded crack.

Case 3
An Elliptical Corner Crack Growing from One Corner of a Through-thickness Hole[9]
The solution to this crack configuration (Fig. 8.8c) incorporates many of the factors discussed in the previous two examples as well as some configurations shown in Fig. 8.7. The maximum stress–intensity condition in this instance is located at A, since

this part of the crack experiences the maximum stress concentration caused by the hole and because A is located at $\beta = 90°$ (see Fig. 8.7j). An approximate solution may be given by

$$K_A \approx 1.12(3\sigma)\sqrt{\pi a/Q} \cdot \sqrt{\sec\frac{\pi a}{2t}} \qquad (8\text{-}27)$$

where
K_A = maximum stress–intensity condition along elliptical surface at A
a = depth of elliptical flaw
c = half width of elliptical flaw
t = plate thickness
1.12 = surface flaw correction at A
3σ = stress concentration effect at A
Q = elliptical flaw correction = $f(a/2c)$

$\sqrt{\sec\dfrac{\pi a}{2t}}$ = finite panel width correction accounting for relatively large a/t ratio (recall Eq. 8-22)

At this point, it is informative to compare the stress–intensity factor K and the stress–concentration factor k_t introduced in Chapter 7. Although k_t accounts for the geometrical variables, crack length and crack–tip radius, the stress–intensity factor K incorporates *both geometrical terms* (the crack length appears explicitly, while the crack–tip radius is assumed to be very sharp) *and the stress level*. As such, the stress–intensity factor provides more information than does the stress–concentration factor.

Once the stress–intensity factor for a given test sample is known, it is then possible to determine the maximum stress–intensity factor that would cause failure. This critical value K_c is described in the literature as the *fracture toughness* of the material.

A useful analogy may be drawn between stress and strength, and the stress–intensity factor and fracture toughness. A component may experience many levels of stress, depending on the magnitude of load applied and the size of the component. However, there is a unique stress level that produces permanent plastic deformation and another stress level that causes failure. These stress levels are defined as the yield *strength* and fracture *strength*. Similarly, the stress–intensity level at the crack tip will vary with crack length and the level of load applied. That unique stress intensity level that causes failure is called the critical stress–intensity level or the *fracture toughness*. *Therefore, stress is to strength as the stress–intensity factor is to fracture toughness.*

Any specimen size and shape may be used to determine the fracture toughness of a given material, provided the stress–intensity–factor calibration is known. Obviously, some samples are more convenient and cheaper to use than others. For example, when the nuclear power plant manufacturers set out to test the steels to be used in nuclear reactors, they chose a small sample like the one shown in Fig. 8.7g so that fracture studies of neutron irradiated samples could be carried out in relatively small environmental chambers. You would use a similar sample in your laboratory if you had a limited amount of material available for the test program or if your testing machine had limited loading capacity. The notched bend bar (Fig. 8.7e) with a long span S also would be an appropriate sample to use when laboratory load capacity is limited.

Of course, this sample would require much more material than the compact tension sample (Fig. 8.7*g*).

8.3 DESIGN PHILOSOPHY

The interaction of material properties, such as the fracture toughness, with the design stress and crack size controls the conditions for fracture in a component. For example, it is seen from Fig. 8.7*a* that the fracture condition for an infinitely large cracked plate would be

$$K = K_c = \sigma \sqrt{\pi a} \qquad (8\text{-}28)$$

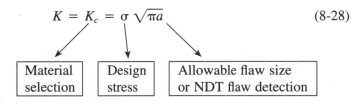

| Material selection | Design stress | Allowable flaw size or NDT flaw detection |

This relation may be used in one of several ways to design against a component failure. For example, if you are to build a system that must withstand the ravages of a liquid metal environment, such as in some nuclear reactors, one of your major concerns is the selection of a suitable corrosion-resistant material. Once done, you have essentially fixed K_c. In addition, if you allow for the presence of a relatively large stable crack—one that can be readily detected and repaired—the design stress is fixed and must be less than $K_c/\sqrt{\pi a}$.

A second example shows another facet of the fracture control design problem. A certain aluminum alloy was chosen for the wing skin of a military aircraft because of its high strength and light weight; hence, K_c was fixed. The design stress on the wing was then set at a high level to increase the aircraft's payload capacity. Having fixed K_c and σ, the allowable flaw size was defined by Eq. 8-28 and beyond the control of the aircraft designers. In one case history, a fatigue crack grew out from a rivet hole in one of the aluminum wing plates and progressed to the point where the conditions of Eq. 8-28 were met. The result—fracture. What was most unfortunate about this particular failure was the fact that the allowable flaw size that could be tolerated by the material under the applied stress was smaller than the diameter of the rivet head covering the hole. Consequently, it was impossible for maintenance and inspection people to know that a crack was growing from the rivet hole until it was too late. This situation could have been avoided in several ways. Had it been recognized beforehand that the wing plate should have tolerated a crack greater than the diameter of the rivet head, the stresses could have been reduced and/or a tougher material selected for the wing plates. It is worth noting that one of the difficulties leading to the early demise of the British Comet jet transport was the selection of a lower toughness 7000 series aluminum alloy for application in critical areas of the aircraft.

The significance of Eq. 8-28 lies in the fact that you must first decide what is most important about your component design: certain material properties, the design stress level as affected by many factors such as weight considerations, or the flaw size that must be tolerated for safe operation of the part. Once such a priority list is established, certain critical decisions can be made. However, once any combination of two or three variables (fracture toughness, stress, and flaw size) is defined, the third factor is fixed.

A leak-before-break condition[16] could develop in a manner shown in Fig. 8.9. Assume that a semielliptical surface flaw with dimensions $2c$ and a is located at the inner surface of a pressure vessel and is oriented normal to the hoop stress direction. Furthermore, we will allow that this crack can grow in the combined presence of a sustained load and aggressive environment (see Chapter 11) or under cyclic loading conditions (see Chapter 13). Recall from Fig. 8.7j and the associated discussion that a semielliptical surface flaw tends to grow more rapidly in a direction parallel to the minor axis of the ellipse (where $\beta = 90°$) until the flaw approaches a semicircular configuration. The crack would then continue to grow as an ever-expanding circle until it breached the vessel's outer wall, thereby allowing fluid to escape. At this point, the crack would break through the remaining unbroken ligament (shaded zone) before assuming the configuration of a through-thickness flaw. Assuming that the crack remained semicircular to the point where breakthrough occurred (at which time a equals the wall thickness), the dimension of the through-thickness flaw would be $2a$ or equivalent to twice the vessel wall thickness, $2t$. Hence, the stress–intensity factor for this crack becomes

$$K = \sigma\sqrt{\pi a} = \sigma\sqrt{\pi t} \qquad (8\text{-}29)$$

Now, if $K < K_c (\text{or } K_{IC})$, then fracture would not take place even though leaking had commenced. In general, then, the leak-before-break condition would exist when a crack of length equal to at least twice the vessel wall thickness could be tolerated (i.e., was stable) under the prevailing stresses.

EXAMPLE 8.2

A 7049-T73 aluminum forging is the material of choice for an 8-cm-internal-diameter hydraulic actuator cylindrical housing that has a wall thickness of 1 cm. After manufacture, each cylinder is subjected to a safety check, involving a single fluid overpressurization that generates a hoop stress no higher than 50% σ_{ys}. The component design calls for an operating internal fluid pressure, corresponding to a hoop stress no higher than 25% σ_{ys}. Prior to overpressurization, a 2-mm-deep semicircular surface flaw that was oriented normal to the hoop stress direction was discovered in one cylinder. Given that $\sigma_{ys} = 460$ MPa and $K_{IC} = 23$ MPa$\sqrt{m}$, would the cylinder have survived the overpressurization test and would the cylinder experience a leak-before-break condition? Also, what were the fluid pressure levels associated with the overpressurization cycle and design stress?

The K level associated with the overpressurization test is given from Figs. 8.7b and 8.7i, where

$$K = 1.1 \left(\frac{2}{\pi}\right) \sigma \sqrt{\pi a}$$

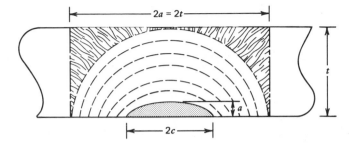

FIGURE 8.9   Diagram showing growth of semielliptical surface flaw to semicircular configuration. At leak condition ($a = t$), unbroken ligaments (shaded areas) break open to form through-thickness crack.

Since the maximum hoop stress level is 50% that of σ_{ys},

$$K = 1.1 \left(\frac{2}{\pi}\right) 230 \times 10^6 \sqrt{\pi \, (2 \times 10^{-3})}$$

$$\therefore K = 12.77 \text{ MPa}\sqrt{m}$$

The prevailing K level is less than K_{IC} and so the cylinder would have survived the proof test. To determine whether the cylinder would experience leak-before-break conditions during normal service conditions (i.e., $\sigma_{design} = 115$ MPa), we use Eq. 8-29 to find

$$K = \sigma\sqrt{\pi \, t}$$

$$\therefore K = 115 \times 10^6 \sqrt{\pi \, (1 \times 10^{-2})} = 20.4 \text{ MPa}\sqrt{m}$$

Since $K < K_{IC}$, leak-before-break conditions would exist, but with a relatively small margin of safety. Finally, we calculate from Eq. 1-63 that the overpressurization and design pressure levels are 57.5 and 28.75 MPa, respectively.

8.4   RELATION BETWEEN ENERGY RATE AND STRESS FIELD APPROACHES

Thus far, two approaches to the relation between stresses, flaw sizes, and material properties in the fracture of materials have been discussed. At this point, it is appropriate to demonstrate the similarity between the two. If Eq. 8-9 is rearranged so that

$$\sigma\sqrt{\pi a} = \sqrt{E\mathscr{G}} \tag{8-30}$$

it is seen from Eq. 8-28 that

$$K = \sqrt{E\mathscr{G}} \quad \text{(plane stress)} \tag{8-31}$$

and

$$K = \sqrt{\frac{E\mathscr{G}}{(1 - v^2)}} \qquad \text{(plane strain)} \qquad (8\text{-}32)$$

for plane strain. This relation between K and $\mathscr{G}$ is not merely fortuitous but, rather, can be shown to be valid based on an analysis credited to Irwin.[17]

Consider the energy needed to reclose part of a crack that had formed in a solid. In reverse manner, then, once this energy was removed the crack should reopen. With the notation shown in Fig. 8.10, the work done per unit area (unit thickness) to close the crack by an amount α is given by

$$\mathscr{G} = \frac{2}{\alpha} \int_0^\alpha \frac{\sigma_y V(x)}{2} \, dx \qquad (8\text{-}33)$$

The constant "2" accounts for the total closure distance, since $V(x)$ is only half of the total crack opening displacement; $1/\alpha$ relates to the average energy released over the total closure distance α; $\sigma_y V(x)/2$ defines the energy under the load deflection curve. From Eq. 8-18 where $\theta = 0$,

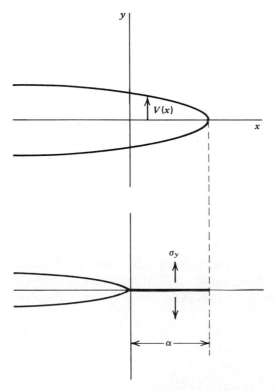

FIGURE 8.10   **Diagram showing partial closure of crack over distance α.**

$$\sigma_y = \frac{K}{\sqrt{2\pi x}} \tag{8-34}$$

while it has been shown[8] that

$$V = \frac{2K}{E}\sqrt{\frac{2(\alpha - x)}{\pi}} \tag{8-35}$$

Combining Eqs. 8-34 and 8-35 with 8-33, note that

$$\mathcal{G} = \frac{2K^2}{\alpha E\pi}\int_0^\alpha \sqrt{\frac{\alpha - x}{x}}\, dx \tag{8-36}$$

Note that $\mathcal{G}$ represents an average value taken over the increment α, while K will vary with α because K is a function of crack length. These difficulties can be minimized by shrinking α to a very small value so as to arrive at a more exact solution for $\mathcal{G}$. Therefore, taking the limit of Eq. 8-36, where α approaches zero, and integrating, it is seen that

$$\mathcal{G} = \frac{K^2}{E} \quad \text{(plane stress)} \tag{8-37}$$

and

$$\mathcal{G} = \frac{K^2}{E}(1 - \nu^2) \quad \text{(plane strain)} \tag{8-38}$$

which is the same result given by Eqs. 8-31 and 8-32.

8.5 CRACK-TIP PLASTIC–ZONE SIZE ESTIMATION

As you know by now, a region of plasticity is developed near the crack tip whenever the stresses described by Eq. 8-18 exceed the yield strength of the material. An estimate of the size of this zone may be obtained in the following manner. First, consider the stresses existing directly ahead of the crack where $\theta = 0$. As seen in Fig. 8.11, the elastic stress $\sigma_y = K/\sqrt{2\pi r}$ will exceed the yield strength at some distance r from the crack tip, thereby truncating the elastic stress at that value. By letting $\sigma_y = \sigma_{ys}$ at the elastic–plastic boundary

$$\sigma_{ys} = \frac{K}{\sqrt{2\pi r}} \tag{8-39}$$

and the plastic–zone size is computed to be $K^2/2\pi\sigma_{ys}^2$. Since the presence of the plastic region makes the material behave as though the crack were slightly longer than

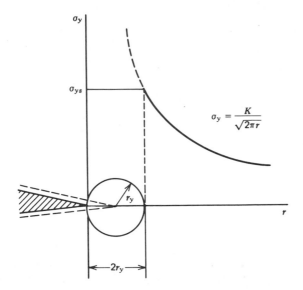

FIGURE 8.11 Onset of plastic deformation at the crack tip. "Effective" crack length taken to be initial crack length plus the plastic zone radius.

actually measured, the "apparent" crack length is assumed to be the actual crack length plus some fraction of the plastic–zone diameter. As a first approximation, Irwin[7] set this increment equal to the plastic–zone radius, so that the apparent crack length is increased by that amount. In effect, the plastic–zone diameter is a little larger than $K^2/2\pi\sigma_{ys}^2$ as a result of load redistributions around the zone and is estimated to be twice that value. Therefore

$$r_y \approx \frac{1}{2\pi} \frac{K^2}{\sigma_{ys}^2} \qquad \text{(plane stress)} \qquad (8\text{-}40)$$

For conditions of plane strain where the triaxial stress field suppresses the plastic–zone size, the plane–strain plastic–zone radius is smaller and has been estimated[18] to be

$$r_y \approx \frac{1}{6\pi} \frac{K^2}{\sigma_{ys}^2} \qquad \text{(plane strain)} \qquad (8\text{-}41)$$

By comparing Eqs. 8-40 and 8-41 it is seen that the size of the plastic zone varies along the crack front, being largest at the two free surfaces and smallest at the midplane.

The reader should recognize that the size of the plastic zone also varies with θ. If the plastic–zone size is determined for the more general case by the distortion energy theory (recall Section 1.6), where σ_x, σ_y, and σ_z are described in terms of r and θ, it can be shown that

$$r_y = \frac{K^2}{2\pi\sigma_{ys}^2} \cos^2\frac{\theta}{2}\left(1 + 3\sin^2\frac{\theta}{2}\right) \qquad \text{(plane stress)} \qquad (8\text{-}42)$$

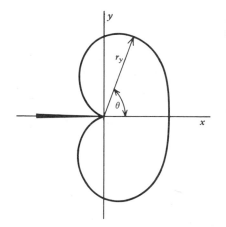

FIGURE 8.12 Crack-tip plastic zone boundary as function of θ.

where the zone assumes a shape as drawn in Fig. 8.12. Hahn and Rosenfield[19] have confirmed this plastic–zone shape by way of etch pit studies in an iron–silicon alloy. Note that when $\theta = 0$, Eq. 8-42 reduces to Eq. 8-40. Equations 8-40 and 8-41 may now be used to determine the effective stress–intensity level K_{eff}, based on the effective or apparent crack length, so that

$$K_{\text{eff}} \approx Y\left(\frac{a + r_y}{W}\right)\sigma\sqrt{a + r_y} \tag{8-43}$$

Since the plastic–zone size is itself dependent on the stress–intensity factor, the value of K_{eff} must be determined by an iterative process that may be truncated at any given level to achieve the desired degree of exactness for the value of K_{eff}. For example, the iteration may be terminated when $(a + r_y)_2 - (a + r_y)_1 \leqslant X$, where X is arbitrarily chosen by the investigator. A special case is an infinite plate with a small central notch, where the stress–intensity factor is defined by Eq. 8-28. Iteration is not necessary in this case, and K_{eff} may be determined directly. Substituting Eq. 8-40 into Eq. 8-28 yields

$$K_{\text{eff}} = \sigma\sqrt{\pi\left(a + \frac{1}{2\pi}\frac{K_{\text{eff}}^2}{\sigma_{ys}^2}\right)} \tag{8-44}$$

Upon rearranging Eq. 8-44, it is seen that

$$K_{\text{eff}} = \frac{\sigma\sqrt{\pi a}}{\left[1 - \frac{1}{2}\left(\frac{\sigma}{\sigma_{ys}}\right)^2\right]^{1/2}} \tag{8-45}$$

so that K_{eff} will always be greater than K_{applied}, although the difference may be very small under low stress conditions.

EXAMPLE 8.3

A plate of steel with a central through-thickness flaw of length 16 mm is subjected to a stress of 350 MPa normal to the crack plane. If the yield strength of the material is 1400 MPa, what is the plastic–zone size and the effective stress–intensity level at the crack tip?

Assuming the plate to be infinitely large, r_y may be determined from Eqs. 8-28 and 8-40 so that

$$r_y \approx \frac{1}{2\pi}\left[\frac{350^2\pi(0.008)}{1400^2}\right] \approx 0.25 \text{ mm}$$

Since r_y/a is very small, it would not be expected that K_{eff} would greatly exceed $K_{applied}$. In fact, from Eq. 8-45

$$K_{eff} = \frac{350\sqrt{\pi(0.008)}}{[1 - \frac{1}{2}(350/1400)^2]^{1/2}} = 56.4 \text{ MPa}\sqrt{m}$$

which is only about 2% greater than $K_{applied}$. When the plastic zone is relatively small in relation to the overall crack length, the plastic–zone correction to the stress–intensity factor is usually ignored in practice. This occurs often under fatigue crack propagation conditions, where the applied stresses are well below the yield strength of the material.

If, on the other hand, a second plate of steel with the same crack size and applied stress level were heat treated to provide a yield strength of 385 MPa, the plasticity correction would be substantially larger. The plastic–zone size would be

$$r_y = \frac{1}{2\pi}\left[\frac{350\sqrt{\pi(0.008)}}{385}\right]^2 = 3.3 \text{ mm}$$

or one-fifth the size of the total crack length. Correspondingly, the effective stress–intensity factor would be considerably greater than the applied level, wherein

$$K_{eff} = \frac{350\sqrt{\pi(0.008)}}{[1 - \frac{1}{2}(350/385)^2]^{1/2}} = 72.4 \text{ MPa}\sqrt{m}$$

which represents a 30% correction. When the computed plastic zone becomes an appreciable fraction of the actual crack length, as found above, and generates a large correction for the stress–intensity level, the entire procedure of applying the plasticity correction becomes increasingly suspect. When such a large plasticity correction is made to the elastic solution, the assumptions of a dominating elastic stress field become tenuous.

8.5.1 Dugdale Plastic Strip Model

Another model of the crack-tip plastic zone has been proposed by Dugdale[20] for the case of plane stress. As shown in Fig. 8.13, Dugdale considered the plastic regions to

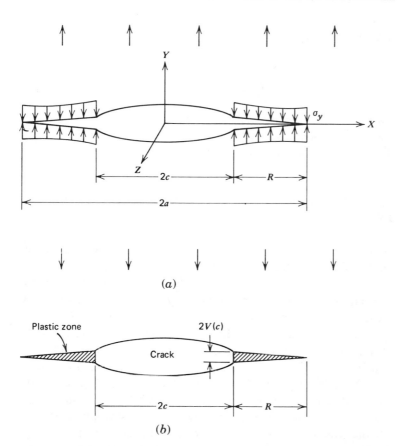

FIGURE 8.13 Dugdale plastic zone strip model for non-strain-hardening solids. Plastic zones *R* extend as thin strips from each end of the crack. (After Hahn and Rosenfield;[19] reprinted with permission of Hahn, *Acta Metall.* **13** (1965), Pergamon Publishing Company.)

take the form of narrow strips extending a distance *R* from each crack tip. For purposes of the mathematical analysis, the internal crack of length $2c$ is allowed to extend elastically to a length $2a$; however, an internal stress is applied in the region $|c| < |x| < |a|$ to reclose the crack. It may be shown that this internal stress must be equal to the yield strength of the material such that $|c| < |x| < |a|$ represents local regions of plasticity. By combining the internal stress field surrounding the plastic enclaves with the external stress field associated with a stress σ acting on the crack, Dugdale demonstrated that

$$c/a = \cos\left(\frac{\pi}{2}\frac{\sigma}{\sigma_{ys}}\right) \qquad (8\text{-}46)$$

or, since $a = c + R$

$$R/c = \sec\left(\frac{\pi}{2}\frac{\sigma}{\sigma_{ys}}\right) - 1 \qquad (8\text{-}47)$$

When the applied stress $\sigma \ll \sigma_{ys}$, Eq. 8-47 reduces to

$$R/c \approx \frac{\pi^2}{8}\left(\frac{\sigma}{\sigma_{ys}}\right)^2 \tag{8-48}$$

By rearranging Eq. 8-48 in the form of $D(K/\sigma_{ys})^2$, it is encouraging to note the reasonably good agreement between Eqs. 8-48 and 8-40 (i.e., $D = \pi/8 \approx 1/\pi$). (Note that 2 r_y is used in Eq. 8-40 for comparison with the Dugdale zone.)

8.6 FRACTURE–MODE TRANSITION: PLANE STRESS VERSUS PLANE STRAIN

As discussed in Section 8.5, the plastic-zone size depends on the state of stress acting at the crack tip. When the sample is thick in a direction parallel to the crack front, a large σ_z stress can be generated that will restrict plastic deformation in that direction. As shown by Eqs. 8-40 and 8-41, the plane–strain plastic–zone size is correspondingly smaller than the plane–stress counterpart. Since the fracture toughness of a material will depend on the volume of material capable of plastically deforming prior to fracture, and since this volume depends on specimen thickness, it follows that the fracture toughness K_c will vary with thickness as shown in Fig. 8.14. When the sample is thin (for example, at t_1) and the degrees of plastic constraint acting at the crack-tip minimal, plane-stress conditions prevail and the material exhibits maximum toughness. (Note that if the sample were made thinner the toughness would gradually decrease because less material would be available for plastic deformation energy absorption.) Alternatively, when the thickness is increased to bring about plastic constraint and plane-strain conditions at the crack tip, the toughness drops sharply to a level that may be one-third (or less) that of the plane-stress value. One very important aspect of this lower level of toughness (the *plane-strain fracture toughness* K_{IC}) is that it does not decrease further with increasing thickness, thereby making this value a conservative lower limit of material toughness in any given engineering application.

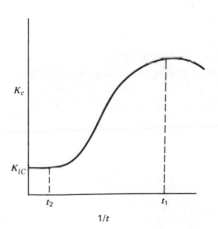

FIGURE 8.14 Variation in fracture toughness with plate thickness.

Once K_{IC} is determined in the laboratory for a given material with a sample at least as thick as t_2 (Fig. 8.14), an engineering component much thicker than t_2 should exhibit the same toughness. To summarize, the plane-stress fracture toughness K_c is related to both metallurgical and specimen geometry, while the plane-strain fracture toughness K_{IC} depends only on metallurgical factors. Consequently, the best way to compare materials of different thickness on the basis of their respective intrinsic fracture-toughness levels should involve a comparison of K_{IC} values, since thickness effects may be avoided.

Since stress-state effects on fracture toughness are affected by the size of the plastic enclave in relation to the sheet thickness, it is informative to consider the change in stress state in terms of the ratio r_y/t, where r_y is computed arbitrarily with the plane-stress plastic-zone size relation as given by Eq. 8-40. Experience has shown that when $r_y/t \geq 1$, plane-stress conditions prevail and toughness is high. At the other extreme, plane-strain conditions will exist when $r_y/t < \frac{1}{10}$. In either case, the necessary thickness to develop a plane-stress or plane-strain condition will depend on the yield strength of the material, since this will control r_y at any given stress-intensity level. Therefore, if the yield strength of a material were increased by a factor of two by some thermomechanical treatment (TMT), the thickness necessary to achieve a plane-strain condition for a given stress-intensity level could be reduced by a factor of four, assuming, of course, that K_{IC} was not altered by the TMT. Clearly, very thin sections can still experience plane-strain conditions in high-yield-strength material, whereas very large sections of low-yield-strength material may never bring about a full plane-strain condition.

Another feature of the fracture-toughness–stress-state dependency is the commonly observed fracture mode transition mentioned in the previous chapter. As shown in Fig. 8.15, the relative degree of flat and slant fracture depends on the crack-tip stress state. When plane-stress conditions prevail and $r_y \geq t$, the fracture plane often assumes

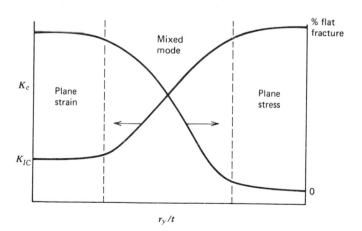

FIGURE 8.15 Effect of relative plastic zone size to plate thickness on fracture toughness and macroscopic fracture surface appearance. Plane-stress state associated with maximum toughness and slant fracture. Plane-strain state associated with minimum toughness and flat fracture.

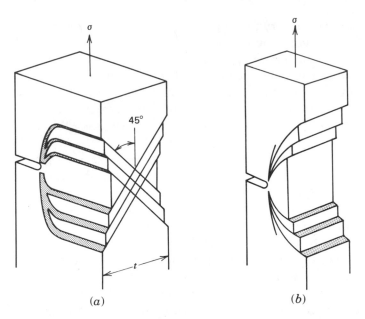

FIGURE 8.16 Crack-tip deformation patterns in (*a*) plane stress, and (*b*) plane strain. (After Hahn and Rosenfield;[19] reprinted with permission of Hahn, *Acta Metall.* **13** (1965), Pergamon Publishing Company.)

a $\pm 45°$ orientation with respect to the load axis and sheet thickness (Fig. 8.16*a*). This may be rationalized in terms of failure occurring on those planes containing the maximum resolved shear stress. (Since $\sigma_z = 0$ in plane stress, a Mohr circle construction will show that the planes of maximum shear will lie along $\pm 45°$ lines in the *YZ* plane.) In plane strain, where $\sigma_z \approx v(\sigma_y + \sigma_x)$ and $r_y \ll t$, the plane of maximum shear is found in the *XY* plane (Fig. 8.16*b*). (σ_y, σ_x, and σ_z may be computed from Eqs. 8-18 and 8-20, where it may be shown, for example, that when $\theta = 60°$, $\sigma_y > \sigma_z > \sigma_x$.) Apparently, the fracture plane under plane-strain conditions lies midway between the two maximum shear planes. This compromise probably also reflects the tendency for the crack to remain in a plane containing the maximum net section stress.

The existence of a fracture-mode transition in many engineering materials such as aluminum, titanium, and steel alloys, and a number of polymers makes it possible to estimate the relative amount of energy absorbed by a component during a fracture process. When the fracture surface is completely flat (Fig. 8.17*c*), plane-strain test conditions probably prevail, and the observed fracture toughness is low. If the fracture is completely of the slant or shear type (Fig. 8.17*a*), plane-stress conditions probably dominate to produce a tougher failure. Obviously, a mixed fracture appearance (Fig. 8.17*b*) would reflect an intermediate toughness condition. By measuring the width of the shear lip and relating it to the size of the plastic-zone radius, it is often possible to estimate the stress intensity factor associated with a particular service failure. The procedure for carrying out this computation is described in Section 14.2. There are exceptions to this stress-state–fracture-mode correlation that should be recognized by the reader. For example, ferritic and pearlitic steels tend to exhibit a smaller shear lip zone than that expected based on estimates of the plane-stress plastic-zone size.[21]

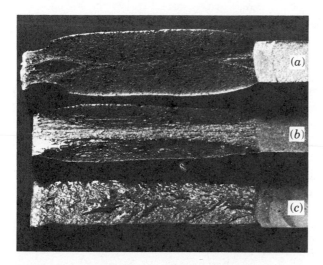

FIGURE 8.17 Fracture-mode transition in alloy steel induced by change in test temperature. (*a*) Slant fracture at high temperature; (*b*) mixed mode at intermediate temperature; (*c*) flat fracture at low temperature; (*d*) double set of shear lips in a steel Charpy sample. Internal shear lips are associated with the formation of center-line delamination. (Photograph courtesy of K. Vecchio.)

From this discussion, it is clear that shear lips form at the free surfaces of the specimen where plane-stress conditions prevail. A unique exception to this pattern is shown in Fig. 8.17*d* where *two* sets of shear lips are found on mating surfaces of a Charpy specimen (i.e., a prenotched bar that is impacted by a falling hammer; see Chapter 9). The external pair of shear lips correspond to the plane-stress condition that normally exists at the specimen's free surfaces. The internal pair of shear lips was produced as a result of the delamination created during the fracture process. (In this instance, the through-thickness stress σ_z (recall Fig. 8.6) acted to split open the Charpy sample along a metallurgical plane of weakness.) Once the delamination was created, two additional free surfaces were created, each corresponding to a plane-stress condition. Consequently, a second set of shear lips was produced in the middle of the specimen—a most unusual location for shear lip development.

8.7 PLANE–STRAIN FRACTURE–TOUGHNESS TESTING

Since the plane-strain fracture toughness K_{IC} is such an important material property in fracture prevention, it is appropriate to consider the procedures by which this property is measured in the laboratory. Accepted test methods have been set forth by the American Society for Testing and Materials under Standard E399-90.[22] Although the reader should examine this standard for precise details, the most important features of K_{IC} testing are summarized in this section.

A recommended test sample is initially fatigue-loaded to extend the machined notch a prescribed amount. (To date, a three-point bend bar (Fig. 8.7*e*), a compact sample (Fig. 8.7*g*), and a C-shaped sample are considered acceptable for a K_{IC} determination. The stress-intensity factor expressions for these three specimen configurations are tabulated in Appendix B.) A clip gage is then placed at the mouth of the crack to monitor its displacement when the specimen load is applied. Typical load-displacement records for a K_{IC} test are shown in Fig. 8.18. From such curves, two important questions should be answered. First, what is the apparent plane-strain fracture-toughness value for the material? Second, is this value *valid* in the sense that a thicker or bigger sample might not produce a lower K_{IC} number for the same material? If a lower toughness level is achieved with a thicker sample, then the value obtained initially is not a valid number. Brown and Strawley[12] examined the fracture toughness of several high-strength alloys and found empirically that a valid plane-strain fracture-toughness test is performed when the specimen thickness and crack length are both greater than a certain minimum value. Specifically

$$t \text{ and } a \geqslant 2.5\left(\frac{K_{IC}}{\sigma_{ys}}\right)^2 \tag{8-49}$$

The ratio $(K_{IC}/\sigma_{ys})^2$ suggests that the required sheet thickness and crack length are related to some measure of the plastic-zone size, since Eqs. 8-40 and 8-41 are of the same form. Using the plastic-zone size determined from Eq. 8-41 and substituting into Eq. 8-49, it is seen that the criteria for plane-strain conditions reflect a condition where

$$a \text{ and } t \geqslant 50 r_y \tag{8-50}$$

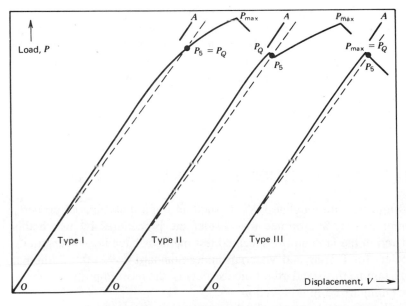

FIGURE 8.18   **Major types of load-displacement records obtained during K_{IC} testing.[22] (Reprinted by permission of the American Society for Testing and Materials from copyright material.)**

Under this condition K_{eff} as defined in Eq. 8-43 would not be significantly different from $K_{applied}$, such that the plastic-zone correction would be unnecessary.

EXAMPLE 8.4

Identical compact tension specimens have been prepared to determine the fracture toughness of the 7178 aluminum alloy, subjected to two different heat treatments. The crack length and thickness of the samples are 4 cm and 1 cm, respectively. From the data shown in Table 8.2 for 7178-T651 and 7178-T7651, would the specimen dimensions, described above, provide valid plane-strain fracture toughness conditions?

From Table 8.2, we see that for the two alloy conditions:

Condition	K_{IC} (MPa$\sqrt{m}$)	σ_{ys} (MPa)
7178-T651	23.1	570
7178-T7651	33	490

From Eq. 8-49, the plane-strain validity criteria require that

$$t \ \& \ a \geq 2.5\left(\frac{K_{IC}}{\sigma_{ys}}\right)^2$$

For alloy 7178-T651,

$$t \ \& \ a \geq 2.5\left(\frac{23.1}{570}\right)^2 = 0.004 \text{ m}$$

Since both crack length (0.04 m) and sheet thickness (0.01 m) are greater than 0.004 m, a valid K_{IC} value could be determined with this specimen.

For alloy 7178-T7651,

$$t \ \& \ a \geq 2.5\left(\frac{33}{490}\right)^2 = 0.011 \text{ m}$$

Since the specimen thickness (0.01 m) is less than that required according to Eq. 8-49 (i.e., 0.011 m), a valid fracture toughness value could not be determined with the specimen described above, even though the crack length dimension (0.04 m) exceeded the size requirement. Therefore, to provide for valid plane-strain conditions for the 7178-T7651 alloy, the specimen thickness would have to be increased to a value >0.011 m. Note that some trial is necessary sometimes before determining the proper specimen dimensions for a valid plane-strain experiment.

To arrive at a valid K_{IC} number, it is necessary to first calculate a tentative number K_Q, based on a graphical construction on the load-displacement test record. If K_Q satisfies the conditions of Eq. 8-49, then $K_Q = K_{IC}$. From ASTM Specification

E-399-90, the graphical construction involves the following procedures.[22] On the load-deflection test record, draw a secant line OP_5 through the origin with a slope that is 5% less than the tangent OA to the initial part of the curve. For the three-point bend bar and the compact tension sample, a 5% reduction in slope is approximately equal to a 2% increase in the effective crack length of the sample, a level reflecting minimal crack extension and plasticity correction.[12] P_5 is defined as the load at the intersection of the secant line OP_5 with the original test record. (Note the similarity between this graphical method and the one described in Chapter 1 for the determination of the 0.2% offset yield strength.) The load P_Q, which will be used to calculate K_Q, is then determined as follows: If the load at every point on the record which precedes P_5 is lower than P_5, then P_Q is P_5 (Fig. 8.18, Type I); if, however, there is a maximum load preceding P_5 which exceeds it, then this maximum load is P_Q (Fig. 8.18, Types II and III).[22] If the ratio of P_{max}/P_Q is less than 1.1, it is then permissible to compute K_Q with the aid of the appropriate K calibration. If K_Q satisfies Eq. 8-49, then K_Q is equal to K_{IC}. If not, then a thicker and/or more deeply cracked sample must be prepared for additional testing so that a valid K_{IC} may be determined.

Another specimen configuration—the short rod or short bar sample—has been standardized to determine K_{IC}.[23] This specimen is either cylindrical or rectangular in shape and contains a deep, machined notch with a chevron configuration, as shown in Fig. 8-19. In the fracture test, a wedge is pushed into the mouth of the chevron-shaped notch. Both the loads and crack-mouth displacements are monitored and analyzed according to methods described by Barker.[24,25] A crack initiates at the tip of the chevron notch with increasing load and grows to the same length at the time of the peak load, provided that conditions of linear elastic fracture mechanics apply. For the case of an ideally elastic material, K_{IC} is given by

$$K_{IC} = A P_c B^{-3/2} \tag{8-51}$$

where A = short rod calibration constant ≈ 22
   P_c = maximum load
   B = short rod diameter

Barker[25] also has modified Eq. 8-51 to account for localized crack-tip plasticity. Fracture toughness values by this method correspond to a slowly advancing steady-state crack that is initiated from a chevron-shaped notch. By contrast, K_{IC} values, obtained from E399 testing requirements, characterize the start of crack extension from a fatigue crack. These differences may cause reported fracture toughness values derived from Standard 1304-89 procedures to be larger in some materials than those corresponding to E399-90 test methods. Accordingly, K_{IC} values determined by these two geometries cannot be assumed to be equivalent.

One obvious advantage of the short rod sample is its small and simple shape. In addition, the fatigue-precracking procedure, necessary in E399-type samples, is not required. This is particularly significant for the case of ceramics, cermets, and brittle alloys in general since these materials are difficult to fatigue precrack. Encouraging reports concerning the use of this test method for ceramics have appeared recently.[26] Toughness levels in brittle ceramics are also being assessed with the use of micro-hardness indentation measurements;[27-34] two such experimental methods have re-

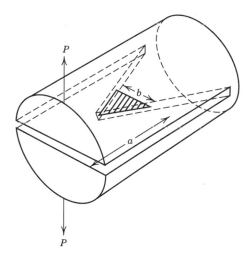

FIGURE 8.19 Short rod specimen containing a deep chevron notch. Wedge loading causes crack to form (shaded area) before maximum load is reached.

ceived considerable attention. In one approach, the sample is loaded with a microhardness indentor such as the equiaxed Vickers type and the lengths of the four resulting cracks are measured from the square corners of the indentation pattern (Fig. 8.20). Only two cracks are typically generated with the elongated Knoop indentor. The mathematical relation between the indentor load P, resulting crack length a, and fracture toughness has been developed and given by

$$K_c = d(E/H)(P/a^{3/2}) \qquad (8\text{-}52)$$

where d = indentor geometry-dependent constant
   E = elastic modulus
   H = indentation hardness value

Marshall[31] concluded that the most consistent results were obtained with the Vickers indentor rather than with the Knoop type. An encouraging correlation has been reported[27] for 12 different ceramic materials between fracture-toughness values obtained from conventional test methods and those derived from Eq. 8-52 (Fig. 8.21a). As such, the microhardness test method holds promise as a useful screening tool for the

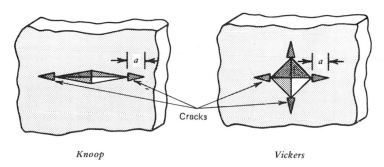

Knoop   Vickers

FIGURE 8.20 Vickers and Knoop indentation and associated crack patterns.

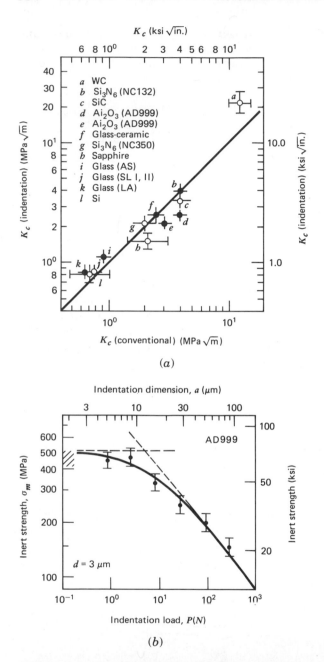

(a)

(b)

FIGURE 8.21   (a) Comparison of fracture-toughness values in different ceramics based on conventional and microhardness test methods.[27] (Reprinted with permission of the American Ceramic Society, G. R. Anstis, P. Chantikul, B. R. Lawn and D. B. Marshall, *J. Amer. Ceram. Soc.*, 64(9), 533 (1981).) (b) Residual beam strength in sintered alumina (99.9% dense) versus indentation load. (Reprinted with permission of the American Ceramic Society, R. F. Cook, B. R. Lawn and C. J. Fairbanks, *J. Amer. Ceram. Soc.*, 68(11), 604 (1985).)

estimation of the fracture toughness in ceramics, based on its limited specimen size requirements and simplicity of test method.

The second microhardness test method makes use of a conventional four-point bend bar. A Knoop hardness impression is first introduced on the expected tensile surface of this sample (with the major axis of the Knoop indentor oriented normal to the flexural stress direction). Following this, the bar is loaded in bending and the flexural stress at fracture noted as a function of indentor load. Typical data for sintered alumina are shown in Fig. 8.21b.[32] The hatched area at the left represents strength levels in bars that contained natural cracks but did not sustain damage through microhardness indentation. The $P^{-1/3}$-dependence of residual strength at large loads conforms with expectations[33] and also mirrors an increase in residual strength with decreasing crack size. At small indentor loads (i.e., small crack sizes), however, residual strength is little changed. Rationalizations for this fairly typical behavior are currently being formulated (e.g., see Refs. 32 and 34). The reader is advised to follow future developments regarding the determination of K_{IC} in such materials based on these microhardness indentation test procedures.

8.8   FRACTURE TOUGHNESS OF ENGINEERING ALLOYS

Typical K_{IC} values for various steel, aluminum, and titanium alloys are listed in Table 8.2 along with associated yield-strength levels. The table also provides a listing of critical flaw sizes for each material, based on a hypothetical service condition involving a through-thickness center notch of length $2a$ embedded in an infinitely large sheet sufficiently thick to develop plane-strain conditions at the crack tip. If it is assumed that the operating design stress is taken to be one-half the yield strength, then the critical crack length would equal $(1/\pi)(K_{IC}/(\sigma_{ys}/2))^2$. One basic data trend becomes immediately obvious: The fracture toughness and allowable flaw size of a given material decreases, often precipitously, when the yield strength is elevated. Consequently, there is a price to pay when one wishes to raise the strength of a material. More will be said about this in Chapter 10.

EXAMPLE 8.5

Assume that a component in the shape of a large sheet is to be fabricated from 0.45–Ni–Cr–Mo steel. It is required that the critical flaw size be greater than 3 mm, the resolution limit of available flaw detection procedures. A design stress level of one-half the tensile strength is indicated. To save weight, an increase in the tensile strength from 1520 to 2070 MPa is suggested. Is such a strength increment allowable? (Assume plane-strain conditions in all computations.)

The answer to this question bears heavily upon the changes in fracture toughness of the material resulting from the increase in tensile strength. At the 1520-MPa strength level, it is found that the K_{IC} value is 66 MPa$\sqrt{m}$, while at 2070 MPa, K_{IC} drops sharply to 33 MPa$\sqrt{m}$ (Fig. 8.22). For a large sheet the stress-intensity factor is

TABLE 8.2   Plane-Strain Fracture Toughness of Selected Engineering Alloys

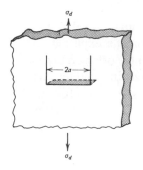

$$\sigma_d = \frac{\sigma_{ys}}{2}; \; K_{IC} = \sigma_d\sqrt{\pi a}$$

$$a_c = \frac{1}{\pi}\left(\frac{K_{IC}}{\sigma_{ys}/2}\right)^2$$

Material	K_{IC}		σ_{ys}		a_c	
	MPa$\sqrt{\text{m}}$	ksi$\sqrt{\text{in.}}$	MPa	ksi	mm	in.
2014-T651	24.2	22	455	66	3.6	0.14
2024-T3	~44.	~40	345	50	~21.	~0.82
2024-T851	26.4	24	455	66	4.3	0.17
7075-T651	24.2	22	495	72	3.0	0.12
7178-T651	23.1	21	570	83	2.1	0.08
7178-T7651	33.	30	490	71	5.8	0.23
Ti-6Al-4V	115.4	105	910	132	20.5	0.81
Ti-6Al-4V	55.	50	1035	150	3.6	0.14
4340	98.9	90	860	125	16.8	0.66
4340	60.4	55	1515	220	2.	0.08
4335 + V	72.5	66	1340	194	3.7	0.15
17-7PH	76.9	70	1435	208	3.6	0.14
15-7Mo	49.5	45	1415	205	1.5	0.06
H-11	38.5	35	1790	260	<0.6	<0.02
H-11	27.5	25	2070	300	0.23	0.009
350 Maraging	55.	50	1550	225	1.6	0.06
350 Maraging	38.5	35	2240	325	<0.4	<0.02
52100	~14.3	~13	2070	300	~0.06	<0.002

determined from Eq. 8-28, where the design stress is $\sigma_{ts}/2$. For the alloy heat treated to the 1520 MPa strength level

$$66 \text{ MPa}\sqrt{\text{m}} = 760 \text{ MPa}\sqrt{\pi a}$$

$$2a = 4.8 \text{ mm}$$

which exceeds the minimum flaw size requirements. At the 2070-MPa strength level, however,

$$33 \text{ MPa}\sqrt{\text{m}} = 1035 \text{ MPa}\sqrt{\pi a}$$
$$2a = 0.65 \text{ mm}$$

which is five times smaller than the minimum flaw size requirement and approximately eight times smaller than the maximum flaw tolerated at the 1520-MPa strength level.

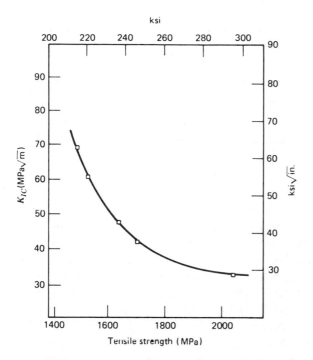

FIGURE 8.22   Fracture toughness in 0.45C–Ni–Cr–Mo steel containing 0.045% S. (After Birkle et al;[34] copyright American Society for Metals, 1966.)

Therefore, it is not possible to raise the strength of the alloy to 2070 MPa to save weight and still meet the minimum flaw size requirement. Furthermore, using the flaw size found in the 1520-MPa material for the 2070-MPa alloy would necessitate a decrease in design stress from 1035 to 380 MPa.

$$\sigma = \frac{33 \text{ MPa}\sqrt{m}}{\sqrt{\pi(0.0024)}} \approx 380 \text{ MPa}$$

Therefore, under similar flaw size conditions, the allowable stress level in the stronger alloy could be only half that in the weaker alloy, resulting in a twofold *increase* in the weight of the component!

The fracture-toughness data for the aluminum alloys deserve additional comment at this time. It is seen that the allowable flaw size in 7075-T651 is only one-eighth that for the 2024-T3. Accordingly, a design engineer would be alerted to the greater propensity for brittle fracture in the 7075-T651 alloy. In actual engineering structures, such as the wing skins of commercial and military aircraft, the relative difference between the two materials is even greater. Since wing skins are about 1 to 2 cm thick, it may be shown by Eq. 8-49 that 7075-T651 would experience approximately plane-strain conditions, while 2024-T3 would be operating in a plane-stress environment, where K_c is about two or three times as large as K_{IC}. Consequently, there may be a

50-fold difference in allowable flaw size between the two materials, taking into account both the metallurgical and stress-state factors.

EXAMPLE 8.6

A plate of Zeusalloy 100, a steel alloy with a yield strength of 415 MPa, has been found to exhibit a K_{IC} value of 132 MPa$\sqrt{m}$. The material is available in various gages up to 250 mm but will be used in very wide 100-mm-thick plates for a given application. If the plate is subjected to a stress of 100 MPa how large can a crack grow from a hole in the middle of the plate before catastrophic failure occurs?

You would be correct in assuming a central crack configuration so that Eq. 8-28 could be used, but would be wrong if you substituted the K_{IC} value of 132 MPa$\sqrt{m}$ into the equation. In fact, not enough information has been provided to answer the question properly. To use K_{IC} in Eq. 8-28 presumes that plane-strain conditions exist in the component. It is seen from Eq. 8-49 that a plate 250 mm thick is required for plane-strain conditions to apply. Since the plate in question is only 100 mm thick, plane-*strain* conditions would not prevail. (In all likelihood, the valid K_{IC} value was obtained from a 250-mm plate). It would be possible to determine K_c for Zeusalloy 100 if data such as shown in Fig. 8.14 were available. In addition, an estimate of K_c may be obtained from the K_{IC} value in the region near plane strain by an empirical relation shown by Irwin,[36] where

$$K_c^2 = K_{IC}^2(1 + 1.4\beta_{IC}^2) \tag{8-53}$$

where $\quad \beta_{IC} = \frac{1}{t}\left(\frac{K_{IC}}{\sigma_{ys}}\right)^2$

$\quad t$ = thickness

As might be expected (recall Section 1.5), K_{IC} values for a particular material may decrease with increasing loading rate and decreasing test temperature. This is particularly true for the case of structural steels. Accordingly, the reader should recognize that the representative K_{IC} values given in Table 8.2 correspond to room temperature experiments conducted at slow loading rates (approximately 10^{-5} s^{-1}). Studies are now being conducted to establish an acceptable standard for the determination of dynamic K_{IC} levels (so-called K_{Id}) at strain rates of about 10 s^{-1} (i.e., impact loading conditions). For additional discussion of K_{IC} dependence on strain rate and temperature, see Section 9.4. Finally, attention is being given to the determination of a crack-arrest toughness parameter, K_{Ia}, associated with the K level where a fast-running unstable crack arrests. (See reports from ASTM Committee E-24.)

8.9 PLANE-STRESS FRACTURE–TOUGHNESS TESTING

The graphical procedures set forth in the last section for the determination of K_{IC} are sometimes confounded by the deformation and fracture response of the material. With rising load conditions, it is possible for the material to experience slow stable crack extension prior to failure, which makes it difficult to determine the maximum stress-intensity level at failure because the final crack length is uncertain. During the early days of fracture-toughness testing, a droplet of recorder ink was placed at the crack tip to follow the course of such slow crack extension and to stain the fracture surface, thereby providing an estimate of the final stable crack length. Since the fracture properties of some high-strength materials are adversely affected by aqueous solutions (see Chapter 11), this laboratory practice has been abandoned. In addition to uncertainties associated with the measurement of the stable crack growth increment Δa, a plastic zone may develop in tougher materials which also must be accounted for in the stress-intensity-factor computation. Consequently, different methods must be employed to determine the fracture-toughness value of a material when the final crack length is not clearly defined but is greater than the initial value.

Irwin[37] proposed that crack instability would occur in accordance with the requirements of the Griffith formulation. That is, failure should occur when the rate of change in the elastic energy-release rate $\partial\mathcal{G}/\partial a$ equals the rate of change in material resistance to such crack growth $\partial R/\partial a$. The material's resistance to fracture R is expected to increase with increasing plastic-zone development and strain hardening; for the case of ceramics, crack resistance increases as a result of strain-induced phase transformations and processes occurring in the wake of the advancing crack front (see Section 10.1.1.) Consequently, both $\mathcal{G}$ and R increase with increasing stress level. The Griffith instability criterion is depicted graphically in Fig. 8.23 as the point of tangency between the $\mathcal{G}$ and R curves when plotted against crack length. Knowing the material's R curve and using the correct stress and crack-length dependence of $\mathcal{G}$ for a given specimen configuration, it would then be possible to determine $\mathcal{G}_c$ or K_c. This fracture-toughness value is generally designated as the *plane-stress* fracture-toughness level, since r_y is no longer much smaller than the sheet thickness and the criterion set forth in Eq. 8-49 is not met.

One should recognize that the fracture-toughness level will vary with the planar dimensions of the specimen. For example, it is seen from Fig. 8.24a that for a given material the fracture-toughness value will depend on the initial crack length, since the tangency point is displaced slightly when the starting crack length is changed. How-

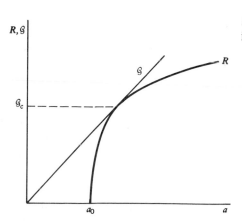

FIGURE 8.23   Instability condition occurs in cracked solid when $\partial\mathcal{G}/\partial a = \partial R/\partial a$.

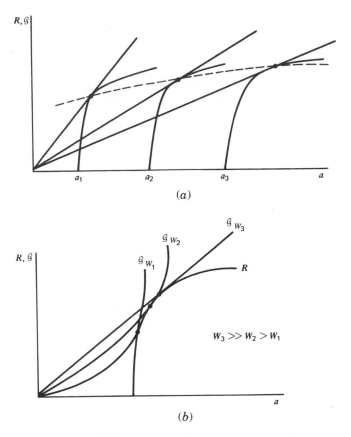

FIGURE 8.24   Effect of (a) initial crack size and (b) plate width on fracture toughness.

ever, this effect is not too large and is minimized when large samples are used, since K_c reaches a limiting value with increasing initial crack length. A considerably larger potential change in K_c values is evident when samples of different planar dimensions are used. Since $\mathcal{G}$ and K depend on specimen configuration, the shape of the $\mathcal{G}$ curve is different for each case. Consequently, the point of tangency and, hence, $\mathcal{G}_c$ will change for a given material. As shown in Fig. 8.24b, the $\mathcal{G}_c$ value increases with increasing sample width and in the limit (the case of an infinitely large panel, where $Y(a/W) = \sqrt{\pi}$ and $\mathcal{G} = \sigma^2 \pi a/E$) reaches a maximum value. Again, it should be fully recognized that the plane-stress fracture toughness of a material is dependent on both metallurgical factors and specimen geometry, while the plane-strain fracture toughness relates only to metallurgical variables.

Although the $\mathcal{G}$ curve can be determined from known analytical relations, such as those provided in Fig. 8.7, the R curve is determined by graphical means. In one accepted procedure the initial step involves the construction of a series of secant lines on the load-displacement record from a test sample (Fig. 8.25a). The compliance values δ/P from these secant lines are used in Fig. 8.25b (the compliance curve for the given specimen configuration) to determine the associated a/W values that reflect the effective crack length $a_{\text{eff}} = a_0 + \Delta a + r_y$. Each a_{eff}/W value is then used to

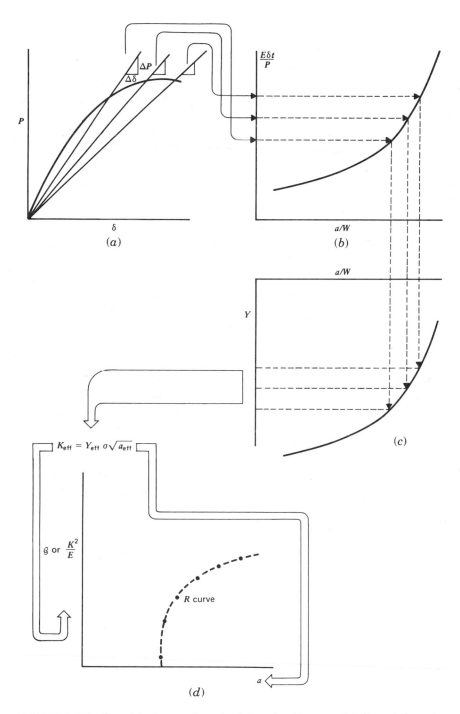

FIGURE 8.25 Graphical procedure to determine R curve. (a) Secant lines drawn on P–δ test record and plotted on compliance curve; (b) procedure to obtain effective a/W values; (c) calibration factors determined from a_{eff}/W and used to calculate K_{eff}^2/E values for (d) R curve construction.

determine a respective K_{eff} value from the calibration curve (Fig. 8.25c). Finally, the $\mathcal{G}$ or K^2/E curve is constructed from these K_{eff} and associated a_{eff} values (Fig. 8.25d). Once the R curve for a given material is determined from this graphical procedure, it becomes part of the information about the material. As such, the same R curve may be used with samples of different crack length and planar dimension, as discussed above and depicted in Figs. 8.24a and 8.24b. Once the R curve is known, the fracture toughness may be determined for an engineering component with any configuration, as long as an analytical expression for $\mathcal{G}$ is known. (Recall that $\mathcal{G}_c$ also may be determined from Eq. 8-17 even when the stress and crack-length dependence of $\mathcal{G}$ is not known.) A recommended practice for the determination of R curves has been developed by Committee E24 of ASTM.[38]

8.10 TOUGHNESS DETERMINATION FROM CRACK-OPENING DISPLACEMENT MEASUREMENT

As demonstrated earlier in this chapter, there is a limit to the extent to which K_{eff} may be adjusted for crack-tip plasticity. Whether one uses Eqs. 8-40 or 8-48 in the computation of r_y, it becomes increasingly inaccurate to determine K_{eff} from Eq. 8-42 when r_y/a becomes large. The allowable loads or stresses in a component rapidly approach a limiting value upon general yielding and, while the associated strains or crack-opening displacements (COD) (i.e., how much the crack opens under load) increase continually to the point of failure, it is more reasonable to monitor the latter in a general yielding situation. Accordingly, the concept of a critical crack-opening displacement near the crack tip has been introduced to provide a fracture criteria and an alternative measure of fracture toughness.[39,40]

By using the Dugdale crack-tip plasticity model[12] for plane stress, it has been possible to conveniently compute the magnitude of the crack-opening displacement $2V(c)$ defined at the elastic-plastic boundary (Fig. 8.13b). Goodier and Field[41] demonstrated that

$$2V(c) = \frac{8\sigma_{ys}c}{\pi E} \ln\left(\sec\frac{\pi\sigma}{2\sigma_{ys}}\right) \tag{8-54}$$

When $\sigma \ll \sigma_{ys}$, $\ln x \approx x - 1$, so that Eqs. 8-47, 8-48, and 8-54 can be combined to yield

$$2V(c) = \frac{K^2}{E\sigma_{ys}} = \frac{\mathcal{G}}{\sigma_{ys}} \tag{8-55}$$

The fracture toughness of a material is then shown to be

$$\mathcal{G}_c \approx 2\sigma_{ys}V^*(c) \tag{8-56}$$

where $V^*(c)$ = critical crack-opening displacement. One important aspect of Eq. 8-55 is that $V(c)$ can be computed for conditions of both elasticity and plasticity, while $\mathcal{G}$ may be defined only in the former situation. The COD concept, therefore, bridges the elastic and plastic fracture conditions. It is important to recognize, however, that the strain fields and crack-opening displacements associated with the crack tip will vary with specimen configuration. Consequently, a single critical crack-opening displacement value for any given material cannot be defined, since it will be affected by the geometry of the specimen used in the test program.[42]

Standard test methods have been developed, which describe the onset of fracture in terms of a critical crack-opening displacement.[43,44] To date, these standards have been used extensively to characterize fracture in structural steel components. Experimental crack-opening values are measured at the specimen surface and then related mathematically to the crack-tip-opening displacement.

The physical significance of Eq. 8-56 warrants additional comment. Since the crack-opening displacement $2V(c)$ is related to the extent of plastic straining in the plastic zone, Eq. 8-56 is analogous to the measurement of toughness from the area under the stress–strain curve in a uniaxial tensile test. As discussed in Chapter 1, maximum toughness is achieved by an optimum combination of stress and strain. Since fracture toughness most often varies *inversely* with yield strength (see Table 8.2 and Chapter 10), Eq. 8-56 would appear to predict toughness–yield-strength trends incorrectly. This is not the case, since an increase in σ_{ys} of an alloy resulting from any thermomechanical treatment is offset normally by a proportionately greater *decrease* in $2V(c)$. The important point to keep in mind is that σ_{ys} and $2V(c)$ are interrelated. As a final note, some uncertainty exists concerning the correct value of yield strength to use in Eq. 8-56. When plane-strain conditions dominate at the crack tip, should not the tensile yield-strength value be elevated to account for crack-tip triaxiality? Certainly, this would be consistent with a similar previous justification to adjust the plastic-zone size estimate for plane-stress and plane-strain conditions (Eqs. 8-40 and 8-41). On this basis, Eq. 8-56 would be modified so that

$$\mathcal{G}_c \approx n\sigma_{ys}2V^*(c) \tag{8-57}$$

where $1 \leq n \leq 1.5 - 2.0$. Consequently, when plane-stress conditions are prevalent, $n = 1$, and n increases with increasing plane strain.

8.11 FRACTURE-TOUGHNESS DETERMINATION AND ELASTIC–PLASTIC ANALYSIS WITH THE J INTEGRAL

Another key parameter has been developed to define the fracture conditions in a component experiencing both elastic and plastic deformation. Using a line integral related to energy in the vicinity of a crack, Rice[45–47] was able to solve two-dimensional crack problems in the presence of plastic deformation. The form of this line integral, the J integral, is given in Eq. 8-58 with failure (crack initiation) occurring when J reaches some critical value.

$$J = \int_C \left(W dy - \mathbf{T} \cdot \frac{\partial \mathbf{u}}{\partial x} ds \right) \tag{8-58}$$

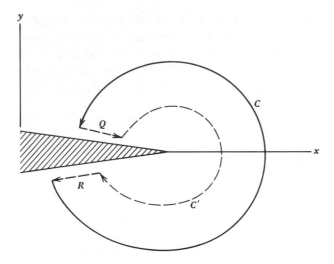

FIGURE 8.26   Line contour surrounding crack tip.

where x, y = rectangular coordinates normal to the crack front (see Fig. 8.26)
   ds = increment along contour C
   $\mathbf{T}$ = stress vector acting on the contour
   $\mathbf{u}$ = displacement vector
   W = strain energy density = $\int \sigma_{ij} d\epsilon_{ij}$

For the case of a closed contour, Rice showed that the line integral is equal to zero. If we examine the contour $C + C' + Q + R$ in Fig. 8.26, we find that $dy = 0$ and and $T_i = 0$ along the contour segments Q and R. Therefore, $J_Q = J_R = 0$, which leads to the fact that $J_C = J_{C'}$. Since the paths along contours C and C' are also opposite in sign, it is concluded that the J integral is path independent and that J can be determined from a stress analysis where σ and ϵ are defined on some arbitrary contour away from and surrounding a crack tip. For example, one could determine J by making use of a finite element analysis to determine σ and ϵ at locations other than at the crack tip.

In establishing the mathematical framework for the J-integral fracture analysis, Rice assumed that deformation theory of plasticity applied; that is, the stresses and strains in a plastic or elastic–plastic body are considered to be the same as for a nonlinear elastic body with the same stress–strain curve. This means that the stress–strain curve is nonlinear and reversible or that almost proportional straining has occurred throughout the body. (Recall the discussion pertaining to the nonlinear elastic stress–strain response of ''whiskers'' (Section 1.2.2.1).) As a result, determination of the crack-tip strain energy using the J integral is valid so long as no unloading occurs. It then follows that the J integral may not be defined rigorously under cyclic loading conditions. As such, the J integral has significance in terms of defining the stress and strain conditions for crack *initiation* under monotonic loading conditions and also in the presence of a limited amount of stable crack extension. (Recall that K_{IC} is used to define the condition for unstable crack *propagation* or the value of K associated with a 2% apparent extension of the crack.) Subsequent studies have shown that this restriction is not as severe as first thought. For example, Hutchinson and Paris[48] presented a theoretical justification for use of the J integral in the analysis of stable crack growth, and Dowling[49] has used the change in J during cyclic loading (ΔJ) to correlate fatigue crack propagation rates.

Of particular significance, Rice[46,47] also presented an alternative and equivalent definition for J with the latter being defined as the pseudopotential energy difference between two identically loaded bodies possessing slightly different crack lengths. The change in strain energy ΔU associated with an incremental crack advance Δa is portrayed in Fig. 8.27 and equal to $J\Delta a$ or $J = (du/da)$, where

$$J = \int_0^P \left(\frac{\partial \delta}{\partial a}\right)_P dP$$

$$J = -\int_0^\delta \left(\frac{\partial P}{\partial a}\right)_\delta d\delta \tag{8-59}$$

From Eq. 8-59, it is seen that J can be determined from load versus load-point-displacement records for specimens containing slightly different crack lengths. Since, for elastic conditions, $\mathcal{G} = (du/da)$, the relation between J and $\mathcal{G}$ is apparent (see Section 8.1.2). For either linear or nonlinear elastic conditions, J is the energy made available at the crack tip per unit crack extension, da. That is, J is equivalent to the crack driving force. Therefore,

$$J = \mathcal{G} = \frac{K^2}{E'} \tag{8-60}$$

where $\quad E' = E \quad$ (plane stress)
$\qquad E' = E/(1 - v^2) \quad$ (plane strain)

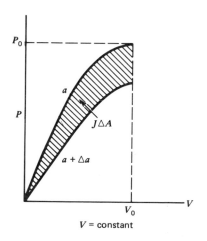 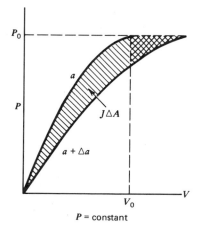

FIGURE 8.27 Determination of J integral based on fixed displacement and fixed load.

For conditions of plasticity, however, energy is dissipated irreversibly during plastic flow within the solid (e.g., through dislocation motion), so that J can no longer represent energy made available at the crack tip for crack extension. However, J can be defined in the same manner for nonlinear elastic and elastic–plastic situations so long as the stress–strain curves are the same. J then becomes a measure of the intensity of the entire elastic–plastic stress–strain field that surrounds the crack tip. It is seen then that J is analogous to K for the case of elastic stress fields. On this basis, J can be used as a failure criterion in that fracture would be expected to occur in two different samples or components if they are subjected to an equal and critical J level.

8.11.1 Determination of J_{IC}

Analogous to K_{IC}, which represents the material's resistance to crack extension, one may define the value J_{IC} which characterizes the toughness of a material near the outset of crack extension. Studies[50,51] have shown that J_{IC} is numerically equal to $\mathcal{G}_{IC}$ values determined from valid plane-strain fracture-toughness specimens for predominantly elastic conditions associated with sudden failure without prior crack extension (Fig. 8.28). For this condition,

$$J_{IC}(\text{plastic test}) = \mathcal{G}_{IC} = \frac{K_{IC}^2}{E}(1 - \nu^2)(\text{elastic test}) \qquad (8\text{-}61)$$

Rice et al.[52] developed a simple method for the determination of the plastic component of J_{IC}. For the case of a plate containing a deep notch and subjected to pure bending, Rice found that

$$J_{pl} = \frac{2}{b}\int_0^{\delta_c} P d\delta_c \qquad (8\text{-}62)$$

where δ_c = displacement of sample containing a crack
P = load/unit thickness
b = remaining unbroken ligament

Equation 8-62 then reduces to

$$J_{pl} = \frac{2A}{Bb} \qquad (8\text{-}63)$$

where B = specimen thickness
A = area under the load versus load-point displacement curve

Equation 8-63 also defines J for a three-point bend specimen with a span/width ratio of four. For the compact specimen, the above relation must be modified to account for the presence of a tensile loading component such that

$$J_{pl} = \frac{\eta A}{Bb} \qquad (8\text{-}64)$$

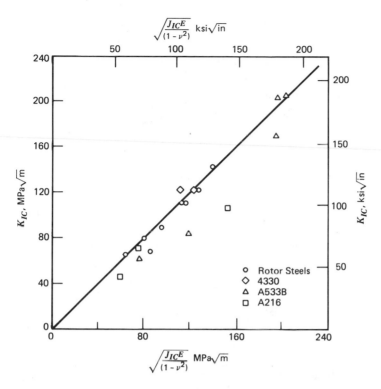

FIGURE 8.28   Comparison of K_{IC} and J_{IC} values for several steel alloys.[53] (Copyright, American Society for Testing and Materials, 1916 Race Street, Philadelphia, PA 19103. Reprinted with permission.)

where $\eta = 2 + 0.522(b/W)$. The new ASTM Standard[54] then defines

$$J_i = J_{el(i)} + J_{pl(i)} \tag{8-65}$$

which defines the J value at a point on the load-displacement line corresponding to a load P_i and displacement δ_l. For the conditions of pure bending and three-point bending with a span/width ratio of four, J is given as

$$J_i = \frac{K_i^2}{E}(1 - \mu^2) + \frac{2A_i}{Bb} \tag{8-66}$$

Alternately,

$$J_i = \frac{K_i^2}{E}(1 - \mu^2) + \frac{A_i\,(2 + 0.522(b/W))}{Bb} \tag{8-67}$$

for the compact specimen design.

EXAMPLE 8.7

An SEN bend bar of a steel alloy was used to conduct a J Integral test. The specimen had dimensions of $B = 10$ mm, $W = 20$ mm, $S = 80$ mm and $a = 10$ mm. The alloy possessed the following mechanical properties: $E = 205$ GPa, and $\mu = 0.25$. The J test developed a load-displacement curve similar to that shown in Fig. 8.29c with the area within the curve $= 5$ Nm^2 and the maximum load $= 15$ kN. Calculate the value of J for this material.

From Eq. 8.66, the value of J is given by

$$J = \frac{K^2}{E}(1 - \mu^2) + \frac{2A}{B(W - a)}$$

Proceeding further, the K relation for this specimen configuration is (see Appendix B-Type 6):

$$K = \frac{PS}{BW^{1.5}}f(a/W)$$

where

$$f(a/W) = \frac{3(a/W)^{0.5}}{2(1 + 2a/W)(1 - a/W)^{1.5}} \times$$

$$[1.99 - (a/W)(1 - a/W)(2.15 - 3.93a/W + 2.7a^2/W^2)]$$

From the dimensional values given above, $f(a/W) = 1.694$ and

$$K = \frac{15,000(80 \times 10^{-3})}{10 \times 10^{-3}(20 \times 10^{-3})^{1.5}}(1.694) = 71.8 \text{ MPa}\sqrt{m}$$

$$\therefore J = \frac{(71.8 \times 10^6)^2}{205 \times 10^9}(1 - \mu^2) + \frac{2(5)}{(0.01)(0.02 - 0.01)}$$

$$J = 25,148 + 100,000 = 125.1 \text{ kN/m}$$

The determination of J_{IC} by this method then involves the measurement of J values from several different samples that have experienced varying amounts of crack extension, Δa.[53] These data are then plotted with J versus Δa and characterized by a best-fit power law relation; as such, this information represents a crack resistance curve R analogous to that described in Section 8.9. (See also ASTM Standard 1152-87 for the determination of J-R curves[55].) J_{IC} is defined at some location on the J–Δa curve corresponding to the onset of cracking. More specifically, ASTM Standard E813-89[54] defines J_{IC} for 0.2-mm crack extension as established from the following recommended test procedure:

1. Load several three-point bend bars or compact specimens to various displacements and then unload. These samples should reveal various amounts of crack extension from the fatigue-precracked starter notch (Fig. 8.29).

2. To record Δa for each sample, the sample is marked. For steel samples, heat tinting at 300°C for 10 min per 25 mm of thickness will discolor the existing fracture surface. For other materials that do not discolor readily after heat tinting (e.g., aluminum alloys), the crack-extension region can be marked by a short amount of fatigue crack growth.

3. The sample is then broken and Δa recorded from that region sandwiched between the fatigue–precrack zone and the marker zone (produced either by heat tinting or cyclic loading). Δa is given by the average of nine readings taken across the crack front from one surface to the other (Fig. 8.26*b*).

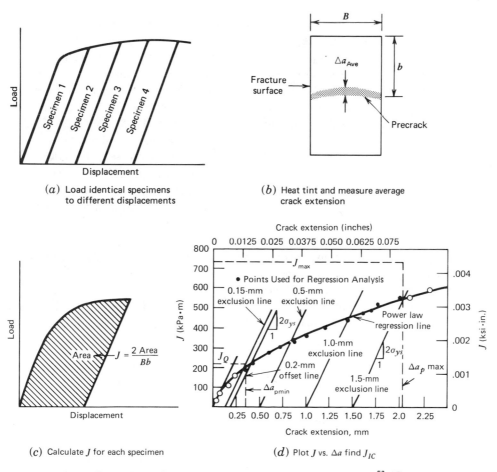

(*a*) Load identical specimens to different displacements

(*b*) Heat tint and measure average crack extension

(*c*) Calculate *J* for each specimen

(*d*) Plot *J* vs. Δa find J_{IC}

FIGURE 8.29 Procedures for multispecimen determination of J_{IC}.[53] (Copyright, American Society for Testing and Materials, 1916 Race Street, Philadelphia, PA 19103. Reprinted with permission.)

4. The area under the $P - \delta$ plot for each sample is measured, and J_{pl} is computed from either Eq. 8-63 or 8-64, depending on the specimen configuration (Fig. 8.29c).

5. To establish the R curve, $J - \Delta a$ data are chosen that conform to certain acceptance criteria based on the construction of a second $J - \Delta a$ line, called the blunting line. The blunting line is conceived from the following considerations: Consider that a sharp crack begins to blunt as the load on the component is increased. Blunting (i.e., *effective* crack extension) then increases until *actual* crack extension occurs. The amount of crack-opening displacement (COD) is given by Eq. 8-55 as $2V(c) = \mathcal{G}/\sigma_{ys}$ or $\mathcal{G} = 2V(c)\sigma_{ys}$. From Fig. 8.30, the *effective* crack extension is approximated by COD/2 where

$$\Delta a = \mathcal{G}/2\sigma_{ys}$$

or

$$\mathcal{G} \approx J \approx 2\sigma_{ys}\,\Delta a \qquad (8\text{-}68)$$

Since Δa values due to blunting are difficult to measure, especially for the case of brittle materials, it has been found more useful to describe the blunting line as $J = 2\sigma_{flow}\Delta a$, where $\sigma_{flow} = (\sigma_{ys} + \sigma_{ts})/2$, which takes into account strain hardening in the material. Additional lines parallel to the blunting line are then constructed with offsets of 0.15, 0.5, 1.0, and 1.0 mm, respectively. When selecting the $J - \Delta a$ data for the R curve, one data point must lie within the 0.15- and 0.5-mm offset lines as well as within the 1.0- and 1.5-mm lines. Several additional points are then taken from the acceptable region of valid data (between 0.15- and 1.5-mm offset lines). The value J_Q is then obtained by noting the intersection of the best-fit power law $J - \Delta a$ curve with a 0.2-mm offset line— the latter corresponding to a fracture-toughness determination for 0.2-mm crack extension. Finally, $J_Q = J_{IC}$ if $B, b > 25J_Q/\sigma_{ys}$. (Other validation criteria are noted in Standard E813-89.)

From the above discussions, it is clear that the determination of J_{IC} involves the testing of numerous specimens, which makes the procedure both tedious and very expensive. Fortunately, another technique has been developed which enables J_{IC} to be determined from multiple loadings of a single sample.[56] After the sample is loaded to a certain load and displacement level, the load is reduced by approximately 10%

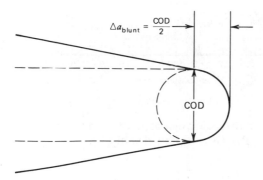

$\Delta a_{blunt} = \dfrac{COD}{2}$

COD

FIGURE 8.30 Effective crack extension approximately given by COD/2.

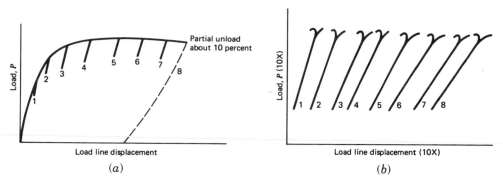

FIGURE 8.31   **Single specimen compliance method for J_{IC} determination. (*a*) Load versus load-line displacement with several partial unloadings; (*b*) amplified segments of (*a*) revealing change in compliance associated with crack extension (ASTM Standard E813-89). (Copyright, American Society for Testing and Materials, 1916 Race Street, Philadelphia, PA 19103. Reprinted with permission.)**

(Fig. 8.31). By measuring the specimen compliance during this slight unloading period, the crack length corresponding to this compliance value can be defined. The crack length could also be inferred with electrical resistance measurements or with an ultrasonic detector. The load is then increased again until another slight unloading event is introduced. Here, again, the new crack length is inferred from any one of the above mentioned techniques. From a number of such load interruptions, Δa values can be determined along with the associated values of J corresponding to the respective locations along the $P - \delta$ plot. This information is then used to obtain an R curve with a $J - \Delta a$ plot. As before, J_{IC} is defined at the intersection point with the computed blunting line. From Fig. 8.32 it is seen that J_{IC}, determined from multiple specimens

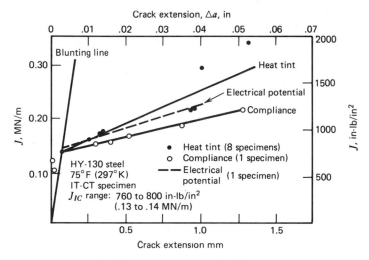

FIGURE 8.32   **Determination of J_{IC} in HY-130 steel based on different test procedures.[53] (Copyright, American Society for Testing and Materials, 1916 Race Street, Philadelphia, PA 19103. Reprinted with permission.)**

using heat tinting, and single specimens, using compliance and electrical potential techniques, are in excellent agreement. Apparently, the slight amount of unloading associated with the compliance technique does not compromise the value of J_{IC} even though the J-integral requirement for deformation plasticity conditions is violated. As noted above, Paris[48,57] has considered this apparent contradiction and concluded that some unloading may be permissible. Furthermore, fatigue crack growth processes may be characterized by a J parameter even though considerable unloading occurs during each loading cycle.[49]

It is instructive to compare specimen size requirements associated with valid J_{IC} and K_{IC} test procedures. Recall from Eq. 8-49 that the relevant specimen dimensions for the K_{IC} test must be

$$\geqslant 2.5 \left(\frac{K_{IC}}{\sigma_{ys}} \right)^2$$

whereas $J_Q = J_{IC}$ if $B, b > 25 \, J_Q/\sigma_{ys}$. Since

$$J_{IC} = \frac{K_{IC}^2}{E/(1 - v^2)}$$

it follows that

$$\frac{K_{IC} \text{ specimen size}}{J_{IC} \text{ specimen size}} = \frac{2.5 \left(\dfrac{K_{IC}}{\sigma_{ys}} \right)^2}{25 \dfrac{K_{IC}^2}{E\sigma_{ys}}} \geqslant 20 \tag{8-69}$$

so that plane-strain fracture-toughness values can be determined by J_{IC} techniques with a much smaller specimen as compared with K_{IC} specimen size requirements.

It is important to recognize that the measurement point for J_{IC} corresponds to minimal (0.2 mm) crack extension while K_{IC} corresponds to 2% apparent crack extension. (The nonlinearity in a $P-\delta$ plot is due to both crack growth and plastic-zone formation at the crack tip.) Depending on the shape of the R curve, the J_{IC} and K_{IC} values may or may not agree closely. Note from Fig. 8.33 that, for the case of a brittle material with a shallow R curve, K_{IC} and J_{IC} should bear good agreement. Conversely, when the R curve is steep, as in the case of a tough material, the agreement between J_{IC} and K_{IC} values would not be as good. In such a case, J_{IC} should be recognized as being a very conservative measurement of toughness, since tough material would be expected to exhibit a finite amount of crack extension at higher loads. In situations where J_{IC} values represent an ultraconservative measure of toughness, other techniques have been attempted to more correctly describe the critical set of conditions for fracture. Paris and co-workers[58–60] have defined a new material parameter, the tearing modulus T, which allows for stable tearing in addition to crack blunting. Paris et al.[58–60] showed that the tearing modulus is given by

$$T = \frac{dJ}{da} \frac{E}{(\sigma_{ys})^2} \tag{8-70}$$

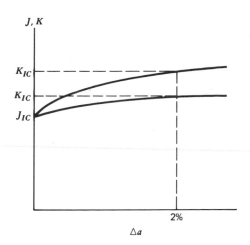

and described a new instability criteria when $T_{applied} > T_{material}$. (Note that (dJ/da) corresponds to the slope of the $J - \Delta a$ curve.) Further studies are in progress to establish the general applicability of this new fracture criterion. (See also ASTM Standard 1152-87, which describes the determination of $J - \Delta a$ curves.[55])

To conclude our discussion of elastic–plastic fracture, we find for the case of ductile materials that fracture may be characterized by a critical crack-opening displacement (δ_c) value or when the J integral reaches the J_{IC} level. Both δ_c and J_{IC} measure the resistance to crack initiation whereas the tearing modulus $T(dJ/da)$ describes the material's resistance to crack extension. When conditions of full-scale plasticity are present, other methods are needed to assess the fracture process; in this regard, the reader is referred to a document entitled "An Engineering Approach to Elastic–Plastic Fracture Analysis."[61]

REFERENCES

1. A. A. Griffith, *Philos. Trans. R. Soc. London, Ser. A* **221,** 163 (1920). (This article has been republished with additional commentary in *Trans. ASM* **61,** 871 (1968).)
2. C. E. Inglis, *Proceedings,* Institute of Naval Architects, Vol. 55, 1913, p. 219.
3. E. Orowan, *Fatigue and Fracture of Metals*, MIT Press, Cambridge, MA, 1950, p. 139.
4. G. R. Irwin, *Fracturing of Metals,* ASM, Cleveland, OH 1949, p. 147.
5. G. R. Irwin and J. A. Kies, *Weld. J. Res. Suppl.* **33,** 193s (1954).
6. H. M. Westergaard, *Trans. ASME, J. Appl. Mech.* **61,** 49 (1939).
7. G. R. Irwin, *Handbuch der Physik*, Vol. 6, Springer, Berlin, 1958, p. 551.
8. P. C. Paris and G. C. M. Sih, ASTM *STP 381*, 1965, p. 30.
9. H. Tada, P. C. Paris, and G. R. Irwin, *The Stress Analysis of Cracks Handbook,* Del Research, Hellertown, PA, 1973.
10. G. C. M. Sih, *Handbook of Stress Intensity Factors*, Lehigh University, Bethlehem, PA, 1973.
11. Y. Murakami, Ed., *Stress Intensity Factors Handbook,* Pergamon Oxford, 1987.
12. W. F. Brown, Jr. and J. E. Srawley, ASTM *STP 410,* 1966.

13. C. Feddersen, ASTM *STP 410*, 1967, p. 77.
14. O. L. Bowie, *J. Math. Phys* **35,** 60 (1956).
15. D. Broek and H. Vliegar, *Int. J. Fract. Mech.* **8,** 353 (1972).
16. G. R. Irwin, *Appl. Mater. Res.* **3,** 65 (1964).
17. G. R. Irwin, *Trans. ASME, J. Appl. Mech.* **24,** 361 (1957).
18. F. A. McClintock and G. R. Irwin, ASTM *STP 381*, 1965, p. 84.
19. G. T. Hahn and A. R. Rosenfield, *Acta Met.* **13,** 293 (1965).
20. D. S. Dugdale, *J. Mech. Phys. Solids* **8,** 100 (1960).
21. R. W. Hertzberg and R. H. Goodenow, *Proceedings, Micro Alloying 75,* Union Carbide, New York, 1977, p. 503.
22. ASTM Standard E399-90, *Annual Book of ASTM Standards*, **03.01,** 509 (1993).
23. ASTM Standard E1304-89, ibid, 962.
24. L. M. Barker, *Eng. Fract. Mech.* **9,** 361 (1977).
25. L. M. Barker and F. I. Baratta, *J. Test. Eval.* **8**(3), 97 (1980).
26. J. J. Mecholsky and L. M. Barker, ASTM *STP 855,* 1984, p. 324.
27. G. R. Anstis, P. Chantikul, B. R. Lawn, and D. B. Marshall, *J. Am. Ceram. Soc.* **64**(9), 533 (1981).
28. A. G. Evans and E. A. Charles, *J. Am. Ceram. Soc.* **59**(7), 371 (1976).
29. B. R. Lawn and D. B. Marshall, *J. Am. Ceram. Soc.* **62**(7), 347 (1979).
30. T. R. Wilshaw and B. R. Lawn, *J. Mater. Sci.* **10**(6), 1049 (1975).
31. D. B. Marshall, *J. Am. Ceram. Soc.* **66**(2), 127 (1983).
32. R. F. Cook, B. R. Lawn, and C. J. Fairbanks, *J. Am. Ceram. Soc.* **68**(11), 604 (1985).
33. B. R. Lawn, in *Fracture Mechanics of Ceramics*, Vol. 5, R. C. Bradt, A. G. Evans, D. P. H. Hasselman, and F. F. Lange, Eds., Plenum, New York, 1983, p. 1.
34. B. R. Lawn, S. W. Frieman, T. L. Baker, D. D. Cobb, and A. C. Gonzalez, *J. Am. Ceram. Soc.* **67**(4), C67 (1984).
35. A. J. Birkle, R. P. Wei, and G. E. Pellissier, *Trans. ASM* **59,** 981 (1966).
36. G. R. Irwin, NRL Report 6598, Nov. 21, 1967.
37. G. R. Irwin, *ASTM Bulletin*, Jan. 1960, p. 29.
38. ASTM Standard E561-92a, *Annual Book of ASTM Standards*, **03.01,** 600 (1993).
39. A. A. Wells, *Brit. Weld. J.* **13,** 2 (1965).
40. A. A. Wells, *Brit. Weld. J.* **15,** 221 (1968).
41. J. N. Goodier and F. A. Field, *Fracture of Solids,* Interscience, New York, 1963, p. 103.
42. D. C. Drucker and J. R. Rice, *Eng. Fract. Mech.* **1,** 577 (1970).
43. British Standard Institution BS5762, BSI, 1979, London.
44. ASTM Standard E1290-93, *Annual Book of ASTM Standards*, **03.01,** 952 (1993).
45. J. R. Rice and E. P. Sorenson, *J. Mech. Phys. Solids* **26,** 163 (1978).
46. J. R. Rice, *J. Appl. Mech.* **35,** 379 (1968).
47. J. R. Rice, *Treatise on Fracture*, Vol. 2, H. Liebowitz, Ed., Academic, New York, 1968, p. 191.
48. J. W. Hutchinson and P. C. Paris, ASTM *STP 668,* 1979, p. 37.
49. N. E. Dowling and J. A. Begley, ASTM *STP 590,* 1976, p. 82.
50. J. A. Begley and J. D. Landes, ASTM *STP 514*, 1972, p. 1.
51. J. A. Begley and J. D. Landes, ASTM *STP 514*, 1972, p. 24.
52. J. R. Rice, P. C. Paris, and J. G. Merkle, ASTM *STP 532*, 1973, p. 231.

53. J. D. Landes and J. A. Begley, ASTM *STP 632*, 1977, p. 57.

54. ASTM Standard E813-89, *Annual Book of ASTM Standards*, **03.01,** 748 (1993).

55. ASTM Standard E1152-87, ibid, 853.

56. G. A. Clarke, W. R. Andrews, P. C. Paris, and D. W. Schmidt, ASTM *STP 590*, 1976, p. 27.

57. P. C. Paris, *Flaw Growth and Fracture*, ASTM *STP 631*, 1977, p. 3.

58. P. C. Paris, H. Tada, A. Zahoor, and H. Ernst, ASTM *STP 668*, 1979, p. 5.

59. P. C. Paris, H. Tada, H. Ernst, and A. Zahoor, ASTM *STP 668*, 1979, p. 251.

60. H. Ernst, P. C. Paris, M. Rossow, and J. W. Hutchinson, ASTM *STP 667*, 1979, p. 581.

61. V. Kumar, M. D. German, and C. F. Shih, EPRI NP-1931, Res. Proj. 1237-1, Elec. Pow. Res. Inst., 1981.

FURTHER READING

J. M. Barsom and S. T. Rolfe, *Fracture and Fatigue Control in Structures*, 2d ed., Prentice-Hall, Englewood Cliffs, NJ, 1987.

D. Broek, *Elementary Engineering Fracture Mechanics,* Sitjhoff & Noordhoff, The Netherlands, 1978.

H. L. Ewalds and R. J. H. Wanhill, *Fracture Mechanics*, Arnold, London, 1984.

J. F. Knott, *Fundamentals of Fracture Mechanics,* Butterworths, London, 1973.

PROBLEMS

8.1   A compact tension test specimen ($H/W = 0.6$), is designed and tested according to the ASTM E399-90 procedure. Accordingly, a Type I load versus displacement (P vs. δ) test record was obtained and a measure of the maximum load P_{max} and a critical load measurement point P_Q were determined. The specimen dimensions were determined as $W = 10$-cm, $t = 5$-cm, $a = 5$-cm, the critical load-point measurement point $P_Q = 100$ kN and $P_{max} = 105$ kN. Assuming that all other E399 requirements regarding the establishment and sharpness of the fatigue starter crack are met, determine the critical value of stress intensity. Does it meet conditions for a valid K_{IC} test if the material yield stress is 700 MPa? If it is 350 MPa?

8.2   An infinitely large sheet is subjected to a gross stress of 350 MPa. There is a central crack $5/\pi$-cm long and the material has a yield strength of 500 MPa.

  (a)   Calculate the stress-intensity factor at the tip of the crack.

  (b)   Calculate the plastic-zone size at the crack tip.

  (c)   Comment upon the validity of this plastic-zone correction factor for the above case.

8.3   A sharp penny-shaped crack with a diameter of 2.5-cm is completely embedded in a solid. Catastrophic fracture occurs when a stress of 700 MPa is applied.

  (a)   What is the fracture toughness for the material? (Assume that this value is for plane-strain conditions.)

(b) If a sheet (0.75-cm thick) of this material is prepared for fracture-toughness testing ($t = 0.75$, $a = 3.75$-cm), would the fracture-toughness value be a valid test number (the yield strength of the material is given to be 1100 MPa)?

(c) What would be a sufficient thickness for valid K_{IC} determination?

8.4 Determine the angle θ (Fig. 8.12) associated with the maximum dimension of the plastic zone.

8.5 Calculate the leak-before-break criterion in terms of flaw size and section thickness.

8.6 Prove to yourself that the ASTM criteria for a valid plane-strain fracture-toughness test (Eq. 8-49) reflects the condition associated with Eq. 8-50.

8.7 For the Ti-6Al-4V alloy test results given in Table 8.2, determine the sizes of the largest elliptical surface flaws ($a/2c \approx 0.2$) that would be stable when the design stress is 75% of σ_{ys}.

8.8 For the same material and design stress described in Problem 8.7, compute the wall thickness of a vessel that contained an elliptical flaw [$(a/2c) = 0.2$] at the beginning of its service life but experienced a leak-before-break condition.

8.9 You are offered an opportunity to earn $10 million by simply hanging from a rope for only one minute. The rope is attached to a glass sheet (300-cm long by 10-cm wide and 0.127-cm thick). Complicating the situation is the fact that:

(a) The glass sheet contains a central crack with total length of 1.62-cm that is oriented parallel to the ground. The fracture toughness of the glass is 0.83 MPa $\sqrt{m}$.

(b) The rope is suspended 3-m above a pit of poisonous snakes. (Recall the good professor from the film ''Raiders of the Lost Ark.'') Would you try for the pot of gold at the end of the rainbow?

8.10 A material possessing a plane-strain fracture-toughness value of 50 MPa$\sqrt{m}$ and a yield strength of 1000 MPa is to be made into a large panel.

(a) If the panel is stressed to a level of 250 MPa, what is the maximum size flaw that can be tolerated before catastrophic failure occurs? (Assume a center notch configuration.)

(b) At the point of fracture, what is the size of the plastic zone at the middle of the panel along the crack front?

(c) If the panel were 2.5-cm thick, would this constitute a valid plane-strain condition?

(d) If the thickness were increased to 10-cm, would there be a change in the critical size of the flaw calculated in part (a)?

8.11 A rather thin steel leaf spring experiences simple bending in one direction and develops a semielliptical flaw ($a/2c = 0.15$) on the tensile surface. As you might expect, the plane of the flaw is oriented normal to the flexural stress direction. Repeated loading causes the crack to expand. Discuss whether the ellipticity ratio ($a/2c$) will increase or decrease.

8.12 A thin-walled pressure vessel 1.25-cm thick originally contained a small semi-circular flaw (radius 0.25-cm) located at the inner surface and oriented normal to the hoop stress direction. Repeated pressure cycling enabled the crack to grow larger. If the fracture toughness of the material is 88 MPa$\sqrt{\text{m}}$, the yield strength equal to 825 MPa, and the hoop stress equal to 275 MPa, would the vessel leak before it ruptured?

8.13 **(a)** Two square steel rods were brazed end to end to produce a rod with dimensions 6.25 × 6.25 × 30-cm. The silver braze is 0.063-cm thick and was produced with material possessing an ultimate strength of 140 MPa. The steel rod sections have yield and tensile strengths of 690 and 825 MPa, respectively, and a plane-strain fracture-toughness value of 83 MPa$\sqrt{\text{m}}$. If the composite rod/braze assembly is loaded in tension, describe how failure will occur if the steel rod contains an elliptical surface flaw 1.25-cm deep and 3.75-cm wide, which is oriented normal to the stress axis (see sketch). With the aid of suitable calculations, determine the stress necessary for failure.

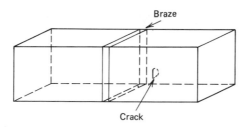

(b) If the same rod had instead contained a through-thickness crack of depth 2.5-cm, describe with suitable equations how failure will occur and at what stress level.

8.14 A 10-cm-diameter aluminum thin-walled pressure vessel, that has a wall thickness of 2-cm, is subjected to a proof pressure cycle of 75 MPa to check on the integrity of the vessel. If the structure withstands this pressure surge, it is then placed in service where a design pressure of 50 MPa is experienced. Assuming that the vessel material has a fracture toughness of 25MPa$\sqrt{\text{m}}$ and contains a 0.5-cm-deep semicircular surface flaw oriented normal to the hoop stress direction, then answer the following questions:

(a) What is the proof stress and the design stress for this vessel?

(b) What is the stress intensity factor associated with the applicaton of the proof stress?

(c) From your answer to the previous question, will the vessel withstand the application of the proof stress, given the presence of the crack?

(d) Determine whether or not the vessel would experience a leak-before-break condition. (Show your calculations).

8.15 A plate of 4335 + V steel contains a semi-elliptical surface flaw, 0.8-mm deep and 2-mm long, that is oriented perpendicular to the design stress direction.

Given that K_{IC} = 72.5 MPa$\sqrt{m}$, σ_{ys} = 1340 MPa, and the operating stress = $0.4\sigma_{ys}$, determine whether the plate merits continued service, based on the requirement that the operative stress intensity level is below 0.5 K_{IC}.

8.16 Is it possible to conduct a valid plane strain fracture toughness test for a CrMoV steel alloy under the following conditions: K_{IC} = 53 MPa$\sqrt{m}$; σ_{ys} = 620 MPa, W = 6-cm and plate thickness, B = 2.5-cm?

8.17 If the plate thickness in the previous problem were 1-cm, would the thickness be sufficient for a J_{IC} test? Assume these material properties: E = 205 GPa, μ = 0.25.

8.18 A rod of soda-lime-silica glass is constrained at 400°K and then cooled rapidly to 300°K.

(a) Would you expect the rod to survive this quench?

(b) If the glass rod contained a 1-mm scratch that was oriented perpendicular to the axis of the rod, would your answer to part **(a)** be the same?

(c) Would failure occur if the temperature drop and the crack size were half that given above? For the glass rod: E = 70 GPa, σ_{ts} = 90 MPa, α = 8×10^{-6}K^{-1}, and K_{IC} = 0.8 MPa$\sqrt{m}$.

CHAPTER 9

TRANSITION TEMPERATURE APPROACH TO FRACTURE CONTROL

9.1 TRANSITION TEMPERATURE PHENOMENON AND THE CHARPY IMPACT SPECIMEN

Before fracture mechanics concepts were developed, engineers sought laboratory-sized samples and suitable test conditions with which to simulate field failures without resorting to the forbidding expense of destructively testing full-scale engineering components. To anticipate the worst possible set of circumstances that might surround a potential failure, these laboratory tests employed experimental conditions that could suppress the capacity of the material to plastically deform by elevating the yield strength: low test temperatures, high strain rates, and a multiaxial stress state caused by the presence of a notch or defect in the sample. Of considerable importance in pressure vessel and bridge and ship structure applications was the fact that in body-centered-cubic metals, such as ferritic alloys, the yield strength is far more sensitive to temperature and strain-rate changes than it is in face-centered-cubic metals such as aluminum, nickel, copper, and austenitic steel alloys. As pointed out in Chapter 2, this increased sensitivity in BCC alloys can be related to the temperature-sensitive Peierls-Nabarro stress contribution to yield strength, which is much larger in BCC metals than in FCC metals.

To a first approximation, the relative notch sensitivity of a given material may be estimated from the yield- to tensile-strength ratio. When the ratio is low, the plastic constraint associated with a biaxial or triaxial stress state at the crack tip will elevate the entire stress–strain curve and allow for a net section stress greater than the smooth bar tensile strength value. Recall from Chapter 7 that a $2\frac{1}{2}$- to 3-fold increase in net section strength is possible in ductile materials that "notch strengthen" (Fig. 9.1a). On the other hand, in materials that have less ability for plastic deformation, the stress concentration at a notch root is not offset by the necessary degree of crack-tip plasticity needed to blunt the crack tip. Consequently, the notch with its multiaxial stress state raises the local stress to a high level and suppresses what little plastic deformation capacity the material possesses, and brittle failure occurs (Fig. 9.1b).

The Charpy specimen (Fig. 9.2a) and associated test procedure provides a relatively severe test of material toughness. The notched sample is loaded at very high strain

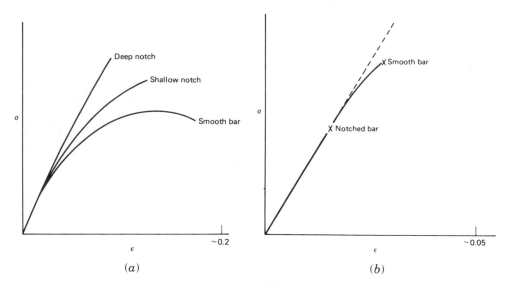

FIGURE 9.1 (*a*) **Plastic constraint resulting from triaxial stresses at notch root produces elevation of flow curve in ductile material.** (*b*) **For material with little intrinsic plastic flow capacity, introduction of sharp crack induces premature brittle failure.**

rates because the material must absorb the impact of a falling pendulum and is tested over a range of temperatures. Considerable data can be obtained from the impact machine reading and from examination of the broken sample.

First, the amount of energy absorbed by the notched Charpy bar can be measured by the maximum height to which the pendulum rises after breaking the sample (Fig. 9.3). If a typical 325-J (240-ft-lb)* machine were used, the extreme final positions of the pendulum would be either at the same height of the pendulum before it was released (indicating no energy loss in breaking the sample), or at the bottom of its travel (indicating that the specimen absorbed the full 325 J of energy). The dial shown in Fig. 9.3 provides a direct readout of the energy absorbed by the sample. Typical impact energy versus test temperature for several metals is plotted in Fig. 9.4. It is clearly evident from this plot that some materials show a marked change in energy absorption when a wide range of temperatures is examined. In fact, this sudden shift or transition in energy absorption with temperature has suggested to engineers the possibility of designing structural components with an operating temperature above which the component would not be expected to fail.

The effect of temperature on the energy to fracture has been related in low-strength ferritic steels to a change in the microscopic fracture mechanism: cleavage at low temperatures and void coalescence at high temperatures. The onset of cleavage and brittle behavior in low-strength ferritic steels is so closely related that "cleavage" and "brittle" often are used synonymously in the fracture literature. This is unfortunate since in Chapter 7 brittle is defined as a low level of fracture *energy* or limited

---

* To convert from foot-pounds to joules, multiply by 1.356.

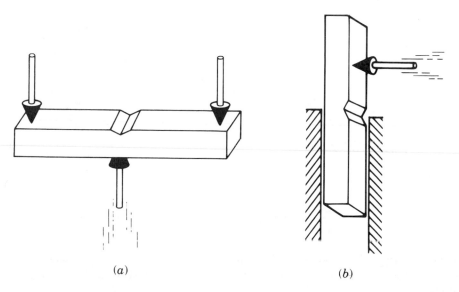

FIGURE 9.2 Flexed-beam impact samples. (*a*) Charpy type[1] (three-point loaded) used extensively with metal alloys. (*b*) Izod type[2] (cantilever-beam loaded) used extensively with polymers. Both samples contain 0.25-mm notch radius.

crack-tip plasticity, while cleavage describes a failure micro-*mechanism*. Confusion arises since brittle behavior can occur without cleavage, as in the fracture of high-strength aluminum alloys; alternatively, you can have 4% elongation (reflecting moderate energy absorption) in a tungsten-25 a/o rhenium alloy specimen and still have a cleavage fracture.[4] Since a direct correlation does not always exist between a given fracture mechanism and the magnitude of fracture energy, it is best to treat the two terms separately.

Unless the fracture energy changes discontinuously at a given temperature, some criterion must be established to *define* the "transition temperature." Should it be

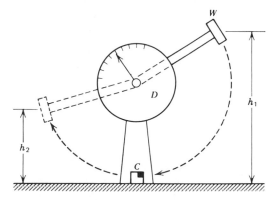

FIGURE 9.3 Diagram showing impact hammer W dropping from height h_1, impacting sample at C and rising to maximum final height h_2. Energy absorbed by sample, related to height differential $h_2 - h_1$, is recorded on dial D.

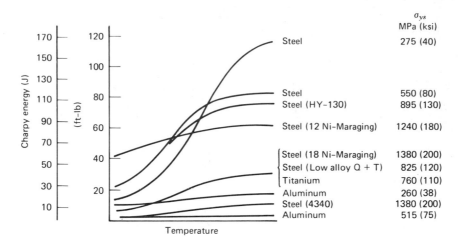

FIGURE 9.4 **Charpy impact energy versus temperature behavior for several engineering alloys.[3] (Reprinted by permission of the American Society for Testing and Materials from copyright material.)**

defined at the 13.5-, 20-, or 27-J (10-, 15-, or 20-ft-lb)* level as it is sometimes done or at some fraction of the maximum or shelf energy? The answer depends on how well the defined transition temperature agrees with the service experience of the structural component under study. For example, Charpy test results for steel plate obtained from failures of Liberty ships revealed that plate failures never occurred at temperatures greater than the 20-J (15-ft-lb) transition temperature. Unfortunately, the transition temperature criterion based on such a specific energy level is not constant but varies with material. Specifically, Gross[5] has found for several steels with strengths in the range of 415 to 965 MPa that the appropriate energy level for the transition temperature criterion should increase with increasing strength.

The same problem arises when the transition temperature is estimated from other measurements. For example, if the amount of lateral expansion on the compression side of the bar is measured (Fig. 9.5), it is found that it, too, undergoes a transition from small values at low temperature to large values at high temperature. (This increase in observed plastic deformation is consistent with the absorbed energy–temperature trend.) Whether the correct transition temperature conforms to an absolute or relative contraction depends on the material. Finally, transitional behavior is found when the amount of fibrous or cleavage fracture on the fracture surface is plotted against temperature. A typical series of fracture surfaces produced at different temperatures is shown in Fig. 9.6a. Here again, the appropriate percentage of cleavage or fibrous fracture (based on comparison with a standard chart such as in Fig. 9.6b or measured directly as in Fig. 9.6c) to use to define the transition temperature will depend on the material as well as other factors. To make matters worse, transition temperatures based on either energy absorption, ductility, or fracture appearance

---

* Dual units are retained for reference since the specific foot-pound energy levels cited above represent long-standing design criteria.

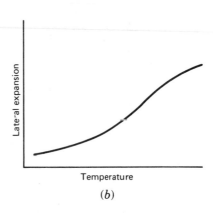

FIGURE 9.5   (*a*) Measurement of lateral expansion at compression side of Charpy bar; (*b*) schema of temperature dependence of lateral expansion revealing transition behavior.

criteria do not agree even for the same material. As shown in Table 9.1, the transition temperatures defined by a 20-J energy criterion and by a 0.38-mm (15-mil) lateral expansion are in reasonably good agreement, but are consistently lower than the 50% fibrous fracture transition temperature. Which transition temperature to use ''is a puzzlement!''

9.2   ADDITIONAL FRACTURE TEST METHODS

In addition to transition temperature determinations from Charpy data, related critical temperatures may be obtained from other laboratory samples, such as the drop-weight and Robertson crack-arrest test procedures. The drop-weight sample[7] (Fig. 9.7) consists of a flat plate, one surface of which contains a notched bead of brittle weld metal. After reaching a desired test temperature, the plate is placed in a holder, weld bead face down, and impacted with a falling weight. Since a crack can begin to run at the base of the brittle weld-bead notch with very little energy requirement, the critical factor is whether the base plate can withstand this advancing crack and not break. According to ASTM Standard E208, the nil ductility temperature (NDT) is defined as that temperature below which the plate ''breaks'' (Fig. 9.7*a*) but above which it does not (Fig. 9.7*b*). (The specimen is considered to be ''broken '' if a crack grows to one or both edges on the tension surface. Cracking on the compression side is not required to establish the ''break'' condition.) Therefore, NDT reflects a go, no-go condition associated with a negligible level of ductility.

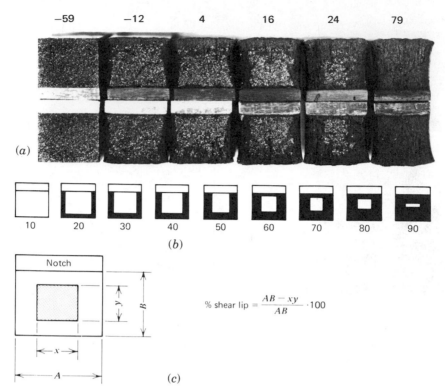

FIGURE 9.6 Transition in fracture surface appearance as function of test temperature. (*a*) Actual fracture series for A36 steel tested in the transverse direction; (*b*) standard comparison chart showing percentage shear lip; (*c*) computation for percentage shear lip.

TABLE 9.1a Transition Temperature Data for Selected Steels[6]

Material	σ_{ys}, MPa σ_{ts}, MPa	Transition Temperature, °C		
		20 J	**0.38 mm**	**50% fibrous**
Hot-rolled C–Mn steel	$\dfrac{210}{442}$	27	17	46
Hot-rolled, low-alloy steel	$\dfrac{385}{570}$	-24	-22	12
Quenched and tempered steel	$\dfrac{618}{688}$	-71	-67	-54

TABLE 9.1b Transition Temperature Data for Selected Steels[6]

Material	σ_{ys}, ksi σ_{ts}, ksi	Transition Temperature, °F		
		15 ft-lb	**15 mil**	**50% fibrous**
Hot-rolled C–Mn steel	$\dfrac{30.5}{64.1}$	80	62	115
Hot-rolled, low-alloy steel	$\dfrac{55.9}{82.6}$	-12	-7	53
Quenched and tempered steel	$\dfrac{89.7}{99.8}$	-95	-88	-66

Break   No break

FIGURE 9.7   Drop-weight test sample with notched weld bead. Used to measure nil ductility temperature (NDT). (*a*) Break; (*b*) no break. (Courtesy of Dr. A. W. Pense, Lehigh University.)

EXAMPLE 9.1

A large-diameter turbine rotor is fabricated from a Cr-Mo-V steel alloy with the yield strength- and fracture toughness-temperature dependence shown in the accompanying graph.* The rotor is designed to operate at 100°C, above both the NDT and 50% FATT, and at a stress 40% that of the material's yield strength at the operating temperature. During a nonroutine shutdown in association with a steam pipe failure, one of the plant technicians discovered a 2.5-cm-deep semicircular surface flaw in the rotor that was oriented normal to the maximum stress direction. Your supervisor decided to delay repair of the rotor until the next scheduled full-scale overhaul. Since the steam pipe repair was time-consuming, he was anxious to get the rotor "on-line" as quickly as possible and urged you to use the "cold" start-up procedure instead of the slower "recommended" start-up procedure. Was your supervisor's judgment correct?

---

* J. M. Barsom and S. T. Rolfe, *Fracture and Fatigue Control in Structures,* 2d ed. Prentice-Hall, Inc. Englewood Cliffs, New Jersey 116 (1987). Reprinted by permission.

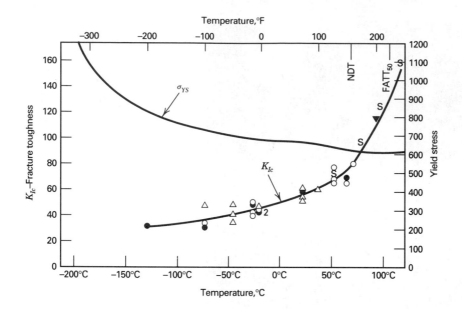

To begin our analysis, we need to describe the two different start-up procedures.

Recommended procedure: "Warm" start-up at 50% of normal rotor design rpm level (ω_{op}) until rotor reaches operational temperature of 100°C. Then increase rpm to 1.15 ω_{op} for 10 minutes to "proof-test" the rotor prior to its return to normal service conditions where $\omega = \omega_{op}$.

Cold start-up procedure: Introduce proof test conditions directly at ambient temperature (10°C) (i.e., $\omega = 1.15 \, \omega_{op}$) and hold for 10 minutes. Then reduce rpm to $\omega = \omega_{op}$ and resume normal service.

From the preceding information and graph, the rotor design stress at the 100°C operational temperature is

$$\sigma_{des} = 0.4\sigma_{ys} = 238 \text{ MPa}$$

Since the operational stress in the rotor varies with ω^2 (e.g., see Section 14.5-Case 3), the "warm-up" stress at $\omega = 0.5\omega_{op}$ is

$$\sigma/\sigma_{des} = (0.5)^2 \qquad \text{so that} \qquad \sigma = 0.25\sigma_{des} \approx 60 \text{ MPa}$$

$$K = \left(\frac{2}{\pi}\right)(1.1) \, \sigma\sqrt{\pi a}$$

$$K = \left(\frac{2}{\pi}\right)(1.1) \, 60\sqrt{\pi(2.5 \times 10^{-2})} = 11.8 \text{ MPa}\sqrt{m}$$

As expected, failure does not occur since $K < K_{IC}$ (53 MPa$\sqrt{m}$ at 10°C).

Once the temperature reaches 100°C, rotor rpm is increased to $1.15\omega_{op}$ so the operating stress increases to 238 $(1.15)^2 = 315$ MPa. Therefore,

$$K = \left(\frac{2}{\pi}\right)(1.1)\,315\sqrt{\pi(2.5 \times 10^{-2})} = 61.8 \text{ MPa}\sqrt{\text{m}}$$

Again, no failure occurs under this "proof-stress" condition since $K < K_{IC}$ (138 MPa$\sqrt{\text{m}}$ at 100°C). (Chapter 13 introduces additional calculations to determine the remaining lifetime of the rotor, given that the 2.5-cm crack will grow with repeated load cycling during continued service operation.)

Under cold start-up conditions, the proof-test stress is applied directly at 10°C. Since the applied stress intensity factor of 61.8 MPa$\sqrt{\text{m}}$ is greater than the fracture toughness of the material at that temperature (i.e., 53 MPa$\sqrt{\text{m}}$) sudden failure would occur. The conclusion: *Use the more cautious start-up procedure and don't be in such a hurry!*

NDT test results have been used in the design of structures made with low-strength ferritic steels. For example, allowable minimum service temperatures (T_{min}) for structures containing sharp cracks have been defined[8] (but may be a function of plate thickness):

1. $T_{min} \geq$ NDT: Permissible when applied stress σ is less than 35–55 MPa
2. $T_{min} \geq$ NDT + 17°C (30°F): Permissible when $\sigma \leq \sigma_{ys}/2$
3. $T_{min} \geq$ NDT + 33°C (60°F): Permissible when $\sigma \leq \sigma_{ys}$
4. $T_{min} \geq$ NDT + 67°C (120°F): Permissible since failure will not occur below the ultimate tensile strength of the material

These criteria represent specific conditions along the curve marked a_5 in Fig. 9.8, corresponding to a large flaw size. For progressively smaller flaw sizes, the allowable

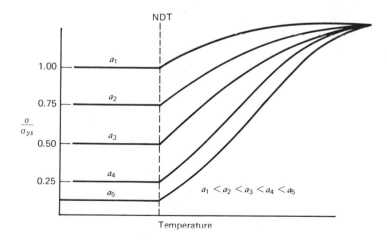

FIGURE 9.8 Fracture analysis diagram revealing allowable stresses as a function of flaw size and operating temperature. Curve a_5 corresponds to crack-arrest temperature as measured from the Robertson test.

stress levels are seen to increase for a given minimum operating temperature. Information of this type has been collected and averaged by Pellini and Puzak[9] and presented in the form of fracture analysis diagrams (FAD) such as the one shown in Fig. 9.8.

Although some success has been achieved by using this diagram for low-strength steel applications, certain inherent deficiencies should be pointed out.[3] First, it is not altogether clear which stress, local, or gross section should be used. Also, the FAD treats all steels of the same strength level alike and ignores the possibility of toughness differences from one steel grade to another. The allowable flaw size to be tolerated by a structure is strongly dependent on the toughness of the material and a reasonable estimate of the stress level; if the latter quantities are not defined, the FAD cannot provide very reliable quantitative information. Nevertheless, these diagrams do represent an attempt to bridge the differences between the transition temperature and fracture mechanics philosophies of fracture control design.

The Robertson sample[10] is designed to measure a crack-arrest condition. As seen in Fig. 9.9, this sample contains a saw cut at one side of the plate and is subjected to a thermal gradient across the plate width such that the starter notch is at the lowest temperature, while the right side of the plate is considerably warmer. After a uniform load is applied normal to the starter crack plane, the plate is impacted on the cold side, causing an unstable crack to grow from the cold starter notch root. The crack will run across the plate until it encounters a plate temperature at which the material offers too much resistance to further crack growth; this is defined as the crack arrest temperature (CAT). From such tests, it has been shown that CAT depends on the material, the magnitude of the applied stress, and the specimen thickness. As a final note, the CAT, when determined over a range of stress levels, provides the information for the construction of the lower curve (marked a_5) in the FAD shown in Fig. 9.8.

Additional specimens and test methods have been developed to evaluate the toughness response of engineering plastics.[11] The Izod sample (Fig. 9.2b) is a notched bar

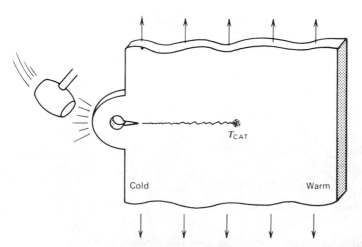

FIGURE 9.9 **Robertson crack-arrest specimen. Sample is uniformly loaded and subjected to temperature gradient. After impact loading, crack propagates from starter notch and is arrested at T_{CAT}.**

that is fixed at one end and impacted on the unsupported section along the side of the bar that contains the notch.[2] Numerous studies have shown that the Izod impact energy of many plastics, usually defined by the energy absorbed per unit area of the net section, possesses a ductile–brittle transition response. Figure 9.10a shows the change in transition temperature for PVC with different notch radii. The large reduction in fracture energy at a given temperature with the presence of a notch is attributed to the high strain rate at the notch root and the virtual elimination of energy necessary to initiate a crack in the sample. A comparison of the room temperature impact resistance for several polymers is given in Fig. 9.10b and reveals the strong notch sensitivity of these materials.[12] Two other impact test methods for engineering plastics are described by the drop-weight and tensile-impact test procedures. The drop-weight method (ASTM D3029[13]) measures the energy to initiate fracture in an unnotched sheet of material; a disk-shaped sample is supported horizontally by a steel ring and struck with different weights that are dropped from a given height. A mean failure energy is then defined on a statistical basis, which corresponds to a 50% failure rate of the disks. The high-speed tensile-impact method (ASTM D1822)[14] makes use of a small tensile specimen that is clamped to a pendulum hammer at one end and attached to a striker plate at the other. The pendulum is released and drops with the attached sample trailing behind. When the striker plate impacts a rigidly mounted stationary anvil, the sample experiences rapid straining in tension to failure. The fracture energy is determined from the differences in pendulum height before and after its release. Impact energies for many engineering plastics are given in the *Modern Plastics Encyclopedia.*[15]

9.3 LIMITATIONS OF THE TRANSITION TEMPERATURE PHILOSOPHY

It is important to recognize some limitations in the application of the transition temperature philosophy to component design. First, the absolute magnitude of the experimentally determined transition temperature, as defined by any of the previously described methods (energy absorbed, ductility, and fracture appearance), depends on the thickness of the specimen used in the test program. This is due to the potential for a plane-strain–plane-stress, stress-state transition when sample thickness is varied. In evaluating this effect, McNicol[6] found that the transition temperature in several steels, based on energy, ductility, and fracture appearance criteria, increased with increasing Charpy bar thickness t. Figure 9.11 shows temperature-related changes in energy absorbed per 2.5-mm sample thickness and percentage shear fracture as a function of sample thickness for A283, a hot-rolled carbon manganese steel. It is clear from this figure that the transition temperature increased with increasing thickness. Moreover, the transition temperature was different for the two criteria. With increasing sample thickness, it would be expected that the transition temperature would rise to some limiting value as full plane-strain conditions were met. This condition is inferred from Fig. 9.12, which shows the transition temperature reaching a maximum level with increasing thickness for three different steel alloys.

It is clear, then, that the defined transition temperature will depend not only on the measurement criteria but also on the thickness of the test bar. Therefore, laboratory

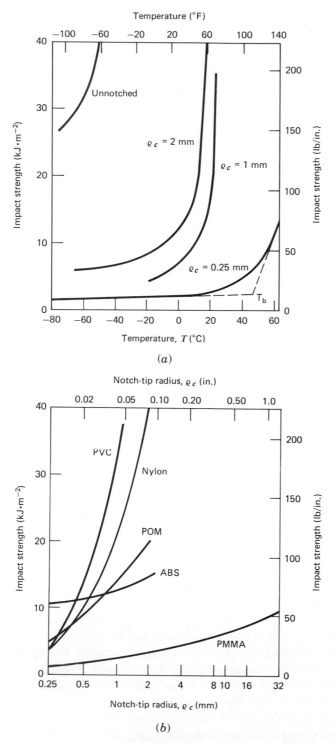

FIGURE 9.10 **Impact strength in (*a*) PVC as a function of temperature for different notch-root radii and (*b*) selected engineering plastics as a function of notch-tip radius.[12] (Reprinted with permission from *Impact Tests and Service Performance of Thermoplastics*, courtesy of Plastics and Rubber Institute.)**

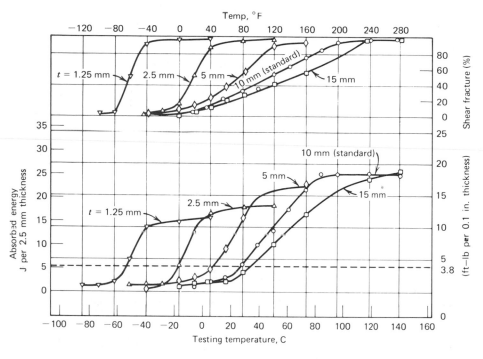

FIGURE 9.11   Adjusted energy-temperature curves and shear fracture–temperature curves for 38-mm-thick plate of A283 steel tested with Charpy V-notch specimens of various thicknesses. Absorbed energy defined at 5.2 J/2.5 mm (3.8 ft-lb/0.1 in.) of specimen thickness.[6] (Reprinted from *Welding Journal* by permission of the American Welding Society.)

results may bear no direct relation to the transition temperature characteristics of the engineering component if the component's thickness is different from that of the test bar. To overcome this difficulty, the dynamic tear test (DT)[8] and drop-weight tear test (DWTT)[16] were developed wherein the sample thickness is increased to the full thickness of the plate. As seen in Fig. 9.13, both tests involve three-point bending of a notched bar. The basic difference between the two is the notch detail: The DWTT contains a shallow notch (5 mm deep) which is pressed into the edge of the sample with a sharp tool, while the DT notch is deeper and embedded within a titanium embrittled electron beam weld. These samples are broken in either pendulum or drop-weight machines that are calibrated to measure the fracture energy of the sample. Hence, energy absorption versus test temperature plots can be obtained in the same manner as with Charpy specimens. As such, the DWTT and DT specimens may be considered to be oversized Charpy samples. The big difference lies in the fact that these samples are much thicker and wider than the Charpy specimen, resulting in much greater plastic constraint at the notch root. As a result, the transition temperature is shifted dramatically to higher temperatures (Fig. 9.14).[17] It is important to note from Fig. 9.14 that the DT, Robertson crack-arrest test, and the drop-weight NDT test results all indicate brittle material response at about −20°C for this material while the Charpy test indicates very tough behavior. Such sharp contrasts in test results are most disturbing when engineering design decisions must be made.

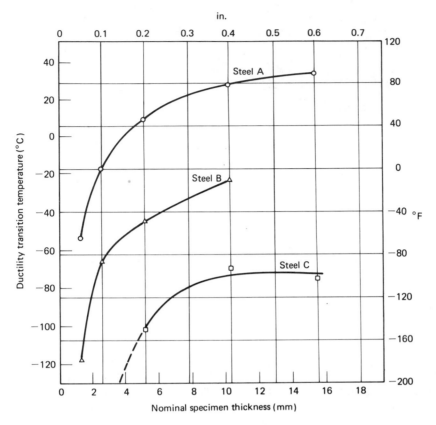

FIGURE 9.12 Effect of specimen thickness on Charpy V-notch ductility transition temperature of steels A, B, and C. The ductility transition temperature was selected with the same relative energy/unit thickness ratio given in Fig. 9.11.[6] (Reprinted from *Welding Journal* by permission of the American Welding Society.)

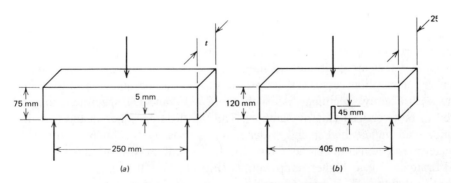

FIGURE 9.13 (*a*) Drop-weight tear test specimen (DWTT) with shallow notch pressed into bar; (*b*) dynamic tear test specimen with machined slot introduced into titanium-embrittled electron beam weld.

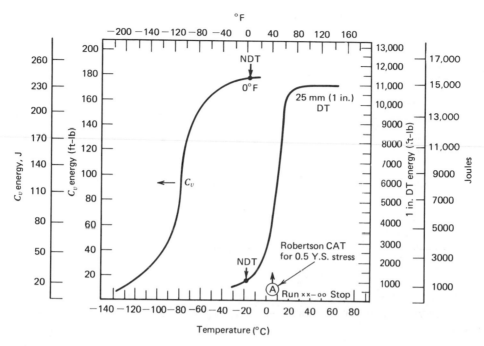

FIGURE 9.14 Comparison of Charpy V-notch, dynamic tear, drop-weight NDT and Robertson crack-arrest test results for A541 (Class 6 steel at 580-MPa yield strength). Note that the C_v test indicates a very high level of toughness at the NDT temperature that, in reality, corresponds to brittle behavior as indicated by the DT and Robertson test result.[17] (Reprinted by permission of the American Society for Testing and Materials from copyright material.)

In addition to transition temperature–thickness effects, there are uncertainties relating to crack-length effects as well. This may be seen by considering implications of a graphical representation of Eq. 8-22 (Fig. 9.15). We see the general relation between flaw size and allowable stress level for a material with a given toughness level. The solid line represents the material toughness K_c, assuming ideally elastic conditions, and the dashed portion of the curve reflects the reality of crack-tip plasticity. On the basis of the energy necessary to break a component, brittle conditions would be associated with the right side of the curve, while tough behavior would be found under conditions associated with the left side of the plot. Consequently, it is seen that a notched bar with crack length a_1 would be brittle at room temperature, but the same material with a crack length a_3 would exhibit tough behavior. If a sample with intermediate crack length, say a_2, were tested, the material also would be tough at room temperature. Since the brittle region of this curve is truncated by the onset of plastic deformation when the applied stress reaches the material yield strength, the brittle domain can be expanded simply by lowering the test temperature. Consequently, if the test temperature were to be reduced from T_1 to T_2, the sample response with crack length a_2 would change from tough to brittle, and the sample with crack length a_3 would still exhibit high toughness. An additional temperature reduction would be necessary for this sample to exhibit brittle behavior.

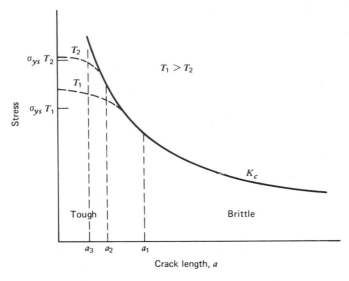

FIGURE 9.15 **Schematic diagram showing relation between allowable stress level and flaw size. Solid line represents material fracture toughness K_c; dashed lines show effect of plasticity.**

From the above discussion, it becomes apparent that a wide range of "transition temperatures" can be obtained simply by changing the specimen thickness and/or the crack length of the test bar. For this reason, transition temperature values obtained in the laboratory bear little relation to the performance of the full-scale component, thereby necessitating a range of correction factors as discussed earlier.

As mentioned above, the onset of brittle fracture is not always accompanied by the occurrence of the cleavage microscopic fracture mechanism. Rather, it should be possible to choose a specimen size for a given material, and tailor both thickness and planar dimensions such that a temperature-induced transition in energy to fracture, amount of lateral contraction, and macroscopic fracture appearance would occur *without the need for a microscopic mechanism transition.* Figure 9.16, from the work of Begley,[18] is offered as proof of this statement. Substandard thickness Charpy bars of 7075-T651 aluminum alloy were tested and shown to exhibit a temperature-induced transition in impact energy and fracture appearance. From Fig. 9.4, no such transition was observed when standard Charpy specimens of an aluminum alloy were broken.

9.4 IMPACT ENERGY—FRACTURE-TOUGHNESS CORRELATIONS

Although handicapped by the inability to bridge the size gap between small laboratory sample and large engineering component, the Charpy test sample method does possess certain advantages, such as ease of preparation, simplicity of test method, speed, low cost in test machinery, and low cost per test. Recognizing these factors, many researchers have attempted to modify the test procedure to extract more fracture information and seek possible correlations between Charpy data and fracture-toughness values obtained from fracture mechanics test samples. In one such approach, Orner

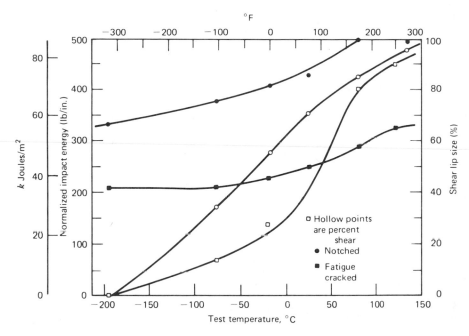

FIGURE 9.16 **Charpy impact data for subsize specimen revealing transition temperature response in 7075-T651 aluminum alloy.[18] (Courtesy James A. Begley.)**

and Hartbower[19] precracked the Charpy sample so that the impact energy for failure represented energy for crack propagation but not energy to initiate the crack.

$$E_T = E_i + E_p \tag{9-1}$$

where E_T = total fracture energy
   E_i = fracture initiation energy
   E_p = fracture propagation energy

They found that a correlation could be made between the fracture toughness of the material $\mathcal{G}_c$ and the quantity W/A, where W is the energy absorbed by the precracked Charpy test piece and A the cross-sectional area broken in the test. Although promising results have been observed for some materials (for example, see Fig. 9.17), the applicability of this test method should be restricted to those materials that exhibit little or no strain-rate sensitivity, since dynamic Charpy data are being compared with static fracture-toughness values. Also, the neglect of kinetic energy absorption by the broken samples as part of the energy-transfer process from the load pendulum to the specimen makes it impossible to develop good data in brittle materials where the kinetic energy component is no longer negligible.[20] Orner and Hartbower did point out, however, that the precracked Charpy sample could be used to measure the strain-rate sensitivity of a given material by conducting tests under both impact and slow bending conditions. Barsom and Rolfe[21] have verified this hypothesis with a direct comparison of static and dynamic test results from precracked Charpy V-notch (CVN) and plane-strain fracture-toughness samples, respectively. First, they established the strain-rate-induced shift in transition temperature for several steel alloys in the strength

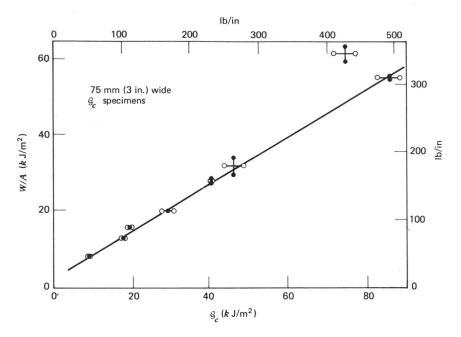

FIGURE 9.17   Relation between fatigue-cracked V-notch Charpy slow bend and $\mathcal{G}_c$ in a variety of 3.2-mm (0.125-in.)-thick aluminum alloys.[19] (Reprinted from *Welding Journal* by permission of the American Welding Society.)

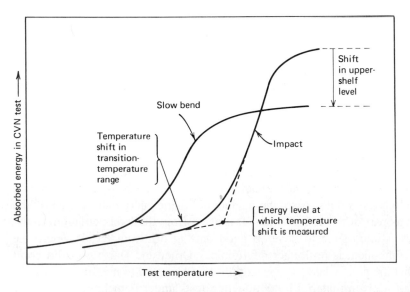

FIGURE 9.18   Diagram of impact energy versus test temperature revealing shift in transition temperature due to change in strain rate. (Note the higher shelf energy resulting from dynamic loading conditions, which may be related to a strain-rate-induced elevation in yield strength.)[21] (Reprinted by permission of the American Society for Testing and Materials from copyright material.)

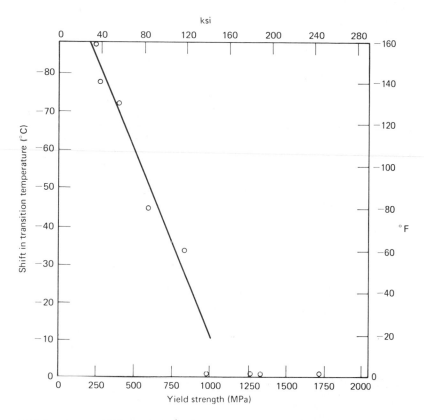

FIGURE 9.19 **Effect of yield strength on shift in transition temperature between impact and slow bend CVN tests.**[21] **(Reprinted by permission of the American Society for Testing and Materials from copyright material.)**

range of 275 to 1725 MPa (Figs. 9.18 and 9.19 and Table 9.2). They noted that the greatest transition temperature shift was found in the low-strength steels and no apparent strain-rate sensitivity was present in alloys with yield strengths in excess of 825 MPa. When these same materials were tested to determine their plane-strain fracture-toughness value, a corresponding shift was noted as a function of strain rate.

Figures 9.20a and 9.20b show static (K_{IC}) and dynamic (K_{ID}) plane-strain fracture-

TABLE 9.2 **Transition Temperature Shift Related to Change in Loading Rate**[21]

Steel	σ_{ys}		Shift in Transition Temperature	
	MPa	(ksi)	°C	(°F)
A36	255	(37)	−89	(−160)
ABS-C	269	(39)	−78	(−140)
A302B	386	(56)	−72	(−130)
HY-80	579	(84)	−44	(−80)
A517-F	814	(118)	−33	(−60)
HY-130	945	(137)	0	(0)
10Ni–Cr–Mo–V	1317	(191)	0	(0)
18Ni (180)	1241	(180)	0	(0)
18Ni (250)	1696	(246)	0	(0)

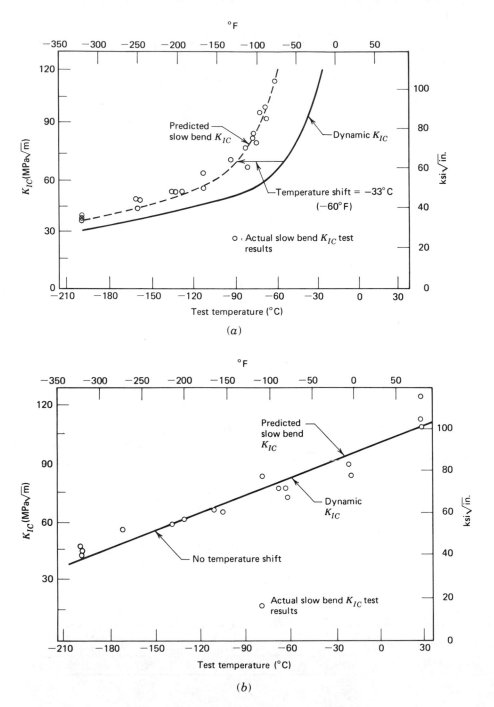

FIGURE 9.20 Use of CVN test results to predict the effect of loading rate on K_{IC}. (*a*) A517-F steel.[21] (*b*) 18Ni-(250) maraging steel. (Reprinted by permission of the American Society for Testing and Materials from copyright material.)

toughness values plotted as a function of test temperature. One additional point should be made with regard to these data. Although K_{IC} increased gradually with temperatures for the high-strength steels, a dramatic transition to higher values was observed for the low- and intermediate-strength alloys. It should be emphasized that this transition was not associated with the plane-strain to plane-stress transition, since all the data reported represented valid plane-strain conditions. A similar transition in plane-strain ductility (measured with a thin, wide sample) occurred in the same temperature region, but no such transition developed in axisymmetric ductility (measured with a conventional round tensile bar). This tentative correlation between the K_{IC} and plane-strain ductility transitions was strengthened with the observation that both transitions were associated with a fracture mechanism transition from cleavage at low temperatures to microvoid coalescence at high test temperatures.[22,23]

It is seen that the toughness levels of both strain-rate sensitive and insensitive materials increased with increasing temperature (Figs. 9.20a and 9.20b). Of significance is the fact that the predicted static K_{IC} values (broken line), obtained by applying the appropriate temperature shift (Fig. 9.19) to the dynamic test results (solid line), were confirmed by experimentation. Since dynamic plane-strain fracture-toughness testing procedures are more complex and beyond the capability of many laboratories, estimation of K_{ID} from more easily determined K_{IC} values represents a potentially greater application of the strain-rate-induced temperature shift in the determination of fracture properties.

Additional efforts have focused on developing empirical relations between impact energy absorbed in DT[17] and Charpy[21] specimens and K_{IC} values. Two such relations are shown in Figures 9.21 and 9.22 with additional correlations given in Table 9.3. It is to be noted that these relations vary as a function of material, the test temperature range, notch acuity, and strain rate. For example, these correlations are different in the upper-shelf energy regime as compared with the transition zone; they depend also on whether the Charpy specimen is precracked and whether it is impacted or tested at slow strain rates. Roberts and Newton[24] examined the accuracy of 15 such relations and concluded that no single correlation could be used with any degree of confidence to encompass all possible test conditions and differences in materials. Furthermore, because of the intrinsic scatter associated with K_{IC} and CVN measurements, the correlations possessed a relatively wide scatter band. Roberts and Newton also pointed out that some of the K_{IC} values used to establish these correlations were invalid with respect to E399-81 test requirements, and that CVN values tended to vary according to the CVN specimen location in the plate.

In addition to these difficulties, certain additional basic problems must not be overlooked. For example, the K_{IC}–CVN correlation implies that you can directly compare data from blunt and sharp notched samples and data from statically and dynamically loaded samples, respectively. The latter difficulty may not be too important for the materials shown in Fig. 9.22 since they all have yield strengths greater than 825 MPa (except A517-F) where strain-rate effects are minimized (see Fig. 9.19 and Table 9.2). The same probably holds true for the DT–K_{IC} data in Fig. 9.21, since only high-strength materials are shown. When the material's fracture properties are sensitive to strain rate, however, a two-step correlation between impact CVN data and K_{IC} values is recommended. First, K_{ID} values are inferred from impact CVN data with the aid of an appropriate correlation (e.g., see Table 9.3). Then K_{IC} is estimated from

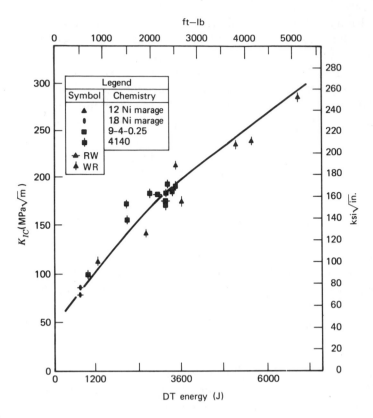

FIGURE 9.21 Relation between 2.5-cm dynamic tear energy and K_{IC} values of various high-strength steels.[17] (Reprinted by permission of the American Society for Testing and Materials from copyright material.)

K_{ID} data through the use of the temperature shift factor (Fig. 9.19). Finally, fracture mechanics—impact energy correlations for engineering plastics have been reported and are reviewed elsewhere.[11]

9.5 INSTRUMENTED CHARPY IMPACT TEST

In recent years, considerable attention has been given to instrumenting the impact hammer in the Charpy machine pendulum so as to provide more information about the load-time history of the sample during the test.[28,29] A typical load-time trace from such a test is shown in Fig. 9.23. A curve of this type can provide information concerning the general yield load, maximum and fracture loads, and time to the onset of brittle fracture. To determine the fracture energy of the sample requires integration of a load-*displacement* record. However, it is possible to calculate the fracture energy from a load-*time* curve if the pendulum velocity is known. Assuming this velocity to be constant throughout the test, the fracture energy is computed to be

$$E_1 = V_0 \int_0^t P\,dt \qquad (9\text{-}2)$$

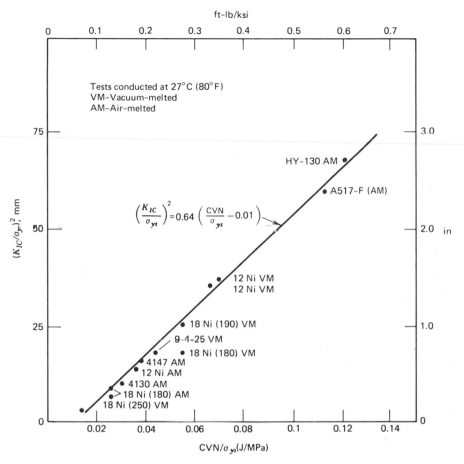

FIGURE 9.22 Relation between K_{IC} and CVN values in the upper shelf region.[21] (Reprinted by permission of the American Society for Testing and Materials from copyright material.)

where E_1 = total fracture energy based on constant pendulum velocity
      V_0 = initial pendulum velocity
      P = instantaneous load
      t = time

In reality, the assumption of a constant pendulum velocity V is not valid. Instead, V decreases in proportion to the instantaneous load on the sample. From the work of Augland,[31] we find that

$$E_t = E_1(1 - \alpha) \tag{9-3}$$

where E_t = total fracture energy

$$E_1 = V_0 \int_0^t P\,dt$$

$$\alpha = \frac{E_1}{4E_0}$$

E_0 = initial pendulum energy

TABLE 9.3 Fracture–Toughness–Charpy Energy Correlations[24]

Material	Notch	Test	Temperature Range	Range of Charpy Results (J)	σ_{ys}(MPa)	Correlation[a]	Reference
A517D 4147 HY130 4130 12Ni–5Cr–3Mo 18Ni–8Co–3Mo	V-Notch	Impact	Upper Shelf	31–121 (23–89 ft-lb)	760–1700 (110–246 ksi)	$\left(\dfrac{K_{Ic}}{\sigma_{ys}}\right)^2 = 0.64\left(\dfrac{CVN}{\sigma_y} - 0.01\right)$ $\left[\left(\dfrac{K_{Id}}{\sigma_{ys}}\right)^2 = 5\left(\dfrac{CVN}{\sigma_{ys}} - 0.05\right)\right]$	21,25
A517F A3202B ABS-C HY-130 18Ni (250) Ni–Cr–Mo–V Cr–Mo–V Ni–Mo–V	V-Notch	Impact	Transition	4–82 (3–60 ft-lb)	270–1700 (39–246 ksi)	$\dfrac{K_{Ic}^2}{E} = 0.22(CVN)^{1.5}$ $\left[\dfrac{K_{Ic}^2}{E} = 2(CVN)^{1.5}\right]$	25
A533B A517F A542	V-Notch	Impact	Transition	7–68 (5–50 ft-lb)	410–480 (60–70 ksi)	$K_{Ic} = 14.6(CVN)^{0.5}$ $[K_{Ic} = 15.5(CVN)^{0.5}]$	26
ABS-C A305-B A517-F	V-Notch	Impact / Slow Bend	Transition	2.7–61 (2–45 ft-lb)	250–345 (36–50 ksi)	$\dfrac{K_{Id}^2}{E} = 0.64\,CVN$ $\left[\dfrac{K_{Id}^2}{E} = 5\,CVN\right]$ $\dfrac{K_{Ic}^2}{E} = 0.64\,CSB$ $\left[\dfrac{K_{Ic}^2}{E} = 5\,CSB\right]$	27

TABLE 9.3 *(continued)*

ABS-C A302-B A517-F	Precracked	Impact	Transition	Slow Bend		27
			2.7–61			
			(2–45 ft-lb)		(36–50 ksi)	250–345

$$\frac{K_{ID}^2}{E} = 0.52 \text{ PCI}$$

$$\left[\frac{K_{ID}^2}{E} = 4 \text{ PCI}\right]$$

$$\frac{K_{IC}^2}{E} = 0.52 \text{ PSB}$$

$$\left[\frac{K_{IC}^2}{E} = 4 \text{ PSB}\right]$$

[a] Correlation in square brackets uses English units.

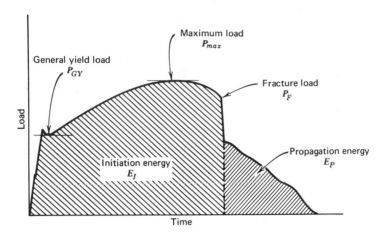

FIGURE 9.23 **Drawing of load-time curve from an instrumented Charpy test.**[30] **(Reprinted by permission of John Wiley & Sons, Inc.)**

When computed total fracture energy values are compared[32] with conventionally determined results based on final pendulum position (direct readout from the impact machine), an almost one-to-one correlation is obtained (Fig. 9.24). On the basis of such good agreement, it has been possible for researchers to use Eq. 9-3 to separately compute the initiation and propagation fracture energies at any given test temperature. The ability to provide such information along with data relating to yielding, maximum, and fracture loads has enabled materials engineers to more clearly identify the various stages in the fracture process. In addition, the instrumented Charpy test provides a relatively inexpensive screening test to compare material properties.

By precracking the Charpy sample and introducing side grooves (to enhance conditions of plane strain at the notch root) it was the hope of some engineers that instrumented Charpy testing could be used to determine $\mathcal{G}_{ID}$ and K_{ID} values. In fact, $\mathcal{G}_{ID}$ could be determined in three separate ways:

1. By relating Charpy energy adsorption, defined by pendulum position, to a $\mathcal{G}_{ID}$ level.
2. By relating Charpy energy absorption, defined from Eq. 9-3, to a $\mathcal{G}_{ID}$ level.
3. Computing $\mathcal{G}_{ID}$ from the appropriate K calibration equation for three-point bending using the maximum load from a load-time trace.

Unfortunately, this is not a realistic goal for many materials, since the Charpy specimen dimensions are generally too small to satisfy the ASTM specimen size requirement for crack length and sample thickness (Eq. 8-49). On the other hand, these dynamic notch-toughness values can be used to rank materials as in the development and evaluation of new materials. Furthermore, instrumented impact test results could be used to establish acceptance and manufacturing quality control specifications and as a test procedure to establish the existence of a temperature-induced transition in dynamic notch-toughness response. To this end, ASTM Committee E24* is attempting

---

* This committee recently has been redesignated as E 08.

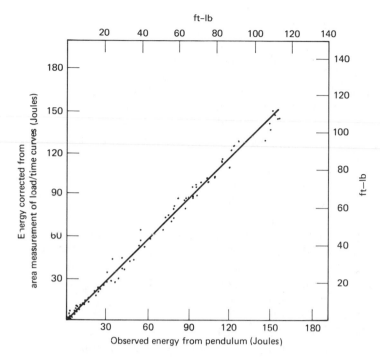

FIGURE 9.24 **Comparison between fracture energy measured from final pendulum height and from load-time record.**[32] **(Reprinted by permission of the American Society for Testing and Materials from copyright material.)**

to establish a recommended test procedure for dynamic notch toughness determination. The interested reader is advised to be alert to further developments in this subject area.

REFERENCES

1. *1981 Annual Book of ASTM Standards,* Part 10, E23-81, p. 273.
2. *1985 Annual Book of ASTM Standards,* Part 08.01, D256, p. 99.
3. W. T. Matthews, ASTM *STP 466,* 1970, p. 3.
4. P. L. Raffo, NASA *TND-4567,* May 1968, Lewis Research Center, Cleveland, OH.
5. J. Gross, ASTM *STP 466,* 1970, p. 21.
6. R. C. McNicol, *Weld. Res. Suppl,* 385s (Sep. 1965).
7. *1981 Annual Book of ASTM Standards,* Part 10, E208-69, p. 416.
8. W. S. Pellini, NRL report 1957, U.S. Naval Research Laboratory, Sept. 23, 1969.
9. W. S. Pellini and P. Puzak, NRL Report 5920, U.S. Naval Research Laboratory, 1963.
10. T. S. Robertson, *Engineering* **172,** 445 (1951).
11. A. J. Kinloch and R. J. Young, *Fracture Behavior of Polymers,* Applied Science, London, 1983.

12. P. I. Vincent, *Impact Tests and Service Performance of Thermoplastics,* Plastics Institute, London, 1971.
13. *1986 Annual Book of ASTM Standards,* Vol. 8.02, D3029-84, p. 761.
14. *1986 Annual Book of ASTM Standards,* Vol. 8.02, D1822-84, p. 138.
15. *Modern Plastics Encyclopedia,* Vol. 62(10A), McGraw-Hill, New York, 1985.
16. *1981 Annual Book of ASTM Standards,* Part 10, E436-74, p. 619.
17. E. A. Lange and F. J. Loss, ASTM *STP 466,* 1970, p. 241.
18. J. A. Begley, *Fracture Transition Phenomena,* Ph.D. Dissertation, Lehigh University, Bethlehem, PA, 1970.
19. G. M. Orner and C. E. Hartbower, *Weld. Res. Suppl.,* 405s (1961).
20. W. F. Brown, Jr. and J. E. Srawley, ASTM *STP 410,* 1966.
21. J. M. Barsom and S. T. Rolfe, ASTM *STP 466,* 1970, p. 281.
22. J. M. Barsom and S. T. Rolfe, *Eng. Fract. Mech.* **2**(4), 341 (1971).
23. J. M. Barsom and J. V. Pellegrino, *Eng. Fract. Mech.* **5**(2), 209 (1973).
24. R. Roberts and C. Newton, Bulletin 265, Welding Research Council, Feb. 1981.
25. S. T. Rolfe and S. R. Novak, ASTM *STP 463,* 1970, p. 124.
26. R. H. Sailors and H. T. Corten, ASTM *STP 514,* Part II, 1972, p. 164.
27. J. M. Barsom, *Eng. Fract. Mech.* **7,** 605 (1975).
28. C. E. Turner, ASTM *STP 466,* 1970, p. 93.
29. R. A. Wullaert, ASTM *STP 466,* 1970, p. 418.
30. T. R. Wilshaw and A. S. Tetelman, *Techniques of Metals Research,* Vol. 5, Part 2, R. F. Bunshah, Ed., Wiley-Interscience, New York, 1971, p. 103.
31. B. Augland, *Brit. Weld. J.* **9**(7), 434 (1962).
32. G. D. Fearnehough and C. J. Hoy, *JISI* **202,** 912 (1964).

PROBLEMS

9.1 Summarize the relative advantages and disadvantages of the transition temperature approach in analyzing the fracture of solids.

9.2 Multiaxial stress conditions may be beneficial or detrimental. Give examples of both situations and discuss the role of material properties.

9.3 Some investigators have established correlations between Charpy CVN and K_{IC} values. Discuss the potential pitfalls of such correlations.

9.4 What would happen to the relative position of the Charpy impact energy curves for the 275-MPa steel and the 1380-MPa 4340 steels shown in Fig. 9.4 if the specimens were tested slowly in bending?

9.5 Speculate as to why some materials exhibit a sharply defined tough–brittle transition temperature while others do not. Consider both macroscopic and microscopic factors in your evaluation.

9.6 Describe the effect of thickness and stress level on the CAT in the Robertson crack-arrest sample (Fig. 9.9).

9.7 Your new assignment as a failure analyst is to evaluate critically the introduction of fracture mechanics concepts in place of current Charpy impact test procedures. Consider both advantages and disadvantages of such a decision in your report to your supervisor.

9.8   The fractographs shown here were obtained from two sets of Charpy samples from two different materials. Unfortunately, the pictures were not labeled, so we do not know from what samples they were obtained. In both materials a tough-to-brittle transition temperature was identified.

   (a)   Assuming that photographs *a* and *b* were taken from Material *X*, speculate on the type of material that was tested and what the relative temperatures were corresponding to photographs *a* and *b*.

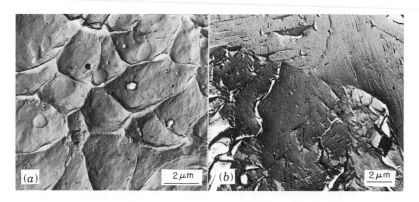

   (b)   Assuming that photograph *a* is representative of the fracture surface in Material *Y* at all test temperatures, speculate on the probable material that was tested.

   (c)   Speculate on the reason or reasons why Materials *X* and *Y* both showed a tough-to-brittle transition temperature.

   (d)   Name the fracture mechanisms shown in photographs *a* and *b* and describe the process for their formation.

9.9   For a bridge steel, it was found that the K_{IC}–CVN correlation in the transition zone was of the form

$$\frac{K_{IC}^2}{E} = 655 \text{ CVN}$$

where K_{IC}, E, and CVN are in units of $\text{Pa}\sqrt{\text{m}}$, Pa, and J, respectively. Compute the fracture toughness of such a material if a transition CVN value of 30 J was recorded.

9.10   Charpy tests are to be performed to evaluate the toughness of a steel plate that is 6 mm thick. The standard Charpy sample is $10 \times 10 \times 55$ mm in size. What difference in impact energy, if any, would you expect if results from this plate are compared with another plate of identical microstructure with the latter plate being 12.5 mm in thickness?

9.11   A 3 cm-diameter penny-shaped slag inclusion is found on the fracture surface of a Ni–Mo–V steel alloy. Would this defect have been responsible for the fracture if the stress acting on the component was 350 MPa? The only fracture toughness data available for this material are Charpy results in the transition temperature regime where impact energy values of 7–10 ft-lb were reported.

CHAPTER 10

MICROSTRUCTURAL ASPECTS OF FRACTURE TOUGHNESS

Lest the reader become too enamored with the continuum mechanics approach to fracture control, it should be noted that the profession of metallurgy predates to a considerable extent the mechanics discipline. To wit: ''And Zillah she also bore Tubal-cain, the forger of every cutting instrument of brass and iron'' (Gen. 4:22).

10.1 SOME USEFUL GENERALITIES

Before considering specific microstructural modifications that improve fracture toughness properties in engineering materials, it is appropriate to cite certain aspects of the material's structure that have a fundamental influence on fracture resistance. It has been pointed out that the deformation and fracture characteristics of a given material will depend on the nature of the electron bond, the crystal structure, and degree of order in the material.[1] The extent of brittle behavior based on these three factors is summarized in Table 10.1 for different types of materials.

It is seen that the more rigidly fixed the valence electrons, the more brittle the material is likely to be. Since covalent bonding involves sharing of valence electrons between an atom and its nearest neighbors only, materials such as diamond, silicon, carbides, nitrides, and silicates tend to be very brittle. Ionic bonding is less restrictive to the location of valence electrons; the electrons are simply transferred from an electropositive anion to an electronegative cation. Furthermore, greater deformation capability is usually found in monovalent rather than multivalent ionic compounds. As mentioned in Chapter 3, plastic flow in ionic materials is also limited by the number of allowable slip systems that do not produce juxtaposition of like ions across the slip plane after a unit displacement. Metallic bonding provides the least restriction to valence electron movement; valence electrons are shared equally by all atoms in the solid. These materials generally have the greatest deformation capability.

As seen in Table 10.1, brittle behavior is more prevalent in materials of low crystal symmetry where slip is more difficult. On the other hand, considerable plastic deformation is possible in close-packed metals of high crystal symmetry. Finally, the ability of a given material to plastically deform generally will decrease as the degree of order of atomic arrangement increases. Consequently, the addition of a solute to a metal lattice will cause greater suppression of plastic flow whenever the resulting solid

TABLE 10.1 **Relation between Basic Structure of Solids and Their Effect on Brittle Behavior**

Basic Characteristic	Increasing Tendency for Brittle Fracture →		
Electron bond	Metallic	Ionic	Covalent
Crystal structure	Close-packed crystals	Low-symmetry crystals	
Degree of order	Random solid solution	Short-range order	Long-range order

solution changes from that of a random distribution to that of short-range order and finally to long-range order.

Additional trends appear when one considers the propensity for brittle fracture based on fundamental engineering properties, such as yield and tensile strength and tensile ductility. Recall from Chapter 1 that toughness was defined by the area under the stress–strain curve; consequently, toughness would be highest when an optimum combination of strength and ductility is developed. Again, in Chapter 8 we saw that fracture toughness of a notched specimen depended on an optimum combination of yield strength and crack opening displacement [$V(c)$ (Eq. 8-56)]. Since $V(c)$ decreases sharply with increasing strength, a basic trend of decreased toughness with increased strength has been identified (see Table 8.2). It is apparent, then, that one is faced with a dilemma: Metallurgists can limit a material's ability to deform by various strengthening procedures that enhance load-bearing capacity, but almost always to the detriment of the fracture toughness. As such, it is not difficult to raise the fracture-toughness level of a material; one only needs to alter the thermomechancal treatment to lower strength and toughness increases as a consequence (Fig. 10.1). This approach is often impractical, however, since material requirements would be increased as a result of the material's lower load-bearing capacity. Of course, one might be satisfied with lower alloy strength to achieve higher toughness if prevention of low-energy fracture were of paramount importance (for example, in the case of nuclear energy generating facilities).

The most desirable approach would involve shifting the curve in Fig. 10.1 up and toward the right so that the material might exhibit both higher strength and toughness

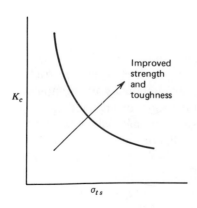

FIGURE 10.1 Inverse relation between fracture toughness and strength. Optimization of alloy properties would involve shifting curve in direction of the arrow.

for a given metallurgical condition. These improvements may be effected in several ways, such as by:

1. Improved alloy chemistry and melting practice to remove or make innocuous undesirable tramp elements that degrade toughness.
2. Development of optimum microstructures and phase distributions to maximize toughness.
3. Microstructural refinement.

It would appear that these variables are extremely important to K_{IC}, since toughness levels can vary widely for the same material and in the same strength range (Fig. 10.2).

10.1.1 Overview of Toughening Mechanisms

A number of mechanisms have been found in engineering solids that enhance energy absorption during the fracture process.[3] Some of these mechanisms are *intrinsic* in nature and reflect basic differences in material ductility and ease of plastic flow. For example, the stabilization of the FCC form of iron rather than the BCC isomorph in a steel alloy contributes to greatly enhanced toughness; the Peierls-Nabarro stress level in the FCC lattice is much reduced (recall Chapter 2) and the low-energy cleavage fracture micromechanism is averted (recall Section 7.7.3.2). A reduction in dispersoid and precipitate particle volume fraction and an increase in particle spacing enhances alloy ductility and represents another intrinsic toughening process.

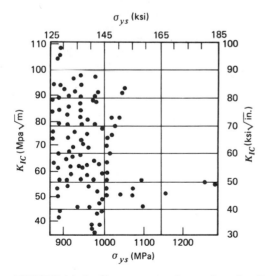

FIGURE 10.2 Fracture toughness data for Ti-6Al-4V alloy. Large scatter in experimental results suggests different microstructures present.[2] (From M. J. Harrigan, *Metals Engineering Quarterly*, May 1974, copyright American Society for Metals.)

A distinctly different set of toughening mechanisms, *extrinsic* in nature, have been identified, which focus attention on events taking place either in the wake of the crack or ahead of the advancing crack front (Fig. 10.3). Each mechanism serves to shield the crack tip from the full impact of the crack driving force, thereby increasing fracture-toughness levels and lowering fatigue crack propagation (FCP) rates (see Section 13.4). A brief description of these mechanisms follows with more detailed discussions of some models deferred to later sections in this chapter.

Fracture toughness can be enhanced through crack deflection and/or meandering of the crack path within the microstructure. The fracture of natural wood products is especially noteworthy in this regard. Energy is also absorbed when hard discrete particles act to temporarily pin the advancing crack front in brittle matrices and attenuate its rate of advance.[4,5] The crack is then forced to move around both sides of the particle before linking up behind the particle. Usually a fracture step is created on the back side of the particle where the cracks rejoin and move forward after particle failure (recall Fig. 7.39).

Several zone-shielding mechanisms have been identified. In one scenario, an unstable phase present in the microstructure transforms under stress to a new crystal form with the associated dissipation of energy and the development of a favorable

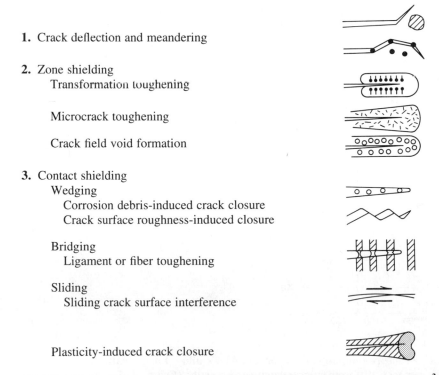

Extrinsic Toughening Mechanisms

1. Crack deflection and meandering

2. Zone shielding
    Transformation toughening

    Microcrack toughening

    Crack field void formation

3. Contact shielding
    Wedging
        Corrosion debris-induced crack closure
        Crack surface roughness-induced closure

    Bridging
        Ligament or fiber toughening

    Sliding
        Sliding crack surface interference

    Plasticity-induced crack closure

FIGURE 10.3 Types of extrinsic crack-tip shielding mechanisms in solids.[3] (With permission, from R. O. Ritchie, *Mechanical Behavior of Materials-5,* Proceedings, 5th Int. Conf., M. G. Yan, S. H. Zhang, Z. M. Zheng, eds., Pergamon, Oxford, 1987.)

compressive residual stress pattern. Zone shielding also occurs through the development of a field of disconnected microcracks or microvoids that serve to relax crack-tip triaxiality and diffuse the intensity of the crack-tip stress singularity.

Another grouping of extrinsic toughening mechanisms involves crack surface contact shielding through, for example, retention of ligaments across the fracture plane. These ligaments may take the form of unbroken fibers in a composite material that eventually pull out from the matrix with the expenditure of considerable amounts of energy. As discussed in Section 13.4, wedging together of oxide debris and/or lateral sliding of adjacent regions on the fracture surface can bring about a significant reduction in FCP rates, below those expected for the prevailing crack driving force. In a related manner, contact shielding can result from the wake of plasticity left behind the advancing crack tip, which affects fatigue resistance. As we will see, these toughening mechanisms are generic in nature and one or more have been identified as occurring in particular metals, ceramics, polymers, and their respective composites.

10.2   TOUGHNESS AND MICROSTRUCTURAL ANISOTROPY

The first approach to toughness improvement that will be discussed is concerned with the means by which cracks are deflected from their normal plane and direction of growth. Such crack deflections can occur at grain boundaries, flow lines, and inclusions that are aligned parallel to a particular processing direction. To better understand the origin of such microstructural alignment, and the associated mechanical anisotropy, we begin by presuming that we have taken a cube of equiaxed polycrystalline material and changed its shape by some mechanical process such as rolling, drawing, or swaging (a combination of drawing and twisting). By the principle of similitude, the conversion of the cube into a thin plate should be reflected by a change in the size and shape of the grains within the solid. That is, the equiaxed grains should be flattened and spread out, as shown in Fig. 10.4a. The alignment of the grain structure in the direction of mechanical working is known as *mechanical fibering* and is exhibited most dramatically in forged products such as the one shown in Fig. 10.5. Here the grains have been molded to parallel the contour of the forging dies. Engineers have found that the fracture resistance of a forged component can be enhanced considerably when the forging *flow lines* are oriented parallel to major stress trajectories and normal to the path of a potential crack. As such, forged parts are considered to be superior to comparable castings because of the benefits derived from the deformation-induced microstructural anisotropy. Japanese armor makers took advantage of this fact in their manufacture of the samurai sword.[6] They heated a billet of iron to an elevated temperature where it was folded back upon itself by repeated blows of forging hammers, and then placed back in the furnace until ready for another forging and folding operation. This was done 10 to 20 times, resulting in a sword blade containing 2^{10} to 2^{20} layers of the original billet. After masterful decoration and a special heat treatment, this aesthetically appealing and structurally sound weapon was ready for its deadly purpose. Of course, when a forged product is used improperly, the flow lines act as readily available paths for easy crack propagation. The reader should note that mechanical fibering involves not only alignment of grains but also alignment of inclu-

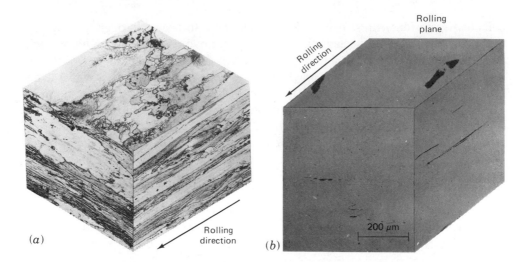

(a)

(b)

FIGURE 10.4 (a) Photomicrograph revealing mechanical fibering associated with rolling of 7075-T651 aluminum plate. (Courtesy J. Staley, Alcoa Aluminum Company.) (b) Alignment of manganese sulfide inclusions on rolling plane in hot-rolled steel plate. (After Heiser and Hertzberg;[7] reprinted with permission of the American Society of Mechanical Engineers.)

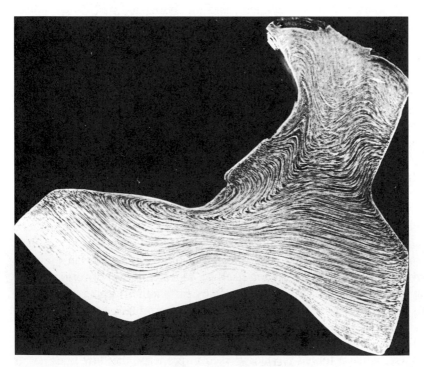

FIGURE 10.5 Flow lines readily visible in forged component. (Courtesy of George Vander Voort, Car Tech Corp.)

sions. For example, in standard steel-making practice, the hot-rolling temperature for billet breakdown exceeds the softening point for manganese sulfide inclusions commonly found in most steels. Consequently, these inclusions are strung out in the rolling direction and flattened in the rolling plane, as shown in Fig. 10.4b. The deleterious nature of these aligned inclusions relative to the fracture properties of steel plate is discussed below, along with recently developed procedures aimed at inclusion shape control.

A drawing operation will convert our reference cube into a long, thin wire or rod. Again, by similitude, the grains are found to be sausage-shaped and elongated in the drawing direction. In a transverse section normal to the rod axis, the grains should appear equiaxed while being highly elongated parallel to the rod axis. For very large draw ratios in BCC metals, however, the grains take on a ribbonlike appearance, because of the nature of the deformation process in the BCC lattice.[8] When a metal is swaged, the elongated grain structure along the rod axis is maintained (Fig. 10.6a), and the transverse section reveals a beautiful spiral nebula pattern, reflecting the twisting action of the rotating dies during the swaging process[9] (Fig. 10.6b).

As we saw from Eqs. 8-18 and 8-20, a triaxial tensile stress state is developed at the crack tip when plane-strain conditions are present. Since fracture toughness was shown to increase with decreasing tensile triaxiality (for example, with thin sections where $\sigma_z \sim 0$), some potential for improved toughness is indicated if ways could be found to reduce the crack-tip-induced σ_x and/or σ_z stresses. One way to reduce the σ_x stress would involve the generation of an internally free surface perpendicular to σ_x and the direction of crack propagation. This can be accomplished by providing moderately weak interfaces perpendicular to the anticipated direction of crack growth, which could be pulled apart by the σ_x stress in advance of the crack tip (Fig. 10.7).[10] Since there can be no stress normal to a free surface, σ_x would be reduced to zero at this interface. In addition to reducing the crack-tip triaxiality by generation of the internally free surface, the crack becomes blunted when it reaches the interface (Fig. 10.7b). Both conditions make it difficult for the crack to reinitiate in the adjacent layer with the result that toughness is improved markedly.

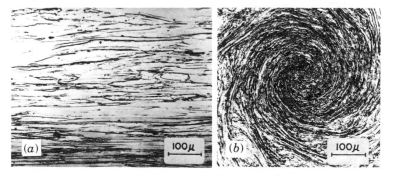

FIGURE 10.6 **Longitudinal and transverse sections of swaged tungsten wire reduced by 87%. (After Peck and Thomas[9]; reprinted with permission of the Metallurgical Society of AIME.)**

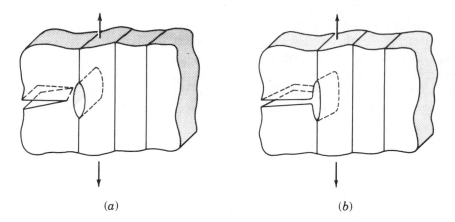

(a) (b)

FIGURE 10.7   **Delamination at relatively weak interface caused by σ_x stress. (a) Delamination ahead of crack-tip reduces tensile triaxiality and (b) reduces crack-tip acuity. Both factors contribute to enhanced fracture resistance.**

Embury et al.[11] conducted laboratory experiments to demonstrate the dramatic improvement in toughness arising from delamination, which can effectively arrest crack propagation. These investigators soldered together a number of thin, mild steel plates to produce a standard-sized Charpy impact specimen with an "arrester" orientation (Fig. 10.8a). As seen in Fig. 10.9a, the transition temperature for the "ar-

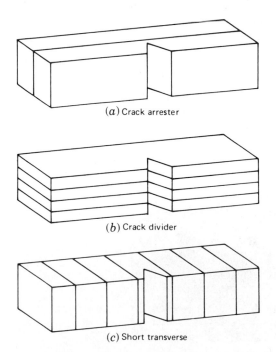

(a) Crack arrester

(b) Crack divider

(c) Short transverse

FIGURE 10.8   **Specimens containing relatively weak interfaces: (a) arrester; (b) divider; and (c) short transverse configurations. (Reprinted with permission from American Society of Mechanical Engineers.)**

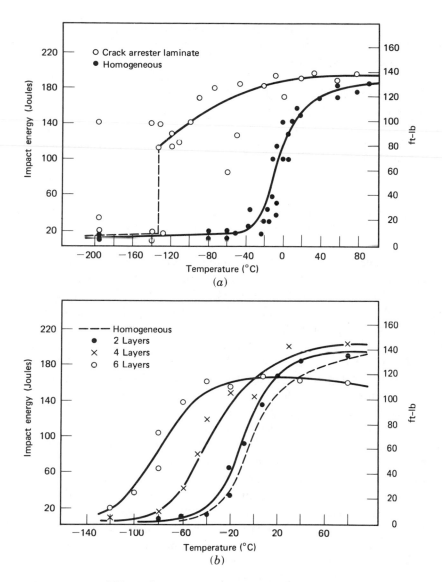

FIGURE 10.9 Effect of (*a*) arrester and (*b*) divider geometry on Charpy impact energy temperature transition.[11] (Reprinted with permission from American Institute of Mining, Metallurgical and Petroleum Engineers.)

rester'' sample was found to be more than 130°C lower than that exhibited by homogeneous samples of the same steel. Additional confirmation of such favorable material response was reported by Leichter[12] who observed 163-J and 326-J Charpy impact energy absorption in ''arrester'' laminates of high-strength titanium and maraging steel alloys, respectively. Such energies are much higher than values expected from homogeneous samples of the same materials (Fig. 9.4).

The benefits of the crack-arrester geometry have been utilized for many years in a number of component designs. For example, one steel fabricator has developed a procedure for on-site construction of large pressure vessels using a number of tightly wrapped and welded concentric shells of relatively thin steel plate. This approach, though expensive to construct, boasts several advantages. First, only a *thin* layer of corrosion-resistant (more expensive) material would be needed (if at all) to contain an aggressive fluid within the vessel, as opposed to a full thickness vessel of the more expensive alloy. Second, the tightly wrapped layers are designed to create a favorable residual compressive stress on the inner layers, thus counteracting the hoop stresses of the pressure vessel. Third, the free surfaces between the layers act as crack arresters to a crack that might otherwise penetrate the vessel thickness. Finally, the metallurgical structure of thin plates (especially for low-hardenability steels) is generally superior to that of thicker sections.

In another example of crack-arrester design, large gun tubes often contain one or more sleeves shrunk fit into the outer jacket of the tube. Here, again, the procedure was originally developed to produce a favorable residual compressive stress in the inner sleeve(s), but serves also to introduce an internal surface for possible delamination and crack arrest. In one actual case history, a fatigue crack was found to have initiated at the inner bore of the sleeve, propagated to the sleeve–jacket interface, and proceeded around the interface, but *not* across the interface into the jacket itself. A similar crack-"arrester" response is found in conventional materials given thermomechanical treatments that produce layered microstructures. McEvily and Bush[13] showed that a Charpy specimen made from ausformed steel (warm rolled above the martensite transformation temperature) completely stopped a 325-J (240-ft-lb) impact hammer when the carbide-embrittled former austenite grain boundaries were oriented normal to the direction of crack propagation (Fig. 10.10).

Triaxiality can also be reduced by relaxing σ_z stresses brought about by delamination of interfaces positioned normal to the thickness direction. When delamination occurs, the effective thickness of the sample is reduced, since σ_z decreases to zero at each delamination. Consequently, the specimen acts like a series of thin-plane stress samples instead of one thick-plane strain sample. For this reason, the resulting shift in transition temperature will depend on the number of weak planes introduced in the specimen—the more planes introduced, the thinner the delaminated segments will be and the greater the tendency for plane-stress response. Embury et al.[11] conducted such tests with laminated samples in the "divider" orientation (Fig. 10.8*b*) and found the transition temperature to decrease with increasing number of weak interfaces (Fig. 10.9*b*). Note that at sufficiently low temperatures all the samples exhibited minimal toughness, indicating that even the thinnest layers were experiencing essentially plane-strain conditions at these low temperatures. Leichter[12] confirmed the beneficial character of the divider orientation in raising toughness. For example, the fracture toughness of a titanium alloy was improved six- to sevenfold for a laminated sample over that of the full-thickness sample made from the same material. It is interesting to note that the homogeneous samples exhibited an extensive amount of flat fracture, while each delaminated layer showed 100% full shear. As we saw in Chapter 8 this difference in fracture mode appearance also reflects the improvement in fracture toughness.

The strength of the interface represents an important parameter in the delamination-induced toughening process. On one hand, the interface should not be so weak that

FIGURE 10.10 Extensive delaminations in ausformed steel with "arrester" orientation. Specimen absorbed 325-J energy.[13] (Copyright American Society for Metals, 1962.)

the sample slides apart like a deck of playing cards. On the other hand, if the interface is too strong, delamination will not occur. In considering this point, Kaufman[14] demonstrated that the toughness of multilayered, adhesive-bonded panels of 7075-T6 aluminum alloy was significantly greater than that shown by homogeneous samples with the same total thickness. Conversely, when similar multilayered panels of the same material were metallurgically bonded (resulting in strong interfaces), no improvement in toughness was observed over that of the homogeneous sample.

We must also consider the third possible orientation of weakened interfaces relative to the stress direction. As one might expect, fracture-toughness properties are lowest in the short-transverse orientation. It is as though all the positive increments in toughness associated with crack "arrester" and "divider" orientations are derived at the expense of short-transverse properties. The spalling fracture of solids, resulting from shock wave–material interactions, often represents a short-transverse fracture event. This arises from the fact that shock waves, produced by the impact of a high-velocity projectile, bounce off the back wall of an object (e.g., armor plate), reverse direction, and return as reflected tensile waves. If the associated tensile stresses of these reflected waves are great enough, nucleation, growth, and coalescence of voids and/cracks will occur[15] Often, a chunk of material (i.e., spall) breaks away from the surface opposite to the impacted surface. Studies have shown that spall formation is nucleated most readily at the interfaces between inclusions and the surrounding matrix. As expected, spall formation is significantly suppressed in steels that contain finer nonmetallic inclusions as a result of an electroslag remelt refining process that in a

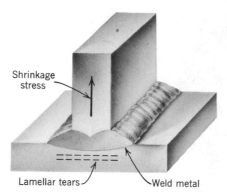

FIGURE 10.11 Lamellar tears generated along rolling planes as a result of weld shrinkage stresses.

conventionally cast steel alloy[16] (see Section 10.3). Since inclusions tend to become aligned parallel to the plane of rolling, a reduction in inclusion content should enhance spall resistance in the short-transverse plane.

A potentially dangerous condition—lamellar tearing—can develop because of the poor short-transverse properties often found in rolled plate. Consider the consequences of a large T-joint weld such as the one shown in Fig. 10.11. After the weld is deposited, large shrinkage stresses are developed that act in the thickness direction of the bottom plate. These stresses can be large enough to cause numerous microfissures at inclusion–matrix interfaces, which were aligned during the rolling operation. Clusters of these short-transverse cracks can seriously degrade the weld joint efficiency and should be minimized if at all possible.

Because of the anisotropy of wrought products, fracture-toughness values may be expected to vary with the type of specimen used to measure K_c or K_{IC}. This is not related to the specimen shape per se in terms of the K calibration but rather to the material anisotropy. For example, a rolled plate containing a surface flaw (arrester orientation) might be expected to exhibit somewhat higher toughness than the same material prepared in the form of an edge-notched plate, where the crack would propagate parallel to the rolling direction. To illustrate this behavior, the fracture-toughness anisotropy in a number of wrought aluminum alloys is shown in Table 10.2 with the fracture-toughness data given as a function of fracture plane orientation (first letter in code) and crack propagation (second letter in code) (Fig. 10.12). Additional data revealing fracture-toughness anisotropy in aluminum, steel, and titanium alloys are given in Table 10.7 at the end of this chapter.

TABLE 10.2a Plane-Strain Fracture-Toughness Anisotropy in Wrought, High-Strength Aluminum Alloys[17]

Alloy and Temper Designation	Product	K_{IC}(MPa$\sqrt{\text{m}}$)		
		L-T	T-L	S-T
2014-T651	127-mm plate	22.9	22.7	20.4
7075-T651	45-mm plate	29.7	24.5	16.3
7079-T651	45-mm plate	29.7	26.3	17.8
7075-T6511	90 × 190-mm extruded bar	34.0	22.9	20.9
7178-T6511	90 × 190-mm extruded bar	25.0	17.2	15.4

TABLE 10.2b   Plane-Strain Fracture-Toughness Anisotropy in Wrought, High-Strength Aluminum Alloys [17]

Alloy and Temper Designation	Product	K_{IC} (ksi$\sqrt{\text{in.}}$)		
		L-T	*T-L*	*S-T*
2014-T651	5-in. plate	20.8	20.6	18.5
7075-T651	1¾-in. plate	27.0	22.3	14.8
7079-T651	1¾-in. plate	27.0	23.9	16.2
7075-T6511	3½ × 7½-in. extruded bar	30.9	20.8	19.0
7178-T6511	3½ × 7½-in. extruded bar	22.7	15.6	14.0

EXAMPLE 10.1

Components were machined without consideration of direction from a plate of 7075-T651 aluminum alloy. If the design stress were set at 40% of the material's yield strength, would all components withstand fracture in the presence of internal 15-mm penny-shaped cracks, regardless of defect orientation?

The mechanical properties of the plate as a function of sample orientation are given in the following table.

	L-T	*T-L*	*S-T*
K_{IC} (MPa$\sqrt{\text{m}}$)	29.7	24.5	16.3
σ_{ys} (MPa)	515	510	460

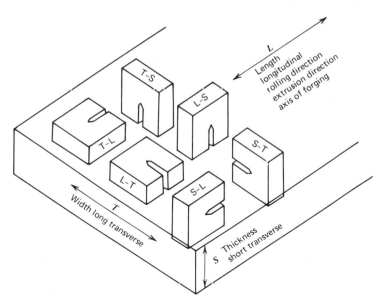

FIGURE 10.12   Code system for specimen orientation and crack propagation direction in plate.

To begin, we will assume that the internal circular crack in each component is oriented normal to the direction of the design stress. Since the stress intensity factor for this crack

$$K = \frac{2}{\pi} \sigma_d \sqrt{\pi a}$$

configuration is given by

$$K = \frac{2}{\pi} (0.4) \, \sigma_{ys} \sqrt{\pi(.0075)}$$

$$K = \frac{2}{\pi} \sigma_d \sqrt{\pi a}$$

where σ_d = design stress and = $0.4\sigma_{ys}$

$$a = \text{disc radius}$$

therefore,

$$K = \frac{2}{\pi} (0.4)\sigma_{ys} \sqrt{\pi(.0075)}$$

By substituting yield strength values in the preceding equation, we see that the stress intensity factor level for the three orientations are 20.1, 19.9, and 18 MPa$\sqrt{\text{m}}$ for the *L-T, T-L,* and *S-T* orientations, respectively. Therefore, components machined in the *L-T* and *T-L* orientations could have sustained the design stress since $K < K_{IC}$, whereas the component prepared in the *S-T* orientation would have fractured abruptly because $K > K_{IC}$. From a design standpoint, it is important that the component have the highest fracture toughness values oriented parallel to the direction of greatest stress.

10.3   IMPROVED ALLOY CLEANLINESS

Although certain elements are added to alloys to develop the best microstructures and properties, other (tramp) elements serve no such useful purpose and are, in fact, often very deleterious. For example, we see from Fig. 10.13 that small amounts of oxygen have a severe embrittling effect on the fracture toughness of diffusion-bonded Ti-6Al-4V alloy.[18] Also, hydrogen in solid solution is known to produce hydrogen embrittlement in a number of high-strength alloys and their weldments. The latter problem is examined in the next chapter. For the moment we focus on those elements that contribute to the formation of undesirable second phases which serve as crack nucleation sites. Edelson and Baldwin[19] demonstrated convincingly that second-phase particles act to reduce alloy ductility (Fig. 10.14). The severe effect of sulfide inclusions

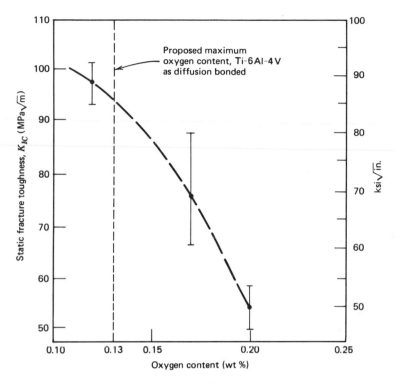

FIGURE 10.13   Effect of oxygen content on the fracture toughness of diffusion bonded Ti-6Al-4V.[18] (Copyright _Aviation Week & Space Technology._)

on toughness in steel is shown in Fig. 10.15, where the Charpy V-notch shelf energy drops appreciably as sulfur content increases. Since the yield strength of this material is greater than 965 MPa, it would be interesting to compute the fracture-toughness level K_{IC} for these alloys with different sulfur content, using the Barsom-Rolfe relation described in Chapter 9 (Fig. 9.22). This will be left to the student as an interesting exercise. By using K_{IC} as the measure of toughness, Birkle et al.[21] demonstrated the deleterious effect of sulfur content at all tempering temperatures in a Ni–Cr–Mo steel (Fig. 10.16).

The task, then, is to remove sulfur, phosphorous, and gaseous elements (such as hydrogen, nitrogen, and oxygen) from the melt before the alloy is processed further. This has been done with a number of more sophisticated melting techniques developed in recent years. For example, melting in a vacuum rather than in air has contributed to a dramatic reduction in inclusion count and in the amount of trapped gases in the solidified ingot. To obtain a still better quality steel, steels are vacuum arc remelted (VAR). In this process, the electrode (the steel to be refined) is remelted from the heat generated by the arc and the molten metal is collected in a water-cooled crucible. Electroslag remelting (ESR) represents a variation of the consumable electrode re-melting process: When the steel electrode is remelted, the molten metal droplets must first filter through a slag blanket floating above the molten metal pool. By carefully

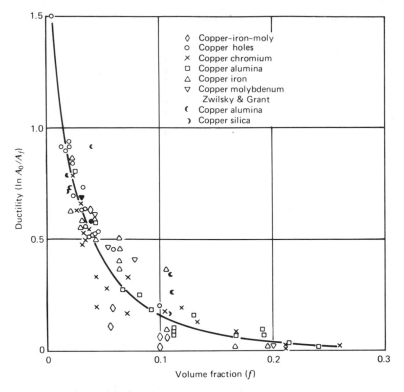

FIGURE 10.14   Effect of second-phase volume fraction on fracture ductility.[19] (Copyright American Society for Metals, 1962.)

controlling the chemistry of the slag layer, various elements contained within the molten drops may be selectively removed.

As one might expect, removal of tramp elements increases the cost of the product. Although these costs are justifiable in terms of improved alloy behavior, the price of the final product may not be competitive in the marketplace. The task is to devise inexpensive means by which the tramp elements are rendered more harmless. One truly excellent example that we may discuss relates to correction of inferior transverse fracture properties in hot-rolled, low-alloy steels. As we saw in the previous section, the alignment of inclusions during rolling develops a considerable anisotropy in fracture toughness. In these alloys, the objectionable particles are manganese sulfide inclusions that become soft at the hot-rolling temperature and consequently, smear out on the rolling plane and in the rolling direction (Fig. 10.17a). The result: very poor transverse fracture properties. Since these alloys are used in automotive designs where components are bent in various directions for both functional and aesthetic reasons, poor transverse bending properties severely restrict the use of these materials. The objective, then, was to suppress the tendency for softening of the sulfide at the hot-rolling temperature and thereby preclude its smearing out on the rolling plane. This was accomplished by very small additions of rare earth metals to the melt. (Certainly rare earth elements are not cheap, but the small amounts necessary result

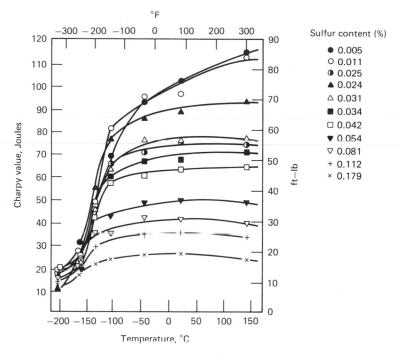

FIGURE 10.15 Effect of sulfur content on Charpy impact energy in steel plate (30R).[20] (Reprinted with permission from American Institute of Mining, Metallurgical, and Petroleum Engineers.)

in limited additional cost per ton of steel produced.) Luyckx et al.[22] found that the manganese in the sulfide was replaced by rare earth elements (mostly cerium), which produced more stable and higher melting point sulfides. Since these did not deform during hot rolling, they maintained their globular shape (Fig. 10.17b), thus giving rise to greater isotropy in fracture properties. This is demonstrated in Fig. 10.18 by noting the rise in transverse Charpy shelf energy with increasing cerium/sulfur ratio. Recent plane-stress fracture-toughness results from a quenched and tempered steel that has no inclusion shape control and from a rare-earth-modified, hot-rolled, low-alloy steel that has inclusion shape control revealed that the K_c anisotropy ratio was much higher for the steel without inclusion shape control.[23] The addition of rare earth metals to steel alloys also enhances lamellar tearing resistance since short-transverse ductility would increase as a result of the elimination of large planar arrays of inclusions.[24]

No one would argue with the desirability of removing sulfides from the microstructure or at least making them more harmless. (It should be noted that sulfur is sometimes deliberately added to certain steel alloys to enhance their machinability.) However, removal of carbides presents a more serious problem, since carbon both in solid solution and in carbides serves as a potent hardening agent in ferrous alloys. In addition, carbon provides the most effective means by which steel hardenability may be raised. And, yet, carbides provide the nucleation sites for many cracks. In a painstaking study designed to identify the origin of microcracks in high-purity iron,

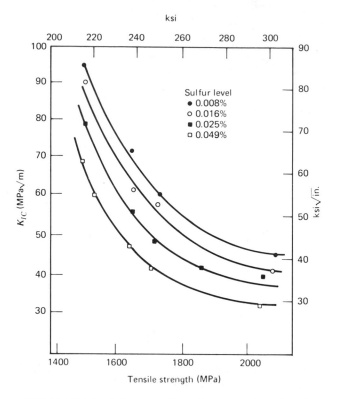

FIGURE 10.16 Influence of tensile strength and sulfur content on plane strain fracture toughness of 0.45C–Ni–Cr–Mo steels. (Copyright American Society for Metals, 1966.)

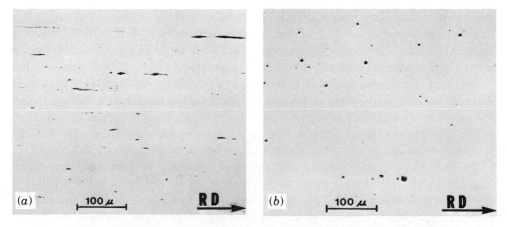

FIGURE 10.17 Longitudinal sections shown: (*a*) elongated manganese sulfide inclusions in quenched and tempered steel without inclusion shape control; (*b*) globular rare earth inclusions found in hot-rolled, low-alloy steel with inclusion shape control.

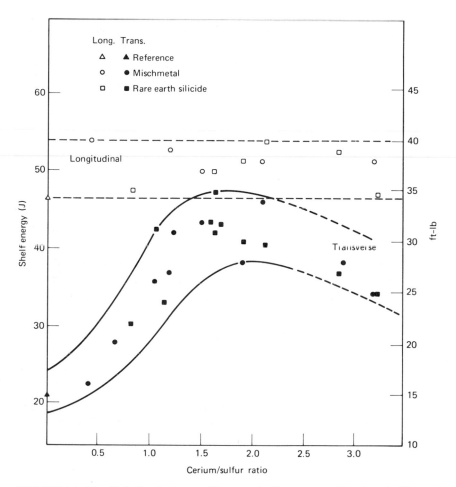

FIGURE 10.18 Relation between Charpy shelf energy and cerium/sulfur ratio in longitudinal and transverse oriented half-sized impact specimens of VAN-80 steel. (Copyright American Society for Metals, 1970.)

McMahon[25] demonstrated that almost every microcrack found could be traced to the fracture of a carbide particle (Table 10.3). This was found true even for the alloy that contained less carbon than the solubility limit. Of particular importance in this study was a critical evaluation of the relative importance of various mechanisms of microcrack nucleation proposed by a number of investigators.

As we see from Fig. 10.19, McMahon found that cracks were much more likely to occur at embrittled grain boundaries and brittle second-phase particles than as a result of twin and/or slip band interactions. Consequently, although dislocation models proposed to account for microcrack formation, such as those shown in Fig. 10.20, appear to be valid for certain ionic materials,[28] they are precluded by other crack nucleation events in metals. Indeed, McMahon stated that "there appears to be no direct evidence of initiation of cleavage by slip band blockage in metals."[25]

TABLE 10.3 **Initiation Sites of Surface Microcracks in Ferrite**[25]

Material	0.035% C	0.035% C	0.005% C
Test temperature	$-140°C$	$-180°C$	$-170°C$
Total microcracks per 10^4 grains	66	43	17
Microcracks originating at cracked carbides	63	42	12
Microcracks probably originating at cracked carbides	3	1	4
Microcracks possibly originating at twin-matrix interface	0	0	1

Because of the negative side effects of carbon solid solution and carbide strengthening in ferrous alloys, attempts have been made to develop alloys that derive their strength instead by precipitation-hardening processes involving various intermetallic compounds.* Maraging steels represent such a class of very low-carbon, high-alloy steels; they are soft upon quenching but harden appreciably after a subsequent aging treatment. It is seen from Fig. 10.21 that for all strength levels the toughness of maraging steels is superior to that of AISI 4340 steel, a conventional quenched and tempered steel. The chemistry of AISI 4340 steel and a typical maraging steel is given in Table 10.4. It is felt that the lower carbon levels in maraging steels are partly

---

* By contrast, a recent review by Lesuer et al. discussed the development of low alloy plain carbon *hypereutectoid* steels containing 1–2.1% carbon; these high carbon alloys exhibit high strength (800–1500 MPa) and generally good ductility (2.2–25%). This attractive combination of properties arises from the development of microstructures containing submicron-sized equiaxed ferrite grains and 15–32 v/o uniformly distributed spheroidized carbide particles. Of further interest, these fine-grained microstructures are amenable to net-shape forming via superplastic methods (recall Section 5.4). (See review by D. R. Lesuer, C. K. Syn, A. Goldberg, J. Wadsworth, and O. D. Sherby, *J. Metals*, **45**(8), 40 (1993)).

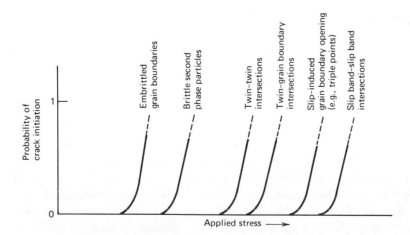

FIGURE 10.19 **Probability as a function of applied stress that a particular microcrack formation mechanism will be operative.**[25] **(Reprinted with permission of Plenum Publishing Corporation, from C. J. McMahon, *Fundamental Phenomena in the Materials Sciences*, Vol. 4, 1967, p. 247.)**

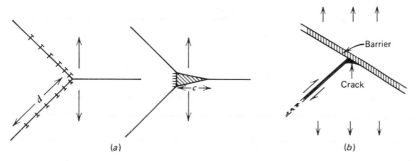

FIGURE 10.20 Dislocation models for crack nucleation: (*a*) Cottrell model.[26] (Reprinted with permission from American Institute of Mining, Metallurgical and Petroleum Engineers). (*b*) Zener model.[27] (Reprinted from *Fracturing of Metals,* copyright, American Society for Metals, 1948.)

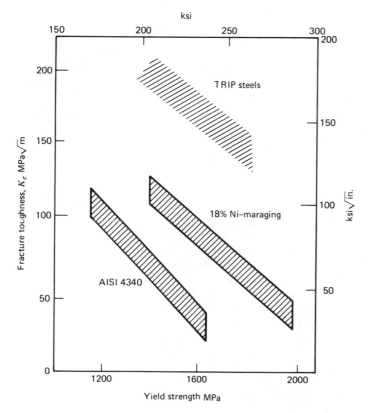

FIGURE 10.21 Fracture-toughness–tensile-strength behavior in AISI 4340, 18% Ni maraging, and TRIP steels.[20] (Reprinted with permission from V. F. Zackay and Elsevier Sequoia S. A.)

TABLE 10.4 Nominal Chemistry of Typical High-Strength Steels

Material	Composition								
	C	Ni	Cr	Mo	Si	Mn	Co	Ti	Al
AISI 4340	0.40	1.65–2.00	0.70–0.90	0.20–0.30	0.2–0.35	0.60–0.80	—	—	—
Maraging steel	0.03 max.	18	—	5	(0.20 max.)	(0.20 max.)	8	0.4	0.1

responsible for their improved toughness and resistance to hydrogen, neutron, and temper embrittlement (see Section 10.6). Additional factors are discussed in Section 10.4.

Striking improvements in the fracture toughness of aluminum alloys have also been achieved by eliminating undesirable second-phase particles. Since precipitation hardening is achieved by dislocation interaction with closely spaced submicronsized particles, the very large, dark inclusions (e.g., Al_7Cu_2Fe, $(Fe,Mn) Al_6$, and Mg_2Si), or secondary microconstituents (depending on your point of view), seen in Fig. 10.22 provide no strengthening increment. Instead, they provide sites for early crack nucleation.

Piper et al.[30] conducted an exhaustive study to determine how various elements affected the strength and toughness properties in 7178-T6 aluminum alloy.* Their results are summarized in Table 10.5. It is seen that some elements (copper and magnesium) provide a solid solution strengthening component to alloy strength, while zinc and magnesium contribute a precipitation-hardening increment. By comparing data from other investigators, Piper et al.[30] determined that strengthening of this alloy caused an expected reduction in fracture toughness. After examining strength and

---

* 7178 aluminum alloy: 1.6–2.4 Cu, 0.70 max. Fe, 0.50 max Si, 0.30 Mn, 2.4–3.1 Mg, 6.3–7.3 Zn, 0.18–0.40 Cr, 0.20 max Ti.

FIGURE 10.22 Metallographic section in 2024-T3 aluminum alloy revealing typically large number of Al_7Cu_2Fe second-phase particles.

TABLE 10.5   Function of Various Elements in 7178-T6 Aluminum Alloy[30]

Element	Function
Zinc	Found in Guinier-Preston zones and subsequently found in $MgZn_2$ precipitates. Element acts as precipitation-hardening agent.
Magnesium	Some Mg_2Si formation but mostly found in $MgZn_2$ precipitates and in solid solution.
Copper	Exists in solid solution, in $CuAl_2$ and Cu–Al–Mg-type precipitates, and in Al_7Cu_2Fe intermetallic compounds.
Iron	Initially reacts to form Al–Fe–Si intermetallic compounds. Copper later replaces Si to form Al_7Cu_2Fe (the large black particles seen in Fig. 10.22).
Silicon	Initially reacts to form Al–Fe–Si compound prior to being replaced by Cu. Also forms Mg_2Si.
Manganese	Exact role not clear
Chromium	Combines with Al and/or Mg to form fine precipitates, which serve to grain refine.

fracture-toughness data in 18 different alloys, all reasonably close to the composition of 7178, they were able to isolate the strength–toughness relation for the major alloying additions. Zinc was found to degrade K_c less for a given strength increment than that associated with the average response of the alloy. This would indicate that zinc is a desirable alloy strengthening addition. Although yield-strength increments associated with copper and magnesium produced about average degradation in K_c, iron was found to degrade fracture toughness by the greatest amount ($3\frac{1}{2}$ times that associated with zinc additions). As expected, reduction in iron content brought about a significant improvement in alloy toughness and an associated reduction in the number of insoluble large particles. More recent results have confirmed the deleterious effect of iron *and* silicon content on fracture toughness (Fig. 10.23).[31–34] This has led to the development of new alloys possessing the same general chemistry as previous ones with the exception that iron and silicon contents are kept to an absolute minimum. Examples of these new materials include 2124 (the counterpart of 2024) and 7475 (the counterpart of 7075), which have the same strength as the older alloys but enhanced toughness (see Fig. 10.23 and Table 10.8).

The role of inclusions in initiating microvoids in a wide variety of aluminum alloys has been examined recently by Broek,[35] who showed that microvoid dimple size was directly related to inclusion spacing (Fig. 10.24). Large particles that fractured at low stress levels allowed for considerable void growth prior to final failure, but smaller particles nucleated and grew spontaneously to failure. Using an analysis similar to that proposed by McClintock[36] and supported by the work of Edelson and Baldwin[19] (Fig. 10.14), Broek suggested that the fracture strain was related to some function of the volume fraction of particles or voids. Hence, $\epsilon_f \propto f(1/V)$, where ϵ_f = fracture strain and V = volume fraction of the second phase. Consequently, the toughness of these alloys would be expected to rise with decreasing particle content. Therefore, one would predict that had Broek examined low iron and silicon alloys in his investigation, he would have found these materials to reveal larger microvoids and fracture strains. Indeed, Kaufman[32] confirmed larger microvoids in the tougher, cleaner aluminum alloys he examined.

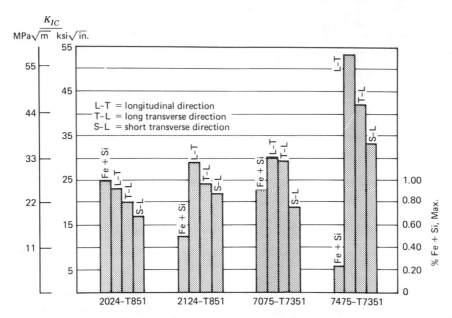

FIGURE 10.23 **High-purity metal (low iron and silicon) and special processing techniques used to optimize toughness of 2xxx and 7xxx aluminum alloys.**[33] **(From R. Seng and E. Spuhler,** *Metal Progress,* **March 1975, copyright American Society for Metals.)**

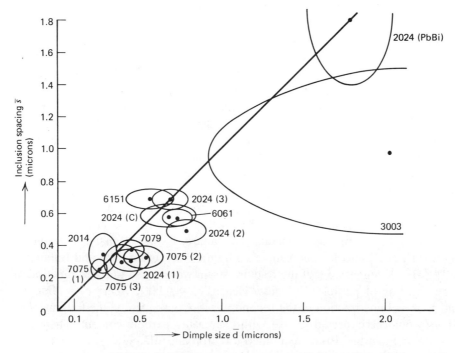

FIGURE 10.24 **Observed relation between microvoid size and inclusion spacing. Numbers represent aluminum alloy designations; ellipses indicate scatter.**[35] **(Reprinted with permission from D. Broek,** *Engineering Fracture Mechanics,* **1973, Pergamon Press.)**

10.4 OPTIMIZING MICROSTRUCTURES FOR MAXIMUM TOUGHNESS

10.4.1 Ferrous Alloys

Numerous studies have been conducted to determine which alloying elements and microstructures provide a given steel alloy with the best combination of strength and toughness. Since these structure–property correlations were established with many different properties (strength, ductility, impact energy, ductile–brittle transition temperature, and fracture toughness), it is difficult to make immediate data comparisons. There are, however, certain general statements that can be made with regard to the role of major alloying elements in optimizing mechanical properties. These are summarized in Table 10.6.

The beneficial role of nickel in improving toughness and lowering the transition temperature (Fig. 10.25) has been recognized for many years and used in the development of new alloys with improved properties. For example, high-nickel steels are presently being evaluated as candidates for cryogenic applications, such as in the construction of liquefied natural gas storage tanks. The explanation for the improved response of nickel steels remains unclear despite the efforts of many researchers. In a recent review of the subject, Leslie[38] examined several theories and found all to be incomplete and/or incapable of rationalizing a number of exceptions to the proposed rule(s). One interesting result of Leslie's review, however, was the identification of platinum as an even better toughening agent than nickel when found in iron solid solutions. Obviously, the exorbitant cost of platinum will preclude its use as an alloying addition in steels.

TABLE 10.6   Role of Major Alloying Elements in Steel Alloys

Element	Function
C	Extremely potent hardenability agent and solid solution strengthener; carbides also provide strengthening but serve to nucleate cracks.
Ni	Extremely potent toughening agent; lowers transition temperature; hardenability agent; austenite stabilizer.
Cr	Provides corrosion resistance in stainless steels; hardenability agent in quenched and tempered steels; solid solution strengthener; strong carbide former.
Mo	Hardenability agent in quenched and tempered steels; suppresses temper embrittlement; solid solution strengthener; strong carbide former.
Si	Deoxidizer; increases σ_{ys} and transition temperature when found in solid solution.
Mn	Deoxidizer; forms MnS, which precludes hot cracking caused by grain-boundary melting of FeS films; lowers transition temperature; hardenability agent.
Co	Used in maraging steels to enhance martensite formation and precipitation hardening kinetics.
Ti	Used in maraging steels for precipitation hardening; carbide and nitride former.
V	Strong carbide and nitride former.
Al	Strong deoxidizer; forms AlN, which pins grain boundaries and keeps ferrite grain size small. AlN formation also serves to remove N from solid solution, thereby lowering lattice resistance to dislocation motion and lowering transition temperature.

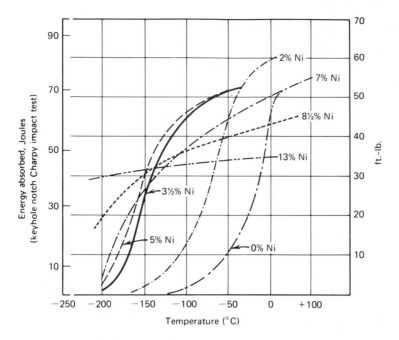

FIGURE 10.25 **Effect of nickel content on transition temperature in steel.[37] (Reprinted with permission from the International Nickel Company, Inc., One New York Plaza, New York.)**

Low[39] examined the effect of typical alloy steel microconstituents on toughness and concluded that the finer ones, namely lower bainite and martensite, provided greater fracture resistance than the coarser high-temperature transformation products such as ferrite, pearlite, and upper bainite. (The question of structural refinement is discussed further in Section 10.5.) More recently, Cox and Low[40] sought explanations for the toughness difference between two types of important commercial steel alloys—quenched and tempered (for example, AISI 4340) and maraging steels (Fig. 10.21). One factor already mentioned (Section 10.3) was the beneficial effect of lower carbon content in the maraging steel. Proceeding further, they found that voids were nucleated in both alloys—nucleated by fracture of Ti (C,N) inclusions in maraging steels and at the interfaces between MnS inclusions and the matrix in AISI 4340 steel. However, a critical difference between the two alloys was noted in the crack growth and coalescence stage. In maraging steels, these voids grew until impingement caused coalescence and final failure. By contrast, the growth of the initial large voids in AISI 4340 was terminated prematurely by the development of void sheets—consisting of small voids—that linked the large voids (Fig. 10.26). This difference was verified by fractographic observations, which revealed uniform and relatively large microvoids in the maraging steel, but a duplex void size distribution in the 4340 steel. The small voids found in AISI 4340 were attributed to fracture of coarse carbide particles found along martensite lath boundaries. Since the strengthening intermetallic precipitates in the maraging steel are much finer and more resistant to fracture than the corresponding carbides in the quenched and tempered steel, Cox and Low concluded that the superior

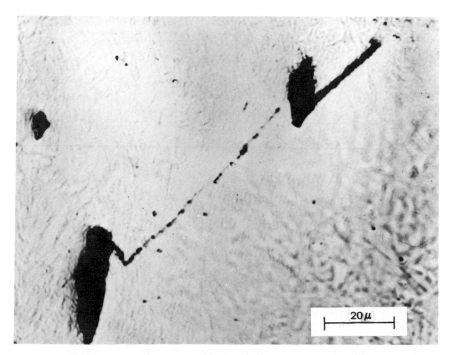

FIGURE 10.26   Large voids in AISI 4340 linked by narrow void sheets consisting of small microvoids.[40] (Copyright © American Society for Metals, 1974.)

toughness shown by the maraging steels could be attributed to their much lower tendency to form void sheets. Correspondingly, they suggested that the quenched and tempered AISI 4340 steel could be toughened by a thermo-mechanical treatment resulting in refinement of the carbides.

Thus far, our discussion has focused on the mechanical properties associated with alloy steel transformation products. Let us now consider the fracture behavior of the parent austenite phase. As we will see, this is both a complex and intriguing task. For one thing, the stability of the austenite phase can be increased through judicious alloying so as to completely stabilize this high-temperature phase at very low cryogenic temperatures or partially stabilize it at room temperature. Low-temperature stability of austenite (γ) is highly beneficial in light of the general observation that austenitic steels are tougher than ferritic (α) or martensitic (α') steels because of the intrinsically tougher austenite FCC crystal structure. In a study of AFC 77, a high-strength steel alloy containing both martensite and austenite microconstituents, Webster[41] showed that the fracture-toughness level increased with an increasing amount of retained austenite in the microstructure (Fig. 10.27). It is believed that the retained austenite phase in this alloy serves as a crack arrester or crack blunter, since it is softer and tougher than the martensite phase. By sharp contrast, retained γ in high-carbon steels can damage overall material response when it undergoes an *ill-timed*, stress-induced transformation to untempered martensite, a much more brittle microconstituent. Several research groups have been experimenting with certain alloys that

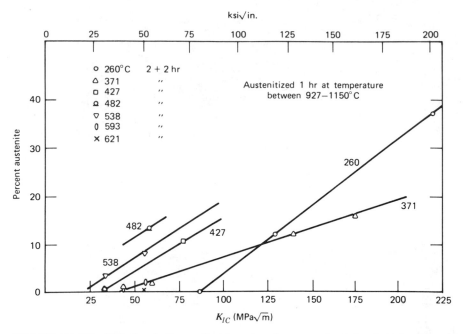

FIGURE 10.27 **Relation between fracture toughness and retained austenite content in AFC 77 high-strength alloy. Retained austenite content varies with austenitizing and tempering temperatures.[41] (Copyright © American Society for Metals, 1968.)**

will undergo (in a carefully controlled manner) this mechanically induced phase transformation. The result has been the development of high-strength steels possessing remarkable ductility and toughness brought about by transformation-induced plasticity in the material. These TRIP[42] (an acronym for *transformation-induced plasticity*) steels compare most favorably with both quenched and tempered and maraging steels (Fig. 10.21).

How can this be? How can you transform a tough phase γ into a brittle phase α′ and produce a tougher alloy? How does this crack-tip zone-shielding mechanism work (recall Fig. 10.3)? Antolovich[43] argued that a considerable amount of energy is absorbed by the system when the γ → α′ transformation takes place. Assuming for the moment that the fracture energy of γ and α′ is the same, the total fracture energy of the system would be the elastic and plastic energies of fracture for each phase plus the energy required for the transformation itself. Since the fracture energy of α′ is lower than that of the γ phase, the toughness of the unstable γ alloy would be greater than that of a stable γ alloy, so long as the energy of transformation more than made up the loss in fracture energy associated with the fracture of α′ rather than the γ phase. Obviously, the toughness of the TRIP steel would be enhanced whenever the toughness difference between the two phases was minimized. An additional rationalization for the TRIP effect has been given, based on the 3% volumetric expansion associated with the γ → α′ transformation. It has been argued[44] that this expansion would provide for some stress relaxation within the region of tensile triaxiality at the crack tip. Bressanelli and Moskowitz[45] pointed out that the *timing* of the transformation was extremely critical to alloy toughness. Transformation to the more brittle α' phase was beneficial only if it occurred during incipient necking. That is, if martensite formed at strains where plastic instability by necking was about to occur, the γ matrix could be strain hardened and, thereby, resist neck formation. If the $\gamma \rightarrow \alpha'$ transformation occurred at lower stress levels prior to necking because the alloy was very unstable, brittle α' would be introduced too soon with the result that the alloy would have lower ductility and toughness. Note in this connection that prestraining these alloys at room temperature would be very detrimental. At the other extreme, if alloy stability were too high, the transformation would not occur and the material would not be provided with the enhanced strain-hardening capacity necessary to suppress plastic necking. Some success has been achieved in relating the fracture toughness of a particular TRIP alloy to the relative stability of the austenite phase.[46,47]

Although the TRIP process offers considerable promise as a means to improve alloy toughness, engineering usage of materials utilizing this mechanism must await further studies to identify optimum alloy chemistry and thermo-mechanical treatments. Equally important will be efforts to reduce the unit cost of these materials to make them more competitive with existing commercial alloys.

10.4.2 Nonferrous Alloys

Microstructural effects are also important when attempting to optimize the toughness of titanium-based alloys. For example, it has been found that toughness depends on the size, shape, and distribution of different phases that are present (Fig. 10.28). We see, for example, that metastable β (BCC phase) alloys possess the highest toughness with α (HCP phase) + β alloys generally being inferior. Furthermore, in these mixed-phase alloys, acicular α rather than equiaxed α within a β matrix is found to provide superior toughness. It is quite probable that the large amount of scatter in K_{IC} values shown in Fig. 10.2 for Ti-6Al-4V (an α + β alloy) was largely a result of variations in the microstructures just described. For further information concerning the effect of compositional and microstructural variables on the fracture behavior of titanium alloys, the reader is referred to the review by Margolin et al.[49]

Let us now reexamine the observation made in Chapter 8 (Table 8.2) that the 2024-T3 aluminum alloy possesses higher toughness than the 7075-T6 sister alloy. Although this is true when comparison is based on the *different* strength levels designated for these alloys, the 7075-T6 alloy actually is the tougher material when compared at the same strength level (Fig. 10.29).

Nock and Hunsicker[51] demonstrated that the superiority of 7000 series alloys was attributable to a relatively small amount of insoluble intermetallic phase and to a reduced precipitate size, which would be less likely to fracture (25 to 50 Å in 7075-T6 versus 500 to 1000 Å in 2024-T86). Recent metallurgical studies have been concerned with optimizing the fracture toughness, strength, and resistance to environmental attack of various aluminum alloys. For example, it has been shown[31,32,48] that while strength decreases, toughness is improved when the material is underaged, with a somewhat smaller improvement being associated with the overaged condition. However, the overaged alloy, with its greater resistance to stress corrosion cracking, is preferred, even though the toughness level is somewhat lower than that obtained in the underaged condition. Attempts are being made to combine mechanical deformation

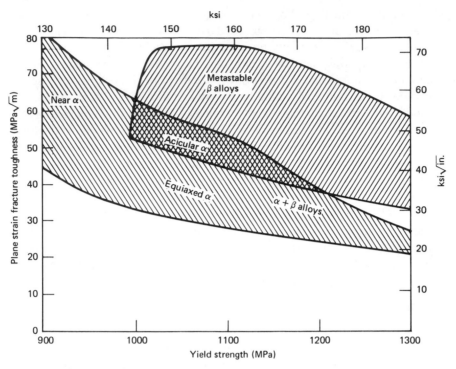

FIGURE 10.28 **Schematic diagram showing effect of alloy strength and microconstituents on toughness in titanium alloys.[48] (Reprinted with permission from A. Rosenfield and A. J. McEvily, Jr., NATO AGARD Report 610, Dec. 1973, p. 23.)**

with variations in aging procedures to optimize material response.[32,52–54] Preliminary results have shown that toughness increases with decreasing size of Al_2CuMg particles and minimization of $Al_{12}Mg_2Cr$ dispersoids.[54] More studies of important aluminum alloys have been reported by Bucci.[55]

Aluminum-lithium alloys were developed to take advantage of their low density and high alloy stiffness. For example, alloys with 3.3% Li are approximately 10% lighter and 15% stiffer than conventional aluminum formulations. Unfortunately, Al-Li alloys typically exhibit lower ductility, toughness, and stress-corrosion resistance than existing Al-Cu-Mg-Zn aerospace alloys. These property deficiencies are attributable to severe planar slip deformation, generated by dislocation cutting of Al_3Li (δ') precipitates[56,57] (recall Fig. 4.12a), and to grain boundary embrittlement, associated with the segregation of alkali metal atoms to grain boundaries.[58–61] (Trace elements of sodium, potassium, cesium, and rubidium are contained in the Li additions to the aluminum; also, these impurities may be picked up from refractory bricks used in the casting process.) For the case of several vacuum-refined experimental heats of the Al-Li 2090 alloy, fracture toughness levels increased dramatically when the Na + K content was reduced from current commercial purity levels of 4–11 ppm Na + K to less than 1 ppm[61] (Fig. 10.30a). There is additional evidence to suggest that Al-Li-based alloy toughness is adversely affected by the diffusion of hydrogen to alkali-metal-contaminated grain boundary regions. Figure 10.30b reveals that increases in

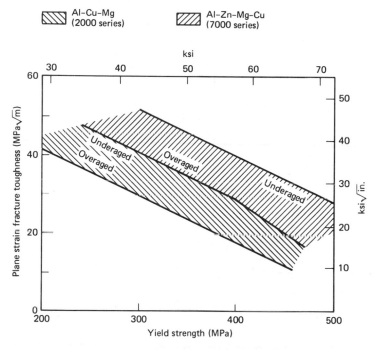

FIGURE 10.29   Toughness versus strength data for 2000 and 7000 series aluminum alloys revealing superior toughness in the latter at any given strength level.[38] (Reprinted with permission from A. Rosenfield and A. J. McEvily, Jr., NATO AGARD Report 610, Dec. 1973, p. 23, based on data from Develay.[50])

toughness are associated with decreases in the number of brittle particles present in the microstructure and, more importantly, to decrease in alkali-metal and hydrogen content.[61] As expected (recall Section 7.7.3), Fig. 10.30b also shows that toughness increases with a shift in observed fracture micromechanisms from cleavage and intergranular separation to transgranular microvoid coalescence.

10.4.3 Ceramics

In most instances, there are fewer than five independent slip systems available in ceramics to allow for arbitrary shape changes in the crystals (recall Section 3.1). In addition, such materials possess ionic or covalent bonds, have low symmetry crystal structures, and exhibit long-range order (see Table 10.1). Consequently, ceramics are typically very brittle and exhibit low fracture toughness. For this reason, their use in engineering components has been limited. Successful efforts to decrease the size and number of defects in the microstructure through improved fabrication procedures have led to improved strength in ceramics. However, these materials remain subject to premature failure in the presence of service-induced defects since the material's intrinsic toughness remains low. To make greater use of these important materials with their exceptional high-temperature capability and wear resistance, much effort has been given to the development of tougher ceramic microstructures, which are less

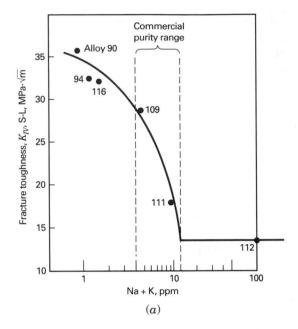

(a)

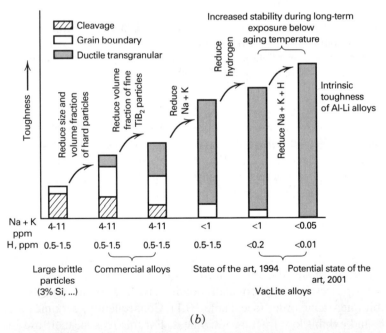

(b)

FIGURE 10.30 Influence of alkali-metal impurities on fracture toughness in Al-Li 2090 alloys. (a) Fracture toughness values given for short transverse orientation and compared after normalization to a yield strength of 435 MPa. Note the sharp increase in toughness at Na + K concentration levels below the commercial purity upper bound; (b) bar graph depicting influence of alkali-metal and hydrogen content on fracture toughness level and operative fracture micromechanisms.[61] (Reprinted from D. Webster, *Advanced Materials and Processes* 145(5), 18 (1994), with permission from ASM International.)

sensitive to the presence of defects.[62–66] For example, Becher[65] reported that when SiC whiskers were added to an alumina matrix, the Weibull modulus of the material's flexural strength increased from 4.6 to 13.4 (Recall from Section 7.3.1 that higher values of the Weibull modulus correspond to decreased variability in the measured property.) To develop materials with greater toughness and reliability, researchers have utilized geometrical and crack-tip shielding extrinsic toughening mechanisms (see Fig. 10.3). Both are responsible for the phenomenological development of R- or T-curve behavior* (recall Section 8.9). For those ceramic materials that exhibit R-curve behavior as a result of crack-tip shielding phenomena, there is no single K_{IC} value to define the toughness of the material. Instead, toughness increases with increasing crack size and crack wake dimension until a steady-state fracture condition is achieved. Furthermore, for stable crack extension associated with crack lengths less than that associated with failure (i.e., a_f, where $\partial K/\partial a \leq \partial T/\partial a$), the material's fracture strength is insensitive to crack size (Fig. 10.31); accordingly, the material's structural reliability is improved. Conversely, non-R-curve materials possess a constant level of toughness and a fracture strength that decreases continuously with crack size. The toughness-dependence on crack size is noted by referring again to Fig. 8.21b. Since the size of the indentation crack varies with indentation load, the relatively constant fracture stress at low loads implies that toughness increases with increasing crack size. Such toughness behavior is related to the crack-tip shielding mechanisms discussed here and summarized in Fig. 10.3.

10.3.4.1 Toughening Mechanisms

Geometrical toughening involves crack deflections arising from crack-tip–grain boundary and/or second phase particle interactions (Figs. 10.3 and 10.7b).[64] For example, with decreasing grain boundary strength and/or increasing grain misorientation, one would expect a crack to become increasingly diverted along a grain boundary path and away from its current path along a cleavage plane within a particular grain. These perturbations in crack plane and directions reduce the local stress intensity factor and lead to a moderate improvement in toughness.[67,68] In addition, crack deflection can result from the interaction of an advancing crack front with a residual stress field, such as one generated by a thermal mismatch between the matrix and reinforcing particles (recall Eq. 7-31). If the difference in coefficient of thermal expansion (CTE) between matrix and spherically shaped second phase particles is positive (i.e., $\alpha_m - \alpha_p > 0$), a compressive radial stress (σ_r) is developed at the particle-matrix boundary along with a tensile tangential stress (σ_t) in the matrix; the crack is then "attracted" to the particle (Fig. 10.32a).[69] Conversely, when $\Delta\alpha < 0$, $\sigma_r > 0$ and $\sigma_t < 0$; in this instance, the crack is "rejected" and passes between the particles (Fig. 10.32b). This more tortuous crack path requires additional fracture

---

* As discussed in Section 8.9, R-curve behavior refers to the crack-dependent change in the material's resistance to fracture (R) (units of energy). For the case of metallic alloys, improved resistance to fracture is derived from the accumulation of plastic deformation. Crack instability occurs when $\partial \mathcal{G}/\partial a = \partial R/\partial a$. The T-curve provides an analogous display of the material's resistance to fracture with toughness (T) (units of stress intensity) plotted versus crack size. Accordingly, instability occurs when $\partial K/\partial a = \partial T/\partial a$.

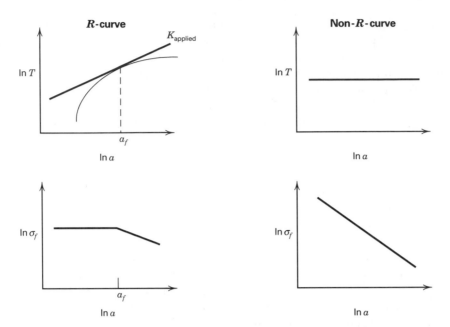

FIGURE 10.31 Toughness- and fracture strength-crack size dependence in R-curve and non-*R*-curve type ceramics. For R-curve materials, toughness increases with crack size up to a_f and fracture strength is insensitive to crack size when $a < a_f$. For non-*R*-curve type materials, toughness is constant and fracture strength decreases with crack size.[86] [Adapted from M. P. Harmer, H. M. Chan, and G. A. Miller, *J. Amer. Ceram. Soc.* 75(7), 1715 (1992). Reprinted by permission.]

energy and should result in greater composite toughness than that associated with crack passage through the particles.

We now consider several *crack-tip shielding* mechanisms that contribute greatly to the toughening of ceramics and their composites. For simplicity, these are distinguished by those mechanisms that occur in the *frontal zone* (ahead of the advancing crack front) and those that act in the *crack wake* (bridging mechanisms) (Fig. 10.33).[64] Shielding zones in structural ceramics are approximately 1–1000 μm in length, whereas those in ceramic and cementitious composites are considerably larger.[64,70] Since dislocation mobility in ceramics is limited due to a large Peierls stress, dislocation cloud formation provides little toughening. Crack-tip stress fields and/or residual stresses can nucleate a cloud of microcracks at weakened microstructural sites such as grain boundaries. Such a cloud can serve to dilate the crack-tip region and reduce the effective stress level. This toughening mechanism is not very important in monophase ceramics; conversely, by increasing the volume fraction of second-phase particles, microcrack density in some two-phase ceramic composites is increased along with toughness.[66,71]

An extremely potent crack-tip zone-shielding mechanism that has attracted considerable attention recently involves the presence of submicron-sized metastable particles of tetragonal zirconia (ZrO_2) that are embedded within a cubic fluorite ZrO_2 matrix

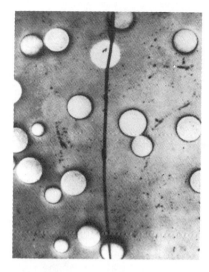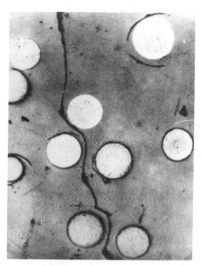

FIGURE 10.32 Crack-path dependence on residual stress state in composites of glass + 10 v/o thoria spheres. (*a*) $\Delta\alpha = 1.8 \times 10^{-6}$ K^{-1}. Crack is drawn toward thoria spheres; (*b*) $\Delta\alpha = -5.1 \times 10^{-6}$ K^{-1}. Crack is deflected away from thoria spheres, resulting in tortuous crack path.[69] (R. W. Davidge and T. J. Green, *Journal of Materials Science*, **3**, 629 (1968). Reprinted by permission.)

or other ceramic matrix (e.g., alumina (Al$_2$O$_3$)).[72-81] The principal feature associated with toughening in these materials is analogous to the transformation-induced plasticity (TRIP) effects associated with certain high-strength steels (recall Section 10.4.1). In the case of partially stabilized zirconia-bearing ceramics, it is possible to bring about, near the crack tip, a stress-induced martensitic transformation of the tetragonal ZrO$_2$ particles to that of the monoclinic polymorph. Figure 10.34 shows a thin-film TEM image of coherent tetragonal particles before and after their stress-induced transformation to the monoclinic form. Note that tetragonal particles away from the crack plane had not transformed. It is generally believed that toughening results from the fact that a portion of the energy available for fracture is dissipated during the stress-induced transformation process. In addition, the transformation process generates a favorable residual compressive stress environment as a result of the 3–5% volume expansion associated with the tetragonal to monoclinic phase change. McMeeking and Evans[82] have computed the contribution of transformation toughening to be

$$K_t = 0.3E\epsilon^T V_f w^{1/2} \tag{10-1}$$

where K_t = toughness contribution due to phase transformation
E = elastic modulus
ϵ^T = unconstrained transformation strain of ZrO$_2$ particles
V_f = volume fraction of ZrO$_2$
w = width of the transformation zone on either side of the crack surface

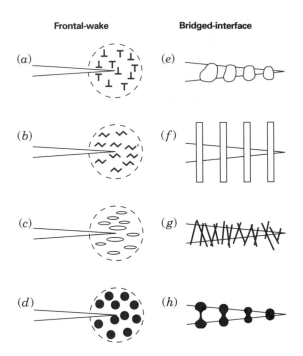

FIGURE 10.33 Crack-tip shielding mechanisms. Frontal zone: (*a*) dislocation cloud; (*b*) microcrack cloud; (*c*) phase transformation; (*d*) ductile second phase. Crack-wake bridging zone: (*e*) grain bridging; (*f*) continuous-fiber bridging; (*g*) short-whisker bridging; (*h*) ductile second phase bridging.[64] [B. Lawn, *Fracture of Brittle Solids, 2d ed.,* Cambridge Solid State Science Series, Cambridge, UK (1993). Reprinted with the permission of Cambridge University Press.]

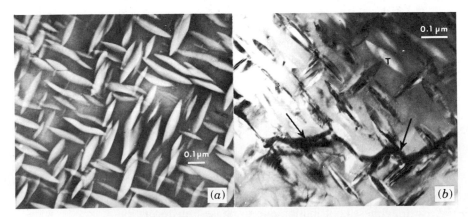

FIGURE 10.34 TEM images of partially stabilized zirconia alloy, containing 8.1 mole percent MgO. (*a*) Coherent tetragonal ZrO_2 particles embedded within a cubic MgO–ZrO_2 matrix; (*b*) ZrO_2 particles near crack plane are transformed from tetragonal to monoclinic form. Note that tetragonal particles away from crack plane are untransformed. (From Porter and Heuer[76] with permission from the American Ceramic Society, Inc.)

Toughness is also enhanced by increasing the density of transformable ZrO_2 particles near the anticipated fracture plane, by choosing chemical systems that enhance the volume change of the transformation process, and by choosing a very rigid matrix (high E) so as to enhance the residual stress field in the transformed zone surrounding the crack.[83] Furthermore, recent studies have noted that the transformation-toughening contribution is greater than that predicted by Eq. 10-1 because of the presence of additional energy-absorbing processes.[83] For additional perspectives on transformation toughening in ceramics, see the recent reviews by Evans[66,78] and a five-paper set by Lange.[84] The ZrO_2 particle size controls the temperatures for spontaneous and stress-induced transformation; the optimum particle size has been found to be 0.1 to 1.0 μm.[73] Examples of the effects of aging time and amount of magnesia content on the fracture toughness of PSZ ceramics are shown in Fig. 10.35. Note that in each case the fracture toughness value can be optimized with appropriate ceramic chemistry or heat treatment. Likewise, since toughening in zirconia-based alloys is achieved by the energy-consuming transformation of tetragonal-ZrO_2 to its monoclinic form, it is important that this phase transformation be inhibited during cooling from the processing temperature and, instead, delayed until the critical loading event; this is accomplished with processing additives such as CaO, MgO, Y_2O_3, and CeO_2. By proper chemistry and heat treatment, zirconia alloys can exhibit toughness levels as high as $20 \text{ MPa}\sqrt{m}$.

Crack bridging with either ductile (metal-toughened ceramics, called *cermets*) or brittle second phase particles (e.g., fibers and whiskers) represents the second major group of toughening mechanisms in brittle ceramics (see Fig. 10.33). Energy is consumed when the interface separates ahead of the advancing crack and the triaxial stress state at the crack tip is relaxed (recall Fig. 10.7). As the crack extends, additional energy is consumed with progressive debonding of the ligaments. These unbroken ligaments produce tractions across the crack wake (i.e., closure forces), which diminish the local crack tip stress level. Eventually, the ligaments fail and pull out of the matrix. (Fig. 10.36a).[64] As such, a steady-state bridging zone of length, L, is developed ahead of the advancing crack tip (Fig. 10.36b). For optimal energy consumption, the fiber-matrix interfacial strength should not be too strong or too weak.[65,87,88] If the bond strength is high, fiber rupture will occur prior to debonding and no bridging will develop. Alternatively, if the interface is too weak, debonding will take place, but little energy will be consumed by the fiber-pull-out process. It has been reported that interfacial debonding and the absorption of high levels of fracture energy will generally occur when the ratio of interface/fiber toughness is < 0.25.[66] Additional efforts are being directed toward the development of optimal interfacial properties through the introduction of a separate interboundary phase.[65,66]

The extent of debonding, fiber pull-out, and associated energy consumption is influenced by the presence of thermally induced residual stresses. For example, when CTE of the matrix is greater than that of the whisker or fiber, the matrix will "clamp" down on the whiskers and generate large frictional forces along the whisker–matrix interface. By reducing the CTE mismatch between the whisker and matrix, the resulting radial compressive stress is reduced and toughness increased by the enhanced ability of the whisker to pull out of the matrix.[65]

It has been shown that *grain bridging*-induced toughening can occur in noncubic monophase ceramics. Here, unbroken grains bridge across the crack plane; these

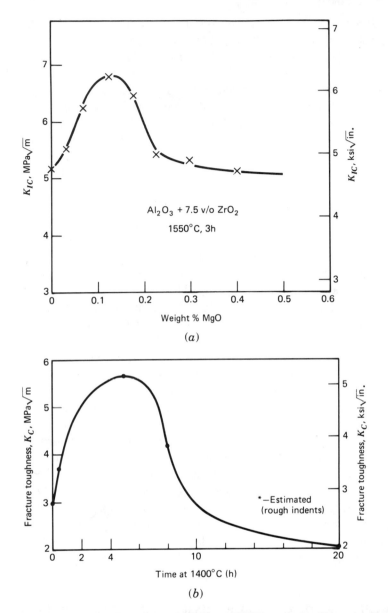

FIGURE 10.35 Fracture toughness of PSZ ceramic alloys. (a) K_{IC} values in Al_2O_3 + 7.5 vol ZrO_2 as a function of MgO content. (From Claussen and Rühle.[73]) (b) K_c values inferred from microhardness indentations in 8.1 mole percent MgO + ZrO_2 as a function of aging time at 1400°C. (From Porter and Heuer[77]; reprinted with permission from the American Ceramic Society, Inc.)

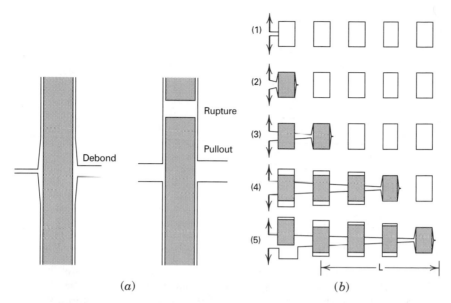

FIGURE 10.36 (*a*) **Debonding and subsequent pull-out of broken fiber from matrix.**[64] **[B. Lawn,** *Fracture of Brittle Solids, 2nd ed.,* **Cambridge Solid State Science Series, Cambridge, UK (1993). Reprinted with the permission of Cambridge University Press.];** (*b*) **sequence of events involving (1) crack deflection, (2) particle debonding, (3) grain pull-out, (4) lengthening of bridging zone, (5) establishment of steady-state bridging zone of length L.**[85] **[N.P. Padture, Ph.D dissertation, Lehigh University, Bethlehem, PA 1991. Reprinted by permission.]**

ligaments are locked in place by thermally induced compressive forces, generated by the anisotropic thermal expansion mismatch between contiguous grains in these materials.[89–91] By contrast, CTE values in cubic ceramics are isotropic and no thermal-stress induced grain bridging is found.[91,92] Other studies show that toughness increases with the introduction of elongated grains[85,93] with the latter simulating the behavior of whiskers and/or short rods. In this instance, the materials are referred to as being ''self-reinforced.''

To this point, our discussions have focused on separate toughening mechanisms. It is striking to note that when multiple mechanisms are active (e.g., whisker bridging and zirconia-based transformation toughening), the combined level of toughening is multiplicitive rather than additive. Such synergism arises from the fact that the crack surface tractions of the bridging ligaments serve to expand the size of the process zone where transformation toughening is occurring.[94] The limit for synergistic material response is found to occur when the ratio of the bridging zone length, L, to the process zone width, h, is $L/h \leq 10$. Figure 10.37 illustrates the combined influence of whisker-reinforcement and transformation toughening mullite composites.[65] The addition of 20 v/o of SiC whiskers doubles the toughness of mullite at 800°C. When a third component is added to this composite (i.e., 20 v/o *monoclinic* zirconia), an additional contribution to overall toughness is achieved. However, the same volume addition of

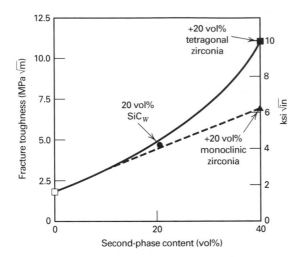

FIGURE 10.37 **Fracture toughness in mullite at 800°K as a function of 20 v/o SiC whisker and 20 v/o zirconia. Superior toughness is associated with the presence of the tetragonal zirconia phase which undergoes a phase transformation as compared with nontranforming monoclinic zirconia.[65] (P. F. Becher, *Journal of the American Ceramic Society*, 74 (2), 255 (1991). Reprinted by permission.)**

tetragonal zirconia leads to a significant increase in toughness, due to the synergistic interaction between whisker bridging and transformation toughening of the metastable tetragonal zirconia phase.

Before concluding this section, it is interesting to reexamine the significance of Eq. 7-37 in light of our subsequent development of the fracture-toughness parameter, K_{IC}. For the case of a semicircular surface flaw, the stress intensity factor at fracture may be given from Eq. 8-26b.

$$K_{IC} = 1.1 \frac{2}{\pi} \sigma \sqrt{\pi a_c} \qquad (8\text{-}26\text{b})$$

where a_c is the depth of the semicircular crack at fracture. (Note that only one surface flaw correction is used in this crack configuration.) Combining Eqs. 8-26b and 7-37, we find

$$M_h = 0.81 \, K_{IC}(r_h/a_c)^{1/2} \qquad (10\text{-}2)$$

where M_h = mirror constant corresponding to the mist–hackle boundary
   K_{IC} = plane-strain fracture toughness
   r_h = distance from crack origin to mist-hackle boundary
   a_c = critical flaw depth of semicircular surface flaw

When available data corresponding to M_h and K_{IC} values for various ceramics are plotted together, the data cluster about a line corresponding to a mist–hackle boundary:

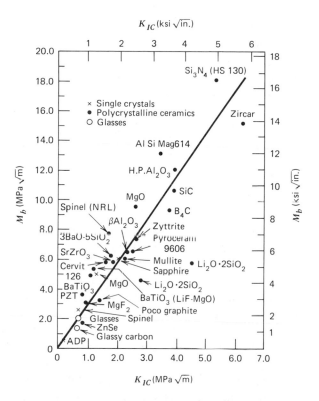

FIGURE 10.38 Mist–hackle constant M_h versus K_{IC} plotted for glassy and crystalline ceramics. Solid line corresponds to a mist–hackle-boundary–critical flaw size ratio of 13 to 1. (From Mecholsky et al.;[95] reproduced with permission from *Journal of Materials Science* 11, 1310 (1976).)

critical flaw size ratio of 13 to 1 (Fig. 10.38). Note that this ratio appears to hold true for both glassy and crystalline ceramics. As such, the critical flaw size in a ceramic can be estimated from Eq. 10-2 if the mist–hackle radius can be measured on the fracture surface. Mecholsky et al.[95] have verified this procedure for a number of glasses and crystalline ceramics.

10.4.4 Polymers

It is generally agreed that crystalline polymers tend to exhibit greater toughness than amorphous polymers owing to their folded chain conformation. There is, however, some controversy regarding the mechanism(s) responsible for whatever toughness amorphous materials possess. According to one proposal,[96,97] toughness should depend on the amount of free volume available for molecule segmental motion. With enhanced motions, toughness should be higher. Litt and Tobolsky[96] found that tough amorphous polymers generally contain a fractional free volume $\bar{f}$ (Eq. 6-6) greater than 0.09. In further support, Petrie[98] found that such polymers, when aged at $T < T_g$, suffered losses in impact energy absorption that were related to corresponding decreases in excess free volume after annealing. In a related finding, it has been shown that the $T < T_g(\beta)$ peak is also identified with impact toughness, the larger and broader the β peak the tougher a material generally tends to be.[97,99,100] For example, the β peak in polycarbonate is much greater than that shown for PMMA, consistent with the much greater toughness associated with the former material.

Finally, it is interesting to note that although a particular *polymer* possessing a low free volume and negligible β peak may be brittle, certain *plastics* based on this polymer may offer good toughness. This improvement in impact resistance is commonly achieved through the use of plasticizers. As discussed in Section 6.2, these high-boiling point, low-MW monomeric liquids serve to separate molecule chains from one another, thus decreasing their intermolecular attraction and providing chain segments with greater mobility. The ductility and toughness of plasticized polymers is found to increase, while their strength and T_g decrease. Care must be taken to add a sufficient amount of plasticizer to a particular polymer so that the material does not actually suffer a loss in toughness. This surprising reversal in material response is referred to as the *antiplasticizer* effect.[101] It has been argued that below a critical plasticizer content, the liquid serves mainly to fill some of the polymer's existing free volume with a concomitant loss in molecular chain segmental mobility.

A brittle polymer may also be toughened with the addition of a finely dispersed rubbery phase in a diameter size range of 0.1 to 10 μm. For a comprehensive treatment of this subject area, the reader is referred to several monographs, conference proceedings, and key papers on the subject.[102-111] Rubber-toughened polymers can be prepared by using several different techniques. An amorphous plastic (typically polystyrene) and an elastomer (polybutadiene) can be mechanically blended using rollers or extruders to produce an amorphous matrix containing relatively large (5–10 μm), irregularly shaped, rubber particles. Such mixtures do not represent an efficient use of the rubbery phase from the standpoint of optimizing the toughness of the blend. Furthermore, the interface between the matrix and rubbery phase is weak and serves as a site for crack nucleation at relatively low stress levels. A far superior blend is achieved by using the solution-graft copolymer technique. In this method, the rubber phase is first dissolved in styrene monomer to a level of 5–10 w/o. When the polymerization process is about one-half completed, the agitated viscous mixture undergoes phase separation with the elastomeric phase separating in the form of discrete spherical particles from the continuous polystyrene matrix. The dissolved styrene monomer in the rubber phase then polymerizes in the form of discrete polystyrene droplets to form a cellular structure, as shown in Fig. 10.39. Some researchers have referred to this morphology as the "salami structure." The dark, skeletal-like structure is the polybutadiene phase that was revealed by staining with osmium tetroxide, to enhance contrast in the image from the transmission electron microscope;[112] the discrete occluded particles and the continuous matrix constitute the polystyrene phase. ABS (acrylonitrile–butadiene–styrene) polymers, a commercially important group of rubber-modified polymers, are produced by an emulsion polymerization method by which the plastic phase is polymerized onto seed-latex particles of the elastomeric phase. Cellular rubber particles are also found in these materials along with solid rubber particles, though the cellular particles in ABS are smaller (usually less than 1-

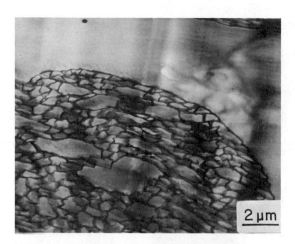

FIGURE 10.39 Photomicrograph revealing duplex structure of toughening phase in HiPS blend. Dark skeletal phase is polybutadiene elastomeric phase. Occluded particles and matrix are polystyrene. (Courtesy Clare Rimnac.)

μm in diameter) than those found in high-impact polystyrene (HiPS). For the case of thermoset resins, successful composite manufacture has been achieved by dissolving the rubber phase in the liquid uncured resin. Upon commencement of the curing reaction, the rubber phase precipitates out from the homogeneous solution as a fine dispersion of rubber particles. More recently, preformed latex-core-hard shell particles have been developed that are added to the resin. Here, the shell material is chosen to effect a strong bond with the matrix phase.

Many studies have demonstrated that amorphous, cross-linked, and semicrystalline matrices can be toughened by various elastomeric additives. Increases of eight- to tenfold in fracture energy have been reported, for example, in polyester[113] and epoxy[105] resins. In such studies, toughness increases with volume fraction of the rubber phase up to a certain point, whereafter toughness remains constant or even decreases. Depending on the matrix phase, the mechanism responsible for toughening has been identified as massive crazing (as in HiPS),[102,103,114,115] or profuse shear banding as reported in ABS, PVC, and epoxy resins.[104,107,108,116–118] Overall, a synergistic interaction exists between the rubbery phase and the matrix. When loading HiPS, for example, a myriad of fine crazes develop at the interface between the glassy matrix and the micron-sized rubber particles (Fig. 10.40). Although craze formation itself is undesirable since it is the precursor of crack formation, the nucleation of many small crazes represents a large sink for strain energy release and serves to diffuse the stress singularity at the main crack front. In addition, the rubber particles act to arrest moving cracks.

Epoxy resins are excellent candidates for use in demanding structural applications, based on their good creep and solvent resistance. Their usage would certainly expand if it were possible to overcome the poor fracture toughness and fatigue crack propagation resistance of these cross-linked polymers. Accordingly, different approaches

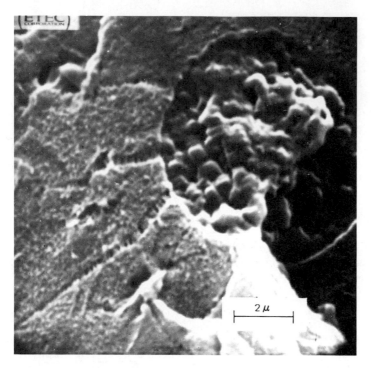

FIGURE 10.40   Matrix crazes emanating from left side of rubber particle as seen on fracture surface in high-impact polystyrene. (Reprinted with permission from John Wiley & Sons, Inc.)

have been examined to improve the toughness of epoxy resins. The most widely studied method involves the addition of rubbery particles to the epoxy matrix, which facilitates three major toughening mechanisms:[104–110,119–123] (1) rubber particle cavitation and associated shear band formation; (2) matrix plastic void growth following rubber cavitation [analogous to microvoid growth in metal alloys in association with the microvoid coalescence fracture process (recall Section 7.3.1)]; and (3) rubber particle bridging (Fig. 10.41a). Several theoretical models[121,122,124,125] have been proposed to assess the relative contribution of these mechanisms to overall toughening. It is theorized that cavitation/shear banding and plastic void growth energy dissipation are proportional to the width of the damage zone and provide *additive* contributions to composite toughness. Since rubber particle bridging influences the size of this zone, it follows that the simultaneous occurrence of these three mechanisms may result in *multiplicative* toughening.[94,125] Huang and Kinloch's additive toughening model[121] successfully predicted their experimental data of an epoxy-rubber composite and noted that particle cavitation and shear band formation contributed a consistently major fraction to toughness over the temperature range from −60°C to 40°C (Table 10.7). By contrast, the toughening contribution by particle bridging decreased with increasing temperature, providing 10% or less of overall composite toughness at temperatures of at least 0°C. Conversely, the toughening contribution associated with plastic void growth increased markedly with increasing temperature.

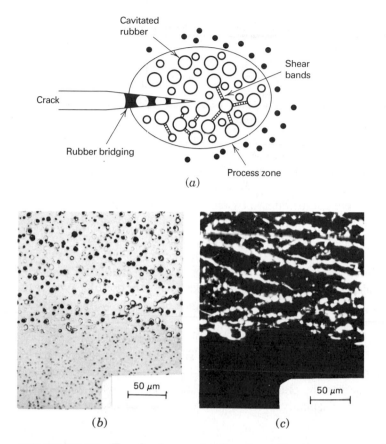

FIGURE 10.41 Toughening mechanisms in rubber-epoxy composites. (*a*) Schematic representation of rubber cavitation and shear banding in the deformation process zone along with crack bridging by rubber particles.[141] (H. Azimi, R. A. Pearson, and R. W. Hertzberg, *J. Mater. Sci. Lett.,* **13,** 1460 (1994). Reprinted by permission. (*b*) Epoxy resin matrix with CTBN rubber particles. Note cavitated rubber particles near fracture surface located near the bottom of the photograph; (*c*) same area as (b) as viewed with cross-polarizers. Note network of shear bands in association with cavitated particles.[108] (Reprinted with permission from R. A. Pearson and A. F. Yee, *J. Mater. Sci.* **21,** (1986) 2475, Chapman & Hall.)

Figure 10.41*b* reveals cavitated rubber particles embedded within the epoxy resin; the extensive shear band network in the same system is made apparent by observing the material between cross-polarizers (Fig. 10.41*c*). There is controversy regarding the sequence of events associated with the toughening process. On one hand, Yee and Pearson argued that rubber particle cavitation relaxes tensile triaxiality at the crack tip, which then facilitates the formation of a network of energy-absorbing shear bands.[104–109] By contrast, Huang and Kinloch[121] concluded that shear bands initiate first at the stress concentration surrounding the rubber particle (cavitated or not) and then terminate at an adjacent particle. Based on a finite element analysis,[123] they determined that the sequence of events between cavitation and shear banding depended

TABLE 10.7 Contribution of Toughening Mechanisms in Rubber-Modified Epoxy [121]

Temperature (°C)	−60	−40	−20	0	23	40
G_{IC} (kJm^{-2})	1.72	1.96	2.53	3.64	5.90	7.23
Rubber bridging (%)	0.36	0.26	0.14	0.11	0.10	0.05
Plastic shear banding (%)	0.64	0.74	0.68	0.60	0.61	0.47
Plastic void growth (%)	0.00	0.00	0.18	0.29	0.29	0.48

on the Young's modulus and Poisson's ratio of the rubber particles. Since the rubbery particles form in the matrix by an in situ process, their mechanical properties depend on composite processing variables. Hence, the sequence of deformation mechanisms for a given composite system can be altered by changing processing conditions.

Yee and Pearson[105] argued that the toughness of a rubber-modified epoxy composite is dominated by the properties of the matrix; for example, they altered the ductility of a thermoset epoxy resin matrix through changes in its cross-link density. As expected, the toughness of the rubber-modified epoxy resin increased markedly with increasing molecular weight between cross-links, M_w^c, whereas the neat resin showed almost no influence of M_w^c (Fig. 10.42). (An increase in M_w^c corresponds to a decrease in cross-link density).

Another approach to epoxy matrix toughening involves the addition of rigid inorganic fillers such as glass spheres and short rods. The addition of such particles enhances the toughness of epoxy-matrix composites by enabling glass spheres and short rods to pin and/or deflect the crack front, facilitates glass particle debonding to enable the epoxy matrix to plastically deform, serves to generate a series of micro-

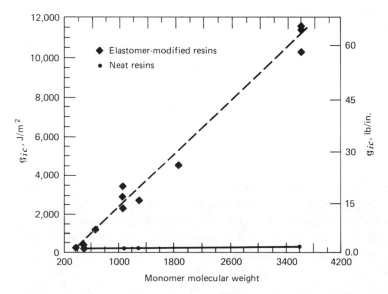

FIGURE 10.42 Fracture toughness in neat and rubber-modified epoxy resin as a function of molecular weight between cross-links.[105] Note dramatic improvement in composite toughness with increasing resin ductility as represented by increased M_w^c. (Reprinted with permission from A. F. Yee and R. A. Pearson, *Toughening in Plastics II*, 2/1 (1985), Plastics and Rubber Institute.)

cracks, and provides bridging elements across the wake of the crack.[126–129] These mechanisms serve to shield the crack tip from the applied stress, thereby reducing the effective crack-tip stress intensity factor (recall Fig. 10.3). For the case of glass spheres, the most important toughening mechanisms are believed to be crack pinning[110,127,128] and matrix plastic deformation in the vicinity of debonded glass spheres.[125] For rod-like particles, the most potent toughening mechanisms involve fiber debonding[130] and pull-out[131–134] (see Section 10.4.5), crack deflection,[125] and fiber bridging[131] (recall Fig. 10.3).

A third approach to the toughening of epoxy resins introduces both rubbery and hard inorganic particles to the epoxy matrix. As noted above, the combination of crack pinning and/or bridging, associated with hard inorganic particles, with energy absorbing process zone mechanisms (e.g., cavitation/shear band formation and plastic void growth), attributable to the rubber particles, provides a multiplicative interaction and synergistic toughening.[94,125] Reports of synergistic toughening in association with hybrid composites are cited in references 136–142. For example, Pearson et al.[139] noted synergistic toughening in an epoxy composite, containing both rubber particles and hollow glass spheres (Fig. 10.43). For this system, synergistic toughening was attributed to the multiplicative interaction between rubber particle cavitation/matrix shear banding and hollow glass sphere-induced microcracking.

10.4.5 Composites

The fracture toughness of fibrous composites depends on the relative contributions of three principal fracture processes. Fracture energy is expected to increase with crack path deflections brought about by fiber–matrix interfacial separation (recall the discussion associated with Fig. 10.3). However, the major toughening contribution is believed to result from fiber pullout. If the fiber length l is less than critical l_c, then the fibers will not fracture but will, instead, pull out of the matrix as the crack progresses across the sample. The energy for this process is given by the shear force at the interface times the distance z over which the force acts. Therefore,

$$\text{energy} = \int_0^z \pi d\tau \, z dz = \frac{\pi d\tau z^2}{2} \qquad (10\text{-}3)$$

where d = fiber diameter
τ = shear stress at interface
z = distance along fiber

From Eq. 10-3, pull-out energy increases with fiber length and is maximized when $z = l_c/2$.[143,144] By combining Eqs. 1-40 and 10-3, we find

$$\text{Energy} = \frac{\pi \, r^3 \sigma_{zz}}{2\tau_{rz}} \qquad (10\text{-}4)$$

where r = fiber radius
σ_{zz} = axial stress along the fiber length
τ_{rz} = shear stress acting along the fiber-matrix interface at the ends of the fiber

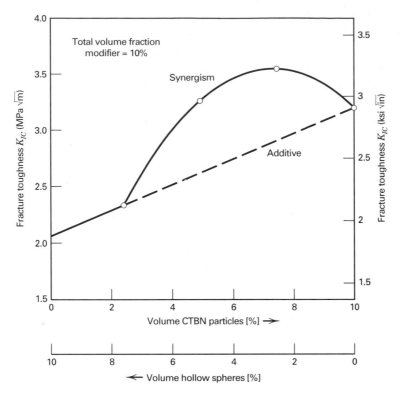

FIGURE 10.43 **Synergistic toughening in CTBN rubber and hollow glass sphere hybrid-epoxy composite, corresponding to 3:1 volume ratio between CTBN rubber and hollow glass spheres.[139] [R. A. Pearson, A. K. Smith, and Y.W. Yee, *Second International Conference on Deformation and Fracture of Composites*, PRI, Manchester, UK, 9-1 (1993). Reprinted by permission.]**

10.5 MICROSTRUCTURAL REFINEMENT

Microstructural refinement represents a unique opportunity by which the material may be *both* strengthened and toughened (Fig. 10.44). This represents a particularly attractive strengthening mechanism in view of the generally observed inverse relation between strength and toughness (Figs. 10.1, 10.16, 10.21, 10.28, 10.29). The toughness and strength superiority of fine-grained materials has been recognized for many years, as evidenced by the well-accepted view that quenched and tempered steel alloy microstructures are superior to those associated with the normalizing process. (Quenched and tempered steels contain the finer transformation products, such as lower bainite and martensite, while normalizing produces coarser aggregates of proeutectoid ferrite and pearlite.) One beneficial effect of grain refinement is revealed by a reduction in the ductile–brittle transition temperature, as shown in Fig. 10.45. In addition to illustrating the beneficial effect of grain refinement on transition temperature, this figure reveals a shift to lower transition temperatures in the ''controlled'' as opposed to the ''standard'' specimens. Kapadia et al.[146] demonstrated that the

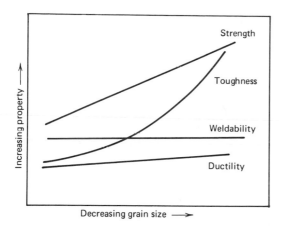

FIGURE 10.44 Simultaneous improvement in alloy strength and toughness with decreasing grain size. Ductility and weldability are not impaired. (Reprinted with permission from American Society of Agricultural Engineers.[145])

superior behavior of the controlled group of samples was attributable to enhanced delamination and associated stress relaxation in divider-type samples which resulted from a thermomechanical treatment designed to accentuate mechanical fibering. More recent data reveal improvements also in K_{IC} levels with reduced grain size. Mravic and Smith[147] reported that a 0.3C–0.9Mn–3.2Ni–1.8Cr–0.8Mo steel with a prior austenite ultrafine grain size of ASTM No. 15 (produced by multiple-cycle rapid austenitizing), exhibited K_{IC} values in the range of 100 to 110 MPa$\sqrt{\text{m}}$ at a strength level of 1930 to 2000 MPa. By comparison, for the same strength level, 4340 steel with a conventional grain size of about ASTM No. 7 exhibits a K_{IC} value of about 55 MPa$\sqrt{\text{m}}$.

To explain the beneficial role of structural refinement on toughness it may be argued that a microcrack will be stopped by an effective barrier (the grain boundary) more often the finer the grain size. As a result, the crack is forced to reinitiate repeatedly, and considerable energy is expended as it alters direction in search of the most likely propagation plane in the contiguous grain. Recall from Chapter 7 that this twisting of the crack front at the boundary gives rise to ''river patterns'' on cleavage fracture surfaces. One may argue, too, that fine-grained structures produce smaller potential flaws, thereby increasing the stress necessary for fracture (Eq. 8-6).

A number of investigators have attempted to describe the role of grain size in cleavage fracture for materials that undergo a temperature-sensitive fracture mechanism transition (see Chapter 9). Cottrell[26,148] and Petch[149] used dislocation theory to independently develop similar relations that could account for the effect temperature and various metallurgical factors have on the likelihood for cleavage failure. By using dislocation models and analyses analogous to those discussed previously with respect to Eqs. 4-8 to 4-11, they found that the fracture stress could be given by

$$\sigma_f \approx \frac{4G\gamma_m}{k_y} d^{-1/2} \qquad (10\text{-}5)$$

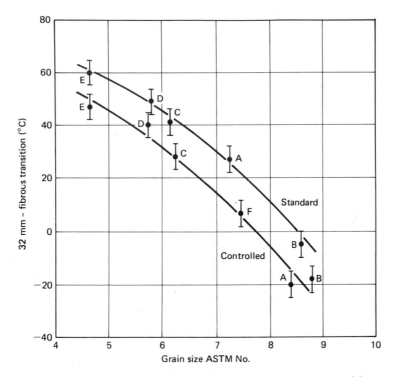

FIGURE 10.45 **Effect of grain refinement in reducing transition temperature in hot-rolled steel. "Controlled" specimens exhibited more delaminations in crack "divider" configuration to account for superior behavior.[146] (Copyright American Society for Metals, 1962.)**

where σ_f = fracture stress
        G = shear modulus
       γ_m = plastic work done around a crack as it moves through the crystal
       k_y = dislocation locking term from Hall-Petch relation (Eq. 4-7)
       d = grain size

The increase in σ_f with decreasing grain size parallels a similar increase in yield strength with grain refinement. The familiar Petch-Hall relation for yield strength is given by

$$\sigma_{ys} = \sigma_i + k_y d^{-1/2} \tag{4-7}$$

where σ_{ys} = yield strength
       σ_i = lattice resistance to dislocation movement resulting from various strengthening mechanisms and intrinsic lattice friction (Peierls stress)
       k_y = dislocation locking term
       d = grain size

As seen in Fig. 10.46, Low[150] demonstrated σ_f to be more sensitive to grain size than the associated yield strength σ_{ys}. There are some important implications to be drawn

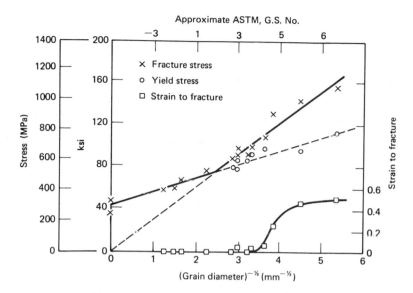

FIGURE 10.46 Yield and fracture strength and fracture strain dependence on grain size in low-carbon steel at −196°C.[150] (Reprinted from *Relation of Properties to Microstructure*, copyright American Society for Metals, 1954.)

from these data. First, the intersection of the yield-strength and fracture-strength curves represents a transition in material response. For large grains (greater than the critical size), failure must await the onset of plastic flow; hence, fracture occurs when $\sigma = \sigma_{ys} = \sigma_f$. For grains smaller than the critical size, yielding occurs first and is followed by eventual failure after a certain amount of plastic flow—the amount increasing with decreasing grain size. The latter situation reflects greater toughness with an increasing ratio σ_f/σ_{ys}. Since σ_f and σ_{ys} are temperature-sensitive properties, the critical grain size for the fracture transition would be expected to vary with test temperature. Consequently, the transition temperature is shown to decrease strongly with decreasing grain size (Fig. 10.47). As such, grain refinement serves to increase yield strength (Eq. 4-7) and fracture strength (Eq. 10-5), while lowering the ductile–brittle transition temperature (Fig. 10.47).

The significance of the terms in Eq. 10-5 has been treated at greater length by Tetelman and McEvily.[151] They argue that γ_m should increase with an increasing number of unpinned dislocation sources, temperature, and decreasing crack velocity. Obviously, the more dislocations that can be generated near the crack tip the more blunting can take place and the tougher the material will be. However, when these sources are pinned by solute interstitials, such as nitrogen and carbon in the case of steel alloys, or are highly immobile, as in ionic or covalent materials, because of a high Peierls stress, γ_m and σ_f are reduced. The beneficial effect of increased test temperature may be traced to a reduction in the Peierls stress and an increase in dislocation velocity. As we saw in Chapter 4, dislocation velocity was found to depend on the applied shear stress.

$$v = \left(\frac{\tau}{D}\right)^m \tag{4-15}$$

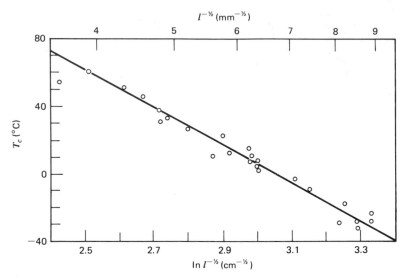

FIGURE 10.47 **Dependence of transition temperature on grain size.**[149] **(Reprinted with permission from MIT Press.)**

where v = dislocation velocity
               τ = applied shear stress
        D,m = material properties

With increasing temperature, D decreases so that for the same applied stress, dislocation velocity will increase, thereby enabling dislocations to move more rapidly to blunt the crack tip. In related fashion, a slower crack velocity will provide more time for dislocations to glide to the crack tip to produce blunting. In short, anything that enhances the number of mobile dislocations, their mobility and speed, and the time allowed for such movement will increase γ_m and σ_f and contribute to improved toughness. By comparing Eqs. 10-5 and 4-7, it should be noted that strengthening mechanisms (such as solid solution strengthening, precipitation hardening, dispersion hardening, and strain hardening) that restrict the number of free dislocations and their mobility contribute toward increasing σ_i, while at the same time reducing the magnitude of γ_m. Therefore, attempts to increase yield strength by increasing σ_i are counterproductive, since γ_m and σ_f decrease. Likewise, k_y can be adjusted to improve σ_f or σ_{ys} but only at the expense of the other. Enhanced dislocation locking will increase σ_{ys} but will decrease σ_f directly (Eq. 10-5) and indirectly (since the number of mobile dislocations and γ_m decrease.). In this regard, the more brittle nature of nitrogen-bearing steel is attributed to its stronger dislocation locking character.

We may summarize this discussion by stating that the only way to improve σ_{ys}, σ_f, and toughness simultaneously is not by changing γ_m, σ_i, or k_y, but rather by decreasing grain size. By using the transition temperature as a measure of toughness (higher toughness corresponding to a lower transition temperature) the diagrams in Fig. 10.48 illustrate that only by grain refinement can you have both high yield strength and toughness. Equations 10-5 and 4-7 are instructive in identifying those parameters

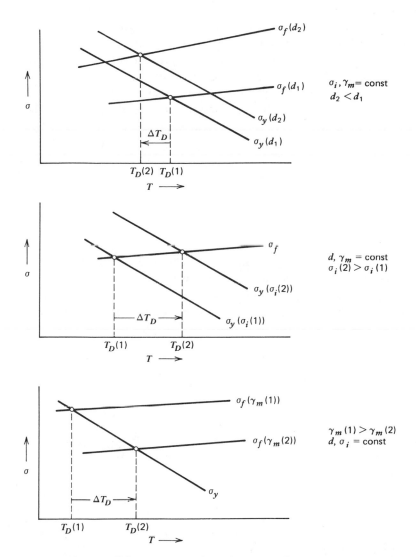

FIGURE 10.48 Diagrams showing effect of $\gamma_m', \sigma_{y'}$ and grain size on $\sigma_{ys'}\sigma_{f'}$ and transition temperature. Only grain refinement produces simultaneous increase in $\sigma_{ys'}\sigma_{f'}$ and reduction in transition temperature. (Reprinted with permission from John Wiley & Sons, Inc.)

that affect the temperature of the fracture mechanism transition (from void coalescence to cleavage). However, the reader should recognize that these relations are not applicable for materials that do not cleave but which, however, may undergo a stress-state-controlled fracture energy transition (Chapter 9).

10.5.1 Other Fracture Models

Since K_{IC} testing is fairly complex and expensive by most standards, it would be valuable to be able to determine K_{IC} on the basis of more readily obtained mechanical properties, such as those associated with a tensile test. Some progress in this regard has been achieved, with failure theories being proposed to account for cracking (e.g., cleavage or interfacial failure) and rupture (e.g., microvoid coalescence) processes.[152–162] For the case of slip-initiated transgranular cleavage, K_{IC} conditions are believed to conform to the attainment of a critical fracture stress σ_f^* over a characteristic distance l_0^* that is related to some microstructural feature.[152–156] The model proposed by Ritchie et al.[155] (Fig. 10.49a) characterizes cleavage fracture by

$$K_{IC} \propto \left[\frac{(\sigma_f^*)^{(1 + n)/2}}{\sigma_{ys}^{(1 - n)/2}} \right] (l_0^*)^{1/2} \tag{10-6}$$

where n = strain-hardening exponent
$\quad\quad\quad \sigma_{ys}$ = yield strength

For the case of low-carbon steels that possess ferrite–pearlite microstructures, l_0^* is found to be equal to several grain diameters and conforms to the spacing between grain boundary carbide particles. Also note that K_{IC} increases with increasing strain-hardening coefficient. Hahn and Rosenfield[163] also noted a direct dependence of K_{IC} on n with

$$K_{IC} \approx n\sqrt{2E\sigma_{ys}\epsilon_f/3} \tag{10-7}$$

where E = elastic modulus
$\quad\quad\quad \epsilon_f$ = true fracture strain in uniaxial tension

Knott has proposed a simpler form of Eq. 10-7 with $K_{IC} = \sqrt{2\pi l_0^*}$; this relation has been examined for different steel microstructures and chemistries.[154]

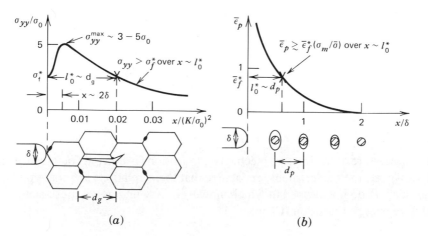

(a)   (b)

FIGURE 10.49 Schematic representations of theoretical models for (a) critical stress-controlled model for cleavage fracture and (b) stress-modified critical strain-controlled model for microvoid coalescence.[153] (Reprinted with permission from R. O. Ritchie and A. W. Thompson, *Metallurgical Transactions*, **16A**, 233 (1985).)

For conditions involving ductile fracture, a stress-modified critical strain criterion has been formulated[156,158,160,161] with fracture defined when a critical strain ϵ_f^* exists over a critical distance l_0^* (Fig. 10.49b):

$$K_{IC} \propto \sqrt{E\sigma_{ys}\epsilon_f^* l_0^*}$$ (10-8)

where ϵ_f^* = critical fracture strain
    l_0^* = multiple of mean distance between microvoid-producing particles

Note that Eqs. 10-7 and 10-8 possess several similar components and that ϵ_f^* increases with the strain-hardening exponent. Chen and Knott[162] also determined for the case of several aluminum alloys that the critical COD value increased with ϵ_f^* and the square of the strain-hardening exponent. Other treatments of ductile fracture, involving microvoid coalescence are discussed in the review by Van Stone et al.[159] The direct dependence of K_{IC} on $\sqrt{\sigma_{ys}}$ (Eqs. 10-7 and 10-8) deserves additional comment since empirical studies (e.g., recall Figs. 8.22 and 10.1) have generally noted an *inverse* dependency of K_{IC} on σ_{ys}. However, as discussed in Section 8.10, increase in σ_{ys} are usually associated with proportionately greater decreases in critical crack opening displacement and critical fracture strain levels.

The models discussed above possess some merit, but it is obvious that further work is necessary before one is able to determine K_{IC} from more readily obtained mechanical properties.

Other attempts have been made to develop K_{IC} data based on certain fractographic measurements in conjunction with the fracture-toughness model that includes crack-opening displacement considerations. From Fig. 10.50, the sharp crack produced by

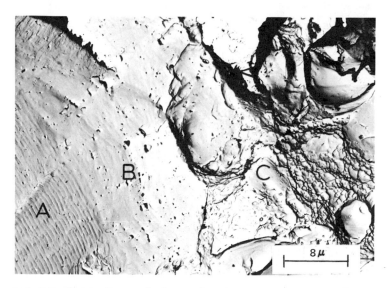

FIGURE 10.50   Precracked sample subsequently loaded to fracture. Region A represents fatigue precracking zone (see Section 13.3); Region B is the "stretch zone" representing crack blunting prior to crack instability; Region C is overload fracture region revealing microvoid coalescence.

cyclic loading (each line represents one load cycle [Region A]) is blunted by a stretching open process (Region B) prior to crack instability where microvoid coalescence occurs (Region C). Some investigators[21,164] have postulated that the width of the "stretch zone" reflects the amount of crack-opening displacement, but Broek[165] claims the depth of this zone to be the more relevant dimension. More recently, Kobayashi and coworkers[166] showed that the stretch zone width (SZW) at instability was related to J_{IC}. They also showed that at J values less than J_{IC} the SZW in several steel, aluminum, and titanium alloys varied with the quantity J/E (Fig. 10.51) in a relation of the form (see also Section 13.4.1)

$$SZW = 95\ J/E \tag{10-9}$$

In a recent study, Klassen et al.[167] characterized the effect of alloying elements on the fracture toughness in high-strength, low-alloy steels and determined J_{IC} values on the basis of stretch zone measurements. Since $J \propto \sigma_{flow} \cdot V(c)$, where SZW is proportional to $V(c)$, they determined that $J_{IC} \propto \sigma_{flow} \cdot SZW$. As expected from our previous discussions, they reported that J_{IC} increased with a decreasing grain size, and volume fraction and size of inclusions. Since stretch zone measurements are difficult to make and vary widely along the crack front, their use in the determination of K_{IC} and J_{IC} should be approached with caution.

10.6 METALLURGICAL EMBRITTLEMENT

Attention will now be focused on several undesirable circumstances that lead to serious loss of a material's fracture toughness. As will be shown, these changes can be brought

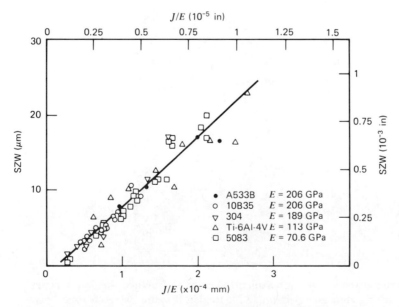

FIGURE 10.51 Stretch zone width (SZW) versus J/E for various steel, aluminum, and titanium alloys. (From Kobayashi et al.[166] with permission from University of Tokyo Press.)

about by alterations in microstructure and/or solute redistribution as produced by improper heat treatment or prolonged exposure to neutron irradiation. For a comprehensive study of this subject, see the review by Briant and Banerji.[168] Regarding the matter of improper heat treatment, two forms of temper embrittlement have been defined (Fig. 10.52). In the first case, high-strength martensitic steels may be embrittled following a short-time temper at low temperatures (in the range of 300 to 350°C). The "350°C embrittlement" is also referred to as "tempered martensite embrittlement" with a single tempering treatment being sufficient to induce embrittlement. The second type of embrittlement is found in lower strength steels and is brought about by a two-step heat treatment, as shown in Fig. 10.52, or by slowly cooling from the initial tempering temperature and through the embrittling temperature range (around 500°C). Note that room temperature embrittlement in the one-step embrittled condition is found when the material is tempered at about 350°C. The fracture energies of samples embrittled by the two-step procedure are shifted to higher test temperatures relative to that of the unembrittled material (i.e., higher ductile–brittle transition temperature).

10.6.1   300 to 350°C or Tempered Martensite Embrittlement

Metallurgists have long recognized the potentially embrittling effects of tempering martensitic steels at about 300 to 350°C. Evidence for embrittlement has been found

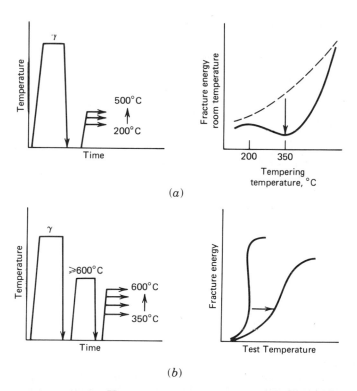

FIGURE 10.52 Heat treatments and associated fracture energies of temper embrittled steels. (a) Tempered martensite embrittlement, and (b) temper embrittlement. (From Briant and Banerji;[168] with permission.)

in this tempering temperature range by noting decreases in notched impact energy, ductility, and tensile strength[169] and a reduction in smooth bar tensile properties when unnotched samples are tested at subzero temperatures[170] (Fig. 10.53). Although precise models to account for all aspects of 300°C embrittlement have not been formulated as yet, certain facts are known. First, the embrittled condition coincides with the onset of cementite precipitation; second, segregation along grain boundaries of impurity elements such as phosphorus (P), sulfur (S), nitrogen (N), antimony (Sb), and tin (Sn) is essential for embrittlement to occur. For example, Capus and Mayer[171] observed no embrittlement trough in high-purity 1.5Ni–Cr–Mo steel when tempered at 300°C, whereas the commercial counterpart of this alloy was embrittled at the same tempering condition (Fig. 10.54). Similar results have been confirmed by Banerji and coworkers

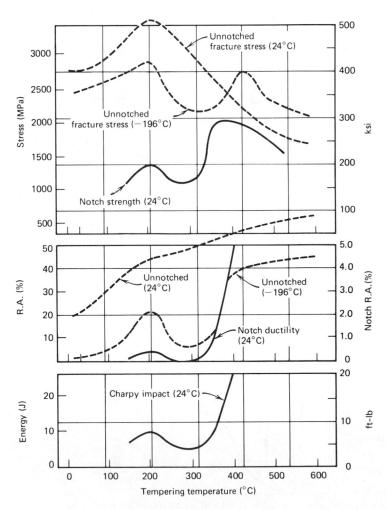

FIGURE 10.53 Notched and unnotched tensile properties at room and low temperatures for SAE 1340 steel, quenched and tempered at various temperatures. Poor properties associated with tempering in range of 300°C.[169] (Reprinted with permission of the American Society for Metals.)

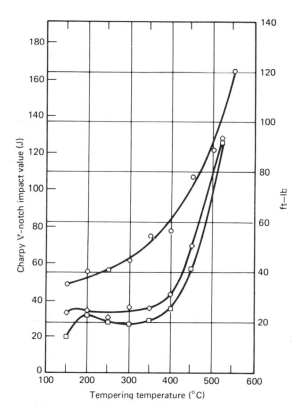

FIGURE 10.54 **Influence of phosphorus and antimony on room temperature impact energy as function of tempering temperature in 1½% Ni–Cr–Mo steel.[171] (From J. M. Capus and G. Mayer, *Metallurgia* 62 (1960); with permission of Industrial Newspapers Ltd.)**

for the case of commercial and high-purity heats of 4340 steel.[172] Tempered martensite embrittlement (TME) represents a problem of intergranular embrittlement brought about as a result of precipitation of carbides along prior austenite grain boundaries that had been embrittled by the segregation of P and S during prior austenization. Indeed, Bandyopadhyay and McMahon[173] observed that the impact energy minima corresponded to a maximum in the amount of intergranular fracture observed. The extent of such P and S segregation is enhanced by the presence of manganese (Mn) and silicon (Si) in the alloy; increased Mn or Si levels were found to increase the fraction of austenite grain boundaries that became embrittled as a result of P or S segregation.[174] Conversely, elimination of Mn and Si from a high-purity NiCrMo alloy restricted impurity segregation at prior austenite grain boundaries, thereby eliminating most of the material's susceptibility to TME.[173] In this context, there exists a basic similarity between tempered martensite embrittlement and temper embrittlement, which will be discussed shortly.

Short of preparing high-purity (but expensive) alloys, the most obvious way to avoid 300°C embrittlement is simply to avoid tempering at that temperature. Usually this involves tempering at a higher temperature but with some sacrifice in strength. However, there are material applications that arise where the higher strengths associated with tempering at 300°C are desired. Fortunately, it has been found possible to obtain the strength levels associated with a 300°C temper while simultaneously suppressing the embrittling kinetics. This optimization of properties has been achieved through the addition of 1.5 to 2% silicon to the alloy steel.[175] Surely, the presence of Si promotes the segregation of P and S to grain boundaries; however, when present in greater amounts (1.5 to 2%), it is believed that silicon suppresses the kinetics of the martensite tempering process with the result that the embrittling reaction shifts to a higher tempering temperature (about 400°C).

10.6.2 Temper Embrittlement

Temper embrittlement (TE) develops in alloy steels when cooled slowly or isothermally heated in the temperature range of 400 to 600°C. The major consequence of TE is found to be an increase in the tough–brittle transition temperature and is associated with intercrystalline failure along prior austenite grain boundaries. Using the change in transition temperature as the measure of TE, the kinetics of the embrittlement process are found to exhibit a C-curve response, with isoembrittlement lines depicting maximum embrittlement in the shortest hold time at intermediate temperatures in the 400 to 600°C range (Fig. 10.55). It is important to note that TE can be largely reversed by reheating the steel above 600°C. Although TE has been recognized for over 85 years (e.g., see the reviews by Holloman,[176] Woodfine,[177] Low,[178] Mc-Mahon,[179] and Briant and Banerji[168]), it is by no means under control. For example, the catastrophic failure in 1969 of two forged alloy steel disks from the Hinkley Point nuclear power station steam turbine rotor offers dramatic proof that additional understanding of the TE process is needed.[180] In this instance, failure was attributed to a combination of two factors: TE resulting from slowly cooling the disks during manufacture through the critical temperature range, and environment-assisted cracking resulting from the entrapment of condensate in the keyways of the disks.

Balajiva and co-workers[181,182] contributed much to our current understanding of temper embrittlement. They demonstrated that TE occurred only in alloy steels of commercial purity but not in comparable alloys of high purity (Fig. 10.56). The most potent embrittling elements were found to be antimony, phosphorus, tin, and arsenic.

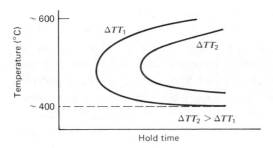

FIGURE 10.55 **Isoembrittlement lines (fixed shift in tough–brittle transition temperature) as function of exposure temperature and hold time.**

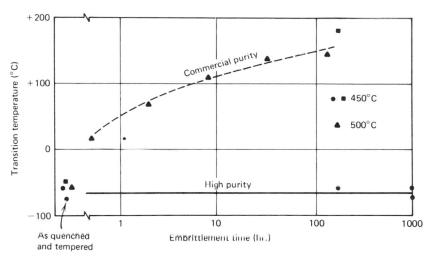

FIGURE 10.56 **Effect of 450 and 500°C exposure on temper embrittlement in commercial and high-purity nickel–chromium steels. (Data from Woodfine and Steven et al.[182,183] Reprinted by permission of the American Society for Testing and Materials from copyrighted work.)**

These results have been verified by others,[184] along with the additional finding that for a given impurity level, Ni–Cr-alloy steels are embrittled more than alloys containing nickel or chromium alone. It has generally been thought that embrittlement resulted from the segregation of impurity elements at prior austenite grain boundaries as a result of exposure to the 400 to 600°C temperature range. This has since been verified using Auger electron spectroscopy[185–187]—a technique by which the chemistry of the first few atomic layers of a material's surface is analyzed. Marcus and Palmberg[185,186] found, in a modified AISI 3340 steel alloy, antimony on the fracture surface (along prior austenite grain boundaries) in amounts exceeding 100 times that of the bulk concentration (0.03 a/o). Furthermore, the high antimony concentration layer was very shallow, extending only one to two atomic layers below the fracture surface.

The severity of embrittlement depends not only on the amount of poisonous elements present such as Sb, Sn, and P (Fig. 10.57), but also on the overall composition of the alloy. Regarding the latter, certain alloying elements may either enhance or suppress grain–boundary segregation of the embrittling species. The respective influence of alloying elements such as chromium (Cr), manganese (Mn), nickel (Ni), titanium (Ti), and molybdenum (Mo) on the temper embrittlement process is reviewed by Briant and Banerji.[168] In another series of studies, the extent of TE in steels was reported to be reduced through the addition of lanthanide metal. In this instance, lanthanides served as scavengers by forming harmless compounds in the matrix with such embrittling impurities as P, As, Sn, and Sb;[189] as a result, the impurity content within the matrix was depleted along with the tendency for TE. (Also see Y. Jingsheng, et al., *J. Metals, 40* (5), 26 [1988].)

The actual mechanism for temper embrittlement remains unclear, though some potentially valuable models are being developed. For example, McMahon and co-workers[172,190–192] have argued that segregation of impurities such as Sb reduces the

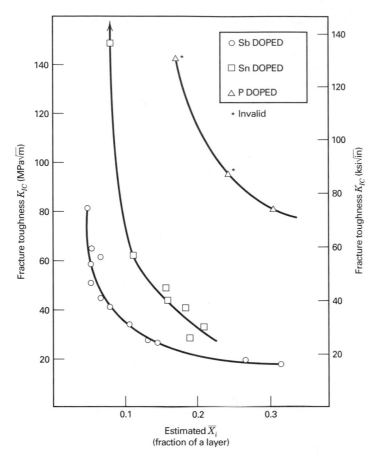

FIGURE 10.57   **Fracture toughness of a Ni–Cr steel doped with antimony (Sb), tin (Sn), and phosphorus (P) versus the average intergranular concentrations $\overline{X}_i$ of each embrittling species. (From J. Kameda;[188] with permission from the Metallurgical Society of AIME.)**

cohesive energy of the grain boundary which, in turn, lowers the local stress necessary to generate an accelerating microcrack. Proceeding further, the lower stress necessary for fracture brings about a sharp drop in the plastic strain rate (i.e., dislocation activity) and associated plastic work term, since the plastic strain rate depends *exponentially* on the applied stress level (recall Eqs. 4-15 and 4-16). McMahon and Vitec[191] concluded for the case of temper embrittled ferritic steels that the relative decrease in plastic strain rate (hence, plastic work) is an order of magnitude larger than the reduction in intergranular cohesive energy. Consequently, even though the ideal cohesive energy term for intergranular fracture represents a small component of the total energy for fracture (recall Eq. 8-8a), the cohesive energy term possesses a disproportionately large influence on the material's fracture toughness through its influence on the plastic work term. On this basis, it is to be expected that temper embrittlement will be controlled by the maximum grain-boundary impurity concentration found in the highly stressed volume of material located near the notch root (Fig. 10.57). This postulate has been verified for the case of Sb-doped Ni–Cr steel.[193]

10.6.3 Neutron-Irradiation Embrittlement

Proper selection of alloy chemistry and heat treatment may provide a material with adequate toughness at the outset of service life for nuclear applications. There exist, however, the potential for mechanical property degradation after exposure to neutron irradiation. For example, we see from Fig. 10.58 a sharp reduction in fracture toughness after neutron irradiation, especially in the region where K_{IC} values normally rise rapidly with test temperature.[194] As a result, the initial K_{IC} level of 200 MPa $\sqrt{m}$ anticipated for this material at room temperature is reduced drastically to about 45 MPa $\sqrt{m}$, representing a 20-fold *decrease* in the allowable flaw size for a given applied stress. As one might expect, considerable attention has been devoted to determine the extent of embrittlement in a number of steels being used or considered for use in reactor pressure vessels. In addition, utilities, reactor manufacturers, and nuclear regulatory agencies throughout the world closely monitor the rate of deterioration in toughness of the containment vessel of the reactor. Of considerable note, it

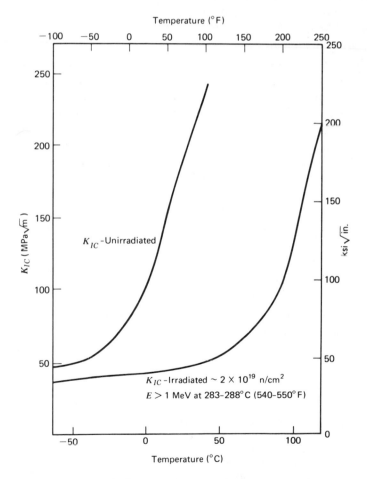

FIGURE 10.58 **Influence of neutron irradiation on fracture toughness as a function of test temperature in A533-B Class I steel.[194] (Reprinted by permission of the American Society for Testing and Materials from copyrighted work.)**

was reported recently that the rate of neutron irradiation embrittlement in 13 different reactors in the United States was occurring "so rapidly . . . that some of the plants may become unsafe to operate by the end of the next year [1982]."[195] In each instance, the service life of the containment vessel was not more than 25% of its design life.

Earlier studies[196,197] had focused on documenting changes in tensile properties (for example, strength and ductility) and Charpy impact energy absorption. To a large extent, these test procedures have continued to the present, along with fracture mechanics studies. For the proceedings of three recent conferences on radiation damage, see references 198 to 200. Researchers have found that neutron irradiation raises the yield strength (mainly as a result of an increase in the lattice friction component σ_i in the Petch-Hall relation) and the tough–brittle transition temperature. At the same time, there is a corresponding decrease in tensile ductility and Charpy impact shelf energy. A diagram showing irradiation-induced changes in Charpy impact response is given in Fig. 10.59, where higher fluence* levels are seen to cause greater embrittlement. Although the cause of this embrittlement is not clearly understood, it is believed to be related to the interaction of dislocations with defect aggregates, such as solute atom-vacancy clusters that are generated by neutron bombardment. As one might expect, the greater the fluence, the greater the number of defect aggregates and the greater the elevation in yield strength and transition temperature.

Studies have shown that the extent of irradiation damage depends strongly on the irradiation temperature, with more damage accompanying low-temperature neutron exposures.[201] It is of some comfort to note that actual service temperatures in current nuclear reactors are in the range of 260 to 288°C (500 to 550°F) where damage is less extensive. Preliminary test results at 307°C show further reductions in the degree of embrittlement. It may be argued that since the defect clusters are probably being annealed out more rapidly at higher temperatures, it would be reasonable to expect minimal irradiation damage at even higher irradiation temperatures, where defect annihilation would be further enhanced.[201] As a matter of fact, neutron irradiation embrittlement in a component can be reversed by annealing the damaged part at a sufficiently high temperature (above the irradiation temperature) to annihilate the defect clusters (e.g., copper-vacancy aggregates).[202,203]

The amount of neutron embrittlement resulting from 288°C irradiation is found to depend strongly on the steel alloy content. For example, Hawthorne et al.[204,205] demonstrated that neutron embrittlement resulting from 288°C exposure could be eliminated completely by careful reduction in residual element content. Although the presence of phosphorus, sulfur, and vanadium in solid solution was identified as being objectionable, *copper* was singled out as the most harmful element. The combined

---

* The rate of neutron irradiation or neutron flux is defined as the number of neutrons crossing a one square centimeter area in one second. The neutron fluence is defined by the product of neutron flux and time. Hence, neutron fluence is given by the number of neutrons/cm². Studies have shown that these neutrons possess a spectrum of widely varying energies, with some capable of much greater damage than others. Since we do not know the amount of damage generated by a neutron with a particular energy level, it has become the interim accepted practice to define neutron fluence as the count of neutrons possessing more than a certain minimum energy.[194] This energy level is usually taken to be 1.6×10^{-13}J (1 MeV). This counting procedure, therefore, assumes that no damage results from neutrons with energies less than this threshold level (since they are not included in the neutron count) and that all neutron energies greater than this value produce the same damage. Certainly, the current neutron counting procedure is not very discriminating but does provide some means for quantitative analysis of damage.

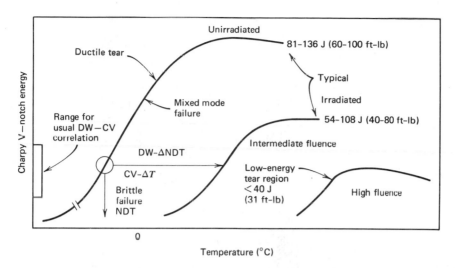

FIGURE 10.59 **Diagram showing transition temperature shift resulting from neutron irradiation of typical reactor-vessel steels.[194] (Reprinted by permission of the American Society for Testing Materials from copyrighted work.)**

effect of irradiation temperature and copper content on the increment in yield strength resulting from a given neutron fluence is shown in Fig. 10.60. Note that low copper levels (<0.03%) or high irradiation temperatures (288°C) cause the smallest increases in yield strength, while low-temperature (232°C) irradiation produces the greatest damage, regardless of the residual element level.

The nature of neutron irradiation is further complicated by size effects other than those associated with stress-state considerations. For one thing, since reactor vessels are fabricated from very thick plate (on the order of 20 cm), through thickness variations in microstructure and associated mechanical properties are expected, with minimum properties anticipated in the midthickness region. Superimposed on this gradient of K_{IC} (unirradiated) values is a separate gradient of irradiation damage that decreases continuously from the inner core surface in association with an attenuation in neutron fluence through the thickness. For example, Loss et al.[206] reported a 20-fold decrease in neutron fluence from 6×10^{19} to 0.3×10^{19} neutrons/cm^2 across a 20-cm-thick pressure vessel wall. From Fig. 10.60 it is seen that the degree of neutron embrittlement decreases markedly with this attenuation in neutron fluence, especially for steels with high copper content.

Although the previous discussion focused on the embrittlement of reactor vessel low-alloy steels, it should be noted that neutron exposure in the temperature range $0.30 < T_h < 0.55$ also causes irradiation damage in stainless steels and other alloys used to contain the nuclear fuel. It is believed that hydrogen and helium—produced by nuclear transmutations—segregate to vacancy clusters and stabilize internal voids.[194] (Neutron irradiation produces both interstitials and vacancies, but there is preferential recombination of interstitials, leaving an excess of vacancies in the lattice.) Since the excess vacancy and gaseous element concentration increases with increasing neutron fluence, it is found that the relative volume change also increases (Fig. 10.61). The particular temperature regime $0.3 < T_h < 0.55$, where void-induced swelling is

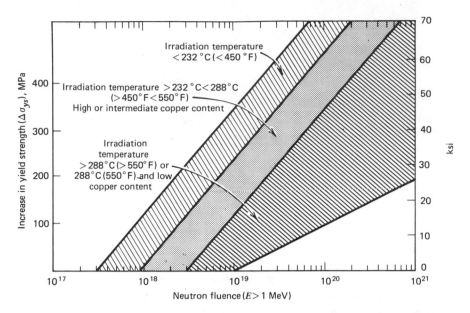

FIGURE 10.60 Effect of copper content and irradiation temperature on neutron-induced increase in room temperature yield strength of A302-B and A533-B steels.[194] (Reprinted by permission of the American Society for Testing and Materials from copyrighted work.)

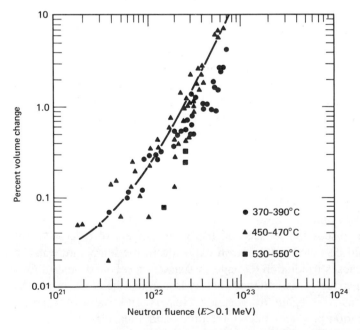

FIGURE 10.61 Effect of neutron fluence on swelling in annealed type 304 stainless steel in the 370 to 550°C temperature range.[194] (Reprinted by permission of the American Society for Testing and Materials from copyrighted work.)

most prevalent, arises from the fact that the kinetics of vacancy condensation are too slow at temperatures below $T_h = 0.3$; above about $T_h = 0.55$, the supersaturation of vacancies is inadequate to sustain the voids.[194] Similarly, it has been found that swelling can be suppressed by addition of refractory elements to the alloy, which attenuate vacancy mobility. In addition to swelling of the fuel-cladding alloy, it is known that the uranium fuel itself undergoes volumetric expansion resulting from the precipitation of krypton and xenon gas bubbles, which form as a result of the fission process.[196] Both swelling of the fuel and cladding are highly undesirable and lead to reductions in the operating life of the fuel and metal cladding.

10.7 ADDITIONAL DATA

Additional K_{IC} data are provided in Table 10.8 to enable the reader to become more familiar with the fracture properties of a number of commercial alloys, ceramics, and polymers. The relatively brittle response of ceramic solids, which was anticipated in Table 10.1, is verified convincingly by the tabulated results.

TABLE 10.8a   Strength and Fracture-Toughness Data for Selected Materials[222]*

Alloy	Material Supply	Specimen Orientation	Test Temperature (°C)	σ_{ys} (MPa)	K_{IC} (MPa$\sqrt{\text{m}}$)
Aluminum Alloys					
2014-T651	Plate	L-T	21–32	435–470	23–27
"	"	T-L	"	435–455	22–25
"	"	S-L	24	380	20
2014-T6	Forging	L-T	"	440	31
"	"	T-L	"	435	18–21
2020-T651	Plate	L-T	21–32	525–540	22–27
"	"	T-L	"	530–540	19
2024-T351	"	L-T	27–29	370–385	31–44
"	"	T-L	"	305–340	30–37
2024-T851	"	L-T	21–32	455	23–28
"	"	T-L	"	440–455	21–24
2124-T851	"	L-T	"	440–460	27–36
"	"	T-L	"	450–460	24–30
2219-T851	"	L-T	"	345–360	36–41
"	"	T-L	"	340–345	28–38
7049-T73	Forging	L-T	"	460–510	31–38
"	"	T-L	"	460–470	21–27
7050-T73651	Plate	L-T	"	460–510	33–41
"	"	T-L	"	450–510	29–38
"	"	S-L	"	430–440	25–28
7075-T651	Plate	L-T	21–32	515–560	27–31
"	"	T-L	"	510–530	25–28
"	"	S-L	"	460–485	16–21
7075-T7351	"	L-T	"	400–455	31–35
"	"	T-L	"	395–405	26–41
7475-T651	"	T-L	"	505–515	33–37
7475-T7351[a]	"	T-L	"	395–420	39–44
7079-T651	"	L-T	"	525–540	29–33
"	"	T-L	"	505–510	24–28

(*continued*)

TABLE 10.8a   (*continued*)

Alloy	Material Supply	Specimen Orientation	Test Temperature (°C)	σ_{ys} (MPa)	K_{IC} (MPa$\sqrt{m}$)
7178–T651	"	L-T	"	560	26–30
"	"	T-L	"	540–560	22–26
"	"	S-L	"	470	17
Ferrous Alloys					
4330V(275°C temper)	Forging	L-T	21	1400	86–94
4330V(425°C temper)	"	L-T	"	1315	103–110
4340(205°C temper)	"	L-T	"	1580–1660	44–66
4340(260°C temper)	Plate	L-T	"	1495–1640	50–63
4340(425°C temper)	Forging	L-T	"	1360–1455	79–91
D6AC(540°C temper)	Plate	L-T	"	1495	102
D6AC(540°C temper)	"	L-T	−54	1570	62
9-4-20(550°C temper)	"	L-T	21	1280–1310	132–154
18 Ni(200)(480°C 6 hr)	Plate	L-T	21	1450	110
18 Ni(250)(480°C 6 hr)	"	L-T	"	1785	88–97
18 Ni(300)(480°C)	"	L-T	"	1905	50–64
18 Ni(300)(480°C 6 hr)	Forging	L-T	"	1930	83–105
AFC77 (425°C temper)	"	L-T	24	1530	79
Titanium Alloys					
Ti6 Al-4V	(Mill anneal plate)	L-T	23	875	123
"	"	T-L	"	820	106
"	(Recryst. anneal plate)	L-T	22	815–835	85–107
"	"	T-L	22	825	77–116
Ceramics[c]					
Mortar[207]	—	—	—		0.13–1.3
Concrete[208]	—	—	—		2–2.3
Al_2O_3[209–211]	—	—	—		3–5.3
SiC[209]	—	—	—		3.4
SiN_4[210]	—	—	—		4.2–5.2
Soda lime silicate glass[210]	—	—	—		0.7–0.8
Electrical porcelain ceramics[212]	—	—	—		1.03–1.25
WC(2.5–3 μm)—3 w/o Co[213]	—	—	—		10.6
WC(2.5–3 μm)—9 w/o Co[213]	—	—	—		12.8
WC(2.5–3.3 μm)—15 w/o Co[213,214]	—	—	—		16.5–18
Indiana limestone[215]	—	—	—		0.99
ZrO_2 (Ca stabilized)[216]	—	—	—		7.6
ZrO_2[216]	—	—	—		6.9
Al_2O_3/SiC whiskers[217]	—	—	—		8.7
SiC/SiC fibers[217]	—	—	—		25
Borosilicate glass/SiC fibers[217]	—	—	—		18.9
Polymers[d]					
PMMA[218]	—	—	—		0.8–1.75[b]
PS[219]	—	—	—		0.8–1.1[b]
Polycarbonate[220]	—	—	—		2.75–3.3[b]

* For additional fracture toughness data for metallic alloys, see C. M. Hudson and S. K. Seward, *Int. J. Fract.* **14**, R151 (1978); **20**, R59 (1982); **39**, R43 (1989). Also, C. M. Hudson and J. J. Ferrainolo, *Int. J. Fract.* **48**, R19 (1991).

[a] Special processing.

[b] K_{IC} is f (crack speed).

[c] For additional K_{IC} data, see reference 221 and Fig. 10.38.

[d] For additional K_{IC} data, see references 139 and 142 and Table 10.7.

TABLE 10.8b Strength and Fracture-Toughness Data for Selected Materials[222]

Alloy	Material Supply	Specimen Orientation	Test Temperature (°F)	σ_{ys} (ksi)	K_{IC} (ksi$\sqrt{\text{in.}}$)
Aluminum Alloys					
2014-T651	Plate	L-T	70–89	63–68	21–24
"	"	T-L	"	63–66	20–22
"	"	S-L	75	55	18
2014-T6	Forging	L-T	"	64	28
"	"	T-L	"	63	16–19
2020-T651	Plate	L-T	70–89	76–78	20–25
"	"	T-L	"	77–78	17–18
2024-T351	"	L-T	80–85	54–56	28–40
"	"	T-L	"	44–49	27–34
2024-T851	"	L-T	70–89	66	21–26
"	"	T-L	"	64–66	19–21
2124-T851	"	L-T	"	64–67	25–33
"	"	T-L	"	65–67	22–27
2219-T851	"	L-T	"	50–52	33–37
"	"	T-L	"	49–50	26–34
7049-T73	Forging	L-T	"	67–74	28–34
"	"	T-L	"	67–68	19–25
7050-T73651	Plate	L-T	"	67–74	30–37
"	"	T-L	"	65–74	26–35
"	"	S-L	"	62–64	22–26
7075-T651	Plate	L-T	70–89	75–81	25–28
"	"	T-L	"	74–77	23–26
"	"	S-L	"	67–70	15–19
7075-T7351	"	L-T	"	58–66	28–32
"	"	T-L	"	57–59	24–37
7475-T651	"	T-L	"	73–75	30–33
7475-T7351[a]	"	T-L	"	57–61	35–40
7079-T651	"	L-T	"	76–78	26–30
"	"	T-L	"	73–74	22–25
7178-T651	"	L-T	"	81	23–27
"	"	T-L	"	78–81	20–23
"	"	S-L	"	68	15
Ferrous Alloys					
4330V(525°F temper)	Forging	L-T	70	203	78–85
4330V(800°F temper)	"	L-T	"	191	94–100
4340(400°F temper)	"	L-T	"	229–241	40–60
4340(500°F temper)	Plate	L-T	"	217–238	45–57
4340(800°F temper)	Forging	L-T	"	197–211	72–83
D6AC(1000°F temper)	Plate	L-T	"	217	93
D6AC(1000°F temper)	"	L-T	−65	228	56
9-4-20(1025°F temper)	"	L-T	70	186–190	120–140
18 Ni(200)(900°F 6 (hr)	"	L-T	70	210	100
18 Ni(250)(900°F 6 hr)	"	L-T	"	259	80–88
18 Ni(300)(900°F)	"	L-T	"	276	45–58
18 Ni(300)(900°F 6 hr)	Forging	L-T	"	280	75–95
AFC77 (800°F temper)	"	L-T	75	222	72
Titanium Alloys					
Ti-6 Al-4V	(Mill anneal plate)	L-T	74	127	112
"	"	T-L	"	119	96
"	(Recryst. anneal plate)	L-T	72	118–121	77–97
"	"	T-L	72	120	70–105

(*continued*)

TABLE 10.8b *(continued)*

Alloy	Material Supply	Specimen Orientation	Test Temperature (°F)	σ_{ys} (ksi)	K_{IC} (ksi$\sqrt{\text{in.}}$)
Ceramics[c]					
Mortar[207]	—	—	—	1.8–2.1
Concrete[208]	—	—	—	0.21–1.30
Al_2O_3[209–211]	—	—	—	2.7–4.8
SiC[209]	—	—	—	3.1
Si_3N_4[210]	—	—	—	3.8–4.7
Soda lime silicate glass[210]	—	—	—	0.64–0.73
Electrical porcelain ceramics[212]	—	—	—	0.94–1.14
WC(2.5–3 μm)—3 w/o Co[213]	—	—	—	9.6
WC(2.5–3 μm)—9 w/o Co[213]	—	—	—	11.6
WC(2.5–3 μm)—15 w/o Co[213,214]	—	—	—	15–16.4
Indiana limestone[215]	—	—	—	0.9
ZrO_2 (Ca stabilized)[216]	—	—	—	6.9
ZrO_2[216]	—	—	—	6.3
Al_2O_3/SiC whiskers[216]	—	—	—	7.9
SiC/SiC fibers[217]	—	—	—	22.7
Borosilicate glass/SiC fibers[217]	—	—	—	17.2
Polymers[d]					
PMMA[218]	—	—	—	0.73–1.6[b]
PS[219]	—	—	—	0.73–1.0[b]
Polycarbonate[220]	—	—	—	2.5–3.0[b]

[a] Special processing.
[b] K_{IC} is f (crack speed).
[c] For additional K_{IC} data, see reference 221 and Fig. 10.38.
[d] For additional K_{IC} data, see references 139 and 142 and Table 10.7.

REFERENCES

1. R. A. Jaffe and G. T. Hahn, *Symposium on Design with Materials That Exhibit Brittle Behavior,* Vol. 1, MAB-175-M, National Materials Advisory Board, Washington, DC, Dec. 1960, p. 126.
2. M. J. Harrigan, *Met. Eng. Quart.,* (May 1974).
3. R. O. Ritchie, *Mechanical Behavior of Materials—5,* Proceedings, 5th International Conference, M. G. Yan, S. H. Zhang, Z. M. Zheng, eds., Pergamon, Oxford 1988.
4. F. F. Lange, *Philos. Mag.* **22,** 983 (1970).
5. A. G. Evans, *Philos. Mag.* **26,** 1327 (1972).
6. C. S. Smith, *A History of Metallography,* University of Chicago Press, Chicago, 1960, p. 40.
7. F. A. Heiser and R. W. Hertzberg, *J. Basic Eng.* **93,** 71 (1971).
8. J. T. Michalak, *Metals Handbook,* Vol. 8, ASM, Metals Park, OH, 1973, p. 220.
9. J. F. Peck and D. A. Thomas, *Trans Met. Soc. AIME,* 1240 (1961).
10. J. Cook and J. E. Gordon, *Proc. R. Soc. London,* **A282,** Ser. *A* 508 (1964).
11. J. D. Embury, N. J. Petch, A. E. Wraith, and E. S. Wright, *Trans. AIME* **239,** 114 (1967).

12. H. I. Leichter, *J. Spacecr. Rockets* **3**(7), 1113 (1966).
13. A. J. McEvily, Jr. and R. H. Bush, *Trans. ASM* **55**, 654 (1962).
14. J. G. Kaufman, *Trans. ASME, J. Basic Eng.* **89**(3), 503 (1967).
15. M. A. Meyers and C. T. Aimone, *Prog. Mater. Sci.* **28**, 1 (1983).
16. M. E. deMorton, R. L. Woodward, and J. M. Yellup, Proceedings, Fourth Tewksburg Symposium, Melbourne, Feb. 1979, p. 11.1.
17. J. G. Kaufman, P. E. Schilling, and F. G. Nelson, *Met. Eng. Quart.* **9**(3), 39 (1969).
18. W. S. Hieronymus, *Aviat. Week Space Tech.,* 42 (July 26, 1971).
19. B. I. Edelson and W. M. Baldwin, Jr., *Trans. ASM* **55**, 230 (1962).
20. J. M. Hodge, R. H. Frazier, and F. W. Boulger, *Trans. AIME* **215**, 745 (1959).
21. A. J. Birkle, R. P. Wei, and G. E. Pellissier, *Trans. ASM* **59**, 981 (1966)
22. L. Luyckx, J. R. Bell, A. McClean, and M. Korchynsky, *Met. Trans.* **1**, 3341 (1970).
23. R. W. Hertzberg and R. Goodenow, *Microalloying 1975,* Oct. 1975, Washington, DC.
24. J. C. M. Farrar, J. A. Charles, and R. Dolby, *Effect of Second Phase Particles on the Mechanical Properties of Steel,* Proc. Conf. 1971, ISI, 171 (1972).
25. C. J. McMahon, Jr., *Fundamental Phenomena in the Material Sciences,* Vol. 4, Plenum, New York, 1967, p. 247.
26. A. H. Cottrell, *Trans. Met. Soc. AIME* **212**, 192 (1958).
27. C. Zener, *Fracturing of Metals,* ASM, Cleveland, 1948, p. 3.
28. T. L. Johnston, R. J. Stokes, and C. H. Li, *Philos. Mag.* **7**, 23 (1962).
29. V. F. Zackay, E. R. Parker, J. W. Morris, Jr., and G. Thomas, *Mater. Sci. Eng.* **16**, 201 (1974).
30. D. E. Piper, W. E. Quist, and W. E. Anderson, *Application of Fracture Toughness Parameters to Structural Metals,* Vol. 31, Metallurgical Society Conference, 1966, p. 227.
31. R. E. Zinkham, H. Liebowitz, and D. Jones, *Mechanical Behavior of Materials,* Vol. 1, Proceedings of the International Conference on Mechanical Behavior of Materials, The Society of Materials Science, Kyoto, Japan, 1972, p. 370.
32. J. G. Kaufman, Agard Meeting of the Structures and Materials Panel, Apr. 15, 1975.
33. R. R. Senz and E. H. Spuhler, *Met. Prog.* **107**(3), 64 (1975).
34. T. Ohira and T. Kishi, *Mater. Sci. Eng.* **78**, 9 (1986).
35. D. Broek, *Eng. Fract. Mech.* **5**, 55 (1973).
36. F. A. McClintock, *Int. J. Fract. Mech.* **2**, 614 (1966).
37. H. E. McGannon, *The Making, Shaping, and Treating of Steel,* 9th ed., United States Steel Corporation, Pittsburgh, 1971.
38. W. C. Leslie, *Met. Trans.* **3**, 5 (1972).
39. J. R. Low, Jr., *Fracture,* Technology Press MIT and Wiley, New York, 1959, p. 68.
40. T. B. Cox and J. R. Low, Jr., *Met. Trans.* **5**, 1457 (1974).
41. D. Webster, *Trans. ASM* **61**, 816 (1968).
42. V. F. Zackay, E. R. Parker, D. Fahr, and R. Busch, *Trans. ASM* **60**, 252 (1967).
43. S. Antolovich, *Trans. Met. Soc. AIME* **242**, 237 (1968).
44. E. R. Parker and V. F. Zackay, *Eng. Fract. Mech.* **5**, 147 (1973).
45. J. P. Bressanelli and A. Moskowitz, *Trans. ASM* **59**, 223 (1968).
46. W. W. Gerberich, G. Thomas, E. R. Parker, and V. F. Zackay, *Proceedings of*

the Second International Conference on the Strength of Metals and Alloys, Asilomar, Calif, 1970, p. 894.

47. D. Bhandarkar, V. F. Zackay, and E. R. Parker, *Met. Trans.* **3,** 2619 (1972).
48. A. R. Rosenfield and A. J. McEvily, Jr., NATO AGARD Report No. 610, Dec. 1973, p. 23.
49. H. Margolin, J. C. Williams, J. C. Chestnutt, and G. Luetjering, *Titanium 80,* H. Kimura and O. Izumi, Eds., AIME, Warrendale, PA, 1980, p. 169.
50. R. Develay, *Met. Mater.,* **6,** 1972, p. 404.
51. J. A. Nock, Jr. and H. Y. Hunsicker, *J. Met.,* 216 (Mar. 1963).
52. D. S. Thompson, S. A. Levy, and D. K. Benson, *Third International Conference on Strength of Metals and Alloys,* Paper 24, 1973, p. 119.
53. N. E. Paton and A. W. Sommer, *Third International Conference on Strength of Metals and Alloys,* Paper 21, 1973, p. 101.
54. J. T. Staley, AIME Spring Meeting, Alcoa Report, May 23, 1974.
55. R. J. Bucci, *Eng. Fract. Mech.* **12,** 407 (1979).
56. E. A. Starke, T. H. Sanders, Jr., and I. G. Palmer, *J. Metals* **33,** 24 (1981).
57. B. Nobel, S. J. Harris, and K. Dinsdale, *Metal Science* **16,** 425 (1982).
58. T. H. Sanders, Jr. NAVAIR Contract No. N62269-76-C-0271, Final Report, June 9, 1979.
59. A. K. Vasudevan, A. C. Miller, and M. M. Kersker, *Proceedings of the Second International Al-Li Conference,* E. A. Starke, Jr. and T. H. Sanders, Jr., Eds., AIME, 181 (1983).
60. D. Webster, *Met Trans.* **18A,** 2181 (1987).
61. D. Webster, *Advanced Materials and Processes* **145**(5), 18 (1994).
62. A. G. Evans, *Ceramic Transactions* Vol 1, *Ceram, Powder Sci.,* G. L. Messing, E. R. Fuller, Jr., and H. Hauser, eds., American Ceramic Society, Westerville, OH, 989 (1989).
63. R. F. Cook, B. R. Lawn, and C. J. Fairbanks, *J. Amer. Ceram. Soc.* **68**(11), 604 (1985).
64. B. Lawn, *Fracture of Brittle Solids,* 2nd ed., Cambridge Solid State Science Series, Cambridge University Press, Cambridge, UK (1993).
65. P. F. Becher, *J Amer. Ceram. Soc.* **74**(2), 255 (1991).
66. A. G. Evans, *J Amer. Ceram. Soc.* **73**(2), 187 (1990).
67. S. Suresh and C. F. Shih, *Int. J. Fracture* **30,** 237 (1986).
68. K. T. Faber and A. G. Evans, *Acta Metall.* **31,** 577 (1983).
69. R. W. Davidge and T. J. Green, *J. Mater. Sci.* **3,** 629 (1968).
70. V. C. Li and T. Hashida, *J. Mater. Sci. Lett.,* **12,** 898 (1993).
71. M. Rühle, *Mater. Sci. Eng.* **A105/106,** 77 (1988).
72. A. H. Heuer and L. W. Hodds, Eds., *Advances in Ceramics,* Vol. 3, *Science and Technology of Zirconia,* ACS, Columbus, OH, 1981.
73. N. Claussen and M. Rühle, *ibid.,* p. 137.
74. E. C. Subbarao, *ibid.,* p. 1.
75. R. C. Garvie, R. H. Hanink, and R. T. Pascoe, *Nature, London* **258**(5537), 703 (1975).
76. D. L. Porter and A. H. Heuer, *J. Am. Ceram. Soc.* **60**(3–4), 183 (1977).
77. D. L. Porter and A. H. Heuer, *J. Am. Ceram. Soc.* **62**(5–6), 298 (1979).
78. A. G. Evans, *Mater. Sci. Eng.* **105/106,** 65 (1988).
79. A. G. Evans and R. M. Cannon, *Acta Metall.* **34**(5), 761 (1986).

80. D. Green, R. H. J. Hannink, and M. V. Swain, *Transformation Toughening of Ceramics*, CRC Press, Boca Raton, FL (1989).
81. A. G. Evans and A. H. Heuer, *J. Amer. Ceram. Soc.* **63**(5–6), 241 (1980).
82. R. M. McMeeking and A. G. Evans, *J. Am. Ceram. Soc.* **65**(5), 242 (1982).
83. S. M. Wiederhorn, *Annu. Rev. Mater. Sci.* **14,** 373 (1984).
84. F. F. Lange, *J. Mater. Sci.* **17,** 225–263 (1982).
85. N. Padture, Ph.D dissertation, Lehigh University, Bethlehem, PA (1991).
86. M. P. Harmer, H. M. Chan, and G. A. Miller, *J. Amer. Ceram. Soc.* **75**(7), 1715 (1992).
87. P. F. Becher, C. H. Hsueh, P. Angelini, and T. N. Tiegs, *J. Amer. Ceram. Soc.* **71**(12), 1050 (1988).
88. S. V. Nair, *J. Amer. Ceram. Soc.* **73**(10), 2839 (1990).
89. P. L. Swanson, C. J. Fairbanks, B. R. Lawn, and Y. W. Mai, *J. Amer.—Ceram. Soc.,* **70**(4), 279 (1987).
90. Y. W. Mai and B. R. Lawn, *J. Amer. Ceram. Soc.* **70**(4), 289 (1987).
91. S. T. Bennison and B. R. Lawn, *Acta Metall,* **37** (10), 2659 (1989).
92. R. F. Cook, C. J. Fairbanks, B. R. Lawn, and Y. W. Mai, *J. Mater, Res.* **2,** 345 (1987).
93. Y. Tajima, K. Urashima, M. Watanabe, and Y. Matsuo, *Ceramic Transactions, Vol. 1, Ceram. Powder Sci.-IIB,* G. L. Messing, E. R. Fuller, Jr., and H. Hauser, Eds., American Ceramic Society, Westerville, OH, 1034 (1988).
94. J. C. Amazigo and B. Budiansky, *J. Mech. Phys. Solids* **36,** 581 (1988).
95. J. J. Mecholsky, S. W. Freiman, and R. W. Rice, *J. Mater. Sci.* **11,** 1310 (1976).
96. M. H. Litt and A. V. Tobolsky, *J. Macromol. Sci. Phys. B* **1**(3), 433 (1967).
97. R. F. Boyer, *Rubber Chem. Tech.* **36,** 1303 (1963).
98. S. E. B. Petrie, *Polymeric Materials,* ASM, Metals Park, OH, 1975, p. 55.
99. J. Heijboer, *J. Polym. Sci.* **16,** 3755 (1968).
100. R. F. Boyer, *Polym. Eng. Sci.* **8**(3), 161 (1968).
101. W. J. Jackson, Jr., and J. R. Caldwell, *J. Appl. Polym. Sci.* **11,** 211 (1967).
102. J. A. Manson and L. H. Sperling, *Polymer Blends and Composites,* Plenum, New York, 1976.
103. C. B. Bucknall, *Toughened Plastics*, Applied Science, Barking, U.K., 1977.
104. *Toughening of Plastics II*, Plast. Rub. Inst., London, 1985.
105. A. F. Yee and R. A. Pearson, in *Toughening of Plastics II*, Rub. Inst., London, 1985, p. 2/1.
106. A. J. Kinloch and D. L. Hunston, in *Toughening of Plastics II,* Plast. Rub. Inst., London, 1985, p. 4/1.
107. A. F. Yee and R. A. Pearson, *J. Mater. Sci.* **21,** 2462 (1986).
108. R. A. Pearson and A. F. Yee, *J. Mater. Sci.* **21,** 2475 (1986).
109. A. F. Yee, ASTM *STP 937*, 1987, p. 383.
110. A. C. Garg and T. W. Mai, *Comp. Sci. and Tech.* **31,** 179 (1988).
111. *Rubber Toughened Engineering Plastics*, A. A. Collyer, Ed., Chapman and Hall, London, 1994.
112. C. E. Hall, *Introduction to Electron Microscopy*, 2nd ed., McGraw-Hill, New York, 1966.
113. G. A. Crosbie and M. G. Phillips, *J. Mater. Sci.* **20,** 182 (1985).
114. C. B. Bucknall and D. G. Street, *Soc. Chem. Ind. Monograph No. 26,* London, 1967, p. 272.

115. C. B. Bucknall, D. Clayton, and W. Keast, *J. Mater. Sci.* **7,** 1443 (1972).
116. H. Brewer, F. Haaf, and J. Stabenow, *J. Macromol. Sci. Phys. B* **14,** 387 (1977).
117. F. Haaf, H. Brewer and J. Stabenow, *Angew Makromol. Chem.* **58/59,** 95 (1977).
118. A. M. Donald and E. J. Kramer, *J. Mater. Sci.* **17**(6), 1765 (1982).
119. J. N. Sultan, R. C. Liable, and F. J. McGarry, *Polym. Symp.* **16,** 127 (1971).
120. J. N. Sultan and F. J. McGarry, *Polym. Eng. Sci.* **13,** 29 (1973).
121. Y. Huang and A. J. Kinloch, *J. Mater. Sci. Lett.* **11,** 484 (1992).
122. A. J. Kinloch, *Rubber Toughened Plastics*, C. K. Riew, Ed., ACS **222,** 67 (1989).
123. Y. Huang and A. J. Kinloch, *Polymer* **33**(24), 5338 (1992).
124. S. Kunz-Douglass, P. W. R. Beaumont, and M. F. Ashby, *J. Mater. Sci.* **15,** 1109 (1980).
125. A. G. Evans, Z. B. Ahmed, D. G. Gilbert, and P. W. R. Beaumont, *Acta Metall.* **34,** 79 (1986).
126. A. G. Evans, S. Williams, and P. W. R. Beaumont, *J. Mater. Sci.* **20,** 3668 (1985).
127. F. F. Lange, *Phil. Mag.* **22,** 983 (1970).
128. A. C. Moloney, H. H. Kausch, T. Kaiser, and H. R. Beer, *J. Mater. Sci.* **22,** 381 (1987).
129. J. Spanoudakis and R. J. Young, *J. Mater. Sci.* **19,** 473 (1984).
130. J. K. Wells and P. W. R. Beaumont, *J. Mater. Sci.* **23,** 1274 (1988).
131. A. G. Evans and R. M. McMeeking, *Acta Met.* **34**(12), 2435 (1986).
132. A. H. Cottrell, *Proc. Roy. Soc. Lond.* **A282,** 2 (1964).
133. A. Kelly and W. R. Tyson, *J. Mech. Phys. Solids* **13,** 329 (1965).
134. M. D. Thouless, O. Sbaizero, L. S. Sigl, and A. G. Evans, *J. Amer. Cer. Soc.* **72**(4), 525 (1989).
135. K. T. Faber and A. G. Evans. *Acta Metall.* **31,** 565 (1983).
136. D. L. Maxwell, R. J. Young, and A. J. Kinloch, *J. Mater. Sci. Lett,* **3,** 9 (1984).
137. A. J. Kinlock, D. L. Maxwell, and R. J. Young, and *J. Mater. Sci.,* **20,** 4169 (1985).
138. I. M. Low, S. Bandyopakhyay, and Y. W. Mai, *Polym. Inter.* **27,** 131 (1992).
139. R. A. Pearson, A. K. Smith, and Y. W. Yee, *Second International Conference on Deformation and Fracture of Comp.,* Manchester, 9-1 (1993).
140. T. Shimizu, M. Kamino, M. Miyagaula, N. Nishiwacki, and S. Kida, *Ninth International Conference on Deformation, Yield and Fracture of Polymers,* Churchill College, Cambridge, U.K., 76/1 April 1994.
141. H. Azimi, R. A. Pearson, and R. W. Hertzberg, *J. Mater. Sci. Lett.,* **13,** 1460 (1994).
142. H. Azimi; PhD dissertation, Lehigh University, Bethlehem, PA (1994).
143. A. Kelly, *Proc. R. Soc. London Ser. A* **319,** 95 (1970).
144. J. O. Outwater and M. C. Murphy, *Proc. 24th SPI/RP Conf.,* Paper 11-B, SPI, New York, 1969.
145. M. Korchynsky, American Society of Agricultural Engineers, Paper No. 70-682, Dec. 1970.
146. B. M. Kapadia, A. T. English, and W. A. Backofen, *Trans. ASM* **55,** 389 (1982).
147. B. Mravic and J. H. Smith, Development and Improved High-Strength Steels for Aircraft Structural Components,'' AFML-TR-71-213, Oct. 1971.

148. A. H. Cottrell, *Fracture*, Technology Press MIT and Wiley, New York, 1959, p. 20.
149. N. J. Petch, *Fracture*, Technology Press MIT and Wiley, New York, 1959, p. 54.
150. J. R. Low, Jr., *Relation of Properties to Microstructure*, ASM, Metals Park, OH, 1954, p. 163.
151. A. S. Tetelman and A. J. McEvily, Jr., *Fracture of Structural Materials*, Wiley, New York, 1967.
152. R. O. Ritchie, *Metals Handbook, Vol. 8, Mechanical Testing*, Metals Park, OH, 1985, p. 465.
153. R. O. Ritchie and A. W. Thompson, *Met. Trans.* **16A**, 233 (1985).
154. J. F. Knott, *Advances in Fracture Research*, Vol. 1, S. R. Valluri, D. M. R. Taplin, P. Rama Rao, J. F. Knott, and R. Dubey, Eds., Pergamon, Oxford, England, 1984, p. 83.
155. R. O. Ritchie, J. F. Knott, and J. R. Rice, *J. Mech. Phys. Solids* **21**, 395 (1973).
156. R. O. Ritchie, W. L. Server, and R. A. Wullaert, *Met. Trans.* **10A**, 1557 (1979).
157. F. A. McClintock, *J. Appl. Mech. Trans. ASME Ser. H* **25**, 363 (1958).
158. R. C. Bates, *Metallurgical Treaties*, J. K. Tien and J. F. Elliott, Eds., AIMI, Warrendale, PA, 1982, p. 551.
159. R. H. Van Stone, T. B. Cox, J. R. Low, and J. A. Psioda, *Int. Met. Rev.* **30**(4), 157 (1985).
160. S. Lee, L. Majno, and R. J. Asaro, *Met. Trans.* **16A**, 1633 (1985).
161. J. R. Rice and M. A. Johnson, *Inelastic Behavior of Solids*, M. F. Kanninen, W. F. Adler, A. R. Rosenfield, and R. I. Jaffee, Eds., McGraw-Hill, New York, 1970, p. 641.
162. C. Q. Chen and J. F. Knott, *Met. Sci.* **15**, 357 (1981).
163. G. T. Hahn and A. R. Rosenfield, ASTM *STP 432*, 1968, p. 5.
164. R. C. Bates and W. G. Clark, Jr., *Trans. ATM* **62**, 380 (1969).
165. D. Broek, *Eng. Fract. Mech.* **6**, 173 (1974).
166. H. Kobayashi, H. Nakamura, and H. Nakazawa, *Recent Research on Mechanical Behavior of Solids*, University of Tokyo Press, 1979, p. 341.
167. R. J. Klassen, M. N. Bassim, M. R. Bayoumi, and H. G. F. Wilsdorf, *Mater. Sci. Eng.* **80**, 25 (1986).
168. C. L. Briant and S. K. Banerji, *Int. Met. Rev.*, Review 232, No. 4, 1978, p. 164.
169. W. F. Brown, Jr., *Trans. ASM* **42**, 452 (1950).
170. E. J. Ripling, *Trans. ASM* **42**, 439 (1950).
171. J. M. Capus and G. Mayer, *Metallurgia* **62**, 133 (1960).
172. S. K. Banerji, H. C. Feng, and C. J. McMahon, Jr., *Met. Trans.* **9A**, 237 (1978).
173. N. Bandyopadhyay and C. J. McMahon, Jr., *Met. Trans.* **14A**, 1313, (1983).
174. J. Yu and C. J. McMahon, Jr., *Met. Trans.* **11A**, 291 (1980).
175. C. H. Shih, B. L. Averbach, and M. Cohen, *Trans. ASM* **48**, 86 (1956).
176. J. H. Hollomon, *Trans. ASM* **36**, 473 (1946).
177. B. C. Woodfine, *JISI* **173**, 229 (1953).
178. J. R. Low, Jr., *Fracture of Engineering Materials*, ASM, Metals Park, OH, 1964, p. 127.
179. C. J. McMahon, Jr., ASTM *STP 407*, 1968, p. 127.
180. D. Kalderon, *Proc. Inst. Mech. Eng.* **186**, 341 (1972).
181. K. Balajiva, R. M. Cook, and D. K. Worn, *Nature London*, **178**, 433 (1956).

182. W. Steven and K. Balajiva, *JISI* **193,** 141 (1959).

183. J. M. Capus, ASTM *STP 407*, 1968, p. 3.

184. J. R. Low, Jr., D. F. Stein, A. M. Turkalo, and R. P. LaForce, *Trans. Met. Soc. AIME* **242,** 14 (1968).

185. H. L. Marcus and P. W. Palmberg, *Trans. Met. Soc. AIME* **245,** 1665 (1969).

186. H. L. Marcus, L. H. Hackett, Jr., and P. W. Palmberg, ASTM *STP 499,* 1972, p. 90.

187. D. F. Stein, A. Joshi, and R. P. LaForce, *Trans. ASM*, **62,** 776 (1969).

188. J. Kameda, *Met. Trans.* **12A,** 2039 (1981).

189. C. I. Garcia, G. A. Ratz, M. G. Burke, and A. J. DeArdo, *J. Met.* **37**(9), 22 (1985).

190. C. J. McMahon, Jr., V. Vitek, and J. Kameda, *Developments in Fracture Mechanics*, Vol. 2, G. G. Chell, Ed., Applied Science, New Jersey, 1981, p. 193.

191. C. J. McMahon, Jr. and V. Vitek, *Acta Met.* **27,** 507 (1979).

192. J. Kameda and C. J. McMahon, Jr., *Met. Trans.* **11A,** 91 (1980).

193. J. Kameda and C. J. McMahon, Jr., *Met. Trans.* **12A,** 31 (1981).

194. S. H. Bush, *J. Test. Eval.* **2**(6), 435 (1974).

195. M. L. Wald, *New York Times*, Sept. 27, 1981, p. 1.

196. A. Tetelman and A. J. McEvily, *Fracture of Structural Materials*, Wiley, New York, 1967.

197. D. McClean, *Mechanical Properties of Metals,* Wiley, New York, 1962.

198. *Irradiation Effects on the Microstructure and Properties of Metals*, ASTM *STP 611*, ASTM, Philadelphia, PA, 1976.

199. N. L. Peterson and S. D. Harkness, Eds., *Radiation Damage in Metals,* ASM, Metals Park, OH, 1976.

200. J. A. Sprague and D. Kramer, Eds., *Effects of Radiation on Structural Materials*, ASTM *STP 683,* 1979.

201. L. E. Steele, *Nucl. Mater.* **16,** 270 (1970).

202. J. A. Spitznagel, R. P. Shogan, and J. H. Phillips, ASTM *STP 611,* 1976, p. 434.

203. J. R. Hawthorne, H. E. Watson, and F. J. Loss, ASTM *STP 683*, 1979, p. 278.

204. U. Potapovs and J. R. Hawthorne, *Nucl. Appl.* **6**(1), 27 (1969).

205. J. R. Hawthorne, ASTM *STP 484*, 1971, p. 96.

206. F. J. Loss, J. R. Hawthorne, C. Z. Serpan, Jr., and P. P. Puzak, *NRC Report 7209*, Mar. 1, 1971.

207. D. J. Nans, G. B. Batson, and J. L. Lott, *Fracture Mechanics of Ceramics,* Vol. 2, R. C. Bradt, D. P. H. Hasselman, and F. F. Lange, Eds., Plenum, New York, 1974, p. 469.

208. R. Rossi, P. Acker, and D. Francois, *Advances in Fracture Research*, Vol. 4, S. R. Valluri, D. M. R. Taplin, R. Ramo Rao, J. F. Knott, and R. Dubey, Eds., Pergamon, Oxford, England, 1984, p. 2833.

209. R. F. Pabst, *ibid.*, p. 555.

210. S. M. Wiederhorn, *ibid.*, p. 613.

211. S. W. Freiman, K. R. McKinney, and H. L. Smith, *ibid.*, p. 659.

212. W. G. Clark, Jr., and W. A. Logsdon, *ibid.,* p. 843.

213. R. C. Lueth, *ibid.*, p. 791.

214. N. Ingelström and H. Nordberg, *Eng. Fract. Mech.* **6,** 597 (1974).

215. R. A. Schmidt, *Closed Loop* **5,** 3 (Nov. 1975).

216. *Guide to Selecting Engineered Materials*, Vol. 2(1), ASM, Metals Park, OH, 1987, p. 83.

217. *Advanced Materials and Processes*, Vol. 2(9), 1986, p. 32.

218. G. P. Marshall and J. G. Williams, *J. Mater. Sci.* **8,** 138 (1973).

219. G. P. Marshall, L. E. Culver, and J. G. Williams, *Int. J. Fract.* **9**(3), 295 (1973).

220. J. C. Radon, *J. Appl. Polym. Sci.* **17,** 3515 (1973).

221. R. C. Bradt, D. P. H. Hasselman, and F. F. Lange, Eds., *Fracture Mechanics of Ceramics*, Vols 1 to 4, Plenum, New York, 1978.

222. J. E. Campbell, W. E. Berry, and C. E. Feddersen, *Damage Tolerant Design Handbook, MCIC-HB 01*, Sept. 1973.

PROBLEMS

10.1   Why would you expect a steel refined by the Bessemer process (air blown through the melt) to exhibit inferior fracture properties to a steel refined in a BOF (oxygen blown through the melt)?

10.2   Discuss three ways in which the toughness of a material may be increased.

10.3   For a stress level of 240 MPa compute the maximum radius of a semicircular surface flaw in 7075-T651 aluminum alloy plate when loaded in the L-T, T-L, and S-L orientations. Assume plane-strain conditions.

10.4   For the room temperature CVN test results shown in Fig. 10.15, estimate the K_{IC} level based on the K_{IC}–CVN upper-shelf correlation given in Fig. 9.22. Assume $\sigma_{ys} = 1000$ MPa. Plot these estimates of K_{IC} versus sulfur content and then superimpose the data from Fig. 10.16. Discuss your results.

10.5   Discuss the overall virtues of fine-grain microstructures with regard to room temperature and elevated temperature behavior.

10.6   An 8-cm-diameter extruded rod of 7178-T651 aluminum alloy is to be machined into a closed-end, hollow cylinder with a 7-cm bore. If a fluid is introduced into the bore and compressed by a piston, calculate the largest semicircular surface flaw (oriented along the axis of the bore) that could withstand a fluid pressure of 50 MPa.

10.7   Two single-edge notch samples were machined from a hot-rolled steel plate according to the manner shown in the accompanying diagram.

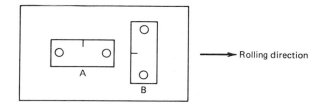

(a)   If we assume that the width of the samples is large compared to the 2.5-cm-long sharp crack in each coupon, compute the apparent fracture toughness exhibited by specimens A and B if the stress to fracture was 325 and 260 MPa, respectively.

(b)   What is the reason for this difference in toughness?

(c)   What could be done to improve the toughness of the material when tested as in specimen B?

10.8   You have been asked to determine K_{IC} for a material that is supplied in the form of extruded bar stock. Compare the K_{IC} values obtained with long bend bar, disk compact, and short rod specimen configurations.

10.9   A rectangularly shaped component is to be fabricated from the least expensive steel available (presumed to be the alloy with the least alloying additions). The final decision is to be made between 4330V (425°C temper) and 9-4-20 (550°C temper). Which alloy would you choose if the component is to experience a stress of half the yield strength in the presence of a circular corner crack with a radius of 10 mm? Would you answer be the same if the design stress were increased to 65% of the yield strength? The properties of these materials are given in Table 10.7.

10.10   A 20-mm-thick pressure vessel that has an internal diameter of 400 mm contains a semicircular crack at the internal surface that extends through 25% of a tank wall. The tank was made from D6AC steel and possesses yield-strength and fracture-toughness values of 1500 MPa and 102 MPa$\sqrt{m}$, respectively. Prove that it would be safe to pressurize the cracked tank to a level of 75 MPa.

10.11   You are given a 10-cm-diameter cylindrically shaped extruded bar (length = 50 cm) of an aluminum alloy.

(a)   What specimen configurations would you use to characterize the degree of anisotropy of the material's fracture toughness?

(b)   If the lowest fracture toughness of the bar was found to be 20 MPa$\sqrt{m}$, what would be the load level needed to achieve this stress intensity level?

(c)   Confirm the presence of plane strain conditions, given that σ_{ys} = 500MPa. (Assume specimen thickness = 1 cm, a/W = 0.5, and bar diameter = 1.2 W.)

10.12   Estimate the fracture stress for a sapphire rod that contains a mirror-mist zone with a mirror boundary located 1 mm from the defect origin.

10.13   A component of A533-B Class I steel was subjected to an accumulated neutron fluence of $\sim 2 \times 10^{19}$ n/cm over a several year period of time. As a result, irradiation damage occurred to reduce the material's plane strain fracture toughness as shown in Fig. 10.58. For purposes of this exercise, assume that the operational temperature of this component is 100° C, the stress is 325 MPa, and the component contains a semicircular surface flaw with a depth of 2 cm.

(a)   Prove that the component can survive normal operational stresses at 100°C, given the presence of the flaw.

(b) If the component is cooled to room temperature and then abruptly re-started at the normal stress level, what is likely to happen?

(c) If the component is replaced with a new one, would it survive the abrupt restarting procedure?

10.14 For lack of a suitable material supply, a thin-walled cylinder is machined from a thick plate of 7178-T651 aluminum alloy such that the cylinder axis is oriented parallel to the rolling direction of the plate. If the cylinder's diameter and wall thickness are 5 cm and 0.5 cm, respectively, determine whether the cylinder could withstand a pressure of 50 MPa in the presence of a 0.2-cm-deep semicircular surface flaw.

10.15 A 100-mm-long rod of Si_3N_4 is tested in three-point bending. If the 10-mm-square rod cross section contains a 0.5-mm-deep semicircular surface flaw at the span midpoint and another one 1.5-mm-deep at the quarter span length position, where would you expect failure to occur?

CHAPTER 11

ENVIRONMENT-ASSISTED CRACKING

The image of stress corrosion I see
Is that of a huge unwanted tree,
Against whose trunk we chop and chop,
But which outgrows the chips that drop;
And from each gash made in its bark
A new branch grows to make more dark
The shade of ignorance around its base,
Where scientists toil with puzzled face.

On Stress Corrosion, S. P. Rideout*

Much attention was given in preceding chapters to the importance of the plane-strain fracture-toughness parameter K_{IC} in material design considerations. It was argued that this value represents the lowest possible material toughness corresponding to the maximum allowable stress-intensity factor that could be applied short of fracture. And, yet, failures are known to occur when the *initial* stress-intensity-factor level is considerably below K_{IC}. How can this be? These failures arise because cracks are able to grow to critical dimensions with the initial stress-intensity level increasing to the point where $K = K_{IC}$ (Eq. 8-28). Such crack extension can occur by a number of processes. Subcritical flaw growth mechanisms involving a cooperative interaction between a static stress and the environment include stress corrosion cracking (SCC), hydrogen embrittlement (HE), and liquid-metal embrittlement (LME). The subject of fatigue and corrosion fatigue is examined in Chapter 13, while SCC, HE, and LME are considered in this chapter. The literature dealing with these topics is as staggering as is the history and significance of the problem.† It is not within the scope of this book to cover this material in depth, especially considering the complexity of the environmentally induced embrittlement phenomena itself. Indeed, as Staehle[1] has pointed out, "A general mechanism for stress corrosion cracking . . . seems to be an unreasonable and unattainable goal. Specific processes appear to operate under specific sets of metallurgical and environmental conditions." More recently, Latanision et al.[2] noted with appropriate sarcasm that "it is no surprise that evidence can be found to contradict virtually every point of view." A montage of some major SCC mechanisms is shown in Fig. 11.1, representing the cumulative results of many researchers. In

---

* Reprinted with permission from *Fundamental Aspects of Stress Corrosion Cracking*, 1969, National Association of Corrosion Engineers.

† A 1978 National Bureau of Standards report places the cost to the U.S. economy for corrosion damage at 70×10^9. (L. H. Bennett, J. Kruger, R. L. Parker, E. Passaglia, C. Reimann, A. W. Ruff, and H. Yakowitz, *Economic Effects of Metallic Corrosion in the United States*, Special Pub. 511-1,2, U.S. Department of Commerce, National Bureau of Standards, Washington, DC, May 1978.)

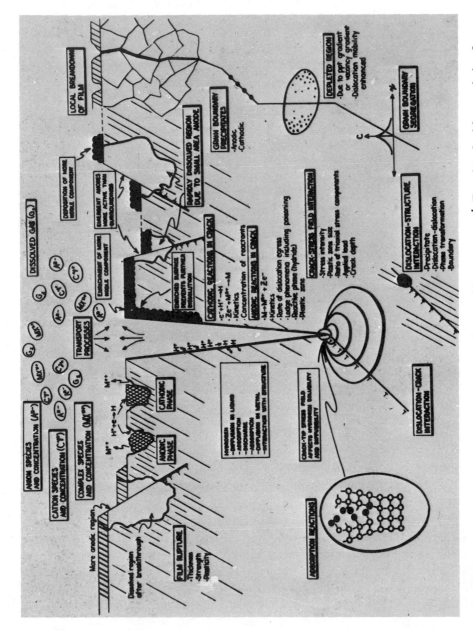

FIGURE 11.1 Montage of important stress corrosion cracking processes.[1] (Reprinted with permission from R. W. Staehle, *Fundamental Aspects of Stress Corrosion Cracking*, 1969, National Association of Corrosion Engineers.)

addition to Ref. 1, the interested reader should find the several dozen other papers in this volume of particular interest regarding the specifics of SCC. In addition, the reader is referred to several comprehensive reviews.[2-10]

In dealing with the problems of SCC, HE, and LME, considerable discussion has surrounded both similarities and differences associated with these processes. For convenience of this discussion, these phenomena are referred to collectively as environment-assisted cracking (EAC). Indeed, Ford[6] has referred to such classifications of SCC, HE, LME, and corrosion fatigue as being "artificial and confining in terms of remedial actions." The objective of this chapter is to provide an overview of EAC that will enable the reader to better appreciate some of the major problems that befall many engineering materials.

Associated with the expanding EAC literature is a growing realization of the complex and interdependent nature of the various cracking processes. For example, Williams and coworkers[11] have advocated that a successful study of EAC requires an integrated interdisciplinary approach involving the participation of fracture mechanics, chemistry, and materials science experts. To illustrate, the interrelated factors associated with hydrogen embrittlement (or hydrogen-assisted-cracking [HAC]) are depicted in Fig. 11.2. Fracture mechanics tests can provide a characterization of the phenomenology of EAC such as the rate of crack advance and the associated crack-velocity dependence on temperature, pressure, and concentration of aggressive species. Surface chemistry and electrochemistry studies are needed to identify the rate-limiting processes, whereas metallurgical investigations are important to identify what alloy compositions and microstructures are susceptible to the cracking process and what fracture micromechanisms are operative (recall Section 7.7.3). A critical point to recognize with regard to Fig. 11.2—or for that matter with a comparable diagram that would describe the sequential processes associated with SCC—is that the slowest process represents the rate controlling step in the embrittlement of the material; this

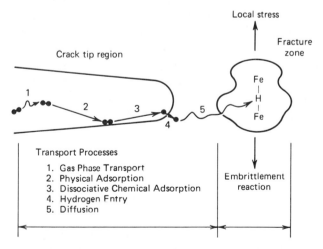

FIGURE 11.2 **Various processes involved in the hydrogen embrittlement of ferrous alloys. (From Williams et al.[11], with permission from the Metallurgical Society of AIME.)**

arises from the fact that these processes are mutually *dependent* on one another (recall Eq. 5-9). By contrast, final fracture can result from several mutually *independent* fracture mechanisms; in this instance, the fastest process will dominate the fracture mode (recall Eq. 5-11).

11.1 EMBRITTLEMENT MODELS

A growing number of models have been proposed to describe various SCC,[2,3,6] HAC,[4,5,12,13] and LME[14,15] fracture processes. The need for so many models attests to the complexity of EAC phenomena. Yet, certain clear similarities and differences in proposed mechanisms are becoming apparent and have led some investigators to conclude that these embrittling processes are often interrelated.[2,4,5] For example, Thompson and Bernstein[4,5] have suggested that SCC may involve both HAC and electrochemical processes that operate in *parallel* (Fig. 11.3). Consequently, EAC may occur by either SCC or HAC processes or by both. The latter condition is illustrated by the iron plus water system; in this instance, the chemical reaction between Fe and H_2O involves the liberation of hydrogen, which then introduces the basis for HAC.

11.1.1 Hydrogen–Embrittlement Models

The flow diagram shown in Fig. 11.4 provides a useful summary of hydrogen sources, transport paths, destinations, and induced fracture micromechanisms. Hydrogen can enter the metal in a number of different ways. Before vacuum degassing techniques were developed, large steel castings were subject to a phenomenon called hydrogen flaking, wherein dissolved hydrogen in the molten metal would form entrapped gas pockets upon solidification. The large, localized pressures associated with these gas pockets generated many sharp cracks which, when located near the casting surface, caused chunks of steel to spall. A service failure of a large rotor forging caused by this type of defect is discussed in Chapter 14. Hydrogen can also be picked up from the electrode cover material or from residual water during welding. After diffusing into the base plate while the weld is hot, embrittlement occurs upon cooling by a process referred to as cold cracking in the weld heat affected zone.[16] Hydrogen may also enter the material as a result of electroplating (i.e., cathodic charging), which contributes to early failure. It is ironic that the electroplating process, designed to protect a material against aqueous environments and SCC, actually undermines fracture resistance of the component by simultaneously introducing another cracking process.*

Hydrogen pickup and associated embrittlement can also be introduced into the metal whenever a sample under stress is exposed to a hydrogen gas atmosphere. It

---

* It is possible to overcome many of the problems associated with cathodic charging by subjecting the electroplated material to a baking treatment. This involves heating the metal to a moderate temperature for a sufficient period of time to drive the hydrogen out of solid solution. Furthermore, weld-related cold cracking is suppressed by preheating the sample. This has the effect of lowering the postweld cooling rate, thereby allowing more time for the hydrogen to diffuse away from the weld zone.

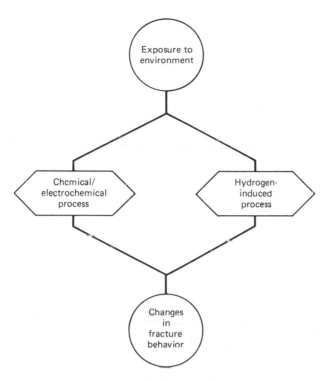

FIGURE 11.3   Parallel processes (SCC and HAC) involved in environment-assisted cracking. (From Thompson[4]; with permission from the Metallurgical Society of AIME.)

should be noted that embrittlement does not occur as a result of prior exposure to hydrogen gas in the absence of stress.[17,18] The dramatic difference in cracking behavior in H-11 steel for oxygen, argon, and hydrogen gases and water is illustrated in Fig. 11.5.[19] Note the severe effect of dry hydrogen and moisture on the cracking rate, while oxygen causes total crack arrest. It is presently believed that molecular hydrogen is dissociated by chemisorption on iron,[20] allowing liberated atomic hydrogen to diffuse internally and embrittle the metal. Likewise, it has been shown that hydrogen is a product of the corrosion reaction between iron and water; this hydrogen then follows the same path as the chemisorped hydrogen to the metal interior.[21] On the basis of this latter observation, it has been suggested that SCC and HE in steels are closely related.[4,5,22] Apparently, oxygen has a greater affinity for iron and forms a protective oxide barrier to block the chemisorption process.[17,19] It is believed that once the oxygen is removed, hydrogen can reduce the oxide layer and thereby react again with a clean iron surface.

Referring again to Fig. 11.4, hydrogen can diffuse rapidly through the lattice because of its small size. Of potentially greater significance, calculations have shown that hydrogen transport rates in association with dislocation motion can be several orders of magnitude greater than that associated with lattice diffusion.[12,23] As such, one cannot rule out hydrogen embrittlement as being a major contributing factor in a cracking process even though the EAC rate is greater than the rate of hydrogen

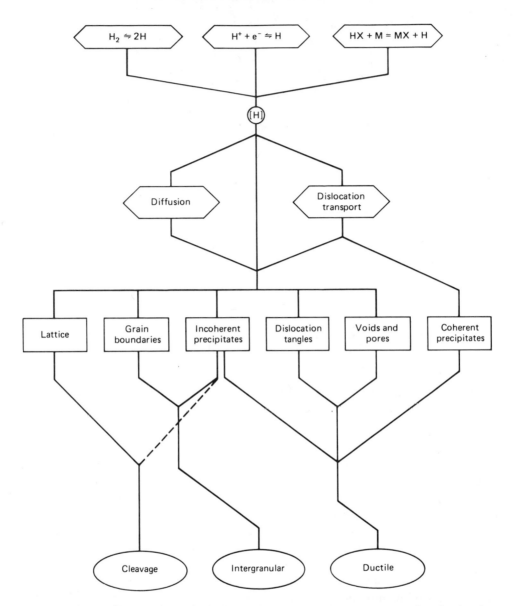

FIGURE 11.4 Flow diagram depicting hydrogen sources, transport paths, destinations, and induced fracture micromechanisms. (From A. W. Thompson and I. M. Bernstein,[5] *Advances in Corrosion Science and Technology,* Vol. 7, 1980, p. 145; with permission from Plenum Publishing Corporation.)

diffusion through the lattice. Furthermore, hydrogen tends to accumulate at grain boundaries, inclusions, voids, dislocation arrays, and solute atoms. To this extent, HAC is controlled by those hydrogen accumulation sites that are most sensitive to fracture. From Fig. 11.4, we see that the cracking process can involve cleavage, intergranular, or ductile (microvoid coalescence) fracture micromechanisms. Beachem[13] reported all three mechanisms in the same steel alloy when tested at different

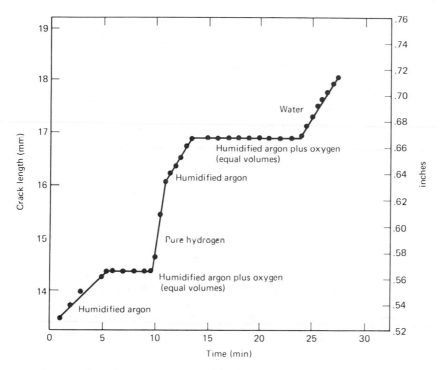

FIGURE 11.5 **Fast crack growth of high-strength steel in water and hydrogen, but crack arrest in oxygen.**[19] **(Reprinted with permission of the American Institute of Mining, Metallurgical, and Petroleum Engineers.)**

stress levels (Fig. 11.6). Consequently, there is no single fracture micromechanism that is characteristic of HAC. For this reason, fractographic information is important to our understanding of HAC, but it does not provide a unique characterization of the degree of embrittlement prevailing at a given time.

The hydrogen–embrittling process, therefore, depends on three major factors: (1) the original location and form of the hydrogen (internally charged versus atmospheric water or gaseous hydrogen); (2) the transport reactions involved in moving the hydrogen from its source to the locations where it reacts with the metal to cause embrittlement; and (3) the embrittling mechanism itself. We may now ask what that embrittling mechanism is. Unfortunately, the answer is not a simple one, as evidenced by the number of theories that have been proposed. According to one model, called the "planar pressure mechanism," the high pressures developed within internal hydrogen gas pores of charged material cause cracking.[24,25] Although this mechanism appears valid for hydrogen-charged steels, it cannot be operative for the embrittlement of steel by low-pressure hydrogen atmospheres. In the latter situation, there would be no thermodynamic reason for a low gas pressure external atmosphere to produce a high gas pressure within the solid. Troiano and coworkers[26,27] have argued that hydrogen diffuses under the influence of a stress gradient to regions of high tensile triaxiality which then interacts with the metal lattice to lower its cohesive strength. A third model to explain HE was proposed by Petch and Stables,[28] who suggested that

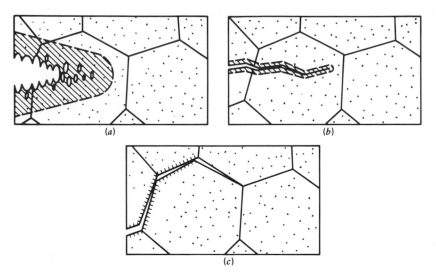

FIGURE 11.6 **Schematic representation of different hydrogen-induced fracture paths as a function of stress level.** (*a*) **High *K* level generates microvoid coalescence;** (*b*) **intermediate *K* level generates transgranular fracture by a quasicleavage mechanism;** (*c*) **low *K* level leads to intergranular fracture path. (From Beachem[13]; with permission from the Metallurgical Society of AIME.)**

hydrogen acts to reduce the surface energy of the metal at internally free surfaces. A significantly different HE model was proposed by Beachem[13] and discussed by Hirth,[12] among others. Beachem suggested that the presence of hydrogen in the metal lattice greatly enhances dislocation mobility at very low applied stress levels. Brittle behavior is then envisioned to occur as a result of extensive but highly localized plastic flow, which can occur at very low shear stress levels. Finally, hydrogen embrittlement may result from the formation of metal hydrides in such materials as titanium, vanadium, and zirconium. In a recent study, Fujita[29] has suggested that hydride-induced embrittlement may occur more frequently than believed previously. Further study of this proposal is needed. Additional references pertaining to the subject of HE are found in the proceedings of relevant conferences.[30,31]

11.1.2   Stress Corrosion Cracking Models

Bursle and Pugh[3] reviewed several different SCC models and concluded that the film-rupture model,[32,33] involving anodic dissolution at the crack tip, was capable of explaining most examples of intergranular SCC (Fig. 11.7). The principal feature of this model is that the protective surface film in the vicinity of the crack tip is ruptured by localized plastic flow. (Note that the protective film away from the crack tip remains intact.) Consequently, an electrolytic cell is created with the bare metal at the crack tip serving as the anode and the unbroken protective surface film serving as the cathode. The exposed bare metal is then subjected to rapid anodic dissolution, thereby allowing the crack to advance. Since the protective film is generally regarded to be passive in character, the rate of anodic dissolution and associated crack extension will

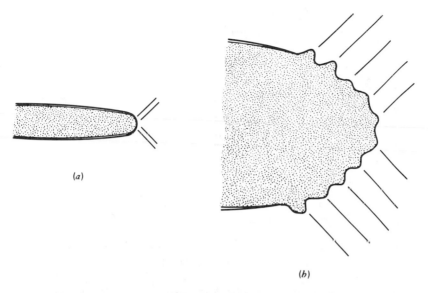

FIGURE 11.7   Diagram showing film-rupture model. Localized plastic flow at crack (*a*) results in numerous film-rupture events associated with transient anodic dissolution (*b*). (From Bursle and Pugh[3]; with permission from the Metallurgical Society of AIME.)

depend, in part, on the repassivation rate (i.e., the rate associated with the formation of a new protective film). It is interesting to note that the SCC process is maximized at intermediate passivation rates.[6] When the passivation rate is low, the crack tip becomes blunt because of excessive dissolution on the crack sides; when the passivation rate is high, the amount of crack-tip penetration per film-rupture event is minimized.

Bursle and Pugh[3] further concluded that transgranular SCC could be described best by a discontinuous cleavage process based on a hydrogen–embrittlement-induced decohesion mechanism. The discontinuous cracking process was characterized by the presence of crack arrest lines on the fracture surface (Fig. 11.8).

11.1.3 Liquid-Metal Embrittlement

When many ductile metals are coated with a micron-thin layer of certain liquid metals and then loaded in tension, the metal's fracture stress and strain are significantly reduced. Fracture times are extremely short, with crack velocities as high as 500 cm/s being reported for aluminum alloys and brass in the presence of liquid mercury (Hg). The reader is referred to several comprehensive review articles on the subject of liquid–metal embrittlement.[15,35–40] A large number of embrittlement systems have been identified (Table 11.1) that are highly specific in that a given liquid metal (e.g., Hg) may embrittle one metal (e.g., Al) but not another (e.g., Mg). Empirical guidelines for the existence of LME are discussed elsewhere.[35,36] It should be noted that embrittlement can also occur when the two metals are in contact and in the *solid* form. In this instance, the vapor phase of the embrittling metal migrates by surface diffusion to the crack tip in the embrittled solid. Important practical examples of the solid-

FIGURE 11.8 Transgranular stress corrosion cracking fracture surface in type 310S stainless steel. Fracture bands were produced in load pulse (10-s spacing) experiments and reflect a discontinuous cracking process. (From Hahn and Pugh[34]; copyright, American Society for Testing and Materials, 1916 Race Street, Philadelphia, PA, 19103. Reprinted with permission.)

metal-induced embrittlement include cadmium–steel couples associated with cadmium-plated steel components and lead–steel couples as found in internally leaded steel alloys.[41,42]

Liquid-metal embrittlement is believed to result from liquid–metal chemisorption-induced reduction in the cohesive strength of atomic bonds in the region of a stress concentration.[39,43,44] From Fig. 11.9, the liquid–metal atom (L) is believed to reduce the interatomic bond strength between solid atoms, S_1 and S_2 at the crack tip, thereby causing bond rupture to occur at reduced stress levels. Once the S_1–S_2 bond is broken, liquid-metal atoms then reduce the strength of the atomic bond between solid atoms S_1 and S_3 with local fracture continuing at a rapid pace. Lynch[45] has proposed an alternative LME mechanism with premature fracture resulting from a reduction in shear strength rather than the cohesive strength of atomic bonds at the crack tip; in this manner, many dislocations can be nucleated at low stress levels that facilitate

TABLE 11.1 Liquid–Metal Embrittlement Systems[40]

	Hg	Ga	Cd	Zn	Sn	Pb	Bi	Li	Na	Cs	In
Aluminum	x	x		x	x				x		x
Bismuth	x										
Cadmium		x			x					x	x
Copper	x	x			x	x	x	x	x		x
Iron	x	x	x	x		x		x			x
Magnesium				x					x		
Silver	x	x									
Tin	x	x									
Titanium	x		x								
Zinc	x	x			x	x					x

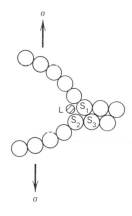

FIGURE 11.9 Model for liquid-metal embrittlement. Liquid-metal atom L reduces interatomic bond strength between atoms S_1 and S_2, S_1 and S_3, and so on.

localized plastic deformation at much reduced stress levels (recall Ref. 12 and 13 as discussed in Section 11.1.1).

11.2 FRACTURE MECHANICS TEST METHODS

As already mentioned, the various manifestations of EAC have been long recognized. Consequently, different approaches to "solving" the problem have been developed, along with "standard" specimen types. For example, stress corrosion cracking studies of engineering materials had often made use of smooth test bars which were stressed in various aggressive environments. Here the *nucleation* kinetics of cracking, as well as its character (transgranular versus intergranular), were examined closely. Most often these studies focused on the nature of anodic dissolution in the vicinity of the crack tip. Recently, more attention has been given to the *propagation* stage, reflecting the more conservative and realistic philosophical viewpoint[46] that defects preexist in engineering components (recall the discussion in Chapter 7). These propagation studies have been aided greatly by the fast-developing discipline of fracture mechanics and are the focus of attention here.

In a dramatic series of experiments, researchers at the Naval Research Laboratories[47–49] showed that certain *precracked*, high-strength titanium alloys failed under load within a matter of *minutes* when exposed to both distilled water and saltwater environments. In all tests, the initial stress-intensity levels were below K_{IC}. Heretofore, it had been felt that these same alloys would represent a new generation of submarine hull materials, based on their resistance to general corrosion, which was vastly superior to steel alloys in these same environments. These initial experiments made use of the very simple loading apparatus shown in Fig. 11.10a. Precracked samples were placed in the environmental chamber and stressed in bending at different initial K levels by a loaded scrub bucket hung from the end of the cantilever beam. Note the strong similarity between the NRL test apparatus and the diagram attributed to Galileo some 400 years earlier (Fig. 11.10b). For each test condition associated with a different initial K value (always less than K_{IC}) the time to failure was recorded. A typical plot of such data is shown in Fig. 11.11 for the environment-sensitive Ti–8Al–1Mo–1V

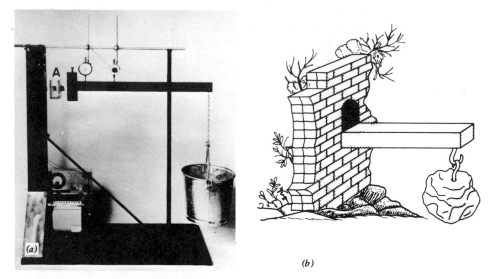

FIGURE 11.10 (*a*) **Environment-assisted cracking test stand. Specimen is placed in environment chamber at A and loaded by weights placed in scrub bucket.[48] (Reprinted with permission from B. F. Brown and C. D. Beachem.** *Corrosion Science* **5 (1965), Pergamon Press.) (*b*) Cantilever beam arrangement (adapted from Galileo).**

alloy. With an apparent fracture toughness level of about 100 MPa$\sqrt{m}$, test failures occurred at initial K levels of only 40 MPa$\sqrt{m}$ after a few minutes of exposure to a 3.5% NaCl solution. At slightly lower K levels, the time to failure increased rapidly, suggesting the existence of a threshold K level, originally designated K_{ISCC},[47] below which stress corrosion cracking would not occur. To be consistent with the philosophical viewpoint expressed in this chapter, K_{ISCC} will be redefined hereafter as K_{IEAC}, where EAC represents environment-assisted cracking; furthermore, K_{IEAC} and K_{EAC} correspond to conditions of plane strain and plane stress, respectively. As a

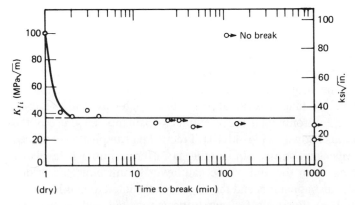

FIGURE 11.11   Initial stress intensity level plotted versus time to break for Ti–8Al–1Mo–1V alloy in 3.5% NaCl solution. Note threshold behavior.[47] (Reprinted by permission of the American Society for Testing and Materials from copyrighted work.)

result, a new safe lower limit of the applied stress-intensity value (time-dependent in this case) was identified with fracture occurring according to the following criteria:

1. $K < K_{IEAC}$ No failure expected even after long exposures under stress to aggressive environments.

2. $K_{IEAC} < K < K_{IC}$ Subcritical flaw growth with fracture occurring after a certain loading period in an aggressive environment.

3. $K < K_{IC}$ Immediate fracture upon initial loading.

The reader must recognize that rigorous and scientific proof for the existence of an environmental threshold is lacking.[50,51] Therefore the use of K_{IEAC} or K_{EAC} data in the design of structural components should be treated with caution. One should keep in mind that for the test conditions associated with Fig. 11.11 (criteria 2, above), stable crack extension causes the initial stress-intensity level to increase to the point where failure occurs when K approaches K_{IC}. Stated differently, the fracture toughness of the material is not affected by the environment; instead, small cracks grow under sustained loads to the point where the critical stress-intensity-factor level is reached (Fig. 11.12).

Determination of K_{IEAC} values is not an easy matter. Since the environmental-threshold level depends on how long one chooses to conduct the test, K_{IEAC} values may vary from one laboratory to another, depending on the patience of the investigator. It may be that K_{IEAC} test times will have to be determined for each material-environment system on an individual basis. For example, a recent ASTM draft of a proposed new standard[52] for the determination of threshold stress intensity factor values for environment-assisted cracking, suggests the following minimum test times for selected structural alloys:

steels ($\sigma_{ys} < 1200$ MPa)	10,000 h
steels ($\sigma_{ys} > 1200$ MPa)	5,000 h
aluminum alloys	10,000 h
titanium alloys	1,000 h

Some materials, such as high-strength steels and titanium alloys, exhibit a rather well-defined K_{IEAC} limit after a reasonable test time period, but in aluminum alloys this does not appear to be the case. Instead, K_{IEAC} values in high-strength aluminum alloys tend to decrease with increasing patience of the investigator. Consequently, K_{IEAC} data must be used carefully, especially in the design of engineering components that will be stressed in an aggressive atmosphere for time periods longer than those associated with generation of the K_{IEAC} data.

During the past few years, different specimen configurations and loading methods have been developed to determine K_{IEAC}. The proposed standard test method,[52] applicable to aqueous and other aggressive environments, utilizes fatigue precracked cantilever beam [SE(B)] or compact fracture [C(T)] specimens (see Appendix B) that are subjected to constant-load testing. For the determination of a material's K_{IEAC} value, it is necessary that SE(B) and/or C(T) specimen dimensions satisfy those corresponding to plane strain conditions (recall Eq. 8-49). It is recommended that

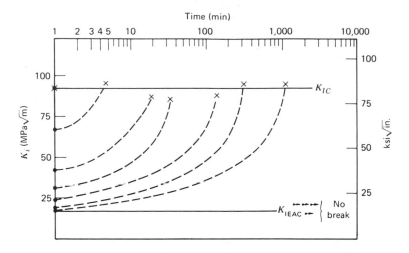

FIGURE 11.12 Change in K level with subcritical flaw growth. Regardless of $K_{initial}$, failure in any sample occurs at $K = K_{IC}$.[48] (Reprinted with permission from B. F. Brown and C. D. Beachem, *Corrosion Science* 5 (1965), Pergamon Press.)

tests be initiated at a minimum of four to six different K levels. K_{IEAC} or K_{EAC} is then determined by bracketing the two loading conditions that produce specimen failure after a relatively long time under load and where specimen failure does not occur after the specimen is loaded for a time interval as previously noted.

EAC data have also been obtained with a modified compact specimen configuration[53] (Fig. 11.13). In this instance, a screw, engaged in the top half of the sample, bears against the bottom crack surface. This produces a crack-opening displacement corresponding to some initial load. In this manner, the specimen is self-stressed and does not require a test machine for application of loads. As the crack extends by environment-assisted cracking, the load and, hence, the K level drop under the prevailing constant displacement condition. The crack finally stops when the K level drops below K_{IEAC}. Consequently, only one specimen is needed to determine K_{IEAC}. Such a test is very easy to conduct and very portable, since the self-stressed sample can be carried to any environment rather than vice versa. All one needs to do is to

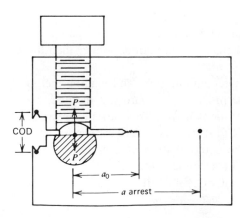

FIGURE 11.13 Modified compact tension sample with threaded bolt bearing on load pin. Initial crack opening displacement determined by extent to which bolt is engaged.

engage the screw thread to produce a given crack-opening displacement and place the specimen in the environment. Samples are examined periodically to determine when the crack stops growing. The K_{IEAC} values is then defined by the residual applied load remaining after the crack has ceased growing and the final crack length as seen on the fracture surface. The major advantages of the modified compact tension sample relative to the precracked cantilever beam are:

1. The need for one sample versus 8 to 10 in determining K_{IEAC}.

2. The specimen is self-stressed and highly portable.

3. The method is less costly.

4. K_{IEAC} is determined directly by the arrest characteristics of the sample because of the continual decrease in K with increasing crack length. By comparison, the K_{IEAC} value determined with the precracked cantilever beam samples represents an interpolated value between the highest K level at which EAC does not occur and the lowest K level where failures still occur.

5. The need for a sharp notch is not as great, since K is initially high, which results in early crack growth. By contrast, a poorly prepared notch in the cantilever beam specimen would involve a considerable period of time for crack initiation, especially at low K levels.

An interlaboratory comparison of K_{IEAC} data obtained with cantilever beam and modified compact samples was recently completed and the results were reported from 16 laboratories.[54] This study found that repeatable and reproducible K_{IEAC} values can be obtained with the use of the constant displacement-modified compact sample. On the other hand, the constant load–cantilever beam sample proved to be less useful because of difficulties with the time for the onset of cracking and the need for subjective judgment in the determination of K_{IEAC}. Problems with the stress analysis of the modified compact sample were also identified since the bolt unloads elastically with crack extension. Consequently, the crack-opening displacement, which was assumed to remain constant, actually decreases slowly with test time. In addition, corrosion reactions involving volumetric expansion of corrosion debris along the crack surface may occur so as to unload the bolt, reduce COD, and, thereby, confound determination of the instantaneous stress-intensity level.

Another K_{IEAC} test procedure that employs a single sample has shown considerable promise. In this method, a conventional fracture-toughness sample is exposed to an aggressive environment and subjected to a rising load but at a loading rate lower than that associated with E399-90 procedures.[55–63] At the load level corresponding to the onset of environment-assisted cracking, the load-displacement trace deviates markedly from that associated with conventional loading rates in air (Fig. 11.14a). Hirano et al.[57] have noted excellent agreement between K_{IEAC} values in 4340 steel that were determined from conventional modified compact samples and from the inflection point in rising load tests so long as the loading rate ($\dot{K}$) in the rising load test was equal to or less than 0.25 MPa$\sqrt{m}$/min (Fig. 11.15). Using this test method, $\dot{K}_{IEAC}$ values can be determined readily after only a few hours of testing.* It should be noted that the

---

* For a recent summary of efforts to standardize slow strain rate test procedures, see reference 63.

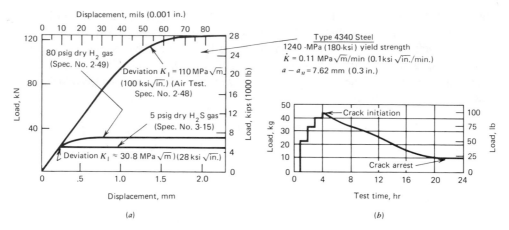

(a)

(b)

FIGURE 11.14 **Rising load test method for the determination of K_{IEAC}. (a) Effect of environment on K_{IEAC} in 4340 steel.[56] (b) Onset of HE-induced cracking in steel based on rising step-load test.[65] (Reprinted with permission. Copyright ASTM)**

influence of $\dot{K}$ on K_{IEAC} depends on the material–environment combination. Finally, Raymond et al.[64,65] have demonstrated that deeply side-grooved Charpy samples can be used to assess the EAC susceptibility of structural materials; the associated test method involves the use of a series of rising step loads with crack initiation noted by a drop in load (Fig. 11.14b).

Although the modified compact tension sample and the rising load test procedure represents improvements in the method by which K_{IEAC} data are obtained, one must

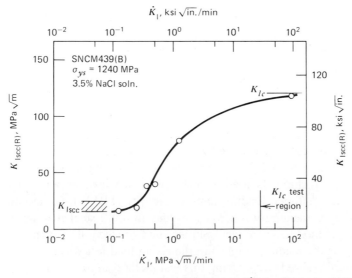

FIGURE 11.15 **Influence of loading rate ($\dot{K}$) on K_{IEAC} in 4340 steel. Data bar corresponds to data from modified compact samples.[57] (Reprinted with permission. Copyright ASTM)**

still contend with the fact that K_{IEAC} may not represent a true material property. As noted above, threshold values are often found to be a function of the length of the test (i.e., the patience of the technician). Furthermore, recent studies have demonstrated that much lower K_{IEAC} values are obtained when a small pulsating load is superimposed on the static load applied during the EAC test[66] (Fig. 11.16). Furthermore, Fessler and Barlo[67] found that K_{IEAC} values decreased with decreasing frequency of the ripple loadings. From the load-time diagram given in Fig. 11.16, the superposition of ripple loading on the static load corresponds to a condition of corrosion fatigue under high mean stress conditions. Pao and Bayles[68] have taken both EAC and fatigue threshold data (see Section 13.4.2) into account to determine the material/environment system susceptibility to ripple loading. They determined that the percentage of degradation of a K_{IEAC} value is given by

$$\% \text{ degradation} = \left[1 - \frac{\Delta K_{th}}{K_{IEAC} (1 - R)} \right] 100 \qquad (11\text{-}1)$$

where $\Delta K_{th} =$ fatigue threshold stress intensity range ($K_{max} - K_{min}$)
(see Section 13.4.2)

$$R = \frac{K_{min}}{K_{max}} = \frac{P_{min}}{P_{max}}$$

In the limit, no ripple loading effects are experienced when $\Delta K_{th}/(1 - R) \geq K_{IEAC}$.

It is believed that such small amplitude cyclic loadings induce oxide-film rupture at the crack tip, which facilitates the cracking process. Since some structures typically experience load fluctuations during their lifetime, it follows that K_{IEAC} values based on traditional static test methods may prove to be nonconservative in assessing the EAC susceptibility of a given material and component. Finally, it has been shown that environment-assisted cracking is influenced by both mechanical and electrochem-

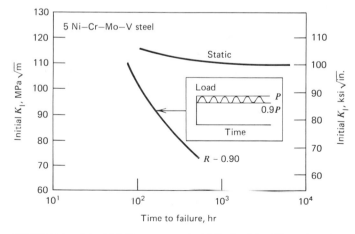

FIGURE 11.16   EAC response in 5Ni–Cr–Mo–V steel under static and ripple loading conditions in salt water at room temperature. $R = 0.9 = K_{min}/K_{max}$. (Adapted from Ref. 66 with permission.)

ical variables. Regarding the latter, Gangloff and Turnbull[69] reported that electro-chemical driving forces are influenced by differences in crack size (especially short cracks), shape, and crack opening.

Engineers and scientists have sought other means of quantifying EAC processes. In this regard, considerable attention has been given to characterize the kinetics of the crack growth rate process by monitoring the rate of crack advance da/dt as a function of the instantaneous stress-intensity level. From the work of Wieder-horn[70–72] on the static crack growth of glass and sapphire in water, a log da/dt–K relation was determined, which took the form shown schematically in Fig. 11.17. Three distinct crack growth regimes are readily identified. In Region I, $(da/dt)_I$ is found to depend strongly on the prevailing stress-intensity level, along with temperature, pressure, and the environment. For some materials, the slope of this part of the curve is so steep as to allow for an alternative definition of K_{IEAC}; that is, the K level below which da/dt becomes vanishingly small. For aluminum alloys that do not appear to exhibit a true threshold level and which have a shallower Stage I slope, a "K_{IEAC}" value can be defined at a specific da/dt level much the same as the yield strength of a material exhibiting Type II behavior (Section 1.2.2.1) is defined by the 0.2% offset method. Environment-assisted crack growth is often relatively independent of the prevailing K level in Region II, but it is still affected strongly by temperature, pressure, and the environment. Finally, Region III reflects a second regime where da/dt varies strongly with K. In the limit, crack growth rates become unstable as K approaches K_{IC}.

In addition to the three *steady-state* crack growth regimes just described, a number of additional *transient* growth regions have been identified followed by a dormant or incubation period prior to steady-state growth.[73] Consequently, the total time to fracture is the summation of transient, incubation, and steady-state cracking periods. Hence

$$t_T = t_{tr} + t_{inc} + t_s \tag{11-2}$$

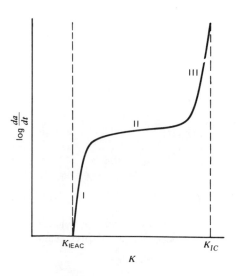

FIGURE 11.17 Diagram showing three stages of environment-assisted cracking under sustained loading in an aggressive atmosphere. Lower and upper K limits of plot determined by K_{IEAC} and K_{IC}, respectively.

where t_T = total time to failure

t_{tr} = total time during transient crack growth

t_{inc} = incubation time

t_s = time of Region I, II, and III steady-state crack growth.

The transient time t_{tr} is usually small relative to t_{inc} and t_s and is often ignored in life computations. The relative importance of the other two regimes in affecting total life is shown schematically in Fig. 11.18. Note that the incubation time decreases rapidly with increasing initial K level.[73–76] (Higher test temperatures also decrease t_{inc}.[75]) Since the initial K level of the bolt-loaded, constant displacement type K_{IEAC} test sample is large, this configuration is preferred over the cantilever beam geometry. Furthermore, in material–environment systems where da/dt is high, incubation times are short and the influence of $\dot{K}$ in rising load tests is limited. As noted previously, the very long times to failure at low K values suggest a threshold condition. It should be recognized, however, that the incubation period represents a large part of the time to failure. As a result, initial crack growth rate readings often are abnormally low, suggesting the existence of an erroneously high K_{IEAC} level. These data, therefore, should be used with extreme caution.

11.2.1 Major Variables Affecting Environment-Assisted Cracking

The degree to which materials are subject to EAC depends on a number of factors, including alloy chemistry and thermomechanical treatment, the environment itself, temperature, and pressure (for the case of gaseous atmospheres). The effect of these important variables on the cracking process will now be considered.

11.2.1.1 Alloy Chemistry and Thermomechanical Treatment

As one might expect, many studies have been conducted to examine the relative EAC propensity of different families of alloys and specific alloys thermomechanically

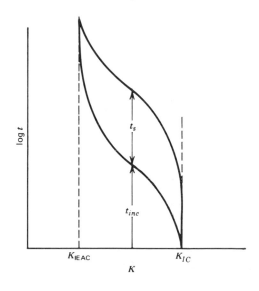

FIGURE 11.18 Diagram showing time for crack incubation and steady-state growth as a function of applied K level. Note smaller incubation time at higher K values.

treated to different specifications. For example, the log *da/dt–K* plot for several high-strength aluminum alloys exposed to alternate immersion in a 3.5% NaCl solution reveals the 7079-T651 alloy to be markedly inferior relative to the response of the other alloys (Fig. 11.19).[77] For example, Stage II cracking in the 7079 alloy occurs at a rate 1000 times greater than in the 7178-T651 alloy and corresponds to a Stage II crack growth rate of greater than 3 cm in 1 hr. No engineering component would be expected to resist final failure for long at that growth rate. During the past 10 years, many investigations have been conducted to improve the EAC resistance of these materials. These studies indicate that overaging is the most effective way to accomplish this objective.[77,78] Coincidentally, toughness is improved while strength decreases as a result of the overaging process. The effect of overaging (denoted by the -T7 temper designation) on 7079 and 7178 aluminum alloys is shown in Fig. 11.20. Although Stage I in the 7079 alloy is shifted markedly to higher *K* levels,

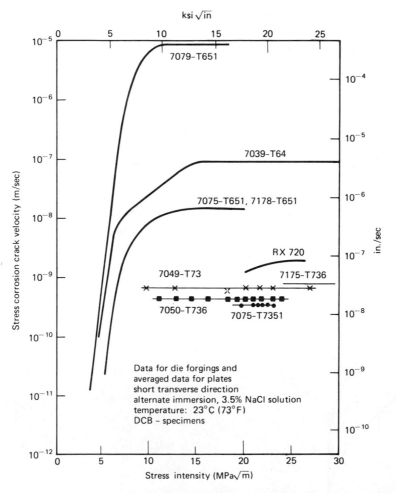

FIGURE 11.19 Environment-assisted cracking in 3.5% NaCl solution for several aluminum alloys and heat treatments.[77] (Reprinted with permission of the American Society for Metals.)

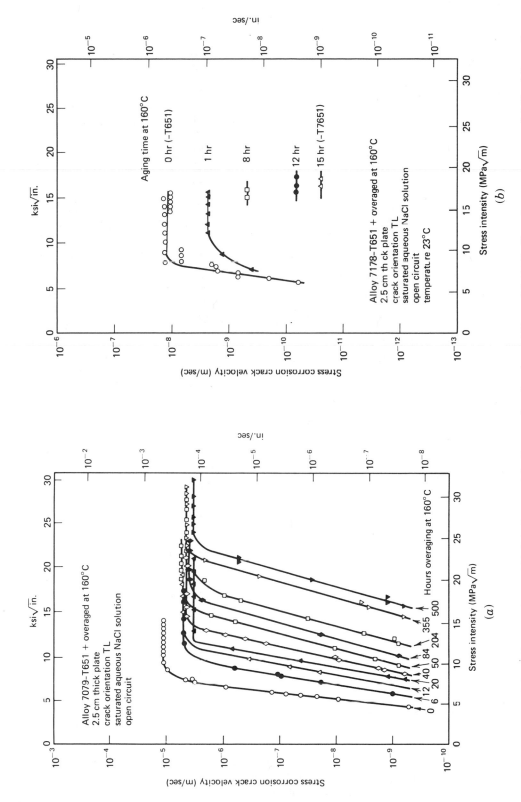

FIGURE 11.20 Effect of overaging on EAC (salt water) in 7xxx series aluminum alloys: (a) 7079 alloy shows pronounced shift of Stage I behavior to higher K levels while $(da/dt)_{\mathrm{II}}$ remains relatively constant; (b) 7178 alloy shows sharp drop in $(da/dt)_{\mathrm{II}}$. (With permission of Markus O. Speidel, Brown Boveri Co.)

reflecting a sharp increase in K_{IEAC}, the growth rates associated with Stage II cracking remain relatively unchanged. Consequently, the major problem of very high Stage II cracking rates in this material remains even after overaging. By contrast, preliminary data for the 7178 alloy show a marked decrease in Stage II crack growth rate with increasing aging time, while Stage I cracking is shifted to a much lesser extent.[77,78] Note the dramatic six order of magnitude difference in Stage II crack growth rates between these two alloys in the overaged condition. Upon reflection, it would be most desirable to have the overaging treatment effect a *simultaneous* lowering of the Stage II cracking rate and a displacement of the Stage I regime to higher K levels. This may prove to be the case in other alloy systems.

In general, K_{IEAC} values tend to be greater in materials possessing higher K_{IC} levels and lower yield strength. For example, we see from Fig. 11.21 that K_{IEAC} values decrease rapidly with increasing yield strength in a 4340 steel.[79] However, Speidel[77,78] has noted exceptions to this rule for the case of aluminum alloys. If the relative degree of susceptibility to environment-assisted cracking is defined by the ratio K_{IEAC}/K_{IC}, the generally observed trend is for K_{IEAC}/K_{IC} to decrease with increasing alloy strength. That is, K_{IEAC} values drop faster than K_{IC} values with increasing strength.

Most often, hydrogen and stress corrosion cracks follow an intergranular path in the important high-strength steel, titanium, and aluminum alloys (see Section 7.7.3.3). This is particularly true for aluminum alloys, which suffer stress corrosion cracking almost exclusively along grain-boundary paths. Consequently, environment-assisted cracking in wrought alloys is usually of greater concern in the short transverse direction than in other orientations. As such, EAC orientation sensitivity parallels K_{IC} orientation dependence as described in Section 10.2.

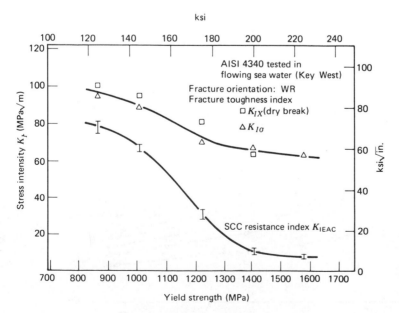

FIGURE 11.21 **Effect of yield strength on K_{IC} and K_{IEAC} (in water) in 4340 steel.[79] (Reprinted with permission from M. H. Peterson, B. F. Brown, R. L. Newbegin, and R. E. Grover, *Corrosion* 23 (1967), National Association of Corrosion Engineers.)**

Takeda and McMahon[80] addressed the controversy regarding the fracture micromechanism associated with hydrogen embrittlement in steel. On the one hand, test conditions have been reported in which the presence of hydrogen allowed plastic flow (i.e., dislocation movement) to occur at abnormally low stress levels and resulted in transgranular fracture; the enhanced level of plastic flow in connection with HAC represents a *strain*-controlled fracture mode. On the other hand, countless examples of hydrogen-induced intergranular failure have been reported in the literature. In this instance, the fracture mode is believed to be *stress*-controlled and influenced by the level of impurity elements segregated at grain boundaries (recall Section 10.6). Takeda and McMahon[80] observed *both* mechanisms in quenched and tempered steel samples possessing virtually the same microstructure and strength. However, the samples were distinguished by slight, but significant, differences in alloy composition. When samples were aged to segregate impurities to the prior austenite grain boundaries, these precracked samples exhibited intergranular failure when loaded in tension in the presence of gaseous hydrogen. In samples that did not contain embrittled prior austenite grain boundaries due to impurity segregation, failure was associated with transgranular fracture (cleavage and microvoid coalescence). Therefore, hydrogen-assisted cracking, associated with an intergranular crack path, was shown to be dependent on the presence of impurity elements at prior austenite grain boundaries, whereas transgranular failure was attributed to the intrinsic enhancement of dislocation motion by the presence of hydrogen in the metal lattice.

K_{IEAC} data for selected materials are listed in Table 11.2. Note the sharp disagreement in results for several aluminum alloys. Since Speidel's values are based on K levels associated with crack velocities less than about 10^{-10} m/s and were conducted over a long time period, they provide a more representative and conservative estimate of the material's environmental sensitivity. Additional K_{IEAC} information is provided in Ref. 81 which also contains numerous log da/dt–K plots for aluminum, steel, and titanium alloys.

11.2.1.2 Environment

As one might expect, the kinetics of crack growth and the threshold K_{IEAC} level depend on the material–environment system. In fact, the reality of this situation is reflected by the formalization of various cracking processes, such as stress corrosion cracking (generally involving aqueous solutions), hydrogen gas embrittlement, and liquid-metal embrittlement. As mentioned at the beginning of this chapter, the complex aspects of the material–environment interaction can be greatly simplified by treating the problem from the phenomenological viewpoint in terms of a single mechanism, environmental-assisted cracking. This concept is supported by Speidel's results shown in Fig. 11.22, which reveal parallel Stage I and II responses for the 7075 aluminum alloy in liquid mercury and aqueous potassium iodide environments.[79] Obviously, the liquid metal represents a more severe environment for this aluminum alloy (some five orders of magnitude difference in Stage II cracking rate), but the phenomenology is the same. Furthermore, we see that the alloy in the overaged condition is more resistant to the liquid-metal EAC, as was the case for the salt solution results discussed above.

Environment-assisted cracking in dry gases does not appear to occur in aluminum alloys.[77] However, with increasing moisture content, cracking develops with increas-

TABLE 11.2 Selected K_{IEAC} Data[81]

Metal	Environment	Test Orientation	Yield Strength MPa	ksi	K_{IC} or (K_{IX})[a] MPa√m	ksi√in.	K_{IEAC} MPa√m	ksi√in.	Test Time (hr)
Aluminum Alloys									
2014-T6	Synth. seawater	S-L	420	61	21	19	18	16	—
2014-T6	NaCl solution	S-L	—	—	—	—	≈8	≈7	≈10,000[b]
2024-T351	3.5% NaCl	S-L	325	47	(55)	(50)	11	10	—
2024-T351	NaCl solution	S-L	—	—	—	—	≈9	≈8	≈10,000[b]
2024-T852	Seawater	S-L	370	54	19	17.6	15	14	≈10,000[b]
2024-T852	NaCl solution	S-L	—	—	—	—	≈17	≈15	≈10,000[b]
2024-T851	Dist. water	L-T	410	59	21	18.6	24	22	—
7075-T6	3.5% NaCl	S-L	505	73	25	23	21	19	—
7075-T6	NaCl solution	S-L	—	—	—	—	≈8	≈7	≈10,000[b]
7075-T7351	3.5% NaCl	S-L	360	52	26	24	23	21	—
7075-T7351	NaCl solution	S-L	—	—	—	—	≤22	≤20	≈10,000[b]
7075-T7351	3.5% NaCl	T-L	365	53	32	29	26	24	—
7175-T66	3.5% NaCl	—	525	76	32	29	≤6.6	≤6	—
7175-T66	NaCl solution	S-L	—	—	—	—	7	6	≈10,000[b]
7175-T736	NaCl solution	—	455	66	27	25	21	19	>1,029
Steel Alloys									
18 Ni(300)-maraging	"	T-L	1960	284	80	72	8	7.5	>150
4340	"	T-S	1335	194	79	72	9	8.5	>333
4340	"	L-T	1690	245	56	51	17	15	>58

4340	Seawater	T-L	1550	225	(69)	(63)	6	5	>20
"	"	"	1380	200	(65)	(59)	11	10	—
"	"	"	1205	175	(83)	(75)	30	27	—
"	"	"	1035	150	(94)	(85)	65	59	—
"	"	"	860	125	(98)	(89)	77	70	—
300M	3.5% NaCl	L-S	1735	252	70	64	22	20	—
"	"	T-L	1725	250	61	56	20	18	—
Titanium Alloys									
Ti-6Al-4V	3.5% NaCl	L-T	890	129	104	95	39 ± 10	35 ± 9	—
"	"	L-S	890	129	99	90	45 ± 8	41 ± 7	—
Ti-8Al-1Mo-1V	"	T-S	825	120	97	88	25	23	—
"	"	"	745	108	123	112	31	28	—
"	Water	T-L	855	124	(105)	(95)	29	26	—
"	Methanol	"	855	124	(105)	(95)	15	14	—
"	CCl₄	"	855	124	(105)	(95)	22	20	—
"	Water ± 21000 ppm Chloride	"	1035	150	(74)	(67)	15	14	—
"	Water ± 100 ppm Chloride	"	1035	150	(65)	(59)	23	21	—
"	Water + 0.1 ppm Chloride	"	1035	150	(65)	(60)	27	24	—

[a] Numbers in parentheses are invalid K_{IC} values which do not satisfy Eq. 8-49.

[b] M. O. Speidel and M. W. Hyatt, *Advances in Corrosion Science and Technology*, Vol. 2, Plenum, New York, 1972, p. 115.

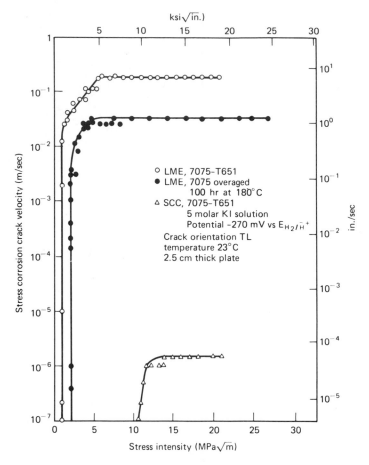

FIGURE 11.22 Environment-assisted cracking with liquid mercury and aqueous io-dide solution in 7075 aluminum alloy.[78] (With permission of Markus O. Speidel, Brown Boveri Co.)

ing speed (Fig. 11.23). Consequently, EAC in aluminum alloys may take the form of stress corrosion cracking and liquid-metal embrittlement but not gaseous hydrogen embrittlement.*

11.2.1.3 Temperature and Pressure

Since EAC processes involve chemical reactions, it is to be expected that temperature and pressure would be important variables. Test results, such as those shown in Fig. 11.24a, for hydrogen cracking in a titanium alloy show the strong effect of temperature on the Stage II cracking rate.[82] These data can be expressed mathematically in the form

---

* There is some debate, however, as to whether aqueous stress corrosion cracking in aluminum is related to hydrogen embrittlement. See M. O. Speidel, *Hydrogen in Metals*, I. M. Bernstein and A. W. Thompson, Eds., ASTM, Metals Park, OH, 1974, p. 249.

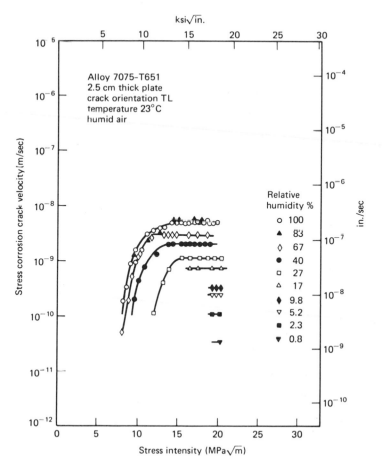

FIGURE 11.23 Effect of humidity on EAC in 7075-T651 aluminum alloy.[78] (With permission of Markus O. Speidel, Brown Boveri Co.)

$$\left(\frac{da}{dt}\right)_{\text{II}} \propto e^{-\Delta H/RT} \qquad (11\text{-}3)$$

where ΔH = activation energy for the rate-controlling process.

The apparent activation energy may then be compared with other data to suggest the nature of the rate-controlling process. Recall from Chapter 5 that at $T_h > 0.5$, ΔH_{creep} was approximately equal to the activation energy for self-diffusion in many materials. In similar fashion, it has been found that the apparent activation energies for the cracking of high-strength steel in water and humidified gas are both about 38 kJ/mol,[3] which corresponds to the activation energy for hydrogen diffusion in the steel lattice.[83] On the other hand, recent studies have shown that the apparent activation energy for Stage II cracking in the presence of gaseous hydrogen is only 16 to 17 kJ/mol.[74,84] Since the embrittling mechanism appears to be the same for the two

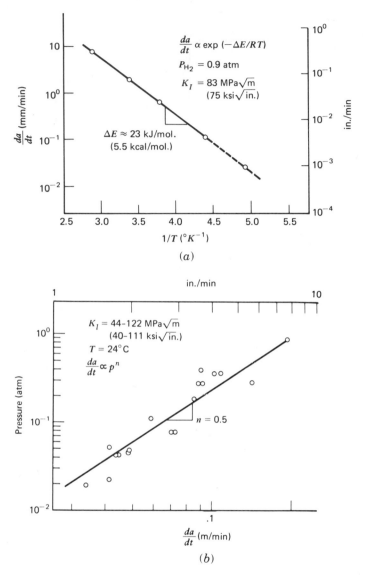

FIGURE 11.24 Effect of (*a*) temperature and (*b*) pressure on hydrogen-induced cracking in Ti–5Al–2.5Sn in Region II.[82] (From D. P. Williams, "A New Criterion for Failure of Materials by Environment-Induced Cracking." *International Journal of Fracture* **9**, 63–74 (1973), published by Noordhoff, Leyden, The Netherlands.)

environments[84] (e.g., the fracture path is intergranular in both cases [see Section 7.7.3.3]), the change in ΔH probably reflects differences in the rate-controlling hydrogen-transport process. In this regard, note that the cracking rate in gaseous hydrogen is higher than that in water (Fig. 11.5).

The increase in Stage II crack growth rate with increasing pressure noted in Fig. 11.24*b* can be described mathematically in the form

$$\left(\frac{da}{dt}\right)_{\mathrm{II}} \propto P^n \qquad (11\text{-}4)$$

In all likelihood, increased pressure enhances hydrogen transport, which in turn increases the cracking rate. At present, there is considerable debate regarding the magnitude of the exponent n.

11.2.2 Environment-Assisted Cracking in Plastics

Engineering plastics are also susceptible to EAC with the extent of structural degradation dependent on the material–environment system, applied stress, and temperature. Examples of such damage include the development of an extensive network of crazes in my friend's drinking glass and the generation of cracks in the shower head used in my home (Fig. 11.25). The extensive array of crazes in the glass resulted from the presence of a residual tensile stress field (generated from thermal cycling in a dishwasher) and simultaneous exposure to alcohol (specifically, a few stiff gin and tonics). Cracking in the shower head was caused by exposure to hot water in the presence of the constant stress produced by tightening the fitting to the water pipe.

Of far greater commercial importance is the premature failure of PVC and polyethylene water and gas transmission pipelines throughout the world. Numerous studies have shown that the rupture life of pipe resin increases with decreasing stress level and temperature (Fig. 11.26). (It is also known that rupture life increases with increasing M_w.) At high applied stress levels (Region A), pressurized pipe samples fail by extensive plastic deformation associated with bulging or ballooning of the pipe's cross section. At lower stress levels (Region B), failure occurs in a brittle manner with little evidence of deformation (recall Section 7.7.1).[85–88] It should be noted that such brittle

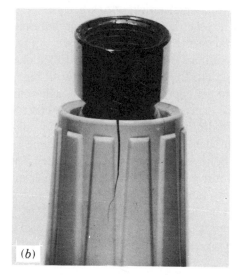

FIGURE 11.25   Environment-assisted cracking in household plastic components. (*a*) **Extensive craze formation in acrylic-based drinking glass (courtesy Elaine Vogel) and** (*b*) **crack formation in plastic showerhead (courtesy Linda Hertzberg).**

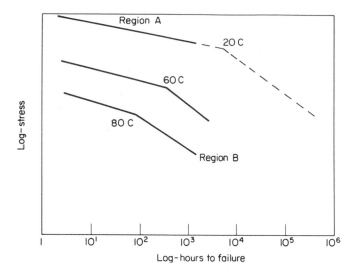

FIGURE 11.26 Environment-assisted cracking lifetime increases with decreasing stress and temperature. Region A corresponds to pipe bulging and Region B is associated with brittle cracking.[88] [Reprinted from R. W. Hertzberg, *Polymer Communications*, **26**, 38 (1985), by permission of the publishers, Butterworth & Co. (Publishers) Ltd. ©.]

fracture is typical of many service failures in gas pipeline systems.[86,87] The locus of failure times at different stress levels is found to be sensitive to the test environment; for example, the addition of a detergent to water exaggerates the level of EAC damage in the pipe resin as compared with that caused by exposure to water alone.

The applicability of fracture mechanics concepts to the study of the kinetics of environmental cracking has been reported.[89,90] So long as the applied K level does not result in extensive crack-tip deformation, da/dt–K plots are typically developed with a form similar to that shown in Fig. 11.17.[90] Furthermore, variation of the test temperature of the EAC test enables one to determine the activation energy of the rate-controlling EAC process (recall Fig. 11.24a). For example, Chan and Williams[89] found the energy of cracking in high-density polyethylene to be approximately equal to the energy associated with the glass transition temperature (α-relaxation peak).

Environment-assisted cracking in polyethylene and polyacetal copolymer results in the development of a tufted fracture surface appearance (Fig. 11.27). These tufts are believed to represent fibril extension and subsequent rupture. For the case of polyethylene, tufting is observed over more than two orders of magnitude of crack growth rate, with the number and length of tufts decreasing at lower crack velocities in association with lower stress intensity levels.[89,91] The identification of this fracture feature is often a clue for the presence of environment-assisted cracking in water pipe systems.

11.3 LIFE AND CRACK-LENGTH CALCULATIONS

Kinetic crack growth data can be integrated to provide estimates of component life and crack length as a function of time.[73,82] For reactions occurring in parallel (Eq.

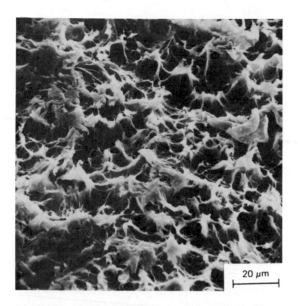

20 µm

FIGURE 11.27 Environment-assisted cracking of polyacetal copolymer under stress in hot water. Tufted fracture surface appearance associated with elongated fibrils that eventually ruptured.

5-9), the effective steady-state cracking rate is controlled by the slowest process acting in Regions I, II, and III. If one ignores the contribution of $(da/dt)_{III}$ ($= \dot{a}_{III}$), then the controlling crack growth rate is given by

$$\frac{1}{\dot{a}_T} = \frac{1}{\dot{a}_I} + \frac{1}{\dot{a}_{II}} \tag{11-5}$$

or

$$\dot{a}_T = \frac{\dot{a}_I \dot{a}_{II}}{\dot{a}_I + \dot{a}_{II}} \tag{11-6}$$

Upon rearrangement of terms, the time devoted to steady-state cracking is

$$t = \int_0^t dt = \int_{a_0}^{a_i} \frac{\dot{a}_I + \dot{a}_{II}}{\dot{a}_I \dot{a}_{II}} \, da \tag{11-7}$$

To solve Eq. 11-7, expressions for $\dot{a}_{I,II}$ are needed in terms of K and the crack length a. In Stage I

$$\dot{a}_I = f(K, T, P, \text{environment}) \tag{11-8}$$

where K = stress intensity factor
       T = temperature
       P = pressure

Since log da/dt–K plots are often linear, Williams[82] has suggested that

$$\dot{a}_{\mathrm{I}} = C_1 e^{mK} \tag{11-9}$$

where C_1 and m are independent of K but may depend on T, P, and environment. For Region II

$$\dot{a}_{\mathrm{II}} = f\,(T,\,P,\,\text{environment}) = C_2 \tag{11-10}$$

Note the lack of K dependence in $\dot{a}_{\mathrm{II}}$ and the fact that C_2 depends on T, P, and environment. From the previous discussion in Section 11.2.1, C_2 can be evaluated by combining Eqs. 11-3 and 11-4, such that

$$\dot{a}_{\mathrm{II}} = C_3 P^n e^{-\Delta H/RT} \tag{11-11}$$

By combining Eqs. 11-9 and 11-11 into Eq. 11-7, it is possible to calculate the length of a crack at any given time, once the various constants are determined from experimental data. The value of Eq. 11-7 lies in its potential to estimate failure times for conditions of T, P, and environments beyond those readily examined in a test program. This life-prediction procedure is now being used in the design of glass components.[92] One additional subtle point should be made regarding the life computation. It should be recognized that the life of a component or test specimen will depend on the rate of change of the stress-intensity factor with crack length dK/da. Consequently, for the same initial K level, the sample with the lowest dK/da characteristic will have the longest life. That is, changing specimen geometry would alter the time to failure. Much work is needed to determine the best method of loading and the best specimen geometry so that a standard EAC test procedure may be established.

REFERENCES

1. R. W. Staehle, Proceedings of Conference, *Fundamental Aspects of Stress Corrosion Cracking,* R. W. Staehle, A. J. Forty, and D. van Rooyen, Eds. NACE, Houston, 1969.
2. R. M. Latanision, O. H. Gastine, and C. R. Conpeau, *Environment-Sensitive Fracture of Engineering Materials,* Z. A. Foroulis, Ed., AIME, Warrendale, PA, 1979, p. 48.
3. A. J. Bursle and E. N. Pugh, *ibid.,* p. 18.
4. A. W. Thompson, *ibid.,* p. 18.
5. A. W. Thompson and I. M. Bernstein, *Advances in Corrosion Science and Technology,* Vol. 7, R. W. Staehle and M. G. Fontana, Eds., Plenum, New York, 1980, p. 53.
6. F. P. Ford, General Electric Report No. 80CRD141, June 1980.
7. R. C. Newman and R. P. M. Proctor, *Brit. Corr. J.* **25,** 259 (1990).
8. W. W. Gerberich, P. Marsh, and H. Huang, *Parkins Symposium on Fundamental Aspects of Stress Corrosion Cracking,* S. M. Bruemmer et al., eds., TMS, Warrendale, PA, 191 (1992).

9. H. K. Birnbaum, *Environment-Induced Fracture of Metals,* R. P. Gangloff and M. B. Ives, eds., NACE, Houston, TX, 21 (1990).
10. R. N. Parkins, *J. Metals* **44**(12), 12 (1992).
11. D. P. Williams III, P. S. Pao, and R. P. Wei, *Environment-Sensitive Fracture of Engineering Materials,* Z. A. Foroulis, Ed., AIME, Warrendale, PA, 1979, p. 3.
12. J. P. Hirth, *Met. Trans.* **11A,** 861 (1980).
13. C. D. Beachem, *Met. Trans.* **3,** 437 (1972).
14. M. G. Nicholas and C. F. Old, *J. Mater. Sci.* **14,** 1 (1979).
15. N. S. Stoloff, *Environment-Sensitive Fracture of Engineering Materials,* Z. A. Foroulis, Ed., AIME, Warrendale, PA, 1979, p. 486.
16. G. E. Linnert, *Welding Metallurgy,* Vol. 2, American Welding Society, New York, 1967.
17. W. Hofmann and W. Rauls, *Weld. J.* **44,** 225s (1965).
18. J. B. Steinman, H. C. VanNess, and G. S. Ansell, *Weld. J.* **44,** 221s (1965).
19. G. G. Hancock and H. H. Johnson, *Trans. Met. Soc. AIME* **236,** 513 (1966).
20. D. O. Hayward and B. M. W. Trapnell, *Chemisorption,* 2d ed., Butterworths, Washington, DC, 1964.
21. F. J. Norton, *J. Appl. Phys.* **11,** 262 (1940).
22. G. L. Hanna, A. R. Troiano, and E. A. Steigerwald, *Trans. Quart. ASM* **57,** 658 (1964).
23. J. K. Tien, A. W. Thompson, I. M. Bernstein, and R. J. Richards, *Met. Trans.* **7A,** 821 (1976).
24. C. A. Zapffe, *JISI* **154,** 123 (1946).
25. A. S. Tetelman and W. D. Robertson, *Trans. AIME* **224,** 775 (1962).
26. A. R. Troiano, *Trans. ASM* **52,** 54 (1960).
27. J. G. Morlet, H. H. Johnson, and A. R. Troiano, *JISI* **189,** 37 (1958).
28. N. J. Petch and P. Stables, *Nature (London)* **169,** 842 (1952).
29. F. E. Fujita, *Hydrogen in Metals,* Vol. 5, paper 2B10, Pergamon, New York, 1977.
30. *Effect of Hydrogen on Behavior of Materials,* AIME, Sept. 7–11, 1975, Moran, Wyoming.
31. I. M. Bernstein and A. W. Thompson, Eds., *Hydrogen Effects in Metals,* AIME, Warrendale, PA, 1980.
32. R. W. Staehle, *The Theory of Stress Corrosion Cracking in Alloys,* NATO, Brussels, 1971, p. 223.
33. R. W. Staehle, *Stress Corrosion Cracking and Hydrogen Embrittlement of Iron-Base Alloys,* NACE, Houston, 1977, p. 180.
34. M. T. Hahn and E. N. Pugh, ASTM, *STP 733,* 1981, p. 413.
35. W. Rostoker, J. M. McCaughey, and M. Markus, *Embrittlement by Liquid Metals,* Van Nostrand-Reinhold, New York, 1960.
36. V. I. Likhtman, E. D. Shchukin, and P. A. Rebinder, *Physico-Chemical Mechanics of Metals,* Acad. Sci. USSR, Moscow, 1962.
37. M. H. Kamdar, *Prog. Mater. Sci.* **15,** 1 (1973).
38. N. J. Kelly and N. S. Stoloff, *Met. Trans.* **6A,** 159 (1975).
39. A. R. C. Westwood, C. M. Preece, and M. H. Kamdar, *Fracture,* Vol. 3, H. Leibowitz, Ed., Academic, New York, 1971, p. 589.
40. M. H. Kamdar, *Treatise on Materials Science and Technology,* Vol. 25, C. L. Briant and S. K. Banerji, Eds., Academic, New York, 1983, p. 362.

41. D. W. Fager and W. F. Spurr, *Corrosion-NACE* **27,** 72 (1971).
42. S. Mostovoy and N. N. Breyer, *Trans. Quart. ASM* **61,** 219 (1968).
43. A. R. C. Westwood and M. H. Kamdar, *Philos. Mag.* **8,** 787 (1963).
44. N. S. Stoloff and T. L. Johnson, *Acta Met.* **11,** 251 (1963).
45. S. P. Lynch, *Acta Met.* **28,** 325 (1981).
46. H. H. Johnson and P. C. Paris, *Eng. Fract. Mech.* **1,** 3 (1967).
47. B. F. Brown, *Mater. Res. Stand.* **6,** 129 (1966).
48. B. F. Brown and C. D. Beachem, *Corr. Sci.* **5,** 745 (1965).
49. B. F. Brown, *Met. Rev.* **13,** 171 (1968).
50. B. F. Brown, *J. Materials,* JMLSA, **5**(4), 786 (1970).
51. R. A. Oriani and P. H. Josephic, *Acta Metall.* **22,** 1065 (1974).
52. *Standard Test Method for Determining Threshold Stress Intensity Factor for Environment-Assisted Cracking of Metallic Materials Under Constant Load,* ASTM, Philadelphia, December 1993.
53. S. R. Novak and S. T. Rolfe, *J. Mater.* **4,** 701 (1969).
54. R. P. Wei and S. R. Novak, *J. Test. Eval.* **15**(1), 38 (1987).
55. P. McIntyre and A. H. Priest, Report MG/31/72, British Steel Corp., London, 1972.
56. W. G. Clark, Jr. and J. D. Landes, ASTM *STP 610,* 1976, p. 108.
57. K. Hirano, S. Ishizaki, H. Kobayashi, and H. Nakazawa, *J. Test. Eval.* **13**(2), 162 (1985).
58. M. Khobaib, AFWAL-TR-4186, WPAFB, Ohio, 1982.
59. R. A. Mayville, T. J. Warren, and P. D. Hilton, *J. Engl. Mater. tech.* **109**(3), 188 (1987).
60. D. R. McIntyre, R. D. Kane, and S. M. Wilhelm, *Corrosion* **44**(12), 920 (1988).
61. *Stress Corrosion Cracking, The Slow Strain-Rate Technique,* G. M. Ugiansky and J. H. Payer, eds., ASTM *STP665* (1979).
62. *Slow Strain Rate Testing for the Evaluation of Environmentally Induced Cracking: Research and Engineering Applications,* R. D. Kane, Ed., ASTM *STP 1210* (1993).
63. R. D. Kane, *ASTM Standardization News* **21**(5), 34 (1993).
64. L. Raymond, *Metals Handbook,* Vol. 13, 9th ed., ASM, Metals Park, OH, 1987, p. 283.
65. D. L. Dull and L. Raymond, ASTM *STP 543,* 1974, p. 20.
66. T. W. Crooker and J. A. Hauser II, NRL Memo Report 5763, April 3, 1986.
67. R. R. Fessler and T. J. Barlo, *ASTM STP 821,* 368 (1984).
68. P. S. Pao and R. A. Bayles, NRL Publication 190-6320 (1991).
69. R. P. Gangloff and A. Turnbull, *Modeling Environmental Effects on Crack Initiation and Propagation,* TMS-AIME, Warrendale, PA, 55 (1990).
70. S. M. Wiederhorn, *Environment-Sensitive Mechanical Behavior,* Vol. 35, Metallurgical Society Conf., A. R. C. Westwood and N. S. Stoloff, Eds., Gordon & Breach, New York, 1966, p. 293.
71. S. M. Wiederhorn, *Int. J. Fract. Mech.* **42,** 171 (1968).
72. S. M. Wiederhorn, *Fracture Mechanics of Ceramics,* Vol. 4, R. C. Bradt, D. P. H. Hasselman, and F. F. Lange, Eds., Plenum, New York, 1978, p. 549.
73. R. P. Wei, S. R. Novak and D. P. Williams, *Mater. Res. Stand.* **12**(9), 25 (1972).
74. S. J. Hudak, Jr., M. S. Thesis, Lehigh University, Bethlehem, PA, 1972.
75. J. D. Landes and R. P. Wei, *Int. J. Fract.* **9**(3), 277 (1973).

76. W. D. Benjamin and E. A. Steigerwald, *Trans. ASM,* **60,** 547 (1967).
77. M. O. Speidel, *Met. Trans.* **6A,** 631 (1975).
78. M. O. Speidel, *The Theory of Stress Corrosion Cracking in Alloys,* J. C. Scully, Ed., NATO, Brussels, Belgium, 1971, p. 289.
79. M. H. Peterson, B. F. Brown, R. L. Newbegin, and R. E. Groover, *Corrosion* **23,** 142 (1967).
80. Y. Takeda and C. J. McMahon, Jr., *Met. Trans.* **12A,** 1255 (1981).
81. *Damage Tolerant Design Handbook,* MCIC-HB-01, Sept. 1973.
82. D. P. Williams, *Int. J. Fract.* **9**(1), 63 (1973).
83. W. Beck, J. O'M. Bockris, J. McBreen, and L. Nanis, *Proc. R. Soc. London Ser. A.* **290,** 221 (1966).
84. D. P. Williams and F. G. Nelson, *Met. Trans.* **1,** 63 (1970).
85. J. B. Price and A. Gray, Proceedings 4th Int. Conf. Plastic Pipes and Fittings, Plast. Rub. Inst., March 1979, p. 20.
86. E. Szpak and F. G. Rice, Proceedings, 6th Plastic Pipe Sym., April 4–6, 1978, Columbus, OH, p. 23.
87. F. Wolter and M. J. Cassady, *ibid.,* p. 40.
88. R. W. Hertzberg, *Polym. Commun.* **26,** 38 (1985).
89. M. K. V. Chan and J. G. Williams, *Polymer* **24,** 234 (1983).
90. K. Tonyali and H. R. Brown, *J. Mater. Sci.* **21,** 3116 (1986).
91. C. S. Lee and M. M. Epstein, *Polym. Eng. Sci.* **22,** 549 (1982).
92. S. M. Wiederhorn, *Ceramics for High Performance Applications,* Brook Hill, Chestnut Hill, MA, 1974, p. 633.

PROBLEMS

11.1 An investigation was made of the rate of crack growth in a 7079-T651 aluminum plate exposed to an aggressive environment under a static stress σ. A large test sample was used with a single-edge notch placed in the T-L orientation. As indicated in the accompanying table, the rate of crack growth under sustained loading was found to vary with the magnitude of the applied stress and the existing crack length. The material exhibits Regions I and II EAC but not Region III.

Cracking Rate (m/sec)	Applied Stress (MPa)	Crack Length (mm)
10^{-9}	35	5
32×10^{-9}	35	10
1×10^{-6}	70	5
1×10^{-6}	70	7.5

If the K_{IC} for the materials is 20 MPa$\sqrt{m}$, how long would it take to break a sample containing an edge crack 5 mm long under a load of 50 MPa? *Hint:* First establish the crack growth rate relations.

11.2 To avoid slowly cooling through the 400 to 600°C temper embrittlement range, one engineer recommended that a thick section component be water quenched from 850°C to room temperature. Are there any problems with this procedure?

11.3   For the 18 Ni (300)-maraging steel listed in Table 11.3, calculate the stress level to cause failure in a center-notched sample containing a crack 5 mm long. What stress level limit would there have to be to ensure that EAC did not occur in a 3.5% NaCl solution?

11.4   How much faster than the room temperature value would a crack grow in a high-strength steel submerged in water if the temperature were raised 100°C?

11.5   A metal plate is found to contain a single-edge notch and is exposed to a static stress in the presence of an aggressive environment. Representative data obtained from crack-growth measurements are given in the following table:

Data	Cracking Rate (m/s)	Applied Stress (MPa)	Crack Length (mm)
1	1×10^{-9}	30	5
2	4.1×10^{-9}	30	8
3	8×10^{-9}	30	10
4	6.4×10^{-8}	60	5
5	6.4×10^{-8}	60	6
6	6.4×10^{-8}	60	7

(a)   What is the growth rate relation among the cracking rate, stress, and crack size?

(b)   Does the relation change? If so, why?

(c)   What was controlling the cracking process in the regime associated with data 4, 5, and 6?

11.6   The type of glass that Griffith used in his classic series of experiments experiences environment-assisted cracking under sustained loads in the presence of moisture. Knowing the usual climate of the British Isles, does this finding suggest any reevaluation of the fracture energy term that Griffith computed?

CHAPTER 12

CYCLIC STRESS AND STRAIN FATIGUE

Daydreamers have two options for supplementary entertainment: doodling or paper clip bending. The doodler is limited by the amount of paper available, while the paper clip bender's amusement is tragically short-lived—the clip breaks after only a few reversals! This simple example describes a most insidious fracture mechanism— failure does not occur when the component is loaded initially; instead, failure occurs after a certain number of similar load fluctuations have been experienced. The author of a book about metal fatigue[1] began his treatise by describing a photograph of his car's steel rear axle, which had failed: "the final fracture occurring at 6:00 AM just after setting out on holiday." Somewhat less expensive damage, but saddening none-theless, was the failure of my son's vehicle (Fig. 12.1). From an examination of the fracture surfaces, it was concluded that this failure originated at several sites and traveled across the section, with occasional arrest periods prior to final separation. Another fatigue failure generated in my household is shown in Fig. 12.2. The failure of this zinc die-cast doorstop nearly destroyed the lovely crystal chandelier in the front hall of my home. Regardless of the material—steel paper clips and car axles, plastic tricycles, zinc doorsteps—fatigue failures will occur when the component experiences cyclic stresses or strains that produce permanent damage. Since the ma-jority of engineering failures involve cyclic loading of one kind of another, it is appropriate to devote considerable attention to this subject in this chapter, and in Chapter 13. The reader is also referred to Chapter 14 for descriptions of case histories of fatigue failures.

12.1 MACROFRACTOGRAPHY OF FATIGUE FAILURES

A macroscopic examination of many service failures generated by cyclic loading reveals distinct fracture surface markings. For one thing, the fracture surface is gen-erally flat, indicating the absence of an appreciable amount of gross plastic defor-mation during service life. In many cases, particularly failures occurring over a long period of time, the fracture surface contains lines referred to in the literature as "clam

FIGURE 12.1 Fatigue fracture of plastic tricycle. (*a*) General location of failure; (*b*) several origins are identified by arrows. Note characteristic fatigue ringlike markings emanating from each origin, which represent periods of growth during life of component. (Courtesy Jason and Michelle Hertzberg.)

shell markings'' arrest lines and/or ''beach markings'' (Fig. 12.3).* These markings have been attributed to different periods of crack extension, such as during one flight or one sequence of maneuvers of an aircraft or the operation of a machine during a factory work shift. It is to be emphasized that these bands reflect *periods* of growth and are not representative of *individual* load excursions. Unique markings associated with the latter are discussed in Chapter 13. It is believed that these alternate crack

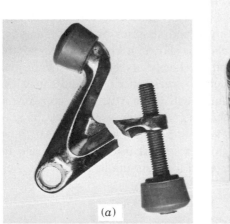 

FIGURE 12.2 Zinc die-cast doorstop fatigue fracture. Fatigue crack grew from corner until reaching a critical size and causing failure. Arrow in (*b*) indicates crack origin.

---

* The use of the expressions ''beach markings'' and ''clamshell'' markings has long been associated with the appearance of certain fatigue fracture surfaces. Clamshell markings on the outer surface of bivalve mollusks represent periods of shell growth and are analogous to macroscopic markings on fatigue fracture surfaces in metals and plastics which correspond to periods of crack grown. No such direct analogy exists for the case of beach markings, which depend on the vagaries of wind and water flow patterns. Nevertheless, reference to ''beach markings'' provides a mental picture of a pattern of parallel lines that typifies the surface of many fatigue fractures.

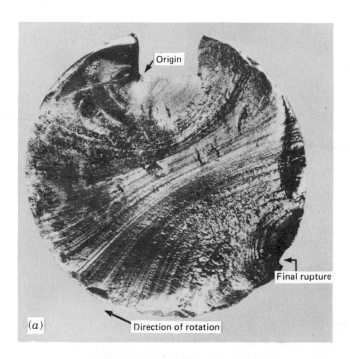

FIGURE 12.3 Fatigue fracture markings. (*a*) Rotating steel shaft. Center of curvature
of earlier "beach markings" locate crack origin at corner of keyway.[2] (By permission
from D. J. Wulpi, *How Components Fail,* copyright American Society for Metals, 1966.)
(*b*) Clam shell markings (C) and ratchet lines (R) in aluminum. Arrows indicate crack
propagation direction (Photo courtesy of R. Jaccard). (*c*) Fatigue bands in high-impact
polystyrene toilet seat. (*d*) Beach markings in South Carolina.

(c)

(d)

FIGURE 12.3 (Continued)

growth and dormant periods cause regions on the fracture surface to be oxidized and/
or corroded by different amounts, resulting in the formation of a fracture surface
containing concentric rings of nonuniform color. Similar bands resulting from variable
amplitude block loading have been found on fracture surfaces (see Sections 13.5.1
and 14.2). Since these "beach markings" often are curved, with the center of curvature
at the origin, they serve as a useful guide to direct the investigator to the fracture
initiation site.

A second set of fracture surface markings are seen in Fig. 12.3*b*. In addition to the set of horizontal clamshell markings (C), one also notes two vertically oriented curved black lines that separate parallel packets of clamshell markings. These lines are called ratchet lines (labeled R) and represent the junction surfaces between the three adjacent crack origins.[2] Since each microcrack is unlikely to form on the same plane, their eventual linkage creates a vertical step on the fracture surface. Once the initial cracks have linked together, the ratchet lines disappears. Hence, the ratchet lines connect contiguous regions where separate cracks had initiated. This point is best illustrated by examination of the fatigue fracture surface in Fig. 12.4. This specimen possessed a polished semicircular notch root from which three fatigue cracks initiated. Note the presence of two small horizontal lines at the edge of the notch root (left side of Fig.

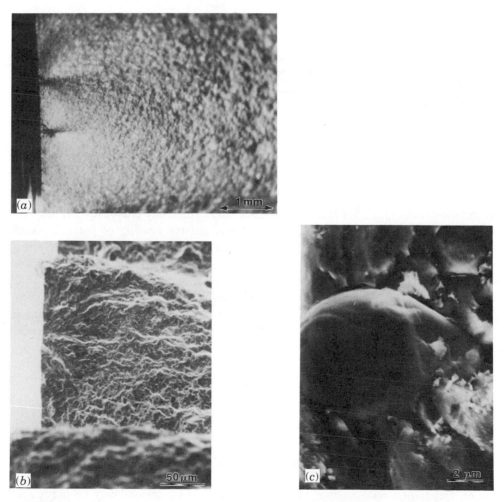

FIGURE 12.4 **Fatigue fracture surface appearance in HSLA steel.** (*a*) **Two ratchet lines separate three crack origins;** (*b*) **SEM image revealing height elevation difference between three cracks;** (*c*) **cerium sulfide inclusion at origin of fatigue crack.** (**From Braglia et al.[7]; copyright, American Society for Testing and Materials, 1916 Race Street, Philadelphia, PA 19103. Reprinted with permission.**)

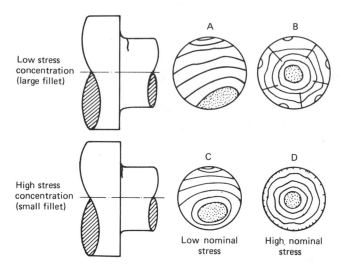

FIGURE 12.5 Diagrams showing typical fatigue fracture surface appearance for varying conditions of stress concentration and stress level.[2] (**By permission from D. J. Wulpi,** *How Components Fail*, **copyright, American Society for Metals, 1966.**)

12.4*a*); these markings are ratchet lines that separate three different cracks that originated on three separate planes (Fig. 12.4*b*). Note the chevronlike pattern on the crack surface in the middle of the photograph. Such markings can be used to locate the crack origin, such as the inclusion that was found at the surface of the polished notch root (Fig. 12.4*c*).

As shown in Fig. 12.5, the fracture surface may exhibit any one of several patterns, depending on such factors as the applied stress and the number of potential crack nucleation sites. For example, we see that as the severity of a design-imposed stress concentration and/or the applied stress increases, the number of nucleation sites and associated ratchet lines increase. Either of these conditions should be avoided if at all possible. In fact, most service failures exhibit only one nucleation site, which eventually causes total failure. The size of this fatigue crack at the point of final failure is related to the applied stress level and the fracture toughness of the material (recall Eq. 8-28).

12.2 CYCLIC STRESS-CONTROLLED FATIGUE

Many engineering components must withstand numerous load or stress reversals during their service lives. Examples of this type of loading include alternating stresses associated with a rotating shaft, pressurizing and depressurizing cycles in an aircraft fuselage at takeoff and landing,* and load fluctuations affecting the wings during flight. Depending on a number of factors, these load excursions may be introduced

---

* For example, recall the recent loss of much of the upper half of the fuselage section from one of Aloha Airlines' Boeing 737 fleet (see *New York Times,* May 1, 1988, p. 1.)

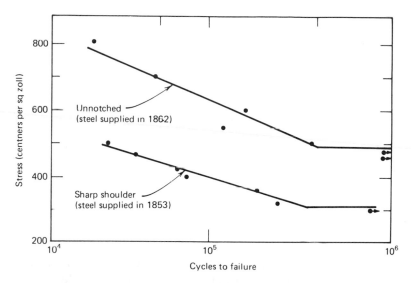

FIGURE 12.6 Wohler's *S-N* curves for Krupp axle steel.[4] (Reprinted by permission of the American Society for Testing and Materials from copyrighted work.)

either between fixed strain or fixed stress limits; hence, the fatigue process in a given situation may be governed by a strain- or stress-controlled condition. Discussion in this section is restricted to stress-controlled fatigue; strain-controlled fatigue is considered in Section 12.3.

One of the earliest investigations of stress-controlled cyclic loading effects on fatigue life was conducted by Wöhler,[3] who studied railroad wheel axles that were plagued by an annoying series of failures. Several important facts emerged from this investigation, as may be seen in the plot of stress versus the number of cycles to failure (a so-called *S-N* diagram) given in Fig. 12.6. First, the cyclic life of an axle increased with decreasing stress level and below a certain stress level, it seemed to possess infinite life—fatigue failure did not occur (at least not before 10^6 cycles). Second, the fatigue life was reduced drastically by the presence of a notch. These observations have led many current investigators to view fatigue as a three-stage process involving initiation, propagation, and final failure stages (Fig. 12.7). When

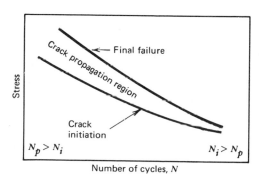

FIGURE 12.7 Fatigue life depends on relative extent of initiation and propagation stages.

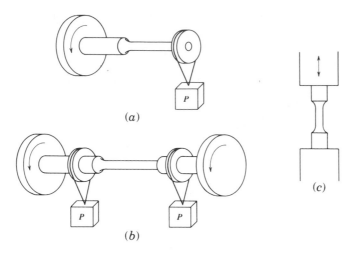

FIGURE 12.8 **Various loading configurations used in fatigue testing. (*a*) Single-point loading, where bending moment increases toward the fixed end; (*b*) beam loading with constant moment applied in gage section of sample; (*c*) pulsating tension or tension–compression axial loading. Test procedures to determine *S-N* diagrams are described in ASTM Standards E466-E468.[5]**

design defects of metallurgical flaws are preexistent, the initiation stage is shortened drastically or completely eliminated, resulting in a reduction in potential cyclic life.

Over the years, laboratory tests have been conducted in bending (rotating or reversed flexure), torsion, pulsating tension, or tension–compression axial loading. Such tests have been conducted under conditions of constant load or moment (to be discussed in this section), constant deflection or strain (Section 12.3), or a constant stress intensity factor (Chapter 13). Examples of different loading conditions are shown in Fig. 12.8. In rotating bending with a single load applied at the end of the cantilevered test bar (Fig. 12.8*a*), the bending moment increases with increasing distance from the applied load point and precipitates failure at the base of the fillet at the end of the gage section. In effect, this represents a notched fatigue test, since the results will depend strongly on fillet geometry. The rotating beam-loaded case (Fig. 12.8*b*) produces a constant moment in the gage section of the test bar that can be used to generate either unnotched or notched test data. Notched test data are obtained by the addition of a circumferential notch in the gage section. Both specimen types generally represent zero mean load conditions or a load ratio of $R = -1 (R \equiv$ minimum load/maximum load). These test specimen configurations and modes of loading may be suitable for evaluating the fatigue characteristics of a component subjected to simple rotating loads. However, it is often more realistic to use the axially loaded specimen (Fig. 12.8*c*) to simulate service conditions that involve direct loading when mean stress is an important variable. Such is the case for aircraft wing loads where fluctuating stresses are superimposed on both a tensile (lower wing skin) and compressive (upper wing skin) mean stress.

Standard definitions regarding key load or stress variables are shown in Fig. 12.9 and defined by

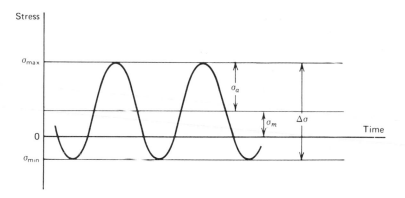

FIGURE 12.9 Nomenclature to describe test parameters involved in cyclic stress testing.

$$\Delta\sigma = \sigma_{max} - \sigma_{min} \tag{12.1}$$

$$\sigma_a = \frac{\sigma_{max} - \sigma_{min}}{2} \tag{12.2}$$

$$\sigma_m = \frac{\sigma_{max} + \sigma_{min}}{2} \tag{12.3}$$

$$R = \frac{\sigma_{min}}{\sigma_{max}} \tag{12.4}$$

Most often, S-N diagrams similar to that shown in Fig. 12.6 are plotted, with the stress amplitude given as half the total stress range. Another example of constant load amplitude fatigue data for 7075-T6 aluminum alloy notched specimens is shown in Fig. 12.10. Note the considerable amount of scatter in fatigue life found among the 10 specimens tested at each stress level. The smaller scatter at high stress levels is believed to result from a much shorter initiation period prior to crack propagation. The existence of scatter in fatigue test results is common and deserving of considerable attention, since engineering design decisions must be based on recognition of the statistical character of the fatigue process. The origins of test scatter are manifold. They include variations in testing environment, preparation of the specimen surface, alignment of the test machine, and a number of metallurgical variables. With regard to the testing apparatus, the least amount of scatter is produced by rotating bending machines, since misalignment is less critical than in axial loading machines.

12.2.1 Effect of Mean Stress on Fatigue Life

As mentioned in the previous section, mean stress can represent an important test variable in the evaluation of a material's fatigue response. It then becomes necessary to portray fatigue life data as a function of two stress variables from the ones defined in Eqs. 12-1 to 12-4. Sometimes this is done by plotting S-N data for a given material

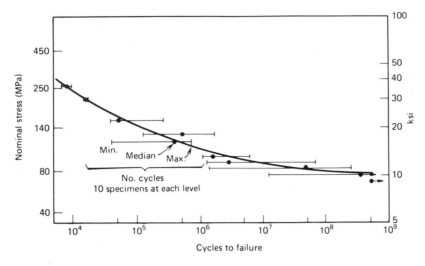

FIGURE 12.10 **Constant load amplitude fatigue data for 7075-T6 aluminum alloy notched specimens (0.25-mm root radius).[6] (Reprinted from Hardrath et al., NASA TN D-210.)**

at different σ_m values, as shown in Fig. 12.11a. Here we see a trend of decreasing cyclic life with increasing σ_m level for a given σ_a level. Alternatively, empirical relations have been developed to account for the effect of mean stress on fatigue life.

$$\text{Goodman relation:} \quad \sigma_a = \sigma_{fat}\left(1 - \frac{\sigma_m}{\sigma_{ts}}\right) \qquad (12\text{-}5)$$

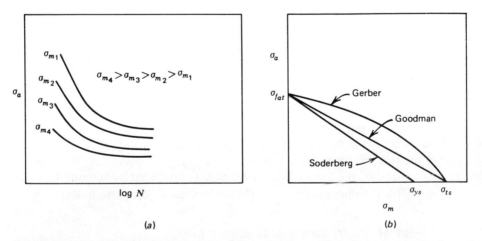

FIGURE 12.11 **Representative plots of data showing effect of stress amplitude and mean stress on fatigue life. (a) Typical S-N diagrams with differing σ_m levels; (b) Gerber, Goodman, and Soderberg diagrams showing combined effect of alternating and mean stress on fatigue endurance.**

$$\text{Gerber relation: } \sigma_a = \sigma_{fat} \left[1 - \left(\frac{\sigma_m}{\sigma_{ts}} \right)^2 \right] \qquad (12\text{-}6)$$

$$\text{Soderberg relation: } \sigma_a = \sigma_{fat} \left(1 - \frac{\sigma_m}{\sigma_{ys}} \right) \qquad (12\text{-}7)$$

where σ_a = fatigue strength in terms of stress amplitude, where $\sigma_m \neq 0$
  σ_m = mean stress
  σ_{fat} = fatigue strength in terms of stress amplitude, where $\sigma_m = 0$
  σ_{ts} = tensile strength
  σ_{ys} = yield strength

These relations are shown in Fig. 12.11b and illustrate the relative importance of σ_a and σ_m on fatigue endurance. Experience has shown that most data lie between the Gerber and Goodman diagrams; the latter, then, represents a more conservative design criteria for mean stress effects.

EXAMPLE 12.1

Suppose that Hertzalloy 100, a certain steel alloy, has an endurance limit and a tensile strength of 700 and 1400 MPa, respectively. Would one expect fatigue failure if a component, manufactured from this alloy, were subjected to repeated loading from 0 to 600 MPa?

From Eq. 12-1, both alternating and mean stress levels are computed to be 300 MPa. The accompanying Goodman diagram shows that this cyclic loading condition (A) lies well within the safe region bounded by the line corresponding to the Goodman relation. Fatigue failure would not be expected. On the other hand, if the component were to possess a residual tensile stress of 700 MPa as a result of a prior welding procedure, the effective mean stress would be 1000 MPa (the sum of the internal residual stress [700 MPa] and the applied external mean stress [300 Mpa]). Under these conditions (B), fatigue failure would be predicted.

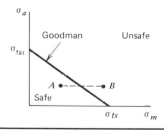

12.2.2 Stress Fluctuation, Cumulative Damage, and Safe-Life Design

Much of the fatigue data discussed thus far were generated from constant amplitude (constant frequency) tests, but these results are not realistic in actual field service conditions. Many structures are subjected to a range of load fluctuations, mean levels, and frequencies. The task, then, is to predict, based on constant amplitude test data, the life of a component subjected to a variable load history. A number of cumulative damage theories, proposed during the past few decades, describe the relative importance of stress interactions and the amount of damage—plastic deformation, crack initiation, and propagation—introduced to a component. For example, if the same amount of damage is done to a component at any stress level as a result of a given fraction of the number of cycles required to cause failure, we see from Fig. 12.12 that $n_1/N_1 + n_2/N_2 + n_3/N_3 = 1$. This may be described in more general form by

$$\sum_{i=1}^{k} \frac{n_i}{N_i} = 1 \tag{12-8}$$

where k = number of stress levels in the block loading spectrum
σ_i = ith stress level
n_i = number of cycles applied at σ_i
N_i = fatigue life at σ_i

Equation 12-8 is the work of Palmgren[8] and Miner[9] and is often referred to as the Palmgren-Miner cumulative damage law. By combining Eq. 12-5 with standard S-N data, one can estimate the total or residual service lifetime of a structural component that experiences multiple load sequences.

EXAMPLE 12.2

A multipurpose traffic bridge has been in service for three years and each day carries a large number of trains, trucks, and automobiles. A subsequent highway analysis reveals a sharp difference between expected and actual traffic patterns that threatens to shorten the useful life span of the bridge. Fortunately, a nearby second bridge was recently completed that can assume all of the train traffic. Given the following fatigue information, estimate the remaining lifetime for the first bridge, assuming that it will carry only truck and automobile traffic.

Vehicle	Fatigue Lifetime	Vehicles / Day
Automobiles	10^8	5000
Trucks	2×10^6	100
Trains	10^5	30

To determine the remaining service lifetime for the bridge, we first need to establish the amount of fatigue damage accumulated during the initial three-year service period.

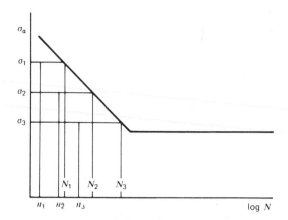

FIGURE 12.12   Component cyclic life determined from $\Sigma\, n_i/N_i = 1$ if damage at σ_i is linear function of n_i and damage is not a function of block sequencing.

For purposes of simplicity, we will assume that there are no load interaction effects corresponding to the three principle stress levels experienced by the bridge members. (A similar assumption is often made by civil engineers, regarding the fatigue life analysis of current bridge decks.) Accordingly, the amount of damage is estimated from the Palmgren-Miner relation where

$$\sum_{i=1}^{k} \frac{n_i}{N_i} = 1$$

For the initial three years of service (1095 days) involving train, truck, and automobile traffic, the amount of bridge damage that incurred, relative to its total design lifetime is estimated to be

$$1095 \left[\frac{5000}{10^8} + \frac{100}{2 \times 10^6} + \frac{30}{10^5} \right] = 0.438$$

Therefore, the remaining fatigue lifetime corresponding to automobile and truck traffic alone will be 56.2% of the combined fatigue lifetime associated with these two stress levels. Therefore,

$$d \left(\frac{5000}{10^8} + \frac{100}{2 \times 10^6} \right) = 0.562$$

where d = remaining days of service.

It follows that an additonal 5620 days or 15.4 years of useful fatigue lifetime remains before the bridge's fatigue design limit is reached. (By comparison, if train traffic were to be continued on this bridge, only 3.85 years of additional useful service life would remain.) At a later date, bridge lifetime could be extended still further by diverting truck traffic to a different route. The residual service life now available for automobile traffic alone could then be calculated in a manner similar to that previously shown, by first adding together the initial increment of life consumed for all traffic (i.e., 0.438) with the lifetime increment corresponding to the second phase of service (i.e., combined automobile and truck traffic). The remaining fraction, based on Eq. 12-8, would then be used to determine the remaining allowable lifetime for automobile traffic alone.

Such computations are used routinely to determine the "safe-life" or durability of numerous engineering components. By further illustration, let us assume that a design engineer assigns a "safe-life" of 1000 cycles to a particular component that experiences a uniform cyclic stress. Based on the S-N data for the component's material, failure after 1000 loading cycles would require an alternating stress of σ_1. The design stress for the component would then be σ_1/F, where F, is the safety factor to account for such variables as: batch-to-batch material property variations, installation-induced differences in component stress level, environmental changes, and existence of adventitious defects (also see Section 12.4). Alternatively, the safety factor could be used to define the allowable cyclic stress necessary for a service lifetime of $F \times 1000$ loading cycles. That is, an allowable stress is determined for a durability of perhaps "two or three" expected component lifetimes. When "safe-life"-designed components reach their lifetime limit, they are removed from service even though they may be defect-free and would surely survive many additional loading cycles. As a result of such conservative design practices, sound parts are often retired from service (or "trashed") with uneconomical consequences. By contrast, the fail-safe design criteria recognizes that cracks can develop in components and provides for a structure that will not fail prior to the time that the defect is discovered and repaired. More will be said about this fatigue design philosophy in Section 13.1.2.

As assumed in Example 12.2, we see that Eq. 12-8 shows no dependence on the order in which the block loads are applied and, as such, represents interaction-free behavior as well. In reality, the Palmgren-Miner law is unrealistic, since the amount of damage accumulated does depend on block sequencing and varies nonlinearly with n_i. For example, if a high load block is followed by a low load block, experimental data in *unnotched* specimens generally indicate $\Sigma n/N < 1$. (The reverse is true for the case of *notched* samples, as is described in Chapter 13; the opposite trend reflects different effects of load interactions on the initiation and propagation stages in the fatigue process.) Since crack propagation begins sooner at the higher stress levels, it is argued that the initial cycles at the second block of lower stress excursions would do more damage than normally anticipated, since the initiation process would have been truncated by the high load block. The deleterious effect of overstressing in unnotched testing is shown in Fig. 12.13. Alternatively, when σ_1/σ_2 is less than unity, $\Sigma n/N > 1$ for some alloys. Such understressing is seen to "coax" the fatigue limit of certain steels that strain age to somewhat higher levels. The use of the Palmgren-Miner cumulative damage law in association with the prediction of fatigue lifetime under random loading conditions is discussed in Section 12.4.

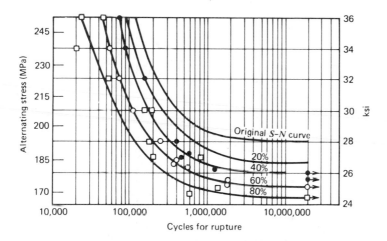

FIGURE 12.13 *S-N* **diagrams showing decreased cyclic life after initial cycling at ±250 MPa for 20, 40, 60, and 80% of anticipated life at that stress for SAE 1020 steel. (Reprinted by permission of the American Society for Testing and Materials from copyright work.)**

12.2.3 Notch Effects and Fatigue Initiation

Fatigue failures almost always initiate at the surface of a component. What are some of the factors contributing to this behavior? First, many stress concentrations, such as surface scratches, dents, machining marks, and fillets, are unique to the surface, as is corrosion attack, which roughens the surface. In addition, cyclic slip causes the formation of surface discontinuities, such as intrusions and extrusions, that are precursors of actual fatigue crack formation. (The processes involved in intrusion and extrusion formation are discussed in Section 12.5.) The data shown in Fig. 12.14 clearly show the serious loss in fatigue limit associated with a deterioration in surface quality. Recall that a similar response was recognized by Wöhler more than 100 years ago (Fig. 12.6).

To quantitatively evaluate the severity of a particular stress concentration, many investigators adopted the stress concentration factor k_t as the comparative key parameter.[4] (From Chapter 8, it is not surprising to find the stress intensity factor also being used in this fashion [see Chapter 13].) Assuming elastic response, the fatigue strength at N cycles in a notched component would be expected to decrease by a factor equal to k_t. For example, if a material exhibits a smooth bar fatigue limit of 210 MPa, the same material would have a fatigue limit of 70 MPa if a theoretical stress concentration factor of 3 were present. In reality, the reduction in fatigue strength at N cycles is less than that predicted by the magnitude of k_t. Rather, the fatigue strength is reduced by a factor k_f, which represents the effective stress concentration factor as affected by plastic flow and by notch root surface area and volume considerations.[12] The relative notch sensitivity q for a given material and notch root detail may be given by[12]

$$q = \frac{1}{1 + p/r} = \frac{k_f - 1}{k_t - 1} \qquad (12\text{-}9)$$

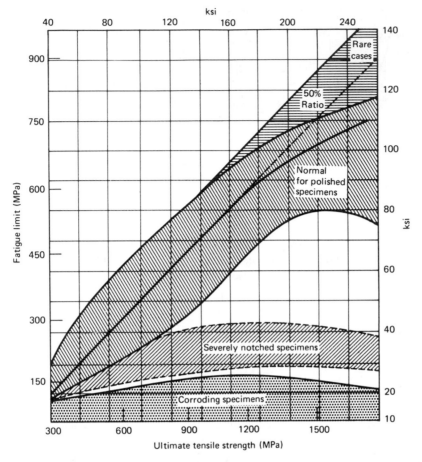

FIGURE 12.14 **Effect of surface condition on fatigue limit in steel alloys.**[11] **(Reprinted with permission from John Wiley and Sons, Inc.)**

where q = notch sensitivity factor, wherein $0 \leqslant q \leqslant 1$
k_t = theoretical stress concentration factor
k_f = effective stress concentration factor
p = characteristic material parameter
r = radius of the notch root

From Fig. 12.15, it is seen that the relative notch sensitivity increases with increasing tensile strength, since high-strength materials usually possess a limited capacity for deformation and crack-tip blunting. Of greater significance, the notch sensitivity factor q decreases markedly with decreasing notch root radii. This results from the fact that k_f increases more slowly than k_t with decreasing notch radius (Eq. 12-9); discrepancies between k_f and k_t as large as a factor of two or three have been noted in some cases. The reason for this apparent paradox (i.e., less severe fatigue damage susceptibility [k_f] as k_t increases) has been attributed to the lack of distinction made between fatigue crack initiation and fatigue crack propagation processes.[14,15] That is,

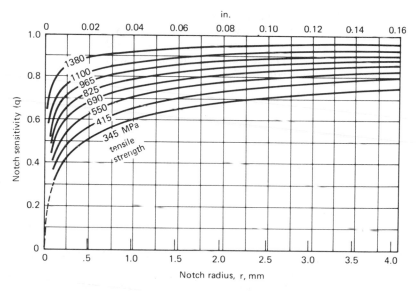

FIGURE 12.15 **Effect of tensile strength and notch acuity on relative notch sensitivity.**[13] (Reprinted with permission from McGraw-Hill Book Co.)

fatigue cracks would be expected to initiate more readily with increasing k_t, but might not always propagate to failure; instead, one might find "nonpropagating cracks" in solids containing stress concentrations beyond some critical value. (See Section 13.4.3 for an expanded discussion of this subject.)

For the present, let us first distinguish between these two stages in the fatigue process. We define initiation life by the number of loading or straining cycles N_i required to develop a crack of some specific size; the propagation stage then corresponds to that portion of the total cyclic life N_p which involves growth of that crack to some critical dimension at fracture. Hence, $N_T = N_i + N_p$ where N_T is the total fatigue life. Does the completion of the crack initiation process correspond to the development of a crack with a length of 1 cm, 1 mm, or 1 μm? Alternatively, should initiation be defined when a newly formed crack attains a length equal to some multiple of the characteristic microstructural unit, such as the grain size in a metal or the spherulitic diameter in a semicrystalline polymer? In general, when does a defect become a crack? These questions have provoked considerable discussion for many years and, yet, no precise definition for crack initiation has been or perhaps can be identified. Part of this difficulty arises from the fact that the fatigue life corresponding to the initiation and growth of a crack to some specified length often depends on the geometry of the test specimen. To illustrate, let us consider the development and growth of a crack from a circular hole in an infinitely wide plate that is subjected to an oscillating tensile stress (Fig. 12.16a). For this discussion, we should recognize that the range of the crack-tip stress intensity factor ΔK, which varies with the cyclic loading history, has a major impact on the rate of growth of the crack. (Chapter 13 deals at considerable length with this subject.) From Section 8.2, the stress intensity factor solution for this crack–hole configuration can be estimated in two distinctly

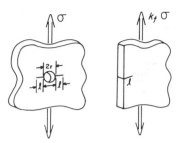

FIGURE 12.16 **Growth of small cracks. (a) From hole with diameter 2r in infinitely wide plate under stress σ. (b) Edge crack in semi-infinite plate with remote stress equal to $k_t\sigma$.**

different ways. At one extreme, when the crack length ℓ is small compared with the hole radius r, the stress intensity factor may be given by

$$K_s = 1.12 k_t \sigma \sqrt{\pi\ell} \tag{8-24}$$

where K_s = stress intensity factor—short-crack solution where the crack tip is embedded within the stress field of the hole

k_t = stress concentration factor for the hole in an infinite plate

σ = remote stress

ℓ = crack length from the surface of the hole

1.12 = surface flaw correction factor

At the other extreme, when $\ell \gg r$, the long-crack stress intensity factor solution is appropriate:

$$K_\ell = \sigma \sqrt{\pi a} \tag{8.25}$$

where K_ℓ = stress intensity factor—long-crack solution where the hole is considered to be part of a long crack

a = crack length = $r + \ell$

r = radius of hole

For the short-crack–hole condition, the stress intensity factor also can be estimated by a single-edge-notched solution where the remote stress is $k_t\sigma$ (i.e., the stress at the surface of the hole; Fig. 12.16b). Hence, if a very small crack or crack nuclei of length ℓ_1 is assumed to exist and we wish to know how many cycles it would take for this defect to grow to a length ℓ_T, both crack configurations would provide the same estimate of cyclic life (Fig. 12.17). However, if one were to define crack initiation when a crack has grown to a length ℓ_2, then different fatigue initiation lives would be found for the two specimen configurations. To avoid such dependence of geometry on the fatigue initiation life, Dowling[14] has recommended that the crack length at initiation be on the order of ℓ_T or less, where ℓ_T corresponds to the transition between short-crack- and long-crack-controlled behavior.

A model for the determination of the transition crack length ℓ_T is suggested in Fig. 12.18. Shown here are the Y calibrations corresponding to Eqs. 8-24 and 8-25 versus the crack length to hole radius ratio ℓ/r. Also shown is the numerical solution for this crack–hole configuration. Note that the short-crack solution closely approximates the

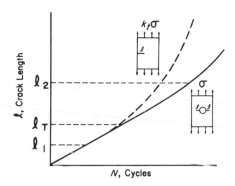

FIGURE 12.17 Difference in crack growth behavior due to specimen configuration when $l > l_T$. (Adapted from Dowling.[14])

numerical solution where the crack length ℓ is small relative to the hole radius r; the long-crack solution provides good agreement when ℓ becomes large compared to ℓ_T. This means that, when $\ell < \ell_T$, fatigue behavior is controlled by the local stress field associated with the notch. Depending on the sharpness of the notch root radius and the magnitude of the applied load, the local stress field could be predominantly elastic or plastic in nature. As a result, the rate of crack growth in this region will be different (see Section 13.4.2 for further discussion). When $\ell > \ell_T$, fatigue behavior is controlled by the remote stress and concepts of linear elastic fracture mechanics. Both local and remote stress fields combine to influence fatigue behavior when $\ell \approx \ell_T$.

The transition point corresponding to the intersection of the curves in Fig. 12.18 depends on the specific stress intensity and stress concentration factors which, in turn, depend on the manner of loading and the crack geometry. For the case of a hole in a large plate, ℓ_T was found to be approximately $r/10$ and was $r/5$ for a sharp notch configuration similar to that of a compact sample. Other values of ℓ_T/r for different crack configurations are given by Dowling.[14]

By estimating the cyclic life of a notched component from unnotched specimen

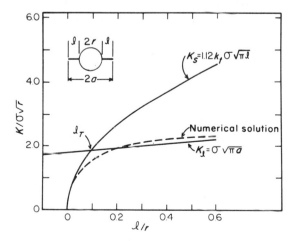

FIGURE 12.18 Y calibrations for short- and long-crack solutions for hole–crack configuration. Dotted curve is numerical solution.[18] (Adapted from Dowling.[14])

test results, the choice of $\ell \leq \ell_T$ as a working definition of fatigue crack initiation avoids the complex problems associated with geometry dependence on initiation life.* Unfortunately, when $\ell < \ell_T$, one may be faced with two additional problems: The size of the crack-tip plastic zone and the microstructural units (e.g., metal alloy grains) may no longer be small compared with ℓ. Both factors compromise the applicability of the stress intensity factor as the controlling parameter for short fatigue crack formation and early growth[16,17] (see Section 13.4.3). On the other hand, linear elastic fracture mechanics would be expected to characterize fatigue crack development when $\ell \gg \ell_T$. Total fatigue life can then be estimated by treating separately the initiation and propagation stages of crack formation. Initiation life can be estimated from notched specimen data characterized in terms of the local stress field $k_t\sigma$ or with the cyclic strain approach (see Section 12.3). In either case, initiation must be defined as the development of a crack whose size is on the order of ℓ_T. Linear elastic fracture mechanics can then be used to calculate the propagation portion of the total life by integrating the crack growth rate–stress intensity factor relation from a crack size of ℓ_T to the critical flaw size at fracture (see Section 13.1.1). The duration of the fatigue crack initiation stage relative to that of propagation depends on the notch root radius (Fig. 12.19). For a given stress level and for the case where the radius is small, total fatigue life is dominated by the propagation stage. On the other hand, when the notch root radius is large, the bulk of the total fatigue life involves initiation of the crack at the notch root.

Another factor that controls fatigue strength at N cycles is the size of the test bar. Although no size effect is observed in *axial* loading of *smooth* bars, a strong size effect is noted in smooth and notched samples subjected to *bending* and in *axially* loaded *notched* bars. In all cases, the section size effect is related to a stress gradient existing in the sample, which in turn controls the volume of material subjected to the highest stress levels (recall Section 7.3.1.1). For the case of bending, the smaller the cross section of the test bar, the higher the stress gradient and the smaller the volume of material experiencing maximum stress. Comparing this situation to that of axially loaded smooth specimens where no stress gradient exists and the entire cross section is stressed equally, one finds bending fatigue strengths to be higher than values obtained from axially loaded samples. From a statistical viewpoint, the larger the volume of material experiencing maximum stress, the greater the probability of finding a weak area that would lead to more rapid failure. As a final note, it is helpful to consider a reappraisal by Findley[20] of the specimen size effect in fatigue testing. If one assumes that fatigue crack initiation will occur when cyclic slip develops over some critical region requiring a minimum driving force σ', then slip can occur only in the outer fibers, where the applied stress is greater than σ'. Since the stress at the outer fiber in bending is

$$\sigma = \frac{Mr}{I} \tag{12-10}$$

---

* It should be recognized that Dowling's working definition of the fatigue initiation stage combines both crack nucleation and early growth to a length ℓ_T. An alternative approach has been proposed wherein the "initiation" stage involves crack nucleation along with a more detailed analysis of the short-crack growth regime (see Section 13.4.3).

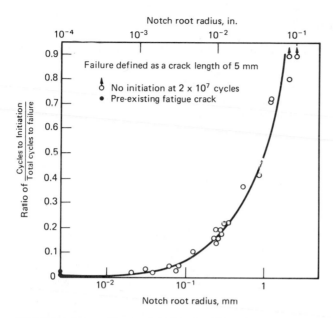

FIGURE 12.19 Ratio of N_i/N_p as a function of notch root radius in 0.23 C, 0.88 Mn, 0.04 Si steel. (From Allery and Birkbeck;[19] Reprinted with permission from *Eng. Fract. Mech.* **4**, 325 (1972). Pergamon Press, Ltd.)

where σ = flexural stress

       M = bending moment

       I = moment of inertia

       r = radius of circular rod or distance from neutral axis to outer fiber

the flexural stress decreases to σ' when one moves a distance Δr from the outer surface, so that

$$\sigma' = \frac{M}{I}(r - \Delta r) \qquad (12\text{-}11)$$

and

$$\sigma = \frac{\sigma'}{1 - \dfrac{\Delta r}{r}} \qquad (12\text{-}12)$$

From Eq. 12-12, we see that no size effect is predicted when uniaxial tension is applied to an unnotched specimen. Here $\sigma = \sigma'$. However, a size effect is anticipated in specimens possessing a large stress gradient (that is, in small specimens subjected to either bending or torsion, and in notched, axially loaded samples). Finally, it is apparent from Eq. 12-12 that the size effect disappears for large samples since $r \gg \Delta r$, whereupon σ' approaches σ.

12.2.4   Material Behavior: Metal Alloys

This section presents an overview of the effect of mechanical properties on material fatigue response. Since detailed discussions of the effect of microstructure and thermo-mechanical treatment on fatigue behavior in various alloy systems would be beyond the scope of this book, the reader is referred to numerous articles in the literature. Books by Forrest,[1] Sines and Waisman,[13] Fuchs and Stephens,[21] and Forsyth[22] should provide an excellent starting point for such an investigation.

Generally, materials tend to exhibit S-N plots of two basic shapes. The plot shows either a well-defined fatigue limit (Fig. 12.6) below which the material would appear to be immune from cyclic damage, or a continually decreasing curve (Fig. 12.10) with no apparent lower stress limit below which the material could be considered completely "safe." (Note the strong resemblance to environment-assisted cracking behavior discussed in the previous chapter.) In materials that possess a "knee" in the S-N curve, the *fatigue limit* is readily determined as the stress associated with the horizontal portion of the S-N curve. It has been found that many steel alloys exhibit this type of behavior. Furthermore, the fatigue limit of steel alloys often is estimated to be one-half the tensile strength of the material (Fig. 12.14). However, it should be noted that the fatigue ratio σ_{fat}/σ_{ts} for such materials can vary between 0.35 and 0.60, as shown in Fig. 12.20a for the case of several carbon and alloy steels. Additional fatigue limit data for other alloys are given in Table 12.1. Initially, it would appear to be good design practice to use a material with as high a tensile strength as possible to maximize fatigue resistance. Unfortunately, this can cause a lot of trouble, since (as shown in Chapters 10 and 11) fracture toughness decreases and environmental sensitivity increases with increasing tensile strength. Since tensile strength and hardness are related, it is possible to estimate the fatigue limit in a number of steels simply by determining the hardness level—a very inexpensive test procedure, indeed. We see from Fig. 12.21 that a good correlation exists up to a hardness level of about $40R_c$. Above $40R_c$, test scatter becomes considerable and the fatigue limit–hardness relation becomes suspect.

The fatigue behavior of nonferrous alloys usually follows the second type of S-N plot, and no clear fatigue limit is defined. Consequently, the "fatigue limit" for any such alloy would have to be defined at some specific cyclic life—usually 10^7 cycles. Examining various aluminum alloys reveals that "fatigue limit"/tensile strength ratios are lower than those found in steel alloys. The fatigue ratio for selected wrought copper alloys is shown in Fig. 12.20b, and fatigue limit data for nonferrous alloys are included in Table 12.1. Many studies have been conducted and theories proposed to account for the relatively poor fatigue response shown by this important group of engineering materials. It is presently felt[24] that extremely fine and atomically ordered precipitates, contained within Al–Cu alloys, are penetrated by dislocations moving back and forth along active slip planes. This action produces an initial strain-hardening response followed by local softening, which serves to concentrate additional deformation in narrow bands and leads to crack initiation. Localized softening is believed to occur by a disordering process resulting from repeated precipitate cutting by the oscillating dislocations. To offset this, it has been suggested that additional platelike particles that are impenetrable by dislocations be added to the microstructure to arrest the mechanically induced disordering process. In this manner, the fine, ordered par-

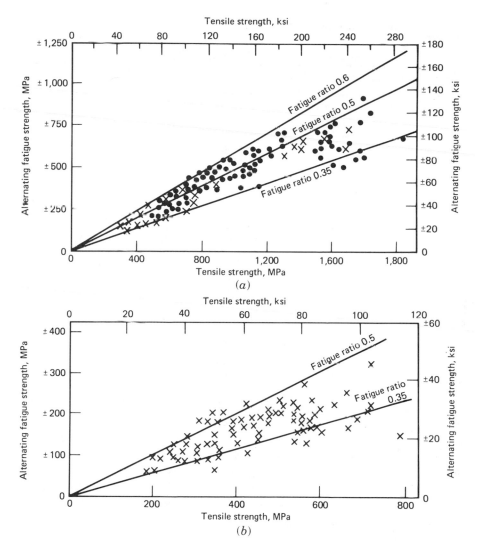

FIGURE 12.20   Relation between rotating, bending unnotched fatigue strength and tensile strength.[1] (*a*) Alloy (●) and carbon (x) steels; (*b*) wrought copper alloys. (P. J. Forrest, *Fatigue of Metals*, Addison-Wesley, Reading, MA, 1962, with permission.)

ticles, penetrable by dislocations, would act as precipitation-hardening agents while relatively larger, platelike particles that are not cut by dislocations, would enhance fatigue behavior.

Since the fatigue limit associated with long cyclic life is strongly dependent on tensile strength, it follows that fatigue behavior should be sensitive to alloy chemistry and thermomechanical treatment. Since large inclusions do not significantly alter tensile strength but do serve as potential crack nucleation sites, their presence in the microstructure is undesirable. By eliminating them through more careful melting practice and stricter alloy chemistry, one finds a reduction in early life failures and a

TABLE 12.1   Fatigue Endurance Limit of Selected Engineering Alloys

Material	Condition	σ_{ts} MPa	(ksi)	σ_{ys} MPa	(ksi)	σ_f MPa	(ksi)
Steel Alloys[a] (Endurance limit based on 10^7 cycles)							
1015	Cold drawn—0%	455	(66)	275	(40)	240	(35)
1015	Cold drawn—60%	710	(102)	605	(88)	350	(51)
1040	Cold drawn—0%	670	(97)	405	(59)	345	(50)
1040	Cold drawn—50%	965	(140)	855	(124)	410	(60)
4340	Annealed	745	(108)	475	(69)	340	(49)
4340	Q & T (204°C)	1950	(283)	1640	(238)	480	(70)
4340	Q & T (427°C)	1530	(222)	1380	(200)	470	(68)
4340	Q & T (538°C)	1260	(183)	1170	(170)	670	(97)
HY140	Q & T (538°C)	1030	(149)	980	(142)	480	(70)
D6AC	Q & T (260°C)	2000	(290)	1720	(250)	690	(100)
9Ni–4Co–0.25C	Q & T (315°C)	1930	(280)	1760	(255)	620	(90)
300M	—	2000	(290)	1670	(242)	800	(116)
Aluminum Alloys[b] (Endurance limit based on 5×10^8 cycles)							
1100-0		90	(13)	34	(5)	34	(5)
2014-T6		483	(70)	414	(60)	124	(18)
2024-T3		483	(70)	345	(50)	138	(20)
6061-T6		310	(45)	276	(40)	97	(14)
7075-T6		572	(83)	503	(73)	159	(23)
Titanium Alloys[c] (Endurance limit based on 10^7 cycles)							
Ti–6Al–4V		1035	(150)	885	(128)	515	(75)
Ti–6Al–2Sn–4Zr–2Mo		895	(130)	825	(120)	485	(70)
Ti–5Al–2Sn–2Zr–4Mo–4Cr		1185	(172)	1130	(164)	675)	(98)
Copper Alloys[c] (Endurance limit based on 10^8 cycles)							
70Cu–30Zn Brass	Hard	524	(76)	435	(63)	145	(21)
90Cu–10Zn	Hard	420	(61)	370	(54)	160	(23)
Magnesium Alloys[c] (Endurance limit based on 10^8 cycles)							
HK31A-T6	—	215	(31)	110	(16)	62–83	(9–12)
AZ91A	—	235	(34)	160	(23)	69–96	(10–14)

[a] *Structural Alloys Handbook,* Mechanical Properties Data Center, Traverse City, MI, 1977.

[b] *Aluminum Standards and Data 1976*, The Aluminum Association, New York, 1976. (See source for restrictions on use of data in design.)

[c] *Materials Engineering* **94**(6) (Dec. 1981), Penton/IPC Publication, Cleveland, OH.

concomitant reduction in the amount of scatter in test results. In this instance, a reduction in inclusion content has a favorable effect on both fatigue behavior and fracture toughness.

Although inclusions and certain other metallurgical microconstituents may have a deleterious effect on unnotched fatigue response, they have an interesting influence on notched fatigue behavior: The notch sensitivity associated with an external notch is lower in a material that already contains a large population of internal flaws. An example of this is seen from a comparison of relative notch sensitivity between flake-graphite gray cast iron and nodular cast iron.[1] The fatigue limit/tensile strength ratio is lowest in the flake-graphite cast iron (0.42 versus 0.48 for the nodular cast iron), which probably reflects the damaging effect of the sharp graphite flakes. Conversely,

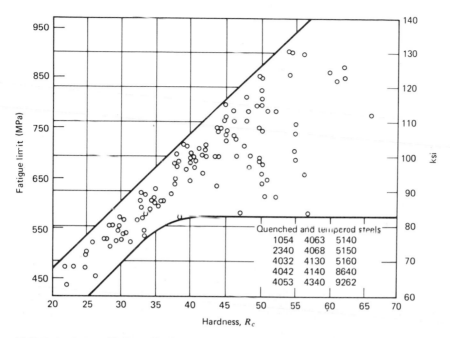

FIGURE 12.21 Fatigue limit of several quenched and tempered steels as a function of hardness level. Considerable uncertainty exists in determining fatigue limits at hardness levels in excess of $40R_c$ (about 1170 MPa).[23] (By permission from *Metals Handbook*, Vol. 1, copyright, American Society for Metals, 1961.)

the notch sensitivity q to an external circumferential V-notch is lowest in the flake-graphite cast iron (0.06 versus 0.25). You might say that with all the graphite flakes present to create a multitude of stress concentrations, one more notch is not that harmful.

12.2.4.1 Surface Treatment

Although changes in overall material properties do influence fatigue behavior (for example, see Fig. 12.21), greater property changes are effected by localized modification of the specimen or component surface, since most fatigue cracks originate in this region. To this end, a number of surface treatments have been developed; they may be classified in three broad categories: mechanical treatments, including shot peening, cold rolling, grinding, and polishing; thermal treatments, such as flame and induction hardening; and surface coatings, such as case hardening, nitriding, and plating. One of the most widely used mechanical treatments involves the use of shot peening. In this process, small, hard particles (shot) about 0.08 to 0.8 mm diameter are blasted onto the surface that is to be treated. This action results in a number of changes to the condition of the material at and near the surface.[25] First, and most importantly, a thin layer of compressive residual stress is developed that penetrates to a depth of about one-quarter to one-half the shot diameter (Fig. 12.22). Since the peening process involves localized plastic deformation, it is believed that the surrounding elastic material forces the permanently strained peening region back toward its original dimensions, thereby inducing a residual compressive stress. Depending on

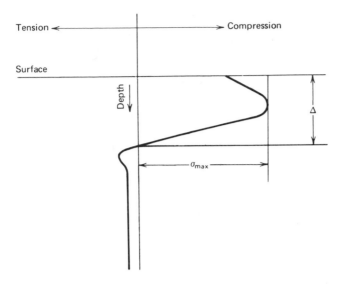

FIGURE 12.22 Diagram showing residual stress distribution after shot peening process. Compressive residual stress extends from surface to a depth Δ.

the type of shot, shot diameter, pressure and velocity of shot stream, and duration of the peening process, the maximum compressive stress can reach about one-half the material yield strength. Consequently, the peening process benefits higher strength alloys more than the weaker ones. Since the peened region has a localized compressive mean stress, it acts to reduce the most damaging tensile portion of the applied alternating stress range (Fig. 12.9), resulting in a substantial improvement in fatigue life.

It should be emphasized that shot peening is most effective in specimens or components that contain a stress concentration or stress gradient; the peening process is also useful in unnotched components that experience a stress gradient (bending or torsion) such as leaf and coil springs, torsion bars, and axles. Shot peening is of limited use when high applied stresses are anticipated (that is, in the low cycle fatigue regime), since large stress excursions, particularly those in the plastic range, cause rapid "fading" of the residual stress pattern. On the other hand, shot peening is very useful in the high-cycle portion of fatigue life associated with lower stress levels.

Another beneficial effect of shot peening, though of secondary importance, is the work-hardening contribution in the peened material that results from plastic deformation. Particularly in cases involving low-strength alloys with high strain-hardening capacity, the material strain hardens, thereby contributing to a higher fatigue strength associated with the higher tensile strength. Finally, the shot peening process alters the surface by producing small "dimples" that, by themselves, would have a deleterious effect on fatigue life by acting as countless local stress concentrations. Fortunately, the negative aspect of this surface roughening is more than counterbalanced by the concurrent favorable residual compressive stress field. To be sure, the fatigue properties of a component may be improved still further if the part is polished after a shot peening treatment.

Surface rolling also produces a favorable residual stress that can penetrate deeper than that produced by shot peening and which does not roughen the component surface. Surface rolling finds extensive use in components possessing surfaces of rotation, such as in the practice of rolling machine threads.

Flame- and induction-hardening heat treatments in certain steel alloys are intended to make the component surface harder and more wear resistant. This is done by heating the surface layers into the austenite phase region and then quenching rapidly to form hard, untempered martensite. Since the tensile strength and hardness of this layer is markedly increased, the fatigue strength likewise is improved [though at the expense of reliability when the hardness exceeds $40R_c$ or about 1170 MPa (Fig. 12.21)]. In addition, since the austenite to martensite phase transformation involves a volume expansion that is resisted by the untransformed core, a favorable compressive residual stress is developed in this layer, which contributes an additional increment to the improved fatigue response of steel alloys heat treated in this manner.

Like flame and induction hardening, case hardening by either carburizing or nitriding is intended primarily to improve wear resistance in steels but simultaneously improves fatigue strength. Components to be carburized are treated in a high-temperature carbonaceous atmosphere to form a carbide-rich layer some 0.8 to 2.5 mm deep (Fig. 12.23a), while nitrided samples are placed in a high-temperature ammonia atmosphere, where nitrogen reacts with nitride-forming elements within the steel alloy to form a nitrided layer about 0.5 mm deep. In both instances, the improvement in fatigue strength results from the intrinsic strength increase within the carburized or nitrided case and also from the favorable residual compressive stress pattern that accompanies the process. The latter factor can be compared to similar residual stress patterns arising from the mechanical and thermal treatments described above, but it is in sharp contrast to the unfavorable residual tensile stresses from nickel and chro-

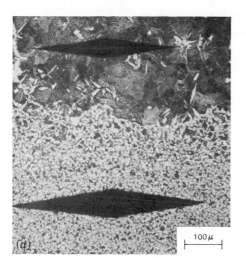 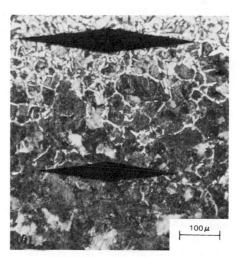

FIGURE 12.23 (a) Photomicrograph showing carburized layer (top) at surface of 1020 steel part. Microhardness impressions reveal considerable hardening in the case. (b) Photomicrograph showing decarburized layer (top) at surface of 1080 steel part. Microhardness impressions reveal softening in the decarburized zone.

mium plating procedures. In these two cases, fatigue resistance is definitely impaired. Such problems are not found with cadmium, zinc, lead, and tin platings, but one must be wary of any electrolytic procedure, since the component may become charged with hydrogen gas and be susceptible to hydrogen-embrittlement-induced premature failure (see Chapter 11).

The improvement in fatigue resistance afforded by case hardening is considerable enough to transfer the fatigue initiation site from the component surface to the case–core boundary region, where (1) the residual stress shifts abruptly to a tensile value, and (2) the intrinsic strength of the core is considerably less than that associated with the case material. As one might expect, case hardening imparts a significant improvement in fatigue resistance to components experiencing a stress gradient, such as those in plane bending or in any notched sample. Here, the applied stress is much lower in the area of the weak link in a case hardened part—the case–core boundary. By contrast, less improvement is anticipated when an axially loaded unnotched part is case hardened, since failure can occur anywhere within the uniformly loaded cross section and will do so at the case–core boundary.

Although case hardening considerably improves fatigue resistance, inadvertent decarburizing in steel alloys during a heat treatment can seriously degrade hardness, strength, and fatigue resistance (Fig. 12.23b). Logically, decarburizing results in a loss of intrinsic alloy strength, since carbon is such a potent strengthening agent in most iron-based alloys (recall Section 4.6). In addition, the propensity for a volumetric contraction in the low carbon surface region, which is restrained by the higher carbon interior regions, may produce an unfavorable residual tensile stress pattern.

From the above, considerable improvement in fatigue properties may be achieved by introducing a favorable residual compressive stress field and avoiding any possibility for decarburization. In fact, Harris[26] showed that when decarburization was avoided and machine threads were rolled rather than cut, the fatigue endurance limit of threaded steel bolts increased by over 400% (Table 12.2).

A number of other conditions may degrade fatigue behavior. These include inadequate quenching, which produces local soft spots that have poorer fatigue resistance; excessive heating during grinding, resulting in reversion of the steel to austenite, which forms a brittle martensite upon quenching; and splatter from welding, which creates local hot spots and causes local metallurgical changes that adversely affect fatigue response.

TABLE 12.2   Fatigue Strength in Threaded^a Bolts[26]

Manufacturing Procedure	Fatigue Strength^b	
	MPa	(ksi)
Thread rolling of unground stock + additional heat treatment	55–125	(8–18)
Machine cut threads	195–220	(28–32)
Thread rolling of ground stock with no subsequent heat treatment	275–305	(40–44)

^{a} K_t 3.5–4.0
^{b} Tensile strength of material = 760–895 MPa (110–130 ksi)

12.2.5   Material Behavior: Polymers

The fatigue failures in polymers may be induced either by large-scale hysteretic heating, resulting in actual polymer melting, or by fatigue crack initiation and propagation to final failure[27] (Fig. 12.24). Over the years, the basic differences and importance of polymer fatigue failures induced by these two processes have become a source of controversy among researchers. Part of this difficulty is due to the nature of the ASTM recommended test procedure (ASTM Standard D671-71).[28]

The major cause of thermal failure is believed to involve the accumulation of hysteretic energy generated during each loading cycle. Since this energy is dissipated largely in the form of heat, an associated temperature rise will occur for every loading cycle when isothermal conditions are not met. As shown in Fig. 12.25, the temperature rise can be great enough to cause the sample to melt, thereby preventing it from carrying any load.[30] Failure is presumed, therefore, to occur by viscous flow, although the occurrence of some bond breakage cannot be excluded.

From the work of Ferry,[32] the energy dissipated in a given cycle may be described by

$$\dot{E} = \pi f J''(f,T)\sigma^2 \qquad (12\text{-}13)$$

where E = energy dissipated
f = frequency
J'' = the loss compliance
σ = the peak stress

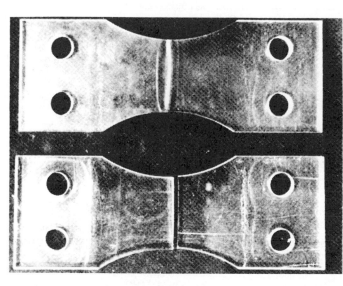

Cyclic thermal softening

True fatigue

FIGURE 12.24   Typical fatigue and cyclic thermal softening failures in poly(methyl methacrylate).[29] (Reproduced by courtesy of The Institution of Mechanical Engineers from an article by I. Constable, J. G. Williams, and D. J. Burns from *JMES* 12, 20 (1970).)

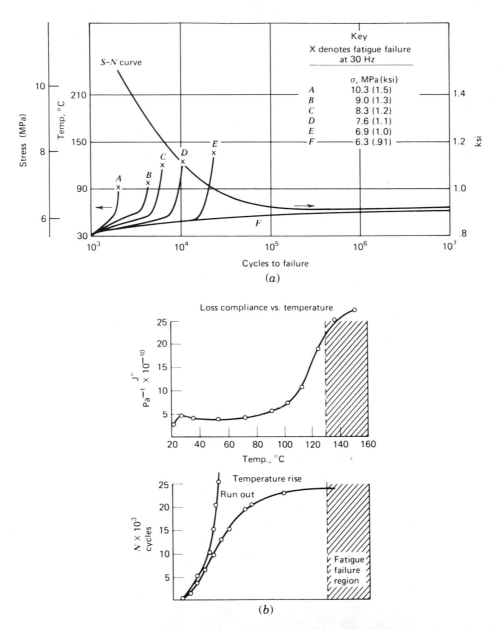

FIGURE 12.25   Effect of temperature rise during fatigue testing. (a) Temperature rise to cause thermal failure at different stress levels (no failure seen in F where sample temperature stabilized);[30] (b) loss compliance and temperature rise.[31] (Reprinted with permission of the Society of Plastics Engineers.)

Neglecting heat losses to the surrounding environment, Eq. 12-13 may be reduced to show the temperature rise per unit time as

$$\Delta\dot{T} = \frac{\pi f J''(f,T)\sigma^2}{\rho c_p}$$ (12-14)

where $\Delta\dot{T}$ = temperature change/unit time
  ρ = density
  c_p = specific heat

Equation 12-14 is useful in identifying the major variables associated with hysteretic heating. For example, the temperature rise is seen to increase rapidly with increasing stress level. With increasing specimen temperature, the elastic modulus is found to decrease. For this reason, larger specimen deflections are required to allow for continuation of the test under constant stress conditions. These larger deflections then contribute to even greater hysteretic energy losses with continued cycling. A point is reached whereby the specimen is no longer capable of supporting the loads introduced by the test machine within the deflection limits of the test apparatus. For such test conditions, ASTM Standard D671-71 defines fatigue failure life (thermal fatigue, in this instance) as the number of loading cycles at a given applied stress range that brings about an "apparent modulus decay to 70% of the original modulus of the specimen determined at the start of the test."[28] Furthermore, the ASTM Standard calls for the investigator to "measure the temperature at failure unless it can be shown that the heat rise is insignificant for the specific material and test condition."[28] Figure 12.25a illustrates a typical curve of stress versus number of cycles to failure for poly(tetrafluoroethylene) (PTFE), along with the superposition of temperature rise curves corresponding to the various stress levels. Note that for all stress levels above the endurance limit (the stress level below which fatigue failure was not observed) the polymer heated to the point of melting (shown by the temperature rise curves A, B, C, D, and E). Evidently, heat was generated faster than it could be dissipated to the surrounding environment. When a stress level less than the endurance limit was applied, the temperature rise became stabilized at a maximum intermediate level below the point where thermal failure was observed (Curve F). These specimens did not fail after 10^7 cycles.

From Eq. 12-14, the rate of temperature rise depends on the magnitude of the loss compliance J'', which is itself a function of temperature. As we see in Fig. 12.25b, the loss compliance rises rapidly in the vicinity of the glass transition temperature after a relatively small change at lower temperatures. Consequently, one would expect the temperature rise in the sample to be moderate during the early stages of fatigue cycling but markedly greater near the final failure time. It is concluded, therefore, that thermal failure describes an event primarily related to the lattermost stages of cyclic life. In further support of this last statement, other tests have been reported wherein intermittent rest periods were interjected during the cyclic history of the sample. In this manner, any temperature rise resulting from adiabatic heating could be dissipated periodically. It would be expected, then, that the fatigue life of specimens allowed intermittent rest periods would be substantially greater than that of uninterrupted test samples. Indeed, several investigators[30,33,34] have shown significant improvement in fatigue life when intermittent rest periods were introduced during testing. On the basis of these test results, it is concluded that linear cumulative damage laws cannot be applied to thermal failures.

Finally, the fatigue lives of polymers subjected to isothermal test conditions are superior to those exhibited by samples examined under adiabatic test conditions, consistent with the absence of hysteretic heating in the former case.[35]

From Eq. 12-14, the fatigue life of a sample should decrease with increasing frequency, since the temperature rise per unit time is proportional to the frequency. Test results have confirmed the anticipated effect of this variable as evidenced by a decrease in endurance limit with increasing frequency.[30] Also, thermal failures are affected by specimen configuration. As mentioned above, the temperature rise resulting from each loading cycle depends on the amount of heat dissipated to the surroundings. Consequently, the fatigue life of a given sample should be dependent on the heat transfer characteristics of the sample and the specimen surface to volume ratio. Indeed, the endurance limit in PTFE has been shown to increase with decreasing specimen thickness.[30] This sensitivity to specimen shape constitutes a major drawback to unnotched specimen tests involving thermal failure, since the test results are a function of specimen geometry and, therefore, do not reflect the intrinsic response of the material being evaluated.

It is clear from the above that thermal fatigue may be suppressed by several factors, such as limiting the applied stress, decreasing test frequency, allowing for periodic rest periods, or cooling the test sample, and by increasing the sample's surface to volume ratio. It is extremely important to recognize, however, that suppressing thermal fatigue by any of the above procedures may not preclude mechanical failure caused by crack initiation and propagation. The corollary is true though: If stresses are reduced to the point where mechanical failure does not occur, this stress level certainly will be low enough such that thermal failure will not occur either.

Although I choose to treat mechanical and thermal fatigue failures as distinctly different events, there are points of common ground. For example, it would be expected that hysteretic heating would take place within the plastic zone at the tip of a crack. Since this heat source is small compared to the much larger heat sink of the surrounding material, it would be expected that any temperature rise would be limited and restricted to the proximity of the crack tip. The influence of such localized heating on fatigue crack propagation in engineering plastics is discussed in Section 13.8.

It would appear then that the likelihood of thermal failure would depend on the size of the heated zone in relation to the overall specimen dimensions. When this ratio is large, as in the case of unnotched test bars, thermal failures are distinctly possible. When the ratio is very small, say, in the case of a notched bar, thermal failures would not be expected.

12.2.6 Material Behavior: Composites

12.2.6.1 Particulate Composites
The fatigue response of composite materials is dependent on the complex interaction between the mechanical properties and volume fraction of the matrix and reinforcing phases, fiber aspect ratio, and the strength of the bond between the two phases; in

addition, the direction of loading (i.e., tensile and/or compressive), environment, temperature, and cyclic frequency also affect fatigue behavior. Within the context of this book, it will be possible to address these factors only in the briefest fashion. We begin this discussion by examining the fatigue response of impact-modified polymers. Consideration of overall fatigue life and the relative effects of rubber on crack initiation and propagation reveals interesting and perhaps unexpected behavior. For example, detailed examinations of several rubber-toughened polymers have shown that the high level of impact strength is not necessarily carried over into fatigue resistance; this is particularly true when compressive stresses are applied. At least with HIPS and ABS tested in tension–compression, the rubbery phase decreases both initiation and propagation lifetimes compared to typical unmodified polystyrenes (PS) or styrene–acrylonitrile copolymers (SAN).[36–38] Evidently, the lower yield strength in the toughened resins greatly facilitates initiation, and compressive loading severely damages craze fibrils during the crack closure part of the load cycle. In fact, if the stress levels in the S-N curves for HIPS and PS are normalized by their respective yield strengths, HIPS appears to be relatively superior to PS. The especially deleterious effect of compressive stresses is consistent with the observation that the fatigue life of HIPS and ABS is increased significantly by switching to tension–tension loading.[37,38]

Figure 12.26 is a three-dimensional plot showing the interaction of M_w and percentage rubber content on fatigue crack initiation lifetime (N_i) in impact-modified poly(vinyl chloride) (PVC).[39] The rubber phase in this blend is methacrylate–butadiene–styrene (MBS) polymer and N_i is defined as the number of loading cycles necessary to nucleate a crack 0.25 mm in length from the root of a polished round hole (radius, 1.59 mm) introduced at the end of the slot in a compact tension sample (recall Fig. 8.7g). Fatigue initiation life increases markedly with increasing molecular weight for all rubbery phase contents. Conversely, in high M_w PVC blends, rubber modification lowers fatigue crack initiation (FCI) resistance, although the addition of MBS in low M_w PVC slightly improves FCI resistance. The generally deleterious influence of rubbery phase on fatigue crack *initiation* resistance contrasts with the superior fatigue crack *propagation* resistance shown for this blend;[40] a 3- to 30-fold reduction in the crack propagation rate was observed when up to 14% MBS was added to the PVC matrix.

Thus, to avoid unpleasant surprises, caution should be used in subjecting rubber-

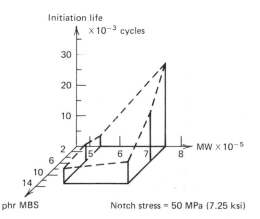

FIGURE 12.26   Three-dimensional plot revealing interactive effects of MW and percent rubber content (MBS) on fatigue crack initiation lifetime in PVC.[39] (Reprinted with permission, *Deformation, Yield, and Fracture*, 1985, The Plastics and Rubber Institute.)

modified plastics to fatigue loading, especially if compressive stresses are involved. Before application in such situations, careful tests should be run to simulate anticipated loading conditions. Furthermore, conclusions based on fatigue initiation studies may not necessarily be extrapolated to the realm of crack propagation.

12.2.6.2   Fibrous Composites

Fatigue failure processes in fibrous composites are complex and include the following:[27,41,42] (1) Damage is progressive, physical integrity may be maintained for many decades of cycles, and criteria for failure may be based arbitrarily on the degradation of a property such as the elastic modulus. (2) Diverse micromechanisms of failure may occur and include fiber deformation and/or brittle fracture, fiber–matrix debonding, delamination of composite plies, and matrix cracking. (3) The balance of micromechanisms depends on such factors as hysteretic heating (modified by the presence of fibers), relative orientation of fiber and stress axes, mode of loading, and the presence and nature of preexisting flaws. (4) In principle, linear elastic fracture mechanics is not applicable to heterogeneous systems, and the concept of a crack requires redefinition in terms of a more diffuse zone of damage. (5) Under some circumstances, fatigue loading can result in effective crack blunting so that fracture toughness may actually increase, at least during part of the fatigue life.[42,43]

   To illustrate the influence of fiber properties, geometry, and volume fraction on composite fatigue life, consider the results shown in Fig. 12.27. Typical S-N curves

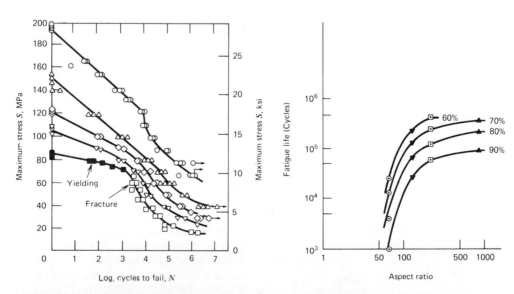

FIGURE 12.27   Fatigue response in polymer–matrix composites. (a) Injection-molded polysulfone matrix composites. $R = 0.1$, frequency = 5–10 Hz. ■□, unreinforced; ▽, 10% glass; ◇, 20% glass; △, 40% glass; ○, 40% carbon.[44] **(*Polymer Composites*, 4, 32 (1983) (b) 50–50 vol% boron-fiber reinforced epoxy. Fatigue life to produce a 20% decrease in elastic modulus as a function of aspect ratio for applied stresses of 60, 70, 80, and 90% of ultimate failure stress.**[45] **(*Polymer Engineering and Science*, 9, 365 (1969). Reprinted with permission from Society of Plastics Engineers.)**

for tensile fatigue are shown in Fig. 12.27a for several injection-molded polysulfone composites.[44] The superiority of all composites to the unreinforced matrix is evident, as is the superiority of increased fiber fraction and carbon relative to glass fibers. Clearly, the greater stiffness and thermal conductivity of carbon constitute significant advantages in lowering the strains at a given stress and minimizing hysteretic heating. The importance of aspect ratio to fatigue performance is illustrated in Fig. 12.27b for a composite containing equal volume fractions of short boron fibers and an epoxy resin matrix.[45] The number of loading cycles required to produce a 20% decrease in elastic modulus at a frequency of 3 Hz (low enough to minimize hysteretic heating) increases sharply with aspect ratio up to a value of $l/d = 200$; little additional improvement in fatigue life is noted with further increases in aspect ratio. The latter is consistent with theoretical predictions as discussed in Chapter 1 (recall Eq. 1-43).

The deleterious influence of compressive loading on continuous carbon fiber-reinforced epoxy resin is shown in Fig. 12.28.[46] Compression-induced effects, such as fiber buckling and delamination, and matrix shear reduce fatigue resistance;[47] cycling in flexure, torsion, and other shear modes is also especially deleterious.[27,41,46] In such cases, the dominance of the fibers in determining fatigue properties decreases, and the matrix and interface play more important roles. Because such severe modes of loading are often more typical of actual service than is axial tension, more testing should be done under these more rigorous conditions.

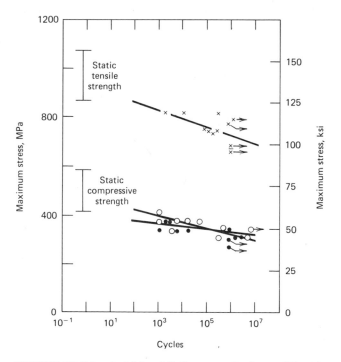

FIGURE 12.28   Axial-load fatigue results for unidirectional, surface-treated, carbon-fiber-reinforced epoxy resin.[46] Volume fraction of carbon fiber = 0.61. Test frequency = 117 Hz. x, zero-tension; ○, zero-compression; ●, fully reversed. Compressive stress plotted positive (lower curve). (Reprinted from *Composite Materials*, 5, 341 (1974).

12.3 CYCLIC STRAIN-CONTROLLED FATIGUE

Localized plastic strains can be generated by loading a component that contains a notch. Regardless of the external mode of loading (cyclic stress or strain controlled), the plasticity near the notch root experiences a strain-controlled condition dictated by the much larger surrounding mass of essentially elastic material. Scientists and engineers from the Society of Automotive Engineers (SAE) and the American Society for Testing and Materials (ASTM) have recognized this phenomenon and have developed strain-controlled test procedures to evaluate cumulative damage in engineering materials.[48–51] These procedures are particularly useful in evaluating component life where notches are present. Indeed, a reasonable assumption can be made that the same number of loading cycles is needed to develop a crack at the notch root of an engineering component and in an unnotched specimen, if the two cracked regions experience the same cyclic stress–strain history. Other examples of strain-controlled cyclic loading include thermal cycling, where component expansions and contractions are dictated by the operating temperature range, and reversed bending between fixed displacements, such as in the reciprocating motion shown in Fig. 12.29.

By monitoring strain and stress during a cyclic loading experiment, the response of the material can be clearly identified. For example, for a material exhibiting Type I stress–strain behavior (Chapter 1) involving only elastic deformation under the applied loads, a hysteresis curve like that shown in Fig. 12.30a is produced. Note that the material's stress–strain response is retraced completely; that is, the elastic strains are completely reversible. For Type II behavior involving elastic–homogeneous plastic flow, the complete load excursion (positive and negative) produces a curve similar to Fig. 12.30b that reflects both elastic and plastic deformation. The area contained within the hysteresis loop represents a measure of plastic deformation work done on the material. Some of this work is stored in the material as cold work and/or associated with configurational changes (entropic changes), such as in polymer chain realignment; the remainder is emitted as heat. From Fig. 12.30b, the elastic strain range in the hysteresis loop is given by

$$\Delta\epsilon_e = XT + QY = \frac{\Delta\sigma}{E} \tag{12-15}$$

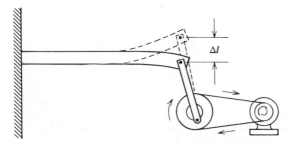

FIGURE 12.29 Reciprocating action produces fixed beam displacements. Compare this case to the stress controlled condition shown in Fig. 12.8a.

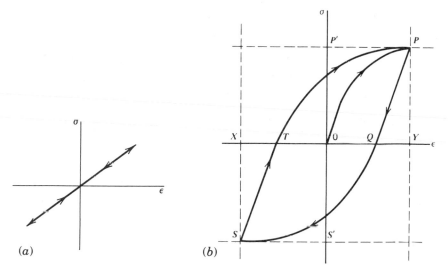

FIGURE 12.30 Hysteresis loops for cyclic loading in (*a*) ideally elastic material and (*b*) material undergoing elastic and plastic deformation.

The plastic strain range is equal to TQ or equal to the total strain range minus the elastic strain range. Hence

$$\Delta\epsilon_p = \Delta\epsilon_T - \frac{\Delta\sigma}{E} \qquad (12\text{-}16)$$

Note that as the amount of plastic strain diminishes to zero, the hysteresis loop in Fig. 12.30*b* shrinks to that shown in Fig. 12.30*a*. Consequently, the elastic strain approaches the total strain. It is important to recognize that *fatigue damage will occur only when cyclic plastic strains are generated*. This basic rule should not be construed as a "security blanket" whenever nominal applied stresses are below the material yield strength, since stress concentrations readily elevate local stresses and associated strains into the plastic range.

12.3.1 Cycle-Dependent Material Response

Cycle-dependent material responses under stress and strain control are shown in Figs. 12.31 and 12.32, respectively, which reflect changes in the shape of the hysteresis loop. It is seen that, in both cases, the material response changes with continued cycling until cyclic stability is reached.* That is, the material becomes either more or less resistant to the applied stresses or strains. Therefore, the material is said to

---

* Although most of our discussions will focus on symmetrical loading about zero, it is important to appreciate what happens to a sample when a nonzero mean stress is superimposed during a cyclic strain experiment. The specimen is found to accumulate strains as a result of each cycle. This accumulation has been termed "cyclic-strain-induced creep" and will contribute to either an extension or contraction of the sample, depending on the sense of the applied mean stress.

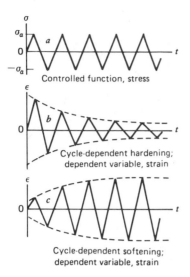

FIGURE 12.31 Cycle-dependent material response under stress control.[52] (Reprinted with permission of the University of Wisconsin Press and the Regents of the University of Wisconsin System.)

cyclically strain harden or strain soften. Referring again to Fig. 12.30b for the case of stress control, where the fatigue test is conducted in a stress range between P' and S', the width of the hysteresis loop TQ (the plastic strain range) contracts when cyclic hardening occurs and expands during cyclic softening. Cyclic softening under stress control is a particularly severe condition because the constant stress range produces a continually increasing strain range response, leading to early fracture (Fig. 12.31). Under cyclic strain conditions within limits of strains X and Y, the hysteresis loop expands above P and below S for cyclic hardening and shrinks below P and above S for cyclic softening. An example of cyclic strain hardening and softening under strain-controlled test conditions is shown in Fig. 12.33.

After cycling a material for a relatively short duration (often less than 100 cycles),

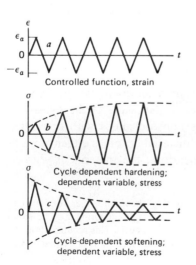

FIGURE 12.32 Cycle-dependent material response under strain control.[52] (Reprinted with permission of the University of Wisconsin Press and the Regents of the University of Wisconsin System.)

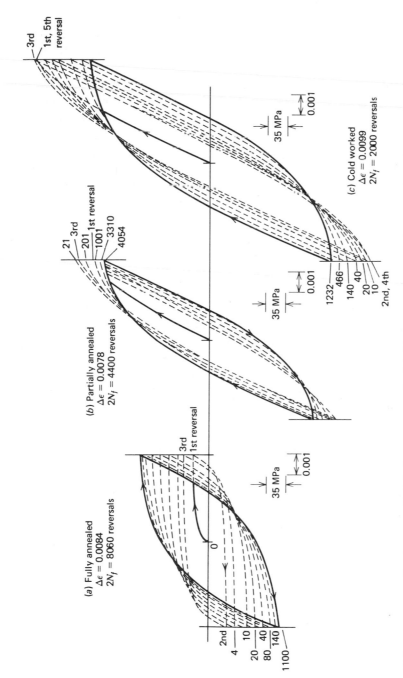

FIGURE 12.33 Strain-controlled fatigue response in OFHC copper. (*a*) Fully annealed sample exhibits cyclic strain hardening; (*b*) partially annealed sample exhibits relative cyclic stability; (*c*) severely cold-worked sample exhibits cyclic strain softening.[53] (Reprinted by permission of the American Society for Testing and Materials from copyrighted work.)

the hysteresis loops generally stabilize and the material achieves an equilibrium condition for the imposed strain limits. The cyclically stabilized stress–strain response of the material may then be quite different from the initial monotonic response. Consequently, cyclically stabilized stress–strain curves are important characterizations of a material's cyclic response. These curves may be obtained in several ways. For example, a series of companion samples may be cycled within various strain limits until the respective hysteresis loops become stabilized. The cyclic stress–strain curve is then determined by fitting a curve through the tips of the various superimposed hysteresis loops, as shown in Fig. 12.34.[54] This procedure involves many samples and is expensive and time-consuming. A faster method for obtaining cyclic stress–strain curves is by multiple step testing, wherein the same sample is subjected to a series of alternating strains in blocks of increasing magnitude. In this manner, one specimen yields several hysteresis loops, which may be used to construct the cyclic stress–strain curve.[55] An even quicker technique involving only one sample has been found to

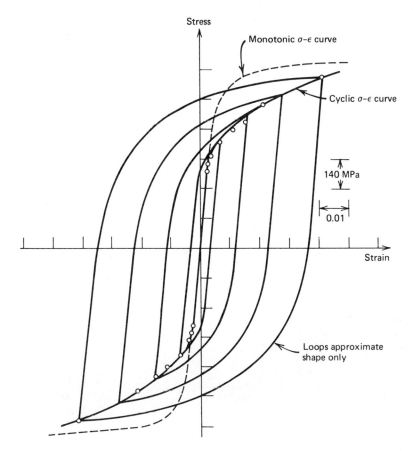

FIGURE 12.34 **Monotonic and cyclic stress–strain curves for SAE 4340 steel. Data points represent tips of stable hysteresis loops from companion specimens.[56] (Reprinted by permission of the American Society for Testing and Materials from copyrighted work.)**

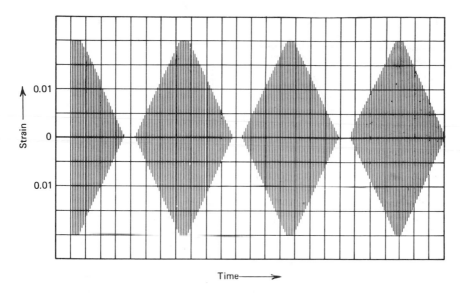

FIGURE 12.35   Incremental step test program showing strain–time plot.

provide excellent results and is used extensively in current cyclic strain testing experiments. As seen in Fig. 12.35, the specimen is subjected to a series of blocks of gradually increasing and then decreasing strain excursions.[55] It has been found that after a relatively few such blocks (the greater the number of cycles within each block, the fewer the number of blocks needed for cyclic stabilization), the material reaches a stabilized condition. At this point, the investigator simply draws a line throught the tips of each hysteresis loop, from the smallest strain range to the largest. As such, each loop contained within the hysteresis envelope represents the cyclically stabilized condition for the material at that particular strain range. By initiating the test with the maximum strain amplitude in the block, the monotonic stress–strain curve is automatically determined for subsequent comparison with the cyclically stabilized curve. In this manner, both the monotonic and cyclic stress–strain curves can be determined from the same sample. Obviously, this method results in savings in test time and money. It should be noted that if a specimen subjected to either multiple or incremental step testing were to be pulled to fracture after cyclic stabilization, the resulting stress–strain curve would be virtually coincident with the one generated by the locus of hysteresis loop tips.

By comparing monotonic and cyclically stabilized stress–strain curves, Landgraf et al.[55] demonstrated that certain engineering alloys will cyclically strain harden and others will soften (Fig. 12.36). From the Holloman relation given by $\sigma = K\epsilon^n$, it is possible to mathematically describe the material's stress–strain response in either the monotonic or cyclically stabilized state. Consequently, one may define the strain-hardening coefficient for both monotonic (n) and cyclic (n') conditions as well as the monotonic yield strength (σ_{ys}) and cyclic (σ'_{ys}) counterpart. Equation 12-17 describes the cyclically stabilized stress–strain curve where K' is the cyclic strength coefficient and $\Delta\sigma$ and $\Delta\epsilon$ are true stress range and true strain range, respectively:

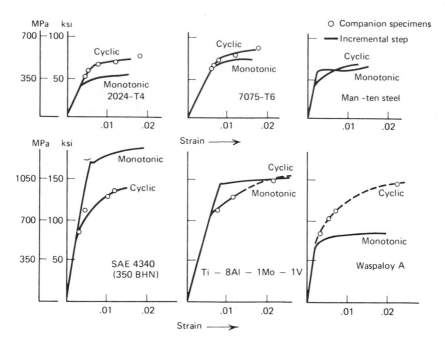

FIGURE 12.36 Monotonic and cyclic stress–strain curves for several engineering alloys.[55] (Reprinted by permission of the American Society for Testing and Materials from copyrighted work.)

$$\frac{\Delta\epsilon}{2} = \frac{\Delta\epsilon_e}{2} + \frac{\Delta\epsilon_p}{2} = \frac{\Delta\sigma}{2E} + \left(\frac{\Delta\sigma}{2K'}\right)^{1/n'} \qquad (12\text{-}17)$$

Typical material property values for a number of metal alloys are given in Table 12.3a,b. Although large changes occur in these properties as a result of cyclic hardening or softening, it is worth noting that, for most metals, n' ranges from 0.1 to 0.2. Beardmore and Rabinowitz[57–59] have conducted an extensive comparison of cyclic and monotonic stress–strain curves for several polymers. For all the materials examined, semicrystalline, amorphous, and two-phase, pronounced cyclic strain softening was observed; by comparison, no cyclic strain hardening took place.

Can one determine in advance which alloys will cyclically harden and which will soften? Manson et al.[54,60] observed that the propensity for cyclic hardening or softening depends on the ratio of monotonic ultimate strength to 0.2% offset yield strength. When $\sigma_{ult}/\sigma_{ys} > 1.4$, the material will harden, but when $\sigma_{ult}/\sigma_{ys} < 1.2$, softening will occur. For ratios between 1.2 and 1.4, forecasting becomes difficult, though a large change in properties is not expected. Also, if $n > 0.20$, the material is likely to strain harden, and softening will occur if $n < 0.10$. Therefore, *initially hard and strong materials will generally cyclically strain soften, and initially soft materials will harden.* Manson's rules makes use of monotonic properties to determine *whether* cyclic hardening or softening will occur. However, the *magnitude* of the cyclically induced change can be determined only by comparison of the monotonic and cyclic stress–strain curves (see Table 12.3a,b and Figs. 12.34 and 12.36).

But why do these materials cyclically harden or soften? The answer to this question

TABLE 12.3a Monotonic and Cyclic Properties of Selected Engineering Alloys[a]

Material	Condition	σ_{ys}/σ'_{ys} (MPa)	n/n'	ϵ_f/ϵ'_f	σ_f/σ'_f (MPa)	b	c
Steel							
SAE 1015	Normalized, 80 BHN	225/240	0.26/0.22	1.14/0.95	725/825	−0.11	−0.64
SAE 950X	As received, 150 BHN	345/335	0.16/0.134	1.06/0.35	750/625	−0.075	−0.54
VAN-80	As received, 225 BHN	565/560	0.13/0.134	1.15/0.21	1220/1055	−0.08	−0.53
SAE 1045	Q + T (650°C), 225 BHN	635/415	0.13/0.18	1.04/1.0	1225/1225	−0.095	−0.66
SAE 1045	Q + T (370°C), 410 BHN	1365/825	0.076/0.146	0.72/0.60	1860/1860	−0.073	−0.70
SAE 1045	Q + T (180°C), 595 BHN	1860/1725	0.071/0.13	0.52/0.07	2725/2725	−0.081	−0.60
AISI 4340	Q + T (425°C), 409 BHN	1370/825	—/0.15	0.48/0.48	1560/2000	−0.091	−0.60
AISI 304 ELC	BHN 160	255/715	—/0.36	1.37/1.02	1570/2415	−0.15	−0.77
AISI 304 ELC	Cold drawn, BHN 327	745/875	—/0.17	1.16/0.89	1695/2275	−0.12	−0.69
AISI 305[b]	0% CW	250/405	—/0.05	—	—	—	—
AISI 305[b]	50% CW	850/710	—/0.11	—	—	—	—
AM 350	Annealed	440/1350	—/0.13	0.74/0.33	2055/2800	−0.14	−0.84
AM 350	Cold drawn 30%, BHN 496	1860/1620	—/0.21	0.23/0.098	2180/2690	−0.102	−0.42
18 Ni maraging	ST(790°C)/1 h + 480°C (4 h), BHN 480	1965/1480	0.015–0.030 / 0.008	0.81/0.60	2240/2240	−0.07	−0.75
Aluminum							
2014-T6	BHN 155	460/415	—/0.16	0.29/0.42	600/850	−0.106	−0.65
2024-T4	—	305/440	0.20/0.08	0.43/0.21	635/1015	−0.11	−0.52
5456	H31, 95 BHN	235/360	—/0.16	0.42/0.46	525/725	−0.11	−0.67
7075-T6	—	470/525	0.113/0.146	0.41/0.19	745/1315	−0.126	−0.52
Copper							
OFHC[c]	Annealed	20/140	0.40/0.16	—	—	—	—
70/30 brass[b]	Annealed	140/240	—/0.08	—	—	—	—
70/30 brass[b]	82% CW	570/475	—/0.11	—	—	—	—
Nickel							
W'aspalloy[c]	—	545/705	0.11/0.17	—	—	—	—
MP35N[b]	0% CW	350/625	—/0.06	—	—	—	—
MP35N[b]	20% CW	7 910/745	—/0.10	—	—	—	—
MP35N[b]	40% CW	1180/1850	—/0.14	—	—	—	—

[a] L. E. Tucker, R. W. Landgraf, and W. R. Brose, SAE Report 740279, Automotive Engineering Congress, Feb. 1974.
[b] Ref. 63.
[c] Ref. 55.

TABLE 12.3b   Monotonic and Cyclic Properties of Selected Engineering Alloys[a]

Material	Condition	σ_{ys}/σ'_{ys} (ksi)	n/n'	ϵ_f/ϵ'_f	σ_f/σ'_f (ksi)	b	c
Steel							
SAE 1015	Normalized, 80 BHN	33/35	0.26/0.22	1.14/0.95	105/120	-0.11	-0.64
SAE 950X	As received, 150 BHN	50/49	0.16/0.134	1.06/0.35	109/91	-0.075	-0.54
VAN-80	As received, 225 BHN	82/81	0.13/0.134	1.15/0.21	177/153	-0.08	-0.53
SAE 1045	Q + T (1200°F), 225 BHN	92/60	0.13/0.18	1.04/1.0	178/178	-0.095	-0.66
SAE 1045	Q + T (700°F), 410 BHN	198/120	0.076/0.146	0.72/0.60	270/270	-0.073	-0.70
SAE 1045	Q + T (360°F), 595 BHN	270/250	0.071/0.13	0.52/0.07	395/395	-0.081	-0.60
AISI 4340	Q + T (800°F), 409 BHN	199/120	—/0.15	0.48/0.48	226/290	-0.091	-0.60
AISI 304 ELC	BHN 160	37/104	—/0.36	1.37/1.02	228/350	-0.15	-0.77
AISI 304 ELC	Cold drawn, BHN 327	108/127	—/0.17	1.16/0.89	246/330	-0.12	-0.69
AISI 305[b]	0% CW	36/59	—/0.05				—
AISI 305[b]	50% CW	123/103	—/0.11				—
AM 350	Annealed	64/196	—/0.13	0.74/0.33	298/406	-0.14	-0.84
AM 350	Cold drawn 30%, BHN 496	270/235	—/0.21	0.23/0.098	316/390	-0.102	-0.42
18 Ni maraging	ST(1450°F/1h) + 900°F (4 h), BHN 480	285/215	$\frac{0.015\text{–}0.030}{0.008}$	0.81/0.60	325/325	-0.07	-0.75
Aluminum							
2014-T6	BHN 155	67/60	—/0.16	0.29/0.42	87/123	-0.106	-0.65
2024-T4	—	44/64	0.20/0.08	0.43/0.21	92/147	-0.11	-0.52
5456	H31, 95 BHN	34/52	—/0.16	0.42/0.46	76/105	-0.11	-0.67
7075-T6	—	68/76	0.113/0.146	0.41/0.19	108/191	-0.126	-0.52
Copper							
OFHC[c]	Annealed	3/20	0.40/0.16	—	—	—	—
70/30 brass[b]	Annealed	20/35	—/0.08	—	—	—	—
70/30 brass[b]	82% CW	83/69	—/0.11	—	—	—	—
Nickel							
Waspalloy[c]	—	79/102	0.11/0.17	—	—	—	—
MP35N[b]	0% CW	51/91	—/0.06	—	—	—	—
MP35N[b]	20% CW	132/108	—/0.10	—	—	—	—
MP35N[b]	40% CW	171/123	—/0.14	—	—	—	—

[a] L. E. Tucker, R. W. Landgraf, and W. R. Brose, SAE Report 740279, Automotive Engineering Congress, Feb. 1974.
[b] Ref. 63.
[c] Ref. 55.

appears to be related to the nature and stability of the dislocation substructure of the material. For an initially soft material, the dislocation density is low. As a result of plastic strain cycling, the dislocation density increases rapidly, contributing to significant strain hardening. At some point, the newly generated dislocations assume a stable configuration for that material and for the magnitude of cyclic strain imposed during the test. When a material is hard initially, subsequent strain cycling causes a rearrangement of dislocations into a new configuration that offers less resistance to deformation—that is, the material strain softens. The processes associated with cyclic strain softening were referred to in earlier technical literature as the Bauschinger effect.[31] To characterize this effect, consider the yielding behavior of a metal alloy (typically strain hardened) that is subjected to a complete loading cycle. After exhibiting a particular strength level associated with initial yielding in tension, the yield-strength level under compressive loading is found to be reduced. Further cycling then leads to additional reductions in both tensile and yield-strength levels (recall Fig. 12.33c). The tendency for cold-worked metals to exhibit cyclic softening (i.e., the Bauschinger effect) is put to good use in certain metal-forming applications.[62] For example, if a strain-hardened metal sheet is passed through a series of roll pairs that are alternately slightly above and below the nominal plane of the workpiece, the alternating tensile and bending stresses associated with this roller-leveling operation induce cyclic strain softening and enhance alloy ductility.

As we saw in Chapters 2 and 4, dislocation mobility that strongly affects dislocation substructure stability depends on the material's stacking fault energy (SFE). Recall that when SFE is high, dislocation mobility is great because of enhanced cross-slip; conversely, cross-slip is restricted in low SFE materials. As a result, some materials cyclically harden or soften more completely than others. For example, in a relatively high SFE material like copper, initially hard samples cyclically strain soften, and initially soft samples cyclically harden; thus, the cyclically stabilized condition is the same *regardless* of the initial state of the material (Fig. 12.37a). In this case, the mechanical properties of the material in the stabilized state are *independent* of prior strain history. This is not true for a low stacking fault energy material, where restricted cross-slip will prevent the development of a common dislocation state from an initially hard and soft condition, respectively. In addition, the low SFE material will harden or soften more slowly than the high SFE alloy. We see from Fig. 12.37b that the material will cyclically soften and harden, but a final stabilized state is never completely achieved and is not equivalent for the two different starting conditions. For such materials, the "final" cyclically stabilized state is *dependent* on prior strain history.

One might then expect to find dislocation substructures in cyclically loaded samples similar to those found as a result of unidirectional loading. In fact, Feltner and Laird[65] observed that "those factors which give rise to certain kinds of dislocation structures in unidirectional deformation affect the cyclic structures in the same way." For example, we see from Fig. 12.38 that at high cyclic strains a cell structure is developed in high SFE alloys. If cyclic straining causes coarsening of a preexistent cell structure, then softening will occur. If the cell structure gets finer, then cyclic straining results in a hardening process. In low SFE alloys, dislocation planar arrays and stacking faults are present. These findings are similar to those discussed in Chapters 2

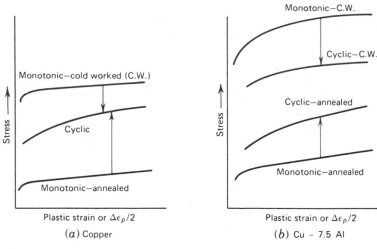

(a) Copper

(b) Cu – 7.5 Al

FIGURE 12.37 **Cyclic response of (a) high stacking fault energy copper and (b) low stacking fault energy Cu–7.5% Al alloy. Cyclically stabilized state in high SFE alloy is path independent.[64] (Reprinted from C. E. Feltner and C. Laird, *Acta Metall.* 15 (1967), with permission of Pergamon Press.)**

for monotonic loading. A parallel condition is found in monotonic and cyclically induced dislocation structures produced at low strains.

12.3.2 Strain Life* Curves

Having identified the response of a solid to cyclic strains, it is now appropriate to consider how cyclically stabilized material properties affect the life of a specimen or engineering component subjected to strain-controlled loading. To accomplish this, it is convenient to begin our analysis by considering the elastic and plastic strain components separately. The elastic component is often described in terms of a relation between the true stress amplitude and number of load reversals

$$\frac{\Delta\epsilon_e E}{2} = \sigma_a = \sigma_f'(2N_f)^b \qquad (12\text{-}18)$$

where $\dfrac{\Delta\epsilon_e}{2}$ = elastic strain amplitude

E = modulus of elasticity

= stress amplitude

fatigue strength coefficient, defined by the stress intercept at one load reversal ($2N_f = 1$)

les to failure

er of load reversals to failure

strength exponent

defined in a number of ways dependent on the ultimate use of the fatigue erial being tested. Some criteria include total fracture, specific changes in p, and the existence of microcracks of a certain size (recall Section 12.2.3).

appears to be related to the nature and stability of the dislocation substructure of the material. For an initially soft material, the dislocation density is low. As a result of plastic strain cycling, the dislocation density increases rapidly, contributing to significant strain hardening. At some point, the newly generated dislocations assume a stable configuration for that material and for the magnitude of cyclic strain imposed during the test. When a material is hard initially, subsequent strain cycling causes a rearrangement of dislocations into a new configuration that offers less resistance to deformation—that is, the material strain softens. The processes associated with cyclic strain softening were referred to in earlier technical literature as the Bauschinger effect.[31] To characterize this effect, consider the yielding behavior of a metal alloy (typically strain hardened) that is subjected to a complete loading cycle. After exhibiting a particular strength level associated with initial yielding in tension, the yield-strength level under compressive loading is found to be reduced. Further cycling then leads to additional reductions in both tensile and yield-strength levels (recall Fig. 12.33c). The tendency for cold-worked metals to exhibit cyclic softening (i.e., the Bauschinger effect) is put to good use in certain metal-forming applications.[62] For example, if a strain-hardened metal sheet is passed through a series of roll pairs that are alternately slightly above and below the nominal plane of the workpiece, the alternating tensile and bending stresses associated with this roller-leveling operation induce cyclic strain softening and enhance alloy ductility.

As we saw in Chapters 2 and 4, dislocation mobility that strongly affects dislocation substructure stability depends on the material's stacking fault energy (SFE). Recall that when SFE is high, dislocation mobility is great because of enhanced cross-slip; conversely, cross-slip is restricted in low SFE materials. As a result, some materials cyclically harden or soften more completely than others. For example, in a relatively high SFE material like copper, initially hard samples cyclically strain soften, and initially soft samples cyclically harden; thus, the cyclically stabilized condition is the same *regardless* of the initial state of the material (Fig. 12.37a). In this case, the mechanical properties of the material in the stabilized state are *independent* of prior strain history. This is not true for a low stacking fault energy material, where restricted cross-slip will prevent the development of a common dislocation state from an initially hard and soft condition, respectively. In addition, the low SFE material will harden or soften more slowly than the high SFE alloy. We see from Fig. 12.37b that the material will cyclically soften and harden, but a final stabilized state is never completely achieved and is not equivalent for the two different starting conditions. For such materials, the "final" cyclically stabilized state is *dependent* on prior strain history.

One might then expect to find dislocation substructures in cyclically loaded samples similar to those found as a result of unidirectional loading. In fact, Feltner and Laird[65] observed that "those factors which give rise to certain kinds of dislocation structures in unidirectional deformation affect the cyclic structures in the same way." For example, we see from Fig. 12.38 that at high cyclic strains a cell structure is developed in high SFE alloys. If cyclic straining causes coarsening of a preexistent cell structure, then softening will occur. If the cell structure gets finer, then cyclic straining results in a hardening process. In low SFE alloys, dislocation planar arrays and stacking faults are present. These findings are similar to those discussed in Chapters 2 and 4

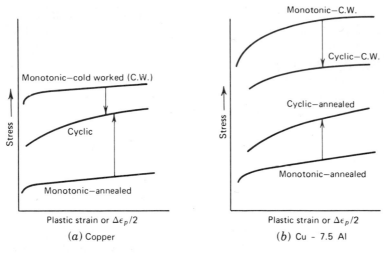

FIGURE 12.37 **Cyclic response of (*a*) high stacking fault energy copper and (*b*) low stacking fault energy Cu–7.5% Al alloy. Cyclically stabilized state in high SFE alloy is path independent.**[64] **(Reprinted from C. E. Feltner and C. Laird,** *Acta Metall.* **15 (1967), with permission of Pergamon Press.)**

for monotonic loading. A parallel condition is found in monotonic and cyclically induced dislocation structures produced at low strains.

12.3.2 Strain Life* Curves

Having identified the response of a solid to cyclic strains, it is now appropriate to consider how cyclically stabilized material properties affect the life of a specimen or engineering component subjected to strain-controlled loading. To accomplish this, it is convenient to begin our analysis by considering the elastic and plastic strain components separately. The elastic component is often described in terms of a relation between the true stress amplitude and number of load reversals

$$\frac{\Delta\epsilon_e E}{2} = \sigma_a = \sigma_f'(2N_f)^b \tag{12-18}$$

where $\dfrac{\Delta\epsilon_e}{2}$ = elastic strain amplitude

E = modulus of elasticity

σ_a = stress amplitude

σ_f' = fatigue strength coefficient, defined by the stress intercept at one load reversal ($2N_f = 1$)

N_f = cycles to failure

$2N_f$ = number of load reversals to failure

b = fatigue strength exponent

---

* "Life" of a test bar can be defined in a number of ways dependent on the ultimate use of the fatigue data and the nature of the material being tested. Some criteria include total fracture, specific changes in the shape of the hysteresis loop, and the existence of microcracks of a certain size (recall Section 12.2.3).

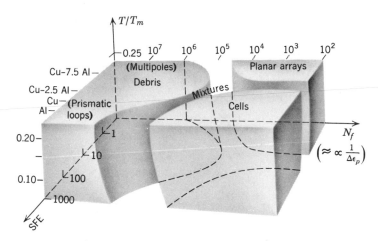

FIGURE 12.38 Schema showing dislocation substructures in FCC metals as a function of stacking fault energy, strain amplitude, and temperature.[65] (Reprinted with permission from the American Institute of Mining, Metallurgical and Petroleum Engineers.)

This relation, which represents an empirical fit of data above the fatigue limit (see Fig. 12.18), is similar in form to that proposed in 1910 by Basquin.[66] A sampling of test results is shown in Fig. 12.39u and fitted to Eq. 12-18. Increased fatigue life is expected with a decreasing fatigue strength exponent b and an increasing fatigue strength coefficient σ_f'. Representative fatigue property data for selected metal alloys are given in Table 12.3a,b.

The plastic component of strain is best described by the Manson-Coffin relation:[54,60,67,68]

$$\frac{\Delta\epsilon_p}{2} = \epsilon_f'(2N_f)^c \qquad (12\text{-}19)$$

where $\dfrac{\Delta\epsilon_p}{2}$ = plastic strain amplitude

ϵ_f' = fatigue ductility coefficient, defined by the strain intercept at one load reversal ($2N_f = 1$)

$2N_f$ = total strain reversals to failure

c = fatigue ductility exponent, a material property in the range -0.5 to -0.7

Data for SAE 4340 steel are plotted in Fig. 12.39b and are fitted to Eq. 12-19. In this instance, improved fatigue life is expected with a decreasing fatigue ductility exponent c and an increasing fatigue ductility coefficient ϵ_f'. Representative values for these qualities are given also in Table 12.3a,b.

Manson et al.[60] argued that the fatigue resistance of a material subjected to a given strain range could be estimated by superposition of the elastic and plastic strain components. Therefore, by combining Eqs. 12-16, 12-18, and 12-19, the total strain amplitude may be given by

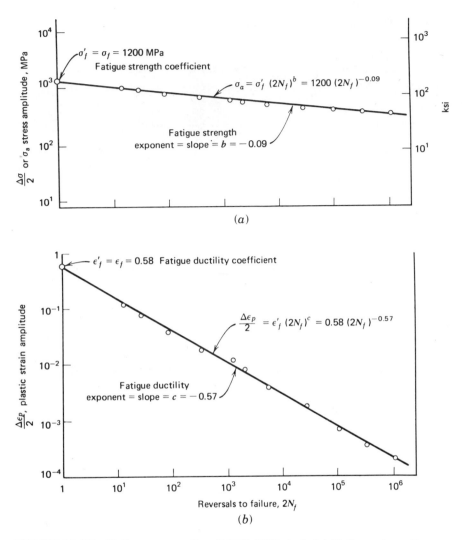

FIGURE 12.39 Fatigue properties of SAE 4340 steel. (*a*) Fatigue strength properties; (*b*) fatigue ductility properties.[53,54] (Reprinted by permission of the American Society for Testing and Materials from copyrighted work.)

$$\frac{\Delta\epsilon_T}{2} = \frac{\Delta\epsilon_e}{2} + \frac{\Delta\epsilon_p}{2} = \frac{\sigma_f'}{E}(2N_f)^b + \epsilon_f'(2N_f)^c \qquad (12\text{-}20)$$

By combining Eqs. 12-17 and 12-20, one finds that $n' = b/c$ and $K' = \sigma_f'/\epsilon_f'^{n'}$. ASTM Standard E606-92 has been prepared to allow for the development of such fatigue properties.[51] The interested reader is referred also to Refs. 48 and 49. It would be expected, then, that the total strain life curve would approach the plastic strain life curve at large strain amplitudes and approach the elastic strain life curve at low total strain amplitudes. This is shown in Fig. 12.40 for a high-strength steel alloy. Finally, when the total strain life plots for strong, tough, and ductile alloys are compared, the

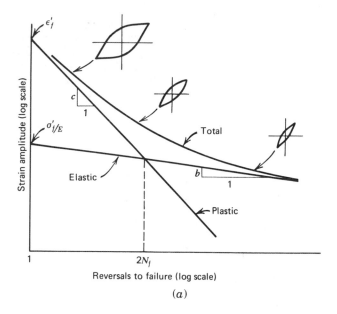

(a)

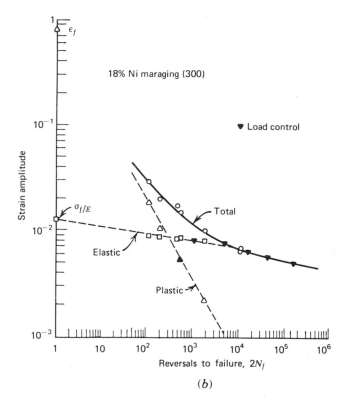

(b)

FIGURE 12.40 Superposition of plastic and elastic strain life curves to produce the total strain life fatigue relation for 18% Ni maraging steel.[56] (Reprinted by permission of the American Society for Testing and Materials from copyrighted work.)

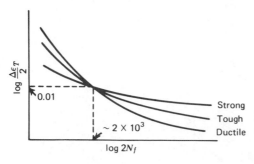

FIGURE 12.41 Schema showing optimum material for given total strain amplitude. Note lack of material property effect for cyclic life of about 10^3 (2×10^3 reversals) corresponding to a total strain range of about 0.01.[56] (Reprinted by permission of the American Society for Testing and Materials from copyrighted work.)

aforementioned trends in material selection are verified. We see from Fig. 12.41 that ductile alloys are best for high cyclic strain applications, and strong alloys are superior in the region of low strains. Note that around 10^3 cycles (2×10^3 reversals), corresponding to a total strain of about 0.01, there appears to be no preferred material. That is to say, *it makes no sense to attempt to optimize material properties for applications if strains of about 0.01 are encountered*—just about any alloy will serve the purpose.

12.4 FATIGUE LIFE ESTIMATIONS FOR NOTCHED COMPONENTS

A number of different procedures dealing with the estimation of component fatigue life have been considered thus far in this chapter. For example, a typical *S-N* curve has been used to establish a "safe life" for an engineering component based on the endurance limit or the fatigue strength associated with a particular cyclic life. For more complex loading histories, Miner's law is applied also to account for the cumulative damage resulting from each block of loads. In this approach, no deliberate attempt is made to distinguish between fatigue crack initiation and propagaton. Since most components contain some type of stress concentration, such as those shown in Fig. 7-6, "safe-life" values must be defined in terms of the local stress at the notch root rather than by the nominal applied stress. Indeed, Dowling and Wilson[69] have shown that specimens and components loaded with the same local stress field (defined by the product of the stress concentration and nominal stress field) exhibit similar crack initiation lives.

Special care is needed to equate the damage resulting from random loading cycles of varying amplitude with strain damage accumulated from constant amplitude loading events (e.g., see Figs. 12.39 and 12.40). To apply the linear damage accumulation concepts pertaining to Miner's law, it is necessary to identify the mean and amplitude values of each loading event so that the appropriate amount of damage can be properly "counted." The difficulty with this task lies in establishing a proper counting procedure that best accounts for the fatigue damage accumulated as a result of the random loading events.

Several counting procedures have been reviewed elsewhere, with the "rainflow method" accorded particular attention.[21,70] With reference to Fig. 12.42a, the random loading excursions result in the cyclic strains at the notch root as given in Fig. 12.42c.

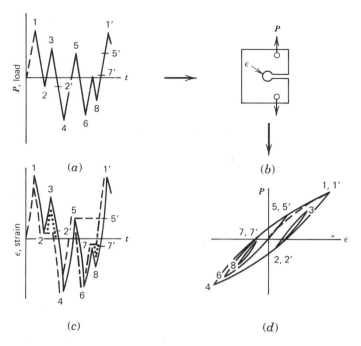

FIGURE 12.42 Random load spectrum applied to a notched component. (*a*) Load–time sequence; (*b*) notched component; (*c*) strain–time sequence at notch root and rainflow analysis procedure; (*d*) hysteresis loops corresponding to rainfall analysis. (Adapted from N. E. Dowling, W. R. Brose, and W. K. Wilson, *Fatigue Under Complex Loading*, SAE, 1977, p. 55, with permission.) (Reprinted with permission, copyright 1977 Society of Automotive Engineers, Inc.)

To identify the effective strain cycles by the rainflow methodology, we begin with the highest peak strain (location 1) and follow the curve to the first strain reversal (at 2). Now proceed horizontally to the next downward segment of the strain–time history (at 2′) (as rainwater would flow off the edge of the roof) and proceed again to the next strain reversal site (at 4). Since a horizontal shift would not encounter another downward segment of the strain–time history sequence, the strain direction would reverse and proceed upward to the next strain inflection point (at 5). Proceeding horizontally again to 5′ and continuing upward, we complete this strain segment at 1′. The initial load–strain hysteresis loop defined from this rainflow path would, therefore, correspond to strain and load limits defined by points 1, 4, and 1′, respectively (Fig. 12.42*d*).

Three additional hysteresis loops can be defined from rainflow analyses of the remaining positions of the original strain–time plot. The largest of these begins with the strain reversal at location 5. Proceeding downward to 6 and reversing direction to 7, we proceed horizontally to 7′ and continue the ascending strain segment until one reaches the location at 5′. The associated load–strain hysteresis loop defined from the second rainflow path corresponds to strain and load limits defined by points 5, 6, and 5′, respectively. Note the nonzero mean strain level corresponding to this hysteresis loop (Fig. 12.42*d*). The third and fourth hysteresis loops derived from the strain–time plot (Fig. 12.42c) correspond to the two remaining segments identified by points 2, 3, and 2′ and 7, 8, and 7′, respectively. Again note that these hysteresis loops (Fig. 12.42d) are not symmetrical about zero strain and involve various amounts of positive and negative cyclic-strain-induced creep. If the random loading sequence shown in Fig. 12.42a were to be repeated n times, the rainflow analysis procedure would characterize all of these random loading events in terms of the four hysteresis loops defined above, each repeated n times. As such, the random loading sequence can be reduced to a series of closed hysteresis loops of the type generated with laboratory samples.

It should also be noted that this simplification of the random loading pattern implies that there are no block sequence effects on component lifetime. This assumption of strain amplitude noninteraction parallels that implicit in Miner's law (recall Section 12.2.2); circumstances where fatigue life depends on load history are discussed in Section 13.5.

As discussed in the previous section, one can make a reasonable assumption that the same number of loading cycles is needed to develop a crack at various notch roots and in unnotched samples, if the cracked regions experience the same cyclic stress–strain history. Estimation of the finite life that involves such cumulative damage has been referred to as the "local stress–strain approach."[48,49] This procedure is not often easy to implement since it is not possible to monitor the elastoplastic stresses and strains near a notch root without resorting to advanced techniques such as elastic–plastic finite element analyses. Instead, estimation procedures have been developed such as the one proposed by Neuber.[71] Neuber's rule states that the *theoretical* stress concentration is equal to the geometric mean of the *actual* stress and strain concentrations (Eq. 12-21):

$$k_t = \sqrt{k_\sigma \cdot k_\epsilon} \tag{12-21}$$

where k_σ = actual stress concentration = $\Delta\sigma/\Delta S$
    $\Delta\sigma$ = local stress range
    ΔS = nominal stress range
    k_ϵ = actual strain concentration = $\Delta\epsilon/\Delta e$
    $\Delta\epsilon$ = local strain range
    Δe = nominal strain range

After substitution of terms, Eq. 12-21 becomes

$$k_t = \sqrt{\frac{\Delta\sigma}{\Delta S} \cdot \frac{\Delta\epsilon}{\Delta e}} \tag{12-22}$$

If the nominal stress ΔS and nominal strain Δe are assumed to be elastic, then Eq. 12-22 can be rearranged to show

$$\Delta\sigma \cdot \Delta\epsilon = (\Delta S \cdot k_t)^2/E \tag{12-23}$$

Topper et al.[72] modified this relation for cyclic loading applications by replacing k_t by the effective stress concentration factor k_f:

$$\Delta\sigma \cdot \Delta\epsilon = (\Delta S \cdot k_f)^2/E \tag{12-24}$$

When Eqs. 12-17 and 12-24 are combined, the notch root stress and strain ranges can be determined as shown in Fig. 12.43. These values can then be used to obtain the cycle life of a notched component from Eq. 12-20 or from actual data (e.g., Fig. 12.10 or 12.40).

If the fatigue data corresponding to Fig. 12-43 reflect cyclic lives necessary to initiate a crack of approximate length l_T, then the fatigue crack initiation lifetime will have been defined (recall Section 12.2.3). Total life can then be given by this initiation time plus the number of cycles necessary to propagate a crack from length l_T to the critical flaw size at final fracture (see Section 13.1.1). If one or more significant flaws are assumed to preexist in a structure, then the crack initiation stage is ignored and total fatigue life is represented only by crack propagation. Such circumstances form the basis of the ''damage tolerant design philosophy,'' which is discussed at considerable length in Chapter 13.

As a final note, let us reexamine the effect of various notches on the fatigue life of a particular material. It is clear from Fig. 12.6 that fatigue life at a *nominal* stress σ is reduced because of the presence of the stress concentration at the notch root. In principle, one should be able to normalize these results by replotting cyclic life versus the *local* stress at the notch root, with the latter value being estimated by the quantity $k_t\sigma$ so long as general yielding does not occur. An example of such a normalized plot is shown in Fig. 12.44a for the case of samples of 4340 steel containing widely varying stress concentrations.[69] Some investigators have used an alternative parameter, $\Delta K/\sqrt{\rho}$, to normalize crack initiation lifetimes where ΔK is the stress intensity factor and ρ is the notch root radius.[7,73,74] In this instance, the computation of ΔK has been based on an idealized geometry in which the notch is equated with a narrow slit (see

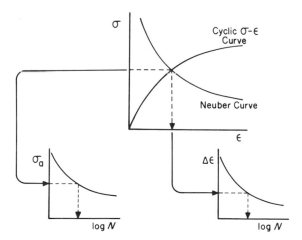

FIGURE 12.43 **Determination of notch root stress and strain from simultaneous solution of Eq. 12-17 (cyclic stress–strain curve) and Eq. 12-24 (Neuber relation). Cyclic life associated with these notch root stress and strain values is determined from the experimental data of unnotched specimens.**

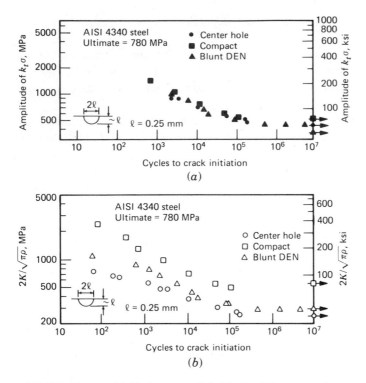

FIGURE 12.44 (*a*) **Fatigue crack initiation life versus the amplitude of the local elastic stress of the notch root, $k_t\sigma$. Specimen shapes include center hole, compact, and double-edge-notched configurations. (After Dowling and Wilson;[69] reprinted with permission of Pergamon Press.) (*b*) Poor correlation in fatigue initiation lives is noted when the $\Delta K/\sqrt{\rho}$ parameter is used for different specimen-notch configurations. (After Dowling;[75] reprinted with permission, copyright 1982, Society of Automotive Engineers, Inc.)**

Eq. 8-25). The normalizing potential of the $\Delta K/\sqrt{\rho}$ term for the case of compact specimens containing a long, narrow notch and different notch root radii is shown in Fig. 12.45 with data for a high-strength low-alloy steel.[7] In this example, the fatigue initiation life was defined when a crack 0.25 mm in length had been developed.

It should be noted that the $\Delta K/\sqrt{\rho}$ parameter does not represent a new approach to our understanding of fatigue fracture.[14,69,75] Instead, it represents an alternative method for estimating the local elastic stress at the notch root. Indeed, this estimation procedure can prove to be inexact when widely varying specimen configurations are compared (Fig. 12.44*b*). This follows from the considerable errors that result when Eq. 8-25 is used to determine the stress intensity factor for the case of blunt notches (e.g., a circular hole). Consequently, the use of the $\Delta K/\sqrt{\rho}$ parameter in normalizing crack initiation data should be restricted to specimens that contain long, narrow notches (e.g., the compact specimen).

12.5 FATIGUE CRACK INITIATION MECHANISMS

Fatigue cracks are initiated at heterogeneous nucleation sites within the material whether they be preexistent (associated with inclusions, gas pores, or local soft spots

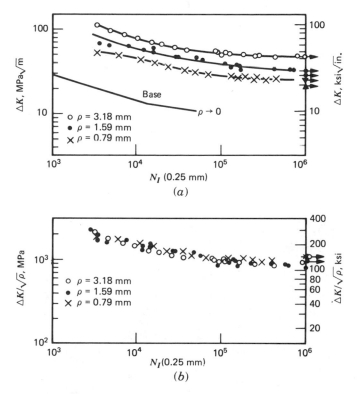

FIGURE 12.45 Fatigue initiation life in high-strength low-alloy steel associated with the development of a crack 0.25 mm in length. (*a*) Cyclic life versus ΔK as a function of notch root radius. (*b*) Cyclic life versus $\Delta K/\sqrt{\rho}$. (After Braglia, Hertzberg, and Roberts;[7] copyright, American Society for Testing and Materials, 1916 Race Street, Philadelphia, PA 19103. Reprinted with permission.)

in the microstructure) or are generated during the cyclic straining process itself (recall Fig. 12.4*c*). As one might expect, elimination of preexistent flaws can result in a substantial improvement in fatigue life. A good illustration of this is found when steels for roller bearings are vacuum melted as opposed to air melted. The much lower inclusion level in the vacuum-melted steel enables these bearings to withstand many more load excursions than the air-melted ones.

The most intriguing aspect of the fatigue crack initiation process relates to the generation of the nucleation sites. Although strains under monotonic loading produce surface offsets that resemble a staircase morphology, cyclic strains produce sharp peaks (extrusions) and troughs (intrusions)[76,77] resulting from nonreversible slip (Fig. 12.46). Many investigators have found that these surface notches serve as fatigue crack nucleation sites. It is probable that these extrusions and intrusions represent the initial stage in microcrack formation. (Recall the discussion of the phenomenology of microcrack formation in Section 12.2.3.) When the surface is periodically polished to remove these offsets, fatigue life is improved.[78] These surface upheavals represent the free surface terminations of dense bands of highly localized slip, the number of which increase with strain range. Careful studies have demonstrated these bands to be softer than the surrounding matrix material,[79] and it is believed that they cyclically

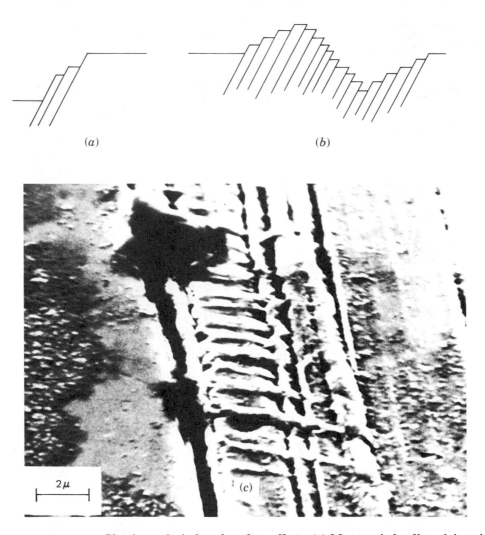

FIGURE 12.46 **Plastic strain-induced surface offsets.** (*a*) **Monotonic loading giving rise to staircase morphology slip offsets;** (*b*) **cyclic loading, which produces sharp peaks (extrusions) and troughs (intrusions);** (*c*) **photomicrograph showing intrusions and extrusions on prepolished surface.**[77] **(Reprinted with permission from W. A. Wood and Academic Press.)**

soften relative to the matrix, resulting in a concentration of plastic strain. These bands are called ''persistent slip bands'' because of two main results. First, when a metallographic section is prepared from a damaged specimen, the deformation bands persist after etching, indicating the presence of local damage. Second, when the surface offsets are removed by polishing and the specimen is cycled again, new surface damage occurs at the same sites. Consequently, although cracking begins at the surface, it is important to recognize that the material within these persistent bands and below the surface is also damaged and will control the location of the surface nucleated cracks.[80] It should also be noted that in polycrystalline metals and alloys, persistent slip bands (PSB) can be arrested at grain boundaries, thereby contributing to crack nucleation at the PSB–grain-boundary junction.

Various dislocation models have been proposed to explain the fatigue crack nucleation process. It is now generally believed that the initial stage of cyclic damage is associated with homogeneous slip and rapid strain hardening. At the point where the cyclic stress–strain curve tends to level out (see Section 12.3.1), slip becomes concentrated along narrow bands (i.e., persistent slip bands) and the band zones become softer. Such localization of plastic strain has been found to be the precursor for fatigue crack initiation in the low strain cycling regime. For a detailed examination of this subject, see the extensive reviews by Laird[81] and Mughrabi.[82]

In a recent review, Kramer[83] has proposed an alternative mechanism for fatigue crack initiation. Based on a wide range of experimental findings, he concluded that the extent of work hardening near the free surface differs markedly from that found at the specimen interior. The development of such a surface layer is believed to influence the stress–strain response of a metal as well as characterize the extent of fatigue damage accumulation. In a number of instances, the dislocation density is considerably higher within a layer extending approximately 100 μm from the free surface than at the specimen interior. It is interesting to note that the improvement in fatigue life by periodic surface polishing, mentioned above, could also be rationalized by the surface layer model; the polishing action would remove the hardened layer, thereby arresting the formation of dislocation pileups believed responsible for the cracking process.

Only recently, has attention been given to an analysis of fatigue initiation mechanisms in engineering plastics. In some amorphous materials and rubber-toughened polymers such as high-impact polystyrene, crazing has been identified as the dominant fatigue damage mechanism.[84,85] Evidence of this effect has been obtained on the basis of volumetric measurements (recall that crazing involves a volume expansion; see Section 6.4) and the progressive change in shape of the cyclic stress–strain hysteresis loops. Regarding the latter, the *tensile* portion of the hysteresis loops tends to flatten

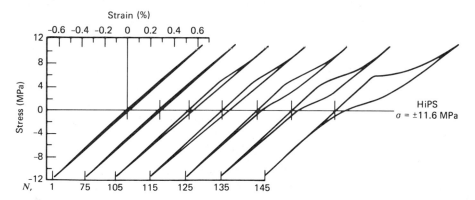

FIGURE 12.47 Hysteresis loops in high-impact polystyrene (HiPS). Note progressive distortion of tensile portion of loop with increasing number of cyclic loads. (After Bucknall and Stevens,[84] from *J. Mater. Sci.* **15**, 295 (1980); with permission from Chapman & Hall.)

and is characteristic of a crazed material (Fig. 12.47). By contrast, fatigue initiation in ABS was traced to a shear yielding process. In the latter material, the hysteresis loops generated during fatigue cycling remained symmetrical until near the end of the fatigue life of the sample. At this point, both shear yielding and crazing occurred.

12.6 AVOIDANCE OF FATIGUE DAMAGE

12.6.1 Favorable Residual Compressive Stresses

To this point, we have examined the influence of mean stress, load spectrum, and surface treatment on fatigue lifetime. Regarding the latter, shot peening and carburizing processes were found to introduce surface compressive stresses that extend the lifetime of service components (recall Section 12.2.4.1). Favorable residual compressive stresses may also be introduced by *mechanical* as well as metallurgical means. For example, cold-rolling fatigue-sensitive areas such as thread roots in threaded fasteners, causes the material to plastically deform and spread laterally in the thread root area; however, such motion is constrained by the bulk elastic substrate, resulting in the development of compressive residual stresses at the notch root. The combination of this favorable residual stress pattern with the reduction in stress concentration brought about by the cold-rolling-induced enlargement of the thread root radius and the development of a favorable grain flow pattern (recall Fig. 10.5) leads to improved fatigue life in cold-rolled fasteners.

Compressive residual stresses can also be introduced around a fastener hole, a classical stress concentration site. In one such commercially successful procedure used extensively in the aircraft industry, an internally lubricated split-sleeve is placed over a mandrel and the assembly inserted into the plate hole (Fig. 12.48a, b).[86,87] When the mandrel is pulled through the split-sleeve, the hole is plastically deformed and expanded since the combined thickness of the split-sleeve and maximum dimension of the mandrel is greater than the hole diameter (Fig. 12.48c). The elastic springback of the plate then creates residual compressive stresses in the plastically deformed region surrounding the hole; the peak residual stress is approximately two thirds of the tensile yield strength of the plate material. Minor distortions of the hole surface, resulting from the expansion process, can be eliminated by a final reaming operation. Since the compressive stress field extends one to two radii from the hole surface (see Fig. 12.49), the reaming operation does little to diminish the beneficial influence of this favorable stress field. Figure 12.50 confirms that fatigue lifetimes may be enhanced three- to tenfold by the split-sleeve cold expansion process. Experience has shown that for best results, holes in aluminum alloy plates should be expanded by at least 3%; in titanium and high strength steel alloys, holes should be expanded by at least $4\frac{1}{2}\%$.[87]

The fatigue resistance of thick-walled cylindrical pressure vessels is also improved by the introduction of a favorable residual compressive stress field at the inner bore surface where applied circumferential stresses are greatest. Three methods for the development of this stress field include: (1) shrinking an outer hollow cylinder (jacket) over the main cylinder (core); (2) constructing the thick-walled cylinder by nesting several prestressed thin sleeves over one another; and (3) a process called *autofrettage* or *self-hooping*. In the first method, an outer jacket—with inner radius slightly smaller

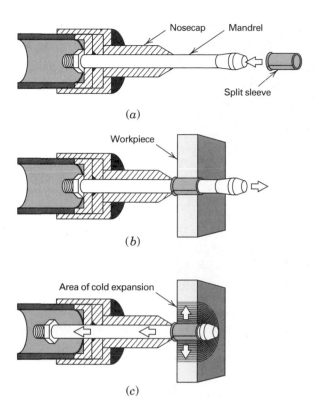

FIGURE 12.48 Split-sleeve cold expansion process. (*a*) Internally lubricated split-sleeve is attached to the mandrel; (*b*) mandrel and split-sleeve are inserted into plate hole; (*c*) mandrel cold expands the hole to generate residual compressive stresses. (Courtesy of Fatigue Technology Inc.)

than the outer radius of the core cylinder—is initially heated and placed over the core cylinder. As the jacket cools, it contracts, thereby generating a compressive circumferential stress field in the core cylinder[88] (Fig. 12.51*a*). Note that the outer jacket experiences a residual tensile stress field and that the magnitude of the circumferential stress in both the core and outer sleeve varies through the wall thickness. (Recall from Section 1.4 that the circumferential stress is essentially constant through the thickness of a *thin*-walled pressure vessel.) When the inner core is pressurized, a circumferential stress distribution is developed as shown in Fig. 12.51*b*. Note that the maximum applied stress at the inner bore surface (location *A*) is reduced from *A'* to *B'* by the superposition of the compressive residual stress at the inner bore. When a nest of thin-walled sleeves is constructed, a resulting residual stress pattern is developed as shown in Fig. 12.51*c*. Notice the development of a quasi-uniform stress field (*E–F*) when the internal-pressure-induced stress field (*C–D*) is superimposed on the latter residual stress pattern (Fig. 12.51*d*).

In the third pressure vessel prestressing procedure (autofrettage or self-hooping), the thick-walled cylinder is internally pressurized to cause yielding, beginning at the inner radius and spreading outward through the wall thickness; the extent of plastic

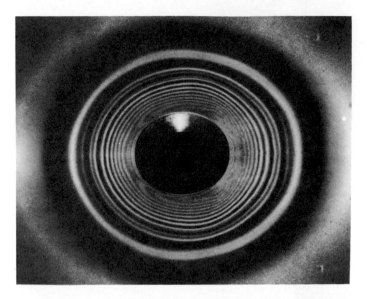

FIGURE 12.49 View of cold-expanded hole with the aid of polarized light through a birefringent plastic coating. Concentric rings correspond to compressive stress field surrounding the hole. (Courtesy of Fatigue Technology Inc.)

deformation depends on the magnitude of the internal pressure. When the pressure is released, the spring-back action of the outer portion of the vessel generates a compressive stress field in the bore region of the vessel as the plastically deformed material is squeezed together. (Note the similarity of the autofrettage process with split-sleeve cold expansion of fasteners holes.) A matching residual tensile stress field is created near the outer radius of the vessel. Since the maximum circumferential stress in a

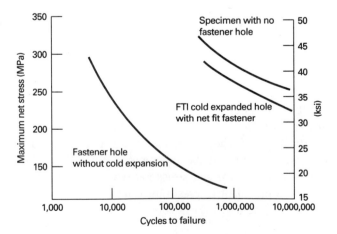

FIGURE 12.50 *S-N* plot of 7075-T6 aluminum specimens revealing the beneficial influence of cold expanded holes on total fatigue lifetme.[86] R. A. Feeler, *Aviat. Equip. Maint.*, 21 (March/April 1964), Reprinted with permission.

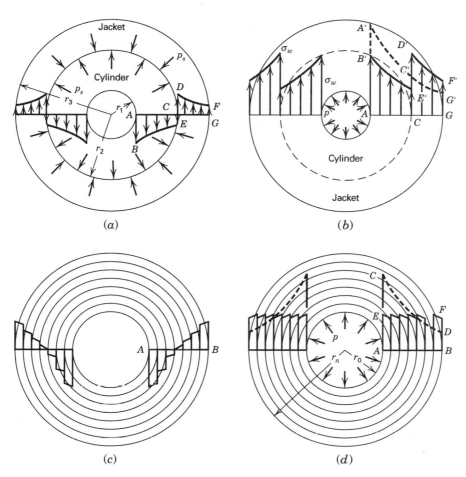

FIGURE 12.51   Introduction of residual stresses in thick-walled pressure vessels. (*a*) Residual stress pattern associated with shrink fitting a jacket onto the inner core hollow cylinder; (*b*) superposition of residual stress pattern from (*a*) and internally applied stress field. Maximum circumferential stress at bore surface is reduced from A' to B'; (*c*) residual stress pattern associated with nested cylinder construction; (*d*) superposition of residual stress pattern from (*c*) and internally applied stress field. Maximum circumferential stress at bore surface is reduced from C to E.[88] (F. B. Seely and J. O. Smith, *Advanced Mechanics of Materials, 2nd Ed.*, John Wiley & Sons, New York (1957). Reprinted by permission of John Wiley & Sons, Inc.)

thick-walled tube is located along the inner bore surface, there is a net positive influence of this residual stress field on the fatigue resistance of the cylinder. An example of the application of this technique to improve the fatigue resistance of a 175-mm gun tube is described in a failure analysis case history (see Section 14.5, Case 7). In this instance, the autofrettage-induced residual compressive stress at the bore surface was greater than the gas pressure-induced circumferential stress generated when the gun was fired!

12.6.2 Pretensioning of Load-Bearing Members

Residual *tensile* stresses also improve the fatigue lifetime of engineering components such as fasteners and associated fastener holes.[89,90] For example, when an interference fit is established between a bolt and the associated plate holes, the elastic stress pattern, shown in Fig. 12.52, is developed in the plate. For the case of an interference-fit bolt with a bolt-plate elastic modulus ratio of 3 (e.g., steel/aluminum), the repeated application of a remote stress σ_a, will result in a local stress in the plate at $y = 0$ and $x = \pm R$ that fluctuates between 1.3 and 1.7 σ_a. (It is assumed here that the initial interference is sufficient such that contact between the bolt and plate is maintained at $x = 0$ and $y = \pm R$ when σ_a is applied.) For the case of a loose-fitting bolt, the maximum local stress where $y = 0$ and $x = \pm R$ is $3\sigma_a$, where 3 represents the stress concentration associated with a hole in a plate. It follows that if the external stress fluctuates between 0 and σ_a, the stress range at the hole in the plate is $3\sigma_a$ as compared with $0.4\sigma_a$ for the case of the interference bolt. Since the fatigue lifetime of a component depends strongly on the applied cyclic stress *range* (recall Section 12.2), the durability of the plate is significantly enhanced with the use of interference-fit fasteners. Note that the introduction of an interference-fit fastener of similar modulus (Fig. 12.52) generates local stresses at the plate hole that fluctuate between 1.0 and 2.0 σ_a. It follows that greater benefits from interference-fit fasteners result from an increased bolt/plate elastic modulii ratio, but that even interference fits with similar materials provide improved fatigue resistance as compared to the case where no interference is present.

Tapered pins also provide the characteristics of interference-fit fasteners. As shown in Fig. 12.53a, tapered bolts (called *taperlock* fasteners) are inserted into tapered holes and then tightened into place.[91] This, again, results in an interference fit and the

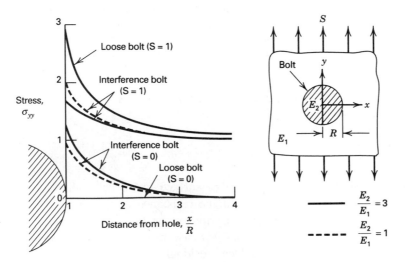

FIGURE 12.52 Elastic stress distribution surrounding plate hole with/without interference-fit bolt.[89] H. F. Hardrath, *J. Test. Eval.* 1 (1), 3 (1972). Copyright ASTM, Reprinted with permission.

Installation sequence

1. Following hole preparation with tapered drill, tapered shank bolt is inserted in hole and firmly seated by hand pressure.

2. Full contact along entire shank of bolt and hole prevents rotation of bolt while tightening washer-nut. During tightening nut spins freely to the locking point; but washer remains stationary and provides a bearing surface against structure.

3. Torquing of washer-nut by conventional wrenching method produces a close tolerance interference fit, seats the bolt head, and creates an evenly balanced prestress condition within the bearing area of the structural joint.

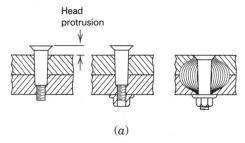

(a)

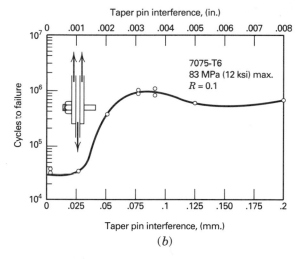

(b)

FIGURE 12.53 (a) **Sequence of installation and geometry of taper pin fasteners;[91] (Excerpted from Assembly Engineering, July, 1967. By permission of the Publisher © 1967. Hitchcock Publishing Co., all rights reserved.)** (b) **Influence of taper pin interference level on fatigue lifetime in 7075-T6 lugs. Applied stress = 83 MPa (12 ksi), $R = 0.1$.[92] (C. R. Smith, *Exp. Mech.*, 5 (8), 19A (1965), with permission.)**

development of a tensile residual stress field in the plate; the latter effectively reduces the local stress range at the hole and increases service lifetime. For example, Fig. 12.53b demonstrates an approximate 20-fold improvement in fatigue lifetime associated with the use of taper pin fasteners.

In related manner, pretensioning of bolts results in their improved fatigue performance. Consider the case where two plates are attached with a bolt and nut (Fig. 12.54). If the nut is finger-tightened, then the entire load acting to separate the plates is carried

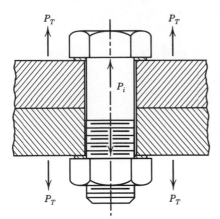

FIGURE 12.54 Preloading of bolt.

by the bolt alone. If, on the other hand, the nut is tightly fastened onto the bolt, the bolt experiences a *tensile* preload, P_i, whereas the plates are squeezed together with a *compressive* preload of the same magnitude. This not only insures a tight and probable leak-proof connection, but the fatigue resistance of the bolt is enhanced, as revealed by the following analysis. From Hooke's law we see that the bolt and plate deflections are

$$\delta_B = \frac{P_B}{k_B} \quad \text{and} \quad \delta_P = \frac{P_P}{k_p} \tag{12-25}$$

where $\delta_{B,P}$ = bolt and plate deflections, respectively
  $P_{B,P}$ = loads carried by bolt and plate, respectively
  $k_{B,P}$ = spring constants for bolt and plate, respectively

Since $\delta_B = \delta_P$, it follows that

$$\frac{P_B}{k_B} = \frac{P_P}{k_P} \tag{12-26}$$

If an external load, P_T, is applied so as to separate the plates, the applied load distribution is

$$P_T = P_B + P_P \tag{12-27}$$

By combining Eqs. 12-26 and 12-27, the total applied load is given by

$$P_T = P_B + \frac{P_B k_P}{k_B} = P_B \left[1 + \frac{k_p}{k_B} \right] \tag{12-28}$$

Alternatively,

$$P_B = \frac{P_T k_B}{k_B + k_P} \tag{12-29}$$

By combining the applied and pretensioning loads, the total load experienced by the bolt is

$$P_B = \frac{P_T k_B}{k_B + k_P} + P_i \qquad (12\text{-}30)$$

Also, the total load experienced by the plate is

$$P_P = \frac{P_T k_P}{k_B + k_P} - P_i \qquad (12\text{-}31)$$

EXAMPLE 12.3

From Fig. 12.54, assume that the stiffness of the plates and washers is four times greater than that of the bolt. If the applied cyclic load, P_T, and preload, P_i, are both 5000 N, find the cyclic loads carried by the bolt and the plate members.

From Eq. 12-30, the maximum load on the bolt is

$$P_{B_{max}} = \frac{5000(1)}{1 + 4} + 5000 = 6000 \text{ N}$$

Therefore, the cyclic load range experienced by the preloaded bolt is $6000 - 5000 = 1000$ N, rather than 5000 N for the case of a non–preloaded bolt. The maximum load experienced by the plates is

$$P_{P_{max}} = \frac{5000(4)}{1 + 4} - 5000 = -1000 \text{ N}$$

with the cyclic load being $-5000 - (-1000) = -4000$ N.

Note that the plates remain in compression even though the applied load is equal to the preload. Furthermore, most of the applied load and the major portion of the load fluctuation is carried by the stiffer plate members and *not* by the bolt. Hence, the fatigue resistance of the bolt should be greatly enhanced. Experimental data given in Table 12.4 confirm this scenario.[91] Therefore, to improve the fatigue life of a bolt, a high tensile preload should be applied (e.g., $P_i \approx 90\% \sigma_{ys}$) and the plates to be joined should be stiff relative to the stiffness of the bolt.* (If a soft gasket or washer is used, the beneficial effect of the preload is significantly diminished.) It should be recognized that the level of bolt preloading can diminish with time due to the extrusion of dirt

---

* It should be noted that tensile preloading of bolts can contribute to premature failure as a result of environmentally assisted crack extension, initiating at the stress concentration between the bolt head and shaft. Fastener failure occurs when the bolt head "pops off." Therefore, environmentally sensitive preloaded bolts should be shielded by the addition of a protective surface treatment, especially at the bolt shaft/head fillet.

TABLE 12.4 Influence of Preload Level on Bolt Fatigue Lifetime [93]

Preload, N(lbs)	Mean Fatigue Lifetime, Cycles
6,315(1420)	6,000
26,330(5920)	36,000
32,115(7220)	215,000
37,450(8420)	5,000,000

or paint particles from the contact surfaces and/or crushing of the joint members such as washers and gaskets. As expected, reduction in the preload level brings about a reduction in fatigue resistance of the bolt.

REFERENCES

1. P. J. Forrest, *Fatigue of Metals*, Addison-Wesley, Reading, MA, 1962.
2. D. J. Wulpi, *How Components Fail*, American Society for Metals, Metals Park, OH, 1966.
3. A. Wöhler, *Zeitschrift für Bauwesen* **10** (1860).
4. R. E. Peterson, Edgar Marbury Lecture, ASTM, 1962, p. 1.
5. ASTM Standards E466-E468, *Annual Book of ASTM Standards*, 1993.
6. H. F. Hardrath, E. C. Utley, and D. E. Guthrie, NASA *TN D-210*, 1959.
7. B. L. Braglia, R. W. Hertzberg, and Richard Roberts, *Crack Initiation in a High-Strength Low-Alloy Steel*, ASTM *STP677*, 1979, p. 290.
8. A. Palmgren, *Bertschrift des Vereines Ingenieure* **58,** 339 (1924).
9. M. A. Miner, *J. Appl. Mech.* **12,** A-159 (1954).
10. J. B. Kommers, *Proc. ASTM* **45,** 532 (1945).
11. D. K. Bullens, *Steel and Its Heat Treatment* **1,** 37 (1938).
12. R. E. Peterson, *Stress Concentration Factors*, Wiley, New York, 1974.
13. G. Sines and J. L. Waisman, *Metal Fatigue*, McGraw-Hill, New York, 1959, p. 298.
14. N. E. Dowling, ASTM *STP677*, 1979, p. 247.
15. S. J. Hudak, Jr., *J. Eng. Mater. Tech.* **103,** 26 (1981).
16. B. Leis, ASTM *STP743*, 1981, p. 100.
17. D. Broek and B. Leis, *Materials Experimentation and Design in Fatigue*, Westbury House, Warwick, Surrey, England, 1981, p. 129.
18. J. C. Newman, Jr., NASA *TND-6376*, Washington, DC, 1971.
19. M. B. P. Allery and G. Birkbeck, *Eng. Fract. Mech.* **4,** 325 (1972).
20. W. N. Findley, *J. Mech. Eng. Sci.* **14** (6), 424 (1972).
21. H. O. Fuchs and R. I. Stephens, *Metal Fatigue in Engineering*, Wiley-Interscience, New York, 1980.
22. P. J. E. Forsyth, *The Physical Basis of Metal Fatigue*, American Elsevier, New York, 1969.
23. *Metals Handbook*, 8th ed., Vol. 1, *Properties and Selection of Metals*, ASM, Novelty, OH, 1961, p. 217.
24. C. Calabrese and C. Laird, *Mater. Sci. Eng.* **13** (2), 141 (1974).

25. F. Sherratt, "The Influence of Shot-Peening and Similar Surface Treatments on the Fatigue Properties of Metals," Part I, S&T Memo 1/66, Ministry of Aviation, U.S. Govt. Report 487487, Feb. 1966.

26. W. J. Harris, "The Influence of Decarburization on the Fatigue Behavior of Steel Bolts," S&T Memo 15/65, Ministry of Aviation, U.S. Govt. Report 473394, August 1965.

27. R. W. Hertzberg and J. A. Manson, *Fatigue of Engineering Plastics*, Academic, New York, 1980.

28. ASTM Standard D671-71, *Annual Book of ASTM Standards*, 1971.

29. I. Constable, J. G. Williams, and D. J. Burns, *JMES* **12**, 20 (1970).

30. M. N. Riddell, G. P. Koo, and J. L. O'Toole, *Polym. Eng. Sci.* **6**, 363 (1966).

31. G. P. Koo, M. N. Riddell, and J. L. O'Toole, *Polym. Eng. Sci.* **7**, 182 (1967).

32. J. P. Berry, *Viscoelastic Properties of Polymers*, Wiley, New York, 1961.

33. L. J. Broutman and S. K. Gaggar, *Proceedings of the Twenty Seventh Annual Technical Conference*, 1972, Society of the Plastics Industry, Inc., Section 9-B, p. 1.

34. A. V. Stinskas and S. B. Ratner, *Plasticheskie Massey* **12**, 49 (1962).

35. L. C. Cessna, J. A. Levens, and J. B. Thomson, *Polym. Eng. Sci.* **9**, 399 (1969).

36. R. L. Thorkildsen, *Engineering Design for Plastics,* E. Baer (Ed.), Van Nostrand-Reinhold, New York, 1964, p. 279.

37. J. A. Sauer and C. C. Chen, *Adv. Polym. Sci.* **52/53,** 169 (1983).

38. J. A. Sauer and C. C. Chen, *Toughening of Plastics*, Plastics and Rubber Institute, London, UK, 1985, paper 26.

39. J. T. Turkanis (Brennock), R. W. Hertzberg, and J. A. Manson, *Deformation, Yield and Fracture of Polymers*, Plastics and Rubber Institute, London, UK, 1985, paper 54.

40. M. D. Skibo, J. A. Manson, R. W. Hertzberg, S. M. Webler, and E. A. Collins, Jr., *Durability of Micromolecular Materials*, R. K. Eby, Ed., ACS Symposium Series, No. 95, American Chemical Society, Washington, DC, 1979, p. 311.

41. R. W. Hertzberg and J. A. Manson, Fracture and Fatigue, *Encyclopedia of Polymer Science and Engineering*, Vol. 7, 2nd ed., Wiley, New York, 1986, p. 378.

42. K. L. Reifsnider, *Int. J. Fract.* **16,** 563 (1980).

43. E. M. Wu, *Composite Materials*, Vol. 5 L. J. Broutman and R. H. Krock, Eds., Academic, New York, 1974, Chapter 3.

44. J. F. Mandell, F. J. McGarry, D. D. Huang, and C. G. Li, *Polym. Compos.* **4**, 32 (1983).

45. R. E. Lavengood and L. E. Gulbransen, *Polym. Eng. Sci.* **9**, 365 (1969).

46. S. Morris, Ph.D. Thesis, University of Nottingham, UK, 1970, quoted in *Composite Materials*, Vol. 5, L. J. Broutman and R. H. Krock, Eds., Academic, New York, 1974, Chapter 8, p. 281.

47. C. C. Chamis, *Composite Materials,* Vol. 5, L. J. Broutman and R. H. Krock, Eds., Academic, New York, 1974, p. 94.

48. R. M. Wetzel (Ed.), *Advances in Engineering*, Vol. 6, *Fatigue under Complex Loading: Analysis and Experiments,* Society of Automotive Engineers, 1977.

49. *Manual on Low Cycle Fatigue Testing,* ASTM *STP465*, 1969.

50. J. Morrow and D. F. Socie, *Materials, Experimentation and Design in Fatigue,*

F. Sherratt and J. B. Sturgeon, Eds., Westbury House, Warwick, Surrey, England, 1981, p. 3.

51. ASTM Standard E606-92, *Annual Book of ASTM Standards,* 1992.
52. B. I. Sandor, *Fundamentals of Cyclic Stress and Strain*, University of Wisconsin Press, Madison, 1972.
53. J. D. Morrow, *Internal Friction, Damping and Cyclic Plasticity* ASTM *STP 378*, 1965, p. 45.
54. R. W. Smith, M. H. Hirschberg, and S. S. Manson, NASA *TN D-1574*, NASA, April 1963.
55. R. W. Landgraf, J. D. Morrow, and T. Endo, *J. Mater, JMLSA* **4**(1), 176 (1969).
56. R. W. Landgraf, *Achievement of High Fatigue Resistance in Metals and Alloys*, ASTM *STP-467,* 1970, p. 3.
57. P. Beardmore and S. Rabinowitz, *Treatise on Materials Science and Technology*, Vol. 6, R. J. Arsenault, Ed., Academic, New York, 1975, p. 267.
58. S. Rabinowitz and P. Beardmore, *J. Mater. Sci.* **9,** 81 (1974).
59. P. Beardmore and S. Rabinowitz, *Polymeric Materials*, ASM, Metals Park, OH, 1975, p. 551.
60. S. S. Manson and M. H. Hirschberg, *Fatigue: An Interdisciplinary Approach,* Syracuse University Press, Syracuse, NY, 1964, p. 133.
61. J. Bauschinger, *Zivilingur* **27,** 289 (1881).
62. S. T. Rolfe, R. P. Haak, and J. H. Gross, *Trans. ASME J. Bas. Eng.* **90,** 403 (1968).
63. J. P. Hickerson and R. W. Hertzberg, *Met. Trans.* **3,** 179 (1972).
64. C. E. Feltner and C. Laird, *Acta Met.* **15,** 1621 (1967).
65. C. E. Feltner and C. Laird, *Trans. AIME* **242,** 1253 (1968).
66. O. H. Basquin, *Proc. ASTM* **10,** Part II, 625 (1910).
67. L. F. Coffin, Jr., *Trans. ASME* **76,** 931 (1954).
68. J. F. Tavernelli and L. F. Coffin, Jr., *Trans. ASM* **51,** 438 (1959).
69. N. E. Dowling and W. K. Wilson, *Advances in Fracture Research*, D. Francois, Ed., Pergamon, Oxford, England, 1981, p. 518.
70. M. Matsuishi and T. Endo, *Fatigue of Metals Subjected to Varying Stress*, paper presented to Japan Society of Mechanical Engineers, Fukuoka, Japan, March 1968.
71. H. Neuber, *Trans. ASME, J. App. Mech.* **8,** 544 (1961).
72. T. H. Topper, R. M. Wetzel, and J. Morrow, *J. Mater, JMSLA* **4**(1), 200 (1969).
73. A. R. Jack and A. T. Price, *Int. J. Fracture Mech.* **6,** 401 (1970).
74. J. M. Barsom and R. C. McNicol, in *Fracture Toughness and Slow-Stable Cracking*, ASTM *STP 559*, 1974, p. 183.
75. N. E. Dowling, Paper presented at SAE Fatigue Conference, April 14–16, 1982, Dearborn, MI.
76. W. A. Wood, *Fracture,* Technology Press of M.I.T. and Wiley, New York, 1959, p. 412.
77. W. A. Wood, *Treatise on Materials Science and Technology*, Vol. 5, H. Herman, Ed., Academic, New York, 1974, p. 129.
78. T. H. Alden and W. A. Backofen, *Acta Met.* **9,** 352 (1961).
79. O. Helgeland, *J. Inst. Met.* **93,** 570 (1965).
80. C. Roberts and A. P. Greenough, *Philos. Mag.*, **12,** 81 (1965).

81. C. Laird, *Metallurgical Treatises*, J. K. Tien and J. F. Elliott, Eds., AIME, Warrendale, PA, 1981, p. 505.

82. H. Mughrabi, *Fifth International Conference on Strength of Metals and Alloys*, P. Haasen, V. Gerold, and G. Kostorz, Eds., Pergamon, Oxford, England, 1980, p. 1615.

83. I. R. Kramer, *Advances in the Mechanics and Physics of Surfaces*, Vol. 3, Gordon & Breech, New York, 1986, p. 109.

84. C. B. Bricknall and W. W. Stevens, *J. Mater. Sci.* **15,** 2950 (1980).

85. M. E. Mackay, T. G. Teng, and J. M. Schultz, *J. Mater. Sci.* **14,** 211 (1979).

86. R. A. Feeler, *Aviation Equipment Maintenance* **21,** March/April 1984.

87. *Fatigue Technology Inc.,* Seattle, Washington, corporate literature.

88. F. B. Seely and J. O. Smith, *Advanced Mechanics of Materials, 2nd Ed.,* John Wiley, New York 1957).

89. H. F. Hardrath, *J. Test. Eval.* **1**(1), 3 (1972).

90. J. H. Crews, *NASA TND-6955,* NASA (August 1972).

91. C. R. Smith, *Assembly Engineering* **10**(7), 18 (1967).

92. C. R. Smith, *Experimental Mechanics* **5**(8), 19A (1965).

93. W. Orthwein, *Machine Component Design*, West Publishing Co., St. Paul, MN, (1990).

PROBLEMS

12.1 Calculate the total strain range that a smooth bar cyclic strain specimen would be able to tolerate before failing after 1000 cycles (2000 load reversals). Show all your computations and assumptions.

$$E = 205 \text{ GPa}$$
$$\sigma_f \text{ (monotonic)} = 1850 \text{ MPa}$$
$$\epsilon_f \text{ (monotonic)} = 0.7$$
$$n' \text{ (cyclic)} = 0.15$$

12.2 Two investigators independently reported fatigue test results for Zeusalloy 300. Both reported their data in the form of $\sigma - N$ curves for notched bars. The basic difference between the two results was that Investigator I reported inferior behavior of the material compared with the results of Investigator II (i.e., lower strength for a given fatigue life) but encountered much less scatter. Can you offer a possible explanation for this observation? Describe the macroscopic fracture surface appearance for the two sets of test bars.

12.3 Two different polymeric materials were evaluated to determine their respective fatigue endurance behavior. Both materials were tested separately in laboratory air and in flowing water. (Water was selected as a suitable liquid test environment since neither polymer was adversely affected by its presence.) Polymer A showed similar *S-N* plots in the two environments whereas Polymer B revealed markedly different results. Speculate as to which environment was associated with the superior fatigue response in Polymer B and characterize the structure and mechanical response of Polymers A and B.

12.4 Calculate the fatigue life of SAE 1015 and 4340 steel tempered at 425°C when the samples experience total strain ranges of 0.05, 0.01, and 0.001. Which alloy is best at each of these applied strain ranges?

12.5 Several samples of a copper alloy (initially strain hardened to varying amounts) were cycled under fixed strain conditions for several hundred cycles. These samples were then tested in tension and found to exhibit the same stress–strain response. Had they been tested in tension before being strain cycled, the results could have been markedly different. Explain what happened to the material during the cyclic loading stage to alter its subsequent mechanical behavior.

12.6 In an effort to determine a material's resistance to fatigue crack initiation, two studies were undertaken. One investigator used a plate sample with a circular hole 1 cm in diameter, and the other investigator used a similar sample with a 4-cm circular hole. In both cases, the definition for fatigue life initiation was taken to be the number of loading cycles necessary to develop a crack 0.25 cm in length. The cyclic lives determined from these two investigations were not in agreement. Why? Based on the results of Dowling, what change in specimen geometry or initiation criteria would you recommend so that both investigators would report similar initiation lives?

12.7 Tensile and fully reversed loading fatigue tests were conducted for a certain steel alloy and revealed the tensile strength and endurance limit to be 1200 and 550 MPa, respectively. If a rod of this material supply were subjected to a static stress of 600 MPa and oscillating stresses whose total range was 700 MPa, would you expect the rod to fail by fatigue processes? *Hint:* Plot a diagram to aid in presenting your answer.

12.8 The fatigue limit for a certain alloy at stress levels of σ_1, σ_2 and σ_3 are 10,000, 50,000, and 500,000 cycles, respectively. If a component of this material is subjected to 2500 cycles of σ_1 and 10,000 cycles of σ_2, estimate the remaining lifetime in association with cyclic stresses at a level of σ_3.

12.9 For a steel alloy with a tensile strength of 1000 MPa, estimate the fatigue strength amplitude for this material when the mean stress is 200 MPa.

CHAPTER 13

FATIGUE CRACK PROPAGATION

As discussed in Chapter 7, a number of engineering system breakdowns can be attributed to preexistent flaws that caused failure when a certain critical stress was applied. In addition, these defects may have grown to critical dimensions prior to failure. Subcritical flaw growth is important to guard against for a number of reasons. First, if a structure or component contains a defect large enough to cause immediate failure upon loading, the defect quite likely could be detected by a number of non-destructive test (NDT) procedures and repaired before damaging loads are applied. If the defect is not detected, the procedure of proof testing (subjecting a structure, such as a pressure vessel, to a preservice simulation test at a stress level equal to or slightly higher than the design stress) would cause the structure to fail, but under controlled conditions with minimum risk to human lives and damage to other equipment in the engineering system. On the other hand, were the crack to be subcritical in size and undetected by NDT, a successful proof test would prove only that a flaw of critical dimensions did not exist at that time. *No guarantee could be given that the flaw would not grow during service to critical dimensions and later precipitate a catastrophic failure.* This chapter is concerned with factors that control the fatigue crack propagation (FCP) process in engineering materials.

13.1 STRESS AND CRACK LENGTH CORRELATIONS WITH FCP

Crack propagation data may be obtained from a number of specimens such as many of those shown in Fig. 8.7. Starting with a mechanically sharpened crack, cyclic loads are applied and the resulting change in crack length is monitored and recorded as a function of the number of load cycles. Many monitoring techniques have been employed, such as the use of a calibrated traveling microscope (Fig. 13.1), eddy current techniques, electropotential measurements, compliance measurements, and acoustic emission detectors. A typical plot of such data is shown in Fig. 13.2, where the crack length is seen to increase with increasing number of loading cycles. The fatigue crack growth rate is determined from such a curve either by graphical procedures or by computation. From these methods, the crack growth rates resulting from a given cyclic load are $(da/dN)_{a_i}$ and $(da/dN)_{a_j}$ when the crack is of lengths a_i and a_j, respectively.

It is important to note that the crack growth rate most often increases with increasing

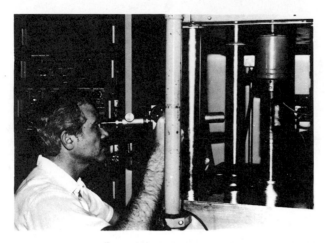

FIGURE 13.1 Monitoring crack length with a traveling microscope.

crack length. (This is generally the case, though not always, as will be discussed below.) It is most significant that the crack becomes longer at an increasingly rapid rate, thereby shortening component life at an alarming rate. An important corollary of this fact is that most of the loading cycles involved in the total life of an engineering component are consumed during the early stages of crack extension when the crack is small and, perhaps, undetected. The other variable that controls the rate of crack propagation is the magnitude of the stress level. It is evident from Fig. 13.2 that FCP rates increase with increasing stress level.

Since many researchers have probed the nature of the fatigue crack propagation process, it is not surprising to find in the literature a number of empirical and theoretical "laws," many of the form

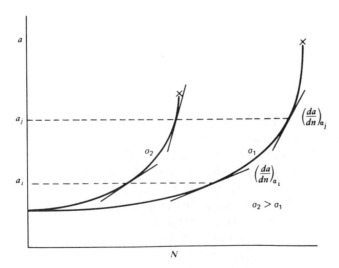

FIGURE 13.2 Crack propagation data showing effect of applied stress level. FCP rate increases with stress and crack length.

$$\frac{da}{dN} \propto f(\sigma, a) \qquad (13\text{-}1)$$

reflecting the importance of the stress level and crack length on FCP rates. Quite often, this function assumes the form of a simple power relation wherein

$$\frac{da}{dN} \propto \sigma^m a^n \qquad (13\text{-}2)$$

where $m \approx 2 - 7$
            $n \approx 1 - 2$

For example, Liu[1] theorized m and n to be 2 and 1, respectively, while Frost[2] found empirically for the materials he tested that $m \approx 3$ and $n \approx 1$. Weibull[3] accounted for the stress and crack length dependence of the crack growth rate by assuming the FCP rate to be dependent on the net section stress in the component. Paris[4] postulated that the stress intensity factor—itself a function of stess and crack length—was the overall controlling factor in the FCP process. This postulate appears reasonable, since one might expect the parameter K, which controlled static fracture (Chapter 8) and environment assisted cracking (Chapter 11) to control dynamic fatigue failures as well. From Chapter 8 the stress intensity factor levels corresponding to the crack growth rates identified in Fig. 13.2 would be $Y_i \sigma_1 \sqrt{a_i}$ and $Y_j \sigma_1 \sqrt{a_j}$ at σ_1 and $Y_i \sigma_2 \sqrt{a_i}$ and $Y_j \sigma_2 \sqrt{a_j}$ at σ_2, respectively. By plotting values of da/dN and ΔK at the associated values of $a_{i,j,\dots,n}$, a strong correlation was observed (Fig. 13.3), which suggested a relation of the form

$$\frac{da}{dN} = A \, \Delta K^m \qquad (13\text{-}3)$$

where $\dfrac{da}{dN}$ = fatigue crack growth rate

   ΔK = stress intensity factor range ($\Delta K = K_{max} - K_{min}$)
   A, m = f (material variables, environment, frequency, temperature, stress ratio)

It is encouraging to note that the FCP response of many materials is correlated with the stress intensity factor range, even though FCC, BCC, and HCP metals all were included in the data base.

Thus, for an interesting period during the early 1960s, the battle of the crack growth rate relations began. Although experimental evidence was mounting in favor of the stress intensity factor approach, it was not until two critical sets of experiments were reported and fully appreciated that the importance of the stress intensity factor in controlling fatigue crack propagation rate was fully accepted. In one paper, Swanson et al.[5] reasoned that if any of the various relations between stress and crack length was truly the critical parameter controlling crack growth rate behavior, then keeping that parameter constant during a fatigue test would cause the crack to grow at a

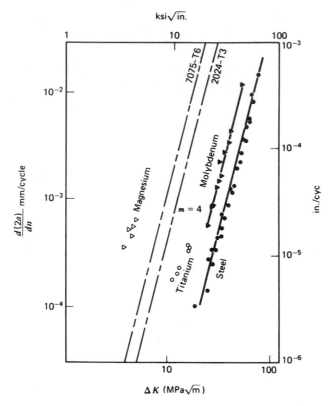

FIGURE 13.3   Fatigue crack propagation for various FCC, BCC, and HCP metals. Data verify power relation between ΔK and da/dN.[4] (With permission of Syracuse University Press.)

constant velocity. In other words, the crack length versus number of cycles curve would appear as a straight line. To achieve this condition, Swanson decreased the cyclic load level incrementally in varying amounts with increasing crack length to maintain constant magnitudes of $\sigma^3 a$, $\sigma^2 a$, σ_{net}, and ΔK, deemed to be the controlling parameters as discussed above. It is demonstrated clearly in Fig. 13.4 for the case of σ_{net} and ΔK that while each parameter when held constant did produce a constant growth rate over a limited range, maintaining a constant stress intensity factor caused the crack to grow at a constant velocity throughout the entire fatigue test life. The stress intensity factor clearly was the key parameter controlling crack propagation. The second set of critical experiments was reported by Paris and Erdogan,[6] who analyzed test data from center cracked panels of a high-strength aluminum alloy. In one instance, the loads were applied uniformly and remote from the crack plane (Fig. 13.5). From Fig. 8.7c, the stress intensity factor for this configuration is found to increase with increasing crack length. Therefore, if a constant load range were applied during the life of the test, the stress intensity level would increase continually because of the increasing crack length. Correspondingly, crack growth rates would be low at first and increase gradually as the crack extended. In the other set of experiments, center notched panels were loaded by concentrated forces acting at the crack surfaces (Fig. 13.5). For this configuration, the stress intensity factor is found to be[7]

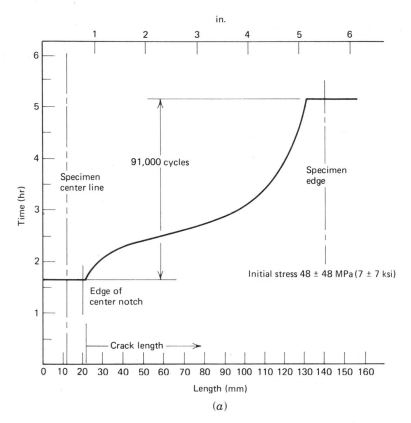

FIGURE 13.4 **Crack length versus time plots revealing crack growth rate behavior in 7079-T6 aluminum alloy. (*a*) Linear load shedding to achieve constant net section stress; (*b*) load shedding to achieve constant Δ*K* conditions. Note constant growth rate over entire specimen life in (*b*).[5] (Reprinted with permission of the American Society for Testing and Materials from copyrighted work.)**

$$K = \frac{P}{\sqrt{\pi a}} \qquad (13\text{-}4)$$

Most interestingly, the stress intensity factor is observed to be large when the crack length is small but decreases with increasing crack length. Consequently, one would predict that if the stress intensity factor did control FCP rates, the crack growth rate would be large initially but would *decrease* with increasing crack length. Such a reversal in FCP behavior would not be predicted if crack growth were controlled by the net section stress, for example, since the latter term would increase with both specimen load configurations. In fact, the experimental results showed the crack growth rate reversal anticipated by the sense of the crack length-dependent change in stress intensity factor *dK/da*. The crack line-loaded sample exhibited the highest growth rates at the outset of the test when the crack was small, with progressively slower growth rates being monitored with increasing crack length. The opposite was true for the grip-loaded sample (Fig. 13.5). Saxena et al.[8] have demonstrated that both increasing and decreasing crack growth rates can be obtained from the same specimen

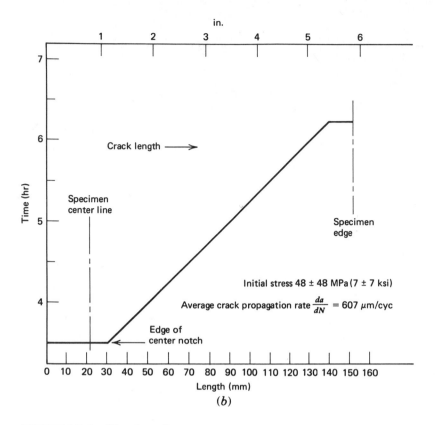

FIGURE 13.4 (*Continued*)

if one part of the test is conducted with a positive dK/da gradient and the other part carried out with a negative dK/da gradient. The desired K and crack growth rate gradients are achieved by programming a computer to control cyclic loads to achieve a particular K gradient according to the relation

$$K_i = K_0 e^{C(a_i - a_0)} \tag{13-5}$$

where K_i = instantaneous K level
K_0 = initial K level

C = normalized K gradient defined as $\dfrac{dK/da}{K}$

a_i = instantaneous crack length
a_0 = initial crack length

Therefore, if C is assigned a negative, zero, or positive value, the associated crack growth rates will decrease, remain the same, or increase, respectively, with increasing crack length. The important point to bear in mind is the fact that, regardless of the K gradient, the associated fatigue crack growth rates fall along the same $da/dN - \Delta K$

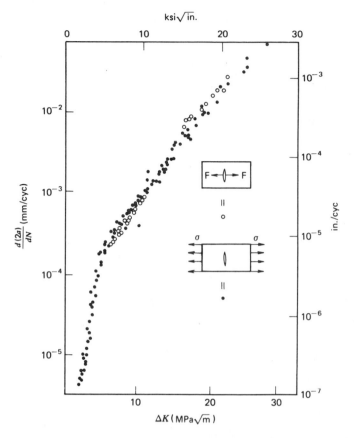

FIGURE 13.5   Fatigue crack propagation behavior in 7075-T6 for remote and crack line loading conditions.[4] (With permission of Syracuse University Press.)

curve, with *da/dN* being controlled by ΔK. The use of computers in the fatigue laboratory is gaining increased importance since unit labor costs per test can be reduced and since the computer allows for more efficient utilization of the fatigue machines and allows for convenient implementation of experiments involving complex loading histories.

Not only did these experiments verify the importance of the stress intensity factor in controlling the fatigue crack propagation process, they illustrated again the interchangeability of specimen geometry and load configuration in the determination of material properties such as crack growth rate or fracture toughness values. As a final note, over the growth range from about 2×10^{-6} to 2×10^{-3} mm/cyc, crack growth rate behavior in relatively inert environments as described by Eq. 13-3 is not strongly dependent on the mean stress intensity level. To a first approximation, crack growth rates are found to double when K_{mean} is doubled. By contrast, when the ΔK level is doubled, the crack would propagate 16 times faster (assuming $m = 4$). Consequently, mean stress effects in this growth rate range are considered to be of secondary importance. (Crack growth rate conditions where mean stress effects are more important are discussed in Section 13.4.) It is to be noted that the secondary importance of mean stress in controlling FCP response is similar to the smaller role played by mean stress in cyclic life tests as portrayed with Gerber and Goodman diagrams (Chapter 12).

Detailed procedures for conducting a fatigue crack propagation test have been established and are to be found in ASTM Standard E647-93.[9] The interested reader is referred to this document and to a collection of papers dealing with various laboratory experiences pertaining to this standard.[10] Furthermore, the reader *must* recognize that data generated in the low crack growth regime, according to conventional test methods described by this standard, may lead to nonconservative estimates of component life. This topic is considered at greater length in Section 13.4.

13.1.1   Fatigue Life Calculations

When conducting a failure analysis, it is often desirable to compute component life for comparison with the actually recorded service life. Alternatively, if one were designing a new part and wished to establish for it a safe operating service life, fatigue life calculations would be required. Such a computation can be performed by integrating Eq. 13-3 (where $\Delta K = Y\Delta\sigma\sqrt{a}$) with the starting and final flaw size as limits of the integration. When the geometrical correction factor Y does not change within the limits of integration (e.g., for the case of a circular flaw where $K = (2/\pi)\sigma\sqrt{\pi a}$ (Fig. 8.7i)), the cyclic life is given by

$$N_f = \frac{2}{(m-2)AY^m\Delta\sigma^m}\left[\frac{1}{a_0^{(m-2/2)}} - \frac{1}{a_f^{(m-2/2)}}\right] \qquad \text{for } m \neq 2 \quad (13\text{-}6)$$

where
$$\begin{aligned}
N_f &= \text{number of cycles of failure} \\
a_0 &= \text{initial crack size} \\
a_f &= \text{final crack size at failure} \\
\Delta\sigma &= \text{stress range} \\
A, m &= \text{material constants} \\
Y &= \text{geometrical correction factor}
\end{aligned}$$

Usually, however, this integration cannot be performed directly, since Y varies with the crack length. Consequently, cyclic life may be estimated by numerical integration procedures by using different values of Y held constant over a number of small crack length increments. It is seen from Eq. 13-6 that when $a_0 \ll a_f$ (the usual circumstance) the computed fatigue life is not sensitive to the final crack length a_f but, instead, is strongly dependent on estimations of the starting crack size a_0.

EXAMPLE 13.1

Compare the differences in fatigue lifetimes for three components that experienced crack extension from 2 to 10 mm, versus where the initial crack length was four times smaller ($a_0 = 0.5$ mm), or where the final crack length was four times larger ($a_f = 40$ mm). Assume that crack growth rates follow a Paris relation where $m = 4$.

We have three scenarios:

Case A: $a_0 = 2$ mm, $a_f = 10$ mm
Case B: $a_0 = 0.5$ mm, $a_f = 10$ mm
Case C: $a_0 = 2$ mm, $a_f = 40$ mm

We are only interested in examining the relative influence of crack length on fatigue lifetimes. Hence, from Eq. 13-6 where $m = 4$,

$$N_f \propto \left[\frac{1}{a_0} - \frac{1}{a_f}\right]$$

Therefore, for Case A (2–10 mm):

$$\left[\frac{1}{0.002} - \frac{1}{0.010}\right] = 500 - 100 = 400$$

For Case B (0.5–10 mm):

$$\left[\frac{1}{0.0005} - \frac{1}{0.010}\right] = 2000 - 100 = 1900$$

For Case C (2–40 mm):

$$\left[\frac{1}{0.002} - \frac{1}{0.040}\right] = 500 - 25 = 475$$

For the case where the initial crack length is smaller, fatigue lifetime increases by 375%.

$$\frac{1900 - 400}{400} \times 100 = 375\%$$

By comparison, when the final crack length is increased by the same relative proportion, fatigue lifetime increases by only 18.75%.

$$\frac{475 - 400}{400} \times 100 = 18.75\%$$

Clearly, fatigue lifetime is far more sensitive to initial rather than final crack size. In a practical sense, one finds that the major portion of fatigue lifetime in a service component is consumed prior to the point where the crack is discovered. Several additional examples[11] demonstrating this integration procedure are now considered.

In the first example, an extruded aluminum alloy is assumed to contain a semielliptical surface flaw with a and c dimensions of 0.15 and 10 mm, respectively (Fig. 13.6a). For a design stress of 128 MPa, a fatigue life computation reveals that the crack would penetrate the back surface after approximately 2,070,000 loading cycles. It should be noted that the numerical integration in this example requires different Y values to be computed for each intermediate crack front configuration and location (i.e., different ellipticity and back face corrections). (If the crack front shape were initially semicircular, the integration procedure would have been greatly simplified.) Several sets of crack growth rate material parameters (A and m values) were used in this computation, which corresponded to different regions of the da/dN versus ΔK curve. Note that most of the fatigue life of this component took place in extending the crack only a short distance from its original contour.

In the second example, the same material is assumed to contain a much deeper semielliptical surface flaw ($a = 1.5$ mm) and is subjected to the same stress level (128 MPa). From Fig. 13.6b we see that only 13,800 loading cycles were needed to cause the crack to penetrate the back face of the more deeply flawed component. As expected, a much shorter life was computed for the component with the larger initial flaw size.

A dramatically different conclusion is reached if life calculations are based on equal starting ΔK conditions (an alternative design assumption). For the first example, the 0.15-mm deep elliptical flaw together with a stress of 128 MPa corresponds to a ΔK level of 3.1 MPa$\sqrt{\mathrm{m}}$. If the component with the 1.5-mm deep elliptical flaw were subjected to the same ΔK level, 22,000,000 loading cycles would be needed to grow the crack through the section. The reason for the 11-fold greater life in the component that contained the much larger flaw (10 times deeper!) is readily apparent when one recognizes the substantial difference in stress range experienced by the two components: The stress range in the part containing the shallow crack was 128 MPa, whereas the same ΔK level in the more deeply flawed component required a stress range of only 40 MPa. From Eq. 13-6, we see that the cyclic life varies inversely with the mth power of the cyclic stress range.

To further illustrate the use of Eq. 13-6 in the computation of fatigue life, let us reconsider the material selection problem described in Chapter 8 (Example 8.5). Let us suppose that the 0.45C–Ni–Cr–Mo steel is available in both the 2070 and 1520 MPa tensile strength levels, and a design stress level of one-half tensile strength is required. It is necessary to estimate the fatigue life of a component manufactured from the material in the two strength conditions. Using the design stress levels, a stress range of 1035 and 760 MPa would be experienced by the 2070 and 1520 MPa materials, respectively. It is immediately obvious from Eq. 13-5 that, all things being equal, the total fatigue life would be lower in the higher strength material because it would experience a higher stress range. Using a value of $m = 2.25$ as found by Barsom et al.[12] for 19 steels, the fatigue life in the stronger material would be reduced by almost a factor of two. This should be considered as a minimum estimate of the reduction in fatigue life, since there is evidence to indicate that the exponent m increases with decreasing fracture toughness.[13] Furthermore, recalling that the critical flaw size in the 2070 MPa-level material was only one-fifth that found in the 1520 MPa alloy, the computed service life in the stronger alloy would be reduced further. This would be true especially if the initial crack was not much smaller than the critical flaw size. Therefore, one concludes that the stronger material is inferior in terms of

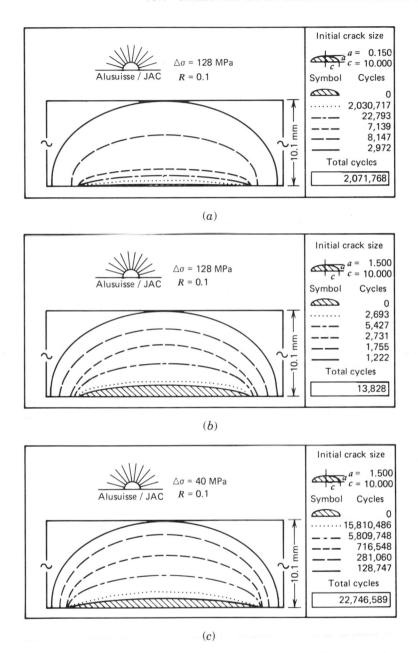

FIGURE 13.6 Crack contour changes associated with fatigue crack propagation in an extruded aluminum alloy. "Cycles" correspond to number of loading cycles needed to grow the crack from one contour to the next. (a) $\Delta\sigma = 128$ MPa, $a = 0.150$ mm, $\Delta K = 3.1$ MPa $\sqrt{m}$; (b) $\Delta\sigma = 128$ MPa, $a = 1.50$ mm, $\Delta K = 10$ MPa $\sqrt{m}$; (c) $\Delta\sigma = 40$ MPa, $a = 1.50$ mm, $\Delta K = 3.1$ MPa $\sqrt{m}$. (Courtesy of Robert Jaccard, Alusuisse Ltd.)

potential fatigue life, critical flaw size, associated fracture toughness, and environment-assisted cracking sensitivity (Fig. 11.17).

13.1.2 Fail-Safe Design and Retirement for Cause

As mentioned in the previous chapter, the ''fail-safe'' design philosophy recognizes that cracks can develop in components but ensures that the structure will not fail prior to the time that the defect is discovered and repaired. Several different design elements have been utilized to provide the structure with ''fail-safe'' characteristics. These include the presence of multiple load paths, which render the structure fail-safe as a result of load shedding from one component to its redundant loading path (e.g., see Section 14.2). Another fail-safe design feature involves the use of crack arresters such as stiffeners (e.g., ''tear straps'') in the fuselage section of aircraft; here, the fatigue crack growth rate of a fuselage crack is attenuated as the crack passes beneath each stiffener[14] (Fig. 13.7). A third fail-safe design feature, already discussed, pertains to the ''leak-before-break'' design of pressure vessels and pipes (recall Section 8.3). It is important to note that the ''fail-safe'' design approach requires periodic inspection of load-bearing components with sufficient flaw detection resolution so as to enable defective parts to be either repaired or replaced in a timely manner.

A specific application of the fail-safe design philosophy involves removal of a particular component only when there is clear evidence for the existence of a defect of critical dimensions. That is, based on a fracture mechanics analysis, the part would be replaced if failure were expected. As such, the component is subject to ''retirement for cause.'' If no defect is found, the part is returned to service with the next inspection interval being determined from a fracture mechanics calculation, based on the exis-

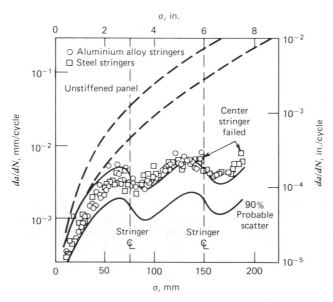

FIGURE 13.7 **Attenuation of FCP rates in aluminum alloy sheet in the presence of stiffeners.[14]**

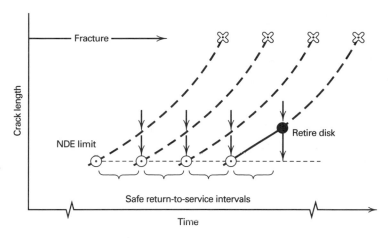

FIGURE 13.8 **Characterization of "retirement for cause" procedures. Components are repeatedly inspected and returned to service until a quantifiable crack is detected, resulting in retirement of the component.**[15]

tence of a crack whose length is just below the inspection resolution limit (Fig. 13.8). In this manner, components that would have been retired, based on attainment of their "safe-life" cyclic limit, would be allowed to continue in service, at considerable savings. To illustrate, the initial safe-life design of the F100 military jet engine involved the replacement of all engine disks after a service life when 1 in 1000 disks would be expected to have initiated a 0.75-mm long crack. Harris[15] concluded that the replacement of safe-life design controls with retirement for cause procedures in 23 F100 military jet engine components would save the U.S. Air Force over $1 billion over a 20-year period of time, thereby representing the greatest savings to date for the Air Force Materials Laboratory on the basis of a technological development.

EXAMPLE 13.2

An NDT examination of a steel component reveals the presence of a 5-mm-long edge crack. If plane-strain conditions prevail (K_{ic} = 75 MPa$\sqrt{m}$), what would be the residual service lifetime of the steel part if it were subjected to repeated stresses of 400 MPa?

Several pieces of information are needed before one can compute the fatigue lifetime of the component. These include the crack growth rate–ΔK relation and the size of the final crack length. Based on experimental test results, $da/dN = 4 \times 10^{-37} \Delta K^4$ where da/dN and ΔK have units of m/cyc and Pa$\sqrt{m}$, respectively. The final crack length is calculated from the instability condition when $K = K_{ic}$. Accordingly,

$$K = K_{ic} = 1.1\sigma\sqrt{\pi a}$$

$$a_f = \left(\frac{75 \times 10^6}{1.1(400 \times 10^6)\sqrt{\pi}}\right)^2; \quad a = 9.2 \text{ mm}$$

From Eq. 13-6

$$N_f = \frac{1}{4 \times 10^{-37} 1.1^4 \pi^2 (400 \times 10^6)^4} \left[\frac{1}{a_0} - \frac{1}{a_f} \right]$$

$$N_f = 617 \text{ cycles.}$$

If the next inspection procedure is scheduled in 6 months' time, should the component be replaced now or retained in service, given that the stresses fluctuate every 4 hours? During the next 180 days, the part would have experienced 1080 loading cycles. Since the computed component lifetime is less than the projected stress cycle count, the part should be replaced now.

13.2 MACROSCOPIC FRACTURE MODES IN FATIGUE

As discussed in Chapter 12, the fatigue fracture process can be separated into three regimes: crack initiation (sometimes obviated by preexistent defects), crack propagation, and final fracture (associated with crack instability). The existence and extent of these stages depends on the applied stress conditions, specimen geometry, flaw size, and the mechanical properties of the material. Stage I, representing the initiation stage, usually extends over only a small percentage of the fracture surface but may require many loading cycles if the nucleation process is slow. Often, Stage I cracks assume an angle of about 45° in the xy plane with respect to the loading direction.[16] After a relatively short distance, the orientation of a Stage I crack shifts to permit the crack to propagate in a direction normal to the loading direction. This transition has been associated with a changeover from single to multiple slip.[17,18] The plane on which the Stage II crack propagates depends on the relative stress state; that is, the extent of plane-strain or plane-stress conditions. When the stress intensity factor range is low (resulting from a low applied stress and/or small crack size), a small plastic zone is developed (Eq. 8-40). When the sheet thickness is large compared to this zone size, plane-strain conditions prevail and flat fracture usually results. With subsequent fatigue crack extension, the stress intensity factor and the plastic zone size increase. When the zone is large compared to specimen thickness, plane-stress conditions and slant fracture are dominant. Therefore, depending on the stress level and crack length, the fractured component will possess varying amounts of flat and slant fracture (see Section 14.2). Consequently, a fatigue crack may start out at 90° to the plate surface but complete its propagation at 45° to the surface (Fig. 13.9). Alternatively, the crack could propagate immediately at 45° if the plastic zone size to plate thickness ratio were high enough, reflecting plane-stress conditions. It is important to recognize that both unstable, fast-moving cracks and stable, slow-moving fatigue cracks may assume flat, slant, or mixed macromorphologies. It should be noted, however, that a unique relation between the stress state and fracture mode was not observed by Vogelesang[19] in 7075-T6 and 2024-T3 aluminum alloys during fatigue crack propagation in aggressive environments. The influence of environment on the fracture mode transition (more flat fracture in corrosive atmosphere than in dry air at the same r_y/t ratio) was

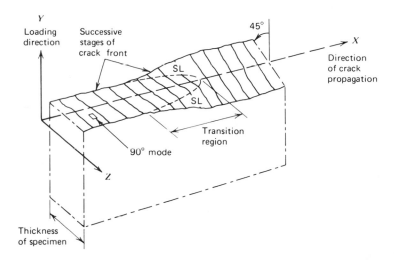

FIGURE 13.9   Diagram showing fracture mode transition from flat to slant fracture appearance. (Adapted from J. Schijve).

believed to be caused by a change in the fracture mechanism. As was just discussed, the plastic zone (defined by Eq. 8-40) can be used to estimate the relative amount of flat and slant fracture surface under both static and cyclic loading conditions. From Fig. 13.10, this plastic zone is developed by the application of a stress intensity factor of magnitude K_1. However, when the latter is reduced by h_k because the direction of loading is reversed, the local stress is reduced to a level corresponding to a stress intensity level of K_2. Since the elastic stress distribution associated with K_1 was truncated at σ_{ys} by local yielding, subtraction of an elastic stress distribution in going from K_1 to K_2 will cause the final crack-tip stress field to drop sharply near the crack tip and even go into compression. At K_2, a smaller plastic zone is formed in which the material undergoes compressive yielding. Paris[4] showed that the size of this smaller plastic zone, which experiences alternate tensile and compressive yielding, may be estimated by substituting h_k for K and $2\sigma_{ys}$ for σ_{ys} in Eq. 8-40. As a result, the size of the reversed plastic zone may be given by

$$r_y = \frac{1}{8\pi}\left(\frac{h_k}{\sigma_{ys}}\right)^2 \tag{13-7}$$

or four times smaller than the comparable monotonic value. Hahn et al.[20] measured the reversed plastic zone dimension using an etch pitting technique and found it to be slightly different and given by

$$r_y^c = 0.033\left(\frac{\Delta K}{\sigma_{ys}}\right)^2 \tag{13-8}$$

where r_y^c is the reversed plastic zone when measured in the y direction from the crack plane to the elastic–plastic boundary.

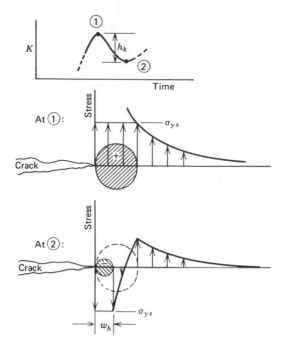

FIGURE 13.10   Monotonic and reversed plastic zone development at tip of advancing fatigue crack.[4] (With permission of Syracuse University Press.)

Since the materials within this smaller plastic zone experiences reversed cyclic straining, it might be expected that cyclic strain hardening or softening would result, depending on the starting condition of the material. This has been borne out by microhardness measurements made by Bathias and Pelloux[21] near the tip of a fatigue crack in two different steels. The significance of material property changes on fatigue crack propagation in this zone is examined further in Section 13.7.

13.3   MICROSCOPIC FRACTURE MECHANISMS

A high-magnification examination of the clamshell markings found on service failure fracture surfaces (Figs. 12.1 and 12.3) reveals the presence of many smaller parallel lines, referred to as fatigue striations (Fig. 13.11). Several important facts are known about these markings. First, they appear on fatigue fracture surfaces in many materials, such as BCC, HCP, and FCC metals, and many engineering polymers, and are oriented parallel to the advancing crack front. In a quantitative sense, Forsyth and Ryder[23] provided critical evidence that each striation represents the incremental advance of the crack front as a result of one loading cycle and that the extent of this advance varies with the stress range. This is shown clearly in Fig. 13.11c, which reveals striations of differing width that results from a random loading pattern.

It is appropriate, then, to emphasize the clear distinction between macroscopically observed clamshell markings, which represent periods of growth during which thousands of loading cycles may have occurred, and microscopic striations, which

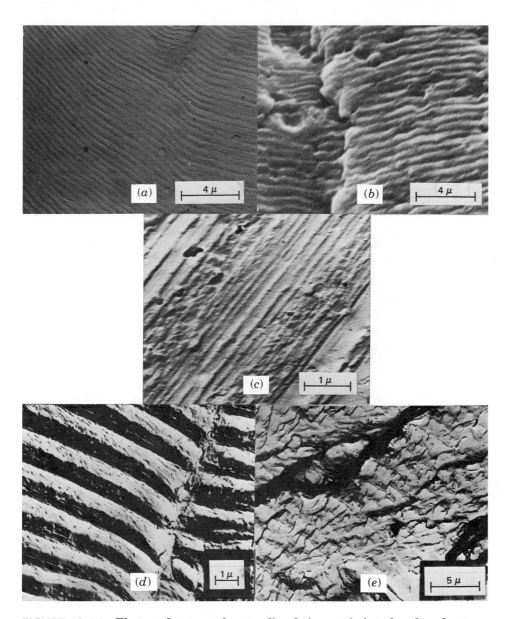

FIGURE 13.11 Electron fractographs revealing fatigue striations found on fracture surface and within macroscopic bands (Figs. 12.1, 12.3, 13.42). (*a*) TEM, constant load range; (*b*) SEM, constant load range; (*c*) TEM, random loading; (*d*) TEM, ductile striations.[22]; (*e*) TEM, brittle striations.[22] (Reprinted with permission of the American Society for Testing and Materials from copyrighted work.)

represent the extension of the crack front during one load excursion. *There can be thousands or even tens of thousands of striations within a single clamshell marking.*

Although these striations provide evidence that fatigue damage was accumulated by the component during its service life, fatigue crack propagation can occur without their formation. Usually, microvoid coalescence occurs at high ΔK levels,[24] and a

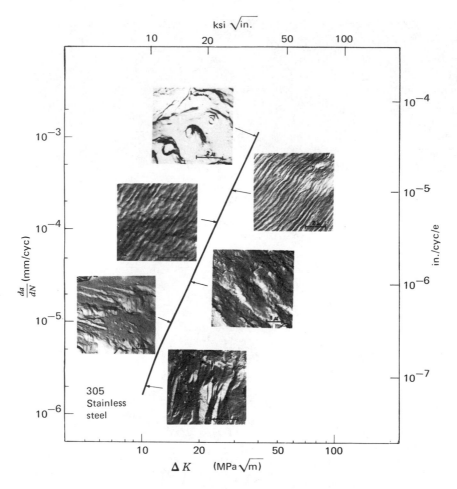

FIGURE 13.12 Change in fracture surface appearance in 305 stainless steel as a function of ΔK level.[18] (Reprinted with permission of the American Society for Testing and Materials from copyrighted work.)

cleavagelike faceted appearance dominates in many materials at very low ΔK levels[18] (Fig. 13.12). It is generally observed that the relative striation density found at intermediate ΔK values seems to vary with stress state and alloy content. Although striations are most clearly observed on flat surfaces associated with plane-strain conditions, elongated dimples and evidence of abrasion are the dominant fractographic features of plane-stress slant fracture surfaces. In terms of metallurgical factors, it is much easier to find striations on fatigue surfaces in aluminum alloys than in high-strength steels. In some cases, it is virtually impossible to identify clearly defined areas of striations in the latter material, thereby making fractographic examinations most difficult.

Fatigue striations can assume many forms, such as the highly three-dimensional or flat ones seen in Fig. 13.11*d, e*. It is not absolutely clear why there are different morphologies, but they are often associated with the test environment during crack propagation. Fatigue striations are relatively flat and assume a cleavagelike appearance when formed in an aggressive environment, but tend to appear more ductile when formed in an inert environment. Although striation morphology may be affected by the service environment, definite and pronounced changes have been observed in the appearance of fatigue striations after exposure to oxidizing or corroding atmospheres. For example, fatigue striations can be completely obliterated as a result of exposure to a high-temperature, oxidizing environment. Even at room temperature, fatigue striations become increasingly more difficult to detect with time. As a result, the amount of fractographic information to be gleaned from a fracture surface decreases with time, particularly in the case of steel alloys; the fracture surface details in aluminum alloys are maintained for a longer period because of the protective nature of the aluminum oxide film that forms quickly on the fracture surface.

The reader should recognize that even when striations are expected to form (Fig. 13.12), they are not always as clearly defined as those in Fig. 13.11. Whether due to environmental and/or metallurgical effects or related to service conditions, such as abrasion of the mating fracture surfaces, striations may appear either continuous or discontinuous, clearly or poorly defined, and straight or curved.

From metallographic sections and electron fractographic examination, three basic interpretations of the morphology of fatigue striations have evolved. The striations are considered to be undulations on the fracture surface with (1) peak-to-peak and valley-to-valley matching of the two mating surfaces, (2) matching crevices separating flat facets, or (3) peak-to-valley matching. Based on these interpretations of striation morphology, different mechanisms have been proposed for striation formation. One mechanism involves plastic blunting processes at the crack tip,[25] which occur regardless of material microscopic slip character; another model takes account of crystallographic considerations, wherein striations are thought to form by sliding off on preferred slip planes.[22,24,26]

The effect of crystallography on striation formation can be supported by both direct and circumstantial evidence. Pelloux[26] demonstrated in elegant fashion with the aid of etch pit studies that striation orientation in an aluminum alloy was sensitive to changes in crystal orientation and that striations tended to form on (111) slip planes and parallel to $\langle 110 \rangle$ directions. The latter is in agreement with theoretical considerations[24,26] (Fig. 13.13) and experimental findings, which show a strong tendency for the macroscopic fracture plane to lie parallel to {100} or {110} planes.[27] It might then be argued that when slip planes are oriented favorably with respect to the maximum resolved shear stresses at the advancing crack tip, a clearly defined striation could be formed. Alternatively, poorly defined striations or none at all might be found when slip planes are unfavorably oriented. It is quite probable that crystallographic considerations dominate striation formation at low ΔK levels where few slip systems are operative, while the plastic blunting model would provide a better picture of events at high ΔK levels.

13.3.1 Correlations with the Stress Intensity Factor

More quantitative information has been obtained from the measurement of fatigue striations than from any other fracture surface detail. Since the striation represents the

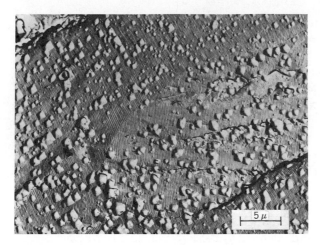

FIGURE 13.13 **Fatigue striations in 2024-T3 aluminum alloy. Note concurrent change in striation and etch pit orientation.**[26] **(Reprinted with permission of the American Society of Metals. Copyright © 1969.)**

position of the crack front after each loading cycle, its width can be used to measure the FCP rate at any given position on the fracture surface. It is not surprising, then, to find reasonable correlation at a given ΔK level between the macroscopically determined growth rate as measured with a traveling microscope and the microscopic growth rate as measured by the width of individual striae[28] (Fig. 13.14). Additional correlations[29] between striation spacing and ΔK have been found for a number of materials (Fig. 13.15). Here, ΔK has been normalized with respect to the elastic modulus of the respective materials examined (see Section 13.7).

The practical significance of the data correlations found in Figs. 13.14 and 13.15 cannot be overemphasized, since such data are very useful in failure analyses.

EXAMPLE 13.3

After a certain period of service, a 15-cm-wide panel of 2024-T3 aluminum alloy was found to contain a 5-cm-long edge crack oriented normal to the stress direction. The crack was found to have nucleated from a small, preexistent flaw at the edge of the panel. The magnitude of the cyclic stress was analyzed to be less than 20% of the yield strength ($\sigma_{ys} \approx 345$ MPa) and was believed to be distributed uniformly along the plane of the crack. Since the crack had reached dangerous proportions, the panel was removed from service and examined fractographically. Average striation widths of 10^{-4} and 10^{-3} mm were found at distances of 1.5 and 3 cm, respectively, from the origin of the crack. Was the premature failure caused by the existence of the surface flaw or related to a much higher cyclic stress level than originally estimated?

For this configuration, the stress intensity factor is given from Fig. 8.7f as

$$K = Y\sigma\sqrt{a}$$

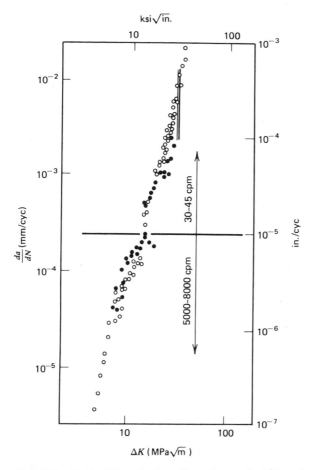

FIGURE 13.14 **Correlation of macroscopic ($\bigcirc$) and microscopic ($\bullet$) crack growth rate with ΔK in 2024-T3 aluminum alloy.**

with $Y(a/W)$ being defined at $Y(1.5/15)$ and $Y(3/15)$ to be 2.1 and 2.43, respectively. From Fig. 13.14, the apparent stress intensity range based on the two striation measurements is found to be 12.7 and 20.9 MPa$\sqrt{m}$, corresponding to crack lengths of 1.5 and 3 cm, respectively. Therefore, two independent estimates of the actual stress range can be obtained directly from the above equation:

$$\Delta\sigma = \frac{\Delta K}{Y\sqrt{a}} = \frac{12.7}{(2.1)\sqrt{0.015}} = 49.4 \text{ MPa}$$

and

$$\frac{20.9}{(2.43)\sqrt{0.03}} = 49.7 \text{ MPa}$$

Since both numbers are self-consistent and are in agreement with the original design estimates, the striation data appear valid. Therefore, it is concluded that premature failure was caused by early crack propagation from the small edge flaw.

Although this procedure is extremely useful, it should be implemented with deliberate caution. First, it is critical to accurately identify the crack length position where the striation spacing measurements were made. The stress level cannot be computed if the crack length is not known. In many service failure reports, striation photographs are presented without identification of the precise location of the region of the fracture surface. Without such information, the photograph serves only to identify the mechanism of failure, but does not enable the examiner to perform any meaningful calculations.

It has been shown repeatedly in laboratory experiments that for constant stress intensity conditions, striation spacings in a local region may vary by a factor of two to four. This results from the fact that striation formation is a highly localized event and is dependent on both the stress intensity factor and metallurgical conditions. In addition, the spacing between adjacent striations depends on the incident angle of the electron beam relative to the orientation of the replica in the transmission electron microscope; that is, the investigator sees a *projected* image of the replica on the viewing screen. Recently, Stofanak et al.[30] concluded that minimal projection errors were found in striation width measurements when the support grid was oriented normal to the electron beam; this resulted from the fact that the carbon replicas had collapsed onto the support grid. The principal reason for the scatter in striation-spacing measurements within a given replica (at a given level of ΔK), therefore, was attributed primarily to metallurgical factors such as grain orientation variations and inclusion distribution. To arrive at a meaningful estimate of crack growth rate at a particular crack length, many measurements of striation spacing should be made. In addition, measurements should be made of different crack length positions, as done in Example 13.3, to serve as a comparative check on the computation.

The prevailing stress intensity factor range could also be estimated with the aid of an empirical correlation identified by Bates and Clark,[29] who showed that

$$\text{striation spacing} \approx 6\left(\frac{\Delta K}{E}\right)^2 \tag{13-9}$$

where ΔK = stress intensity factor range
E = modulus of elasticity

It is particularly intriguing that Eq. 13-9 can be used to estimate ΔK based on fractographic information for any metallic alloy (Fig. 13.15). Since the exponent in Eq. 13-9 is approximately two, while the exponent in the Paris-type Eq. 13-3 depends on material variables, environment, frequency, and temperature, and can vary from about 2 to 7, agreement between macroscopic and microscopic crack growth rates should not be expected in the majority of instances. Consequently, striation spacing measurements should be used in conjunction with Eq. 13-9 or compared with previ-

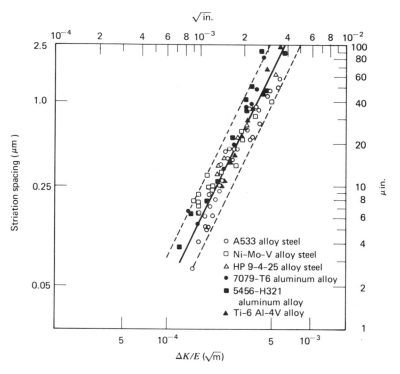

FIGURE 13.15 **Correlation of fatigue striation spacing with Δ*K* normalized with respect to elastic modulus.[29] (Reprinted with permission of the American Society of Metals. Copyright 1969.)**

ously determined fractographic information, rather than macroscopic data, whenever estimations of the prevailing Δ*K* level are desired.

13.4 CRACK GROWTH BEHAVIOR AT Δ*K* EXTREMES

13.4.1 High Δ*K* Levels

Although Eq. 13-3 does provide a simple relation by which crack growth rates may be correlated with the stress intensity factor range, it does not account for crack growth characteristics at both low and high levels of Δ*K*. If enough data are obtained for a given material—say, four to five decades of crack growth rates—the *da/dN* versus Δ*K* curve assumes a sigmoidal shape, as shown in Fig. 13.16. That is, the Δ*K* dependence of crack growth rate increases markedly at both low and high Δ*K* values. At the high growth rate end of the spectrum, part of this deviation sometimes may be accounted for by means of a plasticity correction since the plastic zone becomes large at high Δ*K* levels. This has the effect of increasing ΔK_{eff} (Eq. 8-43) and thus tends to straighten out the curve. Another factor to be considered is that as K_{max} approaches K_c, local crack instabilities occur with increasing frequency, as evidenced by increasing amounts of microvoid coalescence and/or cleavage on the fracture surface. As

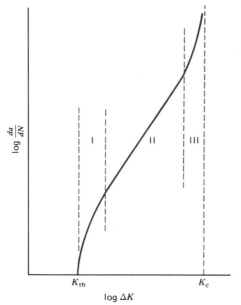

FIGURE 13.16 Diagram showing three regimes of fatigue crack growth response. Region I, crack growth rate decreases rapidly with decreasing ΔK and approaches lower limit at ΔK_{th}; Region II, midrange of crack growth rates where "power law" dependence prevails; Region III, acceleration of crack growth resulting from local fracture as K_{max} approaches K_c.

might be expected, this effect is magnified with increasing mean stress. Characterizing the mean stress level by R, the ratio of minimum to maximum loads, it is seen from Fig. 13.17a that crack growth rates at high ΔK values increase with increasing mean stress, and little mean stress sensitivity is observed at lower ΔK levels. One relation expressing crack growth rates in terms of ΔK, K_c and a measure of K_{mean} was proposed by Forman et al.[32] in the form

$$\frac{da}{dN} = \frac{C \, \Delta K^n}{(1 - R)K_c - \Delta K} \tag{13-10}$$

where $C, n =$ material constants

$\quad\quad K_c =$ fracture toughness

$\quad\quad R =$ load ratio $\left(\dfrac{K_{min}}{K_{max}}\right)$

From Eq. 13-10, we see that the simple power relation (Eq. 13-3) has been modified by the term $[(1 - R)K_c - \Delta K]$, which decreases with increasing load ratio R and decreasing fracture toughness K_c, both of which lead to higher crack growth rates at a given ΔK level. A typical plot of normalized data according to Eq. 13-10 is shown in Fig. 13.17b. Although Eq. 13-10 correctly identifies material FCP response under combinations of high ΔK and K_{mean} conditions, the relation is difficult to apply because of difficulties associated with the determination of the K_c value, which, as shown in Chapter 8, varies with planar and thickness dimensions of the test sample.

Other relations describing mean stress effects on FCP response have taken account of the plastic zones at the crack tip[33] and the plastic deformation process itself. With regard to the latter, Elber[34] proposed that the crack might be partially closed during

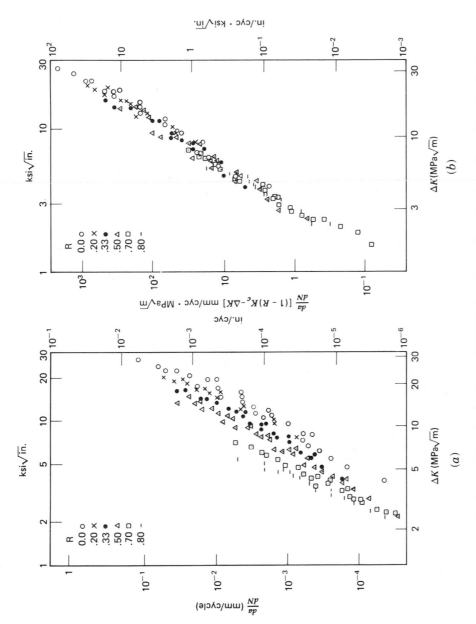

FIGURE 13.17 Fatigue crack propagation in 7075-T6 aluminum alloy showing effect of load ratio R and applicability of Forman, Kearney, and Engle relation. (a) ΔK vs. da/dN; (b) ΔK vs. $[(1 - R) K_c - \Delta K]$ da/dN. Note less scatter in b. (Data from Hudson.)

part of the loading cycle, even when $R > 0$. He argued that residual tensile displacements, resulting from the plastic damage of fatigue crack extension, would interfere along the crack surface in the wake of the advancing crack front and cause the crack to close above the minimum applied load level. This hypothesis was verified with compliance measurements taken from fatigued test panels that showed that an *effective* change in crack length (i.e., change in compliance) occurred prior to any *actual* change in crack length. In other words, the crack was partially closed for a portion of the loading cycle and did not open fully until a certain opening K level, K_{op}, was applied. As a result, the damaging portion of the cyclic load excursion would be restricted to that part of the load cycle that acted on a fully opened crack. From Fig. 13.18, the effective stress intensity factor range ΔK_{eff} would be denoted by the opening level K_{op} to K_{max}, rather than by the applied ΔK level $K_{max} - K_{min}$. In this connection, it is interesting to note that crack growth rates are relatively insensitive to compressive loading excursions where $R < 0$. In fact, a number of investigators[31,35] have shown that the fatigue response of materials subjected to $R < 0$ loading conditions can be approximated by simply ignoring the negative portion of the load excursion, since the crack would be closed. The marked change in importance of mean stress when negative load excursions are encountered is shown in Fig. 13.19. We see in Fig. 13.19a, where $R \geqslant 0$, that FCP rates in 7075-T6 aluminum alloy are affected by changes in mean load level. By contrast, no significant change is found in crack growth rates when $R \leqslant 0$, (Fig. 13.19b). (Note that in the latter figure, ΔK was defined as K_{max}; that is, $K_{min} = 0$.)

Newman[36] has supported the concept of crack closure based on finite element computations, while Katcher and Kaplan[37] and Chu[38] have accounted for mean stress effects in fatigue crack propagation experiments based on closure measurements. Others, however, have shown that closure concepts cannot always explain mean stress effects.[39–42] Part of the controversy surrounding the use of crack closure measurements in FCP tests is due to uncertainties associated with the appropriate measurement technique and differences in closure values determined from alternative detection methods. To date, crack closure has been measured with crack mouth displacement gages, surface mounted strain gages, back face mounted strain gages, potential drop, and other transducer probe methods. For an evaluation of the relative effectiveness of these methods, see the review by Fleck.[43]

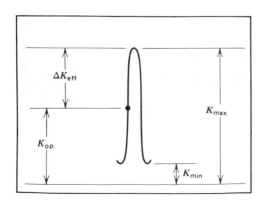

FIGURE 13.18 Crack surface interference results in crack opening K_{op} to be above zero. ΔK_{eff} defined as $K_{max} - K_{op}$.

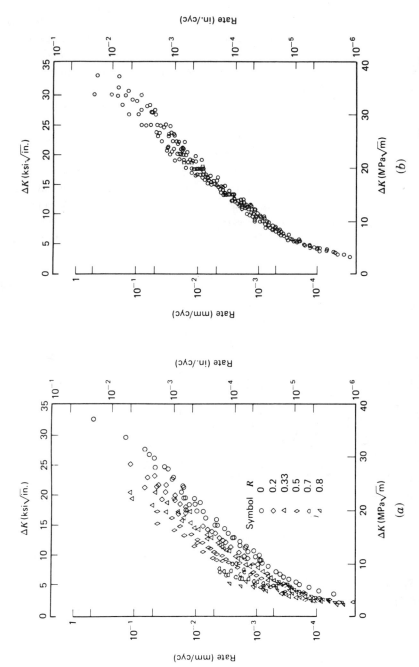

FIGURE 13.19   Variation of fatigue crack growth in 7075-T6 aluminum alloy, where (*a*) *R* ≥ 0[31]; (*b*) *R* < 0.[31] (From C. M. Hudson, NASA TN D-5390, 1969.)

Two examples showing the applicability of the crack closure model to the analysis of fracture surface markings are now given. Hertzberg and von Euw[44] showed that the fracture mode transition (FMT) was dependent on ΔK_{eff}. From Chapter 8, this transition is related to a critical ratio of plastic zone size to panel thickness. Therefore, it was surprising to find the transition occurring instead at a constant crack growth rate[45,46] (Fig. 13.20). From Table 13.1, the FMT did not occur at a specified value of K_{max} or ΔK. However, it did occur at a constant ΔK_{eff} level, which would account for the FMT being observed at a constant growth rate.

For the second example, it is instructive to reexamine the results of Kobayashi et al.[47] pertaining to the correlation between the stretch zone width and J/E (recall Fig. 10.45). Since the stretch zones resulted from the load excursions associated with the various J values needed to determine J_{IC} (recall Fig. 8.27a), these stretch zones can be considered to be equivalent to very large fatigue striations. As such, the width of these stretch bands could be related to $(\Delta K/E)^2$ (Eq. 13-9), where the ΔK values correspond to the loading ranges associated with the J_{IC} test procedure. Also note that J/E for predominantly elastic conditions is equal to $(K/E)^2$, which corresponds to the Bates and Clark relation (Eq. 13-9). When the striation spacings of the fatigue pre-

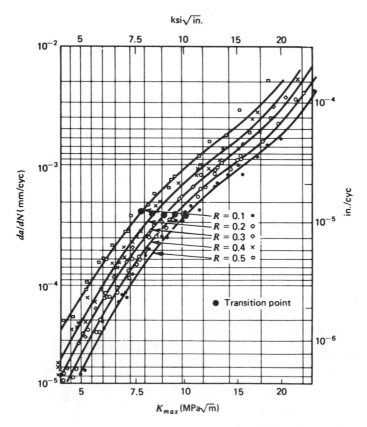

FIGURE 13.20 Crack propagation behavior in 2024-T3 Al clad sheet. Note fracture mode transition (●) at constant crack growth rate.[46] (Reprinted with permission of the American Society for Testing and Materials from copyrighted work.)

TABLE 13.1   Fracture Mode Transition in 2024-T3 Aluminum Alloy[44,45]

	K_{max}		ΔK_{app}		ΔK_{eff}^a	
R	MPa$\sqrt{m}$	kg/mm$^{3/2}$	MPa$\sqrt{m}$	kg/mm$^{3/2}$	MPa$\sqrt{m}$	kg/mm$^{3/2}$
0.1	10.4	33.5	9.4	30.2	5.1	16.3
0.2	11.2	36	8.9	28.8	5.2	16.7
0.3	12.1	39	8.5	27.3	5.2	16.9
0.4	13	42	7.8	25.2	5.1	16.6
0.5	14.3	46	7.1	23	5.0	16.1

^{a} ΔK_{eff} in 2024-T3 aluminum alloy was calculated from the relation[34] $\Delta K_{eff} = \Delta K_{app} (0.5 + 0.4R)$

cracked zones of the J test samples and the stretch zone widths from the J tests were plotted versus $(\Delta K/E)^2$ and J/E, respectively, a similar relation was noted, though the two sets of data were displaced along the horizontal axis (Fig. 13.21). When these two sets of fatigue fracture markings were compared on the basis of ΔK_{eff} (where $\Delta K_{eff} = K_{max} - K_{op}$) rather than ΔK_{app}, the SZW and striation spacing data were in excellent agreement.[48] The normalization of these data not only confirms the importance of using the effective ΔK level to correlate fatigue data such as fatigue fracture

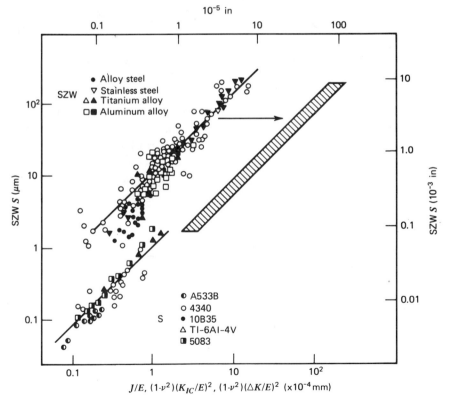

FIGURE 13.21   Relation between stretch zone width (SZW) and striation width (S) with fracture mechanics parameters. Shaded band corresponds to relative shift of SZW data to reflect ΔK_{eff} conditions. (After Kobayashi et al.[47] as modified by Hertzberg[48]; with permission from *Int. J. Fracture* 15, R69 (1979).)

surface markings, but also supports the assumption that the stretch zone is nothing more than a large fatigue striation.

13.4.2   Low ΔK Levels

At the other end of the crack growth rate spectrum, the simple power relation (Eq. 13-3) is violated again for low ΔK conditions, where the FCP rate diminishes rapidly to a vanishingly small level (Fig. 13.16). From such data[49,50] as shown in Fig. 13.22, a limiting stress intensity factor range (the threshold level ΔK_{th}) is defined and represents a service operating limit below which fatigue damage is highly unlikely. In this sense, ΔK_{th} is much like K_{IEAC}, the threshold level for environment-assisted cracking (see Chapter 11). Designing a component such that $\Delta K \leqslant \Delta K_{th}$ would be a highly desirable objective, but it is sometimes not very realistic in the sense that ΔK_{th} for engineering materials often represents only 5 to 10% of anticipated fracture toughness values (Table 13.2). Therefore, to operate under $\Delta K \leqslant \Delta K_{th}$ conditions would require that virtually all defects be eliminated from a component and/or the design stress be extremely low. This is desirable in the design of nuclear power generation equipment where safety is of prime concern; however, designing an aircraft such that $\Delta K \leqslant \Delta K_{th}$ is highly impractical. Theoretically, you could design an airplane that would not fatigue, but the beefed-up structure necessary to reduce the stress level to below the ΔK_{th} level would weigh so much that the plane would not be able to take off! Since many engineering structures do fulfill their intended service life without incident, it is apparent that some components do operate under $\Delta K \leqslant \Delta K_{th}$ conditions.

As seen in Table 13.2 and Fig. 13.22, the effect of K_{mean} (i.e., R ratio) on crack propagation becomes important once again at very low ΔK levels and has been the focus of considerable attention.[51–56] In this crack growth regime, different crack closure mechanisms than residual plasticity have been identified that strongly influence K_{op} levels and associated ΔK_{eff} values. For example, crack-tip zone shielding (recall Fig. 10.3) occurs when an irregular crack path is generated, with the coarse facets on the mating fracture surfaces coming in contact during fatigue cycling.[57–59] With increasing surface roughness, K_{op} levels increase, whereas ΔK_{eff} and the corresponding crack growth rates decrease (Fig. 13.23). It follows that the sensitivity of ΔK_{th} to R ratio for a given material depends on the observed level of crack closure. At one extreme where measured closure levels are minimal, no appreciable change in ΔK_{eff} would occur with increasing R ratio. Indeed, Minakawa et al.[61] reported no R ratio sensitivity on ΔK_{th} in closure-free IN9021-T4 P/M aluminum alloy (Fig. 13.24). Conversely, a significant decrease in ΔK_{th} with increasing R ratio was noted in the the 7090-T6 P/M aluminum alloy that exhibited pronounced roughness-induced closure in the threshold regime; this strong R ratio dependence of ΔK_{th} results from a sharp increase in ΔK_{eff} and associated FCP rates with increasing R ratio as K_{min} rises above K_{op}. Threshold conditions are then met only after the applied ΔK level is reduced.

An alternative crack closure mechanism has been proposed by Suresh et al.[62,63] to account for differences in ΔK_{th} for $2\frac{1}{4}$Cr–1Mo steel when tested in different gaseous atmospheres. The threshold fatigue value in this material decreased when the test atmosphere was changed from air to hydrogen. While it is tempting to rationalize this difference in terms of a hydrogen embrittlement-type argument, these authors pointed

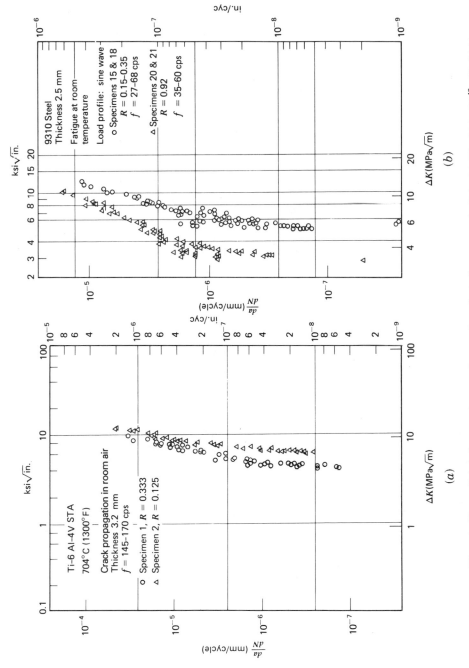

FIGURE 13.22 Threshold condition for crack growth as a function of stress ratio *R* in (*a*) Ti–6A1–4V[49]; (*b*) 9310 steel.[50] (Reprinted with permission of the American Society for Testing and Materials from copyrighted work.)

TABLE 13.2 Threshold Data in Engineering Alloys

Material	R	ΔK_th MPa√m	ksi√in.	Ref.
9310 Steel	0.25	~6.1	~5.5	a
	0.9	~3.3	~3	a
A533B Steel	0.1	8	7.3	b
	0.3	5.7	5.2	b
	0.5	4.8	4.4	b
	0.7	3.1	2.8	b
	0.8	3	2.75	b
A508	0.1	6.7	6.1	b
	0.5	5.6	5.1	b
	0.7	3.1	2.8	b
T-1	0:2	~5.5	~5	c
	0.4	~4.4	~4	c
	0.9	~3.3	~3	c
Ti–6A1–4V	0.15	~6.6	~6	d
	0.33	~4.4	~4	d
18/8 Austenitic steel	0	6.1	5.5	e
	0.33	5.9	5.4	e
	0.62	4.6	4.2	e
	0.74	4.1	3.7	e
Copper	0	2.5	2.3	e
	0.33	1.8	1.6	e
	0.56	1.5	1.4	e
	0.80	1.3	1.2	e
60/40 Brass	0	3.5	3.2	e
	0.33	3.1	2.8	e
	0.51	2.6	2.4	e
	0.72	2.6	2.4	e
Nickel	0	7.9	7.2	e
	0.33	6.5	5.9	e
	0.57	5.2	4.7	e
	0.71	3.6	3.3	e
300-M Steel	0.05	8.5	7.6	f
(650°C temper-oil quench)	0.70	3.7	3.3	f
300-M Steel	0.05	6.2	5.6	f
(650°C temper-step cooled)	0.70	2.7	2.4	f
2024-T3 Aluminum	0.80	1.7	1.5	g
2219-T851 Aluminum	0.1	3.0	2.7	h
	0.5	1.7	1.5	h
A356 Cast aluminum	0.1	6.1	5.5	i
	0.8	2.4	2.1	i
AF42 Cast aluminum	0.5	3.4	3.1	j
	0.8	1.7	1.5	j

[a] P. C. Paris, *MTS Closed Loop Magazine,* **2**(5), 1970.

[b] P. C. Paris, et al., ASTM *STP 513,* 1972, p. 141.

[c] R. J. Bucci et al., *op. cit.,* p. 177.

[d] R. J. Bucci et al., *op. cit.,* p. 125.

[e] L. D. Pook, *op. cit.,* p. 106.

[f] M. F. Carlson and R. O. Ritchie, *Metal Sci.* **11,** 368 (1977).

[g] R. A. Schmidt and P. C. Paris, ASTM *STP 536,* 1973, p. 79.

[h] R. J. Bucci, Alcoa Report No. 57-79-14 (1979).

[i] A. Saxena et al., *J. Test. Eval.* **6,** 167 (1978).

[j] R. J. Stofanak et al., *Eng. Fract. Mech.,* **17**(6), 527 (1983).

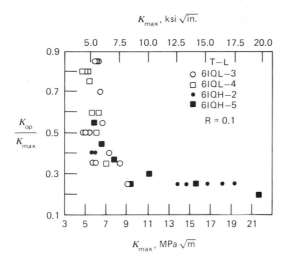

FIGURE 13.23 Increase in crack closure level in threshold regime for extruded aluminum alloy with crack propagation in the T-L orientation.[60] (Reprinted with permission, R. J. Stofanak, R. W. Hertzberg, G. Miller, R. Jaccard, and K. Donald, *Engineering Fracture Mechanics*, 17(6) 527 (1983), Pergamon Journals Ltd.)

out that dry argon also accelerated near-threshold fatigue crack growth rates relative to air; in fact, dry argon behaved like dry hydrogen. To explain the enhanced fatigue resistance of this material when tested in air as compared to dry argon and hydrogen, Suresh et al.[62,63] noted that fatigue testing in air creates an oxide layer on the fracture surface that thickens in the threshold regime as a result of closure-induced fretting (Fig. 13.25). Similar oxide layers in the threshold regime have been reported by others.[52,55,56] The thicker oxide layer, in turn, would be expected to increase K_{op} and bring about a corresponding decrease in ΔK_{eff}, thereby leading to lower crack growth

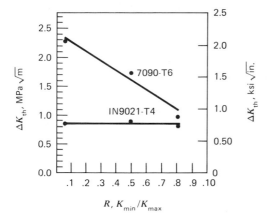

FIGURE 13.24 Influence of R ratio on ΔK_{th} in P/M IN9021-T4 and 7090-T6 aluminum alloys.[61] (Reprinted with permission from K. Minakawa, G. Levan, and A. J. McEvily, *Metallurgical Transactions*, 17A, 1787 (1986).)

(a)   (b)

FIGURE 13.25 Dark bands of oxide debris formed on fracture surfaces of $2\frac{1}{2}$Cr–1Mo steel in region associated with ΔK_{th}. (a) $\Delta K_{th} = 7.7$ MPa $\sqrt{m}$, $R = 0.05$; (b) $\Delta K = 3.1$ MPa $\sqrt{m}$, $R = 0.75$. Note that the oxide band almost disappears at high R ratios. (From Suresh et al.[62]; reprinted with permission from the Metallurgical Society of AIME.)

rates at a given applied ΔK. Tu and Seth[64] have also reported higher ΔK_{th} values in a seemingly more aggressive atmosphere (steam) than in air (Fig. 13.26). In this connection, they found more corrosion products on the steam-atmosphere fracture surfaces in the threshold regime (consistent with the Suresh et al. fretting oxide-induced closure model) than on other parts of the fracture surface. In addition, they observed more crack branching in the specimens tested in the steam atmosphere, which could have further reduced the effective crack-tip stress intensity factor.

It follows from Fig. 13.17 that the influence of such variables as R ratio and environment could be taken into account when FCP rates are compared on the basis of ΔK_{eff}. To accurately determine ΔK_{eff}, however, requires that precise measurements of closure be made throughout the test. Unfortunately, closure is often difficult to measure in a consistent manner and is subject to spurious interpretation, especially in the threshold regime[65–69] (Fig. 13.27). For example, in a recent interlaboratory comparative study, investigators found for the same material supply and test condition that K_{op}/K_{max} ratios varied markedly from 0.1 to 0.5.[65] Consequently, the use of ΔK_{eff} to normalize FCP information is potentially valuable for the purpose of data analysis but suffers, at present, from uncertainties in the precision of closure measurements.

Before concluding this section, it is appropriate to consider the similarity between the threshold stress intensity range, defined from FCP data, and the fatigue limit, determined from stress–cyclic life plots. Furthermore, it is intriguing to consider a possible correlation between these two fatigue parameters. To begin this analysis, consider the propagation of a small through-thickness crack in a large panel, for which $\Delta K = \Delta\sigma\sqrt{\pi a}$. At the threshold level where $\Delta K = \Delta K_{th}$, the corresponding stress

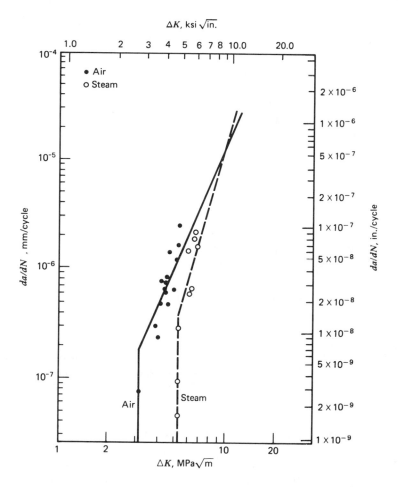

FIGURE 13.26 **Near-threshold fatigue crack growth behavior in Ni–Cr–Mo–V A471 rotor steel-tested at $R = 0.35$, 100 Hz in air, and steam at 100°C. Note higher ΔK_{th} when steel was tested in steam environment. (After Tu and Seth[64]; copyright, American Society for Testing and Materials, 1916 Race St., Philadelphia, PA 19103. Reprinted with permission.)**

range $\Delta\sigma_{th}$ is expected to vary with $1/\sqrt{a}$. A problem arises immediately when one attempts to define $\Delta\sigma_{th}$ where a approaches zero; the computed $\Delta\sigma_{th}$ is surely much larger than the experimentally determined value—the fatigue limit $\Delta\sigma_{fat}$—corresponding to test results from an unnotched sample (recall Section 12.2). In fact, several investigators have observed that when a is very small, $\Delta\sigma_{th}$ approaches an asymptotic limit corresponding to $\Delta\sigma_{fat}$.[70–75] To characterize this behavior, Haddad and his co-workers[72] proposed the existence of an "intrinsic crack length" for a given material, a_0, such that

$$\Delta K = \Delta\sigma\sqrt{\pi(a + a_0)} \qquad (13\text{-}11)$$

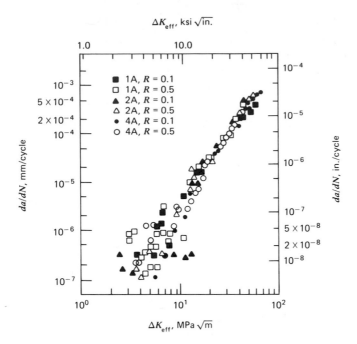

FIGURE 13.27 FCP rates in P/M Astroloy nickel-based alloy at load ratios of 0.1 and 0.5. Note good correlation at high ΔK_{eff} levels but considerable scatter exists in the ΔK_{th} regime. Symbols correspond to different grain sizes: $\square$, 5 μm; $\triangle$, 13 μm; $\bigcirc$, 50 μm.[69]

We see from the above discussion that, when a approaches zero,

$$a_0 = \frac{\Delta K_{th}^2}{\pi \Delta \sigma_{fat}^2} \tag{13-12}$$

where ΔK_{th} = threshold ΔK from long-cracked panel experiments ($a \gg a_0$)
   $\Delta \sigma_{fat}$ = endurance limit from unnotched samples

Haddad et al.[72] gave no physical significance to the intrinsic crack length a_0.* However, Tanaka and coworkers[74,75] developed a similar model along theoretical grounds and concluded that a_0 was related to the combined influence of the materials' grain size and crack closure behavior. Of particular significance, they gathered published data for both ferrous and nonferrous alloys to show a general relation between the threshold stress $\Delta \sigma_{th}$, normalized by the endurance limit $\Delta \sigma_{fat}$, and the crack length a, normalized by the intrinsic crack size a_0 (Fig. 13.28a). Note that when long cracks are present in the test sample, $\Delta \sigma_{th}$ varies with $1/\sqrt{a}$. This is to be expected when crack growth behavior is controlled by linear elastic fracture mechanics (LEFM) considerations. At the other extreme where $a < a_0$, the threshold stress asymptotically approaches a value equal to the endurance limit of an unnotched sample.

---

* It is intriguing to note the analogous form of Eq. 8-40 as compared with Eq. 13-12 given above.

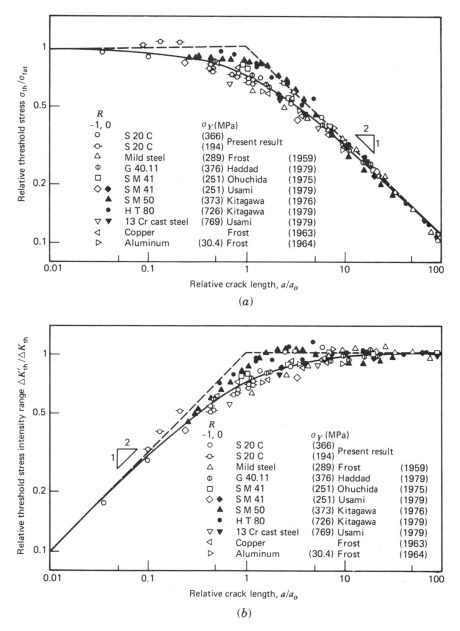

FIGURE 13.28 Normalized threshold behavior versus relative crack size in various materials. (*a*) Relative threshold stress; (*b*) relative stress intensity range. (After Tanaka et al.[75]; with permission from *Int. J. Fract.* **17**, 519 (1981).)

Using their results and those of others,[70,72,76–80] Tanaka et al.[75] showed a similar correlation between $\Delta K'_{\text{th}}$ normalized by the threshold value corresponding to a long-crack sample ΔK_{th}, and the crack length, a, normalized by the intrinsic crack length a_0 (Fig. 13.28*b*). In this instance, $\Delta K'_{\text{th}}$ is found to be independent of crack length when $a > a_0$ (LEFM-control). On the other hand, ΔK_{th} decreases when the crack length is small relative to a_0. That is, very small cracks can grow at ΔK levels previously thought safe (i.e., $\Delta K < \Delta K_{th}$). It follows that ΔK_{th} values associated with long-crack test specimens may lead to nonconservative life estimates of a component that contains very small cracks. For this reason, the data given in Table 13.2 should be used with extreme caution.

Considering the short-crack problem from a different perspective, we see that the concept of crack similitude is clearly violated. Similitude implies that different sized cracks will possess the same plastic zone size, stress and strain distributions, and crack growth rates, if the stress intensity factor is the same.[81–83] The breakdown of similitude can be traced to several factors: (1) when continuum requirements are violated, that is, crack lengths are small compared with the scale of the microstructure; (2) when linear elastic fracture mechanics concepts are violated, that is, the length of the crack is small compared with the dimension of the crack-tip plastic zone; (3) when different crack propagation mechanisms are encountered; and (4) when different closure levels are found at the same applied ΔK level ($K_{max} - K_{min}$) for long and short cracks, respectively. Regarding the latter point, crack-tip shielding is largely absent in small cracks since the crack wake has not yet developed. In the absence of associated closure, ΔK_{eff} in short cracks for a given ΔK_{app} is often significantly higher than that noted in long-crack samples and surely contributes markedly to the much higher growth rates observed in short-crack samples.

13.4.2.1 Estimation of Short-Crack Growth Behavior

Since real structures may initially contain short cracks without a wake zone, it could be argued that component lifetime is dominated by the behavior of such defects, rather than by predictions based on long-crack FCP information. In fact, the large closure levels encountered in the ΔK_{th} regime for long-crack samples may well be a consequence of the constant R ratio ($R^c = 0.1$) ΔK-decreasing test procedure (Standard E647-93a) and not a fundamental characteristic of the material in question.[84,85] That is, long-crack data may provide an overly optimistic assessment of a material's FCP resistance.

Initial efforts to obtain more conservative FCP data for design life calculations have focused on the generation of large quantities of short-crack data.[86,87]* Unfortunately, the characterization of short-crack growth behavior is time-consuming, tedious, and subject to large amounts of experimental scatter.[86–88] For example, errors of up to 50% have been reported[89] in repetitive readings of crack length during the same short-crack fatigue experiment. Furthermore, severalfold differences in FCP rates have been observed at the same nominal ΔK level (Fig. 13.29a). Also note that short-crack growth rates typically exceed those associated with long-crack samples and that crack growth occurs at ΔK levels below ΔK_{th}. Attempts to correlate these short-crack results with corresponding long-crack data on the basis of ΔK_{eff} have met with partial success;[88,90,91] the major problem encountered is that such correlations are based on two experimental data bases—short-crack growth rates and crack closure measurements—that each possess large amounts of scatter.

---

* A large collection of short-crack data is also included in the recent conference proceedings *Fatigue 87,* Vols. 1–3, R. O. Ritchie and E. A. Starke, Jr., Eds., EMAS Ltd., West Midlands, England, 1987.

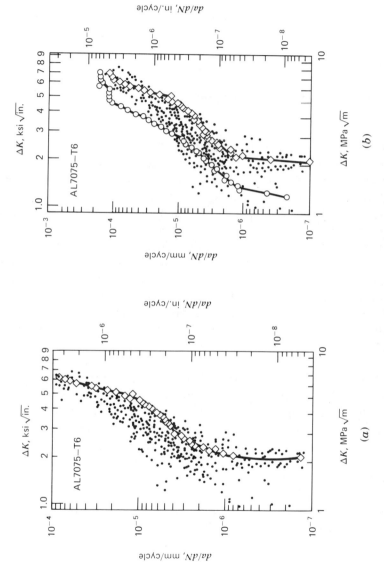

FIGURE 13.29   (*a*) Comparison of $R^c = 0.1$ K-decreasing data (solid line, ◇) with short-crack results (individual datum) for 7075-T6 aluminum alloy. (*b*) Same results from (*a*) along with $K^c_{max} = 10$ MPa $\sqrt{m}$ test results (○). Note that K^c_{max} curve is conservative relative to 85–90% of short-crack data.[95] (See Ref. 95 for sources of short-crack data.) (Reprinted with permission from W. A. Hertzberg, and R. Jaccard, *Fat. & Fract. Engng. Mater. & Struct.*, 11 (4), 303 (1988).)

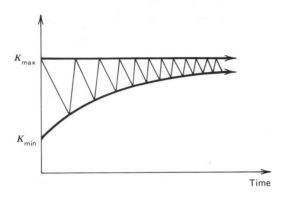

FIGURE 13.30 Schema showing constant K_{max} threshold test procedure.

Clearly, a different long-crack laboratory test method based on standard specimen geometries is needed to better simulate actual service loading conditions. Such a test method has been confirmed and is based on maintaining a constant maximum stress intensity (K^c_{max}) level during the ΔK-decreasing test procdure. By maintaining a constant K_{max} value (Fig. 13.30), mean stress and associated R ratios are found to increase markedly as ΔK decreases.[92–94] The development of such high mean stress and R ratio levels produces a long crack with no associated crack closure, which closely describes the behavior of short cracks.[95–97] The K^c_{max} data shown in Figs. 13.29b and 13.31 reveal compelling results for aluminum, iron, and nickel-based alloys, respectively, that clearly demonstrate the utility of the K^c_{max} test procedure as a method by which a conservative estimate of short-crack growth rates may be obtained.* In sharp contrast, the $R^c = 0.10$ curve anticipated almost none of the accelerated growth characteristics of short cracks.

13.4.2.2 Life Estimation Methodology Involving Short Cracks

As noted in Section 13.1.1, the prediction of a component's service lifetime under cyclic loading conditions depends on the proper use of experimental data that effectively simulates the response of the component under actual service conditions. This author believes that computations based on the combined use of K^c_{max} and R^c data provide a more realistic and conservative estimate of component service lifetime than that obtained solely from the use of an R^c data base (Fig. 13.32). At low ΔK levels associated with the propagation of short cracks, it is recommended that lifetime computations be based on K^c_{max} test results since the latter closely tracks short-crack growth rates in numerous metal alloys (recall Figs. 13.29b and 13.31). As the short crack lengthens, a crack wake is generated in association with increased levels of closure produced by one or more zone shielding mechanisms. To estimate the remaining service lifetime, it is suggested that subsequent calculations utilize the ap-

---

* It has also been shown that by precracking long-crack samples of 7475-T6 aluminum alloy in compression, a closure-free condition is established that generates FCP rates similar to those found under K^c_{max} test conditions. (H. Nowack and R. Marissen, *Fatigue 87,* Vol. 1, R. O. Ritchie and E. A. Starke, Jr., Eds., EMAS Ltd., West Midlands, England, 1987, p. 207.)

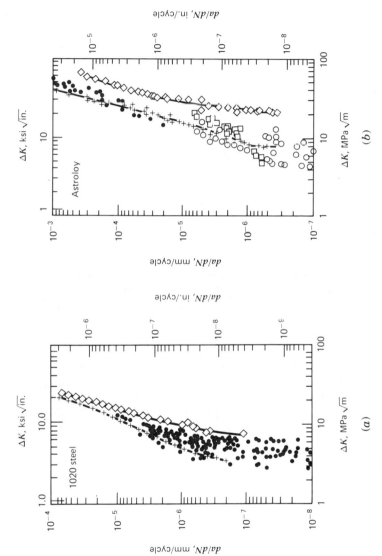

FIGURE 13.31 FCP data revealing the good agreement between K^c_{max} (+) and short-crack results. R^c results ($\diamond$) are nonconservative. See Ref. 95 for sources of short-crack data. (*a*) 1020 steel; (*b*) Astroloy nickel-based superalloy. ($\bullet$ $\bigcirc$ $\square$ short crack data) (Reprinted with permission from W. A. Herman, R. W. Hertzberg, and R. Jaccard, *Fat. & Fract. Engng. Mater. & Struct*, 11 (4), 303 (1988).)

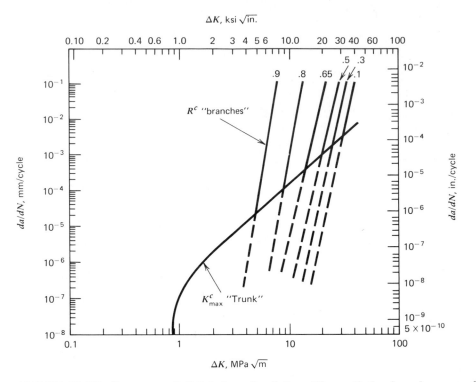

FIGURE 13.32 **Recommended data base for fatigue life prediction based on use of** K_{max}^c **trunk and appropriate** R^c **branch.**[95] **(Reprinted with permission from W. A. Herman, R. W. Hertzberg, and R. Jaccard, *Fat. & Fract. Engng. Mater. & Struct.*, 11 (4), 303 (1988).)**

propriate R^c data base; the latter will account for the presence of high applied mean stress levels as well as residual stress fields.

The results of representative life predictions for AC062/61 (6005) aluminum alloy are shown in Fig. 13.33. The individual data points correspond to tests conducted on butt-welded beams of this material, whereas the solid lines represent stress–cyclic life (*S-N*) simulation curves based on a series of fatigue life calculations.[95,98,99] To initiate this *S-N* simulation, a single datum is chosen corresponding to a low stress level and minimum cyclic life. These stress and cyclic lifetime values are then used to calculate an "effective" or "fictitious" crack size that depends on the assumed flaw shape and magnitude of stress concentration present in the component. Considering this effective flaw size to be common to those test samples that exhibited minimum lifetime at each stress level in the stress–cyclic life data base, the associated cyclic lifetime for each stress level can then be obtained for the component by integration of Eq. 13-6. The computational method employed in this example used both K_{max}^c and R^c data bases. At low ΔK levels, near ΔK_{th}, A and m values were chosen from the K_{max}^c trunk curve. When the R ratio of the K_{max}^c trunk curve decreased to the level corresponding to the combined applied and residual stress levels of the welded samples, latter stages of fatigue life at higher ΔK levels were calculated by using A and m values taken from appropriate R^c curves. Figure 13.33 shows excellent prediction of the lower bound of

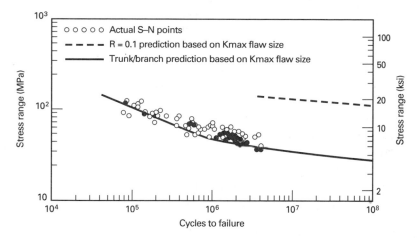

FIGURE 13.33   *S-N* lifetime simulation for 6005 aluminum alloy based on K^c_{max} and R^c test results. (Reprinted from R. W. Hertzberg, W. A. Herman, T. Clark, and R. Jaccard, *ASTM 1149*, J. M. Larsen and J. E. Allison, eds., 197 (1992).)

the *S-N* data based on the use of the K^c_{max} and the appropriate R^c curves corresponding to the different stress levels chosen in the development of the actual *S-N* data set. A poor fit between the data base and predicted curve is observed when the *S-N* prediction is based solely on the $R^c = 0.10$ data base.

Finally, the crack growth behavior of physically short cracks is complicated further when the crack is embedded within the stress field of a notch. Due to rapidly decreasing stresses, crack growth rates may *decrease* initially with *increasing* crack length. Depending on the local stress conditions, the crack could then either grow at an accelerating rate with increasing crack length or arrest[100–102] (Fig. 13.34). The existence of ''nonpropagating cracks'' has been confirmed by Frost,[103] as shown in Fig. 13.35. These data reveal the change in long life fatigue strength of steel samples containing notches of the same length but different notch root radii. Note how the

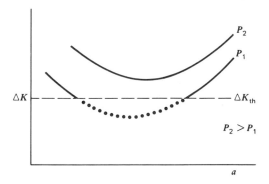

FIGURE 13.34   Effective ΔK level as a function of crack length at two load levels. At P_1, ΔK will decrease below $ΔK_{th}$ as crack moves through notch root zone. Crack arrest will then occur.

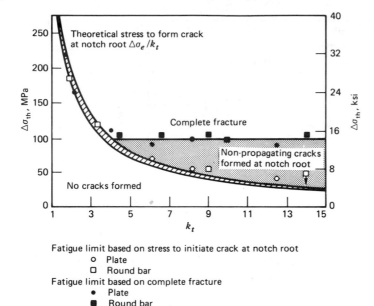

FIGURE 13.35 **Effect of k_t on fatigue strength for crack initiation and complete fracture of mild steel. (After Frost[103]; reprinted by permission of the Council of the Institution of Mechanical Engineers from the *Journal of Mechanical Engineering Science*.)**

fatigue strength for crack initiation decreases continuously with increasing k_t and becomes *independent* of the stress concentration factor beyond some critical value of k_t. That is, beyond some critical k_t value a fatigue crack could initiate but then would not propagate to failure. The presence of these nonpropagating cracks is consistent with the fact that fatigue crack growth essentially ceases when the stress intensity factor decreases below the threshold level ΔK_{th}.

As a final note, the sharp decrease in the fatigue sensitivity factor q with increasing notch acuity (recall Fig. 12.15) can be interpreted as reflecting the development of nonpropagating cracks. That is, a sharp notch root radius would cause a fatigue crack to nucleate and grow a short distance from the notch. If this crack were to stop when $\Delta K < \Delta K_{th}$, then the influence of the notch on fatigue life would be diminished.

13.5 INFLUENCE OF LOAD INTERACTIONS

Much of the FCP data discussed thus far were gathered from specimens subjected to simple loading patterns without regard to load fluctuations. Although this may provide a reasonable simulation condition for components experiencing nonvarying load excursions, constant amplitude testing does not simulate variable load-interaction effects, which, in some cases, can either extend or shorten fatigue life measurably. In the most simple case, involving superposition of single-peak tensile overloads on a regular sinusoidal wave form, laboratory tests[21,85,104–113] have demonstrated significant FCP delay after each overload, with the amount of delay increasing with both magnitude and number of overload cycles.

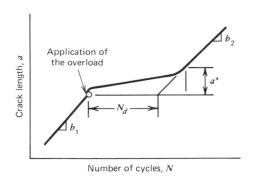

FIGURE 13.36 Crack growth rate plot illustrating effect of single-peak tensile overload. Note cyclic delay N_d and overload affected crack increment $a*$.

The retarding effect of a peak overload is demonstrated clearly in Fig. 13.36 for a constant ΔK loading situation, where the crack growth rate associated with the invariant ΔK level is given by the constant slope b_2. Obviously, the FCP rate is depressed after the overload for a distance $a*$ from the point of the overload. Hertzberg and coworkers[106,109] have shown that this distance corresponds to the plastic zone dimension of the overload. Therefore, once the crack grows through the overload plastic zone, resumption of normal crack propagation is expected. Recent studies have shown that the extent of delay N_d depends on the effective overload ratio as defined by the ratio of the overload ΔK level and the *effective* ΔK base level corresponding to the prevailing closure value (Fig. 13.37). It follows that increased amounts of cyclic delay occur when a given overload ΔK_{OL} is applied in conjunction with conditions associated with large amounts of closure (i.e., low ΔK_{eff} levels). For example, increased amounts of delay were observed in 2024-T3 aluminum alloy with increasing base ΔK level and decreasing sheet thickness (Fig. 13.38a); both factors contribute to enhance plane-stress test conditions and greater levels of plasticity-induced crack closure. It follows, therefore, that the data in Fig. 13.38a can be normalized in terms of the overload plastic zone size/sheet thickness ratio (Fig. 13.38b). Note that the amount of delay increases dramatically as this ratio approaches unity.

Additional experimental findings lend further credence to the crack closure model. It has been shown that when the fracture surfaces in the wake of the fatigue crack are removed with a narrow grinding wheel, the crack growth rate upon subsequent load

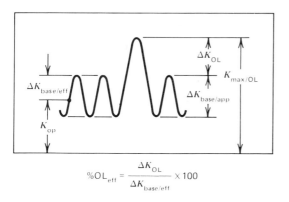

FIGURE 13.37 Definition of terms associated with single-peak tensile overload.

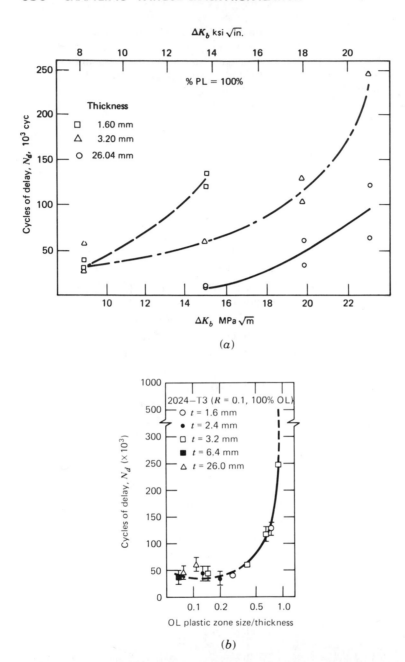

FIGURE 13.38 **Overload induced delay in 2024-T3 aluminum alloy as a function of sheet thickness.** (*a*) **Dependence on base** *K* **level**[108]**; (Reprinted with permission, W. J. Mills and R. W. Hertzberg,** *Engineering Fracture Mechanics,* **7, 705 (1975), Pergamon Journals Ltd.);** (*b*) **dependence on overload plastic zone/sheet thickness ratio.**[111] **(Reprinted with permission, R. S. Vecchio, R. W. Hertzberg, and R. Jaccard,** *Scripta Met.,* **17, 343 (1983), Pergamon Journals Ltd.)**

cycling is higher than before the machining operation.[107,109] Obviously, this effect may be rationalized in terms of the elimination of fracture surface material that was causing interference. It would be expected that if crack surface interference really does occur then some evidence of abrasion should be found on the fracture surface. As shown in Fig. 13.39 for both single-peak overload and high–low block loading sequences, extensive abrasion and obliteration of fracture surface detail is readily apparent. (Note the large striation or stretch band associated with the single-peak overload in Fig. 13.39*a*.)

The importance of large overload cycles in affecting fatigue life of engineering components is illustrated by the finding of Schijve et al.[114] They found that, under aircraft flight simulation conditions involving a random load spectrum, when the highest wind-related gust loads from the laboratory loading spectrum were eliminated, the specimens showed lower test sample fatigue life than did specimens that experienced some of the more severe load excursions. This fascinating load interaction phenomenon has led some investigators to conclude that an aircraft that logged some bad weather flight time could be expected to possess a longer service life than a plane having a better flight weather history.

In sharp contrast to the results shown in Fig. 13.38, the amount of overload-induced cyclic delay *increases* with decreasing base ΔK values at low ΔK levels in conjunction with an increase in oxide- and roughness-induced crack closure levels (Fig. 13.40*a*; also recall Fig. 13.23); here, again, cyclic delay increases with increasing *effective* overload ratio. Since crack closure levels at low K values tend to increase with increasing grain size (see Section 13.7), it was possible to normalize the number of overload-induced delay cycles of several aluminum alloys in terms of the ratio of overload plastic zone to grain dimension (Fig. 13.40*b*). It is seen that trends in overload-induced delay follow a U-shaped curve with delay maxima occurring at both low and high ΔK levels in conjunction with enhanced closure levels. It should be noted, however, that the extensive amount of delay found at low ΔK levels may be

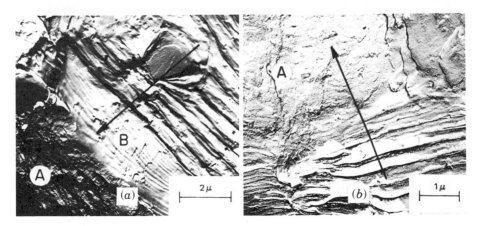

FIGURE 13.39   Abrasion in regions A resulting from overload cycling. (*a*) Single-peak overload. Note stretch zone at B. (*b*) High–low block loading sequence. Arrow indicates crack direction.[106] (Reprinted with permission of the American Society for Testing and Materials from copyrighted work.)

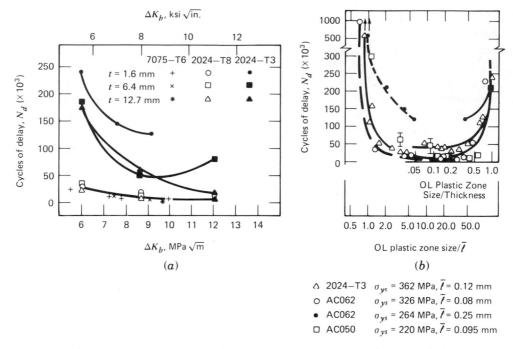

FIGURE 13.40   (*a*) **Effect of sheet thickness and base line *K* on overload induced delay (see Ref. 111 for sources of data). (*b*) U-shaped curve showing tensile overload-induced delay maxima at low and high *K* levels corresponding to small overload plastic zone size/grain size ratios and large overload plastic zone size/sheet thickness ratios, respectively.**[69]

illusory in that it is attributed to the excessive crack closure levels developed in the long-crack test samples used in the overload experiments. Since low ΔK levels in conjunction with short cracks exhibit less closure, the effective overload ratio and associated number of delay cycles in real structures are expected to be considerably smaller.

When attempts are made to estimate the fatigue life of a component that is subjected to variable spectrum loading, it is necessary to consider the influence of *compressive* as well as tensile overloads. Whereas tensile overloads temporarily slow down the rate of crack growth or arrest altogether its advance, compressive overloads tend to accelerate crack growth[85,115–118] (Fig. 13.41). Furthermore, if a tensile overload is followed immediately by a compressive overload, the beneficial effect of the tensile overload may be significantly reduced.[116,119] To underscore the deleterious impact of compressive overloads on overall fatigue life, de Jonge and Nederveen[120] observed a 3.3-fold increase in fatigue life of 2024-T3 aluminum test samples when the ground–air–ground (GAG) cycles were removed from the loading spectrum. (The cycles correspond to the transition from compressive loading on the lower wing skin when the plane is on the ground to tensile loading when the aircraft is in flight.) Since aircraft do land from time to time, fatigue life predictions made without GAG cycles would be definitely unconservative!

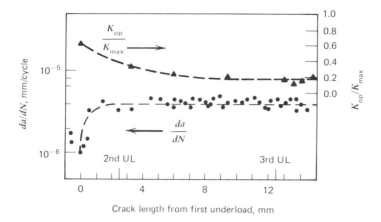

FIGURE 13.41 Influence of compressive underloads applied following a K-decreasing procedure in the threshold regime for an aluminum alloy. Note that FCP rates increase after the initial underload in conjunction with a decrease in crack closure level.[85]

13.5.1 Load Interaction Macroscopic Appearance

Similar to the macroscopic appearance of clamshell markings on many fatigue fracture surfaces (recall Fig. 12.3), macrobands are sometimes found that result from variable amplitude block loading[121] (Fig. 13.42a). The alternating dark and light bands reflect differences in the magnitude of the prevailing ΔK level associated with each loading block. (The relation between the magnitude of ΔK and band brightness is discussed

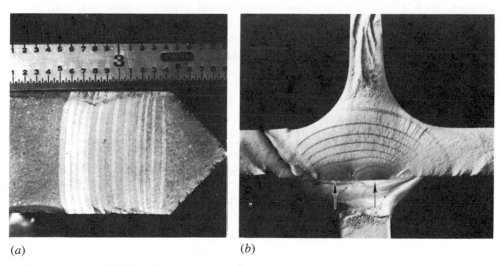

(a)   (b)

FIGURE 13.42 (a) Photograph showing macrobands on fatigue fracture surface in steel alloy resulting from variable amplitude block loading.[121] (Courtesy of H. I. McHenry.); (b) photograph of marker band test on aluminum alloy. Arrows indicate crack origins. All block loads conducted with constant maximum stress. (Courtesy R. Jaccard.)

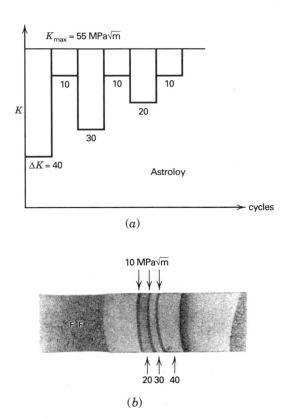

(a)

10 MPa√m

(b)

FIGURE 13.43   (a) **Marker band test profile conducted on Astroloy;** (b) **Macroscopic appearance for** K_{max}^c = **55 MPa√m block load profile. Numbers correspond to ΔK level for each block, 2.7X;** (c) **FCP data for Astroloy nickel-base alloy. Open data points correspond to continuously decreasing ΔK test. Closed points represent block loading data. Note excellent agreement of results.[122] (Reprinted from C. Ragazzo, R. W. Hertzberg, and R. Jaccard,** *J. Test. Eval.* **ASTM, 23(1), 19 (1995).)**

in Section 14.2.) Such marker bands can be used to characterize the size and shape of an advancing crack in a component where crack advance is predominantly internal (Fig. 13.42b). Although such marker bands can be formed by conventional constant low R ratio block loading test procedures, load interaction effects can interfere with the conduct of such a test; for example, crack growth delay can occur in cases where low level ΔK block loads follow high ΔK load excursions (recall the previous section). By contrast, a K_{max}-constant block loading procedure, avoids load interaction effects and, as a result, provides a useful method for documenting the size and shape of a progressing internal flaw.[122] This information, along with crack growth rate data for the component's alloy, can be used to improve the accuracy of fatigue life predictions.

To illustrate, an Astroloy nickel-base alloy test bar was subjected to the constant K_{max} block load test profile shown in Fig. 13.43a.[122] The marker bands from each block load segment were typically 0.25 mm to 0.75 mm wide in the low ΔK regions, and 1 mm or more in width in the higher ΔK regions. Each load block created a

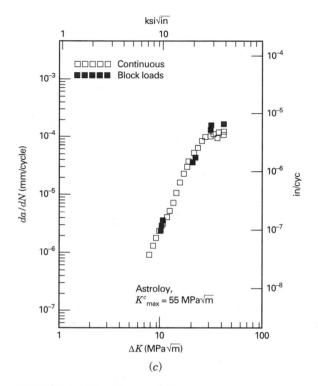

FIGURE 13.43 (*Continued*)

clearly defined marker band (crack growth from right to left), identified by its associated ΔK level (Fig. 13.43*b*). Additional tests[122] revealed that the development of
contrast changes on fatigue fracture surfaces is dependent on the ΔK_{eff} level and is
essentially independent of the K_{max} level. Therefore, changes in fracture surface
brightness are the result of changes in the cyclic plastic zone size and not the monotonic plastic zone size. By contrast, crack growth delay is dependent on changes in
the monotonic plastic zone size (recall Fig. 13.36). As such, when the monotonic
plastic zone is held constant in a block loading test by maintaining K_{max} constant, no
load interaction between load blocks would be expected. Indeed, the *da/dN-ΔK* data
for a continuously decreasing ΔK test are shown in Fig. 13.43*c* along with crack
growth information for the constant-ΔK block segments identified in Fig. 13.43*a*.
Essentially no difference in crack growth rate is found at a given ΔK level, between
continuously varying ΔK and block load sequence test results, even when the magnitude of the ΔK blocks was decreased by factors of two to four. (Similar results were
found for the case of steel and aluminum alloys.)[122] Therefore, it is possible to
periodically characterize the size and shape of growing internal cracks without affecting their subsequent crack advance behavior. Using this technique, it should be
possible for an investigator to more accurately compute cyclic life intervals in structural components based on improved knowledge of crack front profiles during the
period of stable crack extension.

13.6 ENVIRONMENTALLY ENHANCED FCP (CORROSION FATIGUE)

Recalling from Chapter 11 that cracks can grow in many materials as a result of sustained loading conditions in an aggressive environment, it is not surprising to find that fatigue crack propagation rates also are sensitive to environmental influences. The involvement of an aggressive environment in fatigue growth surely depends on a complex interaction between chemical, mechanical, and metallurgical factors. To this extent, corrosion fatigue studies benefit greatly from an interdisciplinary approach to test design and analysis of data.[123] Furthermore, the characteristics of a material–environment–stress system can be either simple or highly complex (Fig. 13.44).

To begin our discussion of the influence of environment on fatigue crack growth behavior, the effect of water partial pressure on fatigue behavior in 7075-T6 aluminum alloy is shown in Fig. 13.45.[125] For the two stress intensity levels examined, a large shift in FCP rates is noted over a small change in water content, resulting in an order of magnitude change in crack growth between very dry and wet test conditions. Tests conducted on several aluminum alloys in other environments, such as wet and dry oxygen, wet and dry argon, and dry hydrogen, indicate that enhanced crack growth in aluminum alloys is due to the presence of moisture.[125–127] This is consistent with static test results reported in Section 11.1.1. This has led investigators to reexamine the relative fatigue behavior of many engineering alloys to determine the relative contribution of environmental effects. For example, earlier test results generated in uncontrolled laboratory environments (Fig. 13.3) indicated a marked superiority in fatigue performance of 2024-T3 versus 7075-T6 aluminum alloys. By conducting tests in these two alloys in both dry and wet argon atmospheres, Hartman[126] and Wei[127] determined that FCP differences in these alloys were minimized greatly by eliminating moisture from the test environment. Consequently, the superiority of 2024-T3 over 7075-T6 in uncontrolled test atmospheres is due mainly to a much greater

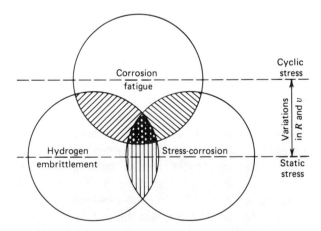

FIGURE 13.44 Schema showing interaction of cracking modes under static and cyclic loads. (After Ford,[124] *Fatigue-Environment and Temperature Effects*, J. J. Burke and V. Weiss, Eds., 1983, with permission from Plenum Publishing Corporation.)

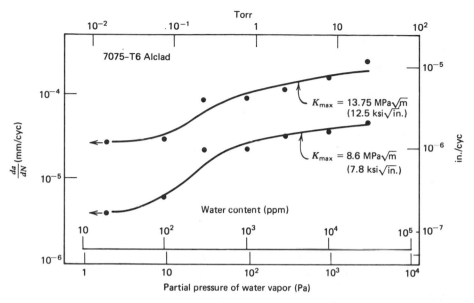

FIGURE 13.45   Fatigue crack growth rate in 7075-T6 aluminum alloy as a function of water vapor content in test atmosphere. Data from Ref. 125 and redrawn by Wei[127] (Reprinted from R. P. Wei. *Engineering Fracture Mechanics,* **1970, Pergamon Press.)**

environmental sensitivity in 7075-T6. This is consistent with the fact that 7075-T6 is more susceptible to environment-assisted cracking than 2024-T3.

During the past few years, many more material–environment systems have been identified as being susceptible to corrosion fatigue. It is found that many aluminum, titanium, and steel alloys are adversely affected during fatigue testing by the presence of water, and titanium and steel alloys (but not aluminum alloys) are affected by dry hydrogen. In these studies, test frequency, load ratio, load profile, and temperature have been identified as major variables affecting FCP response of a material subjected to an aggressive environment. For example, the harmful effects of a 3.5% saline solution on fatigue performance in a titanium alloy are shown in Fig. 13.46 as a function of test frequency. As might be expected, FCP rates increase when more time (i.e., lower frequencies) is allowed for environmental attack during the fatigue process. It should be pointed out that no important frequency effects are found in metals when tested in an inert atmosphere. Also note the negligible environmental effect on FCP at high crack growth rates where the mechanical process of fatigue damage is probably taking place too quickly for chemical effects to be important.

Fatigue crack growth rate sensitivity to environment and test frequency has also been found in ferrous alloys (Fig. 13.47).[129,130] What is most intriguing about these data is the fact that they reveal a significant environmental sensitivity, even though all tests were conducted with K_{max} maintained *below* the K_{IEAC}* level for the material.

---

* Recall from Chapter 11 that the threshold level for stress corrosion cracking, K_{ISCC}, has been redefined by the more general term for environment-assisted cracking, K_{IEAC}.

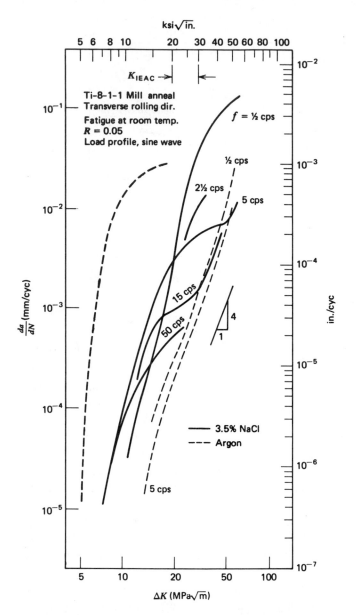

FIGURE 13.46   Effect of frequency on fatigue crack growth in Ti–8Al–1Mo–1V alloy in 3.5% NaCl and argon atmospheres.[128] Dashed curve at left corresponds to testing at 5 Hz in 3.5% NaCl and with $R = 0.75$. (Courtesy R. J. Bucci, Alcoa Research Laboratories.)

Why should there be any environmental effect during fatigue if the tests were conducted below K_{IEAC}? Perhaps a protective film, developed at the crack tip under sustained loading conditions acts to protect the material from the environment but is ruptured by fatigue cycling, thereby permitting the corrodent to reattack the crack-tip region. Such a hypothesis is supported by observations made by Bucci,[128] who showed

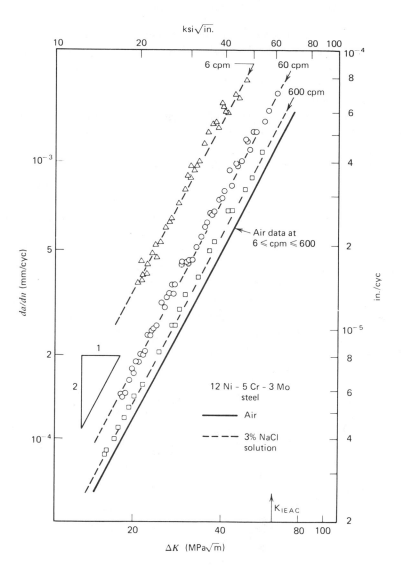

FIGURE 13.47 Corrosion fatigue (3% NaCl solution) crack growth data in 12Ni–5Cr–3Mo steel as a function of test frequency. All tests conducted with $\Delta K < K_{IEAC}$[130] (Reprinted with permission of the American Society for Testing and Materials from copyrighted work.)

that environment-assisted cracking does occur below K_{IEAC} when the arrested crack is subjected to a period of load cycling and then reloaded below the previously established K_{IEAC} level. This would suggest that constant loading conditions used to obtain a K_{IEAC} value for a given material represent a *metastable* condition, easily upset by the imposition of a number of load fluctuations.

Although the effect of mean stress on FCP in the intermediate growth rate regime was found generally to be of secondary importance (Section 13.4), it does become a major variable during corrosion fatigue conditions (Fig. 13.46). From these data, it would appear that high R ratio conditions enhance the corrosion component of crack growth, while low R ratio testing reflects more of the intrinsic fatigue response of the material. The greater importance of mean stress effects during environmentally enhanced fatigue crack propagation may be rationalized with the aid of the superposition model described in the next section.

As might be expected, the other major test variable relating to corrosion fatigue is that of test temperature. Many investigators have found FCP rates to increase with increasing temperature. For many years, a controversy has existed concerning the origin of this FCP temperature sensitivity. Is it due to a creep component or to an environmental component, both of which increase with increasing test temperature? In a series of experiments conducted at elevated temperatures in inert environments and in vacuum,[131-135] it was shown that neither temperature nor frequency had any effect on fatigue crack propagation rates. In fact, test results were comparable to room temperature results. This was the case even when the inert environment was liquid sodium.[134] On the basis of these results, it is concluded that higher FCP rates at higher temperatures mainly result from material–environment interactions, rather than a creep contribution. A similar conclusion was reached by Coffin for the case of unnotched samples, which were tested under constant strain range conditions. Figure 13.48 reveals a plot of cyclic life versus plastic strain range for many ferrous and nonferrous alloys that were tested in room temperature air or high-temperature vacuum or argon.

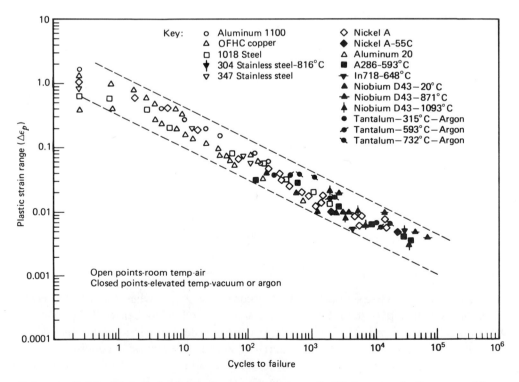

FIGURE 13.48 **Plot of plastic strain range versus cyclic life in several alloys tested in room temperature air or in high-temperature vacuum or argon. (From Coffin[131]; reprinted with permission from the Metallurgical Society of AIME.)**

The general agreement among these results suggests that temperature is not a significant variable when fatigue tests are conducted in inert atmospheres over a range of temperatures.

Before concluding this section, it is important to point out that unusual environmental influences may be associated with the growth of short cracks. Gangloff[136] reported that for the case of 4140 steel (σ_{ys} = 1300 MPa), when exposed to a hydrogen-producing environment, short cracks (0.1 to 0.8 mm deep) grew as much as an order of magnitude faster than 20- to 50-mm-long cracks for comparable ΔK levels and environmental (3% NaCl) conditions. Furthermore, he noted that the corrosion fatigue growth rates could not be described by a single-valued function of ΔK in that crack growth rates were *higher* at lower stress levels. Gangloff postulated that fatigue crack growth of physically short cracks in the presence of an aggressive environment could be described in terms of a purely chemical mechanism based on processes responsible for the control of the hydrogen concentration levels at the crack tip. Surely, additional studies are needed to explore the generality of this short-crack–cyclic load–environment interrelated phenomenon.

13.6.1 Corrosion Fatigue Superposition Model

Wei and Landes[137] and Bucci[128] developed a model that would account for effects of environment, test frequency, wave form, and load ratio on corrosion fatigue crack propagation behavior. They approximated the total crack extension rate under corrosion fatigue conditions by a simple superposition of the intrinsic fatigue crack growth rate (determined in an inert atmosphere) and the crack extension rate due to a sustained load applied in an aggressive environment (determined as the environment-assisted crack growth rate). Therefore

$$\left(\frac{da}{dN}\right)_T = \left(\frac{da}{dN}\right)_{\text{fat}} + \int \frac{da}{dt} K(t)dt \qquad (13\text{-}13)$$

where
$\left(\dfrac{da}{dN}\right)_T$ = total corrosion fatigue crack growth rate

$\left(\dfrac{da}{dN}\right)_{\text{fat}}$ = fatigue crack growth rate defined in an inert atmosphere

$\dfrac{da}{dt}$ = crack growth rate under sustained loading

$K(t)$ = time-dependent change in stress intensity factor

Two important aspects of this model should be emphasized. First, its linear character implies that there is no interaction (or synergism) between the purely mechanical and environmental components. Second, the model also depends on the assumption that the same mechanisms control the fracture process in both environment-assisted cracking and corrosion fatigue. For example, in the case of steels, both processes are believed to be controlled by hydrogen embrittlement. The application of the superposition model is demonstrated with the aid of Fig. 13.49. The FCP of the material in an inert environment $(da/dN)_{\text{fat}}$ is determined first and plotted in Fig. 13.49*a*. Next,

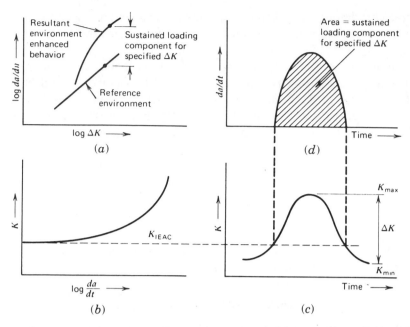

FIGURE 13.49 **Schematic diagram of superposition model. (a) Fatigue behavior in both reference and aggressive environments; (b) sustained loading environment cracking behavior; (c) K versus time for one cycle of loading; (d) sustained loading crack velocity versus time. Note that the model cannot predict environment enhanced fatigue crack propagation below K_{IEAC}.[128] (Courtesy of R. J. Bucci, Alcoa Research Laboratories.)**

the sustained loading crack growth rate component developed during one load cycle is obtained by integrating the product of the environment-assisted cracking rate da/dt and the time-dependent change in stress intensity level $K(t)$ over the time period for one load cycle (Fig. 13.49b–d). This increment then is added to $(da/dN)_{fat}$ to obtain $(da/dN)_T$, the corrosion fatigue crack growth rate.

Studies by Wei and coworkers[138,139] have attempted to account for environmentally enhanced fatigue crack growth behavior where $K_{max} < K_{IEAC}$ (see Fig. 13.47). To this end, a third component has been added to Eq. 13-13, which represents a cyclic-dependent contribution that involves a synergistic interaction between fatigue and environmental damage.

13.7 MICROSTRUCTURAL ASPECTS OF FCP IN METAL ALLOYS

A review of the literature reveals that there is a major influence of metallurgical variables on fatigue crack propagation at both low and high ΔK levels (Fig. 13.50). Conversely, many studies conducted in the intermediate growth rate regime reveal that metallurgical variables such as yield strength, thermomechanical treatment, and preferred orientation do not have a pronounced effect on FCP rates in aluminum and steel alloys; that is, fatigue crack propagation at intermediate ΔK levels is relatively

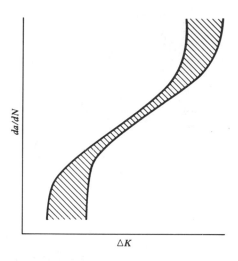

FIGURE 13.50 Major influences of metallurgical variables on fatigue crack growth are observed at both low and high ΔK levels.

structure insensitive. In almost every case, the transition from structure-sensitive to structure-insensitive crack growth behavior is associated with a concomitant transition in fracture mechanisms (recall Fig. 13.12). These fractographic observations strongly suggest the role of microstructural influences in the determination of the operative micromechanisms of fracture. To confirm this hypothesis, let us consider some earlier results pertaining to the fatigue crack propagation response in numerous titanium alloys. Several investigators[140–145] reported that below some critical ΔK level, a highly faceted fracture surface appearance was developed in conjunction with a highly branched crack front (Fig. 13.51a, b). Above this ΔK range, the fracture surface was much smoother overall and covered with fatigue striations. At the same time, the crack front was no longer bifurcated (Fig. 13.51c, d). In each instance, the fracture mechanism transition correlated with the development of a reversed plastic zone size (recall Section 13.2) equal to the grain size of the controlling phase in the alloy microstructure. For example, Yoder et al.[146–151] found that the transition from structure-sensitive to structure-insensitive FCP behavior occurred when the height of the cyclic plastic zone above the Mode I crack plane was equal to the average dimension of the Widmanstätten packet $\overline{\ell}_{wp}$. That is,

$$\overline{\ell}_{wp} = r_y^c = 0.033\left(\frac{\Delta K_T}{\sigma_{ys}}\right)^2 \qquad (13\text{-}14)$$

where　$\overline{\ell}_{wp}$ = average Widmanstätten packet size
　　　　r_y^c = cyclic plastic zone height above Mode I plane
　　ΔK_T = ΔK value at the fracture mechanism transition
　　σ_{ys} = cyclic yield strength (recall Section 12.3.1), but often taken to be the monotonic σ_{ys}

When Eq. 13-19 is rearranged, the stress intensity factor range at the mechanism transition is given by[148]

$$\Delta K_T = 5.5\sigma_{ys}\sqrt{\overline{\ell}} \qquad (13\text{-}15)$$

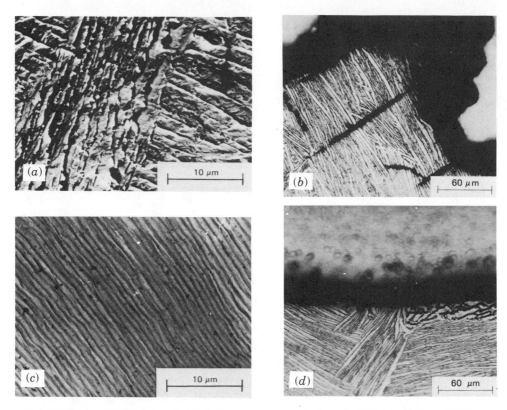

FIGURE 13.51 Fatigue fracture surface micromorphology in Ti-based alloy. (*a*) Faceted fracture surface; (*b*) multiple crack paths at $\Delta K < \Delta K_T$; (*c*) fatigue striations; (*d*) unperturbed crack profile at $\Delta K > \Delta K_T$. [(*a*) and (*c*) from Yoder et al[148]; reprinted with permission from the Metallurgical Society of AIME. (*b*) and (*d*) from Yoder et al.[149]; reprinted with permission from *Eng. Fract. Mech.* **11,** 805 (1979), Pergamon Press, Ltd.]

where $\overline{\ell}$ is generalized to correspond to the controlling alloy phase associated with the fracture mechanism transition.

The transition from structure-sensitive to structure-insensitive behavior also influences the dependence of the macroscopic growth rate on the applied ΔK level as noted by the slope change drawn in Fig. 13.52. It follows from Eq. 13-15 and Fig. 13.52 that, with increasing grain size of the relevant phase, the transition to structure-sensitive behavior (i.e., steeper slope of the $da/dN - \Delta K$ curve) should occur at higher ΔK levels.[148–151] Consequently, the FCP rate of these materials should decrease in the regime below ΔK_T with *increasing* grain size. An example of this behavior is shown in Fig. 13.53. Note the increase in ΔK_T with increasing Widmanstätten packet size and the associated shift to lower fatigue crack growth rates at a given ΔK level. Yoder et al.[151] compared such results with those from other investigations and confirmed the increasing influence of grain size on fatigue crack growth rates with decreasing ΔK levels.

From the previous discussions, one would expect that metallurgical variables such as grain size also should influence the Region I fatigue threshold value. This follows

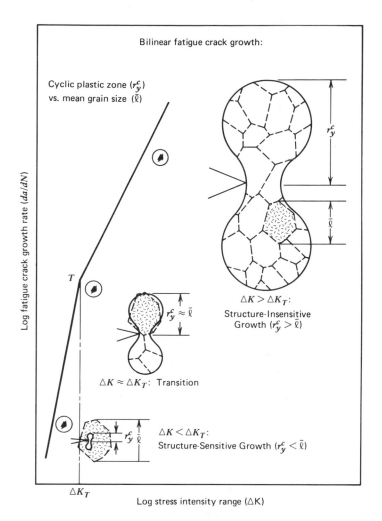

FIGURE 13.52 Bilinear fatigue crack growth behavior. Structure-sensitive behavior observed when reversed plastic zone < mean grain size. Structure-insensitive mode occurs when reversed plastic zone > mean grain size. (After Yoder et al[151]; reprinted with permission from the Metallurgical Society of AIME.)

from the fact that ΔK_{th} and ΔK_T do not differ greatly since the slope of the ΔK versus the da/dN curve in this regime is so steep.[152] Indeed, results reported for ferritic and pearlitic steels have shown that ΔK_{th} increases with the square root of grain size with a relation of the form

$$\Delta K_{th} = A + B\sqrt{d} \tag{13-16}$$

where A and B are material constants and d is the ferrite grain size.[73,74,153,154] Tanaka and coworkers[73,74] have proposed a model to explain this behavior in the materials they examined. They suggested that threshold conditions are established when the slip band at the crack tip is unable to traverse the nearby grain boundary. This should

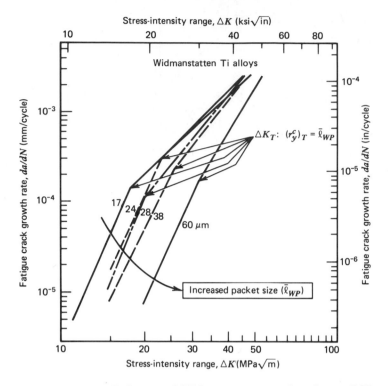

FIGURE 13.53 **Influence of Widmanstätten packet size on FCP response in Ti alloy.** **(After Yoder et al**[149]**; reprinted with permission from** *Eng. Fract. Mech.* **11, 805 (1979), Pergamon Press, Ltd.)**

occur when the slip band size or the cyclic plastic zone dimension is approximately equal to the average grain diameter. Since the plastic zone size varies with the square of the stress intensity factor, it follows that the stress intensity level associated with the threshold condition should increase with increasing grain size.

It is important to recognize the similar dependence of ΔK_T and ΔK_{th} on structural size (Eqs. 13-15 and 13-16) and to recall that $\overline{\ell}$ corresponds to the controlling alloy phase or structural unit responsible for the fracture mechanism transition. Similarly, the term d in Eq. 13-16 should correspond to the dimension of the critical structural unit, which acts as the effective barrier to slip. Depending on the material, this barrier may correspond to the grain size, Widmanstätten packet dimension as in Ti alloys, subgrain size, or the dislocation cell dimension.

Since ΔK_{th} values in certain steel and titanium alloys increase with *increasing* grain size, one is faced with a design dilemma. Recall from earlier discussions in Sections 4.2, 10.5, and 12.2.4 that the material's yield strength, fracture toughness, and smooth bar fatigue endurance limits, respectively, should increase with *decreasing* grain size. Consequently, some compromises become necessary when attempts are made to optimize simultaneously these four mechanical properties. This difficulty becomes readily apparent when one sets out to optimize the fatigue life of an engineering component. If one assumes that the component does not contain an initial flaw, then the fatigue life should depend strongly on the initiation stage of fatigue damage. As such, the fatigue life should increase with an increase in the endurance limit, σ_{fat}. Since σ_{fat} increases with increasing tensile strength (Fig. 12.14), then a reduction in grain size would be expected to enhance the material's fatigue resistance. On the other hand, if the component contains a preexistent flaw that can subsequently grow to failure, then the fatigue life can be improved by lowering crack growth rates at a given ΔK level (particularly at low ΔK levels) and/or by increasing ΔK_{th}. Consequently, if one changes the grain size to optimize σ_{fat}, the associated ΔK_{th} value will have been reduced, and vice versa. These conflicting trends can be shown in schematic form by superimposing the influence of grain size on σ_{fat} and $\Delta\sigma_{th}$, the latter being computed from the stress intensity factor formulation (Fig. 13.54). Note that the characteristic crack length parameter a_0, as described by Haddad et al.,[72] increases with increasing grain size.

Since fatigue crack initiation and propagation processes usually occur at component surfaces and in interior regions, respectively, the development of a duplex grain structure may lead to an optimization of fatigue performance. For example, if a component is thermomechanically treated so as to develop a fatigue-initiation resistant, fine-grained surface zone, coupled with a crack-propagation resistant, coarse-grained interior, then the component would be expected to exhibit superior overall fatigue resistance. Indeed, by shot peening and locally recrystallizing the surface zone of a Ti-8 Al alloy, a duplex grain structure was developed with fine grains located at the surface and coarse grains developed within the sample interior (Fig. 13.55a).[155] Preliminary S-N data attest to the superior fatigue response of this duplex-grain microstructure as compared with that of a completely coarse grained material (Fig. 13.55b).

The strong influence of grain size and slip character on ΔK_{th} may well reflect their impact on the development of roughness-induced crack closure; large grains and extended planar slip behavior would be expected to enhance crack surface interference and reduce ΔK_{eff}, thereby promoting higher ΔK_{th} levels as determined with conventional E647-93 test procedures. The emerging class of aluminum–lithium alloys are

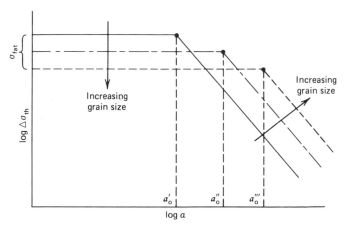

FIGURE 13.54 Influence of grain size on the determination of the characteristic crack length parameter a_0. Note that increasing grain size decreases σ_{fat} and increases σ_{th}.

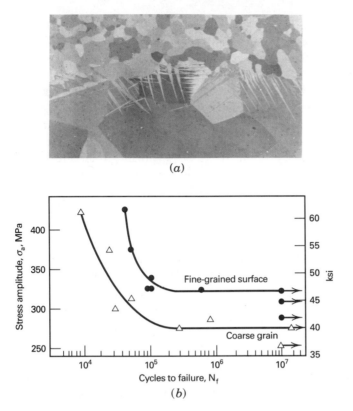

(a)

(b)

FIGURE 13.55 **Shot peening-induced surface layer recrystallization in Ti-8Al.** (*a*) **Microstructure revealing fine grains in surface layer and coarse grains in substrate (250X);** (*b*) **S-N plot revealing superior fatigue response of material containing duplex grain-size microstructure.**[155] **(Reprinted from L. Wagner and J. K. Gregory,** *Advanced Materials and Processes,* **146 (1), 50 HH (1994), with permission from ASM International.)**

particularly noteworthy in this regard.[156,157] However, the beneficial influence of microstructure in the threshold regime is significantly reduced in the relatively closure-free environment associated with the growth of short cracks.[97,157] In fact, preliminary results suggest that the difference between short- and long-crack FCP rates is greatest in those alloys that exhibit the highest long-crack closure levels; hence, it is ironic that the most nonconservative estimates of fatigue life could be associated with the use of long-crack data from those alloys that "look the best," according to low-R^c threshold test procedures.

Recalling Fig. 13.12, the observed transition from faceted growth to striation formation in close-packed alloys could very well reflect the structure-sensitive to structure-insensitive transition discussed above. For the case of BCC steels, a similar set of fracture micromechanism transitions is observed. To wit, microvoids are found at the highest ΔK levels and gradually yield to fatigue striation formation with decreasing ΔK values. In most instances, the spacing between adjacent striations is found to vary with the second power of ΔK, consistent with the Bates and Clark relation (Eq. 13-9). At very low ΔK levels, the fracture surface micromorphology in

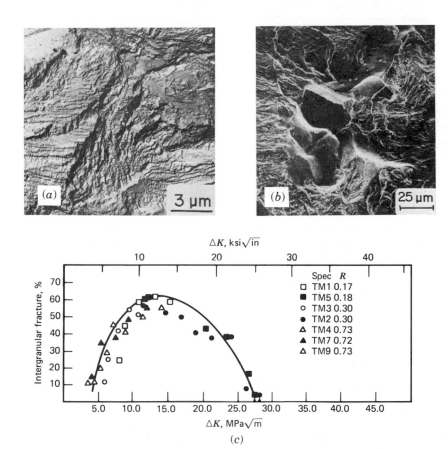

FIGURE 13.56   (*a*) Fractograph of AISI 9310 steel alloy (double-vacuum melted) revealing crisp faceted appearance in fatigue threshold region. $\Delta K \approx 8$ MPa $\sqrt{\mathrm{m}}$. (*b*) A471 steel revealing localized evidence of intergranular failure as a result of fatigue loading. $\Delta K \approx 7$ MPa $\sqrt{\mathrm{m}}$. Tested in water at 160 Hz. (*c*) Percentage of intergranular fracture in a medium carbon steel versus ΔK.[56] Tested in air. (Reprinted with permission from *Eng. Fract. Mech.* **7**, 69 (1975), Pergamon Press, Ltd.)

certain steels has been shown to possess a highly faceted texture, similar to that found in FCC alloys (Fig. 13.56*a*). Overlapping the striation formation-faceted growth transition in BCC alloys is yet another fracture mechanism; in the ΔK range between 5 and 25 MPa$\sqrt{\mathrm{m}}$, many investigators have reported varying amounts of intergranular fracture (Fig. 13.56*b*).[55,56,158,167] The presence of intergranular fracture regions has been reported in low[160,166] and high-carbon steels,[55] high-strength steels,[158,162–165] and in microstructures composed of ferrite, bainite, martensite, and austenite. In some instances, investigators[55,56,140,159,160,166,168] have found that intergranular fracture takes place when the size of the reversed plastic zone is comparable to the relevant microstructural dimension, such as the prior austenite grain size. As such, there should be a maximum amount of intergranular fracture at some intermediate ΔK level with decreasing amounts of grain-boundary failure being observed at both higher and lower ΔK levels (Fig. 13.56*c*). Although this correlation is encouraging, other investigators[169] have questioned whether the maximum amount of intergranular failure should correspond instead to some critical K_{max} value. Furthermore, for the case of 4340 steel tested in air, Cheruvu[170] reported no correlation between the prior austenite grain size (20 to 200-μm) and the reverse plastic zone dimension at ΔK values corresponding to a maximum incidence of intergranular fracture. Surely more studies are needed to more fully identify the processes responsible for this fracture mechanism transition.

In a number of related studies, the amount of intergranular fracture observed tended to increase with the aggressiveness of the test environment.[161,164] Researchers have also found that the incidence of intergranular fracture can sometimes be traced to the presence of an embrittling grain-boundary film or solute segregation at certain grain boundaries.[161,162] Ritchie[158] found the amount of intergranular failure in a high-strength steel to be considerably greater when the material was heat treated to bring about a temper-embrittled condition. It appears, therefore, that intergranular facets are produced in various materials as a result of the *combined* influence of environmental and microstructural factors. In turn, the extent of the environmental sensitivity to intergranular fracture was found to be dependent on test frequency, with more inter-granular facets being observed in 4340 steel (in 585 Pa water vapor) at 1 Hz than at 10 Hz.[171]

Metallurgical factors also affect fatigue crack propagation rates at high ΔK levels. This is because as ΔK becomes very large, K_{max} approaches K_{IC} or K_c where local fractures occur with increasing frequency and produce accelerated growth (recall Eq. 13-10). Since tougher materials are typically cleaner and will exhibit fewer local instabilities, their crack growth rates should be lower at high K levels. This is consistent with the well-established rule of thumb regarding low cycle fatigue (analogous to high FCP rates); low cycle fatigue resistance is enhanced by improvements in toughness and/or ductility. A number of investigators have verified this relation and have rationalized differences in macroscopic and microscopic crack growth rates. For example, in the high ΔK regime, FCP rates in a banded steel (consisting of alternate layers of ferrite and pearlite (Fig. 13.57a)) increased when tested in the arrester, divider, and short transverse directions, respectively (Fig. 13.57b).[172] Although little difference in fatigue response was found at low ΔK levels among the three orientations tested, a 40-fold difference in macroscopic crack growth rate was observed in going from the least to the most damaging loading direction. It is interesting to speculate whether the relative fatigue resistance associated with the three test orientations in the banded steel would be reversed at growth rates below 10^{-5} mm/cyc (the point where the FCP resistance was comparable for each test direction). Indeed, Mayes and Baker[173] reported that the highest ΔK_{th} value of a high-sulfur semifree machining steel was associated with the short transverse orientation. They interpreted these results in terms of increased fracture surface roughness associated with the rupture of non-metallic inclusions; the enhanced roughness, in turn, resulted in a greater degree of crack closure.

It is important to note that FCP anisotropy was also found in homogenized samples of the banded steel when the layered microstructure was eliminated (but not the alignment of sulfide particles [see Fig. 10.4b]). Since the *microscopic* growth rate (i.e., fatigue striation spacings) was the same in the three crack plane orientations, it was concluded that the anisotropy in *macroscopic* FCP was related to different amounts of sulfide inclusion fracture in the three orientations. Consequently, macro-

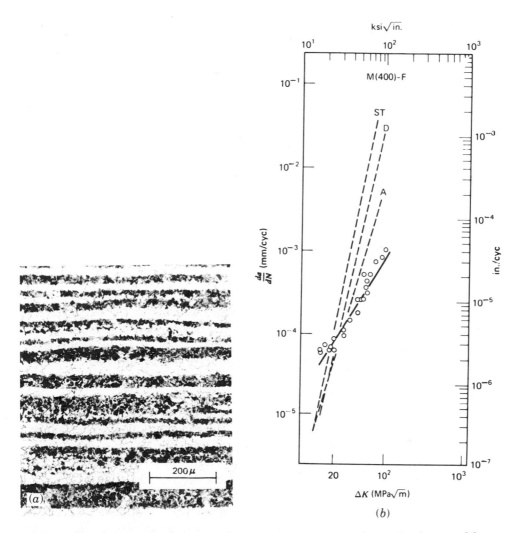

FIGURE 13.57 (*a*) Microstructure in banded steel revealing alternating layers of ferrite and pearlite. (*b*) Fatigue crack propagation in banded steel (layers of ferrite and martensite) as a function of specimen orientation. (ST, short transverse; D, divider; A, arrester geometry.) Striation spacing (data points and solid line) is seen to be independent of specimen orientation. (From F. Heiser and R. W. Hertzberg, *Transactions of the ASME*, 1971, with permission of the American Society of Mechanical Engineers.)

scopic FCP is considered to be the summation of several fracture mechanisms, the most important being striation formation and local fracture of brittle microconstituents. From this, macroscopic growth rates may be described by

$$\left(\frac{da}{dN}\right)_{\text{macro}} = \sum Af(K)_{\substack{\text{striation} \\ \text{mechanism}}} + Bf'(K)_{\substack{\text{void} \\ \text{coalescence}}}$$
$$+ \; Cf''(K)_{\text{cleavage}} + Df'''(K)_{\substack{\text{corrosion} \\ \text{component}}} + \cdots \qquad (13\text{-}17)$$

From Eq. 13-18, we know at least that $A \approx 6/E^2$ and $f(K) \approx \Delta K^2$.

In another attempt to correlate macroscopic and microscopic crack growth rates above the crossover point, Bates[174] adjusted Eq. 13-9 so that

$$\left(\frac{da}{dN}\right)_{macro} \approx \frac{6}{f_s}\left(\frac{\Delta K}{E}\right)^2 \tag{13-18}$$

where f_s is the percentage of striated area on the fracture surface. This relation is consistent with observations by Broek[175] and Pelloux et al.,[176] who also found increasing amounts of particle rupture and associated void coalescence on the fracture surface with increasing stress intensity levels. The latter investigators noted a marked increased in particle rupture when the plastic zone dimension grew to a size comparable to the particle spacing. Although Eq. 13-18 provides a rationale for differences in macroscopic and microscopic FCP rates, it may be too impractical to use because an extensive amount of fractographic information is required.

At intermediate ΔK levels, metallurgical factors do not appear to influence fatigue crack growth rates to a significant degree.[177] For example, only modest shifts in the slopes of log ΔK − log da/dN plots were found in studies of brass and stainless steel alloys in both cold-worked and annealed conditions where 4- to 10-fold differences in monotonic yield strength were reported.[27] Fairly strong crystallographic textures were developed in both the cold-worked and recrystallized conditions, but, again, little effect was noted on FCP response for both brass and steel specimens oriented so as to present maximum densities of {111}, {110}, and {100} crystallographic planes, respectively, on the anticipated crack plane. It was noted, however, that the actual fracture plane and crack direction were affected strongly by crystallographic texture in that the gross crack plane avoided a {111} orientation,[27,178] consistent with expectations of the striation formation model discussed in Section 13.3.

It has been suggested that FCP does not depend on typical tensile properties because monotonic properties are not the controlling parameter. Instead, cyclically stabilized properties may hold the key to fatigue crack propagation behavior. Starting or monotonic properties between two given alloys may differ widely, but their final or cyclically stabilized properties would not. For example, soft alloys would strain harden and hard ones would strain soften; as a result, the materials would be more similar in their final state than at the outset of testing. Consequently, if fatigue crack propagation response were dependent on cyclically stabilized properties, smaller differences in FCP behavior would be expected than that based on a comparison of monotonic values. A number of studies[179–183] have been conducted to establish correlations between cyclic strain and FCP data. For example, the slope m of the da/dN − ΔK plot is seen to decrease with increasing cyclic yield strength σ'_{ys} and cyclic strain-hardening exponent n'.[180] Furthermore, it has been shown that log A varies inversely with m (Eq. 13-3). Although it is encouraging to find such correlations, more work is needed before it will be possible to predict FCP rates from cyclic strain data. For one thing, it is not clear from Fig. 13.58 whether a high or low slope is desirable for optimum fatigue performance.[184] Obviously, the intercept A from Eq. 13-3 is equally important in this determination. For example, alloy A would be better than alloy C but alloy D, which has the same slope as A, would be worse than C. Furthermore, the choice between alloy B and C would depend on the anticipated crack growth rate

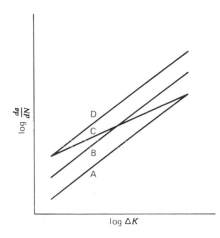

Figure 13.58 Diagram showing relative fatigue crack propagation behavior of several materials (A, B, C, D).[184] **(Reprinted with permission of the American Society for Metals, copyright © 1974.)**

regime for the engineering component. If many fatigue cycles were anticipated, the designer should opt for alloy B, since fatigue cracks would propagate more slowly over most of the component service life and allow for much greater fatigue life. In a low cycle fatigue situation, representative of conditions to the right of the crossover point, alloy C would be preferred.

In summary, fatigue crack propagation response in metal alloy systems is sensitive to structural variables at both low- and high-stress intensity value extremes. Both crack growth rates and fracture surface micromechanisms change with grain size, inclusion content, and mechanical properties such as yield strength. At low ΔK levels, structural sensitivity is observed when the crack-tip plastic zone is small as compared with the critical microstructural dimension. At high ΔK levels, fatigue crack growth rates are accelerated as K_{max} approaches K_c.

13.7.1 Normalization and Calculation of FCP Data

To this point, we have seen that microstructurally and/or crack length-induced differences in FCP data for a given alloy may be normalized by comparing crack growth rates as a function of ΔK_{eff}. It is interesting to note that differences in FCP rates between various metal alloy systems are dependent on the modulus of elasticity—a structure-insensitive property.[185] To illustrate, Ohta et al.[186] demonstrated excellent correlation of FCP data for aluminum, stainless steel, and low-carbon steel alloys when ΔK values were normalized by E (Fig. 13.59a); the minimal degree of scatter in Fig. 13.59b is attributed to their use of P_{max}^c-test conditions, which eliminated crack closure.

It is obvious, from earlier discussions, that the intrinsic fatigue crack propagation resistance of a metallic alloy may be characterized by its closure-free behavior with K_{max}^c data providing a convenient estimate of the latter. Regardless of the test method used (i.e., R^c or K_{max}^c), ASTM Standard 647-93 provides an operative definition of the ΔK threshold value (ΔK_{th}) at a crack growth rate of 10^{-10} m/cycle.[9] Based on the slip characteristics of a crystalline metal alloy, it may be reasonable to define a closely related ΔK value as that driving force corresponding to a growth rate of a single

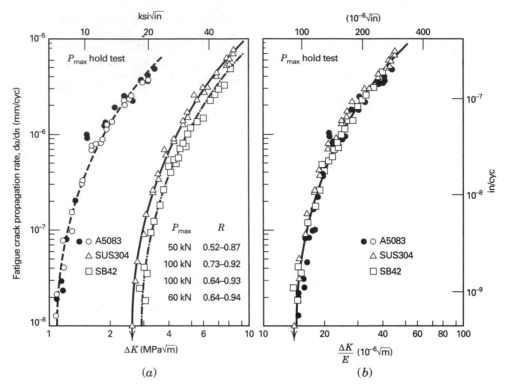

FIGURE 13.59 **Fatigue crack propagation rates in 5083 aluminum, 304 stainless and SB42 steel alloys[186], plotted as a function of: (a) ΔK; and (b) $\Delta K/E$. (Reprinted from A. Ohta, N. Suzuki, and T. Mawari, *Int. J. Fat.*, 14(4), 224 (1992), by permission of the publishers, Butterworth Heinemann, Ltd. ©.)**

Burgers vector. For purposes of identification, we may define this stress intensity factor driving force as ΔK_b. (For example, ΔK_b in steel and aluminum alloys would be identified at fatigue crack growth rates of 2.48 and 2.86 $\times$ 10^{-10} m, respectively.) One may then define ΔK_b as the limit of continuous damage accumulation with crack growth increments, $n\mathbf{b}$ ($n \geq 1$), occurring when $\Delta K \geq \Delta K_b$. Any growth increment less than the minimum unit of deformation (i.e., $\mathbf{b}$) would correspond to discontinuous crack extension.

If one examines the closure-free intrinsic fatigue crack propagation response for aluminum, nickel, titanium, and steel alloys, it is remarkable to note that $(\Delta K_b/E)^2$ values (units of m) correspond to the Burgers vector for each material[187,188] (Eq. 13-19).

$$\Delta K_b = E\sqrt{\mathbf{b}} \tag{13-19}$$

where E = modulus of elasticity
$\qquad\mathbf{b}$ = Burgers vector

Hence, a datum for the fatigue crack growth response of a crystalline metal in the near threshold regime would correspond to $da/dN = \mathbf{b}/cyc$ and $\Delta K_b = E\sqrt{\mathbf{b}}$.* This datum would identify the point above which crack growth would be continuous and occur in multiples of the Burgers vector and below which crack growth would be discontinuous in nature. Strong agreement was found between this computed datum (Point A) and K_{max}^c- or ΔK_{eff}-based experimental test results for several aluminum, steel, nickel, and titanium alloys[187,188] (e.g., see Fig. 13.60a, b).

Proceeding further, it was found for a wide range of metallic alloys that K_{max}^c-generated da/dN-ΔK plots tended to assume a log–log relation[187,188] with crack growth rates being approximately dependent on ΔK^3. As such, additional data points could be calculated directly at FCP rates above $\mathbf{b}/cyc$ wherein

$$da/dN = \mathbf{b}(\Delta K/\Delta K_b)^3 \tag{13-20}$$

where $\mathbf{b}$ = Burgers vector
            ΔK = closure-free stress intensity factor range
            ΔK_b = closure-free ΔK level associated with $da/dN = \mathbf{b}/cyc$

By combining Eqs. 13-19 and 13-20, it follows that

$$da/dN = \mathbf{b}(\Delta K/E\sqrt{\mathbf{b}})^3 = (\Delta K/E)^3(1/\sqrt{\mathbf{b}}) \tag{13-21}$$

The dashed lines shown in Figs. 13.60a, b represent data points calculated from Eq. 13-21 within an arbitrarily defined ΔK range from ΔK_b to $10\Delta K_b$ (Point B). The agreement between experimental and computed data points is most encouraging. Finally, one of the steel alloys shown in Figure 13.60 is replotted along with associated short-crack test data (Fig. 13.60c). One may readily conclude that both experimental and calculated closure-free curves provide an upper bound estimate of short crack behavior (recall Section 13.4.2.1).

Finally, an attempt was made to predict ΔK_b and associated ΔK data, corresponding to Eq. 13-21, for the case of a material (Cu-Be alloy 25) for which no closure-free data were known to the author. Using values of E and $\mathbf{b}$ of 125 GPa and 2.55×10^{-10} m, respectively, and assuming that closure-free crack growth rates are dependent on ΔK^3, Points A and B were computed (Fig. 13.60d). Two K_{max}^c fatigue tests were then performed at 35 and 45 MPa$\sqrt{m}$, respectively. Excellent agreement is seen between the experimental and computed data. (Note that the slopes of the $K_{max}^c = 35$ and 45 MPa$\sqrt{m}$ data plots were 2.97 and 3.03, respectively, in excellent agreement with the assumed $da/dN/\Delta K^3$ dependence.) In summary, this simple model provides a means by which closure-free FCP data in monolithic metal alloys may be predicted in the ΔK range near ΔK_{th} and above.

---

* Note the agreement between the form of the $\Delta K_b = E\sqrt{\mathbf{b}}$ relation and the theoretical models by Sadananda and Shahinian[189] and Yu and Yan[190] for ΔK_{th} where the latter were found to be proportional to $G\sqrt{\mathbf{b}}$ and $Ee_f\sqrt{\mathbf{b}}$, respectively.

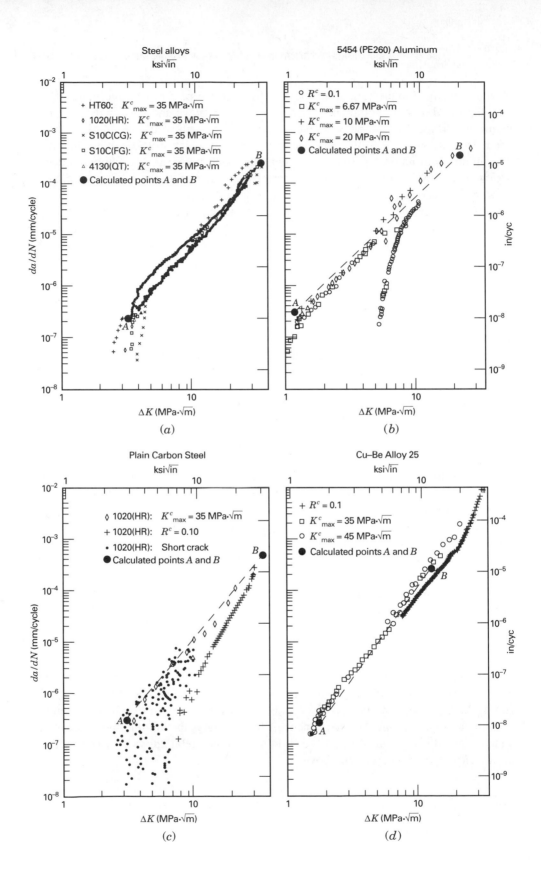

EXAMPLE 13.4

A 30-cm diameter and 1.5-cm-thick, cylindrically shaped steel pressure vessel contains a small surface crack, oriented normal to the hoop stress direction. (a) Confirm that a leak-before-break condition exists, given that the vessel experiences a cyclic gas pressure between 40 and 50 MPa and the material's fracture toughness is 180 MPa$\sqrt{m}$. (b) At the point where the crack breached the wall thickness, estimate the rate of fatigue crack advance.

The maximum and minimum hoop stresses are given from Eq. 1-63 where

$$\sigma = \frac{P \cdot r}{t}$$

$$\sigma = \frac{50 \cdot 15}{1.5} = 500 \text{ MPa and } \frac{40 \cdot 15}{1.5} = 400 \text{ MPa}$$

From Eq. 8-29 the leak-before-break condition will occur when

$$K_c > K = \sigma\sqrt{\pi a} = \sigma\sqrt{\pi t}$$

In this instance,

$$K = 500 \cdot 10^6 \cdot \sqrt{\pi \cdot 0.015} = 108.5 \text{ MPa}\sqrt{m} < 180 \text{ MPa}\sqrt{m}$$

This vessel satisfies the leak-before-break condition. At this stage of damage development, the vessel contains a through-thickness crack with a total length of 3 mm. Since the cyclic hoop stress varies from 400 to 500 MPa (i.e., $R = 0.8$), little or no crack closure would be expected. Accordingly, crack growth rates can be estimated from Eq. 13-21. Since $E = 205$ GPa and $b = 2.48 \times 10^{-10}$ m, the crack growth rate is given by

$$da/dN = (\Delta K/E)^3 (1/\sqrt{b}) \tag{13-21}$$

$$da/dN = \left(\frac{(500 - 400)\sqrt{\pi 0.015}}{E} \cdot 10^6\right)^3 \cdot \left(1/\sqrt{2.48 \times 10^{-10}}\right)$$

$$da/dN = 7.54 \times 10^{-5} \text{ mm/cyc.}$$

which is in very good agreement with the experimental data shown in Fig. 13.60a.

FIGURE 13.60 Experimental and calculated (line A-B) FCP data for several metallic alloys.[187,188] *(a)* several steel alloys (R. W. Hertzberg, *Int. Journal of Fracture*, **64** (3), 135 (1993). Reprinted by permission of Kluwer Academic Publishers.); *(b)* 5454 aluminum; *(c)* long- and short-crack data in hot-rolled 1020 steel; *(d)* Cu-Be Alloy 25. (R. W. Hertzberg, *Matls. Sci. Eng.*, A190, 25 (1995). Reprinted by permission.)

13.8 FATIGUE CRACK PROPAGATION IN ENGINEERING PLASTICS

A growing number of studies concerning the FCP behavior of engineering plastics have been conducted and information is now available for more than two dozen different materials.[191–198] With such a body of data, certain conclusions and generalities may be drawn. As in metals, the FCP rates of polymers are strongly dependent on the magnitude of the stress intensity factor range, regardless of polymer chemistry or molecular arrangement (Fig. 13.61). Note the data correlation for both amorphous, semicrystalline, and rubber-modified polymers on the same plot of ΔK versus da/dN. In a sense, this is analogous to plots of data from metal alloys possessing various crystal structures (Fig. 13.3). Figure 13.61 also shows the relative ranking of the fatigue resistance of metals and plastics when compared as a function of ΔK and $\Delta K/E$. It is clear that plastics will exhibit superior or inferior FCP resistance as compared with metal alloys, depending on whether cycling is conducted under strain-controlled or stress-controlled conditions, respectively.

On the basis of these results, it is concluded that superior FCP resistance is exhibited by semicrystalline polymers (e.g., nylon 66, ST801 (rubber modified nylon 66); nylon 6, poly(vinylidene fluoride), (PVDF); polyacetal (PA); and poly(ethylene terapthalate), (PET)). In all likelihood, the superior fatigue resistance of crystalline polymers relative to that associated with amorphous structures is not fortuitous. Koo et al.[199,200] and Meinel and Peterlin[201] have pointed out that crystalline polymers not only can dissipate energy when crystallites are deformed, but they can also apparently reform a crystalline structure that is extremely strong. To further illustrate this point, the remarkably superior FCP resistance of amorphous PET was traced to strain-induced crystallization that took place within the plastic zone ahead of the crack tip.[202] Furthermore, the percent crystallinity increased with increasing ΔK level (i.e., increasing plastic zone size); for a given level of crystallinity, FCP resistance was greater in thermal treatments that lead to enhanced resistance to cyclically induced polymer chain disentanglement through the development of higher tie molecule densities.[194,195]

The adverse influence of cross-linking and the beneficial role of rubbery additions on FCP resistance have been examined and are described elsewhere.[192,193] For the present discussion, it is relevant to note that the fatigue crack propagation resistance of engineering plastics increases directly with the material's fracture toughness. Indeed, a striking correlation is evident in a large number of polymeric systems between values of ΔK^* (the value of ΔK required to drive a crack at a constant value of da/dN) and K_{cf} (the maximum value of K during the fatigue test preceding unstable crack extension) (Fig. 13.62). Since K_{cf} represents some measure of the material's fracture toughness, it is seen that the greater the toughness of a polymer, the greater the driving force required to advance the crack at a constant speed. A similar correlation is found for the case of rubber and hollow glass sphere-toughened hybrid composites.[197] Evidence for synergistic toughening at intermediate combinations of CTBN rubber particles and hollow glass spheres was shown in Fig. 10.43; such behavior was attributed to the multiplicitive interaction between rubber particle cavitation/matrix shear banding and hollow glass sphere-induced microcracking. Fig. 13.63a reveals that the fatigue crack propagation response of these blends possesses a similar synergistic response at the same rubber/glass ratio. Azimi[197] concluded that this behavior was

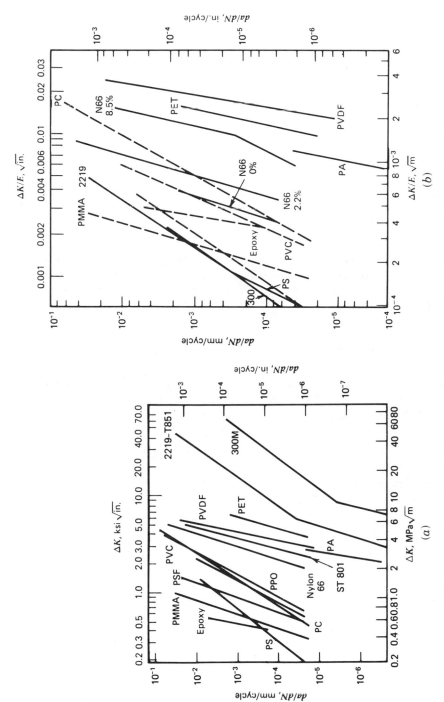

FIGURE 13.61 FCP behavior in various metal and polymeric materials. (*a*) Data plotted versus ΔK; (*b*) data plotted versus $\Delta K/E$. (Adapted from J. A. Manson et al., *Advances in Fracture Research*, D. Francois et al., Eds., Pergamon Press, Oxford, 1980.)

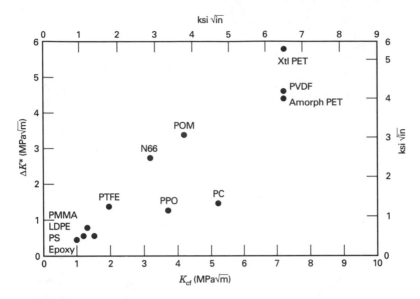

FIGURE 13.62 Relation between ΔK^* (the ΔK driving force to generate a crack velocity of 5×10^{-4} mm/cycle) and K_{cf} (the maximum value of K during the fatigue test preceding unstable crack extension) for selected amorphous and semicrystalline polymers. (Reprinted from Pecorini).[194]

attributable to the development of an enlarged plastic zone consisting of cavitated rubber particles with associated shear banding and distinct secondary deformation branches generated by the interaction between the elastic stress fields of the advancing crack front and the hollow glass spheres (Fig. 13.63b). Finally, it is worth noting that the fatigue crack propagation resistance of rubber-modified epoxy blends is relatively insensitive to modifier content at low ΔK levels (Fig. 13.64).[197,198] Above the transition level (referred to as ΔK_T, where the plastic zone size is on the order of the rubber particle diameter), rubber-modified blends reveal superior FCP resistance.* For example, Azimi determined that when the plastic zone dimension is smaller than the rubber particle size (i.e., where $\Delta K < \Delta K_T$), little influence of rubber particles on FCP response is expected. Conversely, when the size of the plastic zone is large enough to engulf many rubber particles, cavitation/shear banding mechanisms are activated, thereby leading to improved FCP resistance (Fig. 13.65). Note the lack of rubber cavitation and matrix dilation when $\Delta K = 0.5$ MPa$\sqrt{m} < \Delta K_T$ and the correspondingly pronounced activity of these two deformation mechanisms when $\Delta K = 2.5$ MPa$\sqrt{m} > \Delta K_T$.

Perhaps the greatest change in FCP resistance occurs when the molecular weight is modified. For example, Rimnac et al.[203] found a thousandfold decrease in FCP rates when the molecular weight (MW) of poly(vinyl chloride) (PVC) was increased by little more than a factor of three (Fig. 13.66). Similarly, major improvements in FCP resistance with increasing MW have been found in polyacetal,[204] polycarbonate,[205]

---

* Recall an earlier discussion of ΔK-dependent transitional behavior (see Fig. 13.52).

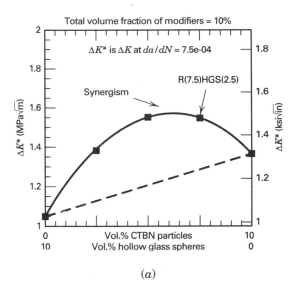

(a)

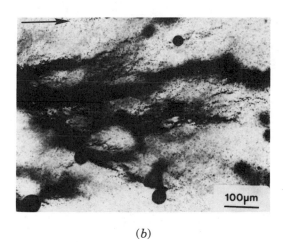

(b)

FIGURE 13.63 Synergistic fatigue response of CTBN rubber particle-hollow glass sphere hybrid epoxy composites. (a) Maximum FCP resistance associated with epoxy blend containing 7.5 v/o CTBN rubbery particles and 2.5 v/o hollow glass spheres. Maximum fracture toughness also associated with same blend composition (recall Fig. 10.43); (b) thin-section of same hybrid epoxy resin composite, viewed under transmitted and bright field conditions. Plastic zone contains rubber cavitation/matrix shear banding and secondary branching due to interaction of elastic stress fields between crack tip and hollow glass spheres. (Reprinted from H. Azimi.[197])

polyethylene,[206] nylon 66,[204] and poly(methyl methacrylate).[207] It is suggested that cyclic loading tends to disentangle whatever molecular network exists, and that this disentanglement is easier at lower MW. In addition, there may be positive contributions from enhanced orientation hardening with the higher MW species. Other investigations have shown a strong influence of molecular weight distribution (MWD) on

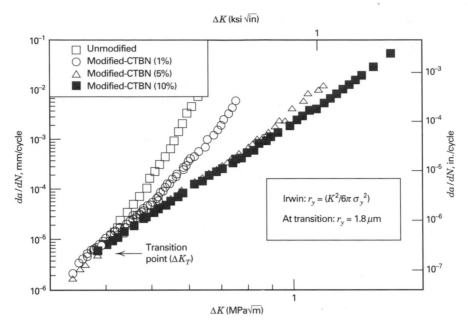

FIGURE 13.64 Influence of rubber particle additions on FCP resistance in epoxy resins. Above ΔK_T, the plastic zone size is large compared with rubber particle size, thereby resulting in improved fatigue performance. (Reprinted from H. Azimi et al[198].)

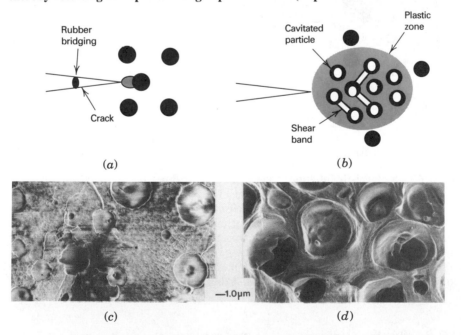

FIGURE 13.65 Mechanism associated with improved FCP resistance in epoxy-CTBN rubber particle blends. Little cavitation and matrix dilation occurs when $\Delta K < \Delta K_T$ and plastic zone is small compared with rubber particle diameter (see a and c). When $\Delta K > \Delta K_T$, plastic zone is large compared with particle dimension and much cavitation and matrix dilation occurs (see b and d). (Reprinted from Azimi.[197])

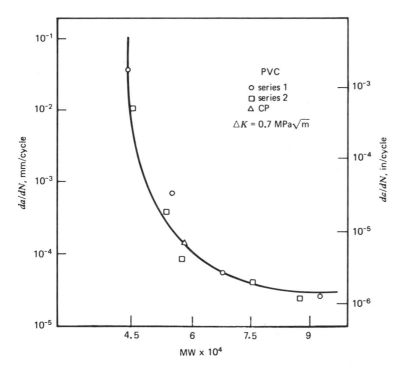

FIGURE 13.66 Fatigue crack propagation rates in PVC at $\Delta K = 0.7$ MPa $\sqrt{m}$ as a function of MW (From Rimnac et al.[203]; reprinted from *J. Macromol. Sci. Phys. B* **19**, 351, (1981), by courtesy of Marcel-Dekker, Inc.)

FCP resistance.[208] Michel et al.[209–211] developed a theoretical model to show that the strong sensitivity of FCP rate to MW and MWD is related to the fraction of molecules that can form effective entanglement networks. It follows that longer chains lead to the development of more fracture-resistant entanglement networks.

13.8.1 Polymer FCP Frequency Sensitivity

One is faced with an interesting challenge when trying to explain the effect of test frequency on polymer fatigue performance. Although hysteretic heating arguments appear sufficient to explain a *diminution* of fatigue resistance with increasing cyclic frequency in *unnotched* polymer test samples (recall Section 12.2.5), the fatigue resistance of several polymers in the *notched* condition is *enhanced* with increasing cyclic frequency. Note the pronounced decrease in FCP rate with increasing test frequency for a given ΔK level in PVC (Fig. 13.67a). Similar attenuation of FCP rates with increasing test frequency has been reported in several other polymeric solids.[192,213,215–217] Other polymers, such as polycarbonate (PC) and polysulfone (PSF), showed no apparent sensitivity of FCP rate to test frequency (Fig. 13.67b). An intriguing correlation has been found between the relative FCP frequency sensitivity in polymers and the frequency of movement of main chain segments responsible for generating the β transition peak (see Chapter 6) at room temperature.[213] Data for

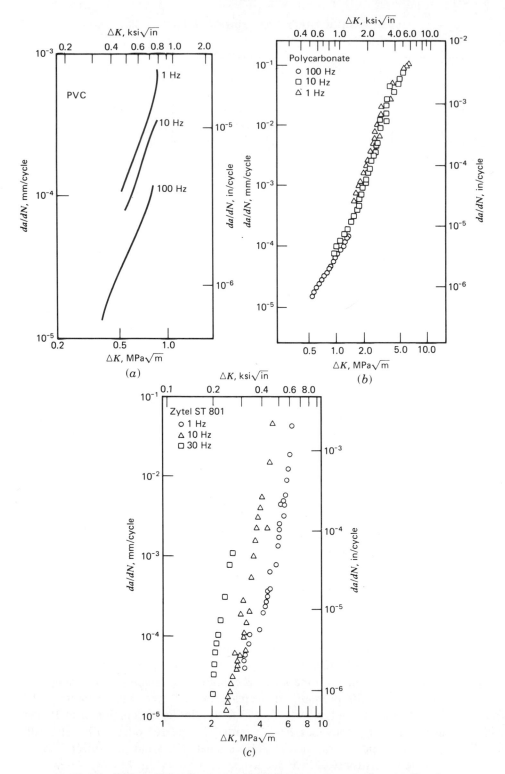

FIGURE 13.67 Effect of test frequency on fatigue crack growth rate. Crack growth can decrease, increase, or remain unchanged with increasing test frequency. (*a*) PVC (Skibo[212]); (*b*) polycarbonate (Hertzberg et al.[213]; with permission from the Society of Plastics Engineers, Inc.); (*c*) impact-modified nylon 66 (Hertzberg et al.[214]; copyright ©, American Society for Testing and Materials, 1916 Race Street, Philadelphia, PA 19103. Reprinted with permission.)

several polymers are shown in Fig. 13.68, along with the fatigue test frequency range. Note the greatest frequency sensitivity in the material that revealed its β peak at a frequency comparable to the fatigue test frequency range. This resonance condition suggests the possibility that localized crack-tip heating may be responsible for polymer FCP frequency sensitivity. One may then speculate whether other materials that were not FCP frequency sensitive at room temperature might be made so at other temperatures, if the necessary segmental motion jump frequency were comparable to the mechanical test frequency. Indeed, this has been verified for PC and PSF under low-temperature test conditions; conversely, the FCP response of PMMA was found to be *less* frequency sensitive at −50°C than at room temperature, which is consistent with expectations.[215,216] Of great significance, the overall frequency sensitivity for all the engineering plastics tested thus far has been shown to be dependent on $T - T_\beta$ (Fig. 13.69). This latter term represents the difference between the test temperature and the temperature corresponding to the β damping peak within the appropriate test frequency range. Experiments with PVC have confirmed a similar relation for fatigue tests conducted in the vicinity of the glass transition temperature (T_α).[218]

Recent studies have verified that the resonance condition, noted above, contributes to localized heating at the crack tip.[219] When the temperature increases, yielding processes in the vicinity of the crack tip are enhanced and lead to an increase in the crack-tip radius. A larger radius of curvature at the crack tip should result in a lower effective ΔK; fatigue crack growth rates should decrease accordingly with increasing test frequency. On the other hand, if the amount of specimen heating becomes *generalized* rather than *localized*, higher FCP rates would be expected at high test frequencies. This special condition was found to exist in impact-modified nylon 66, a material possessing a high degree of internal damping (Fig. 13.67c). Temperature

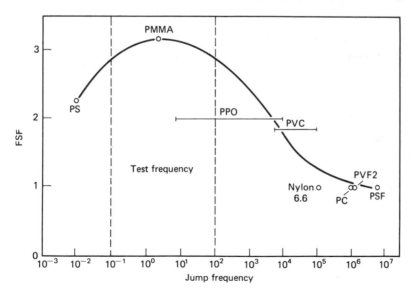

FIGURE 13.68   Relation between FCP frequency sensitivity and the room temperature jump frequency for several polymers.[213] (Reprinted with permission of the Society of Plastics Engineers.)

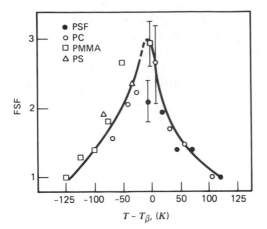

FIGURE 13.69 Frequency sensitivity factor (FSF) relative to normalized β-transition temperature $T - T_\beta$. (From Hertzberg et al.[216]; reproduced from R. W. Hertzberg et al., *Polymer* 19, 359 (1978), by permission of the publishers, Butterworth & Co. Ltd.)

measurements obtained with an infrared microscope and thermocouples revealed crack-tip temperatures in the range of 130°C and substantial heating throughout the specimen's unbroken ligament.[214,219] Such major temperature rises in the specimen decrease the specimen stiffness and enhance damage accumulation (recall Section 12.2.5). It is seen then that the antipodal behavior of rubber-toughened nylon 66 with that of PMMA, PVC, or polystyrene reflects a different balance between gross hysteretic heating (which lowers the elastic modulus overall) and localized crack-tip heating (which involves crack-tip blunting). Materials like PC and PSF, which do not reveal FCP rate frequency sensitivity over a large ΔK range at room temperature, exhibited no significant localized crack-tip heating.[220]

13.8.2 Fracture Surface Micromorphology

At least two distinctly different sets of parallel markings have been found on the fatigue fracture surfaces of amorphous plastics, such as PMMA, PS, and PC. At relatively large ΔK levels, striations are found that correspond to the incremental advance of the crack as a result of one load cycle[221] (Fig. 13.70a). Similar markings

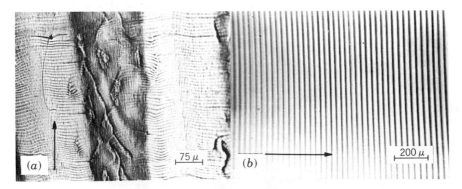

FIGURE 13.70 Fatigue fracture surface markings in amorphous plastics. (*a*) Striations associated with crack advance during one load cycle; (*b*) discontinuous growth bands equal in size to crack-tip plastic zone. Arrow indicates crack direction.

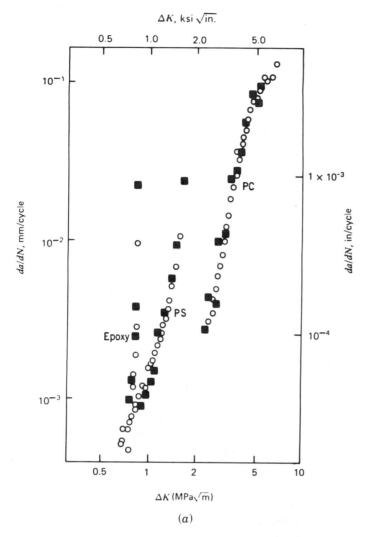

(a)

FIGURE 13.71   (a) Macroscopic ($\bigcirc$) and microscopic ($\blacksquare$) fatigue crack growth in epoxy, polystyrene, and polycarbonate.[221] (b) Size of discontinuous growth bands as function of ΔK in polystyrene.[223]

have been reported for rubber.[222] The dependence of fatigue striation spacing on the stress intensity factor range and the excellent correlation with associated macroscopic crack growth rates in epoxy, PS, and PC may be seen in Fig. 13.71a.[221] The essentially exact correlation between macroscopic and microscopic growth rates reflects the fact that 100% of the fracture surface in this ΔK regime is striated; that is, only one micromechanism is operative. Contrast this with the results for metals, where several micromechanisms operate simultaneously (Eq. 13-17). In the latter case, the two measurements of crack growth rate do not always agree (see Fig. 13.57b).

The other sets of parallel fatigue markings have been found at low ΔK levels and at high test frequencies in PS, PC, PSF, PMMA, and at all stress intensity levels in

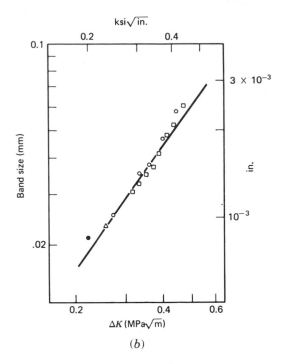

FIGURE 13.71 (*Continued*)

PVC.[192,193,221–227] These bands (Fig. 13.70*b*) are too large to be caused by the incremental extension of the crack during one loading cycle. Instead, they correspond to discontinuous crack advance following several hundred loading cycles during which the crack tip remains stationary. The fatigue fracture sequence that produces these markings is shown in Fig. 13.72*a*. The plastic zone—actually a long, thin craze—is seen to grow continuously, although it is characterized by a decreasing rate with increasing craze length. When the craze reaches a critical length, the crack advances abruptly across the entire craze and arrests. With further cycling, a new craze is developed and the process is repeated. The sequence involving continuous craze growth and discontinuous crack growth is modeled in Fig. 13.72*b*. Close examination of the fracture surface reveals that the growth bands consist of equiaxed dimples, which decrease in diameter from the beginning to the end of each band (Fig. 13.73). The variable dimple size is believed to reflect the void size distribution within the craze prior to crack instability; it also parallels the extent of crack opening displacement with increasing distance from the crack tip.[224]

The size of these bands increases with ΔK (Fig. 13.71*b*) and corresponds to the dimension of the crack-tip plastic zone[223,225,226] as computed by the Dugdale plastic strip model

$$R \approx \frac{\pi}{8} \frac{K^2}{\sigma_{ys}^2}$$

(8-48)

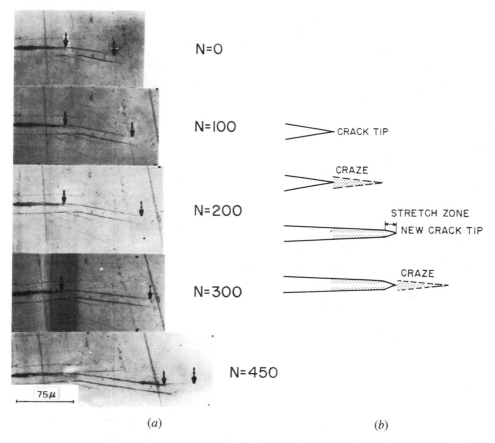

N=0

N=100

N=200

N=300

N=450

75μ

CRACK TIP

CRAZE

STRETCH ZONE

NEW CRACK TIP

CRAZE

(a)   (b)

FIGURE 13.72 (a) **Composite micrograph revealing position of craze ($\downarrow$) and crack ($\downarrow$) tip after fixed cyclic increments in PVC.**[214] **(Reproduced with permission from *Journal of Materials Science* 8 (1973).) (b) Schema showing model of discontinuous cracking process.**

This calculation can be reversed to compute the yield strength that controls the crack-tip deformation process. By setting R equal to the band width for a given ΔK level, an inferred yield strength was determined in PS and other materials and found equal to the craze stress for that material. The number of loading cycles required for craze development and sudden breakdown is estimated by dividing the bandwidth dimension by the corresponding macroscopic crack growth rate; depending on the material and ΔK level, band stability can extend from 10^2 to 10^5 loading cycles (Fig. 13.74).

In summary, two different sets of fatigue markings may be found on the fracture surfaces of the same plastic, each band corresponding to either one or as many as 100,000 load cycles. Without accurate macroscopic growth rate data or fractographic analysis, it is difficult to distinguish between them. To do so is highly desirable, since the markings would provide valuable information concerning the number of cycles associated with the fatigue crack propagation process.

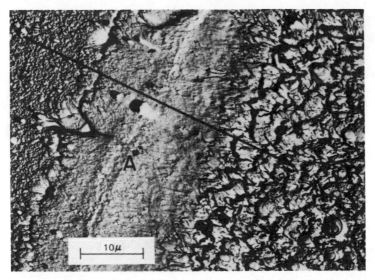

FIGURE 13.73 Transmission electron fractograph showing variable equiaxed dimple size in PVC discontinuous growth bands. Largest dimples are found near beginning of band. Region A is the arrest line between bands. Arrow indicates crack propagation direction.[214] (Reproduced with permission from *Journal of Materials Science* **8** (1973).)

13.9 FATIGUE CRACK PROPAGATION IN CERAMICS

Scientific thought, including that expressed in earlier editions of this text, once held that without an aggressive environment, ceramic materials were not subject to pure cyclic damage. This view was based mainly on the negligible amount of plastic deformation capacity believed to be available at the crack tip within simple ceramic microstructures. Recent studies,[229–241] however, have demonstrated clearly that fatigue damage does occur under both cyclic tensile and compressive loading conditions and that stable FCP takes place. Figure 13.75 shows representative crack growth data for transformation-toughened, whisker-reinforced, and single-phase ceramics and composites; typical data for high-strength aluminum and steel alloys are added for comparative purposes. For these and other reported results, brittle solids reveal a trend toward higher ΔK_{th} values with increasing K_c (i.e., $\Delta K_{th}/K_c \approx 0.6$) (see Table 13.3). Furthermore, a growing literature in ceramic and ceramic composites reveals a Paris-type relation (Eq. 13-3) with a crack growth rate-ΔK dependence between 15 and 42 (see Table 13.3) as compared with 2–4 for the case of monolithic metal alloys. If one were to compute the change in fatigue lifetime for a metal versus ceramic component in association with a twofold increase in nominal stress level (recall Eq. 13-6), one would find lifetime decreasing by factors of 16 and 4×10^{12}, respectively!* Alternatively, a 16-fold decrease in fatigue lifetime in a ceramic (the same as that for the

---

* Values of $m = 50$–100 have also been reported. (See R. H. Dauskardt, R. O. Ritchie, and B. N. Cox, *Advanced Materials & Processes,* **144**(2), 30 (1993). Accordingly, a twofold increase in stress would decrease projected lifetime by as much as 30 orders of magnitude!

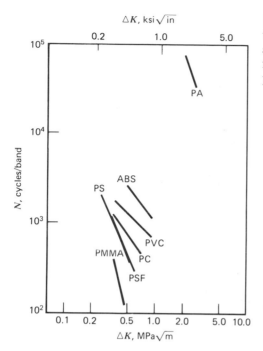

FIGURE 13.74   Cyclic stability of discontinuous growth bands in several polymers as function of ΔK. (From Hertzberg et al.[228] from *J. Mater. Sci.* 13, 1038 (1978); with permission from Chapman and Hall.)

metal alloy but assuming $m = 42$) would correspond to less than a 7% increase in stress level, hardly discernible within the range of residual stresses present in ceramic components. As a result, one must anticipate highly unreliable fatigue life predictions in ceramic components.

When assessing component lifetime, it is important to recognize that cyclic-loading induced damage accumulation in brittle solids can lead to much higher crack growth rates (i.e., shorter lifetimes) than those associated with environmental cracking under sustained loading conditions (i.e., static fatigue); for example, when pyrolitic-carbon coated graphite is tested in Ringer's solution, crack growth rates are much greater when cyclic rather than static loads are applied (Fig. 13.76a). Figure 13.76b shows a time-corrected comparison of cyclic versus static fatigue crack growth in Al_2O_3-28v/oSiC$_w$ in ambient air. Note that cyclic fatigue rates far exceed those of the statically loaded composite but are slower than static fatigue rates in monolithic aluminum oxide.

The crack growth resistance in ceramics and their composites depends strongly on crack-tip shielding mechanisms (e.g., transformation toughening in partially stabilized zirconia systems and whisker/fiber bridging in reinforced composites). To illustrate, Steffen et al.[232] accounted for the transformation-toughened shielding contribution on ΔK_{tip} ($K - K_s$), where K_s is found from Eq. 10-1; as a result, they were able to significantly normalize FCP data in a Mg-PSZ alloy subjected to different heat treatments[232] (Fig. 13.77).

Given the presence of assorted shielding mechanisms in toughened and/or reinforced ceramics, one may expect that short crack behavior will occur.[232,233] As discussed in Section 13.4.2.1, this arises from the limited wake in small cracks, which

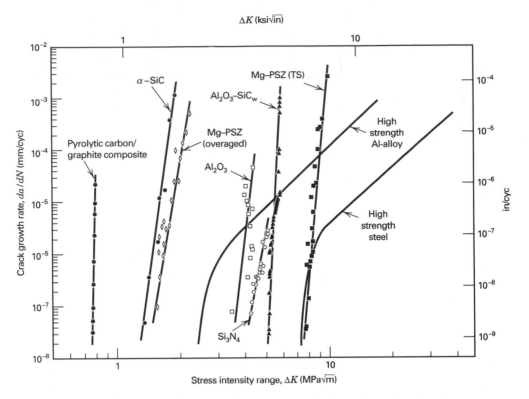

FIGURE 13.75 Fatigue crack propagation response for transformation-toughened, whisker-reinforced, and single phase ceramics and composites.[230] Data for high strength aluminum and steel alloys are included for comparative purposes. Apart from the threshold region, note that *da/dN*-dependence on ΔK is much greater in ceramics compared to metallic alloys. Reprinted from *Acta Metallurgica et Materialia*, **41** (9), R. H. Dauskardt, 2765 (1993) with kind permission from Elsevier Science Ltd., The Boulevard, Langford Lane, Kidlington OX5 1GB, UK.

minimizes the crack closure level. An example of differences in long versus short crack behavior is shown in Fig. 13.78 for the case of the Al_2O_3-SiC_w composite. Note that small cracks, introduced from microindentations, grew at ΔK levels about 2–3 times lower than ΔK_{th} values associated with long crack test results. In addition, short crack growth rates tended to *decrease* with increasing stress intensity level. These differences in growth rate behavior between long and short cracks in this ceramic composite were largely attributed to residual stresses introduced into the short crack sample during the indent process. When the crack-tip K level was corrected for residual stresses, the data were normalized; the latter suggests that a stabilized level of shielding was achieved after a limited advance of the indent crack.

Recent studies[229,231] have identified a large number of crack growth mechanisms in ceramics and composites that relate to both intrinsic crack-tip-microstructural interactions and extrinsic crack-tip shielding mechanisms (Fig. 13.79) (also recall Section 10.1.1). A growing number of studies (e.g., see references 230 and 231) have

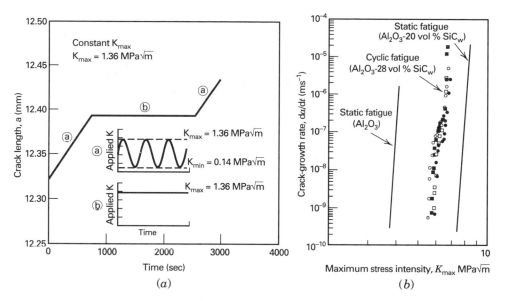

FIGURE 13.76 Comparison of crack propagation rates under sustained and cyclic loading conditions. (*a*) Sustained- vs. cyclic-loading (R=0.1 (50 Hz)) response in pyrolytic-carbon coated graphite tested in Ringer's solution at 37° C. Note much higher rates under cyclic loading conditions (region a) than under sustained loading conditions (region b).[229] (R.O. Ritchie and R. H. Dauskardt, *J. Ceram. Soc. Japan*, 99(10), 1047 (1991). Reprinted with permission of Fuji Technology Press, Ltd.); (*b*) time-corrected comparison of static vs. cyclic loading rates in Al_2O_3 and Al_2O_3-SiC composites. Composite with cyclic loading exhibits higher crack growth rates than with statically loaded conditions but lower growth rates than with unreinforced Al_2O_3.[234] (R. H. Dauskardt, B. J. Dalgleish, D. Yao, R. O. Ritchie, and P. F. Becker, *Journal of Materials Science*, 28, 3258 (1993). Reprinted by permission.)

shown cyclic-induced frictional wear of bridging zones and crushing of asperities on interlocking interfaces to be responsible for the major portion of the degradation process (see Fig. 13.79a2). With progressive wear, crack tip shielding is attenuated, whereas the effective stress intensity factor increases along with the crack growth rate.

Suresh and coworkers[236,237,241] have also confirmed the existence of crack growth in polycrystalline alumina under cyclic compression. It is believed that residual tensile stresses, induced at the notch tip by the applied compressive loads, contributes to the accumulation of damage associated with grain boundary failure. Crack growth rates were found to decrease with increasing crack length before arresting (Fig. 13.80). An examination of the crack wake revealed the development of debris that enhanced closure and contact between the mating fracture surfaces. When the debris was periodically removed by ultrasonic cleaning, crack growth rates increased (Curve *A*) relative to that associated with the retention of crack wake debris (Curve *B*).

The existence of fatigue damage accumulation in ceramic and ceramic composites under cyclic compressive loading raises considerable concern regarding life prediction for such brittle solids. That is, brittle solids are up to ten times stronger in uniaxial compression than in tension and, therefore, are believed to be much safer when loaded

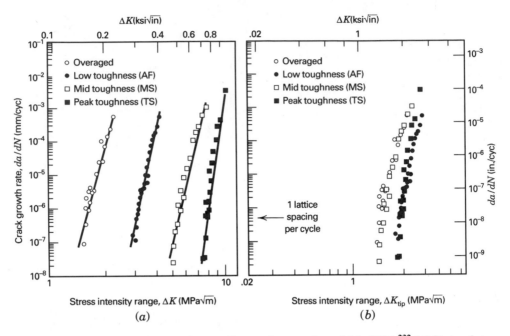

FIGURE 13.77 FCP response in transformation-toughened Mg-PSZ.[232] (*a*) Data plotted vs. ΔK; (*b*) data plotted vs. ΔK_{tip} ($K_{max} - K_s$). Note that FCP resistance increases with magnitude of transformation toughening. (A. A. Steffen, R. H. Dauskardt, and R. O. Ritchie, *J. Amer. Ceram. Soc.*, 76(6), 1259 (1991) Reprinted by permission.)

in compression than in tension. However, it has been shown[236–241] that subcritical crack growth occurs in the presence of stress concentrations with cyclic compressive stress levels far below the material's compressive strength; such results have been reported for the case of monolithic, transformable, and reinforced ceramics. Therefore, a reliable design methodology for ceramics and ceramic composites must recognize the potential for subcritical crack growth under *both* cyclic compression and tensile loading as well as static loading conditions.

13.10 FATIGUE CRACK PROPAGATION IN COMPOSITES

As might be expected, the addition of reinforcing fibers and whiskers leads to a reduction in FCP rates for a given composite material. This results from the reduction in cyclic strain within the matrix (characterized by $\Delta K/E$ (recall Fig. 13.59*b*)) and the transfer of cyclic loads to the fibers. Furthermore, as cracking proceeds within the matrix, unbroken fibers remain behind the advancing crack front and restrict crack opening. This crack-tip shielding mechanism, involving fiber bridging (recall Fig. 10.3), leads to vastly reduced FCP rates as demonstrated in Fig. 13.81 for the case of the ARALL hybrid composite[242–244] (recall Fig. 4.21*a*). The conventional FCP behavior of 7075-T6 is contrasted with the behavior of a 3/2-ARALL hybrid composite

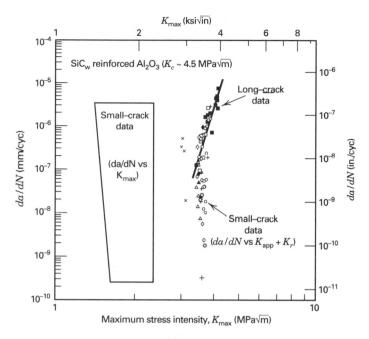

FIGURE 13.78 Short- and long-crack growth rate data in Al_2O_3-SiC composite.[233] Short crack growth data in box corresponds to applied ΔK test conditions whereas individual short crack datum refer to ΔK values corrected for residual stresses generated during the indentation process. (R. H. Dauskardt, M. R. James, J. R. Porter, and R. O. Ritchie, *J. Amer. Ceram. Soc.* **75**(4), 759 (1992) Reprinted by permission.)

containing three layers of 7075-T6 aluminum alloy separated by two layers of an adhesive/aramid fiber composite. The crack growth resistance of prestrained ARALL is of special interest in that a 1000-fold reduction in crack growth rates is achieved from that of the unreinforced aluminum alloy matrix. In this instance, when the composite is prestrained into the elastic–plastic regime (elastic aramid fiber response versus plastic deformation in the aluminum alloy) a significant residual compressive stress is developed in the aluminum layers upon removal of the prestrain load. As a result, crack initiation and propagation resistance in the composite increases dramatically. In fact, crack growth rates are found to *decrease* with increasing crack length in the aluminum layers as more aramid fibers bridge the crack wake. The overall efficiency of this crack bridging mechanism depends on the rate of damage accumulation within the adhesive/aramid fiber layers, resulting from adhesive shear deformation and delamination; thinner layers and a greater number of laminates per unit thickness will reduce damage accumulation by these processes and suppress FCP rates. Recent studies have shown that the length of the crack-tip bridging zone is on the order of 3–5 mm.[245] Since ARALL composites are lighter than aluminum sheets of the same thickness and possess vastly improved FCP resistance while maintaining the bending, milling, drilling, riveting, and bolting characteristics of conventional

Mechanisms of Cyclic Fatigue Crack Growth in Ceramics

a) Extrinsic mechanisms

1. Degradation of transformation toughening

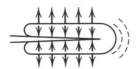

- degree of reversability of transformation
- cyclic accomodation of transformation strain
- cyclic modification of zone morphology

2. Damage to bridging zone
   - fraction and wear degradation of:

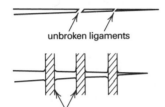

unbroken ligaments

whisker/fiber reinforcements

- crushing of asperities and interlocking zones

3. Fatigue of ductile reinforcing phase

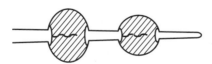

b) Intrinsic mechanisms

1. Accumulated (damage) localized microplasticity/microcracking

2. Mode II and III crack propagation on unloading

3. Crack tip blunting/resharpening

   a) Continuum

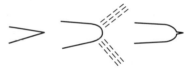

   b) Alternating shear

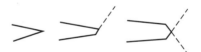

4. Relaxation of residual stresses

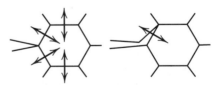

FIGURE 13.79 Examples of proposed fatigue crack advance mechanisms in polycrystalline ceramics and composites. (R. O. Ritchie, R. H. Dauskardt, W. Yu, and A. M. Brendzel, *J. Biomedical Materials*, **24**, 189 (1990). Reprinted by permission of John Wiley & Sons, Inc.)

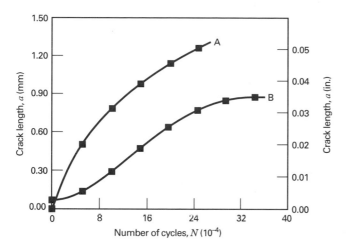

FIGURE 13.80 Fatigue crack growth in polycrystalline alumina under cyclic compression. Crack wake retained (Curve B) and removed (Curve A) to reveal influence of crack closure on crack extension.[236] (Data adapted from L. Ewart and S. Suresh, *J. Mater. Sci. Lett.*, **5**, 774 (1986).)

monolithic aluminum alloys, these hybrid composites are being considered for use in fatigue critical airplane components.

The action of these aramid fibers as numerous *microscopic* stiffeners within the laminate may be compared with the action of conventional *macroscopic* stiffeners used in large panel structures.[14] These stiffeners serve to take up the load from the main panel, as the crack passes, thereby reducing the ΔK level at the crack tip. The associated reduction in crack growth rate is readily apparent each time the crack in the main panel passes beneath a panel stiffener (recall Fig. 13.7).

Chan and Davidson[246] recently completed a review of the fatigue response of continuous fiber-reinforced metal matrix composites and showed that the prevailing fatigue fracture micromechanisms were strongly dependent on the respective properties of the matrix, interface, and fibers (Fig, 13.82). For the case of strong interfaces and weak fibers, failure is typically dominated by fiber and matrix fracture events. Conversely, when strong fibers are weakly bonded to the matrix, crack branching occurs as the crack moves through the matrix. Furthermore, crack bridging occurs when the latter scenario is present along with residual compressive stresses at fiber-matrix interfaces. As discussed previously, crack branching and crack bridging lead to enhanced FCP resistance.

Studies of fatigue damage accumulation processes in short-fiber reinforced plastics have identified the need to redefine the crack length and the prevailing ΔK level at the crack tip.[247–250] First, the crack is not a simple entity that can be directly characterized in terms of its length. Instead, the crack should be viewed as a main crack surrounded by many secondary cracks that lie away from the plane of the main crack; the crack-tip region, therefore, contains many microcracks with overall crack exten-

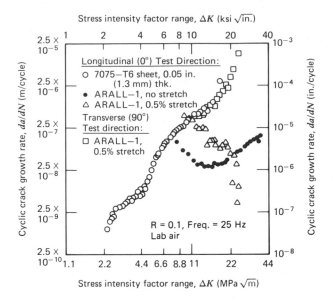

FIGURE 13.81   FCP response in prestrained and unprestrained ARALL hybrid composite as compared with that of 7075-T6 base metal. Note attenuated crack growth due to presence of aramid fibers. (With permission from R. J. Bucci, Alcoa Technology Center.)

sion characterized by the development and growth of this diffuse damage zone. In addition to microcrack-induced crack-tip zone shielding, the crack tip is also influenced by the bridging of fibers across the crack surfaces. Both microcracking and fiber bridging in the crack-tip region shield the crack tip from the full influence of the applied loads (recall Fig. 10.3), thereby reducing the effective ΔK level. The prevailing ΔK level is also profoundly affected by out-of-plane growth of the crack (e.g., crack deflection parallel to the fiber axis and stress direction).

The improvement in FCP resistance of injection-molded nylon 66 with the addition of short glass fibers is shown clearly in Fig. 13.83[248] and may be traced to the following: (1) load transfer from the matrix to the much stronger fibers along with the overall increase in specimen stiffness; (2) additional energy dissipation mechanisms associated with fiber debonding and pullout, and local plastic deformation in the matrix around the fibers; and (3) reduction in the effective ΔK level in associaton with microcracking and fiber bridging–crack-tip shielding.

The influence of fiber orientation on FCP rates deserves additional comment. One would typically expect crack growth resistance to be greater in samples containing fibers oriented parallel to the stress axis (i.e., perpendicular to the anticipated crack plane) than in the plane of the notch. The situation is complex for the case of injection-molded aparts since fibers are oriented parallel to the injection-molding direction along the mold surfaces and transverse to the molding direction in the core region because of 'the divergent flow pattern in the middle of the component (recall Fig. 6.37). Depending on the thickness of the two skin layers of the injection-molded

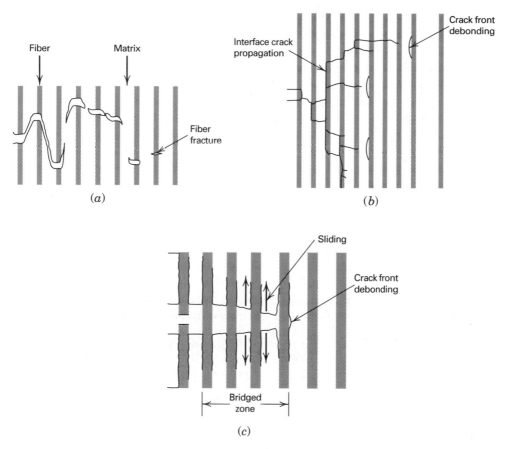

FIGURE 13.82   **Fatigue fracture micromechanisms in metal-matrix composites associated with conditions of:** (*a*) **strong interfaces and weak fibers—fiber fracture dominated;** (*b*) **strong fibers and weak interfaces—interfaces decohesion dominated; and** (*c*) **strong fibers and weak interfaces, coupled with residual compressive stresses at fiber-matrix interfaces—fiber bridging dominated.**[246] **(K. S. Chan and D. L. Davidson,** *Proc. of Engineering Foundation, International Conference on Fatigue of Advanced Materials.* **R. O. Ritchie, R. H. Dauskardt, and B. N. Cox, eds., 1991. Reprinted by permission.)**

plaque relative to that of the core region, FCP resistance of the material may be superior when loaded either parallel or perpendicular to the injection molding direction.[247–250] For example, the same FCP rates in longitudinal- and transverse-oriented samples containing 18 v/o glass fibers implies a microstructural balance, whereas 31 v/o glass fiber FCP results imply domination of the skin layers in the determination of FCP resistance of this composite (Fig. 13.79). Furthermore, the influence of orientation on FCP behavior depends on the thickness of the injection-molded plaque in that the relative influence of the surface layers will vary[251]; the FCP resistance in the L and T directions should decrease and increase, respectively, with increasing plaque thickness.

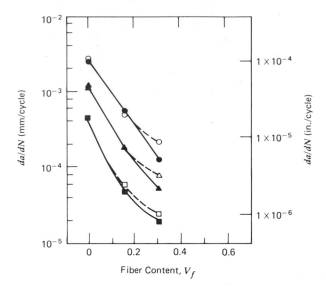

FIGURE 13.83 **Effect of fiber volume fraction v_f on fatigue crack growth rates in sgfr Nylon 66 at various levels of ΔK for two specimen orientations (frequency = 10 Hz, minimum/maximum load ratio = 0.1). ΔK = 4.5 MPa $\sqrt{\text{m}}$: ●, L direction; ○, T direction. ΔK = 4.0 MPa $\sqrt{\text{m}}$: ▲, L direction; △, T direction. ΔK = 3.5 MPa $\sqrt{\text{m}}$: ■, L direction; □, T direction, (Reprinted with permission, R. W. Lang, J. A. Manson, and R. W. Hertzberg in *Polymer Blends and Composites in Multiphase Systems*, C. D. Hans, ed., *Advances in Chemistry Series, No. 206*, 1984, p. 261.)**

REFERENCES

1. H. W. Liu, *Appl. Mater. Res.* **3**(4), 229 (1964).
2. N. E. Frost, *J. Mech. Eng. Sci.* **1**(2), 151 (1959).
3. W. Weibull, *Acta Met.* **11**(7), 725 (1963).
4. P. C. Paris, *Fatigue—An Interdisciplinary Approach,* Proceedings, 10th Sagamore Conference, Syracuse University Press, Syracuse, NY, 1964, p. 107.
5. S. R. Swanson, F. Cicci, and W. Hoppe, ASTM *STP 415,* 1967, p. 312.
6. P. C. Paris and F. Erdogan, *J. Basic Eng. Trans. ASME, Series D* **85**(4), 528 (1963).
7. P. C. Paris and G. C. M. Sih, ASTM *STP 381,* 1965, p. 30.
8. A. Saxena, S. J. Hudak, Jr., J. K. Donald, and D. W. Schmidt, *J. Test Eval.* **6,** 167 (1978).
9. ASTM Standard E647-93, *1993 Annual Book of ASTM Standards,* Philadelphia, PA, 1993, p. 765.
10. *Fatigue Crack Growth Measurement and Data Analysis,* ASTM *STP 738,* S. J. Hudak, Jr., and R. J. Bucci, Eds., ASTM, Philadelphia, PA, 1981.
11. R. Jaccard, *Fatigue Life Prediction of Aluminum Structures Based on SN-Curve Simulation,* 2nd Int. Conf. on Aluminum Weldments, Aluminum-Verlag, Dusseldorf, Germany (1982).
12. J. M. Barsom, E. J. Imhof, and S. T. Rolfe, *Engl. Fract. Mech.* **2**(4), 301 (1971).
13. G. Miller, *Trans. Quart. ASM* **61,** 442 (1968).

14. C. C. Poe, AFFDL-TR-70-144, 207 (1970).
15. J. A. Harris, Jr., *Engine Component Retirement for Cause, Vol. I. Executive Summary,* AFWAL-TR-87-4069, Wright-Patterson Air Force Base, Ohio (1987).
16. P. J. E. Forsyth, *Acta Met.* **11**(7), 703 (1963).
17. D. O. Swenson, *J. Appl. Phys.* **40**, 3467 (1969).
18. R. W. Hertzberg and W. J. Mills, ASTM *STP 600,* 1976, p. 220.
19. L. B. Vogelesang, *Report LR-286,* Delft, The Netherlands, 1979.
20. G. T. Hahn, R. G. Hoagland, and A. R. Rosenfield, *Met. Trans.* **3**, 1189 (1972).
21. C. Bathias and R. M. Pelloux, *Met. Trans.* **4**(5), 1265, (1973).
22. J. C. McMillan and R. W. Hertzberg, ASTM *STP 436,* 1968, p. 89.
23. P. J. E. Forsyth and D. A. Ryder, *Metallurgia* **63**, 117 (1961).
24. R. W. Hertzberg, ASTM *STP 415,* 1967, p. 205.
25. C. Laird and G. C. Smith, *Philos. Mag.* **7**, 847 (1962).
26. R. M. Pelloux, *Trans. Quart. ASM* **62**(1), 281 (1969).
27. J. H. Weber and R. W. Hertzberg, *Met. Trans.* **4**, 595 (1973).
28. R. W. Hertzberg and P. C. Paris, Proceedings, International Fracture Conference, Sendai, Japan, Vol. 1, 1965, p. 459.
29. R. C. Bates and W. G. Clark, Jr., *Trans. Quart. ASM* **62**(2), 380 (1969).
30. R. J. Stofanak, R. W. Hertzberg, and R. Jaccard, *Int. J. Fract.* **20**, R145 (1982).
31. C. M. Hudson, NASA Tech. Note *D-5390,* 1969.
32. R. G. Forman, V. E. Kearney, and R. M. Engle, *J. Basic Eng. Trans. ASME* **89**, 459 (1967).
33. R. Roberts and F. Erdogan, *J. Basic Eng. Trans. ASME* **89**, 885 (1967).
34. W. Elber, ASTM *STP 486,* 1971, p. 230.
35. W. Illg and A. J. McEvily, Jr., NASA *TND-52,* 1959.
36. J. C. Newman, Jr., Ph.D. dissertation, VPI, Blacksburg, VA, May 1974.
37. M. Katcher and M. Kaplan, ASTM *STP 559,* 1974, p. 264.
38. H. P. Chu, *Trans. Quart. ASM* **62**(2), 380 (1974).
39. N. F. Boutle and W. D. Dover, *Fracture 1977* **2**, 1065 (1977).
40. A. Clerivet and C. Bathias, *Eng. Fract. Mech.* **12**, 599 (1979).
41. R. D. Brown and J. Weertman, *Eng. Fract. Mech.* **10**, 757 (1978).
42. T. C. Lindley and C. E. Richards, *Mater. Sci. Eng.* **14**, 281 (1974).
43. N. Fleck, Cambridge University, Report TR.89, January 1982.
44. R. W. Hertzberg and E. F. J. vonEuw, *Int. J. Fract. Mech.* **7**, 349 (1971).
45. D. Broek and J. Schijve, National Aeronautical and Astronautical Research Institute, Amsterdam, Holland, NLR-TR-M 2.111, 1963.
46. J. Schijve, ASTM *STP 415,* 1967, p. 415.
47. H. Kobayashi, H. Nakamura, and H. Nakazawa, *Recent Research on Mechanical Behavior of Solids,* Univ. Tokyo Press, Tokyo, 1979, p. 341.
48. R. W. Hertzberg, *Int. J. Fract.* **15**, R69 (1979).
49. R. J. Bucci, P. C. Paris, R. W. Hertzberg, R. A. Schmidt, and A. F. Anderson, ASTM *STP 513,* 1972, p. 125.
50. R. J. Bucci, W. G. Clark, Jr., and P. C. Paris, ASTM *STP 513,* 1972, p. 177.
51. M. Klesnil and P. Lukas, *Mater. Sci. Eng.* **9**, 231 (1972).
52. R. J. Stofanak, R. W. Hertzberg, G. Miller, R. Jacard, and J. K. Donald, *Eng. Fract. Mech.* **17**(6), 527 (1983).
53. R. O. Ritchie, *J. Eng. Mater. Tec.* **99**, 195 (1977).

54. J. L. Robinson and C. J. Beevers, *Met. Sci.* **7,** 153, (1973).
55. R. J. Cooke and C. J. Beevers, *Mater. Sci. Eng.* **13,** 201 (1974).
56. R. J. Cooke, P. E. Irving, C. S. Booth, and C. J. Beevers, *Eng. Fract. Mech.* **7,** 69 (1975).
57. N. Walker and C. J. Beevers, *Fat. Eng. Mater. Struct.* **1,** 135 (1979).
58. K. Minakawa and A. J. McEvily, *Scripta Met.* **15,** 663 (1981).
59. S. Suresh and R. O. Ritchie, *Met. Trans.* **13A,** 1627 (1982).
60. R. J. Stofanak, R. W. Hertzberg, G. Miller, R. Jaccard, and K. Donald, *Eng. Fract. Mech.* **17**(6), 527 (1983).
61. K. Minakawa, G. Levan, and A. J. McEvily, *Met. Trans.* **17A,** 1787 (1986).
62. S. Suresh, G. F. Zamiski, and R. O. Ritchie, *Met. Trans.* **12A,** 1435 (1981).
63. S. Suresh, D. M. Parks, and R. O. Ritchie, *Fatigue Thresholds,* J. Bäcklund, A. F. Blom, and C. J. Beevers, Eds., EMAS Publ. Ltd., Warley, England, 1982.
64. L. K. L. Tu and B. B. Seth, *J. Test. Eval.* **6**(1), 66 (1978).
65. ASTM Committee E-24 Round-Robin Study on Closure Measurements, Bal Harbour, Florida, November 10, 1987.
66. R. S. Vecchio, J. Crompton, and R. W. Hertzberg, *Int. J. Fract.* **31**(2), R29 (1986).
67. C. H. Newton and R. W. Hertzberg, ASTM *STP 982,* 139 (1988).
68. J. K. Donald, *op cit.,* p. 222.
69. R. S. Vecchio, Ph.D. dissertation, Lehigh University, Bethlehem, PA, 1985.
70. H. Kitagawa and S. Takahashi, *Trans. Japan Soc. Mech. Eng.* **45,** 1289 (1979).
71. H. Ohuchida, S. Usami, and A. Nishioka, *Trans. Japan Soc. Mech. Eng.* **41,** 703 (1975).
72. M. H. El Haddad, K. N. Smith, and T. H. Topper, *Trans. ASME* **101H,** 42 (1979).
73. S. Taira, K. Tanaka, and M. Hoshina, ASTM *STP 675,* 1979, p. 135.
74. Y. Nakai and K. Tanaka, Proc. 23rd Japan Cong. Mater. Res., 1980, p. 106.
75. K. Tanaka, Y. Nakai, and M. Yamashita, *Int. J. Fract.* **17,** 519 (1981).
76. H. Kitagawa and S. Takahashi, Proc. 2nd. Int. Conf. on Mech. Beh. Maters., ASM, Metals Park, Ohio, 1976, p. 627.
77. M. H. El Haddad, T. H. Topper, and K. N. Smith, *Eng. Fract. Mech.* **11,** 573 (1979).
78. N. E. Dowling, ASTM *STP 637,* 1977, p. 637.
79. N. E. Frost, *J. Mech. Eng. Sci.* **5,** 15 (1963).
80. N. E. Frost, *J. Mech. Eng. Sci.* **6,** 203 (1964).
81. D. Broek and B. N. Leis, *Materials, Experimentation and Design in Fatigue,* F. Sherratt and J. B. Sturgeon, Eds., Westbury House, Guildford, 1981, p. 129.
82. J. Schijve, *Fatigue Thresholds,* Vol. 2, EMAS Ltd., West Midlands, England, 1982, p. 881.
83. R. O. Ritchie and S. Suresh, *Mater. Sci. Eng.* **57,** 1983, L27.
84. J. C. Newman, AGARD Conf. Proc. No. 328, paper 6, 1983.
85. C. H. Newton, Ph.D. Dissertation, Lehigh University, Bethlehem, PA, 1984.
86. R. O. Ritchie and J. Lankford, Eds., *Small Fatigue Cracks,* AIME, Warrendale, PA, 1986.
87. K. J. Miller and E. R. de los Rios, Eds., *The Behavior of Short Fatigue Cracks,* Mechanical Engineering Publications Ltd., London, 1986.
88. S. Suresh and R. O. Ritchie, *Int. Metals Rev.* **29**(6), 445 (1984).
89. B. N. Leis and T. P. Forte, ASTM *STP 743,* 1981, p. 100.

90. E. R. de los Rios, E. Z. Tang, and K. J. Miller, *Fat. Eng. Mater. Struct.* **7**(2), 97 (1984).
91. P. K. Liaw and W. A. Logsdon, *Eng. Fract. Mech.* 22(1), 115 (1984).
92. D. E. Castro, G. Marci, and D. Munz, *Fat. Fract. Eng. Mater.,* in press.
93. H. Doker and M. Peters, *Fatigue 84,* C. Beevers, Eds., EMAS Ltd., West Midlands, England, 1984, p. 275.
94. H. Doker and G. Marci, *Int. J. Fat.* **5**(4), 187 (1983).
95. W. A. Herman, R. W. Hertzberg, and R. Jaccard, *J. Fat. Fract. Eng. Mater. Struct.,* **11**(4), 303 (1988).
96. W. A. Herman, R. W. Hertzberg, C. H. Newton, and R. Jaccard, *Fatigue 87,* R. O. Ritchie and E. A. Starke, Jr., Eds., 2, EMAS Ltd., West Midlands, England, 1987, p. 819.
97. R. W. Hertzberg, W. A. Herman, T. R. Clark, and R. Jaccard, *ASTM 1149,* J. M. Larsen and J. E. Allison, Eds., ASTM Philadelphia (1992) 197.
98. R. Jaccard, 3rd Int. Conf. Alum. Weld., Munich, FRG, 1985.
99. R. Jaccard, 2nd Int. Conf. Alum. Weld., Dusseldorf, FRG, 1982.
100. M. H. El Haddad, K. N. Smith, and T. H. Topper, ASTM *STP 677,* 1979, p. 274.
101. R. A. Smith and K. J. Miller, *Int. J. Mech. Sci.* **20,** 201 (1978).
102. M. M. Hammouda, R. A. Smith, and K. J. Miller, *Fat. Eng. Mater. Struct.* **2,** 139 (1979).
103. N. E. Frost, *J. Mech. Eng. Sci.* **2**(2), 109 (1960).
104. J. Schijve and D. Broek, *Aircr. Eng.* **34,** 314 (1962).
105. J. Schijve, D. Broek, and P. deRijk, NLR Report *M2094,* Jan. 1962.
106. E. F. J. vonEuw, R. W. Hertzberg, and R. Roberts, ASTM *STP 513,* 1972, p. 230.
107. V. W. Trebules, Jr., R. Roberts, and R. W. Hertzberg, ASTM *STP 536,* 1973, p. 115.
108. W. J. Mills and R. W. Hertzberg, *Eng. Fract. Mech.* **7,** 705 (1975).
109. W. J. Mills and R. W. Hertzberg, *Eng. Fract. Mech.* **8,** 657 (1976).
110. W. J. Mills, R. W. Hertzberg, and R. Roberts, ASTM *STP 637,* 1977, p. 192.
111. R. S. Vecchio, R. W. Hertzberg, and R. Jaccard, *Scripta Met.* **17,** 343 (1983).
112. R. S. Vecchio, R. W. Hertzberg, and R. Jaccard, *Fat. Eng. Mater. Struct.* **7**(3), 181 (1984).
113. C. H. Newton, R. S. Vecchio, R. W. Hertzberg, and R. Jaccard, *Fatigue Crack Growth Threshold Concepts,* AIME, Warrendale, PA, 1984, p. 379.
114. J. Schijve, F. A. Jacobs, and P. J. Tromp, NLR *TR 69050 U,* June 1969.
115. R. I. Stephens, D. K. Chen, and B. W. Hom, ASTM *STP 595,* 1976, p. 27.
116. B. M. Hillberry, W. X. Alzos, and A. C. Skat, AFFDL-TR-75-96, Aug. 1975.
117. M. K. Himmelein and B. M. Hillberry, ASTM *STP 590,* 1975, p. 321.
118. E. Zaiken and R. O. Ritchie, *Eng. Fract. Mech.* **22**(1), 35 (1985).
119. W. S. Johnson, ASTM *STP 748,* 1981, p. 85.
120. J. B. deJonge and A. Nederveen, ASTM *STP 714,* 1980, p. 170.
121. H. I. McHenry, Ph.D. dissertation, Lehigh University, Bethlehem, PA (1970).
122. C. Ragazzo, R. W. Hertzberg, and R. Jaccard, ASTM *J. Test. Eval.* **23**(1), 19 (1995).
123. D. P. Williams III, P. S. Pao, and R. P. Wei, *Environment-Sensitive Fracture of Engineering Materials,* Z. A. Foroulis, Ed., AIME, Warrendale, PA, 1979, p. 3.

124. F. P. Ford, *Fatigue-Environment and Temperature Effects,* J. J. Burke and V. Weiss, Eds., Plenum, New York, 1983.
125. A. Hartman, F. J. Jacobs, A. Nederveen, and P. deRijk, NLR *TN/M 2182,* 1967.
126. A. Hartman, *Int. J. Fract. Mech.* **1**(3), 167 (1965).
127. R. P. Wei, *Eng. Fract. Mech.* **1,** 633 (1970).
128. R. J. Bucci, Ph.D. dissertation, Lehigh University, Bethlehem, PA, 1970.
129. J. M. Barsom, *Eng. Fract. Mech.* **3**(1), 15 (1971).
130. E. J. Imhof and J. M. Barsom, ASTM *STP 536,* 1973, p. 182.
131. L. F. Coffin, Jr., *Met. Trans.* **3,** 1777 (1972).
132. H. D. Solomon and L. F. Coffin, ASTM *STP 520,* 1973, p. 112.
133. M. W. Mahoney and N. E. Paton, *Nucl. Tech.* **23,** 290 (1974).
134. L. A. James and R. L. Knecht, *Met. Trans.* **6A,** 109 (1975).
135. H. W. Liu, and J. J. McGowan, AFWAL-TR-81-4036, June 1981.
136. R. P. Gangloff, *Res. Mech. Let.* **1,** 299 (1981).
137. R. P. Wei and J. D. Landes, *Mater. Res. Stand.* **9,** 25 (1969).
138. T. W. Weir, G. W. Simmons, R. G. Hart, and R. P. Wei, *Scripta Met.* **14,** 357 (1980).
139. R. P. Wei and G. W. Simmons, *Int. J. Fract.* **17,** 235 (1981).
140. P. E. Irving and C. J. Beevers, *Mater. Sci. Eng.* **14,** 229 (1974).
141. J. L. Robinson and C. J. Beevers, 2nd Inter. Conf. Titanium, Cambridge, England, 1972.
142. A. Yuen, S. W. Hopkins, G. R. Leverant, and C. A. Rau, *Met. Trans.* **5,** 1833 (1974).
143. M. F. Amateau, W. D. Hanna, and E. G. Kendall, *Mechanical Behavior, Proceedings of the International Conference on Mechanical Behavior of Materials,* Vol. 2, Soc. Mater. Sci., Japan, 1972, p. 77.
144. M. J. Harrigan, M. P. Kaplan, and A. W. Sommer, *Fracture Prevention and Control,* D. W. Hoeppner, Ed., Vol. 3, ASM, 1974, p. 225.
145. J. C. Chestnutt, C. G. Rhodes, and J. C. Williams, ASTM, *STP 600,* 1976, p. 99.
146. G. R. Yoder, L. A. Cooley, and T. W. Crooker, *Proc. 2nd Int. Conf. Mech. Beh. of Mater.,* ASM, Metals Park, OH, 1976, p. 1010.
147. G. R. Yoder, L. A. Cooley, and T. W. Crooker, *J. Eng. Mater. Tech.* **99,** 313 (1977).
148. G. R. Yoder, L. A. Cooley, and T. W. Crooker, *Met. Trans.* **8A,** 1737 (1977).
149. G. R. Yoder, L. A. Cooley, and T. W. Crooker, *Eng. Fract. mech.* **11,** 805 (1979).
150. G. R. Yoder, L. A. Cooley, and T. W. Crooker, *Trans. ASME* **101,** 86 (1979).
151. G. R. Yoder, L. A. Cooley, and T. W. Crooker, *Titanium 80,* H. Kimura and O. Izumi, Eds., Vol. 3, AIME, Warrendale, PA, 1980, p. 1865.
152. G. R. Yoder, L. A. Cooley, and T. W. Crooker, ASTM *STP 791,* 1983, p. I-348.
153. J. Masounave and J. P. Bailon, *Proc. 2nd Int. Conf. Mech. Beh. Master.,* ASM, Boston, 636 (1976).
154. J. Masounave and J. P. Bailon, *Scripta Met.* **10,** 165 (1976).
155. L. Wagner and J. K. Gregory, *Advanced Materials and Processes,* **146**(1), 50HH (1994).
156. K. T. Venkateswara Rao, W. Yu and R. O. Ritchie, submitted to *Met. Trans A.,* 1987.

157. K. T. Venkateswara Rao, W. Yu and R. O. Ritchie, submitted to *Met. Trans A.,* 1987.
158. R. O. Ritchie, *Metal Sci.* **11,** 368 (1977).
159. E. J. Prittle, *Fracture 1977,* **2** ICF 4, p. 1249.
160. G. Birkbeck, A. E. Inckle, and G. W. J. Waldron, *J. Mater. Sci.* **6,** 319 (1971).
161. T. C. Lindley, C. E. Richards, and R. O. Ritchie, *Metallurgia and Metal Forming,* 268 (Sept. 1976).
162. M. W. Lui and I. LeMay, *IMS Proceedings,* 1971, p. 227.
163. R. O. Ritchie, *Met. Trans.* **8A,** 1131 (July 1977).
164. R. D. Zipp and G. H. Walter, *Metallography,* **7,** 77 (1974).
165. I. LeMay and M. W. Lui, *Metallography,* **8,** 249 (1975).
166. G. W. J. Waldron, A. E. Inckle, and P. Fox, 3rd Scanning Electron Microscope Symposium, 1970, p. 299.
167. P. R. V. Evans, N. B. Owen, and B. E. Hopkins, *Eng. Frac. Mech.* **3,** 463 (1971).
168. P. E. Irving and C. J. Beevers, *Met. Trans.* **5,** 391 (1974).
169. G. Clark, A. C. Pickard, and J. F. Knott, *Engineering Fracture Mechanics,* Vol. 8, Pergamon, Oxford, England, 1976, pp. 449–451.
170. N. S. Cheruvu, *ASTM Symposium on Fractography in Failure Analysis of Metals and Ceramics,* ASTM, April 1982, Philadelphia.
171. P. S. Pao, W. Wei, and R. P. Wei, *Environment-Sensitive Fracture of Engineering Material,* Z. A. Fouroulis, Ed., AIME, Warrendale, PA, 1979, p. 3.
172. F. A. Heiser and R. W. Hertzberg, *J. Basic Eng. Trans. ASME* **93,** 71 (1971).
173. I. C. Mayes and T. J. Baker, *Fat. Eng. Mater. Struct.* **4**(1), 79 (1981).
174. R. C. Bates, Westinghouse Scientific Paper *69-1D9-RDAFC-P2,* 1969.
175. D. Broek, Paper 66, Second International Fracture Conference, Brighton, 1969, p. 754.
176. S. M. El-Soudani and R. M. Pelloux, *Met. Trans.* **4,** 519 (1973).
177. J. M. Barsom, *J. Eng. Ind., ASME, Series B* **92**(4), 1190 (1971).
178. J. H. Weber and R. W. Hertzberg, *Met. Trans.,* **2,** 3498 (1971).
179. B. Tomkins, *Philos. Mag.* **18,** 1041 (1968).
180. J. P. Hickerson, Jr., and R. W. Hertzberg, *Met. Trans.* **3,** 179 (1972).
181. S. Majumdar and J. D. Morrow, ASTM *STP 559,* 1974, p. 159.
182. S. D. Antolovich, A. Saxena, and G. R. Chanani, *Eng. Fract. Mech.* **7,** 649 (1975).
183. A. Saxena and S. D. Antolovich, *Met. Trans.* **6A,** 1809 (1975).
184. R. W. Hertzberg, *Met. Trans.* **5,** 306 (1974).
185. S. Pearson, *Nature* (*London*) **211,** 1077 (1966).
186. A. Ohta, N. Suzuki, and T. Mawari, *Int. J. Fat.,* **14**(4), 224 (1992).
187. R. W. Hertzberg, *Int. J. Frac.* **64**(3), 135 (1993).
188. R. W. Hertzberg, *J. Mater. Sci. Eng.,* **A190,** 25 (1995).
189. K. Sadananda and P. Shahinian, *Int. J. Fract.,* **13,** 585 (1977).
190. C. Yu and M. Yan, *Fat. Engng. Mater. Struct.* **3,** 189 (1980).
191. J. A. Manson and R. W. Hertzberg, *CRC Crit. Rev. Macromol. Sci.* **1**(4), 433 (1973).
192. R. W. Hertzberg and J. A. Manson, *Fatigue of Engineering Plastics,* Academic, New York, 1980.
193. R. W. Hertzberg and J. A. Manson, *Encyl. Polym. Sci. Eng.,* Vol. 7, 2d ed., Wiley, New York, 1986, p. 378.

194. T. Pecorini, Ph.D. Dissertation, Lehigh University, Bethlehem, PA 1992.
195. T. Pecorini and R. W. Hertzberg, *Polymer,* **34**(24), 5053 (1993).
196. T. Clark, Ph.D. Dissertation, Lehigh University, Bethlehem, PA 1993.
197. H. R. Azimi, Ph.D. Dissertation, Lehigh University, Bethlehem, PA 1994.
198. H. R. Azimi, R. A. Pearson, and R. W. Hertzberg, *J. Mater. Sci. Lett.,* **13,** 1460 (1994).
199. G. P. Koo, *Fluoropolymers, High Polymers,* Vol. 25, L. A. Wall, Ed., Wiley-Interscience, New York, 1972, p. 507.
200. G. P. Koo and L. G. Roldan, *J. Polym. Sci. Polym. Lett. Ed.* **10,** 1145 (1972).
201. G. Meinel and A. Peterlin, *J. Polym. Sci. Polym. Lett. Ed.* **9,** 67 (1971).
202. A. Ramirez, Ph.D. dissertation, Lehigh University, Bethlehem, PA, 1982.
203. C. M. Rimnac, J. A. Manson, R. W. Hertzberg, S. M. Webler, and M. D. Skibo, *J. Macromol. Sci., Phys.* **B19**(3), 351 (1981).
204. P. E. Bretz, R. W. Hertzberg, and J. A. Manson, *J. Appl. Polym. Sci.* **27,** 1707 (1982).
205. G. Pitman and I. M. Ward, *J. Mater. Sci.* **15,** 635 (1980).
206. F. X. de Charentenay, F. Laghouati, and J. Dewas, *Deformation Yield and Fracture of Polymers,* Plastics and Rubber Institute, Cambridge, England, 1979, p. 6.1.
207. S. L. Kim, M. D. Skibo, J. A. Manson, and R. W. Hertzberg, *Polym. Eng. Sci.* **17**(3), 194 (1977).
208. S. L. Kim, J. Janiszewski, M. D. Skibo, J. A. Manson, and R. W. Hertzberg, *ACS Org. Coatings Plast. Chem.* **38**(1), 317 (1978).
209. J. C. Michel, J. A. Manson, and R. W. Hertzberg, *Org. Coatings Plast. Chem.* **45,** 622 (1981).
210. J. C. Michel, J. A. Manson, and R. W. Hertzberg, *Polym. Prepr. Am. Chem. Soc. Div. Polym. Chem.* **26**(2), 141 (1985).
211. J. C. Michel, J. A. Manson, and R. W. Hertzberg, *Polymer* **25,** 1657 (1984).
212. M. D. Skibo, Ph.D. dissertation, Lehigh University, Bethlehem, PA, 1977.
213. R. W. Hertzberg, J. A. Manson, and M. D. Skibo, *Polym. Eng. Sci.* **15,** 252 (1975).
214. R. W. Hertzberg, M. D. Skibo, and J. A. Manson, ASTM *STP 700,* 1980, p. 49.
215. M. D. Skibo, R. W. Hertzberg, and J. A. Manson, *Fracture 1977,* **3,** 1127 (1977).
216. R. W. Hertzberg, J. A. Manson, and M. D. Skibo, *Polymer* **19,** 359 (1978).
217. R. W. Hertzberg, M. D. Skibo, J. A. Manson, and J. K. Donald, *J. Mater. Sci.* **14,** 1754 (1979).
218. J. D. Phillips, R. W. Hertzberg, and J. A. Manson, manuscript in preparation.
219. M. T. Hahn, R. W. Hertzberg, R. W. Lang, J. A. Manson, J. C. Michel, A. Ramirez, C. M. Rimnac, and S. M. Webler, *Deformation, Yield and Fracture of Polymers,* Plastics and Rubber Institute, Cambridge, England, 1982, p. 19.1.
220. R. W. Lang, unpublished research.
221. R. W. Hertzberg, M. D. Skibo, and J. A. Manson, ASTM *STP 675,* 1979, p. 471.
222. E. H. Andrews, *J. Appl. Phys.* **32**(3), 542 (1961).
223. M. D. Skibo, R. W. Hertzberg, and J. A. Manson, *J. Mater. Sci.* **11,** 479 (1976).
224. R. W. Hertzberg and J. A. Manson, *J. Mater. Sci.* **8,** 1554 (1973).

225. J. P. Elinck, J. C. Bauwens, and G. Homes, *Int. J. Fract. Mech.* **7**(3), 227 (1971).
226. C. M. Rimnac, R. W. Hertzberg, and J. A. Manson, ASTM *STP 733,* 1981, p. 291.
227. J. Janiszewski, R. W. Hertzberg, and J. A. Manson, ASTM *STP 743,* 1981, p. 125.
228. R. W. Hertzberg, M. D. Skibo, and J. A. Manson, *J. Mater. Sci.* **13,** 1038 (1978).
229. R. O. Ritchie and R. H. Dauskardt, *J. Ceram. Soc. Japan,* **99**(10), 1047 (1991).
230. R. H. Dauskardt, *Acta Metall. Mater.* **41**(9), 2765 (1993).
231. S. Lathabai, J. Rödel, and B. R. Lawn, *J. Amer. Ceram. Soc.* **74,** 1340 (1991).
232. A. A. Steffen, R. H. Dauskardt, and R. O. Ritchie, *J. Amer. Ceram. Soc.* **76**(6), 1259 (1991).
233. R. H. Dauskardt, M. R. James, J. R. Porter, and R. O. Ritchie, *J. Amer. Ceram. Soc.* **75**(4), 759 (1992).
234. R. H. Dauskardt, B. J. Dalgleish, D. Yao, R. O. Ritchie, and P. F. Becher, *J. Mater. Sci.* **28,** 3258 (1993).
235. T. Hoshide, T. Ohara, and T. Yamada, *Int. J. Fract.* **37,** 47 (1988).
236. L. Ewart and S. Suresh, *J. Mater, Sci, Lett.* **5,** 774 (1986).
237. L. Ewart and S. Suresh, *J. Mater, Sci,* **22,** 1173 (1987).
238. S. Suresh and J. R. Brockenbrough, *Act Met,* **36,** 1455 (1988).
239. S. Suresh, *Int. J. Fract.,* **42,** 41 (1990).
240. S. Suresh, *Fatigue of Materials,* Cambridge Solid State Science Series, Cambridge University Press, Cambridge, England, Chapter 13 (1991).
241. L. A. Sylva and S. Suresh, *J. Mater. Sci.* **24,** 1729 (1989).
242. R. Marissen, *Eng. Fract. Mech.* **19**(2), 261 (1984).
243. R. Marissen, *Int. Conf. Aero. Sci.,* Vol. 2.6.2, 1986, p. 801.
244. R. J. Bucci, L. N. Mueller, R. W. Schultz and J. L. Prohaska, 32nd Int. SAMPE Symp., Anaheim, CA, April 6, 1987.
245. R. O. Ritchie, W. Yu, and R. J. Bucci, *Engineering Fracture Mechanics* **32**(3), 361 (1989).
246. K. S. Chan and D. L. Davidson, *Proc. of Engineering Foundation, International Conference, Fatigue of Advanced Materials,* R. O. Ritchie, R. H. Dauskardt, and B. N. Cox, Eds., Santa Barbara, CA, 325 (1991).
247. R. W. Lang, J. A. Manson, and R. W. Hertzberg, *Polym. Eng. Sci.* **22**(15), 982 (1982).
248. R. W. Lang, J. A. Manson, and R. W. Hertzberg, *Polymer Blends and Composites in Multiphase Systems,* C. D. Han, Ed., ACS Adv. Chem. Ser. No. 206, ACS, New York, 1984, p. 261.
249. K. Friedrich, *Colloid. Polym. Sci.* **259,** 808 (1981).
250. K. Friedrich, *Deformation, Yield and Fracture of Polymers,* Plast. Rub. Inst., London, 1982, p. 26.1.
251. K. Friedrich, private communication, 1982.

FURTHER READING

S. Suresh, *Fatigue of Materials,* Cambridge Solid State Science Series, Cambridge University Press, Cambridge, U.K. (1991).

J. A. Bannantine, J. J. Comer, and J. L. Handrock, *Fundamental of Metal Fatigue Analysis,* Prentice Hall, Englewood Cliffs, New Jersey (1990).
H. O. Fuchs and R. I. Stephens, *Metal Fatigue in Engineering,* Wiley-Interscience, John Wiley & Sons, New York (1980).

PROBLEMS

13.1 During the course of a simple sinusoidal wave form fatigue test, the machine command signal was changed for one cycle at position A. The resulting data are shown below. Describe what changes were made in the machine signal and the corresponding effect on the specimen.

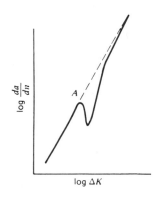

13.2 **(a)** A 10-cm-square, 20-cm-long extruded bar of 7075-T6511 is hollowed out to form a thin-walled cylinder (closed at one end), 20 cm long with an outer diameter of 9 cm. The cylinder is fitted with a 7-cm-diameter piston designed to increase pressure within the cylinder to 55 MPa. On one occasion, a malfunction in the system caused an unanticipated pressure surge of unknown magnitude, and the cylinder burst. Examination of the fracture surface revealed a metallurgical defect in the form of an elliptical flaw 0.45 cm long at the inner diameter wall and 0.15 cm deep. This flaw was oriented normal to the hoop stress of the cylinder. Compute the magnitude of the pressure surge responsible for failure. (For mechanical property date see Tables 10.2 and 10.8.)

(b) Assume that another cylinder had a similarly oriented surface flaw but with a semicircular ($a = 0.15$ cm). How many pressure cycles could the cylinder withstand before failure? Assume normal operating conditions for this cylinder and that the material obeys a fatigue crack propagation relation

$$\frac{da}{dn} = 5 \times 10^{-39}(\Delta K)^4$$

where *da/dn* and ΔK have the units of m/cyc and Pa$\sqrt{\text{m}}$, respectively.

13.3 A large steel plate is used in an engineering structure. A radical metallurgy graduate student intent on destroying this component decides to cut a very sharp notch in the edge of the plate (perpendicular to the applied loading direction). If he walks away from the scene of his dastardly deed at a rate of 5 km/h, how far away will he get by the time his plan succeeds? Here are hallowed hints for the hunter:

  (a) The plate is cyclically loaded uniformly from zero to 80 kN at a frequency of 25 Hz.

  (b) The steel plate is 20 cm wide and 0.3 cm thick.

  (c) The yield strength is 1400 MPa and the plane-strain fracture toughness is 48 MPa$\sqrt{m}$.

  (d) The misled metallurgist's mutilating mark was measured to be 1 cm long (through thickness).

  (e) A janitor noted, in subsequent eyewitness testimony, that the crack was propagating at a velocity proportional to the square of the crack-tip plastic zone size. (The janitor had just completed a correspondence course entitled "Relevant Observations on the Facts of Life" and was alerted to the need for such critical observations.)

  (f) Post-failure fractographic examination revealed the presence of fatigue striations 2.5×10^{-4} mm in width where the crack was 2.5 cm long.

13.4 If the plate in the previous problem had been 0.15 or 0.6 cm thick, respectively, would the villain have been able to get farther away before his plan succeeded? (Assume that the load on the plate was also adjusted so as to maintain a constant stress.)

13.5 Estimate the stress intensity factor range corresponding to an observed striation spacing of 10^{-4} mm/cyc in the steel alloy shown in Fig. 13.57b. Compare the results you would get when ΔK is determined from the striation data and the *macroscopic* data in Fig. 13.57b. Also, compute ΔK from Eq. 13-9.

13.6 The Liberty Bell, like most other bells, contains a large percentage of tin (18–25 w/o) in a copper-based alloy. From the copper–tin phase diagram (see any number of metallurgy texts or handbooks), describe the probable mechanical properties for this alloy. From this, decide whether a deliberate overload would have prolonged the bell's service life.

13.7 Many years ago, a crack was discovered in "Big Ben," the bell located in the Parliament building in London. To avoid its catastrophic failure or complete replacement, it was decided to replace the clapper with a smaller one and to rotate the bell to change the point of clapper impact. Using fracture mechanics concepts, explain how this alteration has succeeded to this day in prolonging "Big Ben's" life.

13.8 **(a)** A material with a plane-strain fracture toughness of $K_{IC} = 55$ MPa$\sqrt{m}$ has a central crack in a very wide panel. If $\sigma_{ys} = 1380$ MPa and the design stress is limited to 50% of that value, compute the maximum allowable fatigue flaw size that can grow during cyclic loading. (Assume that plane-strain conditions prevail.)

(b)   If the initial crack had a total crack length of 2.5 mm, how many loading cycles (from zero to the design stress) could the panel endure? Assume that fatigue crack growth rates varied with the stress intensity factor range raised to the fourth power. The proportionality constant may be taken to be 1.1×10^{-39}.

13.9   A thin-walled cylinder of a high-strength aluminum alloy ($K_{IC} = 24\,\text{MPa}\sqrt{\text{m}}$) has the following dimensions: length = 20 cm; outer diameter = 9 cm; inner diameter = 7 cm. A semicircular crack of depth $a = 0.25$ cm is discovered on the inner diameter and oriented along a line parallel to the cylinder axis. If the cylinder is repeatedly pressurized, how many pressure cycles could the cylinder withstand before failure? The pressure within the cylinder reaches 75 MPa, and the material obeys a fatigue crack propagation relation of the form

$$\frac{da}{dN} = 5 \times 10^{-39}(\Delta K)^4$$

where da/dN and ΔK have the units of m/cycle and $\text{Pa}\sqrt{\text{m}}$, respectively.

13.10   A steel plate ($K_{IC} = 54\,\text{MPa}\sqrt{\text{m}}$) contains a central crack 0.2 mm in length that is oriented normal to the stress axis. The plate is subjected to an alternating stress $\Delta\sigma = 180$ MPa with a mean stress of 90 MPa. Laboratory experiments have shown that this material experiences fatigue crack growth under these conditions, with the growth rate da/dN varying with ΔK according to the relation

$$da/dN = 4 \times 10^{-37}\Delta K^m$$

where da/dN and ΔK are given in units of m/cycle and $\text{Pa}\sqrt{\text{m}}$. The value of m was never computed, but the investigator noted that the crack growth rate varied directly with the square of the plastic zone dimension at the crack tip. Determine the number of loading cycles that can be withstood by the plate before final failure.

13.11   A wide panel contains a central crack that has a total length of 2.0 cm. Ten thousand loading cycles resulted in crack growth from 2.0 cm to a total length of 4.0 cm. What was the magnitude of the applied stress range? The material has an elastic modulus of 210 GPa and a power-law fatigue crack growth relation of

$$da/dN = 4 \times 10^{-37}\Delta K^4$$

where da/dN and ΔK possess units of m/cyc and $\text{Pa}\sqrt{\text{m}}$, respectively. For the point where the crack is 2.0 cm long, what would be the striation spacing?

13.12   If the starting K level for the two plates shown below is the same, would the fatigue lifetime be the same for the two components? In giving your answer, describe whether the rate of crack growth would differ between components

A and B as the crack lengthened and how you would characterize the crack driving force?

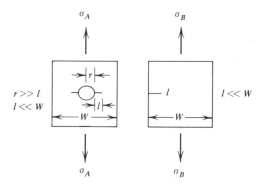

13.13 A certain steel alloy was chosen for use in a fatigue limiting service application. Experimental test results provide the following mechanical properties for this material:

$K_{ic} = 50$ MPa$\sqrt{\text{m}}$

$\Delta K_{\text{th}} = 4$ MPa$\sqrt{\text{m}}$

$\sigma_{ts} = 1000$ MPa

$da/dN = 4 \times 10^{-37} \Delta K^4$

A 2-m-wide plate of this material experiences a cyclic stress range of 200 MPa ($R = 0.1$) during component operation. As part of a routine maintenance sequence, no defect was discovered as a result of NDT inspection, capable of detecting cracks as small as 1 mm.

(a) Is it safe to say that one need not worry about either sudden or progressive failure of this plate?

(b) If fatigue failure is a possibility, estimate the minimum service lifetime for this plate.

13.14 Assume that the rotor described in Example 9.1 is returned to service, following the "recommended" start-up procedure. After a certain period of time in service, the rotor experienced a progressive surge in rpm that reached 1.3 ω_{op} at which point catastrophic failure occurred.

(a) What was the final crack length at the time of fracture?

(b) How many days had elapsed from the time the rotor was returned to service to the instant of final fracture?

To answer this problem it is necessary to know that the rotor operated at a rotational frequency of 3600 rpm with a stress that fluctuated between 218 and 238 MPa.

13.15 A 2-cm-long through thickness crack is discovered in a steel plate. If the plate experiences a stress of 50 MPa that is repeated at a frequency of 30 cpm, how long would it take to grow a crack, corresponding to the design limit where

$K_{\text{limit}} = K_{\text{Ic}}/3$. Assume that $K_{\text{Ic}} = 90$ MPa$\sqrt{m}$ and the material possesses a growth rate relation where $da/dN = 4 \times 10^{-37} \Delta K^4$, with da/dN and ΔK being given in units of m/cycle and Pa$\sqrt{m}$, respectively.

13.16 An 8-cm-square bar of steel is found to contain a 1-mm corner crack, oriented perpendicular to the length of the bar. If an axial stress is applied from 0 to 420 MPa at a frequency of once every 10 minutes, how long will it take for the rod to fracture? The properties of the bar are: $K_{\text{Ic}} = 70$ MPa$\sqrt{m}$, $\sigma_{ys} = 1500$ MPa, and the crack growth relation is $da/dN = 2 \times 10^{-37} \Delta K^4$, with da/dN and ΔK being given in units of m/cycle and Pa$\sqrt{m}$, respectively.

13.17 An aluminum plate is inspected for the presence of cracks after every 50,000 loading cycles. The NDT procedure that is routinely employed possesses a resolution limit of 1 mm. Through a mix-up, the new inspection team calibrated the instrument to yield a crack resolution limit of 1 cm. No crack was found on this occasion but unstable fracture took place following 34,945 additional loading cycles in association with the development of an edge crack, oriented normal to the major stress direction. Are there grounds for a lawsuit based on improper inspection procedures? The key stress level fluctuates between 50 and 100 MPa. The properties of the alloy are: $K_{\text{Ic}} = 30$ MPa$\sqrt{m}$, $\sigma_{ys} = 550$ MPa, $E = 70$ GPa, and the crack growth rate relation is given by $da/dN = 5 \times 10^{-35} \Delta K^4$, with da/dN and ΔK being given in units of m/cycle and Pa$\sqrt{m}$, respectively.

13.18 The presence of striations on the fatigue fracture surface of an aluminum alloy is used to determine the magnitude of an overload cycle. Striations immediately before the overload have a width of 2×10^{-4} mm, corresponding to 50% crack closure loading conditions; the overload cycle produced a striation width of 10^{-3} mm. What was the magnitude of the overload cycle?

CHAPTER 14

ANALYSES OF
ENGINEERING FAILURES

We have come now to the moment of truth—we must now use our knowledge and understanding of fracture mechanics and the relation between mechanical properties and microstructure to analyze actual service failures. However, before discussing recommended procedures for failure analyses and the details of several case histories, it is best to stand back for a moment and view component failures in a broader sense. To begin, we must ask who bears responsibility for these failures? Is it the company or individual that manufactured the component or engineering system, or the company or person that operated it when it failed? Such is the basis for debate in many product liability lawsuits. For example, opposing lawyers might ask of manufacturer and user the following questions:

1. Were engineering factors such as stress, potential flaw size, material, and environment considered in the design of the part?
2. Was the part underdesigned?
3. Was a proper material selection made for the manufacture of the part?
4. Was the part manufactured properly?
5. What limits were placed upon the use of the part and what, if any, service life was guaranteed?
6. Were these limits conservative or unconservative?
7. Were these limits respected during the operation of the part?

A product liability case often becomes entangled in a number of ambiguities arising from incomplete or unsatisfactory answers to these questions. As such, it is important for the practicing engineer called in to analyze a failure and, perhaps, testify in court, to identify the major variables pertaining to the design and service life of the component. Because an individual from one field may be reluctant to challenge the conclusions drawn by an expert in another discipline, it becomes difficult to reconcile the two points of view without an overview of the facts involved. In many cases, these differences contrast the importance of the continuum versus the microstructural approach to the understanding of the component response (or failure). The most valuable expert witness is one who can appreciate and evaluate the input from different disciplines and educate the court as to their respective significance in the case under study. On the basis of such expert testimony, the courts are able to draw conclusions and render judgments. A delightful statement, made by the auditor for the 1919 Boston molasses tank law suit, relates to this decision-making process:

> *Weeks and months were devoted to evidence of stress and strain, of the strength of materials, of the force of high explosives, of the bursting power of gas and of similar technical problems. . . . I have listened to a demonstration that piece "A" could have been carried into the playground only by the force of a high explosive. I have thereafter heard an equally forcible demonstration that the same result could be and in this case was produced by the pressure caused by the weight of the molasses alone. I have heard that the presence of Neumann bands* in the steel herein considered along the line of fracture proved an explosion. I have heard that Neumann bands proved nothing. I have listened to men upon the faith of whose judgment any capitalist might well rely in the expenditure of millions in structural steel, swear that the secondary stresses in a structure of this kind were negligible and I have heard from equally authoritative sources that these same secondary stresses were undoubtedly the cause of the accident. Amid this swirl of polemical scientific waters it is not strange that the auditor has at times felt that the only rock to which he could safely cling was the obvious fact that at least one-half the scientists must be wrong. By degrees, however, what seem to be the material points in the case have emerged.[1]*

A more recent service failure has had an even greater impact on our understanding of fracture and has led to the development of design procedures to guard against such future accidents. In this instance, a key structural member in the wing assembly of an F-111 fighter-bomber fractured, thereby leading to the loss of the plane and the death of the two pilots.[2] A postfracture examination of the broken wing section revealed that a large crack, suspected of having been introduced during the heat-treatment procedure, had gone undetected during the various stages of fabrication and assembly of the wing component. The darkened appearance of the elliptically shaped defect on the fracture surface was believed to represent oxidation, which occurred during the normal heat treatment cycle of this forged part (Fig. 14.1). This preexistent flaw was surrounded by a narrow, shiny band which represented the extent of fatigue crack extension during the 109-h service life of the component. Beyond this point, fracture proceeded in an unstable fashion. Based on this aircraft accident and its associated failure analysis, the United States Air Force changed procedures regarding the safe design of aircraft; this led to the development of military specification MIL-A-83444.[3] Embodied within this document are requirements for the *damage tolerance* of a given component. That is, damage is assumed to exist in each component (e.g., a crack located at a rivet hole) and the structure is designed to ensure that the crack will not grow to a critical size within a specified period of time (recall Section 13.1.2). To perform the computations needed for this design procedure, it is necessary to know

---

* Deformation twin bands in BCC iron.

FIGURE 14.1 Fracture surface of F-111 wingbox area. Dark, semielliptical surface flaw preexisted the flight service. Smooth bright band at boundary of dark flaw represents fatigue crack propagation zone prior to unstable fracture. (After Wood[2]; reprinted with permission from *Eng. Fract. Mech. 7,* 557 (1975), Pergamon Press, Ltd.)

how the crack growth behavior of the specified material depends on such variables as the stress intensity factor range, load ratio, test temperature and frequency, environment, and complex load interactions. To this end, data described in Chapter 13 assume great importance in the implementation of this military specification.

In the following sections, attention is given to the identification of typical defects, consideration of fracture surface examination techniques, and identification of data needed for a successful failure analysis. A discussion of numerous service failures then follows.

14.1 TYPICAL DEFECTS

A wide variety of defects can be found in a given engineering component.[4] These flaws may result from such sources as material imperfections, defects generated during service, and defects introduced as a result of faulty design practice. Regarding the first source mentioned, defects can be found within the original material supply or can be introduced during the manufacturing process. Typical material defects include porosity, shrinkage cavities, and quench cracks. Other microstructural features can trigger crack formation if the applied stresses exceed some critical level. These include nonmetallic inclusions, unfavorably oriented forging flow lines (recall Section 10.2), brittle second phases, grain-boundary films, and microstructural features resulting from 300°C and temper embrittlement. The list of manufacturing defects includes machining, grinding and stamping marks (such as gouges, burns, tears, scratches, and cracks), laps, seams, delaminations, decarburization, improper case hardening, and defects due to welding (e.g., porosity, hot cracking, cold cracking, lack of penetration, and poor weld bead profile).

Defects can be introduced into the component during service conditions as a result of excessive fretting and wear. Environmental attack can also cause material degradation as a result of general corrosion damage, liquid metal and hydrogen embrittlement, stress corrosion cracking, and corrosion fatigue. Surely, cyclic loading can initiate fatigue damage without an aggressive environment and may lead to serious cracking of a component.

Finally, defects can be introduced into a component through faulty design. These human errors include the presence of severe stress concentrations, improper selection of material properties and surface treatments, failure to take remedial actions (such as baking a steel part after it has been cadmium plated to remove charged hydrogen gas), inadequate or inaccurate stress analysis to identify stress fields in the component, and improper attention to important load and environmental service conditions as they relate to material performance.

14.2 MACROSCOPIC FRACTURE SURFACE EXAMINATION

The functions of a macroscopic fracture surface examination are to locate the crack origin, determine its size and shape, characterize the texture of the fracture surface, and note any gross markings suggestive of a particular fracture mechanism. To begin, one should attempt to identify whether there are one or more crack origins, since this may provide an indication of the magnitude of stress in the critical region. In general, the number of crack nuclei increases with increasing applied stress and magnitude of an existing stress concentration factor (Fig. 12.5). Even when one crack grows to critical dimensions, secondary cracks can develop before final failure because of load adjustments that may accommodate the presence of the primary defect.

Whether there are one or more fracture nucleation sites, it is of utmost importance to locate them and identify precisely the reason for their existence. When the fracture mechanism(s) responsible for growth of the initial defect to critical proportions is known, the engineer can recommend ''fixes'' or changes in component design.

The task at this point is to find the origin. This was not difficult in the case of the tricycle and doorstop failures mentioned in Chapter 12, but one can well imagine the difficulty of sifting through the wreckage of a molasses tank, ship, or bridge failure (see Chapter 7) to find their respective fracture origins. For these situations, there could be literally thousands of linear meters of fracture surface to examine. Where does one begin? Once begun, how does one know the direction in which to proceed to locate the origin? As discussed in Section 7.7.2, the fracture surface often reveals contour lines that point back to the crack origin. These features, referred to as chevron markings, are found on the fracture surfaces in many engineering solids and aid the investigator in locating the region where the crack had formed or preexisted. The microscopist is then able to focus attention on the micromorphological features of the origin and gain insight into the cause of failure. Sometimes, however, these markings may be obscured by other fracture markings, such as by secondary fractures in anisotropic materials. A crack ''divider'' orientation Charpy specimen of banded steel (Fig. 13.62) reveals many fracture surface delaminations caused by σ_z stresses acting parallel to the crack front, but they cloud the expected ''chevron'' pattern (Fig. 14.2).

As one follows the path of the crack, shear lips are often found that represent the regions on the fracture surface that correspond to plane-stress conditions (recall Section 8.6). As such, it is tempting to relate the size of the plastic zone to the amount of shear lip found on the fracture surface.[5] Since $r_y \approx (1/2\pi)(K^2/\sigma_{ys}^2)$ at the surface of the plate and the shear lips form on $\pm 45°$ bands to the sheet thickness, it is seen

FIGURE 14.2 Crack "divider" fracture surface. Delaminations obscure antici-pated chevron pattern.[6] (Reprinted with permission from Metals Society, *JISI* 209, 975 (1971).)

from Fig. 14.3 that the depth D of the shear lips can be approximated by the plastic zone radius. Hence

$$D \approx r_y \approx \frac{1}{2\pi}\left(\frac{K}{\sigma_{ys}}\right)^2 \tag{14-1}$$

Combining Eqs. 8-28, and 14-1, we find

$$\text{shear lip} \approx \frac{1}{2\pi}\frac{Y^2\sigma^2 a}{\sigma_{ys}^2} \tag{14-2}$$

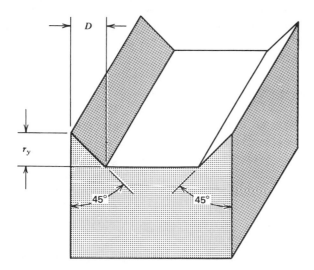

FIGURE 14.3 Schema showing relation between shear lip depth and estimated plane stress plastic zone size.

The geometrical correction factor Y for the component and the crack length where the shear lip was measured can be used with Eq. 14-2 to estimate the prevailing stress level. This approach to determining the stress level is highly empirical and appears to work satisfactorily only for certain materials such as high-strength aluminum and certain alloy steels, but not for lower strength steels. When the correlation does not hold, it usually results in too low an estimate of K from Eq. 14-1; that is, the shear lip is smaller than what would be expected from the actual plastic zone. Because Eq. 14-2 is a highly empirical determinant of stress level, computed values must be considered tentative until corroborated by additional findings. The use of shear lip measurements to determine the stress level is discussed in Case History 1 in Section 14.5.

It is possible that a crack may initiate by one mechanism and propagate by one or more different ones. For example, a crack may begin at a metallurgical defect, propagate for a certain distance by a fatigue process, and then continue growing by a combination of fatigue crack propagation and environment-assisted cracking when the stress intensity factor exceeds K_{IEAC}. Such mechanism changes may be identified by changes in texture of the fracture surface. For example, the fracture surface shown in Fig. 14.4a reveals the different textures associated with fatigue and stress corrosion subcritical flaw growth. In the broken wing strut (Fig. 14.4b), we find the regions of FCP (shiny areas) interrupted by two separate localized crack instabilities (dull areas), which probably were caused by two high load excursions during the random loading life history of the strut. Another example of a plane-strain "pop-in" is found on the fracture surface of a fracture-toughness test sample that exhibited Type II (Fig. 8.18) load-deflection response (Fig. 14.4c).

"Pop-in" can also result from the presence of local residual stresses. If a crack is embedded within a localized stress concentration region, the application of a moderate load could develop a stress intensity level (magnified by the local stress concentration) equal to K_{IC} or K_c (depending on the prevailing stress state), which would cause the crack to run unstably through the component. However, the crack would soon run out of the region of high stress concentration and arrest (and produce a fracture surface marking), since the moderate load without the stress concentration does not possess the necessary driving force to sustain crack growth.

Other macroscopic arrest lines, such as the fatigue crack propagation "beach mark" and load block band, were discussed in Chapter 12 (Fig. 12.3a) and Chapter 13 (Figs. 13.42a, b and 3.43b). A more critical examination of Fig. 12.3a reveals a striking feature. As expected, the crack initiation site in the steel shaft is located in the vicinity of a stress concentration at the base of the shaft key way. The exceptional aspect of this fracture is the extent to which the fatigue crack was able to grow prior to the onset of unstable fracture. Indeed, the crack is seen to have grown more than 90% across the component width. The fatigue fracture in a magnesium housing provides another example of this unusual pattern (Fig. 14.5).[8] In this instance, multiple cracks had initiated (note the ratchet lines) at the center hole and propagated across more than 95% of the section width. For cracks to grow to the extent noted in Figs. 12.3a and 14.5a, one of two scenarios must exist: either the fracture toughness of the material must be extremely high or the stress level must be extremely low in association with a normal value of fracture toughness. The first scenario can be dismissed since the

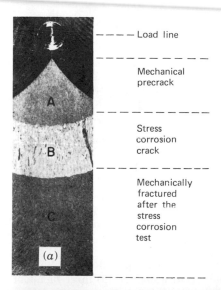

- - - - Load line

- - - - - - - -
Mechanical
precrack

- - - - - - - -
Stress
corrosion
crack

- - - - - - - -
Mechanically
fractured
after the
stress
corrosion
test

(a)

- - - - - - - -

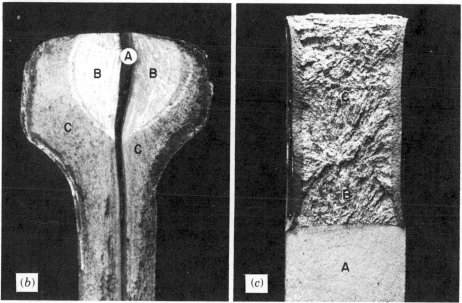

FIGURE 14.4 Macrofractographs revealing fracture mechanism transitions. (a) Transition from fatigue A to stress corrosion cracking B to fast fracture C[7]; (b) wing strut with metallurgical delamination A, fatigue B, and static fracture C; (c) fracture-toughness sample revealing fatigue precracking zone A, pop-in instability B, and fast fracture C. (Fig. 14.4a reprinted with permission of Markus O. Speidel, Brown Boveri Co.)

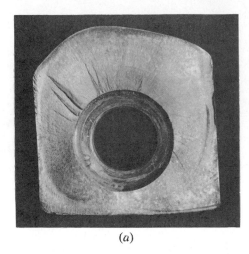

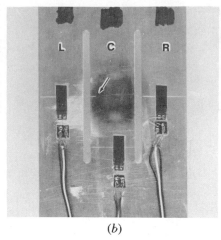

(*a*) (*b*)

FIGURE 14.5 (*a*) **Fatigue fracture surface in a magnesium helicopter housing. Note that fatigue markings are observed over 90% of the fracture surface. (*b*) three-ligament specimen [left (L), center (C), and right (R)] containing small crack (see arrow) in central (C) load path.[8] (R. W. Hertzberg and T. J. Pecorini,** *Int. T. Fat.,* **15, 509 (1993), by permission of the publishers, Butterworth Heinemann, Ltd. ©.)**

steel and magnesium alloys shown do not possess unusual fracture properties. Therefore, fracture in both instances must have taken place in association with a very low stress level. A dichotomy is immediately apparent. If the stress level was so low, then the resultant cyclic stress intensity levels at nascent cracks would be below the threshold stress intensity for the material, and any preexisting crack would not grow: hence, failure would never have occurred.

The rationalization for this dilemma is that the load level was, indeed, high enough to cause the crack to initiate but dropped progressively as the crack lengthened. There are two ways in which the load level can drop. First, loads can shed if the component is loaded under fixed displacement conditions (e.g., recall the bolt-loaded compact sample used to generate stress corrosion cracking data (Fig. 11.3). Alternatively, load shedding can occur when there is a redundant load path such that the load in the cracked segment would be transferred to the unbroken ligament(s), much as load transfer occurs in a composite material from a low-stiffness to a high-stiffness component. That is, as the crack grows in one load path, the stiffness of that segment decreases relative to the unbroken load path(s) and the load will shed to the stiffer members. As the number of redundant paths increases, the response of the component approaches that of a component being loaded under fixed displacement conditions.

To confirm a load shedding scenario, a three-ligament 2024-T3 aluminum alloy specimen (Fig. 14.5*b*) was tested in fatigue with a crack introduced into the 23-mm-wide center load path.[8] Figure 14.6*a* shows the relation between load and crack length for the actual data measured from the cracked central path of the three-ligament specimen, as well as the theoretical curves based on constant load and the three-ligament loading configurations. Clearly, the measured decrease in the actual load in the cracked ligament is in reasonable agreement with theoretical expectations. Figure

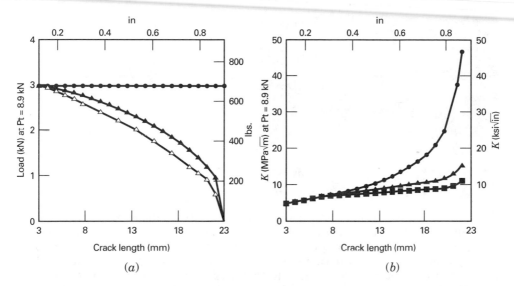

FIGURE 14.6 (*a*) **Load versus crack length within the 23-mm-wide central ligament: ●, without load shedding (theoretical); ▲, with load shedding (theoretical); △, measured during test (*b*) theoretical relationship between stress intensity and crack length within a 23-mm-wide central ligament: ●, without load shedding; ▲, with load shedding; ■, constant displacement.**[9] **(R. W. Hertzberg and T. J. Pecorini,** *Int. J. Fat.,* **15, 509 (1993), by permission of the publishers, Butterworth Heinemann, Ltd. ©.)**

14.6*b* depicts the theoretical K levels for conditions of constant load, constant displacement, and multiple-path load shedding, respectively. Load shedding greatly reduces the stress intensity in this specimen geometry below that for a constant load condition. Indeed, the stress intensity attenuation produced in the three-ligament specimen is almost as low as the attenuation that would be produced under constant displacement. Also note that the K level, corresponding to an a/W level of 0.95, remains beneath the fracture toughness value for the 2024-T3 aluminum alloy three-legged specimen.

The fracture surface of the cracked central ligament progressed entirely across its width in a flat fracture mode, corresponding to plane-strain conditions associated with minimal escalation in the K level with increasing crack length (Fig. 14.7). This appearance is strikingly similar to load-shedding service failures shown in Fig. 12.3*a* and 14.5*a*. (No clam shell markings are observed in the present sample since periodic load fluctuations were not introduced.) By comparison, a crack in a single-ligament sample (with no load shedding) developed shear lips at $a/W \approx 0.25$ and subsequently ruptured abruptly (see arrow); the latter fracture surface mode reflects a significant increase in stress intensity with increasing crack length (Fig. 14.6*b*).

We now conclude our consideration of the macroscopic appearance of fatigue fracture surfaces by determining the cause of contrast differences between marker bands generated by different ΔK levels. Recalling Fig. 13.43*b*, we see that fatigue marker bands become darker with decreasing ΔK level, though the fast fracture region

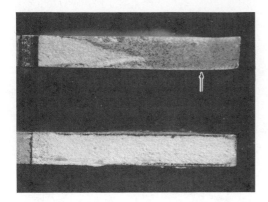

FIGURE 14.7   **Macroscopic appearance of fatigue failures in single-ligament (top) and three-ligament (bottom) samples. Arrow indicates onset of fast fracture in the single-load path sample. Crack growth direction is from left to right. Note lack of shear-lip development in three-ligament sample (bottom) in association with modest increase in ΔK level with increasing crack length.[8] R. W. Hertzberg and T. J. Pecorini, *Int. J. Fat.*, 15, 509 (1993), by permission of the publishers, Butterworth Heinemann, Ltd. ©.**

(Region FF), corresponding to K_c conditions, is also dark in appearance. Therefore, there must be a progressive darkening in the contrast of the fracture surface at ΔK levels above those associated with the block loading test described in Section 13.5.1. Indeed, Fig. 14.8 reveals the fracture surface brightness-ΔK relation in Astroloy for a $K_{\max}^c = 85$ MPa$\sqrt{\text{m}}$ test that shows the fracture surface to be dark at low and high ΔK levels, and relatively bright at intermediate ΔK levels.[9]

It is found that such contrast differences are related to ΔK-induced changes in fracture micromechanisms and their associated influenced on fracture surface roughness. In the low ΔK regime, faceted growth dominates and a rough, dark surface is

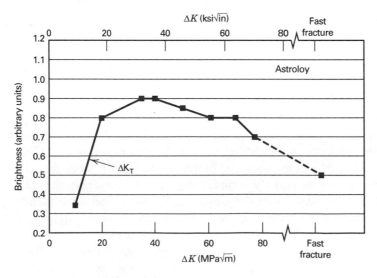

FIGURE 14.8   **Effect of ΔK level on fracture surface brightness in Astroloy nickel-base alloy, corresponding to test at $K_{max}^c = 85$ MPa $\sqrt{\text{m}}$.[9]**

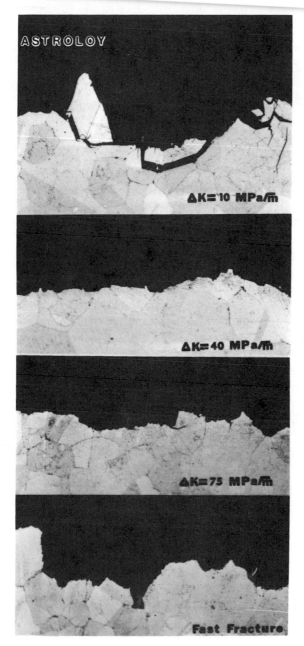

FIGURE 14.9   Crack profiles in Astroloy corresponding to ΔK levels of (*a*) 10 (*b*) 40 (*c*) 75MPa$\sqrt{\mathrm{m}}$, (*d*) fast fracture.[9] (400×)

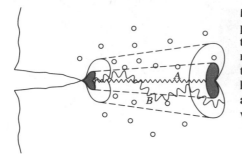

FIGURE 14.10 Dependence of crack-particle encounters on the size of the plastic zone. High ΔK conditions produces macroscopically rougher fracture path (B) than that generated at intermediate ΔK levels (A). (Rough appearance will again appear at low ΔK levels in association with faceted growth.)

generated (Figs. 14.8 and 14.9a). Above the micromechanism transition point (recall Eq. 13-14), a change occurs from the rough faceted mechanism to the relatively flat striation mode of fatigue crack advance (Fig. 14.9b). At progressively higher ΔK levels, the associated plastic zone size "sees" many second phase particles within the microstructure, which enables the crack to wander along a tortuous path (Path B) as compared with that corresponding to lower ΔK levels (Path A) where the plastic zone does not see the particles (Fig. 14.10). As such, the plastic zone acts as a "filter"; when the filter is large, many weak particles are encountered and the fracture surface is relatively rough (Fig. 14.9c, d) and dark in appearance (Fig. 14.8). It is the increased roughness at both low and high ΔK levels (associated with different micromechanisms) that causes the fracture surface to be relatively dark, whereas the flatter, striated region at intermediate ΔK levels generates a brighter fracture surface appearance (Fig. 14.11; also review Fig. 13.51).

14.3 METALLOGRAPHIC AND FRACTOGRAPHIC EXAMINATION

Having located the crack origin, a typical failure analysis would proceed normally with two main interim objectives: (1) identification of the micromechanism(s) of

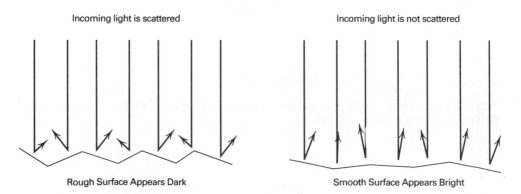

FIGURE 14.11 Schematic diagram illustrating the influence of fracture surface roughness on brightness level.[9]

subcritical flaw growth, and (2) estimation of the stress intensity conditions prevailing at the crack tip when failure occurred. As was discussed in Chapter 7, metallographic techniques have often been used to determine the crack path relative to the component microstructure. In addition to identifying the microscopic fracture path (transgranular versus intergranular), metallographic sections are useful in establishing the metallurgical condition of the material. Grain size and shape offer important clues to the thermomechanical history of the component (see Chapter 10). For example, a coarse-grained structure is indicative of a very high temperature annealing process, while an elongated grain structure indicates not only the application of a deformation process in the history of the material, such as rolling, forging, and drawing, but also the deformation direction. Such information allows one to anticipate the presence of anisotropic mechanical properties that must be identified in relation to the predominant stress direction. Examples of mechanical property anisotropy were given in Tables 10.2 and 10.8.

Determination of the microstructural constituents enables the examiner to determine whether a component has been heat treated properly. Identification of a possible grain-boundary phase, for example, can explain the occurrence of an intercrystalline fracture. Finally, with the aid of an inclusion count, the relative cleanliness of the metallurgical structure can be determined. Although it is not possible to express the fracture toughness of a material in terms of some measure of inclusion content [recently, some progress has been achieved in this regard (see Chapter 10)], it is known that fracture toughness decreases with increasing inclusion content. Hence, a trained metallographer may ascertain from metallographic examination whether the material in question is representative of good or bad stock.

A fractographic study with either a scanning or transmission electron microscope reveals the microscopic character of the fracture surface. Since this topic has been discussed at considerable length in earlier chapters, the reader is referred specifically to Chapters 7 and 13 and Appendix A before proceeding further.

14.4   COMPONENT FAILURE ANALYSIS DATA

Having been introduced to the fundamentals of fracture mechanics analysis, stress intensity analysis of cracks, macroscopic and microscopic features of the fracture surface, and the pertinent mechanical property data necessary to adequately characterize the performance of a given material, the reader should be able to synthesize this information and solve a given service failure problem. The following checklist will assist the investigator in this task. The checklist, which takes into account the component geometry, stress state, flaw characterization, fractographic observations, metallurgical information (including component manufacture and other service information), summarizes the raw data desirable for a complete failure analysis of a fractured component. Experience has shown, however, that in the majority of instances, certain facts are never determined, and educated guesses or estimates (such as the K calibration estimations described in Section 8.2) must be introduced to complete the analysis. The objective of a checklist is to minimize the amount of guesswork while maximizing the opportunity for firm quantitative analysis.

SUGGESTED CHECKLIST OF DATA DESIRABLE FOR COMPLETE FAILURE ANALYSIS

I. Description of Component Size, Shape, and Use _____

_____

    A. Specify areas of design stress concentrations _____

_____

        1. Magnitude of stress concentration at failure site _____

_____

II. Stress State for Component

    A. Type of stresses

        1. Magnitude of design stress levels

            a. Mean stress _____

            b Stress range _____

        2. Type of stress (e.g., Mode I, II, III, or combinations) _____

_____

        3. Presence of stress gradient _____

    B. State of stress: plane strain vs. plane stress

        1. Fracture surface appearance: percent shear lip _____

_____

        2. Estimation from calculated plastic zone size to thickness ratio ____

_____

    C. Nature of load variations

        1. Hours of component operation _____

        2. Load cycle frequency _____

        3. Type of loading pattern

            a. Random loading _____

            b. Existence of overloads resulting from abnormal service life events _____

III. Details of Critical Flaw

    A. Date of previous inspection _____

        1. Findings of previous inspection _____

_____

_____

B.   Nature of critical flaw leading to fracture (make use of clearly labeled sketches and/or macrophotographs with accurate magnifications)

    1.   Location of critical flaw by macroscopic examination   _____

    _____

    _____

    2.   Critical flaw size, shape, and orientation at instability   _____

    _____

    _____

    3.   Surface or imbedded flaw   _____

    4.   Direction of crack propagation as determined by

       Chevron markings   _____   Pop-in   _____

           Beach markings   _____

       Direction   _____

C.   Manufacturing flaws related to crack initiation

       Scratches   _____   Misfits   _____

       Undercuts   _____   Others   _____

       Weld defects   _____

D.   Metallurgical flaws related to crack initiation

       Inclusions   _____   Voids   _____

       Second-phase particles   _____   Weak interfaces   _____

       Entrapped slag   _____   Others   _____

E.   Fractographic observations

    1.   Qualitative observations

       Dimpled rupture   _____   Fatigue striations   _____

       Cleavage   _____   Corrosion   _____

       Intercrystalline   _____   Fretting   _____

    2.   Quantitative observations

       a.   Striation spacings at known crack length positions   _____

       _____

       b.   Striation spacing evidence of uniform or random loading   _____

       _____

       c   Stretch zone width at onset of unstable crack extension   _____

       _____

IV. Component Material Specifications

A. Alloy designation _____

B. Mechanical properties

	σ_{ys}	σ_{ts}	% elong.	% R.A.	K_{IC}	K_{IEAC}	Fatigue Characterization
Specified							
Actual							

C. Alloy chemistry

	Elements						
	A	B	C	D	E	F	G
Specified							
Actual							

D. Melting practice

Air melted _____

Vacuum melted _____

Other _____

E. Ingot breakdown

Hot rolled _____

Cold rolled _____

Cross rolled _____

F. Thermomechanical treatment

1. Annealing or solution treatment condition _____

2. Tempering or aging treatment _____

3. Intermediate mechanical working _____

G. Component manufacture

1. Forged _____   Spun _____

   Cast _____   Extruded _____

   Machined _____   Other _____

2.   Joint detail

Welded _____   Bolted _____

Brazed _____   Other _____

Adhesive bonded _____

H.   Surface treatment

Shot peened _____   Flame or induction
                                        hardened

Cold rolled _____
                                        _____
Carburized _____   Plated _____

Nitrided _____   Pickled _____

                                        Other _____

I.   Component microstructure

1.   Presence of mechanical fibering and/or banding from chemical
segregation _____

2.   Grain size and shape _____

a.   Elongated with respect to stress axis _____

b.   Grain run-out in forgings _____

3.   Inclusion count and classification _____

14.5   CASE HISTORIES

In this section ten actual case histories are discussed; all involve the application of fracture mechanics principles to failure analysis. Although not all component failures require a fracture mechanics analysis, the latter approach is emphasized here to demonstrate the applicability of the fundamental concepts introduced in Chapters 7 through 13. The interested reader is referred to compilations of more than 100 additional failure analyses, which have been reported elsewhere.[10–13] The recent collection of papers[13] is of particular interest because fracture mechanics analyses are used extensively. In one material selection case history, Reid and Baikie[14] analyzed the static and cyclic loading service conditions for high-pressure water pipes and concluded that one particular steel alloy was preferred over another, even though the latter was approximately 50% cheaper than the alloy chosen. It was found that the considerably greater potential service life and greater margin of safety associated with the steel alloy selected more than compensated for its relatively higher unit cost. The reader should recognize that the total price tag for a given component or engineering system depends on both initial material and fabrication costs, along with expenses associated with maintenance and repair. Often, initial material costs are of secondary importance and are subordinate to other financial considerations. The paper by Pearson and Dooman[15] is of interest because the fracture mechanics analysis demonstrated that failure of a truck-mounted propane tank was more complicated than first assumed. A fractographic analysis revealed that a girth weld crack had existed for several years prior to failure and was initially thought to be responsible for the explosion of the tank. However, a fracture mechanics analysis revealed that failure could not have occurred unless the preexisting crack had experienced internal gas pressures far in excess of the allowable level. Further investigation confirmed that such high gas pressures were present: The relief valve of the tank was badly corroded and unable to open, therefore producing excessive gas pressures within the tank. Also contributing to higher gas pressures were solar heating and the proximity of the tank to the truck's hot tailpipe.

CASE 1
Analysis of Crack Development during Structural Fatigue Test[5]

This failure analysis reported by Paris[5] represents an excellent, well-documented example of the use of several different and independent fracture mechanics procedures in the solution of a fracture problem. A program load fatigue test was conducted on a 1.78-cm-thick plate of D6AC steel that had been tempered to a yield strength of 1500 MPa. Fracture of the plate occurred when fatigue cracks that had developed on both sides of a drilled hole grew into a semicircular configuration, as shown in Fig. 14.12. Note the growth rings within the two corner cracks produced by fatigue block

FIGURE 14.12   Two corner cracks emanating from through-thickness hole, revealing fatigue growth bands and shear lips.[5] (From R. J. Gran, F. D. Orazio, Jr., P. C. Paris, G. I. Irwin, and R. W. Hertzberg, AFFDL-TR-70-149, March 1971.)

loading conditions. These may be compared with similar markings shown in Fig. 13.42a. The stress at failure was reported to be 830 MPa, and the maximum and minimum stresses in each of the load blocks were also known.

The stress intensity factor at fracture was determined by three separate methods. First, the stress intensity factor solution for the given crack configuration was estimated in two ways. The actual hole–crack combination was approximated by a semi-circular surface flaw with a radius of 0.86 cm and by a through thickness flaw with a total length of 1.73 cm. These estimates reflect lower and upper bound solutions, respectively, since the former solution does not account for the hole passing through the entire plate thickness, and the latter solution indicates more fatigue crack growth than was actually observed. The lower bound of the stress intensity factor may be given by[16]

$$K_L = \left[1 + 0.12\left(1 - \frac{a}{c}\right)\right]\sigma\sqrt{\frac{\pi a}{Q}}\sqrt{\sec\frac{\pi a}{2t}} \tag{14-3}$$

where K_L = lower bound stress intensity solution
   a = crack depth, 0.86 cm or 0.0086 m
   c = half-flaw width, 0.86 cm
   σ = applied stress, 830 MPa
   Q = elliptical flaw correction, 2.5
   t = plate thickness, 1.78 cm

$$K_L = \left[1 + 0.12\left(1 - \frac{0.86}{0.86}\right)\right][830]\sqrt{\frac{\pi(0.0086)}{2.5}}\sqrt{\sec\frac{\pi(0.86)}{2(1.78)}}$$

$$K_L = 101.3 \ \text{MPa}\sqrt{\text{m}}$$

The upper bound solution is given by

$$K_U = \sigma\sqrt{\pi a}$$

$$= 830\sqrt{\pi(0.0086)}$$

$$= 136 \ \text{MPa}\sqrt{\text{m}}$$

From these results, the actual K level at fracture may be bracketed by

$$101 < K_c < 136 \ \text{MPa}\sqrt{\text{m}}$$

with the correct value being more closely given by the lower bound solution because of a smaller error in this estimation. Consequently, K_c (or K_{IC}) $\approx$ 110 MPa$\sqrt{\text{m}}$.

The stress intensity factor was then estimated by measurement of the shear lip depth (about 0.8 mm) along the surface of the hole (Fig. 14.12). From Eq. 14-1

$$\text{shear lip depth} \approx \frac{1}{2\pi} \frac{K^2}{\sigma_{ys}^2}$$

$$8 \times 10^{-4} \approx \frac{1}{2\pi}\left(\frac{K}{1500}\right)^2$$

$$K \approx 106 \text{ MPa}\sqrt{m}$$

which agrees extremely well with the previous estimate of 110 MPa$\sqrt{m}$.

Two additional estimates of the critical stress intensity factor were obtained by using measurements of the fatigue growth bands. It was known that the last band was produced by 15 load fluctuations between stress levels of 137 and 895 MPa. This growth band measured 0.32 mm, and the average crack growth rate was found to be

$$\frac{da}{dn} \approx \frac{\Delta a}{\Delta n} \approx 3.2 \times 10^{-4}/15 \approx 2.1 \times 10^{-5} \text{ m/cyc}$$

From the fatigue crack growth rate data of Carmen and Katlin[17] the corresponding ΔK level was determined to be about 77 MPa$\sqrt{m}$. The maximum K level was then given by

$$K_{max} = \Delta K\left(\frac{\sigma_{max}}{\Delta\sigma}\right) \tag{14-4}$$

$$K_{max} = 77\left(\frac{895}{758}\right)$$

$$K_{max} = 91 \text{ MPa}\sqrt{m}$$

A similar calculation was made for the next to last band where

$$\Delta n = 2$$

$$\Delta a = 0.16 \text{ mm}$$

$$\sigma_{min} = 138 \text{ MPa}$$

$$\sigma_{max} = 992 \text{ MPa}$$

$$\frac{da}{dn} \approx 1.6 \times 10^{-4}/2 \approx 8 \times 10^{-5} \text{ m/cyc}$$

From Carmen and Katlin's results, the ΔK level corresponding to this crack growth rate was found to be 82.5 MPa$\sqrt{m}$. Again using Eq. 14-4

$$K_{max} = 82.5\left(\frac{992}{854}\right) = 95.8 \text{ MPa}\sqrt{m}$$

In both instances, estimates of K_c from fatigue growth bands were in excellent agreement with values based on estimates of the prevailing stress intensity factor and shear lip measurements. Finally, the average critical stress intensity factor (101 MPa$\sqrt{m}$) is almost identical with the known K_{IC} level for this material (see Table 10.8). To summarize, the analysis of this laboratory failure clearly demonstrates a number of different and *independent* approaches based on fracture mechanics concepts that one can employ in solving a service failure. Ideally, one should use a number of these procedures to provide cross-checks for each computation.

CASE 2
Analysis of Aileron Power Control Cylinder Service Failure[5]

Several failures of an aileron hydraulic power control unit were experienced by a certain fighter aircraft. These units consisted of four parallel chambers, pressurized by two separate pumps. Failures occurred by cracking through either the inner or the outer chamber walls. In either case, the resulting loss of pressure contributed to an aircraft malfunction. Test results indicated the normal mean pressure in these chambers to be about 10.3 MPa, with fluctuations between 5.2 and 15.5 MPa caused by aerodynamic loading fluctuations. Furthermore, during an in-flight aileron maneuver, the pressure was found to rise sometimes to 20.7 MPa, with transient pulses as high as 31 MPa resulting from hydraulic surge conditions associated with rapid commands for aileron repositioning. In one particular case, an elliptical surface flaw grew from the inner bore of one cylinder toward the bore of the adjacent cylinder. A series of concentric markings suggested the initial fracture mode to be fatigue. At this point, the crack had grown to be 0.64 cm deep and 1.42 cm long. Subsequently, the crack appeared to propagate by a different mechanism (macroscopic observation) until it became a through-thickness flaw 2.7 cm long, at which time unstable fracture occurred. It was considered likely that the latter stage of subcritical flaw growth was controlled by an environment-assisted cracking process that would account for the change in fracture surface appearance, similar to that shown in Fig. 14.4a. The component was made from 2014-T6 aluminum alloy and was manufactured in such a way that the hoop stress within each chamber acted perpendicular to the short transverse direction of the original forging. From the *Damage Tolerant Design Handbook,*[18] the yield strength and fracture toughness of the material in this direction are given as 385 MPa and 19.8 MPa$\sqrt{m}$, respectively.

Additional data concerning the geometry of the power control unit are given below:

$$\text{chamber wall thickness } (t) = 0.84 \text{ cm}$$
$$\text{elliptical crack depth } (a) = 0.64 \text{ cm}$$
$$\text{elliptical crack length } (2c) = 1.42 \text{ cm}$$
$$a/2c = 0.445$$
$$\text{elliptical flaw correction factor } (Q) \cong 2.2$$
$$\text{bore diameter } (D) = 5.56 \text{ cm}$$
$$\text{through thickness crack length } (2a_1) \cong 2.7 \text{ cm}$$

To use the plane-strain fracture-toughness value in subsequent fracture calculations, it is necessary to verify that t and $a \geqslant 2.5(K_{IC}/\sigma_{ys})^2$. This condition is met for this case history and supported by the observation that the fracture surface was completely flat. The stress necessary to fracture the unit may be computed by the formula for a through-thickness flaw where

$$K_{IC} = \sigma\sqrt{\pi a}$$

Setting $K_{IC} = 19.8$ MPa$\sqrt{m}$ and $a = 1.35$ cm

$$19.8 = \sigma\sqrt{\pi(1.35 \times 10^{-2})}$$

$$\sigma = 96.1 \text{ MPa}$$

The chambers have a large diameter-to-thickness ratio so that pressurization could be analyzed in terms of a thin-walled cylinder formulation. Since both cylinders are pressurized, the hoop stress between cylinder bores is estimated to be

$$\sigma_{\text{hoop}} = \frac{2PD}{2t}$$

where $P = $ internal fluid pressure.

Using the component dimensions and the calculated stress level at fracture (i.e., 96.1 MPa), the pressure level at fracture P is calculated to be

$$96.1 = \frac{2P(5.56 \times 10^{-2})}{2(8.4 \times 10^{-3})}$$

$$P = 14.5 \text{ MPa}$$

Since the normal mean pressure in the cylinder bores is about 10.3 MPa and reaches a maximum of about 15.5 MPa, unstable fracture could have occurred during either normal pressurization or during pressure buildups associated with an aileron repositioning maneuver.

As mentioned above, the change in fracture mechanism when the elliptical crack reached a depth and length of 0.64 and 1.42 cm, respectively, could have been due to the onset of static environment-assisted cracking at a stress intensity where the cracking rate became independent of the K level (i.e., Stage II behavior). For such an elliptical flaw

$$K^2 = \left[1 + 0.12\left(1 - \frac{a}{c}\right)\right]^2 \sigma^2 \frac{\pi a}{Q}\left(\sec\frac{\pi a}{2t}\right) \qquad (14\text{-}3)$$

with the result that

$$K^2 = \left[1 + 0.12\left(1 - \frac{0.64}{0.71}\right)\right]^2 \sigma^2 \frac{\pi(6.4 \times 10^{-3})}{2.2}\left[\sec\frac{\pi(0.64)}{2(0.84)}\right]$$

$$K = 0.14\sigma$$

Assuming that the major stresses associated with static environment-assisted cracking were those associated with the mean pressure level of about 10.3 MPa, the associated hoop stress is calculated to be

$$\sigma_{hoop} = \frac{2(10.3)(5.56 \times 10^{-2})}{2(8.4 \times 10^{-3})}$$
$$= 68.2 \text{ MPa}$$

Using this stress level, the stress intensity level for the onset of static environment-assisted cracking is estimated to be

$$K = 0.14\sigma$$
$$= 0.14(68.2)$$
$$= 9.5 \text{ MPa}\sqrt{m}$$

Unfortunately, no environment-assisted cracking (EAC) data for this material–environment system are available to check whether the number computed above is reasonable. It is known, however, that EAC rates in this alloy become appreciable in a saltwater environment when the stress intensity level approaches 11 MPa$\sqrt{m}$. Further material evaluations would be needed to determine whether hydraulic fluid has a similar effect on the cracking response of this alloy at stress intensity levels of about 11 MPa$\sqrt{m}$.

CASE 3
Failure of Arizona Generator Rotor Forging[19–21]

This case history does not describe a true service failure, since the rotor failed during a routine balancing test *before* it was placed in service and at an operating speed *less* than that for design operation. The forged rotor, manufactured more than 20 years ago, did not possess benefits derived from current vacuum degassing melting practices as described in Chapter 10; consequently, a large amount of hydrogen gas was trapped in the ingot as it solidified. With time, the hydrogen precipitated from the solid to form hydrogen flakes, evidenced by disk-shaped internal flaws such as the one shown in Fig. 14.13. Investigators[19,20] concluded that these 2.5- to 3.8-cm-diameter circular defects existed before the balancing test and were responsible for its failure, although no specific hydrogen flake could be identified as the critical nucleation site.

The forging material contained 0.3C, 2.5Ni, 0.5Mo, and 0.1V, exhibited room temperature tensile yield and ultimate strengths of 570 and 690 MPa, respectively, and a Charpy V-notch impact energy at the fracture temperature (27°C) of 5.4 to 16.3 J. The rotor contained a central hole along its entire bore. This was done to remove the central section of the original ingot, which normally contains a relatively high percentage of inclusions and low melting point micro-constituents, and to permit a more thorough examination of the rotor for evidence of any defects.[19] By introducing the bore hole, the centrifugal tangential stresses at the innermost part of the rotor are doubled according to Eqs. 14-5 and 14-6, even when the inner bore diameter is very small:

FIGURE 14.13 Hydrogen flake (dark circle) that contributed to fracture of Arizona turbine rotor.[19] (Reprinted with permission from Academic Press.)

$$\sigma_{\max_{(\text{solid cylinder})}} = \frac{3 + v}{8} \rho\omega^2 R_2^2 \qquad (14\text{-}5)$$

$$\sigma_{\max_{(\text{hollow cylinder})}} = \frac{3 + v}{4} \rho\omega^2 \left(R_2^2 + \frac{1 - v}{3 + v} R_1^2 \right) \qquad (14\text{-}6)$$

where v = Poisson's ratio
ρ = mass density
ω = rotational speed
R_1 = inner radius
R_2 = outer radius

Although one would normally try to keep stresses as low as possible, the higher stress levels associated with introduction of the bore hole are justified for the reasons cited above. Using these equations, Yukawa et al.[19] determined the maximum bore tangential stress to be 350 MPa at the fracture speed (3400 rpm).

From the above description of the Arizona rotor failure, the most reasonable stress intensity factor calibration would appear to be that associated with an internal circular flaw.[21] Assuming the worst condition, where the flaw is oriented normal to the maximum bore tangential stress, we have from Fig. 8.7

$$K_{IC} = \frac{2}{\pi} \sigma\sqrt{\pi a}$$

TABLE 14.1 K_{IC}–CVN Correlations

	Estimated K_{IC}	
CVN, J (ft-lb)	Barsom–Rolfe[22] MPa$\sqrt{m}$ (ksi$\sqrt{in.}$)	Sailors–Corten[23] MPa$\sqrt{m}$ (ksi$\sqrt{in.}$)
5.4–16.3	24–55	34–59
(4–12)	(22–50)	(31–54)

Using the K_{IC}–CVN relations proposed by Barsom and Rolfe[22] and Sailors and Corten[23] (see Chapter 9) for the transition temperature regime where

$$\frac{K_{IC}^2}{E} = 2(CVN)^{3/2}* \quad \text{(Barsom–Rolfe)} \tag{14-7}$$

$$\frac{K^2{}_{IC}}{E} = 8(CVN)* \quad \text{(Sailors–Corten)} \tag{14-8}$$

estimates of the K_{IC} value for the rotor material were obtained and are summarized in Table 14.1. These values must be considered as first-order approximations in view of normal test scatter in Charpy energy measurements and the empirical nature of both Eqs. 14-7 and 14-8, but they do provide a starting point from which critical flaw sizes may be computed and compared with experimentally observed hydrogen flake sizes. (Obviously it would have been more desirable to have actual fracture toughness values to use in these computations.) For example, using the K_{IC} values derived from the Sailors–Corten relation in Eq. 14-8, the critical flaw size range is calculated to be

$$34 \text{ to } 59 = \frac{2}{\pi}(350)(\sqrt{\pi a})$$

$$a = 0.74 \text{ to } 2.2 \text{ cm}$$

or a hydrogen flake diameter range of about 1.5 to 4.3 cm, in excellent agreement with the observed size of these preexistent flaws. The reader should take comfort in the knowledge that hydrogen flakes have been eliminated from current large forgings by vacuum degassing techniques, and overall toughness levels of newer steels have been increased measurably.

CASE 4
Failure of Pittsburgh Station Generator Rotor Forging[19,21]
The Pittsburgh rotor was similar in design and material selection to the Arizona rotor described in the previous case history except that it did not contain a bore hole. Consequently, the stresses were computed from Eq. 14-5 to be roughly half those found in the Arizona rotor. On the other hand, the lack of the bore hole increased the likelihood of finding potentially damaging microconstituents along the rotor center

---

* English units.

line. As we will see, the latter potential condition was realized and did contribute to the fracture. The Pittsburgh rotor failed on March 18, 1956 during an overspeed check. (Overspeed checks were conducted routinely after a shutdown period and before the rotor was returned to service.) The rotor was designed for 3600 rpm service and failed when being checked at 3920 rpm. It is important to note that on 10 previous occasions during its two-year life the rotor satisfactorily endured similar overspeed checks above 3920 rpm. Surely, failure during the eleventh check must have come as a rude shock to the plant engineers. One may conclude, therefore, that some subcritical flaw growth must have taken place during the two-year service life to cause the rotor to fail during the eleventh overspeed test but not during any of the other 10 tests, even though these tests were conducted at higher stress levels. Macrofractographic examination revealed the probable initiation site to be an array of nonmetallic inclusions in the shape of an ellipse 5 × 12.5 cm and located nearly on the rotor center line (Fig. 14.14).[19] The maximum bore tangential stress at burst speed was found to be 165 MPa and the temperature at burst equal to 29°C. The tensile properties of the rotor material were given as 510 and 690 MPa for the yield and tensile strength, respectively, with the room temperature Charpy impact energy equal to 9.5 J.

If we take the critical flaw to be equivalent to a 5 × 12.5-cm elliptical crack— assuming that all the inclusions had linked up prior to catastrophic failure (possibly as a result of subcritical flaw growth)—the stress intensity factor at fracture could be given by

$$K = \sigma\sqrt{\pi a/Q}$$

FIGURE 14.14 **Cluster of inclusions contributing to fracture of Pittsburgh turbine rotor.[19] (Reprinted with permission from Academic Press.)**

The elliptical flaw shape factor Q for the condition where $a/2c = 2.5/12.5 = 0.2$ and $\sigma/\sigma_{ys} = 165/510 = 0.32$ is found from Fig. 8.7h to be 1.28. The fracture toughness of the material is then calculated to be

$$K_{IC} = 165 \sqrt{\frac{\pi(2.5 \times 10^{-2})}{1.28}}$$

$$K_{IC} \approx 41 \text{ MPa}\sqrt{m}$$

This result compares very favorably with K_{IC} estimates based on the Barsom–Rolfe[22] and Sailors–Corten[23] K_{IC}–CVN correlations (Eqs. 14-7 and 14-8), where values of 37 and 45 MPa$\sqrt{m}$ may be computed, respectively.

Although the estimated K_{IC} value derived from the crack configuration and stress information was remarkably close to the values determined from the empirical K_{IC}–CVN correlations, it must be kept in mind that the latter values represent only a crude approximation of K_{IC}. Such derived values can vary widely because of the considerable scatter associated with Charpy energy measurements. Nevertheless, the basic merits of using the fracture mechanics approach to analyze this failure have been clearly demonstrated.

CASE 5
Stress Corrosion Cracking Failure of the Point Pleasant Bridge[24]
The failure of the Point Pleasant, West Virginia, bridge in December 1967 occurred without warning, resulting in the loss of 46 lives. Several studies were conducted immediately afterward to determine the cause(s) of failure, since the collapse caused considerable anxiety about the safety of an almost identical bridge built around the same time and possessing a similar design and structural steel. Failure was attributed to brittle fracture of an eyebar (Fig. 14.15) that was about 17 m long, 5.1 cm thick, and 30.5 cm wide in the shank section. The ends of the bar were 70 cm in diameter and contained 29.2-cm-diameter holes. It was determined that a crack had traversed one of the ligaments (the one on the top in Fig. 14.15) of the eye (along the transverse center line) with little apparent energy absorption (the fracture surface was very flat with little shear lip formation). The ligament on the opposite side of the hole suffered extensive plastic deformation before it failed, probably as a result of a bending overload. After removing the rust from the fracture surface, investigators[24] found two discolored regions covered with an adherent oxide layer. These regions were contiguous and in the shape of two elliptical surface flaws (Fig. 14.16). The size of the large flaw was

$$a = 0.3 \text{ cm}$$

$$2c = 0.71 \text{ cm}$$

$$a/2c = 0.43$$

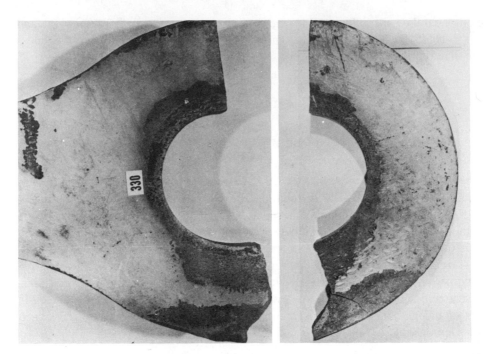

FIGURE 14.15 Fractured eyebar responsible for failure of Point Pleasant Bridge.[24] (Reprinted from *Journal of Testing and Evaluation* with permission from American Society for Testing and Materials.)

FIGURE 14.16 Fracture surface of broken eyebar from Point Pleasant Bridge showing two elliptical surface flaws.[24] (Reprinted from *Journal of Testing and Evaluation* with permission from American Society for Testing and Materials.)

The smaller flaw had the dimensions

$$a \approx 0.1 \text{ cm}$$

$$2c = 0.51 \text{ cm}$$

$$a/2c \approx 0.2$$

Portions of the hole surface were heavily corroded, and some secondary cracks were parallel to the main fracture surface but initiated only in those regions where corrosion damage was extensive. These findings suggested the strong possibility that stress corrosion and/or corrosion fatigue mechanism(s) were involved in the fracture process. The hypothesis was further substantiated by metallographic sections which showed that the secondary cracks contained corrosion products and propagated in an irregular pattern from corrosion pits at the hole surface. Furthermore, some of these secondary cracks were opened in the laboratory, examined in the SEM and electron microprobe, and found to contain high concentrations of sulfur near the crack origin.[25] The presence of sulfur on the fracture surface was believed to be from H_2S in the air near the bridge rather than associated with manganese sulfide inclusions (commonly found in this material). The sensitivity of the bridge steel to H_2S stress corrosion cracking was verified by several tests performed on notched specimens. Fatigue crack propagation data were also obtained and used to examine the possibility that the two surface flaws had propagated instead by corrosion fatigue. Taking the maximum alternating stress on the bridge to be ± 100 MPa, Bennett and Mindlin[24] estimated that it would require over half a million load cycles to propagate a crack from a depth of 0.05 cm to one 0.25 cm deep. Since this was considered to be an unrealistically large number, it was concluded that the actual fracture mechanism was stress corrosion cracking.

Attention was then given to an evaluation of the steel's fracture-toughness capacity. Using both Charpy V-notch and fracture-toughness test procedures, the SAE 1060 steel (0.61 C, 0.65 Mn, 0.03 S), which had been austenitized, water quenched, and tempered for 2 h at 640°C, was shown to be brittle. For example, the material was found to exhibit an average plane-strain fracture toughness level of 51 MPa$\sqrt{m}$ at 0°C, the temperature of fracture. This low value is consistent with the fact that the material displayed a strong stress corrosion cracking tendency—something usually found only in more brittle engineering alloys (see Chapter 11). Based on a measured yield strength of 550 MPa, these results were found to reflect valid plane-strain test conditions for the specimen dimensions chosen.

Estimating the stress intensity level by

$$K = 1.1\sigma\sqrt{\pi a/Q}$$

Bennett and Mindlin computed the stress level at fracture by considering only the larger surface flaw:

$$K = 1.1\sigma\sqrt{\pi a/Q}$$

$$= 1.1\sigma\sqrt{\frac{\pi(3 \times 10^{-3})}{1.92}}$$

$$= 7.7 \times 10^{-2}\sigma$$

or

$$\sigma = 13\ K$$

Using the range of experimentally determined K_{IC} values (47.3 to 56.1 MPa$\sqrt{m}$, the stress level at fracture was found to be

$$\sigma = 615 - 730\ \text{MPa}$$

This represents an upper bound range of the fracture stress, since allowance was not made for the presence of the smaller contiguous elliptical flaw. If one assumes the crack to be elliptical with a maximum depth of 0.3 cm but with $2c = 1.6$ cm, then $a/2c \approx 0.19$ and $Q = 1.05$. This assumption should lead to a slight underestimate of the stress level:

$$K = 1.1\sigma\sqrt{\frac{\pi(3 \times 10^{-3})}{1.05}}$$

$$\sigma = 9.6K$$

Again using the K_{IC} range of 47.3 to 56.1 MPa$\sqrt{m}$, a lower stress range is found to be

$$\sigma = 455 - 540\ \text{MPa}$$

It is concluded that the actual stress range for failure was

$$455 - 540 < \sigma_{\text{actual}} \ll 615 - 730\ \text{MPa}$$

It is seen that the failure stress is approximately equal to the material yield strength. Since the shank section of the eyebar was recommended for a design stress of 345 MPa, Bennett and Mindlin concluded that stresses on the order of the yield strength could exist at the considerable stress concentration associated with this region.

On the basis of this detailed examination, it was concluded that the critical flaw was developed within a region of high stress concentration and progressed by a stress corrosion cracking mechanism to a depth of only 0.3 cm before fracture occurred. Consequently, the hostile environment, the inability to adequately paint the eyebar and thus protect it from atmospheric attack, the low fracture toughness of the material, and the high design stress all were seen to contribute to failure of the bridge. It should come as no surprise that the combination of low toughness and high stress would result in a small critical flaw size (see Eq. 8-28).

CASE 6
Weld Cold Crack-Induced Failure of Kings Bridge, Melbourne, Australia[26]

On a cold winter morning in July 1962, while a loaded truck with a total weight of 445 kN was crossing the bridge, a section of this 700-m-long elevated four-lane freeway fractured, causing a portion of the bridge to drop 46 cm. Examination of the four main support girders that broke revealed that all had suffered some cracking *prior* to installation (Fig. 14.17). Indeed, subsequent welding tests established that a combination of poor detail design of the girder flange cover plate, poor weldability of the steel, poor welding procedure, and failure to properly dry low-hydrogen electrodes before use contributed to the formation of weld cold cracks located at the toe of transverse welds at the ends of the cover plates. In three of these girders, 10-cm-long through-thickness cracks had developed before erection but none were ever discovered during inspection. In addition, it was determined that girder W14-2 was almost completely broken before the span failed. (The crack in this girder extended across the bottom flange and 1.12 m up the web.)

The collapse of the span was traced to failure of girder W14-3, which contained a T-shaped crack extending 12.5 cm across the bottom flange and 10 cm up the web (Fig. 14.17). Madison[26] postulated that the stress intensity condition at instability could be approximated by the superposition of two major components. One major K component was attributed to uniform bending loads acting along the flange and perpendicular to the 12.5-cm-long flange crack. Accordingly

$$K = \sigma\sqrt{\pi a}\sqrt{\sec \pi a/W} \qquad (8\text{-}22)$$

where $\sqrt{\sec \pi a/W}$ = finite width correction

σ = bending stress, 83 MPa

a = 6.25 cm

W = 41 cm

$K = 83\sqrt{\pi(6.25 \times 10^{-2})\ \sec \pi(6.25/41)}$

    = 39 MPa$\sqrt{m}$

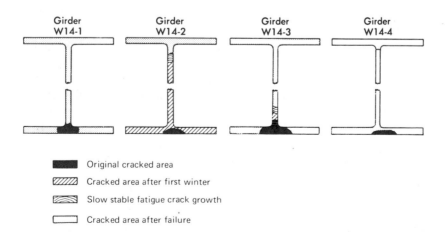

Original cracked area

Cracked area after first winter

Slow stable fatigue crack growth

Cracked area after failure

FIGURE 14.17   **Diagram showing extent of cracking of girders from Kings Bridge, Australia[26] (Courtesy of Dr. Ronald Madison.)**

The second K component was related to load transfer from the web, which produced wedge force loads extending 10 cm along both sides of the flange crack. These loads reflect residual stresses generated by the flange to web welds. For this configuration the K calibration is[16,27]

$$K = \frac{\sigma\sqrt{a}}{\sqrt{\pi}}\left[\sin^{-1}\frac{c}{a} - \left(1 - \frac{c^2}{a^2}\right)^{1/2} + 1\right] \tag{14-9}$$

where $2a$ = crack length, 12.5 cm
σ = wedge force, 262 MPa
$2c$ = length of wedge force, 10 cm

$$K = \frac{262\sqrt{6.25 \times 10^{-2}}}{\sqrt{\pi}}\left\{\sin^{-1}\frac{5}{6.25} - \left[1 - \left(\frac{5}{6.25}\right)^2\right]^{1/2} + 1\right\}$$

$$K = 49 \text{ MPa}\sqrt{m}$$

Therefore, $K_T = 39 + 49 = 88$ MPa$\sqrt{m}$. Note the significant contribution of the residual stresses. This value was found to be in reasonably good agreement with the dynamic fracture toughness of samples prepared from the bridge steel.

CASE 7
Failure Analysis of 175-mm Gun Tube[28]

In April 1966, U.S. Army gun tube No. 733 failed catastrophically after a crack located near the breech end of the tube reached critical proportions. Brittle fracture was suspected since little evidence could be found for plastic deformation. The gun barrel, manufactured from a high-strength steel alloy, broke into 29 pieces that were hurled over distances up to 1.25 km from the firing site (Fig. 14.18). Davidson and coworkers[28] reported this to be the first such brittle fracture of the 175-mm gun tube. Previously, large-caliber gun tubes manufactured from medium-strength, high-toughness steel had been reported typically to fail by excessive wear and erosion of the barrel bore, with such wear resulting in a loss of projectile accuracy.[29-31] Since gun tube No. 733 had been manufactured to a higher strength but lower toughness specification, these latter properties were immediately called into question as being responsible for the catastrophic failure.

To analyze the cause of this fracture, we follow the outline of the "Checklist" and define the component configuration, the prevailing stresses prior to and at the time of the fracture, the details of the critical flaw, and the material properties. For stress analysis purposes, the gun barrel can be thought of as being a thick-walled tube, 10.5 m in length, with outer and inner diameters of 37.3 and 17.8 cm, respectively. At the time of failure, the gun was being fired at two-minute intervals, with the final round generating a nominal pressure of 345 MPa. Altogether, the gun tube experienced 373 rounds at a nominal peak pressure of 345 MPa and 227 rounds at a pressure of 152 MPa. The fracture surfaces of the many broken segments revealed a predominantly flat-fracture appearance, indicative of plane-strain fracture conditions. The critical flaw was found to be semielliptical in shape, as denoted by its darkened appearance

FIGURE 14.18   Fragments from exploded 175-mm gun tube. (After Davidson et al.[28])

(presumably a result of the deposition of combustion products during firing), with half-minor axis and major axis dimensions of 0.94 and 2.79 cm, respectively (Fig. 14.19a). The material was a forged AISI 4335 steel, modified with respect to the overall Cr and Mo content and by the addition of 0.14% V.[28] Selected tensile and fracture properties of this material are shown in Table 14.2.

Davidson and coworkers[28] initially considered the possibility that failure had occurred as a result of higher than expected pressure during firing; it was thought that this condition would account for the early development and growth of the critical flaw and its small final dimensions. Subsequent examination of the gun tube fragments, however, revealed no evidence of overpressure. Furthermore, nothing abnormal was found when tests were conducted of the ammunition being fired at the time of the failure. The possibility of the environment-assisted cracking under sustained loading conditions was also ruled out since the time under service load (during actual firing) was too short (about 20×10^{-3} s) and the magnitude of residual tensile stresses in the tube was too low. Finally, loading rate effects on the material fracture toughness were not considered to be of any consequence for this high-strength steel (recall Section 9.4).

After further analysis of the fracture surface markings, the character of the steel's microstructure, and the prevailing stress intensity levels, the following fracture scenario was identified. Crack initiation was believed to have occurred on the inner bore of the gun tube from a thermally induced cracking process known as "heat checking." This results in the development of a random network of cracks that typically penetrate

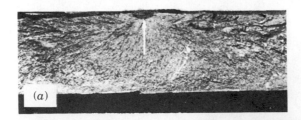

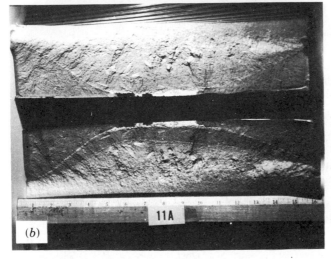

FIGURE 14.19   Fracture surfaces of broken 175-mm gun tubes. (*a*) Fracture surface of gun tube No. 733. Note small semielliptical surface flaw representing critical crack size. (*b*) Fracture surface of autofrettaged gun tube revealing leak-before-break condition. (After Davidson et al.[28])

up to 0.13 cm below the inner bore surface, which is in contact with the hot combustion gases. For the firing conditions associated with this gun tube, the heat checking pattern was found to be fully developed after only ten rounds of ammunition were fired. The total life of the gun tube was then assumed to reflect only fatigue crack propagation (one-round = one loading cycle) during which time the crack grew from a presumed depth of 0.13 cm to the 0.94×2.79-cm semielliptical configuration at fracture.

TABLE 14.2   Mechanical Properties of Gun Tube No. 733

Property	Undefined	Near Failure	Toward Muzzle	Toward Breech
Yield strength, MPa (ksi)	1180 (171)			
Tensile strength, MPa (ksi)	1385 (201)			
Elongation, %	10			
21°C reduction area, %		9–28	17–22	18–34
−40°C Charpy energy, J (ft-lb)		6.1–8.8 (4.5–6.5)	10.2–11.5 (7.5–8.5)	5.4–11.5 (4.0–8.5)
21°C fracture toughness, MPa$\sqrt{m}$ (ksi$\sqrt{in.}$)		89–91 (81–83)	74–99 (67–90)	—

TABLE 14.3 Fracture Data for 175-mm Gun Tubes with 170–190 ksi Yield Strength

Tube No.	Total Cycles to Failure	σ_{ys} MPa	(ksi)	Charpy J	(ft-lb)	K_{IC} MPa$\sqrt{m}$	(ksi$\sqrt{in.}$)	Critical Flaw cm	(in.)
733	373	1180	(171)	8.1	(6)	88	(80)	0.94	(0.37)
863	1011	1270	(184)	12.2	(9)	103	(94)	4.3	(1.7)
1131	9322	1255	(182)	19	(14)	142	(129)	4.3	(1.7)
1382	1411	1275	(185)	14.9	(11)	108	(98)	3.8	(1.5)
1386	4697	1250	(181)	19	(14)	116	(106)	4.6	(1.8)
Typical values for 35 tubes	4000	1240	(180)	16.3	(12)	121	(110)	3.8	(1.5)

Judging from the low fracture-toughness properties of the steel near the failure site (Table 14.3) and evidence for intergranular and cleavage micromechanisms on the fatigue fracture surfaces, Davidson et al.[28] concluded that a condition of temper embrittlement had contributed to both accelerated fatigue crack growth and premature final fracture of tube No. 733. A study of other gun tubes confirmed the relation between gun tube life and material fracture properties. Note in Table 14.3 that the total cycles to failure (at 345 MPa) and the final flaw depth increased with increasing Charpy energy and fracture toughness.

The stress intensity factor in an internally pressurized thick-walled tube containing a long, straight surface flaw located in the inner bore is given by Bowie and Freese[32] in the form

$$K = f(a/W, r_2/r_1)P\sqrt{\pi a} \tag{14-10}$$

where a/W = crack depth to tube thickness ratio
r_2/r_1 = outer-to-inner radius ratio
P = internal pressure
a = crack depth with crack plane being normal to hoop stress direction

At final failure, where $a = 0.94$ cm,

$$K = 2.7P\sqrt{\pi a} \tag{14-11}$$

Since the crack shape at fracture was semielliptical, Eq. 14-11 was modified[33] for the appropriate a/W and $a/2c$ values such that

$$K = 1.7P\sqrt{\pi a} \tag{14-12}$$

The stress intensity factor at fracture in association with $P = 345$ MPa and $a = 0.94$ cm is therefore computed to be 99 MPa$\sqrt{m}$. This value is in fairly good agreement with the reported toughness for the tube material (Table 14.2). To estimate the service life of gun tube No. 733, the crack growth rate expression in Eq. 14-13 was integrated

$$\frac{da}{dN} = 6.49 \times 10^{-12}\Delta K^3 \tag{14-13}$$

where *da/dN* is m/cycle. (This relation was derived from laboratory tests conducted on a material with 50% higher toughness.) Since the calibration factor Y for the changing crack front configuration in the tube varied with the crack length, the integration should most properly be carried out numerically or in parts where Y is held constant over the various intervals of integration. As a first approximation, the integration was performed assuming that Y possesses a constant value of 2.2, corresponding to a simple average between the values of 2.7 and 1.7 in Eqs. 14-11 and 14-12, respectively. The computed life, assuming only stress fluctuations with a range of 345 MPa, was found to be 2070 cycles, between 5 and 6 times greater than the number of 345 MPa stress fluctuations experienced by the gun tube prior to failure.

Several reasons can be given to show that the actual and computed gun tube lives are actually in much closer agreement. A more realistic determination of Eq. 14-13 should reflect the temper-embrittled nature of the material. For example, Ritchie[34] reported FCP rates 2.5 times greater in a temper-embrittled 43XX type steel than in properly treated samples of the same material. Also, the low fracture toughness of the material in gun tube No. 733 would be expected to result in higher crack growth rates at a given ΔK level (recall Eq. 13-10). Finally, Eq. 14-13 was based on test results from laboratory air-test conditions and not from experiments conducted in the presence of more aggressive hot combustion gas products. Taken together, these factors would all be expected to lower the estimated fatigue life below the 2070-cycle value initially computed. Furthermore, the effective service life is most likely greater than 373 cycles at a nominal pressure of 345 MPa since no damage was attributed to the 227 rounds fired at a pressure of 152 MPa. (It is estimated that the life of gun tube No. 733 would have been about 10% greater in the absence of the 227 lower stress rounds.)

The failure analysis report contained additional information pertaining to the avoidance of future gun tube fractures. As a short-range interim procedure, all gun tubes possessing a Charpy impact energy less than 13.5 J were immediately withdrawn from the field. Other tubes were assigned a reduced service life of 300 rounds at 345 MPa instead of the original 800 rounds. Following these changes, no additional field failures occurred. Gun tubes currently in the manufacturing process were heat treated to a lower strength level so that both impact and fracture properties could be increased. Indeed, the cyclic life of these gun tubes increased to about 10,000 rounds, while the final crack depth at fracture was twice that shown in Table 14.3. To further minimize the risk of brittle fracture, gun tubes were subsequently heat treated to a lower yield strength in the range of 965 to 1100 MPa and given an autofrettage treatment. In the autofrettage treatment used in this case (recall Section 12.6.1), the gun tube is subjected to a hydrostatic internal pressure sufficiently high to produce plastic deformation about halfway across the tube thickness. When this pressure is removed, the yielded zone experiences a compressive residual stress gradient with the highest compressive stress located at the inner surface of the gun tube (Fig. 14.20). Note that the compressive residual stress is numerically greater than the hoop stress at the inner bore. As a result, the fatigue life should increase appreciably. To wit, autofrettaged tubes withstood more than 20,000 firing cycles at 345 MPa, representing a 50-fold improvement in fatigue life over that experienced by gun tube No. 733! Associated with this vast improvement in the fatigue life of the gun tube was a trend toward stable fatigue crack propagation completely through the tube wall (Fig. 14.19*b*); hence,

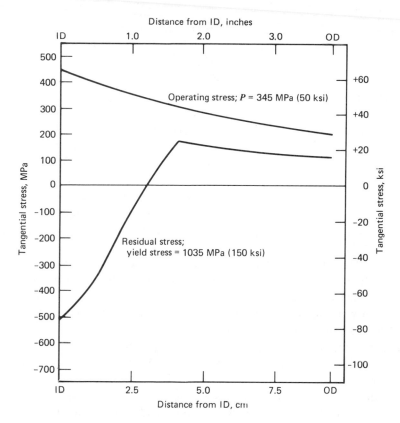

FIGURE 14.20   Operating hoop stress gradient in gun tube versus residual stress profile resulting from autofrettage treatment. Note overall compressive stress at inner wall of tube. (After Davidson et al.[28])

the combination of an increase in fracture toughness, because of a reduction in yield strength, and the development of a favorable residual compressive stress, a result of the autofrettage treatment, created a leak-before-break failure condition (recall Section 8.3).

CASE 8
Hydrotest Failure of a 660-cm-Diameter Rocket Motor Casing[35]

This failure analysis describes the catastrophic rupture of a 660-cm rocket motor casing that fractured prematurely during a hydrotest at an internal pressure of only 56% that of the planned value. Experiments of this type were being performed to demonstrate the feasibility of designing solid-propellant rocket casings with a thrust capacity of 27×10^6 N. This particular case had been fabricated by welding together many sections of a 250-grade air-melted maraging steel. Nominal yield- and tensile-strength values for the base plate (1.85 cm thick at the fracture origin) were 1585 and 1725 MPa, respectively, and the weld efficiency was assumed to be 90%. Initially, approximately 300 m of longitudinal and circumferential welds were prepared by a submerged arc process. Subsequent nondestructive inspections revealed the presence of numerous weld defects that were removed by grinding and repaired using a manual TIG welding process. In turn, some of these weld repairs were found to be defective and in need of repair. Altogether, approximately 100 m of weld repairs and re-repairs were required.

It was planned that the hydrotest be extended to a water pressure of 6.6 MPa, 10% above the maximum expected operating pressure of the rocket motor casing. Instead, failure occurred when the internal water pressure had reached only 3.7 MPa. During the course of the test, a number of stress waves were detected with the aid of several accelerometers and strain gages that had been mounted onto the casing. Although some of these waves may have reflected the relative motion of motor casing components, such as bolts within bolt holes, other stress waves, including the ones associated with final fracture, most likely were associated with subcritical crack growth.

On fracturing, the rocket motor casing broke into a large number of pieces that were subsequently collected and reassembled as shown in Fig. 14.21. The relative locations of the fracture segments and the local directions of crack propagation were determined by noting the chevron markings on the fracture surfaces (recall Section 7.7.2). In addition, the chevron pattern revealed that fracture had originated from two preexistent flaws, which were fairly close to one another. Both of these cracks were located within the heat-affected zone (HAZ) beneath the TIG weld repair and within the coarse-grained heat-affected zone of the submerged arc weld. Electron fractographic studies revealed that the fracture had progressed in an intergranular fashion through these coarse-grained regions. For this reason, it is quite possible that these defects had been produced by cold cracking in the HAZ. The primary flaw had a clean appearance, was elliptical in shape (3.6×0.25 cm), and was oriented parallel

FIGURE 14.21 **Reassembled fragments from ruptured 660-cm rocket motor casing. (From Srawley and Esgar.[35])**

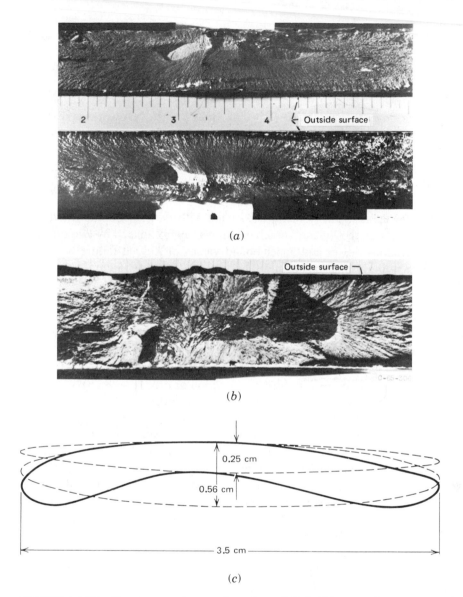

FIGURE 14.22 Fracture surface regions from failed 660-cm rocket motor casing. (*a*) Mating fracture surface of primary flaw (3.6 × 0.25 cm). (*b*) Secondary weld flaw. Note oxide-induced discoloration of fracture surface. (*c*) Approximate configuration of primary flaw. (From Srawley and Esgar.[35])

to a longitudinal weld centerline (Fig. 14.22*a*). The other defect was oriented perpendicular to the longitudinal weld centerline and measured roughly 4 × 0.5 cm. Furthermore, part of the fracture surface of this defect was black and extended to the surface of the casing (Fig. 14.22*b*). X-ray diffraction analysis determined that the black debris was a combination of Fe_2O_3 and Fe_3O_4; presumably, this preexistent defect, which extended to the free surface, became oxidized during the 450°C aging treatment.

Investigators concluded from an examination of the chevron pattern on the fracture surface that final failure had initiated at the "clean" defect. Furthermore, by triangulating the stress wave signals from the accelerometer devices, the stress wave origin was located much closer to the "clean" flaw than to the "black" flaw. It was reasoned that, after the "clean" crack began to propagate unstably, the changing stress distribution resulted in independent growth from the "black" origin. It is interesting to note that had the "clean" defect been detected and removed prior to the hydrotest, premature failure still would have occurred; in this instance, the "black" defect would have provided the initiation site.

To estimate the stress intensity factor associated with the catastrophic fracture of this rocket motor casing, the "clean" crack origin was approximated by two different elliptical configurations; lower and upper bound values of K_{max} at failure were estimated by assuming the minor axis of the flaw to be 0.25 and 0.56 cm, respectively (Fig. 14.22c). In both instances,

$$K = \sigma\sqrt{\pi a/Q} \cdot f(a/t) \tag{14-14}$$

where σ = design stress, approximated by the hoop stress
        a = one-half minor axis of the elliptical flaw
        Q = elliptical correction factor, a function of $a/2c$ (Fig. 8.7h)
$f(a/t)$ = finite width correction factor, $\sqrt{\sec \pi a/t}$ (Eq. 8-22)

The applied stress σ, neglecting any additional stress component from manufacturing-induced residual stresses, was computed to be approximately 690 MPa. (Since two separate investigative teams reached different conclusions regarding the possible existence and sense of a residual stress pattern, no attempt was made to include such a stress component in the stress estimate.) From Eq. 14-14, the lower- and upper-bound estimates of K_{max} are 44 and 65 MPa$\sqrt{m}$, respectively. These estimates are then compared with measured values of the material's fracture toughness, corresponding to the microstructure surrounding the "clean" crack. The most accurate estimate of K_{IC} in this region would have required testing a sample with the crack tip embedded in the overlapping heat-affected zones of the submerged arc and the TIG manual repair welds. Unfortunately, no fracture-toughness specimens were prepared from such a location. Instead, K_{IC} was based on values obtained from specimens located in the HAZ of a submerged arc weld that had not been TIG-weld repaired. For this location, K_{IC} was measured to be 85 MPa$\sqrt{m}$, which is higher than the computed estimates of the maximum stress intensity factor at fracture. It should be noted, however, that investigators estimated the fracture toughness in the overlapping heat-affected zones to be considerably less than 85 MPa$\sqrt{m}$, based on microstructural and fractographic evidence. As such, one would have expected much better agreement between the computed stress intensity level at fracture and the material's fracture toughness, had the latter been determined in the relevant region of the microstructure.

One of the major conclusions drawn from the analysis of this fracture was the fact that the NDT techniques used in the manufacture of this rocket motor casing (dye penetrant, ultrasonic, radiographic, and visual) were much less sensitive and reliable than had been expected. (Recall the similar circumstances surrounding the F-111 aircraft accident that were discussed at the beginning of this chapter.) In addition to the two overlooked defects already discussed, 11 other defects were discovered during a postfailure reinspection of the welds. To be sure, it is uncertain how many of these defects initiated and/or grew to detectable dimensions during the hydrotest. At any rate, it is instructive to compare the loading conditions necessary to fracture a test specimen containing one of these defects with that value based on a fracture mechanics computation. To this end, a section of the casing containing a weld defect was removed and tested in tension to fracture (σ_{max} = 793 MPa). The fracture surface revealed a defect, 1.5 cm long × 0.3 cm wide, with the long dimension oriented parallel to the plate surface. This defect was located in the heat-affected zone of a TIG repair weld, which was embedded, in turn, in the center of submerged arc weld metal. For a fracture-toughness specimen containing a crack in the submerged arc weld metal zone, the best estimate of K_{IC} was 54.1 MPa$\sqrt{m}$. Using Eq. 14-14, the maximum stress intensity factor associated with fracture of the section containing the 1.5 cm × 0.3 cm flaw is 53.2 MPa$\sqrt{m}$, in excellent agreement with the material's intrinsic resistance to fracture.

CASE 9
Premature Fracture of Powder-Pressing Die[36]

Sintered metal powder rods were to be compacted in the die shown schematically in Fig. 14.23. After the powder was placed in the die, the charge was compressed from both ends with two movable plungers. During the die's initial compaction cycle, sudden failure occurred at an applied stress level roughly 20% *below* the rated value. Why?

On close examination, the fracture origin was traced to a small semicircular flaw located at the surface of the inner bore near one end of the cylinder (Fig. 14.24). This defect experienced the full effect of the hoop stress σ_t in the thick-walled cylinder as given by

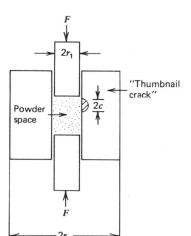

FIGURE 14.23 Schematic drawing of powder-pressing die. (From Ashby and Jones[36]; with permission from Pergamon Press.)

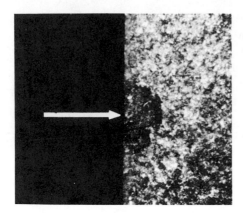

FIGURE 14.24 Photograph showing semicircular surface flaw at inner bore surface of die. (From Ashby and Jones[36]; with permission from Pergamon Press.)

$$\sigma_t = P\frac{\left(\dfrac{1}{r^2} + \dfrac{1}{r_o^2}\right)}{\left(\dfrac{1}{r_i^2} - \dfrac{1}{r_o^2}\right)} \qquad (14\text{-}15)$$

where r_i = inner radius of thick-walled cylinder
r_o = outer radius of thick-walled cylinder
r = radius
P = internal pressure
σ_t = hoop stress (tangential)

For this die, the relevant dimensions of the cylinder and initial flaw size are

$$r_i = 6.4 \text{ mm}$$

$$r_o = 38 \text{ mm}$$

$$a = 1.2 \text{ mm}$$

$$2c = 2.4 \text{ mm}$$

Neglecting the presence of the flaw, the tangential stress acting at the surface of the inner bore is found to be 1.06 P. The maximum allowable pressure for this die, based on a failure criterion of incipient plastic deformation in conjunction with a safety factor of three, is then given by

$$P = \frac{\sigma_{ys}}{1.06 \times 3} = 630 \text{ MPa}$$

where $\sigma_{ys} = 2000 \text{ MPa}$.

Instead of experiencing plastic deformation, the die failed in a brittle fashion associated with a flat fracture appearance. For this specimen–crack configuration, the prevailing stress intensity factor is given by

$$K = 1.1 \frac{2}{\pi} \sigma \sqrt{\pi a} \qquad (14\text{-}16)$$

where σ = tangential stress = 1.06 P

1.1 = surface flaw correction

$\frac{2}{\pi}$ = a semicircular flaw correction factor

(Note the similarity between this relation and Eq. 8-26.) From independent studies, the fracture toughness of this medium carbon chromium steel was determined to be 22 MPa$\sqrt{m}$. Therefore, from Eq. 14-16, the internal pressure at fracture was found to be 512 MPa, roughly 20% below the allowable pressure level.

Surely, the combined effects of low fracture toughness and very high strength level for die material contributed to the premature failure. In addition, evidence of hydrogen cracking was also reported, which could account for the presence of the preexistent flaw. An improved design would involve either the use of this material at a lower strength level and associated higher toughness or the use of another alloy with superior toughness at comparable strength levels.

CASE 10
A Laboratory Analysis of a Lavatory Failure[37]

The reader will certainly agree that the sudden collapse of a toilet seat is enough to distract one's concentration! To think, the thoughts of the day on one's mind and the pieces of a broken polymer literally under one's behind! A segment of such a broken seat was shown previously (Fig. 7.2b). Note the clear evidence for clamshell markings (Fig. 14.25a) indicative of repeated loadings and unloadings as well as variable hold-time periods associated with irregular biological functioning and/or sustained periods of thought or literature review. An analysis of the fractured toilet seat reveals the origin at a surface flaw (see arrow Fig. 12.3c) and the development of a semielliptical fatigue corner crack, which extended to a depth of 0.95 cm and 1.9 cm along the bottom edge of the seat. The last person to use the seat weighed approximately 980 Newtons (220 lb) with an estimated 20–25% of that amount being supported by the individual's legs. From a careful reassessment of the load on the toilet seat at the time of fracture, it was ascertained that the total buttock and thigh loads were applied along only one of the two leaves of the toilet seat. The latter body force distribution was associated with repositioning of the skeletal frame in conjunction with the use of sanitary tissue. Anatomical considerations provide an approximation of the force distribution on this leaf as a linearly decreasing load between the back support and the front tab. The flexural stress at the bottom (tensile) surface of the seat was estimated to be 11 MPa.[37] The stress intensity factor at final failure was estimated from the relationship given in Eq. 14-17:

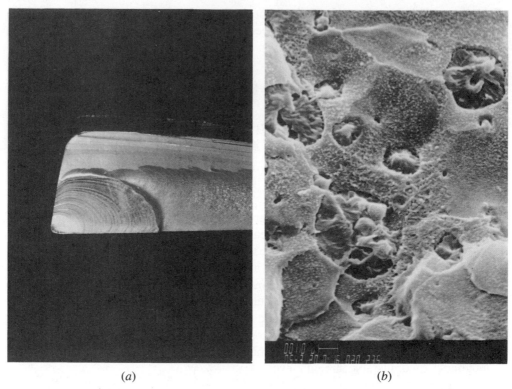

$$(a) \qquad\qquad (b)$$

FIGURE 14.25 Failed toilet seat. (*a*) Fracture surface showing clamshell markings associated with loading and unloading; (*b*) typical fracture surface micromorphology, consistent with appearance of high impact polystyrene (HiPS).[37] R. W. Hertzberg, M. T. Hahn, C. M. Rimnac, J. A. Manson, and P. C. Paris, *Int. Journal of Fracture*, **22**, R57 (1983). Reprinted by permission of Kluwer Academic Publishers.

$$K = 1.1 \, M_B \, \sigma \, \sqrt{\pi a/Q} \cdot \sqrt{\sec \pi a/2t} \qquad (14\text{-}17)$$

where M_B = correction factor for the flexural stress at the crack front away from the bottom surface of the seat[38] (0.9).

σ = flexural stress at bottom surface of seat (11 MPa)

a = crack depth 0.95 cm.

Q = ellipticity correction factor (1.3)

$\sqrt{\sec \dfrac{\pi a}{2t}}$ = finite width correction factor[39] (1.19)

Substitution of the appropriate values into Eq. 14-17 reveals $K_{\max} \approx 2.2 \, \text{MPa}\sqrt{\text{m}}$.

Laboratory studies were then conducted to identify the material used in the manufacture of the toilet seat. First, a fragment of the seat was burned; the resultant trail of smoke was consistent with the presence of a styrene-based polymer. The fractured surface was then suitably coated and examined in the SEM. Figure 14.25*b* reveals a representative region typical of the appearance of high-impact polystyrene (HiPS).[40]

This rubber-modified polymer is a blend of spherical polybutadiene-polystyrene particles embedded within a polystyrene matrix.

From an earlier study,[41] the maximum K level associated with the fatigue fracture of HiPS was found to be approximately 2.4 MPa$\sqrt{m}$ in good agreement with the estimated K_{max} of 2.2 MPa$\sqrt{m}$ for the toilet seat. Finally, the last few fatigue bands seen on the fracture surface were approximately 10 to 20 μm in width and consistent with the macroscopic growth rate in HiPS near final failure. As such, we conclude that these last fracture bands are fatigue striations resulting from individual load excursions associated with attempts at the fulfillment of certain biological functioning. Whether subcritical flaw growth was enhanced by the prevailing aqueous and gaseous environments remains a topic for future study. In conclusion, it is suggested that the reader be mindful of the potential for subcritical flaw growth and premature fracture of toilet seat leaves. A simple nondestructive visual examination is recommended, time permitting.

14.5.1 Additional Comments

Before concluding this section, it is important to comment further on the general problem of fatigue and fracture in welded bridges. Since these structures usually are very complicated because many cover plates, stiffeners, attachments, and splices are added to the basic beam, it is important to recognize the potential danger associated with a particular weld detail. Fisher et al.[42–46] have conducted an extensive study of this problem and proposed several categories of relative attachment detail severity as shown schematically in Fig. 14.26. Categories E and A represent the potentially most damaging and least damaging weld details, respectively. More recently,[46] two additional categories, designated B' and E', were identified which correspond to more severe conditions than those shown in Fig. 14.26. Category B' represents a partial-penetration longitudinal groove weld and category E' represents thicker coverplates [greater than 25 mm (1 in.)] than those associated with category E details.

One important function of these diagrams is to direct the attention of the field engineer to the most critical details in the bridge design so that no time is wasted in examining those areas experiencing a lower stress concentration. Also, the differences in stress concentration associated with categories A, B, C, D, and E have been used by the American Association of State Highway and Transportation Officials to arrive at allowable stress ranges for each detail.[42] For example, a category E detail is allowed only one-third the stress range of a category A region when a cyclic life of up to 2×10^6 cycles is anticipated and only about one-fifth of that value when more than 2×10^6 cycles are desired.

Based on more than 2000 test results, fatigue design curves were defined for each of the welded steel bridge details described above and are shown in Fig. 14.27.[46] Each design curve corresponds to 95% confidence limits for 95% survival based on regression analysis of each respective database. Furthermore, it was concluded that straight-line extensions at a slope of -3.0 to low stress levels of each S-N plot represented a more conservative lower bound estimate of fatigue life for these welded details than one based on discrete endurance limits (dotted horizontal lines) for each

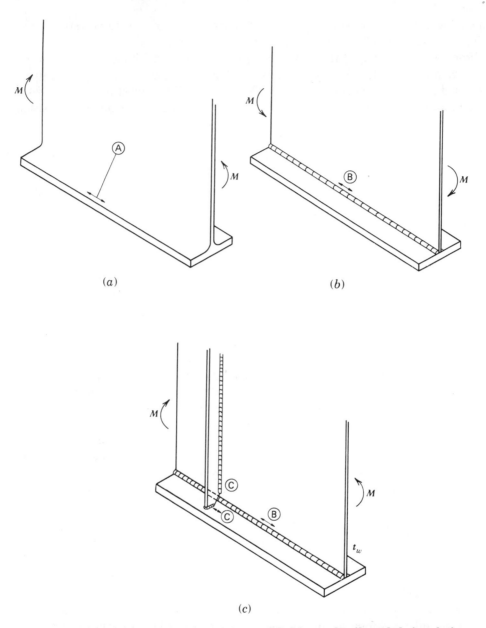

FIGURE 14.26 Drawings showing various welded beam details and their relative stress concentration severity (increasing from category A to E).[42] (Reprinted from *Research Results Digest 59*, with permission of the National Cooperative Highway Research Program.)

detail. Note that Fig. 14.27 essentially shows *S-N* plots for welded details containing different stress concentration factors. (Recall Fig. 12.45*a*).

As a final note, the engineer should remember the subtle but important difference between load- and displacement-controlled conditions governing the behavior of a given structure. For example, for the case of the two fixed-ended beams shown in

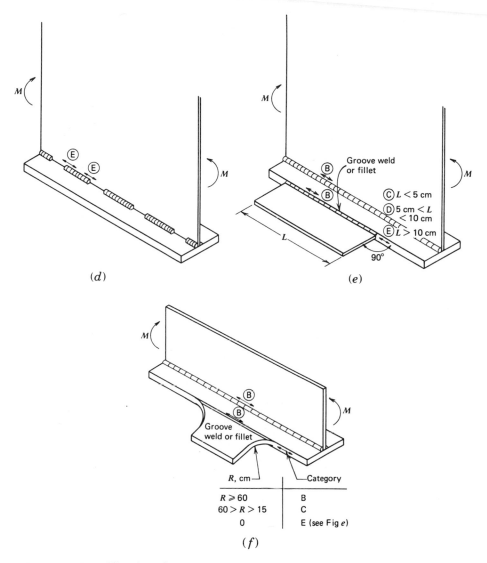

FIGURE 14.26 (*Continued*)

Fig. 14.28, the maximum flexural stress for load-induced and displacement-induced conditions are

$$\text{load-induced: } \sigma_{\text{max}} = \frac{PLc}{8I} \tag{14-18}$$

where σ_{max} = maximum flexural stress
    P = point load
    L = beam length
    c = distance to outermost fiber
    I = moment of inertia

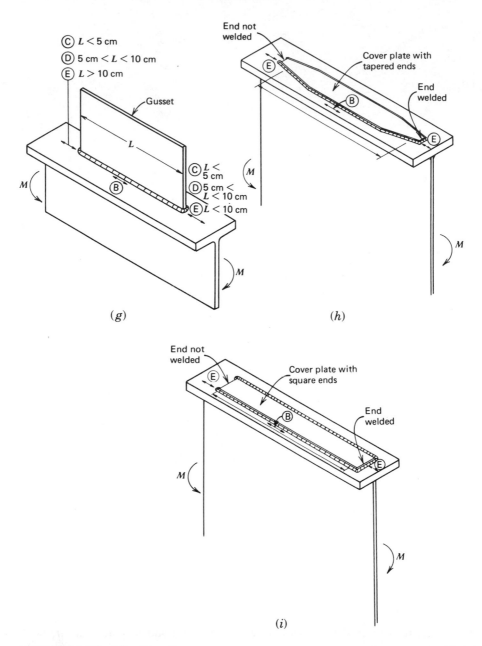

FIGURE 14.26 *(Continued)*

and

$$\text{displacement-induced: } \sigma_{max} = \frac{6Ec\Delta}{L^2} \qquad (14\text{-}19)$$

where E = modulus of elasticity
Δ = displacement

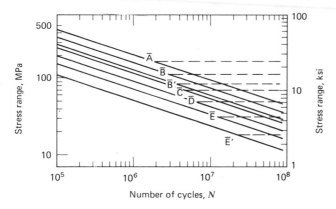

FIGURE 14.27 Proposed American Association of State Highway and Transportation Officials (AASHTO) fatigue design curves[46] (With permission.)

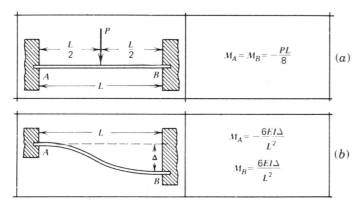

FIGURE 14.28 Beams fixed at both ends. (*a*) Load controlled; (*b*) displacement controlled.

The passage of a fully loaded truck across a bridge might represent a situation that could be analyzed by Eq. 14-18, but if one of the bridge foundations had settled by an amount Δ then Eq. 14-19 would prove more correct. Identification of the importance of load or displacement in controlling the bridge response is critical, since the design changes one would make to improve either load or displacement resistance of a given beam are mutually incompatible. For instance, we see from Eq. 14-18 that for a given load P, the load-bearing capacity of a beam is enhanced by *decreasing* its unsupported length and/or *increasing* its moment of inertia or rigidity. By sharp contrast, we see from Eq. 14-19 that for a given displacement Δ, the flexural stress may be reduced by *increasing* the beam length and/or *decreasing* its rigidity. For example, adding a cover plate to a flange experiencing load control would enhance its fatigue life but would prove to be counterproductive if the beam were displacement controlled. A discussion of such a dilemma with regard to the Lehigh Canal Bridge is given by Fisher et al.[43]

REFERENCES

1. *Engineering News-Record* **94**(5), 188 (Jan. 29, 1925).
2. H. A. Wood, *Eng. Fract. Mech.* **7,** 557 (1975).
3. Military Specification, Mil-A-83444, *Damage Tolerance Design Requirements for Aircraft Structures,* USAF, July 1974.
4. T. J. Dolan, *Met. Eng. Quart.* 32 (Nov. 1972).
5. R. J. Gran, F. D. Orazio, Jr., P. C. Paris, G. R. Irwin, and R. W. Hertzberg, AFFDL-TR-70-149, Mar. 1971.
6. F. A. Heiser and R. W. Hertzberg, *JISI* **209,** 975 (1971).
7. M. O. Speidel, *The Theory of Stress Corrosion Cracking in Alloys,* J. C. Scully, Ed., NATO, Brussels, 1971.
8. R. W. Hertzberg and T. J. Pecorini, *Int. J. Fatigue* **15**(6), 509 (1993).
9. C. Ragazzo, R. W. Hertzberg, and R. Jaccard, *J. Test. Eval.* **23**(1), 19 (1995).
10. *Metals Handbook,* Vol. 10, ASM, Metals Park, OH, 1975.
11. *Source Book in Failure Analysis,* American Society for Metals, Metals Park, OH, Oct. 1974.
12. T. P. Rich and D. J. Cartwright, Eds., *Case Histories in Fracture Mechanics,* AMMRC MS 77-5, 1977, p. 3.9.1.
13. C. M. Hudson and T. P. Rich, Eds., *Case Histories Involving Fatigue and Fracture Mechanics,* ASTM, STP 918, Philadelphia, PA, 1986.
14. C. N. Reid and B. L. Baikie, *op. cit.,* p. 102.
15. H. S. Pearson and R. G. Dooman, *op. cit.,* p. 65.
16. P. C. Paris and G. C. M. Sih, ASTM *STP 381,* 1965, p. 30.
17. C. M. Carmen and J. M. Katlin, ASME Paper No. 66-Met-3, *J. Basic Eng.,* 1966.
18. J. E. Cambell, W. E. Berry, C. E. Feddersen, *Damage Tolerant Design Handbook,* MCIC-HB-01, Sept. 1973.
19. S. Yukawa, D. P. Timio, and A. Rubio, *Fracture,* Vol. 5, H. Liebowitz, Ed., Academic, New York, 1969, p. 65.
20. C. Schabtach, E. L. Fogelman, A. W. Rankin, and D. H. Winnie, *Trans. ASME* **78,** 1567 (1956).
21. R. J. Bucci and P. C. Paris, Del Research Corporation, Hellertown, PA, Oct. 23, 1973.
22. J. M. Barsom and S. T. Rolfe, *Eng. Fract. Mech.* **2**(4), 341 (1971).
23. R. H. Sailors and H. T. Corten, ASTM *STP 514,* Part II, 1972, p. 164.
24. J. A. Bennett and H. Mindlin, *J. Test. Eval.* **1**(2), 152 (1973).
25. D. B. Ballard and H. Yakowitz, *Scanning Electron Microscope 1970,* IITRI, Chicago, April 1970, p. 321.
26. R. Madison, Ph.D. dissertation, Lehigh University, Bethlehem, PA, 1969.
27. H. Tada, P. C. Paris, and G. R. Irwin, *The Stress Analysis of Cracks Handbook,* Del Research Corporation, Hellertown, PA, 1973.
28. T. E. Davidson, J. F. Throop, and J. H. Underwood, *Case Histories in Fracture Mechanics,* T. P. Rich and D. J. Cartwright, Eds., AMMRC MS 77-5, 1977, p. 3.9.1.
29. J. C. Ritter and M. E. deMorton, *J. Austral. Inst. Met.* **22**(1), 51 (1977).
30. R. S. Montgomery, *Wear* **33**(2), 359 (1975).
31. R. B. Griffin et al., *Metallography* **8,** 453 (1975).

32. O. L. Bowie and C. E. Freese, *Eng. Fract. Mech.* **4**(2), 315 (1972).
33. I. S. Raju and J. C. Newman, Jr., *Eng. Fract. Mech.* **11**(4), 817 (1979).
34. R. O. Ritchie, *Int. Metals Rev. Nos. 5 and 6,* Review 245, 1979, p. 205.
35. J. E. Srawley and J. B. Esgar, *NASA RM-X-1194,* 1967.
36. M. F. Ashby and D. R. H. Jones, *Engineering Materials,* Pergamon, Oxford, England, 1980.
37. R. W. Hertzberg, M. T. Hahn, C. M. Rimnac, J. A. Manson, and P. C. Paris, *Int. J. Fract., 22,* R57 (1983).
38. B. J. Gross and J. E. Srawley, NASA *TND-2603* (1965).
39. C. E. Feddersen, ASTM *STP 410,* p. 77 (1967).
40. J. A. Manson and R. W. Hertzberg, *J. Polym. Sci.* (Phys.) **11,** 2483 (1973).
41. R. W. Hertzberg, J. A. Manson, and W. C. Wu, ASTM *STP 536,* p. 391 (1973).
42. J. W. Fisher, *NCHRP Research Results Digest* **59** (Mar. 1974).
43. J. W. Fisher, B. T. Yen, and N. V. Marchica, Fritz Engineering Laboratory Report No. 386.1, Lehigh University, Bethlehem, PA, Nov. 1974.
44. J. W. Fisher, K. H. Frank, M. A. Hirt, and B. M. McNamee, NCHRP *Report 102,* 1970.
45. J. W. Fisher, P. A. Albrecht, B. T. Yen, D. J. Klingerman, and B. M. McNamee, NCHRP *Report 147,* 1974.
46. P. B. Keating and J. W. Fisher, NCHRP *Report 286,* 1986.

FRACTURE SURFACE PRESERVATION, CLEANING AND REPLICATION TECHNIQUES, AND IMAGE INTERPRETATION

A.1 FRACTURE SURFACE PRESERVATION

The first step to be taken prior to a fractographic examination—whether by SEM or TEM imaging—involves preservation of the fracture surface; if the fine details on the fracture surface are mechanically or chemically attacked, a meaningful fractographic analysis becomes suspect, if not impossible. Therefore, unless a fresh fracture surface is examined immediately, some successful method of surface retention must be employed. For example, if specimens are sufficiently small, they may be stored in a desiccator or vacuum storage jar. Alternatively, the fresh fracture surfaces can be protected with a lacquer spray that dries to form a transparent protective layer. In this manner, one is able to perform a macroscopic examination of the various fracture features even though the surface is protected. Field fractures can be protected by applying a coating of fresh oil or axle grease so long as these oil-based products do not react chemically with the metal surface. Boardman et al.[1] have also reported success in protecting fracture surfaces with a commercially available petroleum-base compound (Tectyl 506). To be sure, these coatings would have to be removed prior to both macroscopic and microscopic examination of the fracture surfaces. Finally, protection can be achieved by applying a strip of cellulose acetate replicating tape to the fracture surface. The tape is initially softened with acetone and then pressed onto the surface and permitted to dry. Unfortunately, the dried tape does not always adhere well to the fracture and has a tendency to pop off or peel away with time and with any significant degree of handling.

A.2 FRACTURE SURFACE CLEANING

In many instances, broken components are received in the laboratory in a condition that precludes their immediate examination in the electron microscope. Fracture sur-

TABLE A.1   Cleaning Procedures for Typical Surface Debris[2]

Cleaning Method	Matter Removed
1. Dry air blast; soft natural bristle brush	Loosely adhering particles
2. Organic solvent cleaning[a]	Hard and viscous debris
Toluene, xylene	Oils, greases
Ketones (acetone, methylethylketone)	Preformed lacquers, varnishes, gums
Alcohols (methyl, ethyl, isopropyl)	Dyes, fatty acids
3. Repeated stripping of successive cellulose acetate replicas	Debris, oxidation and corrosion products
4. Water-based detergent cleaning in conjunction with ultrasonic agitation[b]	Oxidation and corrosion products
5. Cathodic cleaning[c] (surface to be cleaned is cathode with inert anode of carbon or platinum to prevent electroplating on fracture surface)	Oxidation and corrosion products
6. Chemical etch cleaning	Very adherent oxidation and corrosion products

[a] A general purpose organic solvent would contain 40% toluene, 40% acetone, and 20% denatured alcohol.

[b] Suggested cleaning solution[2]: 15 g Alconox powder + 350 mL water. Solution is heated to 95°C and agitated ultrasonically for 15 to 30 min, but not to the point where fracture surface becomes etched (R. S. Vecchio and R. W. Hertzberg, ASTM *STP 827,* 1984, p. 267).

[c] For suggested electrolytes, see Refs. 3, 4, and 5.

face details may have been obliterated by repeated rubbing together of the mating pieces or with other hard objects. Fires resulting from the component fracture can also seriously damage fracture surfaces and even lead to localized melting of surface detail. In most other situations, fracture surfaces are found to be covered with loose foreign particles and grease or a layer of corrosion or oxidation product. Depending on the tenacity of the debris and the electrochemical layer, various cleaning methods are suggested as outlined in Table A.1. It is suggested that cleaning methods 1 to 3 be attempted initially since they are easy to perform and do not degrade the fracture surface. On the other hand, methods 4 to 6, when carelessly employed, can result in permanent damage to the fracture surface. When using method 3—the repeated replica stripping technique—it is suggested that the first and, perhaps, the second replica be kept for possible further study. For example, the fracture surface debris retained on the plastic strip can be analyzed chemically using an energy dispersive X-ray device. (For this purpose, the tape must first be coated with a conductive layer prior to insertion

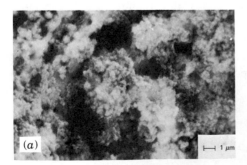

FIGURE A.1   SEM images of fatigue fracture surface in steel alloy. (*a*) **Before cleaning.** (*b*) **After ultrasonic cleaning in heated Alconox solution for 30 minutes.[2]** (R. D. Zipp, *Scanning Electron Microscopy I,* **1979, p. 355, with permission.)**

in the SEM.) An example showing the results of method 4 is shown in Fig. A.1. The fatigue fracture surface of a steel alloy was initially sprayed with a saltwater solution that resulted in the accumulation of corrosion debris on the fracture surface (Fig. A.1a). Individual fatigue striations and occasional secondary fissures (the widely separated black lines) were made visible in the SEM after the specimen was ultrasonically cleaned in a heated Alconox solution for 30 minutes (Fig. A.1b).

Since the formation of an oxide or corrosion layer involves atoms from the base metal, subsequent removal of this layer during any cleaning procedure simultaneously removes some detail from the metal substrate. The extent of this damage depends on the thickness of the electrochemical layer relative to the depth of the fracture surface marking. For example, when the thickness of the corrosive layer is small compared to the depth of the fracture surface contours, removal of the corrosive layer should not significantly impair the fractographic analysis. On the other hand, when fracto-

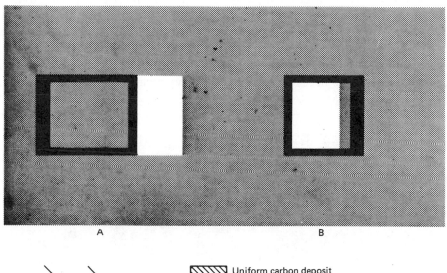

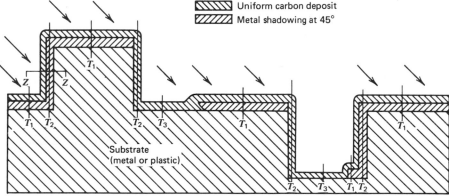

FIGURE A.2   Schema revealing stages in preparation of a fracture surface replica. Replica appears opaque to electron beam when effective thickness is T_2. Brightest image occurs in "shadow" where replica has thickness T_3.[1] (Courtesy of Cedric D. Beachem, Naval Research Laboratory.)

graphic features are themselves shallow, such as fatigue striations in a high-strength steel alloy, removal of the electrochemical layer will seriously reduce the overall depth of those fracture markings to the point where they become extremely difficult to discern.

A.3 REPLICA PREPARATION AND IMAGE INTERPRETATION

Several replication procedures have been developed in the metallurgical laboratory. In the one-step process, a carbon film is vacuum deposited directly onto the fracture surface and subsequently floated free by placing the sample in an acid bath to dissolve the sample. Although this technique produces a high-resolution replica, it is not employed often since the specimen is destroyed in the replication process. The most commonly used technique is a nondestructive, two-stage process that produces a replica with reasonably good resolution. A presoftened strip of cellulose acetate is pressed onto the fracture surface and allowed to dry. (To produce replicas from the fracture surfaces of polymeric mateials that would dissolve in acetone, the author has found a water slurry of polyacrylic acid to be an effective replicating media.) When it is stripped from the specimen, the tape carries an impression of the fracture surface topography. Since this tape is opaque to the electron beam, further steps in the replication procedure are necessary. A layer of heavy metal is deposited at an acute

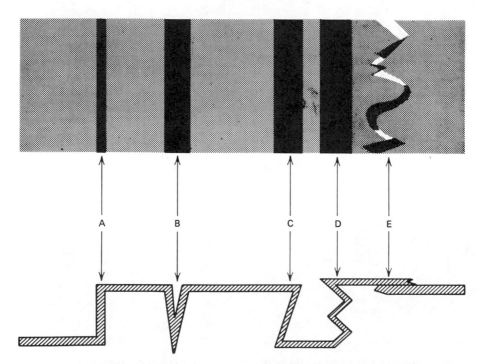

FIGURE A.3 Schema showing different interpretations of opaque black bands.[1] (Courtesy of Cedric D. Beachem, Naval Research Laboratory.)

angle on the side of the tape bearing the fracture impression. This is done to improve the eventual contrast of the replica. Finally, a thin layer of carbon is vapor deposited onto the tape. The plastic tape, heavy metal, carbon composite is then placed in a bath of acetone where the plastic is dissolved. (The polyacrylic acid replica would be dissolved in a water bath.) In the final step, the heavy metal, carbon replica is removed from the acetone bath and placed on mesh screens for viewing in the electron microscope. Since the viewing screens are only 3 mm in diameter, selection of the critical region(s) for examination is most important. This factor emphasizes the need for a carefully conducted macroscopic examination that should direct the examiner to the primary fracture site. By adhering to the recommended procedures for the preparation of replicas, little difficulty should be encountered. For a more extensive discussion of replica preparation techniques see Refs. 36 and 37 in Chapter 7.

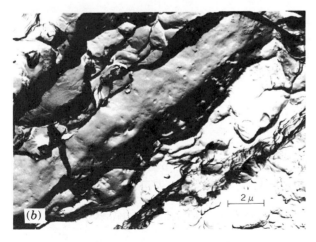

FIGURE A.4   Electron fractographs revealing microvoid coalescence and a large elevation. (a) Replica viewed at zero tilt, note large opaque band; (b) same area as a but replica tilted 24 degrees about horizontal axis. Note fracture details on fracture wall. (Courtesy of R. Korastinsky, Lehigh University.)

An important part of electron fractography is the interpretation of electron images in terms of actual fracture mechanisms. In addition to the references just cited, the reader is referred to the work of Beachem.[6,7] These results are summarized briefly. The fluorescent screen of the microscope (and the photographic film) will react to those electrons that are able to penetrate the replica, that is, the more electrons that penetrate the replica, the brighter the image and vice versa. We see from Fig. A.2 that the vertical walls of the raised and depressed regions at A and B, respectively, show up as black lines, since the electrons cannot penetrate the replica segment with thickness T_2. On the other hand, "shadows" are produced in the absence of the heavy metal (more resistant to electron penetration), which had been deposited at an angle to the fracture surface. These regions appear as white areas *behind* a raised particle and *within* a depressed region associated with a replica thickness T_3. The *location* of the shadow helps to determine whether the region being examined is above or below the general fracture plane. At this point, it is crucial to recognize the image reversal between one- and two-stage replicas. If the substrate in Fig. A.2 were the actual fracture surface, then region A would be correctly identified as lying above the fracture plane while region B would lie below this surface. If the substrate were the plastic, however, the previous conclusion would be completely incorrect. That is, an elevated region on the replica really represents a depressed region on the actual fracture surface. Therefore, when two-stage replicas are prepared, the images produced should be interpreted in reverse fashion—*everything that looks up is really down and vice versa.*

Finally, care should be exercised when attempting to determine the meaning of black bands on the photographs. For the example shown in Fig. A.2, these bands represent simple vertical elevations associated with regions A and B. As seen in Fig. A.3, other more complex explanations are possible. In any case, these opaque bands reflect an elevation on the fracture surface that should reveal fracture mechanism details if reoriented with respect to the electron beam (Fig. A.4).

REFERENCES

1. B. E. Boardman, R. Zipp, and W. A. Goering, *SAE Automotive Engineering Meeting,* Detroit, Michigan, Oct. 1975, Paper No. 750967.
2. R. D. Zipp, *Scanning Electron Microscopy,* Scanning Electron Microscopy, Inc., Chicago, 1979, p. 355.
3. H. DeLeiris, E. Mencarelli, and P. A. Jacquet, *Mem. Scient. Rev. Met.* **63**(5), 463 (1966).
4. P. M. Yuzawich and C. W. Hughes, *Prac. Metall.* **15,** 184 (1978).
5. R. Löhberg, A. Gräder, and J. Hickling, *Microstructural Sci.* **9,** 421 (1981).
6. C. D. Beachem, *NRL Report,* **6360,** U. S. Naval Research Laboratory, Washington, DC, Jan. 21, 1966.
7. C. D. Beachem, *Fracture,* Vol. 1, H. Liebowitz, Ed., Academic, New York, 1968, p. 243.

K CALIBRATIONS FOR TYPICAL FRACTURE TOUGHNESS AND FATIGUE CRACK PROPAGATION TEST SPECIMENS

Type	Stress Intensity Formulation	Configuration
1. Compact specimen[a] C(T)	$K = \dfrac{P}{BW^{1/2}} f(a/W)$ where $f(a/W) = \dfrac{(2 + a/W)}{(1 - a/W)^{3/2}} [0.886 + 4.64a/W$ $- 13.32(a/W)^2 + 14.72(a/W)^3 - 5.6(a/W)^4]$	
2. Disk-shaped compact specimen[a] DC(T)	$K = \dfrac{P}{BW^{1/2}} f(a/W)$ where $f(a/W) = \dfrac{(2 + a/W)}{(1 - a/W)^{3/2}} [0.76 + 4.8a/W$ $- 11.58(a/W)^2 + 11.43(a/W)^3 - 4.08(a/W)^4]$	
3. Wedge opening loaded specimen[b] (WOL)	$K = \dfrac{P}{BW^{1/2}} f(a/W)$ where $f(a/W) = \dfrac{(2 + a/W)}{(1 - a/W)^{3/2}} [0.8072 + 8.858(a/W)$ $- 30.23(a/W)^2 + 41.088(a/W)^3 - 24.15(a/W)^4$ $+ 4.951(a/W)^5]$	

4. Center-cracked tension specimen[c] (CCT)

$$K = \frac{P\sqrt{\pi a}}{BW} f(a/W)$$

where $f(a/W) = \sqrt{\sec \frac{\pi a}{W}}$

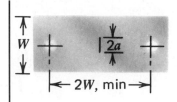

5. Arc-shaped specimen A(T)

$$K = \frac{P}{BW^{1/2}} \left(3\frac{X}{W} + 1.9 + 1.1a/W\right)$$
$$\times [1 + 0.25(1 - a/W)^2(1 - r_1/r_2)] f(a/W)$$
where $f(a/W) = [(a/W)^{1/2}/(1 - a/W)^{3/2}]$
$$\times [3.74 - 6.3(a/W) + 6.32(a/W)^2 - 2.43(a/W)^3]$$

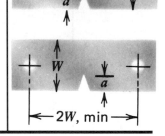

6. Bend specimen[a] SE(B)

$$K = \frac{PS}{BW^{3/2}} f(a/W)$$

where $f(a/W) = \dfrac{3(a/W)^{1/2}}{2(1 + 2a/W)(1 - a/W)^{3/2}}$
$$\times [1.99 - (a/W)(1 - a/W)(2.15 - 3.93a/W + 2.7a^2/W^2)]$$

7. Single edge-notched specimen[d] SE(T)

$$K = \sigma\sqrt{a}[1.99 - 0.41(a/W) + 18.7(a/W)^2 - 38.48(a/W)^3 + 53.85(a/W)^4]$$

[a] ASTM Standard E 399-81, *Annual Book of ASTM Standards,* Part 10, 1981.

[b] A. Saxena and S. Hudak, *Int. J. Fract.* **14,** 453 (1978).

[c] C. E. Feddersen, ASTM *STP 410,* 1976, p. 77.

[d] B. Gross, J. Srawley, and W. F. Brown, Jr., *NASA Tech. Note D-2395,* NASA, Aug. 1964.

MATERIALS INDEX